Whiz-kid MATHS3

Teen Edition

Higher GCSE UK

For Secondary Schools

Grades 5 - 9

Achieve Your Grade

Arinze Oranye

FIRST CONTACT
Iconic Concepts Limited
176 Milwards
Harlow Essex
CM194SJ
United Kingdom
www.iconicconcepts.co.uk
edward@iconicconcepts.co.uk

ACKNOWLEDGEMENTS

The author would like to thank the following organisations for permission to reproduce photographs.

1) www.123rf.com for most of the pictures used in this book.

2) Wikipedia – Picture of Fibonacci

Every effort was made to contact copyright owners/holders of materials reproduced in this book. If through oversight, any omissions will be rectified in future printings of this book if notice is given to the copyright holder.

I humbly submit to The Almighty God, for His Mercies, Guidance, Love and Grace from inception to the completion of this book.

CONTENTS

1 Numbers and Indices

This section covers the following topics:

- Multiples Factors & Primes
- LCM and HCF
- Test for divisibility
- Squares and Cubes
- Indices
- Standard forms

LEARNING OBJECTIVES

By the end of this unit, you should be able to:

a) Understand the meaning of multiples
b) Understand and be able to write factors of numbers
c) Understand and identify prime numbers
d) Find the Highest Common Factor (HCF) of two or more numbers
e) Find the Lowest Common Multiple (LCM) of two or more numbers
f) Write or express numbers as a product of its factors
g) I can test if a number is divisible by 2, 3, 4, 5, 6, 7, 8, 9 and 10
h) Understand powers and roots
i) Use index laws for calculating indices
j) Understand and work in Standard Form

KEYWORDS

- Factors & Multiples
- Prime numbers & prime factors
- Product of prime factors
- Highest common factor and lowest common multiple
- Divisibility
- Squares, square roots, Cubes, cube roots
- Standard form

1.1 MULTIPLES

By now, you must be familiar with the necessary times tables as it will be vital in setting out multiples of numbers.

Multiples of a number are all the numbers in the multiplication table of that number. The most important fact to remember is that the first multiple of a number is the **number itself.**

Also, multiples of a number go on forever (infinite).

The multiples of 3 are the answers to
(1×3), (2×3), (3×3), (4×3), (5×3), (6×3), (7×3) ....

Therefore, the multiples of 3 are **3, 6, 9, 12, 15, 18, 21.........**

EXERCISE 1A

1) Write down the first four multiples of:

a) 1
b) 2
c) 5
d) 7
e) 9
f) 10
g) 13
h) 17
i) 20

2) Look at the list of numbers below.

2	5	7	12	16	22	30	36	49	54	70

a) Which numbers are multiples of 2?

b) Which numbers are multiples of 3?

c) Which numbers are multiples of 4?

d) Which numbers are multiples of 5?

e) Which numbers are multiples of 7?

3) Write down a number that is odd and a multiple of 3

4) Write down a number that is odd and a multiple of 2

5) Write down a number that is even and a multiple of 5?

6) Write down a number that is a square number and a multiple of 7

7) Write down two multiples of 4 with a sum of 28

8) Look at the list of numbers below.

a) Which numbers in the list are multiples of 11?

b) Which numbers in the list are multiples of 9?

c) Which number in the list is a square number?

9) Is 18 a multiple of 6? Explain fully.

10) Sanusi says "the first five multiples of 4 are 8, 12, 16, 20, 24."
Is Sanusi correct? Explain fully.

11) Write down:

a) the first five multiples of 7

b) the first ten multiples of 3

c) a common multiple of 3 and 7

12) Find two multiples of 6 that have a difference 18 but with a product of 360.

13) Three multiples of 9 add up to 117. Find the three numbers.

1.1.1 LOWEST COMMON MULTIPLE (LCM)

The lowest common multiple of two or more numbers is the smallest (first) of their common multiples.

Example 1: Find the LCM of 3 and 5

List some multiples of 3 and 5 as follows:

3: 3 6 9 12 15 18 21.......

5: 5 10 15 20 25 30 35.......

The lowest number which is in both lists is 15.

Therefore, **15** is the LCM of 3 and 5.

Example 2: Find the LCM of 4, 5 and 6

List some multiples of 4, 5 and 6 as follows:

4: 4 8 12 16 20 24 28 32 36 40 44 48 52 56 60 64......

5: 5 10 15 20 25 30 35 40 45 50 55 60 65.......

6: 6 12 18 24 30 36 42 48 54 60 66........

The lowest number which is on all the lists is 60.

Therefore, **60** is the LCM of 4, 5 and 6.

USING PRODUCT OF PRIME FACTORS TO WORK OUT THE LOWEST COMMON MULTIPLE (LCM)

When listing multiples of numbers to find LCM and it becomes cumbersome, it is advisable to use the prime factor method.

Example 1: Find the LCM of 16 and 20

Find all the prime factors of 16 and 20 (refer to section 4.5)

16: $2 \times 2 \times 2 \times 2 \longrightarrow 2^4$
20: $2 \times 2 \times 5 \longrightarrow 2^2 \times 5$

In this system, the LCM contains the highest powers of **each** factor/number.

The highest power of 2 is $\mathbf{2^4}$
The highest power of 5 is $5^1 = \mathbf{5}$

The LCM is the product of all the highest powers

$2^4 \times 5 = 2 \times 2 \times 2 \times 2 \times 5 = \mathbf{80}$ ✓

Example 2: Find the LCM of 20, 24 and 30 using the product of prime method.

Find all the prime factors of 20, 24 and 30 using the factor tree method (Refer to section 4.6)

20: $2 \times 2 \times 5 \longrightarrow 2^2 \times 5$

24: $2 \times 2 \times 2 \times 3 \longrightarrow 2^3 \times 3$

30: $2 \times 3 \times 5 \longrightarrow 2 \times 3 \times 5$

The LCM contains the highest powers of **each** number/factor.

The highest power of 2 is $\mathbf{2^3}$
The highest power of 3 is $\mathbf{3^1 = 3}$
The highest power of 5 is **5**

Therefore, the LCM is the product of all the highest factors
$2^3 \times 3 \times 5 = 2 \times 2 \times 2 \times 3 \times 5 = \mathbf{120}$

USING THE VENN DIAGRAM TO WORK OUT THE LOWEST COMMON MULTIPLE (LCM)

Example 1: Find the LCM of 16 and 20 using the Venn diagram.

Find the prime factors of 16 and 20 (refer to section 4.5)

16: $2 \times 2 \times 2 \times 2$

20: $2 \times 2 \times 5$

Draw a Venn diagram as shown below. The common prime factors are listed in the intersection part.

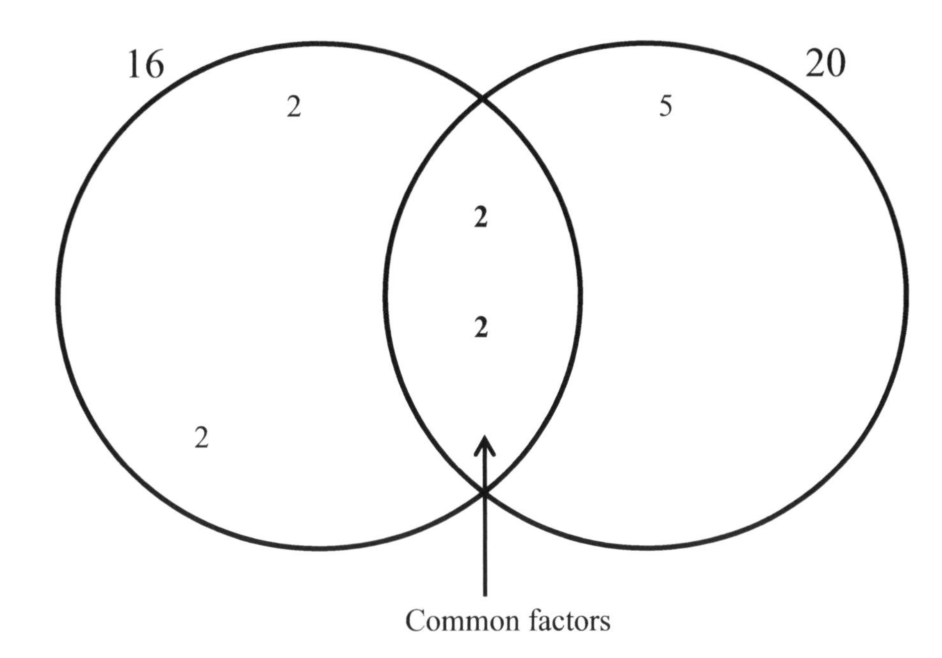

The lowest common multiple is the product of **all** the prime factors in the diagram.

LCM = $2 \times 2 \times 2 \times 2 \times 5 =$ **80** ✓

In chapter 4.6, we shall learn how to use the Venn diagram to find the highest common factor (HCF) of two or more numbers. It is simply performing the steps above and multiplying only the common factors in the intersection part above which is $2 \times 2 = 4$. Therefore, the HCF of 16 and 20 is **4**.

EXERCISE 1B

1) Find the LCM of:

a) 2 and 4

b) 3 and 6

c) 5 and 7

d) 7 and 10

e) 6 and 8

f) 10 and 15

g) 15 and 20

h) 4 and 5

2) Find the LCM of these sets of numbers.

a) 2, 3 and 4

b) 5, 6 and 7

c) 10, 12 and 16

d) 2, 7 and 8

e) 5, 7 and 9

f) 4, 6 and 7

g) 30 and 45

h) 50 and 60

3) Tony and Emma took part in a jogging experiment. Tony took 40 seconds to jog around the athletics track once and Emma 45 seconds. If they both start from the same point at the same time, how long will it take in minutes before they cross the start line together?

4) By leaving your answers in prime factors, find the LCM of the following:

a) $3 \times 3 \times 3 \times 2 \times 2$ and $2 \times 3 \times 3$

b) $2 \times 2 \times 3 \times 3 \times 5$ and $2 \times 2 \times 5 \times 5$

c) $2 \times 2 \times 3 \times 3 \times 3$ and $5 \times 5 \times 2 \times 2 \times 2 \times 3$

d) $2 \times 3 \times 7 \times 7$ and $2 \times 2 \times 2 \times 7 \times 7 \times 7$

e) $5 \times 5 \times 7 \times 11 \times 11$ and $7 \times 7 \times 7 \times 11$

f) $2 \times 3 \times 3 \times 3 \times 3 \times 4 \times 4 \times 5 \times 5 \times 5$ and $2 \times 2 \times 3 \times 3 \times 5 \times 5$

5) Jude says, "The LCM of two numbers is always the product of the numbers."
Is Jude correct?
Explain fully with an example.

6) Find the LCM of 70 and 80

7) Find the LCM of 10, 20 and 40

8) Okoye and Chidi have the same number of drinking cups. Okoye placed his cups into four equal parts while Chidi arranged his own in 2 equal parts. Find the least number of cups they could each have.

1.2 FACTORS

Like in multiples, the knowledge of multiplications is important.

A factor of a number will divide **exactly** into that number. *Exactly* means there will be no remainder(s).

A fact about factors is that every number has at least two factors: **1 and itself**.

Any number that divides exactly into 6 is a factor of 6.

The factors of 6 are: **1, 2, 3** and **6**

We can also find factor as pairs. Factor pairs of a number are the factors of that number.

What are the factors of 20?

Using factor pairs: (1 and 20) because $1 \times 20 = 20$
(2 and 10) because $2 \times 10 = 20$
(4 and 5) because $4 \times 5 = 20$

Putting them in order 1 2 4 5 10 20

Therefore, the factors of 20 are 1 2 4 5 10 and 20

Example 1: Find all the factors of 12.

To start with, 1 and the number itself are factors. So, 1 and 12 are factors of 12.

2 and 6 are also factors because $2 \times 6 = 12$

3 and 4 are also factors because $3 \times 4 = 12$

Putting all in order, the factors of 12 are **1 2 3 4 6 and 12**
Note: Factors of a number start with **1** and end with the number itself.

EXERCISE 1C

1) Find all the factors of:

a) 3
b) 4
c) 14
d) 24
e) 30
f) 35
g) 36
h) 48
i) 56
j) 84
k) 100
l) 120

2) Look at the following list of numbers:

3 5 6 7 9

From the list above, choose a number or numbers that are factors of the following:

a) 48
b) 21
c) 18
d) 27
e) 30
f) 108

3) Write down a number that is odd and a factor of 40

4) Write down even numbers that are factors of 18

5) Write down three factors of 50 that have a sum of 40

6) Write down two factors of 35 that have a sum of 6

7) Write down two factors of 72 that have a difference of 14

8) Anthony Joshua says that 7 is a factor of 56.
Is Anthony correct?
Explain fully

9) Add together all the factors of 30

1.3 PRIME NUMBERS

A prime number is a number that has only **two** factors, **1** and **itself**. This statement means that once a whole number has more than two factors, the number is **not** a prime number.

Example 1: a) list all the factors of 1, 2, 3, 4, 5, 6, 7, 8 and 9

b) Which of the numbers are prime numbers?

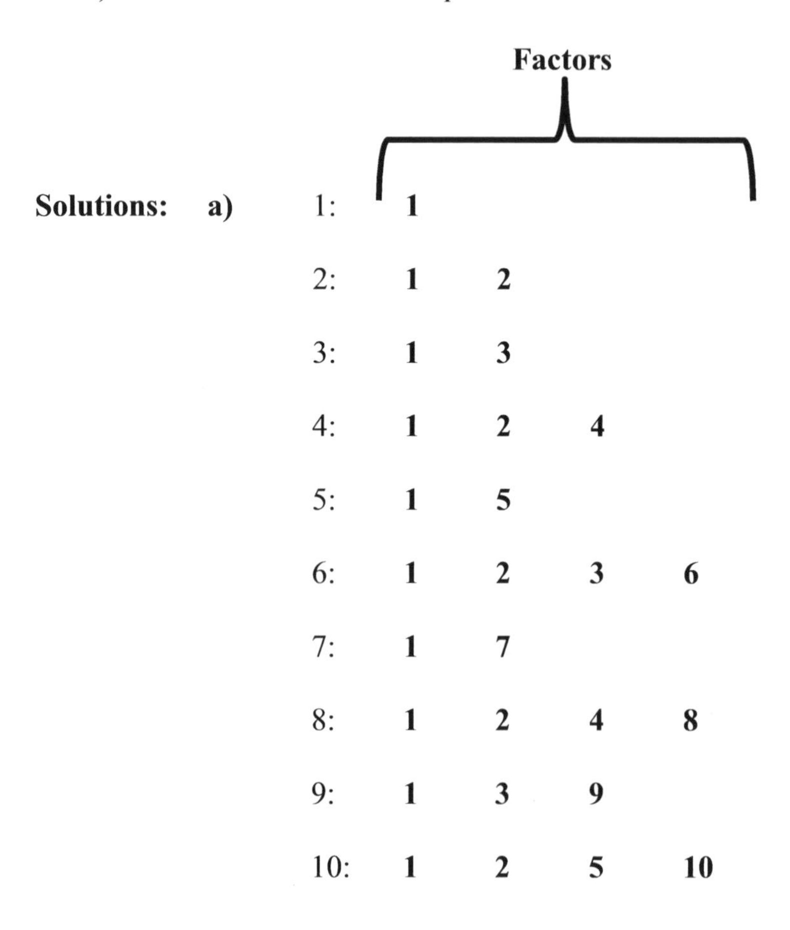

b) Since prime numbers have only two factors, the prime numbers from the list above are **2, 3, 5,** and **7**

Notice that the number **1** is **not** a prime number as it has only **one** factor.

Also, the number **2** is the **only even prime number.**

Example 2: Ifeanyi says: "**9** is a prime number because it is an odd number." Is he correct? Explain fully

Solution: List all the factors of 9. They are 1, 3 and 9.
However, a prime number is a number that has only two factors. 9 has three factors and cannot be a prime number.

Therefore, **9 is not** a prime number because it has **more** than two factors, 1, 3 and 9 and a prime number must have only two factors.

The first 20 prime numbers are:

2	3	5	7	11	13	17	19	23	29
31	37	41	43	47	53	59	61	67	71

1.4 PRIME FACTORS

Prime factors of a number are the factors of that number that are prime numbers.

Example: The factors of 20 are: 1 (2) 4 (5) 10 20

From the factors, only 2 and 5 are prime numbers.

Therefore, **2** and **5** are the **prime factors** of 20.

POINTS TO NOTE

- A number is prime if the only factors are 1 and itself
- 1 **is not** a prime number since it has only one factor, 1
- 2 is the only even prime number
- Prime factors of a number are the prime numbers from factors of that number

EXERCISE 1D

1) Identify all the prime numbers from the numbers below:

a) 1	3	7	15	19	27
b) 2	5	8	13	18	33
c) 6	9	14	28	31	41
d) 10	20	25	37	59	97
e) 4	11	17	60	78	121

2) Find the prime factors of

a) 4   b) 12   c) 34   d) 50

3) Identify a prime number that is also a factor of 6

4) Identify a prime number that is also a factor of 24

5) a) Copy and complete the 7 by 7 square grid following the existing pattern.

1	2	3	4	5	6	7
8	9	10	11	12	13	14

b) Colour in all the prime numbers in the table above.

c) Add up all the prime numbers in the 5th column.

d) Is the number obtained in **part c** above a prime number? Explain fully.

1.5 PRODUCT OF PRIME FACTORS

In section 4.4, prime factors were discussed. Please revisit if you need extra help.

Any whole number which is not a prime number can be broken down into prime factors. It is called prime factor decomposition. The product of the prime factors obtained will always give the original number.

There are different ways to write numbers as the product of prime factors:

Method 1: Factor tree method

Example 1: Write 20 as the product of prime factors.

Step1: Find any two numbers that multiply to give 20 →
If any of the factors are prime, circle it.

Step 2: For the factors **not circled, in this case, 4**, repeat step 1.

Step 3: Since all the numbers are circled, it means **no more decomposition** (breaking down). At this point, you have decomposed the number, 20 into its prime factors.

Multiply all the circled numbers $2 \times 2 \times 5$

Therefore, as a product of prime factors, $20 = 2 \times 2 \times 5$ or $\mathbf{2^2 \times 5}$ in index forms.

Example 2: Write 320 as the product of prime factors.

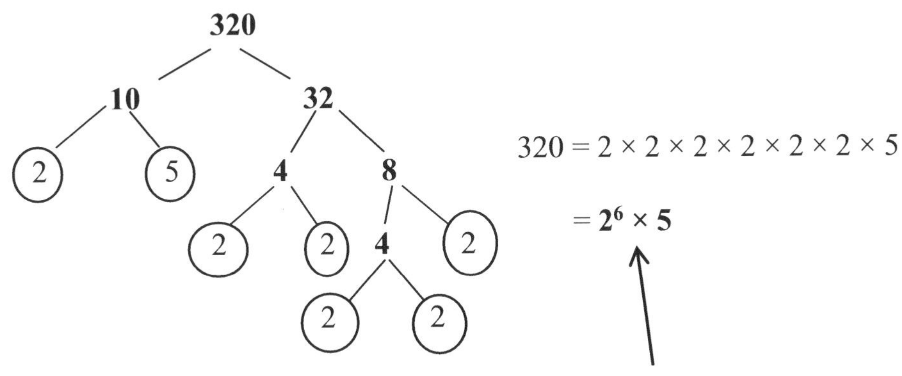

This is the index form

Method 2 of finding the product of prime factors

This is a method of dividing the number by the smallest (first) prime number that divides into the number. The process is repeated until the answer becomes 1.

Example 3: Express 60 as the product of prime factors

The first prime number that divides into 60 is 2. Use 2 to divide. Repeat the process until the answer becomes 1.

2	60
2	30
3	15
5	5
	1

Therefore, 60 written as the product of prime factors is $2 \times 2 \times 3 \times 5 = \mathbf{2^2 \times 3 \times 5}$

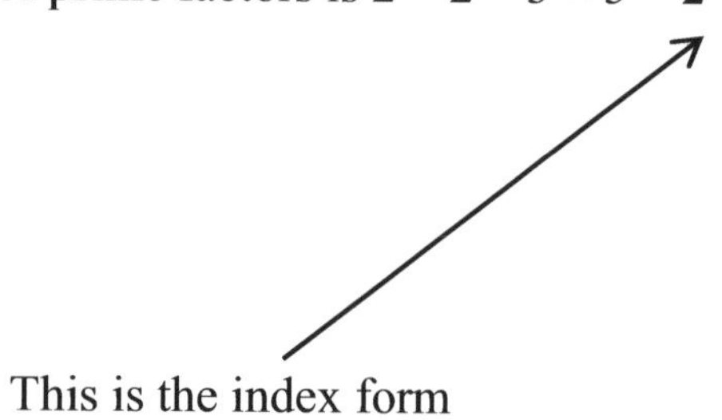

This is the index form

EXERCISE 1E

1) Express each of the numbers below as a product of prime factors

a) 30 c) 70 e) 720

b) 55 d) 144 f) 940

2) a) Complete the factor tree.

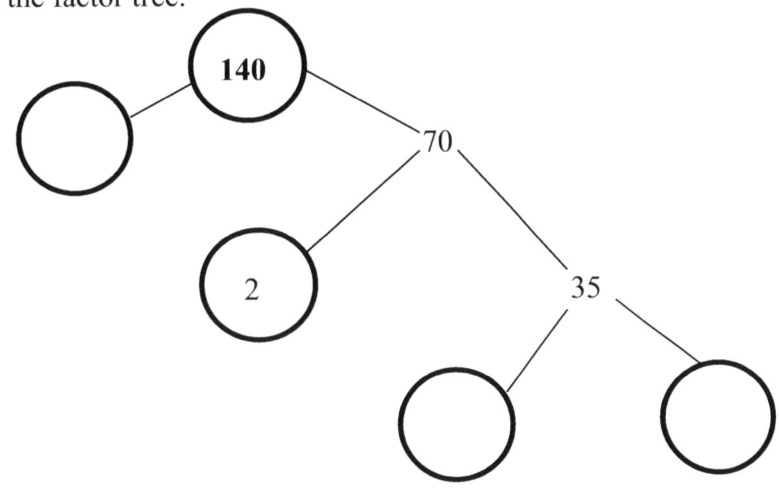

b) Write down the prime factors of 140 in index form.

3) Write each number as a product of prime factors in index form.

a) 150 b) 500 c) 510

4) Write the following as the product of two prime factors.

a) 6 b) 21 143

5) Copy and complete the factor trees.

a)

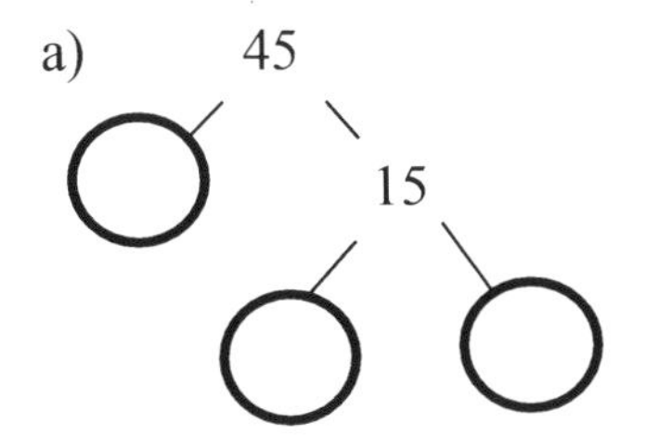

b)

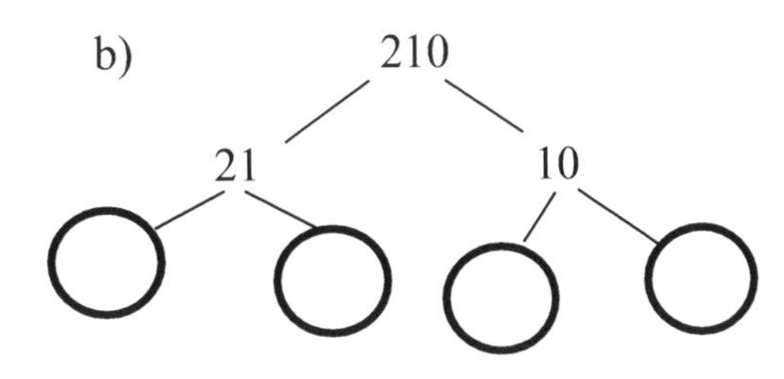

6) Write in index form.

a) $2 \times 2 \times 2 \times 3 \times 3 \times 3$

b) $2 \times 2 \times 5 \times 7 \times 7 \times 7 \times 7$

c) $3 \times 3 \times 3 \times 3 \times 5 \times 5$

7) Anthony says, "Prime factors will be different if you chose a different factor pair."
Is he correct?

Explain fully

8) Write each number as a product of prime factors in index form.

a) 3300

b) 1540

1.6 HIGHEST COMMON FACTOR (HCF)

The highest common factor of two or more numbers can be obtained. The first method is to find all the factors of the numbers and identify the **common factors**. The highest number of the common factors will be the HCF of the numbers.

Refer to **section 1.2** for finding factors of numbers.

Example 1: Find the HCF of 4 and 8

Identify all the factors of 4 and 8 as follows:

4:	**1**	**2**	**4**	
8:	**1**	**2**	**4**	8

The common factors are 1, 2 and 4.

However, 4 is the largest number of the common factors. Therefore, **4** is the highest common factor (HCF) of 4 and 8.

It is also the highest number that can divide into both numbers.

Example 2: Find the HCF of 24 and 36

Identify all the factors of 24 and 36.

24:	**1**	**2**	**3**	**4**	**6**	8	**12**	24	
36:	**1**	**2**	**3**	**4**	**6**	9	**12**	18	36

The common factors are 1, 2, 3, 4, 6 and 12.

However, the highest number common to both numbers is 12.

Therefore, **12** is the highest common factor (HCF) of 24 and 36.

It is the highest number that can divide into both numbers.

USING PRIME FACTORS TO FIND HCF

When listing factors of numbers in other to find HCF becomes cumbersome, it is advisable to use the prime factor method.

Example 1: Find the HCF of 24 and 36

Find all the prime factors of 24 and 36

24: $2 \times 2 \times 2 \times 3 \longrightarrow 2^3 \times 3$

36: $2 \times 2 \times 3 \times 3 \longrightarrow 2^2 \times 3^2$

In this system, find the pairs of common prime factors and multiply them.

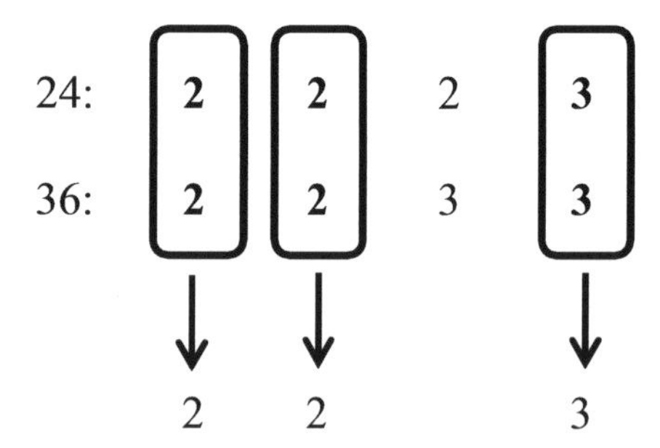

The HCF would be $2 \times 2 \times 3 =$ **12** 

ALTERNATIVELY

After finding the product of prime factors, we multiply the lowest power of **each common** prime factor to give the HCF.

$$24: 2 \times 2 \times 2 \times 3 = 2^3 \times 3$$

$$36: 2 \times 2 \times 3 \times 3 = 2^2 \times 3^2$$

LOWEST POWER OF 2 is $\mathbf{2^2}$ while the LOWEST POWER OF 3 is **3.**

Therefore, HCF of 24 and 36 $= 2^2 \times 3 = 2 \times 2 \times 3 =$ **12** ✓

Example 2: Find the HCF of $3 \times 3 \times 3 \times 3 \times 5 \times 5 \times 7 \times 7 \times 7$ **and** $3 \times 3 \times 3 \times 7 \times 7$
Leave your answer in prime factors in index form.

$3 \times 3 \times 3 \times 3 \times 5 \times 5 \times 7 \times 7 \times 7 = 3^4 \times 5^2 \times 7^3$ and

$3 \times 3 \times 3 \times 7 \times 7 = \mathbf{3^3} \times \mathbf{7^2}$

The lowest power of 3 contained in the two numbers is 3^3

The lowest power of 7 contained in the two numbers is 7^2

* *Notice that the number* ***5*** *is* ***not*** *common to both numbers and as such is* ***not included****.*

Therefore, in prime factor index form, the HCF is $\mathbf{3^3 \times 7^2}$ ✓

Please note: The above index answer is acceptable if the question says so.
If not, work out the real answers to $3^3 \times 7^2$ as detailed below.

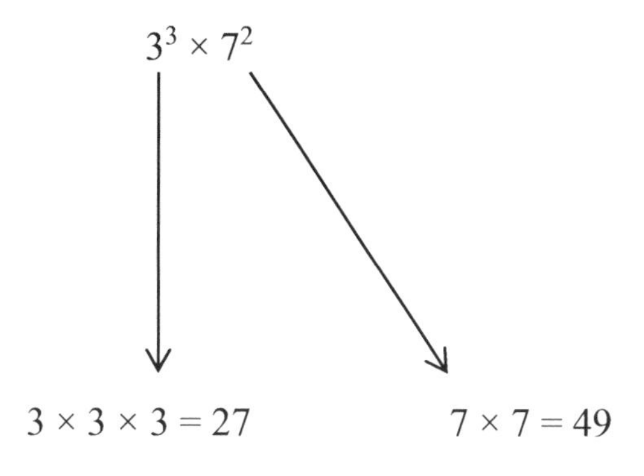

Highest common factor (HCF) $= 27 \times 49$

$= \mathbf{1323}$

USING THE VENN DIAGRAM TO WORK OUT THE HIGHEST COMMON FACTOR (HCF)

Example 1: Find the HCF of 16 and 20 using the Venn diagram.

Find the prime factors of 16 and 20 (refer to section 4.5)

16: $2 \times 2 \times 2 \times 2$

20: $2 \times 2 \times 5$

Draw a Venn diagram as shown below. The common prime factors are listed in the intersection part.

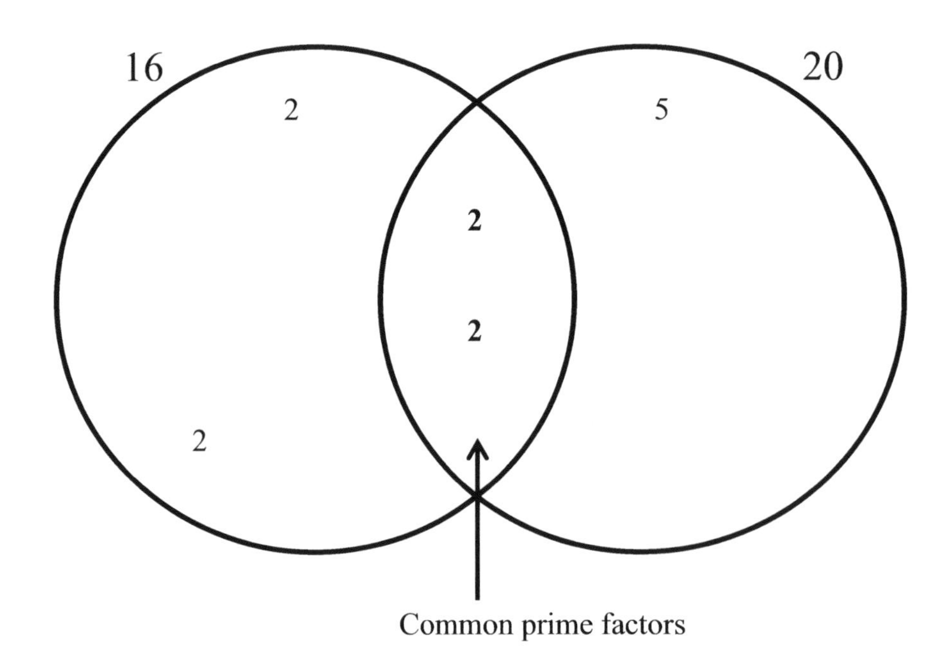

The highest common factor is obtained by multiplying the numbers in the intersection part.

$$2 \times 2 = 4$$

Therefore, the HCF of 16 and 20 is **4**.

In all, you may use any method of your choice to find HCF unless told otherwise.

EXERCISE 1F

1) i) Write down all the factors of the numbers below.

   ii) Circle the common factors of each pair.

   iii) Pick out the highest common factor.

   a) 6 and 10

   b) 12 and 18

   c) 5 and 25

   d) 27 and 49

   e) 30 and 76

   f) 36 and 48

2) Find the highest common factor using the prime factor method for the following pairs of numbers.

   a) 40 and 60

   b) 120 and 150

   c) 420 and 700

3) Find the HCF of the following numbers, leaving your answers in prime factor forms using index notation.

   a) $3 \times 3 \times 5 \times 5 \times 5 \times 5 \times 7 \times 7 \times 7$ and $3 \times 5 \times 5 \times 5 \times 7 \times 7$

   b) $2 \times 2 \times 2 \times 2 \times 3 \times 3 \times 3 \times 3 \times 3 \times 3 \times 11\ 11 \times 11$ and $2 \times 2 \times 3 \times 3 \times 3$

   c) $5 \times 5 \times 7 \times 7 \times 7 \times 13$ and $5 \times 7 \times 13$

4) Find the HCF of the following numbers using the Venn diagram method.

   a) 18 and 60

   b) 20 and 75

5) Using the index notation, find the HCF of the numbers below.

a) 8 and 40

b) 63 and 270

c) 30, 60 and 90

d) 25, 70 and 130

6) Write out the common factors of

a) 4 and 16

b) 12 and 20

c) 45 and 63

d) 8, 20 and 52

7) David needed to cut out congruent squares for making envelopes from a piece of rectangular brown cardboard paper as shown.

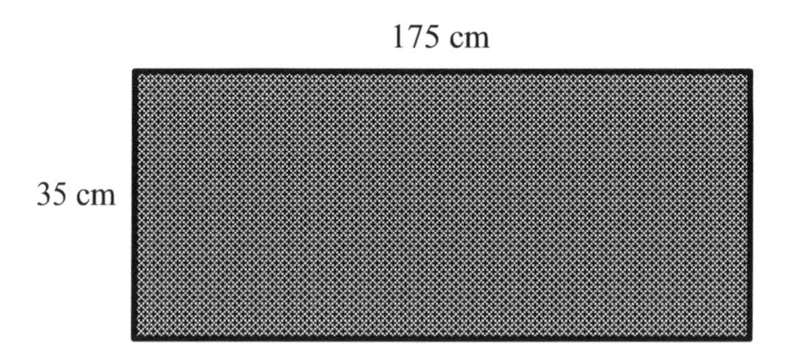

Without wasting any cardboard paper, what is the maximum square size David can use for the envelopes?

8)

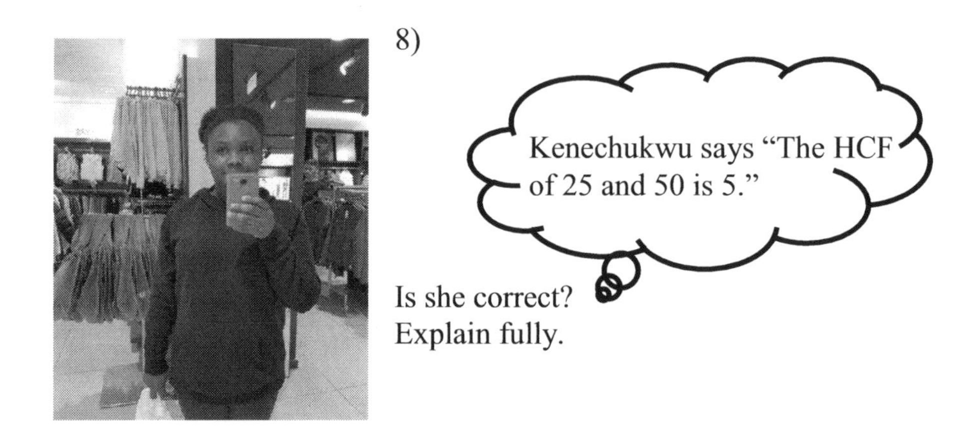

Is she correct? Explain fully.

1.7 DIVISIBILITY TEST

To test if a whole number is divisible by 2, 3, 4, 5, 6, 7, 8, 9 or 10, use the divisibility rules which are set out below.

DIVISIBILITY BY 2

A number is divisible by 2 if the last digit is 0 or even.

Look at the numbers below.

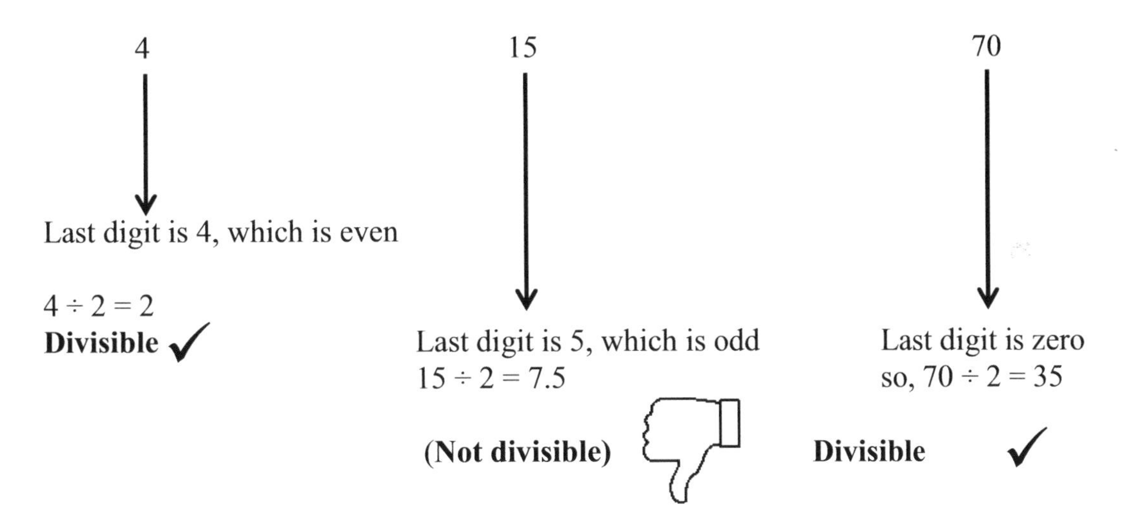

DIVISIBILITY BY 3

To find out if a number is divisible by 3, add up all the digits. If the sum is divisible by 3, then the number is divisible by 3.

If you cannot figure out if the added numbers are divisible by 3, add together the digits of the added numbers and divide by 3 to check.

Example 1: Check if 453 is divisible by 3.

$4 + 5 + 3 = 12$
Clearly, 12 is divisible by 3 because $12 \div 3 = 4$
Therefore, 453 **is divisible** by 3

$453 \div 3 = 151$ ✓

Example 2: Is 79796223 divisible by 3?

$7 + 9 + 7 + 9 + 6 + 2 + 2 + 3 = 45$

If you are familiar with your 3 times table, then 45 is divisible by 3 since $15 \times 3 = 45$

Yes, 79796223 **is divisible** by 3. ✓

You may use a calculator to check. $79796223 \div 3 = 26\ 598\ 741$

Example 3: Is 265 divisible by 3?

$2 + 6 + 5 = 13$

$13 \div 3 = 4.3333333.....$ (**Not** a whole number)

13 is **not** divisible by 3. Therefore 265 **is not divisible** by 3.

DIVISIBILITY BY 4

If the **last two digits** of a number are divisible by 4, then the number is divisible by 4

Example 1: Is 459 divisible by 4?

The last two digits are 59

$59 \div 4 = 14.75$ (**Not** a whole number)

No, 459 **is not divisible** by 4

Example 2: Is 2792 divisible by 4?

The last two digits are 92

$92 \div 4 = 23$ (Whole number)

Yes, 2792 **is divisible** by 4 ✓

DIVISIBILITY BY 5

A number is divisible by 5 if the last digit is 0 or 5.

Example 1: Is 345 divisible by 5?

The last digit is 5; therefore, it **is divisible** by 5. 

345 ÷ 5 = 69 (whole number)

Example 2: Is 677 divisible by 5?

The last digit is 7; therefore, it **is not divisible** by 5

677 ÷ 5 = 135.4 (**Not** a whole number)

Example 3: Is 120 divisible by 5?

The last digit is 0; therefore, it **is divisible** by 5 ✓

120 ÷ 5 = 24 (whole number)

DIVISIBILITY BY 6

A number is divisible by 6 if it is divisible by both 2 and 3. Check the rules above!

Example 1: Is 156 divisible by 6?

Step 1: Check with divisibility by 2 rules:
It is even; therefore, it can be divided by 2 (Yes)

Step 2: Check with divisibility by 3 rules:
1 + 5 + 6 = 12
…and 12 is divisible by 3;
therefore, it can be divided by 3 (Yes)

Since 156 is divisible by 2 and 3, it is divisible by 6 

Example 2: Is 86 divisible by 6?

Step 1: 86 is even, so it is divisible by 2 (Yes)

Step 2: Check for divisibility by 3.

$8 + 6 = 14$
$(14 \div 3 = 4.666666\ldots)$ (X)

14 is not divisible by 3. Therefore, 86 is not divisible by 3 as well.

Since 86 failed divisibility by 3 rules, it cannot be divided by 6.

DIVISIBILITY BY 7

For a number to be divisible by 7, double the last digit and subtract it from a number made by the other digits. If the outcome/result is divisible by 7, then the number itself is also divisible by 7.

Example 1: Is 238 divisible by 7?

Double the last digit $8 \times 2 =$ **16**

Subtract 16 from the remaining numbers $23 - 16 =$ **7**

7 is divisible by 7 $(7 \div 7 = 1)$ .....(Whole number)

Therefore, 238 **is divisible** by 7 ✓

Example 2: Is 957 divisible by 7?

Double the last digit $7 \times 2 = 14$

Subtract 14 from the remaining numbers $95 - 14 = 81$

$81 \div 7 = 11.57\ldots.$ (Not a whole number)

Therefore, 957 is **not divisible** by 7

DIVISIBILITY BY 8

If the **last three** digits are divisible by 8, then the number is divisible by 8

Example 1: Is 1 547 divisible by 8?

The last three numbers are 547.
$547 \div 8 = 68.375$ (Not a whole number)

Therefore, 1547 is **not divisible** by 8

Example 2: Is 32 984 divisible by 8?

The last three numbers are 984
$984 \div 8 = 123$ (Whole number)

Therefore, 32 984 **is divisible** by 8 ✓

Points to note: *At times the last three numbers might be a problem to divide. Halve three times, and if the result is still a whole number, then that number is divisible by 8.*

DIVISIBILITY BY 9

A number is divisible by 9 if the sum of the digits is divisible by 9.

Example 1: Is 154 divisible by 9?

$1 + 5 + 4 = 10$

$10 \div 9 = 1.1111111$ (**Not** a whole number)

Therefore, 154 is **not divisible** by 9

Example 2: Is 322 866 divisible by 9?

$3 + 2 + 2 + 8 + 6 + 6 = 27$

$27 \div 9 = 3$ (Whole number)

Therefore, 322 866 **is divisible** by 9 

DIVISIBILITY BY 10

A number is divisible by 10 is the last digit is zero (if the number ends in zero (0).

Example 1: Is 4 567 divisible by 10?

The last number is 7

4 567 **is not divisible** by 10 because the last number is not zero

Example 2: Is 56 430 divisible by 10?

The last number (digit) is 0

Therefore, 56 430 **is divisible** by 10 ✓

EXERCISE 1G

1) Are these numbers divisible by **2**? Give a reason.

a) 12
b) 26
c) 27
d) 69
e) 780
f) 7865

2) Are these numbers divisible by 3? Give a reason using divisibility tests.

a) 12
b) 21
c) 34
d) 46
e) 180
f) 654

3) Obinna says "287945 is divisible by 7." Is he correct? Explain using the divisibility rule.

4) Are these numbers divisible by 9? Perform divisibility tests to check.

a) 81
b) 108
c) 1 553
d) 2 345
e) 5 166
f) 142 911

5) Is 65432 divisible by 5? Explain fully.

6) Is 7893451 divisible by 8? Explain fully

7) Is 6754320 divisible by 2 and 10? Explain fully.

8) Using the divisibility test, check whether the following numbers are divisible by 4.

a) 876
b) 1258

1.8 POWERS AND ROOTS

Powers are often called **indices**. Repeated multiplications by the same number can be shown using **index notations.**

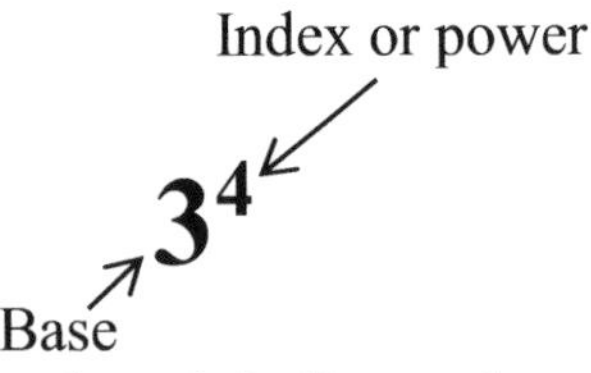

The number, 4, indicates the number of times the base number, 3, is multiplied.
So, $3^4 = 3 \times 3 \times 3 \times 3 = 81$
It is read as 3 to the power of 4.

Note: 3^4 **is not** 3×4 as this would give 12, which is not the answer.

SQUARE NUMBERS

When a number is multiplied by itself, we say the number has been **squared.**
5 squared = $5^2 = 5 \times 5 = \mathbf{25}$
7 squared = $7^2 = 7 \times 7 = \mathbf{49}$
9 squared = $9^2 = 9 \times 9 = \mathbf{81}$

25, 49 and 81 are square numbers.
Similarly,
$2.4^2 = 2.4 \times 2.4 = \mathbf{5.76}$
$1.6^2 = 1.6 \times 1.6 = \mathbf{2.56}$

Therefore, when a number is multiplied by itself, the result or outcome is called a **square number**.

A common *misconception* is thinking that to square a number is simply to multiply by 2. It is wrong.

SQUARE OF NEGATIVE NUMBERS

When two negative numbers are multiplied together, the answer is always a **positive** number.

$(-2)^2 = -2 \times -2 = 4$
$(-5)^2 = -5 \times -5 = 25$

Therefore, the square of a negative number is a positive number.

SQUARE ROOT (√)

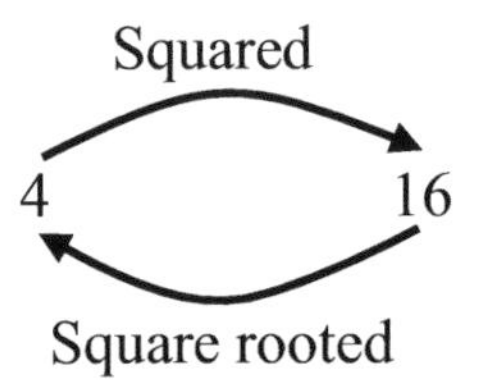

Example 1
$\sqrt{16} = \mathbf{4}$..........because $4 \times 4 = 16$
$\sqrt{49} = \mathbf{7}$ ....because $7 \times 7 = 49$

Remember, the square root of a number could be positive or negative. So,

$\sqrt{16} = \pm 4$ which means **4** or **-4**
Proof: $4 \times 4 = 16$ and $-4 \times -4 = 16$

Example 2: The area of a square is 144m^2. Work out the length of the sides.

Solution:
$\sqrt{144} = 12$
Therefore, the length of each side $\boldsymbol{x} = \mathbf{12}$ **m**.
Check: 12 × 12 = 144

	144 m²
x	
	x

SQUARE ROOT OF BIG NUMBERS

At times, we may be required to work out the square root of big numbers without a calculator.
The knowledge of prime factors is important.

Example 1: Work out the square root of 3136.

Using the knowledge of prime factor decomposition,

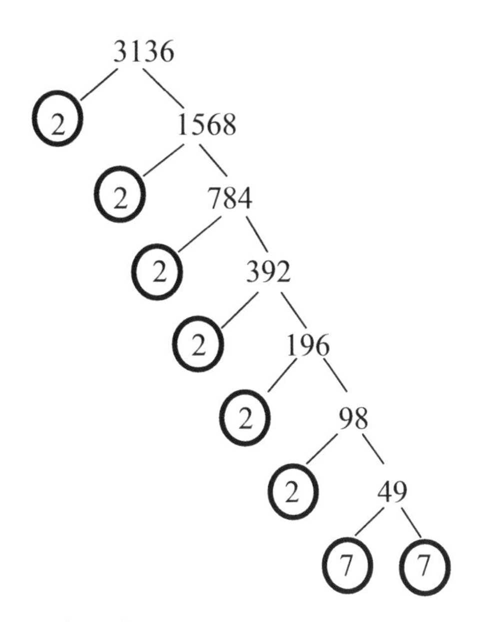

$= 2^6 \times 7^2$

Pair the prime numbers as follows:

$(2 \times 2) \times (2 \times 2) \times (2 \times 2) \times (7 \times 7)$

$2 \times 2 \times 2 \times 7 = 56$

Therefore, $\sqrt{3136} = \mathbf{56}$ ✓

Example 2: Work out $\sqrt{4761}$

As a product of prime factors,
$4761 = 3^2 \times 23^2$.
Writing in pairs gives $(3 \times 3) \times (23 \times 23)$

Therefore, $\sqrt{4761} = 3 \times 23 = \mathbf{69}$ ✓

Check: $69 \times 69 = 4761$

Example 3: Work out the values of

a) $\sqrt{\frac{49}{81}}$ b) $\sqrt{1\frac{9}{16}}$ c) $\sqrt{\frac{128}{162}}$

Solutions:

a) $\sqrt{\frac{49}{81}} = \frac{\sqrt{49}}{\sqrt{81}} = \frac{\mathbf{7}}{\mathbf{9}}$

b) $\sqrt{1\frac{9}{16}} = \sqrt{\frac{25}{16}} = \frac{\sqrt{25}}{\sqrt{16}} = \frac{5}{4} = \mathbf{1\frac{1}{4}}$

From converting $1\frac{9}{16}$ to improper fraction

c) Divide both numbers by 2 to get square numbers.

$\sqrt{\frac{128}{162}} = \sqrt{\frac{64}{81}} = \frac{\sqrt{64}}{\sqrt{81}} = \frac{\mathbf{8}}{\mathbf{9}}$

CUBE NUMBERS

The number 27 can be written as $3 \times 3 \times 3$ or 3^3. We say ***three cubed.***

Examples

$1^3 = 1 \times 1 \times 1 = \mathbf{1}$
$2^3 = 2 \times 2 \times 2 = \mathbf{8}$
$3^3 = 3 \times 3 \times 3 = \mathbf{27}$

The numbers 1, 8 and 27 are all **cube numbers**.

Cube numbers are formed when a number is multiplied by itself three times.

Note: 4^3 is not $4 \times 3 = 12$  

It is $4 \times 4 \times 4 = 64$ ✓

CUBE ROOTS ($\sqrt[3]{\ }$)

Finding the cube root of a number is the opposite of finding the cube of that number.

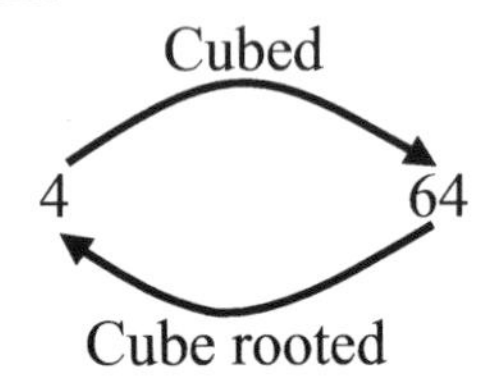

Hence, the cube root of 64 is ($\sqrt[3]{64}$) = **4**, because $4 \times 4 \times 4 = 64$

$\sqrt[3]{8} = 2$, because $2 \times 2 \times 2 = 8$

Likewise,
$\sqrt[3]{-8}$ = **-2,** because $-2 \times -2 \times -2 = -8$

EXERCISE 1H

1) From the list of numbers below, write down
a) a square number b) a cube number
c) the square root of 169
d) the cube root of 27.

1	**3**	**13**	**18**	**64**
125	**338**	**500**		

2) Write down the sum of the first five square numbers.

3) Calculate
a) 9^2 b) 1.5^2 c) 13^2 d) $(-11)^2$

4) Work out
a) $\sqrt{225}$ b) $\sqrt{289}$ c) $\sqrt{441}$ d) $\sqrt{6.25}$

5) Is it possible to work out $\sqrt{-49}$? Explain fully.

6) Work out the value of $(-3)^2 + (-10)^2$

7) Find the difference between $(-6)^2$ and 11^2

8) Work out
a) 5^3 b) $(-4)^3$ c) $(\frac{1}{2})^3$ d) $(\frac{2}{5})^3$

9) Work out
a) $\sqrt[3]{64}$ b) $\sqrt[3]{512}$ c) 7^3 d) 1.2^3

10) Copy and complete.
a) $\sqrt{?} = 11$
b) $8 - \sqrt{1} = ?$
c) $25 + \sqrt{?} = 30$
d) $1002 + \sqrt{400} = ?$

11) Work out
a) $6^2 + 3^3 + 1^2$
b) $\sqrt{484}$
c) $\sqrt{9604}$
d) $\sqrt{4225}$
e) $\sqrt{\frac{400}{100}}$
f) $\sqrt{2\frac{14}{25}}$
g) $\sqrt{\frac{121}{625}}$
h) $\sqrt[3]{343}$
i) $4.5^2 - 2.3^2$
j) $\sqrt{3^2 + 4^2}$
k) Work out the negative square root of
a) 4 b) 196 c) 900

1.9 LAWS OF INDICES

We already know that $5^2 = 5 \times 5 = 25$. In index notation, $3 \times 3 \times 3 = 3^3$ but the value is 27.

1) Multiplication Rule: To multiply powers of the same numbers or letters, add the index numbers.

$$C^x \times C^y = C^{x+y}$$

The base numbers or letters must be the same for multiplication rule to work.

Example 1: Simplify $4^2 \times 4^3$ and write your answer in index form.

$4^2 \times 4^3 = 4^{2+3} = \mathbf{4^5}$

Example 2: Simplify $t^{10} \times t^{12}$

$t^{10} \times t^{12} = t^{10+12} = \mathbf{t^{22}}$

Example 3: Simplify $6^5 \times 6^{-2}$

$6^5 \times 6^{-2} = 6^{5+-2} = \mathbf{6^3}$

Example 4: Simplify $5d^6 \times 2d^3$
First multiply the numbers: $5 \times 2 = 10$
Then, $d^6 \times d^3 = d^{6+3} = d^9$
Therefore, $5d^6 \times 2d^3 = \mathbf{10d^9}$

2) Division rule: To divide powers of the same numbers or letters, subtract the index numbers.

$$C^x \div C^y = C^{x-y}$$

Just like multiplication rule, the base numbers or letters must be the same before applying division rule/law.

Example 5: Simplify $12^7 \div 12^5$
$12^7 \div 12^5 = 12^{7-5} = \mathbf{12^2}$

Example 6: Simplify $x^7 \div x^3$
$x^7 \div x^3 = x^{7-3} = \mathbf{x^4}$

Example 7: Simplify $20c^7 \div 4c^2$
$= 20 \div 4 = 5, \quad c^7 \div c^2 = \mathbf{c^5}$
Therefore, $20c^7 \div 4c^2 = \mathbf{5c^5}$

Example 8: Simplify $\dfrac{5^8 \times 5^7}{5^4 \div 5^2}$

Numerator: $5^8 \times 5^7 = 5^{8+7} = 5^{15}$
Denominator: $5^4 \div 5^2 = 5^{4-2} = 5^2$

$$= \frac{5^{15}}{5^2} = 5^{15-2} = \mathbf{5^{13}}$$

NEGATIVE INDICES RULE

$$x^{-n} = \frac{1}{x^n}$$

Example 9: $4^{-2} = \dfrac{1}{4^2} = \mathbf{\dfrac{1}{16}}$

Example 10: $c^5 \div c^7$
$= c^{5-7} = c^{-2} = \mathbf{\dfrac{1}{c^2}}$

Example 11: $e^{-4} \div e^{-6}$
$= e^{-4--6} = \mathbf{e^2}$

POWERS OF ZERO

Any number raised to the power zero is equal to 1. $d^5 \div d^5 = d^{5-5} = d^0 = \mathbf{1}$ since any number divided by itself is **1.**

Therefore, $5^0 = 1$, $35^0 = 1$, $567^0 = 1$

RAISED POWERS

Example 1: Simplify $(3^2)^3$

This is the same as $3^2 \times 3^2 \times 3^2$
$= 3^{2+2+2} = \mathbf{3^6}$

Generally, for one power raised to another power, **multiply the indices**.
$(3^2)^3 = 3^{2 \times 3} = \mathbf{3^6}$

Example 2: Simplify $(5^{-2})^3$
This is $5^{-2 \times 3} = \mathbf{5^{-6}}$

Example 3: Simplify $(w^4)^5 \times w^7$
This is $w^{4 \times 5} \times w^7 = w20 \times w^7 = \mathbf{w^{27}}$

EXERCISE 1I

1) Write the following using index notation.

a) $5 \times 5 \times 5$
b) $6 \times 6 \times 6 \times 6 \times 6$
c) $13 \times 13 \times 13 \times 13 \times 13 \times 13 \times 13$
d) $4 \times 4 \times 4 \times 7 \times 7$

2) Evaluate these powers without using a calculator.

a) 1^2
b) 2^3
c) 17^2
d) 10^3
e) 20^0
f) 50^3
g) 22^2
h) 30^3

3) Evaluate the following.

a) $6^2 + 2^2$
b) $7^2 - 3^2$
c) $11^2 + 3^2 + 5^3$
d) $(6 - 2)^2$
e) $1^3 + 3^3 + 7^2$
f) $3^2 + 2^3 - 6^0$
g) $9^0 - 8^2$
h) $15^2 - 12^2$

4) Simplify these and leave your answers in index form.

a) $2^2 \times 2^6$
b) $4^2 \times 4^5$
c) $10^4 \times 10^5$
d) $17^2 \times 17^9$
e) $3^1 \times 3^0$
f) $9^8 \times 9$
g) $e^3 \times e^5$
h) $y^7 \times y^9$
i) $w \times w^4$
j) $3d^2 \times 2d^8$
k) $7h^3 \times 15h^5$
l) $y \times y \times y \times y$

5) Simplify these and leave your answers in index form.
a) $4^8 \div 4^5$
b) $3^9 \div 3^2$
c) $n^5 \div n^3$
d) $c^{13} \div c^7$
e) $y^9 \div y^3$
f) $12w^8 \div 4w^3$
g) $5^6 \div 5^2$
h) $15w^9 \div 5w$
i) $3d^2 \div d$
j) $(5^{-3})^2$

6) Write as an ordinary number.
a) 4^{-1} b) 3^{-1} c) 5^{-2} d) 13^{-3}

7) Simplify and write your answers in index form.
a) $5^3 \div 5^5$
b) $5^{-3} \div 5^{-2}$
c) $10^6 \times 10^{-4}$
d) $5^3 \times 5^4 \div 5^2$
e) $\dfrac{n^{12} \div n^7}{n^8 \div n^4}$
f) $\dfrac{8^4 \times 8^4}{8^5 \div 8^2}$
g) $\dfrac{11^{15}}{11^3 \times 11^6}$
h) $(15^{-3})^4$

8) Write **true** or **false**.
a) $5^2 \times 5^3 = 5^6$
b) $8^7 \div 8^2 = 8^5$
c) $12^{-1} = \frac{1}{12}$
d) $175^0 = 175$
e) $y^{12} \times y^0 = y^{12}$
f) $4^{-2} \div 4^{-4} = y^{-6}$

9) Simplify fully

a) $6x^2 \times 2x^4$
b) $25a^6 \div 5a^2$
c) $4xy \times 4xy$
d) $(5x)^2$
e) $\dfrac{88p^4y^2}{11p^2y}$

1.10 STANDARD FORM

Standard form is another way of writing very large or tiny numbers. However, most numbers can be written using standard form.

The standard form is based on our decimal system which is based on the powers of ten. Index notations are used when writing powers of ten.

Examples:
$10^1 = 10$
$10^2 = 10 \times 10 = 100$
$10^3 = 10 \times 10 \times 10 = 1000$
$10^4 = 10000$ ….and so on

Remember, $10^0 = 1$

Negative powers of 10 can also be used when writing decimals.
Examples:

$$10^{-1} = \frac{1}{10^1} = \frac{1}{10} = 0.1$$

$$10^{-3} = \frac{1}{10^3} = \frac{1}{1000} = 0.001$$

If we use a calculator to work out the answer to 600000 × 50000, the calculator may display the answer as

3E^{10} OR **3×10^{10}** OR **3^{10}**

Look at the number 8×10^{11}, the *11* means that 8 is followed by 11 zeros. This could be seen as 800 000 000 000.

Therefore, 8×10^{11}, 1.3×10^{-8}, 2×10^3… are all examples of numbers in standard form.

A number is in standard form if it is written in the form $\boldsymbol{a} \times \mathbf{10}^{\boldsymbol{n}}$ where $\boldsymbol{a}$ is a number between 1 and 10 and $\boldsymbol{n}$ is a positive or negative whole number.

In other words, numbers written in standard form must have **two parts**.

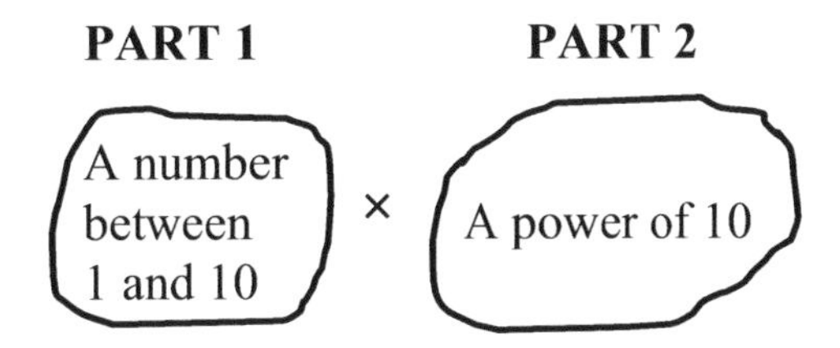

Examples
1) 2×10^3 is in standard form
2) 0.2×10^5 is **NOT** in standard form
3) $6.5^2 \times 10^{-7}$ is in standard form
4) 12×10^5 is **NOT** in standard form

Example 5: Write 235 in standard form.

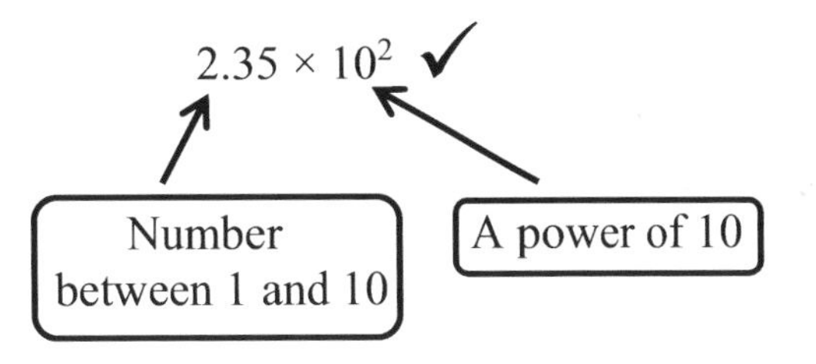

Example 6: Write the following numbers in standard form.

a) 35 b) 6000 c) 27500 d) 862.37

Answers:

a) 35 = $\mathbf{3.5 \times 10^1}$
b) 6000 = $\mathbf{6 \times 10^3}$
c) 27500 = $\mathbf{2.75 \times 10^4}$
d) 862.37 = $\mathbf{8.6237 \times 10^2}$

Example 7: Write these numbers
a) 2×10^3 b) 6.3×10^5 c) 3.261×10^9
as ordinary numbers.

Answers:
a) $2 \times 10^3 = 2 \times 1000 =$ **2000**
b) $6.3 \times 10^5 = 6.3 \times 100000 =$ **630000**
c) $3.261 \times 10^9 = 3.261 \times 1000000000$
$=$ **3261000000**

Example 8: Write these numbers in standard form. a) 0.002 b) 0.000023
c) 0.00405 d) 0.04007

Answers:
a) $0.002 = 2 \times 10^{-3}$
b) $0.000023 = 2.3 \times 10^{-5}$
c) $0.00405 = 4.05 \times 10^{-3}$
d) $0.04007 = 4.007 \times 10^{-2}$

Example 9: Write these numbers as ordinary numbers (decimal fractions)
a) 7×10^{-3} b) 9.5×10^{-5}
c) 6.203×10^{-7} d) 4.257×10^{-6}

Answers:
a) $7 \times 10^{-3} = \frac{7}{1000} = 0.007$
b) $9.5 \times 10^{-5} = \frac{9.5}{100000} = 0.000095$
c) $6.203 \times 10^{-7} = 0.0000006203$
d) $4.257 \times 10^{-6} = 0.000004257$

EXERCISE 1J

Write the following numbers in standard form.

1) 200
2) 500
3) 4000
4) 120000
5) 650
6) 2345
7) 3037
8) 10000
9)340000
10) 39000
11) 40
12) 54
13) 989
14) 1345
15) three thousand
16) sixty thousand
17) 1 billion
18) 7893000
19) 9050
20) 9999

21) The population of Nigeria is estimated to be 210 000 000. Write this in standard form.

22) In chemistry, a hydrogen atom weighs about 0.000 000 000 000 000 000 000 00168 g. Write this number in standard form.

23) The speed of a car is 200 km/s. Write this in **m/s** and standard form.

24) Which of the numbers below are in standard form?

a) 24×10^3 b) 3.5×10^{-2} c) 485×10^{-3}

25) Write 4 trillion in standard form.

EXERCISE 1K

Write the following numbers in standard form.

1) 0.02
2) 0.003
3) 0.00005
4) 0.000123
5) 0.0809
6) 0.000007
7) 0.00064
8) 0.0000444
9) 0.012
10) 0.045
11) 0.00323
12) 0.009
13) 0.000084
14) 0.06128

15) Write these numbers as ordinary numbers (decimal fractions).
a) 1.8×10^{-5}
b) 6.7×10^{-3}
c) 2.05×10^{-4}
d) 7.3×10^{-2}
e) 8.9×10^{-1}
f) 5.02×10^{-3}

ORDERING NUMBERS IN STANDARD FORM

To order numbers means to write them in order of size (smallest to highest or highest to lowest). However, it could be difficult when dealing with negative powers (tiny numbers).

RULE 1
Start by comparing the powers of 10. Know that 10^{-7} is smaller than 10^{-6} and 10^{-5} is lower than 10^{-4}, and so on.

Therefore, 3.5×10^{-7} is smaller than 2.4×10^{-5}.

RULE 2
If the numbers have the same powers of 10, compare the number part of the standard form.

6.7×10^{-3} is bigger than 5.2×10^{-3} and 2.3×10^{-11} is bigger than 1.2×10^{-11} since they have the same powers of 10.

Example 1: Write these numbers in order from smallest to highest.
4.3×10^{-2}, 6.2×10^{-7}, 2.6×10^{-5}, 3.4×10^{-2}

Solution:
10^{-7} is the smallest, so 6.2×10^{-7} is the smallest number. This is then followed by 2.6×10^{-5}, then 3.4×10^{-2} and finally 4.3×10^{-2}.

From smallest to highest, the numbers are **6.2×10^{-7}, 2.6×10^{-5}, 3.4×10^{-2} and 4.3×10^{-2}**

CALCULATIONS IN STANDARD FORM

To multiply numbers in standard form, the ***number*** parts are multiplied together and then the powers of ten after.

Example 1:
Work out $(3 \times 10^4) \times (4 \times 10^5)$ and leave your answer in standard form.

$10^4 \times 10^5 = 10^9$

$(3 \times 10^4) \times (4 \times 10^5)$

$3 \times 4 = 12$

This becomes 12×10^9.
However, 12×10^9 is **not** in standard form since the first part must be a number between 1 and 10 (exclusive) and 12 is more than 10.
In standard form, $12 \times 10^9 = \mathbf{1.2 \times 10^{10}}$

Example 2:
Work out $(3.5 \times 10^3) \times (2.5 \times 10^{11})$ and leave your answer in standard form.

$(3.5 \times 2.5) \times (10^3 \times 10^{11}) = \mathbf{(8.75 \times 10^{14})}$

Example 3:
Work out $(2.4 \times 10^{-5}) \div (4.8 \times 10^6)$

$$\left(\frac{2.4}{4.8}\right) \frac{\times}{\times} \left(\frac{10^{-5}}{10^6}\right)$$

$= (2.4 \div 4.8) \times (10^{-5} \div 10^6)$
$= 0.5 \times 10^{-11}$

In standard form, $0.5 \times 10^{-11} = \mathbf{5 \times 10^{-12}}$

Example 4: Work out $(3.5 \times 10^{-5})^2$ and leave your answer in standard form.

Remember: $(3.5 \times 10^{-5})^2$
$= (3.5 \times 10^{-5}) \times (3.5 \times 10^{-5})$
$= (3.5 \times 3.5) \times (10^{-5} \times 10^{-5})$
$= 12.25 \times 10^{-10}$
In standard form,
$12.25 \times 10^{-10} =$ **1.225×10^{-9}**

Example 5: Express the answer to 2 700 × 3 000 000 in standard form

First convert to standard form.
$2\ 700 = 2.7 \times 10^3$, $3\ 000\ 000 = 3 \times 10^6$
It becomes $(2.7 \times 3) \times (10^3 \times 10^6)$
= **8.1×10^9**

ADDING AND SUBTRACTING IN STANDARD FORM

When adding or subtracting in standard form, work out the calculations in the brackets and then add or subtract.

Example 6
Work out $(3.2 \times 10^3) + (4.1 \times 10^2)$

↙ ↓

3200 + 410

$3200 + 410 = 3610$
In standard form, **3.61×10^3**

Example 7
Work out $(5.2 \times 10^{-3}) - (2.7 \times 10^{-4})$

↙ ↓

0.0052 - 0.00027

$0.0052 - 0.00027 = 0.00493$
In standard form = **4.93×10^{-3}**

EXERCISE 1L

1) Calculate the following and give your answer in standard form.

a) 10×200
b) 400×7
c) 35×10
d) 12×5
e) 500×50
f) $(40)^2$
g) $(1.3 \times 10^5) \times (2.2 \times 10^3)$
h) $(7.5 \times 10^{13}) \times (3 \times 10^{-4})$
i) $(20)^3$
j) $(2.3 \times 10^{-5}) \times (1.2 \times 10^{-6})$

2) If $c = 3 \times 10^6$ and $d = 2 \times 10^{11}$, find
a) cd b) $c \div d$ c) $c + d$ d) $d - c$

3) These numbers are in standard form; arrange them in order of size (smallest first)

a) 2.3×10^5, 8.6×10^4, 1.5×10^8
b) 6.5×10^{-3}, 2.5×10^{-6}, 3.7×10^{-3}, 6.9×10^{-3}
c) 6.1×10^{-9}, 7.1×10^{-9}, 4.5×10^{-6}, 3.4×10^{-6}

4) Work out the following and write your answer in standard form.

a) $40000 \div 20$
b) $6000 \div 12$
c) $(4.8 \times 10^4) \div (1.2 \times 10^6)$
d) $(3 \times 10^{-5}) \div (1.5 \times 10^4)$
e) $(9 \times 10^{-2}) \div (3 \times 10^{-3})$
f) $(2 \times 10^{-5}) \times 10^{-7}$
g) $(4 \times 10^{-4}) + (3.5 \times 10^{-6})$
h) $(8.8 \times 10^{-3}) - (2.2 \times 10^{-5})$

5) The radius of the sun is 1.39×10^6 km approximately.
a) Find the diameter of the sun.
b) Find the surface area if the sun is considered a sphere. 2.43×10^{13} km^2

6) A computer part is rectangular.
Calculate the area of the part.

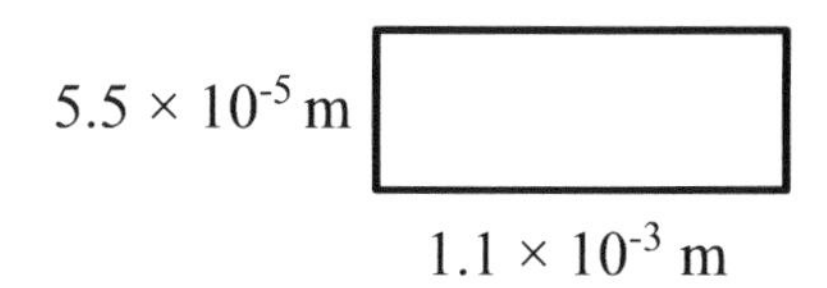

7) The mass of the Earth is 5 973 600,000 000 000 000 000 000 kg.

a) Write this in standard form

b) The mass of Venus is 4.87×10^{24} kg, work out the difference in mass between Venus and Earth.
Leave your answer in standard form.

c) If the radius of Venus is 6 052 km, work out the surface area if it is spherical.

d) If the mass of the Sun is 333000 times mass of the Earth, calculate the mass of the Sun.

8)

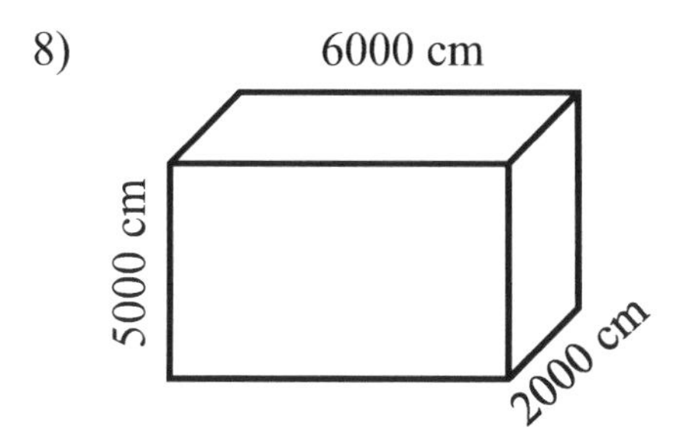

a) Calculate the volume of the cuboid.

b) Write your answer to part **a** in standard form.

c) Calculate the area of the base of the cuboid and leave your answer in standard form.

Chapter 1 Review Section
Assessment

1) Write down the first five multiples of

a) 3 .... **1 mark**

b) 4 ....**1 mark**

c) 14 .... **1 mark**

2) Write down all the factors of

a) 13 ....**1 mark**

b) 36 .... **1 mark**

c) 104 .... **1 mark**

3) From the cloud, write down the number(s) that are

a) factors of 35

b) factors of 63

c) multiples of 6

d) prime numbers

e) multiples of 92

.... **5 marks**

4) Write 250 as a product of prime factors in index form. .... **3 marks**

5) Work out the HCF of

a) 27 and 63 .... **2 marks**

b) 400 and 500 .... **3 marks**

6) Work out the LCM of

a) 2 and 7 .... **2 marks**

b) 35 and 50 .... **2 marks**

c) 182 and 420 .... **3 marks**

7) Using the Venn diagram method, find the LCM and HCF of

a) 66 and 132 .... **5 marks**

b) 420 and 560 .... **5 marks**

8) Find the HCF of the following and leave your answers in prime factors and index notation.

a) $2 \times 2 \times 3 \times 3 \times 3 \times 5 \times 5 \times 5$ and $2 \times 3 \times 3 \times 3 \times 3 \times 5$ .... **1 mark**

b) $7 \times 7 \times 7 \times 7 \times 11 \times 11$, $7 \times 7 \times 13 \times 13 \times 13 \times 13$ and

$7 \times 11 \times 13 \times 13$ .... **1 mark**

9) Azubuike says "The Lowest common multiple of two numbers cannot be one of the numbers."

Is Azubuike correct? Explain fully.

.... **2 marks**

10) The Lowest common multiple of 8 and another number is 72. If the other number is more than 8 but less than 15, what is the other number?

………. **2 marks**

11) Tolu, Nnamdi and Okechukwu started at the same time to ring a school bell.

Tolu rings a bell every 3 seconds.

Nnamdi rings a bell every 5 seconds.

Okechukwu rings a bell every 11 seconds.

How long will it take before they ring the bell at the same time?

……… **3 marks**

12) a) Write 54 as a product of its prime factors. ……… **2 marks**

b) Express your answer in index form ……… **1 mark**

c) Find the HCF of 54 and 126 ……… **2 marks**

13)

Dictionaries are sold in packs of 12.

Scrabbles are sold in packs of 20.

A shop wants to buy the same number of dictionaries and scrabbles.

What is the lowest number of packs of dictionaries and scrabbles they could buy?

……… **3 marks**

14) Henry says "1 is a prime number and 2 is not."
Is Henry correct?

Explain fully. .... **3 marks**

15) Write down the next **five** prime numbers larger than 36.

.... **2 marks**

16) Draw a factor tree for the following numbers.

a) 16 .... **2 marks**

b) 144 .... **2 marks**

17) Is the number 189357 divisible by 3?

Explain your answer.

.... **3 marks**

18) Write down the prime factor of 750. .... **2 marks**

19) From these set of numbers

12 **16** **32** **33** **47** **54**

Write down

a) a prime number .... **1 mark**

b) a multiple of 9 .... **1 mark**

c) a multiple of 8 .... **1 mark**

d) a factor of 144 .... **1 mark**

e) a square number .... **1 mark**

20) Copy and complete the table.

x	1	4	9	16	25	144
x^2						
$\sqrt{x}$						
$-\sqrt{x}$						

.... **18 marks**

21) Which number in each pair is greater?

a) 2^2 or $(-3)^2$ .... **1 mark**
b) 7^2 or $(-7)^2$ .... **1 mark**
c) $\sqrt{144}$ or 4^2 .... **1 mark**
d) $\sqrt[3]{216}$ or 2^3 .... **1 mark**
e) 0.3^2 or 0.3^3 .... **1 mark**

22) Simplify these and leave your answer in index form.

a) $7^2 \times 7^8$ .... **1 mark**
b) $3^6 \times 3^2$ .... **1 mark**
c) $12^3 \div 12^7$ .... **1 mark**

d) $\dfrac{(4^5)^2 \div (4^2)^5}{(4^3)^2 \times 4}$ .... **3 marks**

e) $a^{-5} \times a^{-11}$ .... **1 mark**

23) Write the following in standard form.

a) 345
b) 28000
c) 8980000000
d) 0.0012
e) 0.0000098
f) 0.405
g) 8732
h) 6 billion
i) 0.0000915

.... **9 marks**

24) Write as decimal numbers.

a) 6×10^3
b) 4.2×10^{-1}
c) 7×10^{-5}
d) 6.6×10^{-6}
e) 9.4×10^{-4}
f) 1.4×10^{-2}

.... **6 marks**

25) Work out the following and write your answer in standard form.

a) $(7.6 \times 10^5) \times (2 \times 10^3)$
b) $(4.3 \times 10^{-2}) \times (3.5 \times 10^{-4})$
c) $(8.4 \times 10^4) \div (4.2 \times 10^6)$
d) $(5.5 \times 10^{-7}) \div (1.1 \times 10^{-3})$
e) $(9.2 \times 10^8) + (4 \times 10^7)$
f) $(5.9 \times 10^4) - (2.3 \times 10^3)$

.... **12 marks**

2 Fractions

In this section, we shall consider the following:

- Types of fractions
- Equivalent fractions
- Simplifying fractions
- Adding and subtraction fractions
- Multiplying and dividing fractions
- The fraction of an amount

LEARNING OBJECTIVES

By the end of this unit, you should be able to

a) understand fractions

b) work out equivalent fractions

c) simplify fractions to their lowest form

d) add and subtract fractions including mixed numbers

e) multiply and divide fractions including mixed numbers

f) work out a fraction of an amount

KEYWORDS

- Fractions
- Numerator
- Denominator
- Add and Subtract
- Mixed number
- Simplify
- Divide and multiply

2.1 UNDERSTANDING FRACTIONS

Fractions usually have two parts, the **numerator** and **denominator**.

The numerator is the top number while the denominator is the bottom number in a fraction.

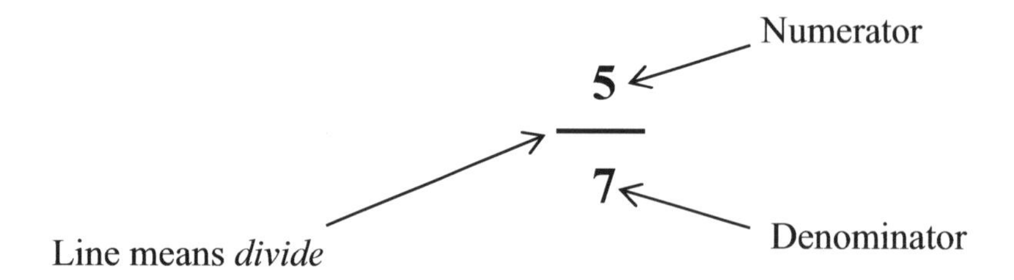

When a whole number is divided into equal parts, each of the parts is a fraction of the whole. Fractions could also be written with a slash (/) instead of a horizontal line (—) like 2/3, 4/7, 5/7.

2.2 TYPES OF FRACTIONS

There are three types of fractions at this level.

a) Proper Fractions

b) Improper fractions

c) Mixed Numbers

PROPER FRACTIONS

In a proper fraction, the numerator (top number) is less than the denominator (bottom number).

Examples of proper fractions: $\frac{2}{3}$ $\frac{4}{7}$ $\frac{12}{15}$

IMPROPER FRACTIONS

In an improper fraction, the numerator is bigger than the denominator.

Examples of improper fractions: $\frac{4}{3}$ $\frac{15}{4}$ $\frac{24}{7}$

MIXED NUMBERS

A whole number combined with a fraction is called a mixed number.

Examples of mixed numbers: $1\frac{2}{3}$ $7\frac{3}{4}$ $11\frac{1}{7}$

2.3 CONVERSION FROM MIXED NUMBERS TO IMPROPER FRACTIONS

Example 1: Convert $2\frac{3}{4}$ to an improper fraction.

- Multiply the denominator (bottom number) by the whole number
- Add the outcome to the numerator
- Finally, divide the result by the denominator **(which stays the same)**

$4 \times 2 = 8$

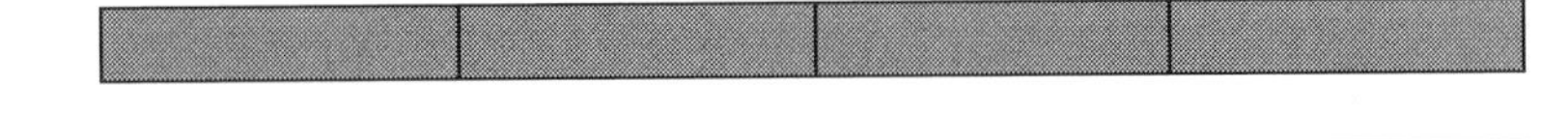

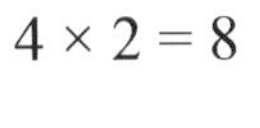

$8 + 3 = 11$

$\frac{11}{4}$ ✓

Example 2: Change $5\frac{1}{2}$ to an improper fraction.

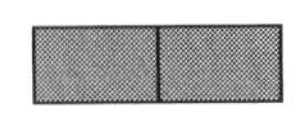

$2 \times 5 = 10$

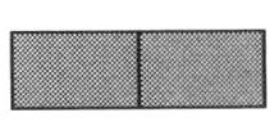

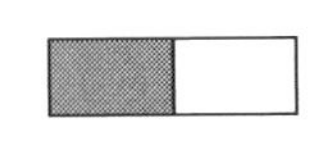

$10 + 1 = 11$

$\frac{11}{2}$ ✓

2.4 CONVERSION FROM IMPROPER FRACTION TO A MIXED NUMBER

Example 1: Convert $\frac{9}{4}$ to a mixed number.

This is a normal division with the remainder on top as the numerator.

Four (4) goes into 9 two (**2**) times, remainder one (**1**) ⟶ $2\frac{1}{4}$ ✓

Example 2: Convert $\frac{29}{8}$ to a mixed number.

8 goes into 29 three (**3**) times remainder five (**5**) ⟶ $3\frac{5}{8}$ ✓

2.5 SHADED FRACTIONS

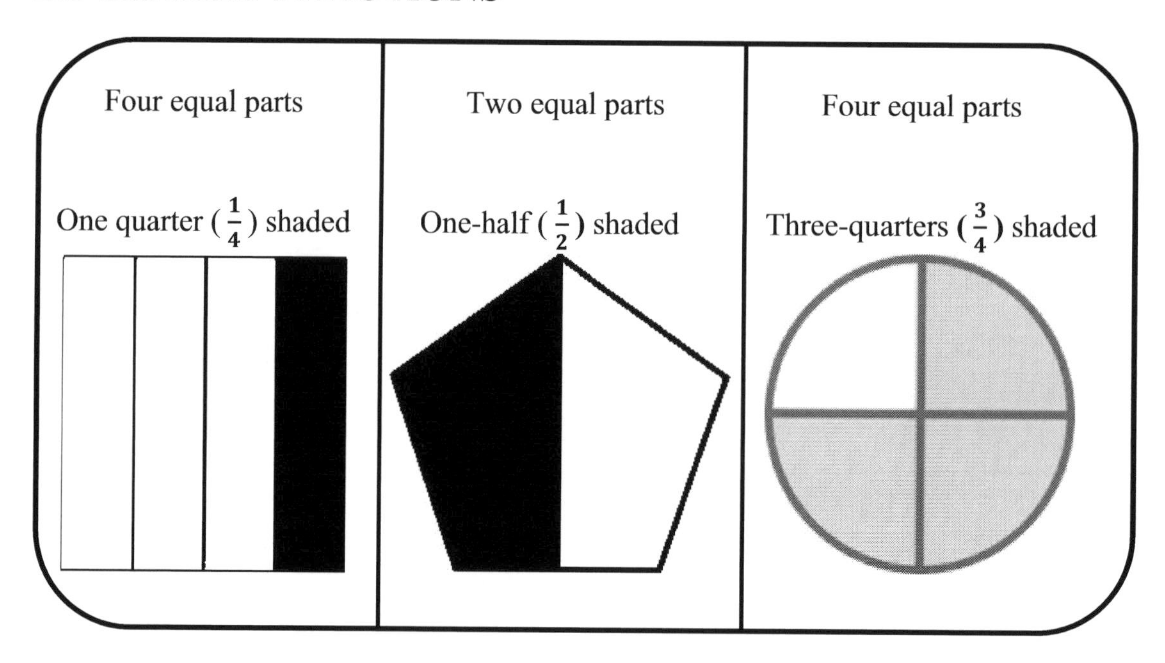

2.6 EQUIVALENT FRACTIONS

These are fractions that have the same value though they may look different.
If **the same number multiplies both the numerators and denominators**, the value of the fraction remains the same provided that the number is not zero (0).

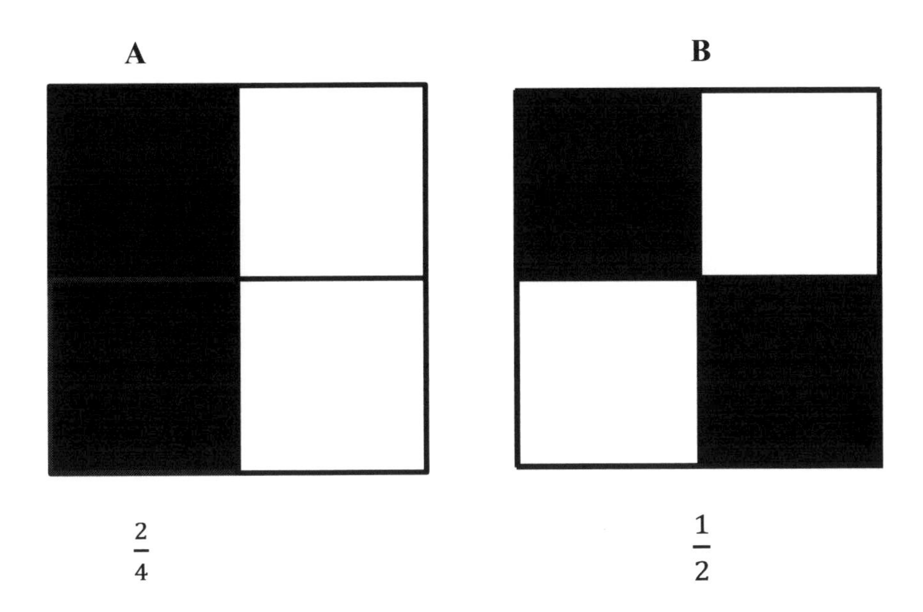

Shapes A and B are the same and are called equivalent fractions. They look different but are the same.

How to form equivalent fractions

$\frac{2}{3}$ is equivalent to $\frac{2}{3} \times \frac{4}{4} = \frac{8}{12}$ .... **Example 1**

$\frac{5}{11}$ is equivalent to $\frac{5}{11} \times \frac{3}{3} = \frac{15}{33}$ .... **Example 2**

Also, equivalent fractions can be obtained by **dividing** the numerator and the denominator by the same number provided that number is not zero (0).

$\frac{16}{20}$ is equivalent to $\frac{16 \div 4}{20 \div 4} = \frac{4}{5}$ .... **Example 3**

Example 4: Draw a pictorial diagram to show that $\frac{3}{5} = \frac{9}{15}$

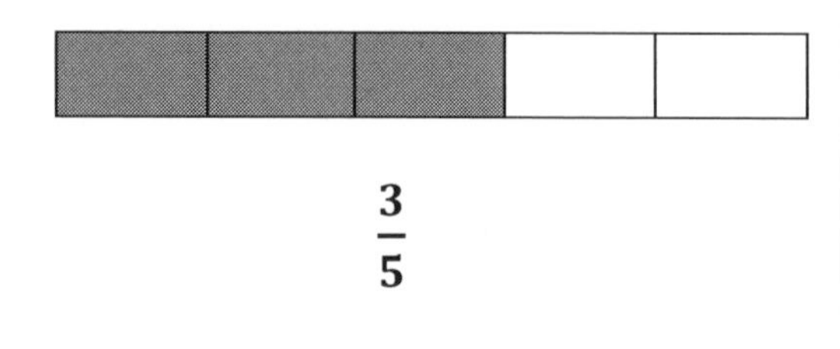

$\frac{3}{5}$

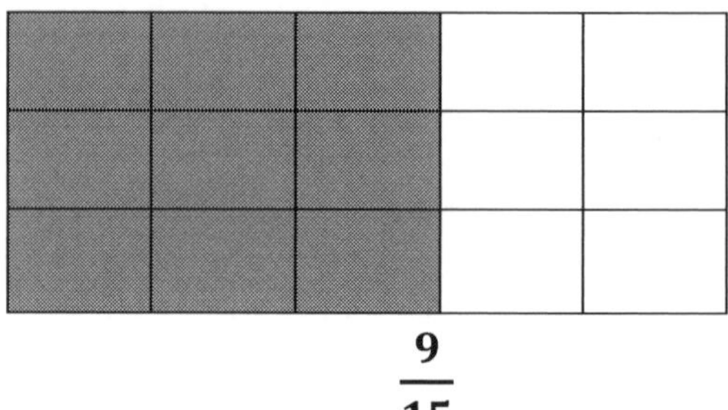

$\frac{9}{15}$

Example 5: Fill in the missing numbers.

a) $\frac{1}{3} = \frac{5}{\square}$ b) $\frac{3}{7} = \frac{\square}{21}$ c) $\frac{12}{18} = \frac{\square}{3}$

Solutions

a) $\frac{1}{3} = \frac{5}{\boxed{15}}$ (numerator × 5, denominator × 5)

b) $\frac{3}{7} = \frac{\boxed{9}}{21}$ (numerator × 3, denominator × 3)

c) $\frac{12}{18} = \frac{\boxed{2}}{3}$ (numerator ÷ 6, denominator ÷ 6)

Note: Whatever you do to the top, you must do to the bottom. In equivalent fractions, × or ÷ is used to find any missing number.

Example 6: Arrange the following fractions in order of size, smallest first.

$$\frac{2}{3}, \frac{1}{5}, \frac{5}{6} \text{ and } \frac{7}{10}$$

The easiest way to compare fractions is to make the denominators the same using equivalent fractions. We also use our knowledge of the lowest common multiple.

The LCM of 3, 5 and 6 is 30.
Therefore, we find equivalent fractions with 30 as the denominator.

$$\frac{2 \times \mathbf{10}}{3 \times \mathbf{10}} = \frac{\mathbf{20}}{\mathbf{30}}$$

$$\frac{1 \times \mathbf{6}}{5 \times \mathbf{6}} = \frac{\mathbf{6}}{\mathbf{30}}$$

$$\frac{5 \times \mathbf{5}}{6 \times \mathbf{5}} = \frac{\mathbf{25}}{\mathbf{30}}$$

$$\frac{7 \times \mathbf{3}}{10 \times \mathbf{3}} = \frac{\mathbf{21}}{\mathbf{30}}$$

Since denominators of the equivalent fractions are now 30, the fraction with the lowest numerator is the lowest fraction. Likewise, the fraction with the highest numerator is the highest fraction.

Therefore, the order of the size of smallest to biggest is $\frac{1}{5}, \frac{2}{3}, \frac{7}{10}$ and $\frac{5}{6}$.

EXERCISE 2A

In questions **1** to **6**, write down the fractions shaded.

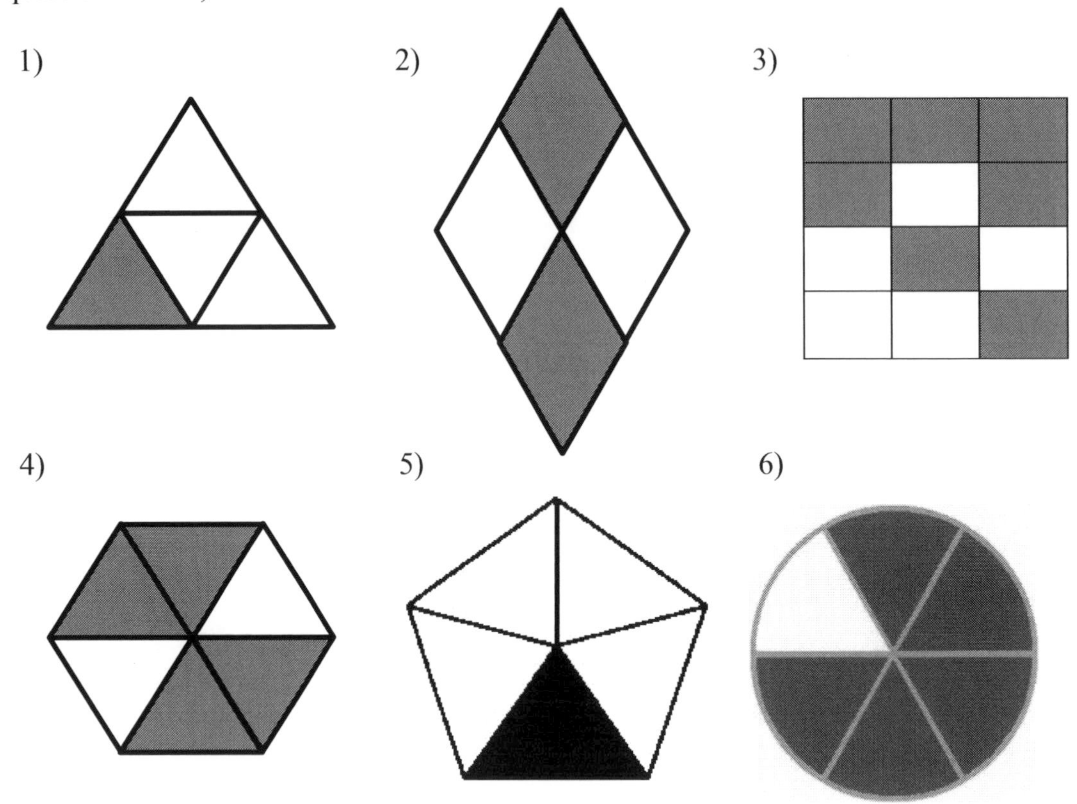

7) Copy and shade $\frac{1}{3}$ of the shape below.

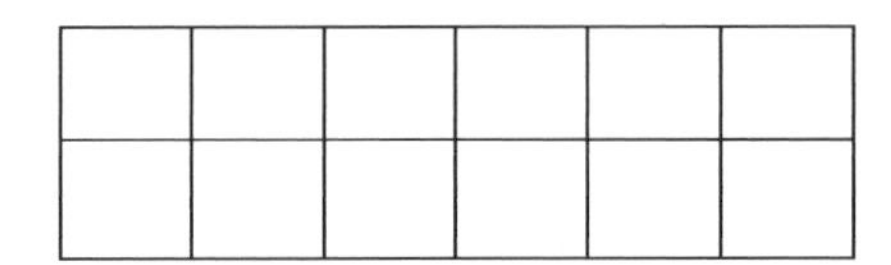

8) Copy and shade $\frac{1}{4}$ of the shape in number 7 above.

9) Copy and shade $\frac{3}{4}$ of the shape in question 7 above.

Change each of these improper fractions to mixed numbers.

10) $\frac{3}{2}$ 11) $\frac{14}{6}$ 12) $\frac{35}{11}$ 13) $\frac{122}{40}$

Change each of these mixed numbers to improper fractions.

14) $1\frac{3}{5}$ 15) $2\frac{1}{3}$ 16) $6\frac{2}{5}$ 17) $9\frac{5}{7}$

In questions 18 to 28, copy and complete the equivalent fractions.

18) $\frac{3}{4} = \frac{9}{?}$

19) $\frac{1}{2} = \frac{7}{?}$

20) $\frac{2}{5} = \frac{?}{15}$

21) $\frac{7}{9} = \frac{?}{18}$

22) $\frac{7}{10} = \frac{?}{30}$

23) $1 = \frac{?}{5}$

24) $\frac{3}{5} = \frac{21}{?}$

25) $\frac{30}{48} = \frac{?}{8}$

26) $\frac{7}{35} = \frac{?}{5}$

27) $\frac{1}{2} = \frac{?}{4} = \frac{?}{8}$

28) $\frac{1}{3} = \frac{?}{9} = \frac{?}{12} = \frac{?}{15}$

2.7 SIMPLIFYING FRACTIONS

When it is ***impossible*** to find a number which is not 1, that will divide exactly into the numerator (top number) and denominator (bottom number), the fraction is said to be in its **simplest form or lowest term.**

The fractions $\frac{2}{3}, \frac{7}{11}, \frac{12}{13}$ are in their simplest forms or lowest terms because no number can divide both the numerator and denominator exactly.

Advice: To simplify a fraction to its lowest form, always divide the fraction by their **highest common factor** (HCF).

Example 1:

Is the fraction $\frac{4}{8}$ in its lowest form?

NO, because 2 or 4 can divide both numbers. In its simplest form,

$$\frac{4 \div 4}{8 \div 4} = \frac{1}{2}$$

$\frac{1}{2}$ is the lowest form of $\frac{4}{8}$

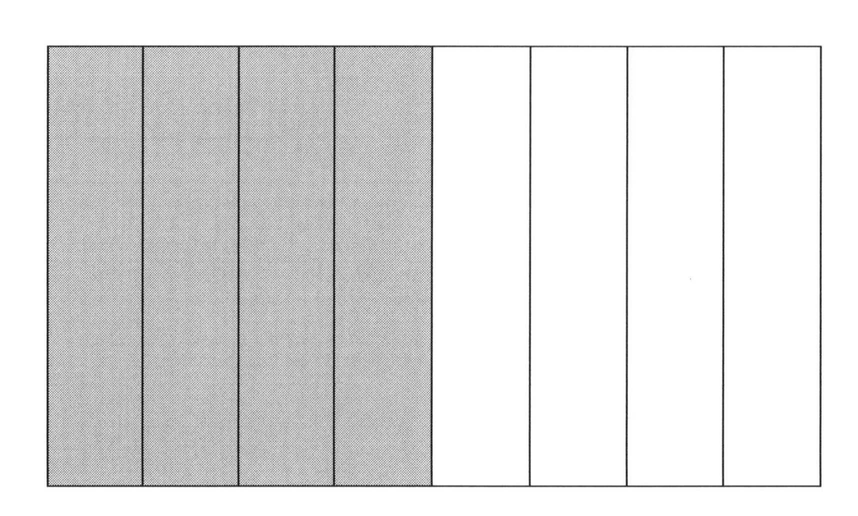

Example 2: Reduce $\frac{25}{30}$ to its lowest form.

Divide both numbers by 5.

$$\frac{25 \div \mathbf{5}}{30 \div \mathbf{5}} = \frac{\mathbf{5}}{\mathbf{6}}$$

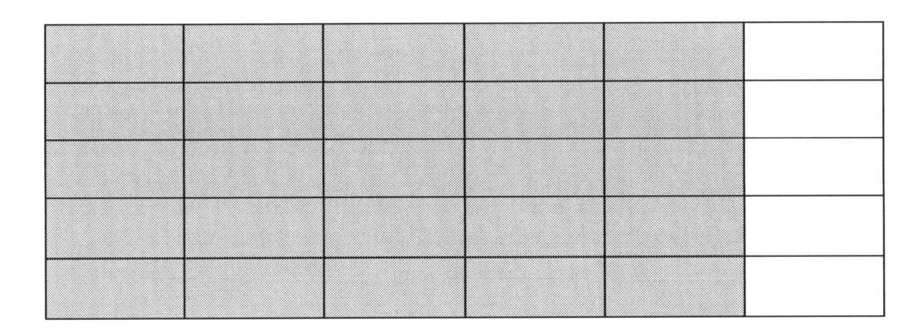

Example 3:

Write $\frac{16}{40}$ in its lowest form.

2, 4 and 8 are the common factors of 16 and 40. We may use any of the common factors to divide.

$\frac{16 \div \mathbf{2}}{40 \div \mathbf{2}} = \frac{8}{20}$ ↑ *Not in the simplest form*

$\frac{8 \div \mathbf{2}}{20 \div \mathbf{2}} = \frac{4}{10}$ ↑ *not in the simplest form*

$\frac{4 \div \mathbf{2}}{10 \div \mathbf{2}} = \frac{\mathbf{2}}{\mathbf{5}}$ ✓ ↑ *in the simplest form*

It took three calculations to get to the fraction in its lowest form.

However, it is advisable to use the **highest common factor** when dividing both numbers. The answer will be obtained quicker in that way. Dividing by 2 or 4 will still give a correct answer, but more calculations are needed.

The highest common factor, in this case, is 8.

Use 8 to divide both numbers.

$$\frac{16 \div 8}{40 \div 8} = \frac{2}{5} \checkmark$$

In one simple calculation, $\frac{16}{40}$ to its lowest form is $\frac{2}{5}$

2.8 ADDING AND SUBTRACTING FRACTIONS

Before fractions can be added or subtracted, the **denominators** (bottom numbers) must be the same. If they are the same, simply add the numerators and divide by one of the denominators (since they are the same). Remember, do not add the denominators.

Example 1: $\frac{2}{7} + \frac{4}{7} = \frac{2+4}{7} = \frac{6}{7}$ ✓

(the Same denominator, so add the numerators)

Also, when subtracting, $\frac{2}{5} - \frac{1}{5} = \frac{2-1}{5} = \frac{1}{5}$ ✓

(the Same denominator, so subtract the numerators)

WHEN DENOMINATORS ARE NOT THE SAME

When fractions have different denominators, we **must** make them the same by finding a common denominator, preferably, the lowest common multiple (LCM). Then write each fraction as an equivalent fraction and perform the given calculation(s).

See sections 4.1.1 for LCM and 5.6 for equivalent fractions.

Example 2: $\frac{1}{3} + \frac{2}{5}$

Since the denominators 3 and 5 are not the same, we must make them the same before adding. The LCM of 3 and 5 is 15, so we make the denominators equal 15.

Find the equivalent fractions using 15 as the denominator.

$$\frac{1 \times \mathbf{5}}{3 \times \mathbf{5}} + \frac{2 \times \mathbf{3}}{5 \times \mathbf{3}}$$

$$= \frac{5}{15} + \frac{6}{15}$$

$$= \frac{5 + 6}{15}$$

$$= \mathbf{\frac{11}{15}} \checkmark$$

Example 3: $\frac{3}{4} - \frac{3}{5}$

$$\frac{3 \times \mathbf{5}}{4 \times \mathbf{5}} - \frac{3 \times \mathbf{4}}{5 \times \mathbf{4}}$$

$$= \frac{15}{20} - \frac{12}{20}$$

$$= \mathbf{\frac{3}{20}} \checkmark$$

EXERCISE 2B

1) Write each fraction in its simplest form.

a) $\frac{10}{20}$

b) $\frac{14}{21}$

c) $\frac{6}{18}$

d) $\frac{6}{33}$

e) $\frac{10}{40}$

f) $\frac{4}{18}$

g) $\frac{10}{48}$

h) $\frac{55}{88}$

i) $\frac{5}{15}$

j) $\frac{12}{40}$

k) $\frac{20}{75}$

l) $\frac{44}{80}$

m) $\frac{50}{250}$

n) $\frac{200}{500}$

o) $\frac{44}{44}$

2) Make both fractions below, so they have the same denominator.

a) $\frac{4}{5}$ and $\frac{2}{3}$

b) $\frac{1}{6}$ and $\frac{4}{5}$

c) $\frac{3}{4}$ and $\frac{1}{2}$

3) Write each of these fractions as a mixed number.

a) Eight fifths

b) Twenty-nine sevenths

c) Fifty-five thirds

d) Eleven sixths

e) $\frac{9}{5}$

f) $\frac{17}{4}$

g) $\frac{142}{13}$

h) $\frac{200}{15}$

4) Work out each of the following calculations in its simplest form. Write as a mixed number when necessary.

a) $\frac{1}{9} + \frac{3}{9}$

b) $\frac{2}{7} + \frac{1}{7}$

c) $\frac{4}{6} + \frac{1}{6}$

d) $\frac{1}{10} - \frac{1}{10}$

e) $\frac{1}{3} + \frac{1}{4}$

f) $\frac{1}{5} + \frac{1}{9}$

g) $\frac{2}{3} + \frac{1}{5}$

h) $\frac{1}{9} - \frac{1}{9}$

i) $\frac{7}{9} - \frac{1}{5}$

j) $\frac{2}{8} + \frac{3}{5}$

k) $\frac{1}{2} + \frac{1}{7}$

l) $\frac{4}{7} + \frac{1}{10}$

m) $\frac{1}{2} + \frac{1}{3} + \frac{1}{4}$

n) $\frac{2}{3} + \frac{3}{5} - \frac{1}{6}$

o) $\frac{4}{5} - \frac{1}{4} + \frac{3}{5}$

5) Copy and complete

a) $\frac{1}{3} + \square = 1$

b) $\frac{1}{9} + \square = 1$

c) $1 - \frac{2}{3} = \square$

d) $\frac{15}{17} + \square = 1$

e) $\square + \frac{6}{13} = 1$

f) $1 - \square = \frac{15}{19}$

6) Write the fractions below in the order of size, largest first.

a) $\frac{1}{2}, \frac{2}{3}, \frac{5}{6}$

b) $\frac{7}{8}, \frac{1}{4}, \frac{1}{2}$

c) $2\frac{1}{3}, 3\frac{2}{3}, 2\frac{2}{3}$

7) What fraction is shaded

a) red

b) green

c) yellow

d) blue?

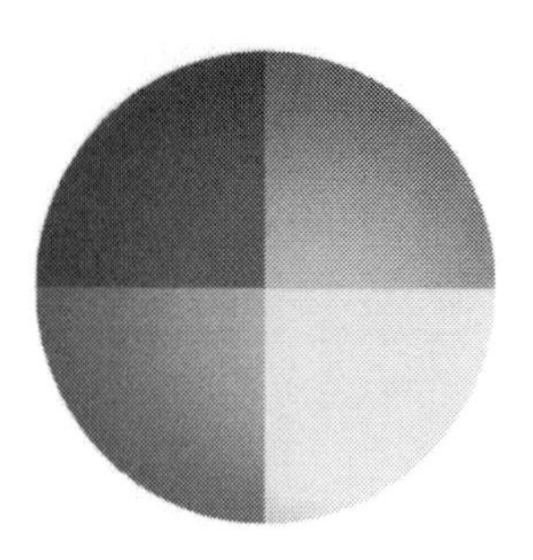

8) Funmi gave $\frac{1}{3}$ of her books to Abubakar and $\frac{1}{5}$ of her books to Chichi. What fraction of Funmi's book does she still have?

9) Five mathematics books are placed side by side and the width shown.

a) What is the total width, ***b metre*** of the five books?

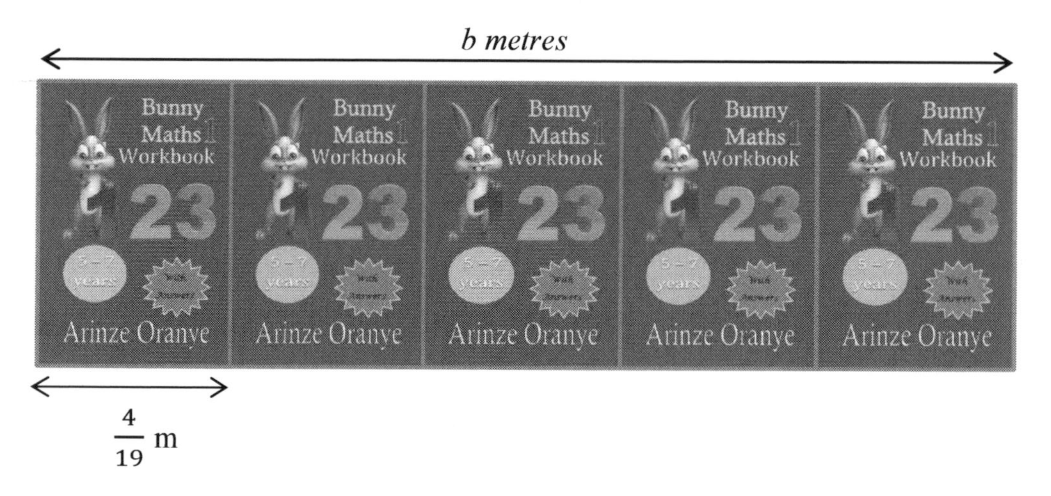

b) A student bought two of the books. What is the combined width of the remaining books?

10) All are equivalent fractions apart from one. Pick the odd one out and give a reason.

$$\frac{4}{15} \qquad \frac{8}{30} \qquad \frac{5}{45} \qquad \frac{20}{75} \qquad \frac{40}{150}$$

11) Complete the missing numbers in the boxes.

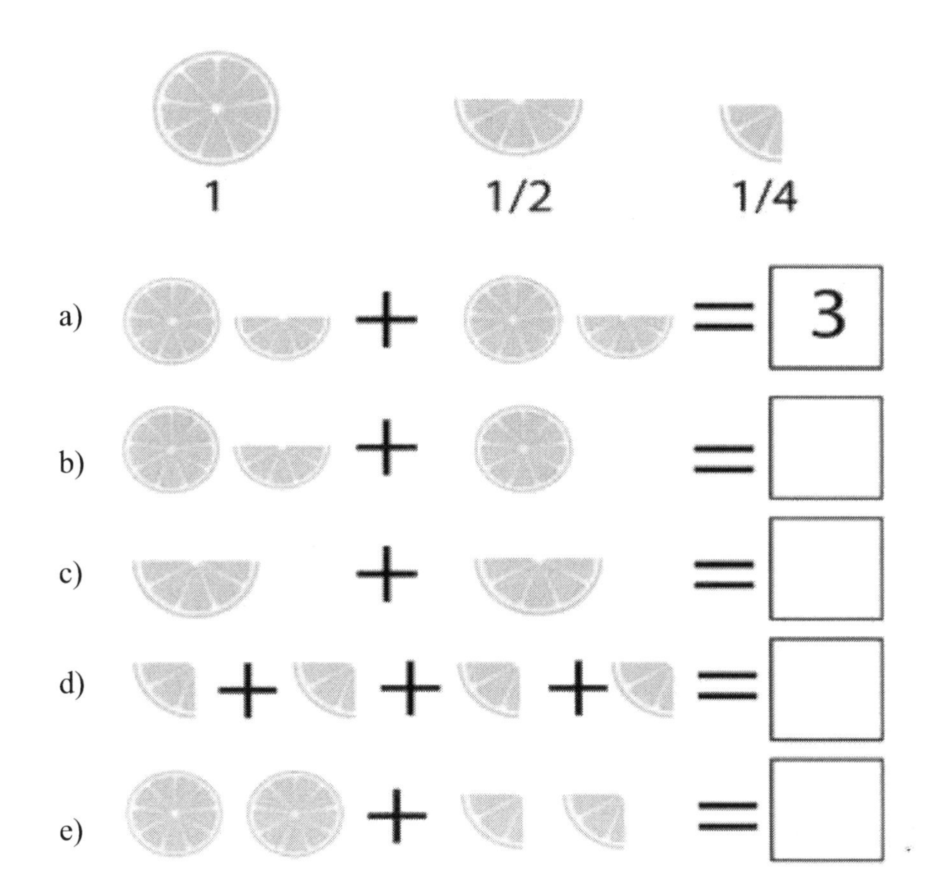

ADDING AND SUBTRACTING MIXED NUMBERS

Example 1: Work out $2\frac{1}{2} + 3\frac{1}{3}$

Solution: Change the mixed numbers to improper fractions (Refer to section 5.3).

For mixed number $2\frac{1}{2}$, change to $\frac{(2 \times 2) + 1}{2} = \frac{5}{2}$

For mixed number $3\frac{1}{3}$, change to $\frac{(3 \times 3) + 1}{3} = \frac{10}{3}$

Therefore, $\frac{5}{2} + \frac{10}{3}$

The denominators are not the same, so we make them the same.

$$= \frac{5 \times \mathbf{3}}{2 \times \mathbf{3}} + \frac{10 \times \mathbf{2}}{3 \times \mathbf{2}}$$

$$= \frac{15}{6} + \frac{20}{6} = \frac{35}{6}$$

However, the answer is top heavy (improper fraction) and should be written as a mixed number (refer to section 2.4).

Therefore, $\frac{35}{6} = 5\frac{5}{6}$

$2\frac{1}{2} + 3\frac{1}{3} = \mathbf{5\frac{5}{6}}$ ✓

A quicker method is to add up the whole numbers and fractions separately.

$2 + 3 = \mathbf{5}$

$$\frac{1}{2} + \frac{1}{3} = \frac{1 \times \mathbf{3}}{2 \times \mathbf{3}} + \frac{1 \times \mathbf{2}}{3 \times \mathbf{2}} = \frac{3}{6} + \frac{2}{6} = \mathbf{\frac{5}{6}}$$

Therefore, 5 and $\frac{5}{6}$ gives $\mathbf{5\frac{5}{6}}$ ✓ ....same answer.

Example 2: Work out $4\frac{1}{2}$ - $3\frac{1}{3}$

Solution: Change the mixed numbers to improper fractions (Refer to section 5.3).

For mixed number $4\frac{1}{2}$, change to $\frac{(2 \times 4) + 1}{2} = \frac{9}{2}$

For mixed number $3\frac{1}{3}$, change to $\frac{(3 \times 3) + 1}{3} = \frac{10}{3}$

Therefore, $\frac{9}{2} - \frac{10}{3}$

The denominators are not the same, so we make them the same.

$$= \frac{9 \times \mathbf{3}}{2 \times \mathbf{3}} - \frac{10 \times \mathbf{2}}{3 \times \mathbf{2}}$$

$$= \frac{27}{6} - \frac{20}{6}$$

$$= \frac{7}{6}$$

However, the answer is top heavy (improper fraction) and should be written as a mixed number (refer to section 5.4).

Therefore, $\frac{7}{6} = 1\frac{1}{6}$

$4\frac{1}{2} - 3\frac{1}{3} = \mathbf{1\frac{1}{6}}$ ✓

EXERCISE 2C

Work out the following and leave your answers in their simplest form.

1) $1\frac{2}{3} + 1\frac{2}{3}$

2) $2\frac{2}{3} + 1\frac{2}{5}$

3) $4\frac{2}{3} + 3\frac{1}{3}$

4) $5\frac{2}{3} - 2\frac{2}{3}$

5) $7\frac{1}{3} - 1\frac{1}{4}$

6) $1\frac{2}{3} + 1\frac{2}{3} + 1\frac{2}{3}$

7) $1\frac{2}{3} - \frac{3}{5} + \frac{2}{7}$

8) $3\frac{2}{7} + \frac{1}{3}$

9) $1\frac{2}{3} - 1\frac{2}{3}$

10) $8\frac{2}{3} - 3\frac{4}{5}$

11) On Tuesday, Edward cycled $9\frac{2}{3}$ kilometres to work from his house.

On his way back, he cycled $4\frac{1}{4}$ km and his bike had a puncture. Edward had a lift back to his house by a friend, Andrew.

How far did Andrew travel to get Edward back to his house?

2.9 MULTIPLYING FRACTIONS

Multiplying fractions is the easiest of all the fraction calculations. To multiply fractions, multiply the numerators (top) numbers and multiply the denominators (bottom numbers). Cancel down when possible.

Example 1: Work out $\frac{1}{4} \times \frac{2}{5}$

Multiply the numerators: $1 \times 2 = 2$

Multiply the denominators: $4 \times 5 = 20$

Therefore, $\frac{1}{4} \times \frac{2}{5} = \mathbf{\frac{2}{20}}$

A more structured approach would be

$$\frac{1}{4} \times \frac{2}{5} = \frac{1 \times 2}{4 \times 5} = \mathbf{\frac{2}{20}}$$

But $\mathbf{\frac{2}{20}}$ is not in its simplest form, so we cancel down

$$\frac{2 \div \mathbf{2}}{20 \div \mathbf{2}} = \mathbf{\frac{1}{10}} \checkmark$$

Example 2: Work out $\frac{3}{7} \times \frac{5}{8}$

$$= \frac{3 \times 5}{7 \times 8}$$

$$= \mathbf{\frac{15}{56}} \checkmark$$

Example 3: The diagram shows a square of side **1 metre** which is divided into Q, R, S, T rectangles.

a) Work out the area of **each** rectangle Q, R, S and T.

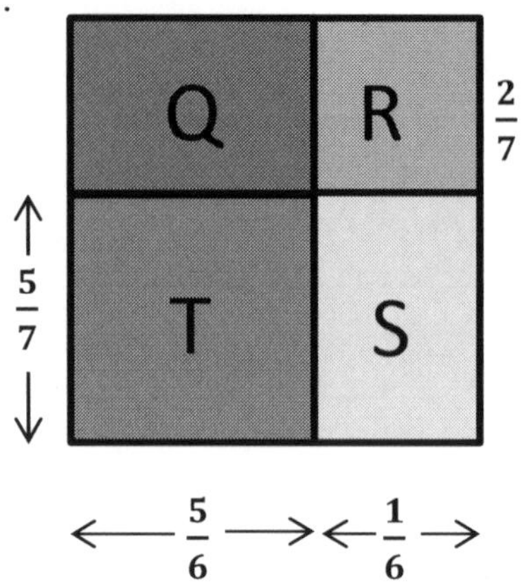

b) Show that the total area of the square is 1m^2

Solution:

Area of a rectangle = length × width

Area of **Q** $= \frac{2}{7} \times \frac{5}{6} = \frac{10}{42} = \frac{\mathbf{5}}{\mathbf{21}}\ \text{m}^2$

Area of **R** $= \frac{2}{7} \times \frac{1}{6} = \frac{2}{42} = \frac{\mathbf{1}}{\mathbf{21}}\ \text{m}^2$

Area of **S** $= \frac{5}{7} \times \frac{1}{6} = \frac{\mathbf{5}}{\mathbf{42}}\ \text{m}^2$

Area of **T** $= \frac{5}{7} \times \frac{5}{6} = \frac{\mathbf{25}}{\mathbf{42}}\ \text{m}^2$

For question b, the total area of the square is the entire area of rectangles Q, R, S and T.

$$\frac{5 \times \mathbf{2}}{21 \times \mathbf{2}} + \frac{1 \times \mathbf{2}}{21 \times \mathbf{2}} + \frac{5}{42} + \frac{25}{42}$$

$$= \frac{10}{42} + \frac{2}{42} + \frac{5}{42} + \frac{25}{42}$$

$$= \frac{10 + 2 + 5 + 25}{42}$$

$$= \frac{42}{42}$$

$$= 1$$

Therefore, the total area of the square is $\mathbf{1\ m^2}$ ✓

EXERCISE 2D

1) Work out and simplify your answers where possible.

a) $\frac{1}{5} \times \frac{2}{3}$

b) $\frac{2}{5} \times \frac{2}{7}$

c) $\frac{3}{4} \times \frac{6}{7}$

d) $\frac{5}{6} \times \frac{2}{9}$

e) $\frac{3}{5} \times \frac{3}{5}$

f) $\frac{1}{4} \times \frac{7}{8}$

g) $\frac{5}{7} \times \frac{8}{9}$

h) $\frac{3}{12} \times \frac{1}{8}$

i) $\frac{2}{9} \times \frac{6}{7}$

j) $\frac{12}{15} \times \frac{1}{7}$

k) $\frac{8}{10} \times \frac{3}{4}$

l) $\left(\frac{2}{20}\right)^2$

2) The height of a Nigerian flag is $\frac{6}{7}$ m.
If fourteen flags are placed on top of each other, what would be the total length of all the flags?

3) Work out the area of the rectangular field.

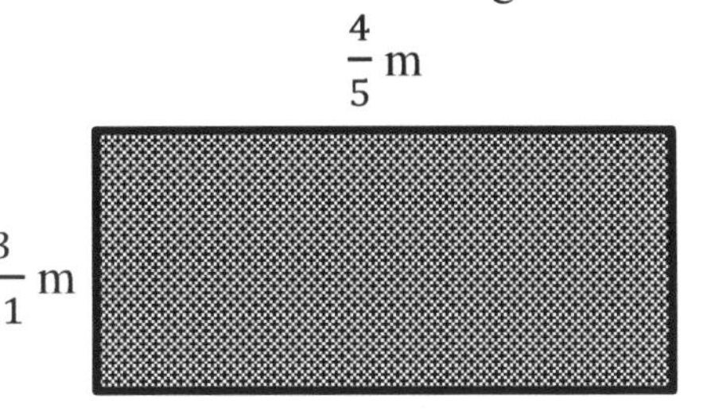

4) Work out the area of the triangle.

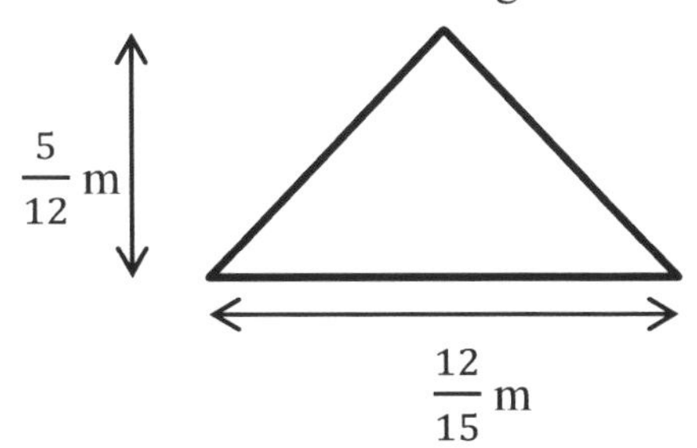

5) A rectangle is divided into four parts as shown below. Work out the area of **each** part of the rectangle. All lengths are in metres.

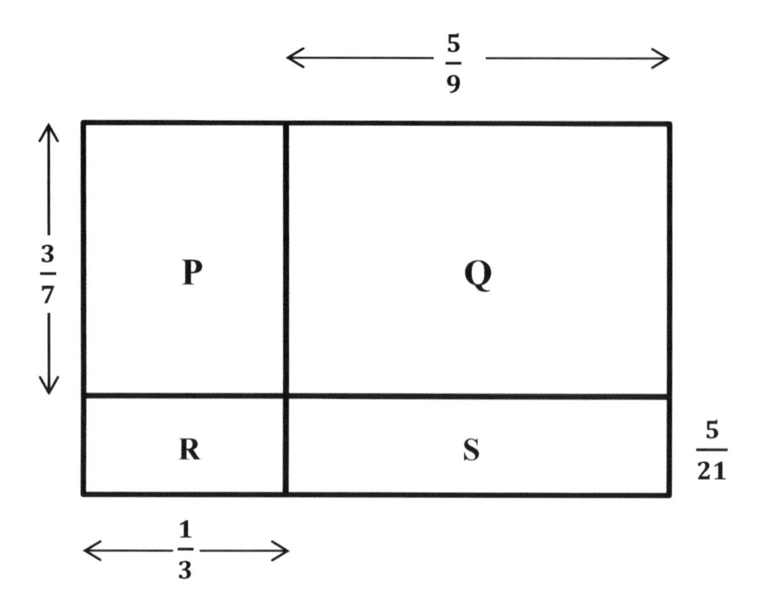

6) Work out the following.

a) $\frac{1}{3} \times \frac{1}{3} \times \frac{1}{3}$

b) $\frac{2}{3} \times \frac{3}{4} \times \frac{1}{5}$

c) $\frac{4}{11} \times \frac{9}{13}$

7) Nnaemeka says "$\frac{5}{6} \times \frac{1}{3} = \frac{6}{9}$"

a) Is he correct? Explain fully

b) If not, what did Nnaemeka do wrong?

FRACTION OF AN AMOUNT

Example 1: Work out $\frac{3}{5} \times 20$

Remember, every whole number can be written as a fraction by dividing by 1.

20 is the same as $\frac{20}{1}$

$$\frac{3}{5} \times \frac{20}{1}$$

$$= \frac{3 \times 20}{5 \times 1}$$

$$= \frac{60}{5} = \mathbf{12} \checkmark$$

Alternatively, you may consider dividing by the denominator and multiplying by the numerator. It is the same thing as cancelling down. 20 ÷ 5 = 4. Then, 4 × 3 = 12.

$$\frac{3}{\cancel{5}_{\mathbf{1}}} \times \frac{\cancel{20}^{\mathbf{4}}}{1} = 3 \times 4 = \mathbf{12}$$

Example 2: Work out $\frac{3}{5}$ of ₦100

In mathematics, **'Of'** in this instance is multiplication (×).

It is the same method of working out fractions as in examples 1 and 2 above.

$$\frac{3}{5} \text{ of ₦}100 = \frac{3}{5} \times \text{₦}100$$

$$= \frac{3}{\cancel{5}_{\mathbf{1}}} \times \cancel{\text{₦}100}^{\mathbf{₦20}}$$

$$= 3 \times \text{₦}20$$

$$= \mathbf{₦60} \checkmark$$

Dividing by the denominator and multiplying by the numerator

Alternatively; $\frac{3}{5}$ of ₦100

$$= \frac{3}{5} \times \frac{100}{1}$$

$$= \frac{3 \times 100}{5 \times 1}$$

$$= \frac{300}{5}$$

= **₦60**

Example 3: $450 \times \frac{3}{9}$

Method 1: Multiply 450 by 3, and then divide by 9

$450 \times 3 = 1350$

$1350 \div 9 =$ **150** ✓

Method 2: Divide 450 by 9, and then multiply by 3

$450 \div 9 = 50$

$50 \times 3 =$ **150** ✓

Cancelling down method

$$\overset{\mathbf{50}}{\cancel{450}} \times \frac{3}{\underset{\mathbf{1}}{\cancel{9}}}$$

$\frac{50 \times 3}{1} =$ **150** ✓

EXERCISE 2E

1) Work out the following.

a) $\frac{1}{5} \times 5$

b) $\frac{2}{5} \times 10$

c) $\frac{3}{4} \times 12$

d) $\frac{7}{8} \times 24$

e) $40 \times \frac{1}{5}$

f) $45 \times \frac{2}{9}$

g) $120 \times \frac{5}{12}$

h) $300 \times \frac{3}{60}$

i) $\frac{1}{5} \times 200$

j) $275 \times \frac{1}{5}$

k) $2000 \times \frac{7}{40}$

l) $\frac{1}{13} \times 13$

2) Work out the following.

a) $\frac{1}{5}$ of ₦200

b) $\frac{1}{3}$ of ₦150

c) $\frac{1}{8}$ of \$400

d) $\frac{1}{9}$ of 54 kg

e) $\frac{5}{8}$ of 64

f) $\frac{4}{5}$ of ₦1500

g) $\frac{1}{5}$ of £77.50

h) $\frac{17}{35}$ of 2100 kg

3)

Emeka paid $\frac{2}{3}$ of the cost of Obiora's jacket for a similar black jacket.

a) How much did Emeka pay for the jacket?

b) Arinze says "$\frac{4}{5}$ of what Emeka paid is more than half the cost of Obiora's jacket."

Is Arinze correct? Explain fully.

MULTIPLYING WITH MIXED NUMBERS

To multiply mixed numbers, it is advisable to change them to improper fractions (Refer to section 2.3)

Example 1: Work out $3\frac{2}{5} \times 7\frac{1}{2}$

Change to improper fractions

$3\frac{2}{5}$ becomes $\frac{(5 \times 3) + 2}{5} = \frac{\mathbf{17}}{\mathbf{5}}$ and $7\frac{1}{2}$ becomes $\frac{(2 \times 7) + 1}{2} = \frac{\mathbf{15}}{\mathbf{2}}$

Multiplying the two improper fractions: $\frac{17}{5} \times \frac{15}{2} = \frac{17 \times 15}{5 \times 2}$

$$= \frac{255}{10} = 25.5$$

$$= \mathbf{25\frac{1}{2}} \checkmark$$

Example 2: Work out $5\frac{1}{4} \times \frac{2}{7}$

$5\frac{1}{4}$ becomes $\frac{(4 \times 5) + 1}{4} = \frac{21}{4}$

Multiplying both fractions gives $\frac{21}{4} \times \frac{2}{7}$

$$= \frac{21 \times 2}{4 \times 7}$$

$= \frac{42}{28}$ (Cancel down by dividing both numbers by 2)

$= \frac{21}{14}$ (Cancel down by dividing both numbers by 7)

$= \frac{3}{2}$ (Change back to a mixed number)

$$= \mathbf{1\frac{1}{2}} \checkmark$$

Example 3: Multiply $4 \times 2\frac{3}{5}$

Change $2\frac{3}{5}$ to an improper fraction. $\frac{(5 \times 2) + 3}{5} = \frac{13}{5}$

Multiplying gives $4 \times \frac{13}{5}$

$$= \frac{4 \times 13}{5}$$

$$= \frac{52}{5}$$

$$= \mathbf{10\frac{2}{5}}$$ ✓

EXERCISE 2F

1) Write the following mixed numbers to improper fractions.

a) $1\frac{3}{4}$ b) $5\frac{3}{5}$ c) $11\frac{4}{9}$ d) $17\frac{1}{3}$

2) Work out the following but give your answer as a mixed number when applicable.

a) $2\frac{3}{5} \times \frac{3}{5}$ b) $1\frac{3}{5} \times 2\frac{3}{7}$ c) $7\frac{3}{5} \times \frac{3}{5}$ d) $3\frac{1}{5} \times 1\frac{4}{10}$

e) $3\frac{3}{4} \times 1\frac{2}{5}$ f) $2\frac{4}{6} \times 6\frac{3}{7}$ g) $12\frac{3}{5} \times 8\frac{3}{4}$ h) $(2\frac{3}{5})^2$

3) Find the product of these numbers. Write as mixed numbers where possible and simplify.

a) $4\frac{3}{7}$ and $\frac{1}{3}$ b) $10\frac{1}{5}$ and $\frac{3}{5}$ c) 2 and $8\frac{5}{6}$ d) 9 and $6\frac{3}{4}$

4) If it takes $\frac{1}{5}$ of a minute to fill a bucket with cold water, what fraction of a minute will it take to fill $15\frac{1}{2}$ buckets with cold water?

5)

A tin of sweetcorn weighs $\frac{1}{5}$kg.
137 of the tins are packed in a bag.

a) What is the total weight of the sweetcorn?

b) If the bag used in packaging weighs $1\frac{3}{5}$ kg, what is the total weight of the sweet corns and bag?

2.10 DIVIDING FRACTIONS

Our mantra would be "**Keep, Change, Flip**."

Keep the first fraction, **change** the division to multiplication (×) and **flip** the second fraction. Once we have successfully applied the mantra, we then multiply out the fractions (See section 2.9).

Example 1: Work out $\frac{5}{6} \div \frac{3}{4}$

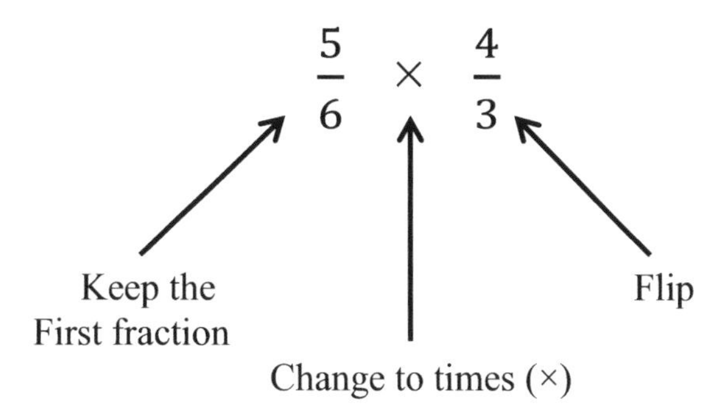

Multiply the two fractions

$$\frac{5 \times 4}{6 \times 3} = \frac{20}{18}$$

Since $\frac{20}{18}$ is not in its simplest form, we cancel down by dividing by 2.

We then have $\frac{20 \div 2}{18 \div 2} = \frac{10}{9}$

Also, the answer is an improper fraction, we then change back to a mixed number (refer to section 5.4)

Therefore, $\frac{10}{9} = \mathbf{1\frac{1}{9}}$ ✓

The same system applies to dividing mixed numbers. However, we must change to improper fractions before applying the *keep, change flip*. See example 2.

Example 2: Work out $3\frac{2}{5} \div 1\frac{3}{4}$

Change the mixed numbers to improper fractions (refer to section 5.3).

$3\frac{2}{5}$ to an improper fraction is $\frac{(5 \times 3) + 2}{5} = \frac{17}{5}$

$1\frac{3}{4}$ to an improper fraction is $\frac{(4 \times 1) + 3}{5} = \frac{7}{4}$

Now, rewrite the fractions using the improper fractions as $\frac{17}{5} \div \frac{7}{4}$

Remember **Keep, change, flip**

$$= \frac{17}{5} \times \frac{4}{7}$$

$$= \frac{17 \times 4}{5 \times 7} = \frac{68}{35} = \mathbf{1\frac{33}{35}} \checkmark$$

Example 3: Work out $10 \div 3\frac{2}{7}$

Change $3\frac{2}{7}$ to improper fraction which will equal $\frac{(7 \times 3 + 2)}{7} = \frac{23}{7}$

Now, rewrite the fractions using the improper fraction as $10 \div \frac{23}{7}$

Remember Keep, change, flip

$$= 10 \times \frac{7}{23}$$

$$= \frac{10 \times 7}{23} = \frac{70}{23} = \mathbf{3\frac{1}{23}} \checkmark$$

Example 4: How many quarters are there in 6?

This means $6 \div \frac{1}{4}$

$$= 6 \times \frac{4}{1}$$

$$= \frac{6 \times 4}{1} = \mathbf{24} \checkmark$$

EXERCISE 2G

1) Work out the fractions and give your answers as a mixed number when possible.

a) $\frac{1}{2} \div \frac{1}{4}$ d) $\frac{6}{7} \div \frac{1}{5}$ g) $10\frac{1}{2} \div \frac{1}{2}$ j) $2\frac{7}{9} \div 1\frac{7}{10}$

b) $\frac{4}{5} \div \frac{2}{3}$ e) $7 \div \frac{1}{2}$ h) $6\frac{2}{9} \div \frac{5}{6}$ k) $5\frac{2}{3} \div 2\frac{7}{10}$

c) $\frac{7}{10} \div \frac{3}{5}$ f) $\frac{1}{3} \div \frac{1}{5}$ i) $20 \div 1\frac{4}{5}$ l) $3\frac{3}{7} \div 1\frac{1}{10}$

2) How many thirds are there in

a) 5 b) 7 c) 9 d) 15?

3) How many tenths are there in

a) 6 b) 8 c) 11 d) 20?

4) The length of the top of a rectangular table tennis table is $2\frac{7}{10}$ m. The area of the top is $4\frac{1}{20}$ m^2.
Work out the width of the top of the table top. Leave your answer as a mixed number.

5) A perimeter fence has seven panels. The width of one panel is $1\frac{8}{10}$ m long. What is the total width of the panels?

6) Work out $15\frac{3}{4} \div 1\frac{3}{4}$, give your answer as a mixed number

7) Work out $12\frac{2}{5} \div 5\frac{7}{8}$, give your answer as a mixed number.

Chapter 2 Review Section

Assessment

1) Look at the picture. What fraction of these students are:

a) males

.... **1 mark**

b) not males

.... **1 mark**

c) wearing white shirts?

.... **1 mark**

2) What fraction of the months of the year starts with the letter J?

.... **1 mark**

3) What fraction of an hour is 20 minutes?

.... **1 mark**

4) Write the following as equivalent fractions with a denominator of 40.

a) $\frac{1}{4}$ b) $\frac{3}{8}$ c) $\frac{4}{5}$ d) $\frac{9}{10}$

.... **4 marks**

5) Change to improper fractions.

a) $2\frac{1}{7}$ b) $1\frac{5}{8}$ c) $12\frac{1}{3}$ d) $20\frac{9}{11}$

.... **4 marks**

6) Add $\frac{4}{7}$ to:

a) $\frac{1}{7}$ b) $\frac{2}{3}$ c) $\frac{5}{6}$ d) $\frac{3}{11}$

.... **8 marks**

7) Simplify each fraction.

a) $\frac{4}{8}$ b) $\frac{9}{27}$ c) $\frac{18}{30}$ d) $\frac{144}{168}$

………….. **8 marks**

8) Copy and complete the pairs of equivalent fractions below.

a) $\frac{4}{7} = \frac{?}{21}$ b) $\frac{12}{20} = \frac{?}{100}$ c) $\frac{6}{8} = \frac{54}{?}$ d) $\frac{?}{15} = \frac{24}{30}$

………….. **4 marks**

9) Change to mixed numbers.

a) $\frac{7}{3}$ b) $\frac{23}{4}$ c) $\frac{59}{7}$ d) $\frac{204}{20}$

………….. **4 marks**

10) Work out and leave your answers as a mixed number where possible. Also, leave your answers in its lowest form where possible.

a) $\frac{4}{17} + \frac{4}{17}$ b) $\frac{8}{9} - \frac{1}{3}$ c) $1\frac{4}{5} + 4\frac{4}{5}$ d) $\frac{2}{3} \div \frac{1}{5}$

e) $3\frac{1}{3} - 1\frac{7}{9}$ f) $3\frac{1}{3} \times \frac{1}{4}$ g) $2\frac{1}{6} \times 2\frac{5}{9}$ h) $8\frac{1}{3} \div 5\frac{2}{5}$

………… **16 marks**

11) Work out

a) $\frac{2}{5}$ of 30 b) $\frac{3}{4}$ of 36 kg c) $\frac{5}{7}$ of ₦420 d) $160 \times \frac{7}{8}$

………….. **8 marks**

12) The diagram is divided into rectangles as shown.

a) Work out the area of **each** rectangle.

b) Work out the **total** area of Q, R, S and T

c) What is the mathematical name of the shape formed by Q, R S and T? Give a reason for your answer.

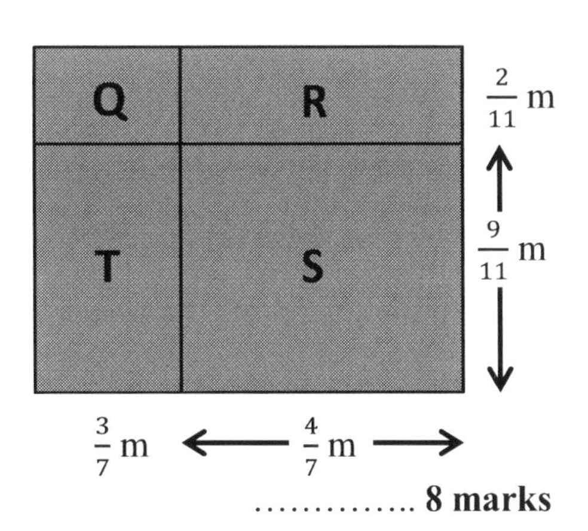

………….. **8 marks**

13) Simplify the following.

a) $\frac{16}{20} \div \frac{4}{10}$

b) $$\frac{\frac{1}{2} \times \frac{6}{7}}{\frac{1}{7}}$$

c) $$\frac{5\frac{3}{5} + \frac{4}{7}}{\frac{2}{3}}$$

14)

The weight of each pack of rice is 2kg.
9 packets of rice are packed into a travelling bag of weight $1\frac{5}{6}$ kg.

a) What is the total weight of the bag when loaded with the 9 packets of rice?

b) Kola bought 21 of such travelling bags. What is the total weight of the travelling bags?

c) Kola gave three travelling bags away to his friend, Tayo. What is the combined weight of Tayo's bags?

d) If a travelling bag costs ₦5 500, how much did Kola pay for the travelling bags?

3 Percentages 1

This section covers the following topics:

- Fractions, decimals and percentages
- Percentage of a quantity
- Chapter Review Section

LEARNING OBJECTIVES

By the end of this unit, you should be able to:

a) Understand the word 'percent.'
b) Change fractions to decimals
c) Change fractions and decimals into percentages
d) Change percentages into fractions and decimals
e) Write one quantity as a fraction of another

KEYWORDS

- Percentage
- Percent
- Quantity
- Fraction
- Decimal

3.1 UNDERSTANDING PERCENTAGES (%)

The word percent means **'out of a hundred.'** A percentage is a special type of fraction.

1% means 1 part per hundred. This can be written as $\frac{1}{100}$.

1% can also be written as a decimal **0.01**.

Likewise,

2% means $\frac{2}{100}$ or 0.02 as a decimal

10% means $\frac{10}{100}$ or 0.1 as a decimal

47% means $\frac{47}{100}$ or 0.47 as a decimal

100% means $\frac{100}{100}$ which is equal to **one whole** (1)

Percentages are equal to fractions with the denominator (bottom number) equal to 100.

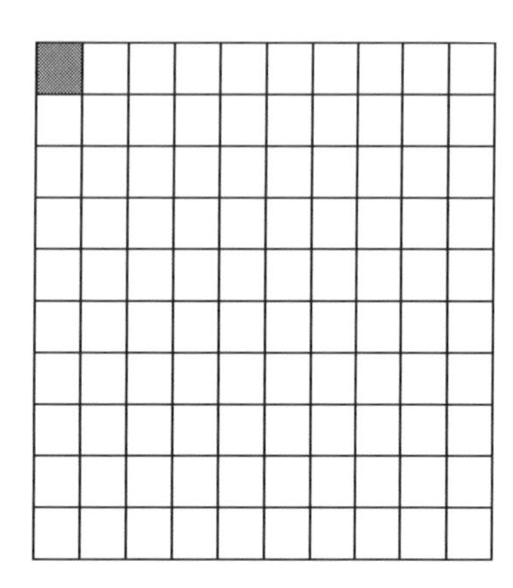

One percent is shaded

$\frac{1}{100}$

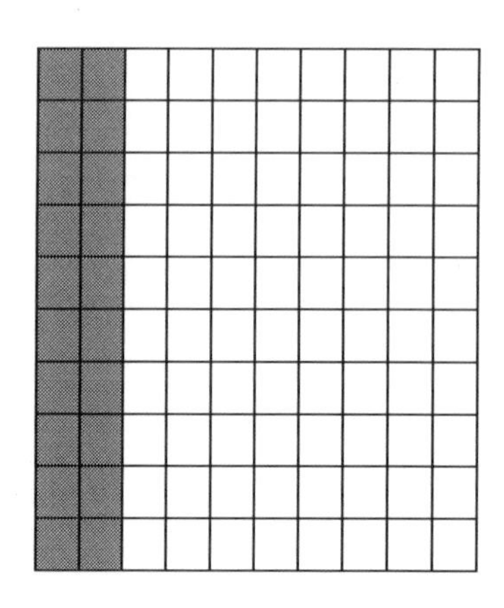

Twenty percent is shaded

$\frac{20}{100}$ or $\frac{1}{5}$ or $\frac{2}{10}$

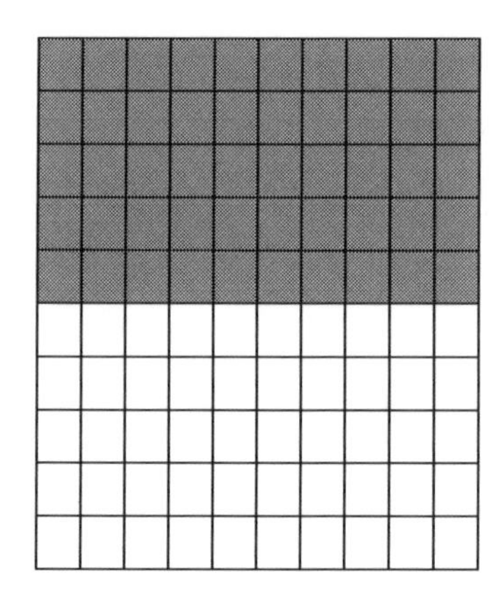

Fifty percent is shaded

$\frac{1}{2}$ or $\frac{50}{100}$ or $\frac{5}{10}$ or $\frac{25}{50}$

Example 1: What percentage of each box has been shaded?

$\frac{1}{4} =$

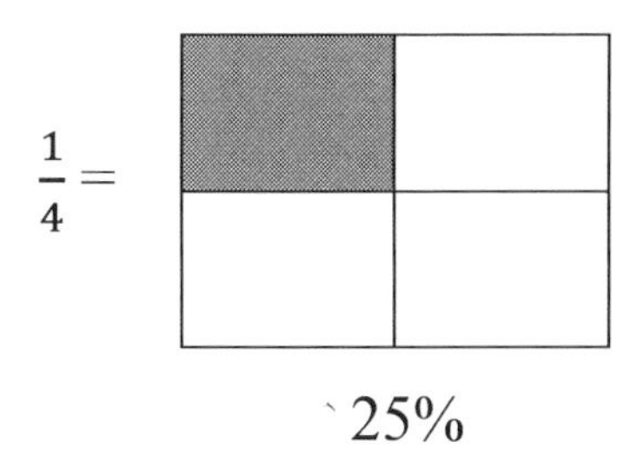

25%

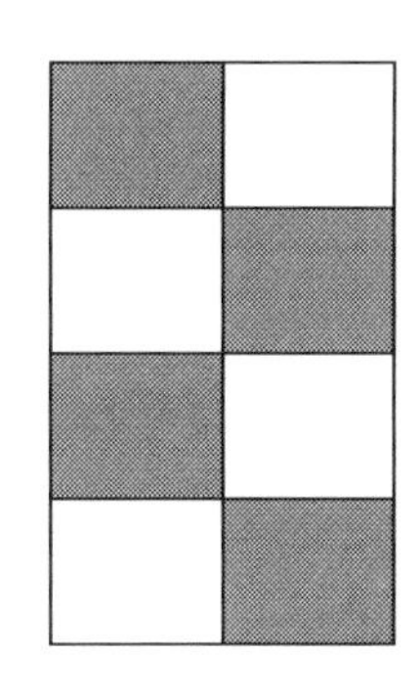

$\frac{4}{8} = \frac{1}{2} = 50\%$

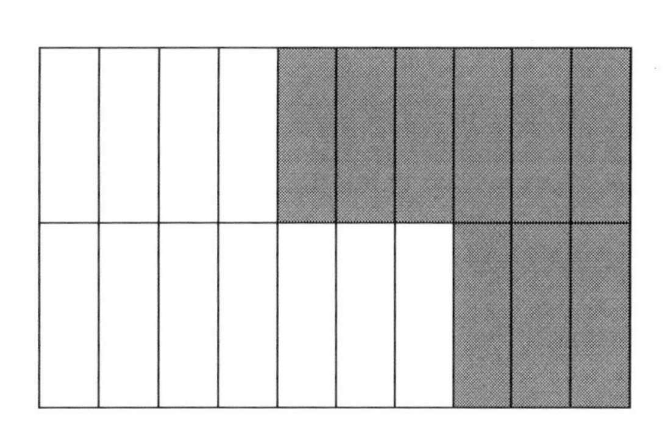

$\frac{9}{20} = 45\%$

Some key percentages to memorise

Fractions	Percentages	
$\frac{1}{4}$	25%	
$\frac{1}{2}$	50%	
$\frac{3}{4}$	75%	
A whole (1)	100%	

Also note:

$\frac{1}{8} = 12\frac{1}{2}\%$

3.2 FRACTIONS TO DECIMALS

A fraction is merely a division. The knowledge in section 3.3 is vital. If you can divide a number, you have successfully converted a fraction to a decimal.

However, the knowledge of equivalent fractions is essential in changing fractions to decimals.

Example 1: Change $\frac{2}{5}$ to a decimal.

Write as an equivalent fraction with a denominator of 100.

$$\frac{2 \times \mathbf{20}}{5 \times \mathbf{20}} = \frac{40}{100} = \mathbf{0.4}$$

Not all numbers will have an equivalent fraction of 100. If that is the case, use other methods, like dividing decimals or change the denominator to an equivalent fraction which you may multiply or divide by a number to make 100 or multiples of 10 (if it is easier). You may also cancel down (reducing to its simplest form) and the find an equivalent fraction with a denominator of 100.

Example 2: Convert $\frac{9}{12}$ to a decimal.

As you can see, 100 is not a multiple of 12, and it will not be possible to make an equivalent fraction with a denominator of 100.

Cancelling down gives $\frac{9 \div \mathbf{3}}{12 \div \mathbf{3}} = \frac{3}{4}$

It now possible to make an equivalent fraction with a denominator of 100

$$\frac{3 \times \mathbf{25}}{4 \times \mathbf{25}} = \frac{75}{100} = \mathbf{0.75}$$

Example 3: Change $\frac{13}{20}$ to a decimal.

$$\frac{13 \times \mathbf{5}}{20 \times \mathbf{5}} = \frac{65}{100} = \mathbf{0.65}$$

3.3 PERCENTAGES TO FRACTIONS

To change a percentage to a fraction, write the percentage as a fraction of a hundred and reduce the fraction to its lowest terms when possible.

Remember, 3% means 3 out of a hundred or $\frac{3}{100}$

4% means 4 out of a hundred or $\frac{4}{100}$

27% means 27 out of a hundred or $\frac{27}{100}$

Example 1: Change 5% to a percentage.

Write as a fraction of 100.

$$5\% = \frac{5}{100}$$

However, $\frac{5}{100}$ is not in its lowest term, so we cancel down.

$$\frac{5 \div 5}{100 \div 5} = \mathbf{\frac{1}{20}}$$

Example 2: Express 38% as a fraction in its lowest form.

Write as a fraction of hundred.

$$38\% = \frac{38}{100}$$

However, $\frac{38}{100}$ is not in its lowest term, so we cancel down.

$$\frac{38 \div 2}{100 \div 2} = \mathbf{\frac{19}{50}}$$

EXERCISE 3A

1) Convert the following fractions to decimals.

a) $\frac{3}{5}$ f) $\frac{1}{8}$ k) $\frac{30}{40}$ p) $\frac{50}{100}$

b) $\frac{4}{5}$ g) $\frac{2}{8}$ l) $\frac{40}{50}$ q) $\frac{2}{20}$

c) $\frac{3}{10}$ h) $\frac{5}{8}$ m) $\frac{7}{20}$ r) $\frac{7}{8}$

d) $\frac{2}{20}$ i) $\frac{3}{50}$ n) $\frac{12}{25}$ s) $\frac{1}{50}$

e) $\frac{7}{25}$ j) $\frac{19}{25}$ o) $\frac{6}{10}$ t) $\frac{16}{20}$

2) In their lowest terms, write the following percentages as fractions.

a) 2% f) 20% k) 50% p) 79%

b) 3% g) 25% l) 55% q) 80%

c) 10% h) 30% m) 60% r) 81%

d) 15% i) 35% n) 70% s) 90%

e) 18% j) 46% o) 75% t) 98%

3.4 FRACTIONS TO PERCENTAGES

Multiplying a fraction or a decimal by **100** changes it to a percentage

Example 1: Express $\frac{1}{4}$ as a percentage.

This means $\frac{1}{4} \times 100$ .... (Refer to section 5.9 on how to multiply fractions and the fraction of an amount.)

$$\frac{1 \times 100}{4} = \frac{100}{4} = 25$$

Therefore, $\frac{1}{4}$ as a percentage is **25%**

Example 2: Write $\frac{2}{20}$ as a percentage.

$$= \frac{2}{20} \times 100$$

$$= \frac{2 \times 100}{20} = \frac{200}{20} = \mathbf{10\%}$$

Example 3: Change $\frac{5}{8}$ to a percentage.

$$= \frac{5}{8} \times 100$$

$$= \frac{5 \times 100}{8} = \frac{500}{8}$$

Refer to section 2.6 on dividing whole numbers

You may also reduce to its simplest form and then change to a mixed number.

$$\frac{500 \div 2}{8 \div 2} = \frac{250}{4} = \frac{250 \div 2}{4 \div 2} = \frac{125}{2} = \mathbf{62\frac{1}{2}\ \%}$$

EXERCISE 3B

1) Write the following fractions as percentages.

a) $\frac{1}{4}$ f) $\frac{3}{20}$ k) $\frac{16}{50}$ p) $\frac{4}{40}$

b) $\frac{1}{5}$ g) $\frac{6}{10}$ l) $\frac{30}{50}$ q) $\frac{7}{14}$

c) $\frac{2}{5}$ h) $\frac{3}{8}$ m) $\frac{6}{25}$ r) $\frac{16}{32}$

d) $\frac{3}{10}$ i) $\frac{7}{20}$ n) $\frac{16}{25}$ s) $\frac{13}{52}$

e) $\frac{4}{5}$ j) $\frac{9}{25}$ o) $\frac{13}{20}$ t) $\frac{16}{320}$

2) Draw each shape below and answer the question for each shape.

a)

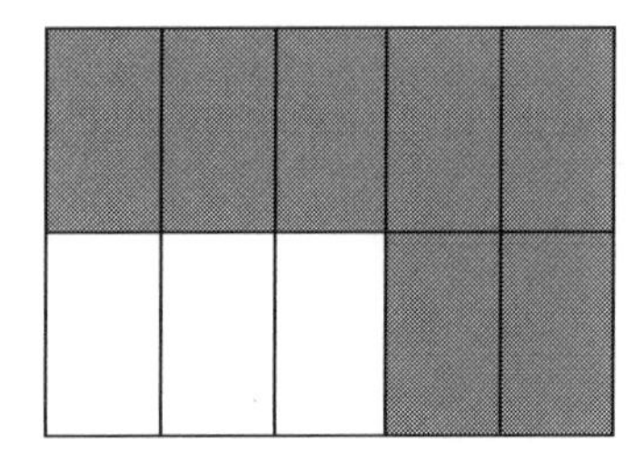

b)

c)

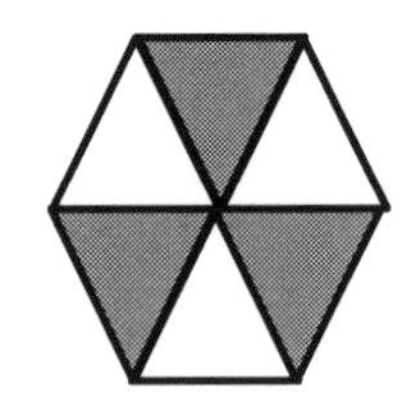

i) What fraction is shaded?
ii) What percentage is shaded?
iii) What fraction is unshaded?
iv) What percentage is unshaded?

3) Copy and complete.

a) $40\% = \frac{?}{10}$

b) $2\% = \frac{?}{100}$

c) $\frac{2}{5} = \square\ \%$

d) $\frac{7}{20} = \square\ \%$

d) $70\% = \frac{?}{10}$

e) $60\% = \frac{?}{5}$

3.5 WRITING ONE NUMBER AS A FRACTION OF ANOTHER

Example 1: Express 4 as a fraction of 10.

This is written as $\frac{4}{10}$

Reduce the fraction to its lowest term.

$\frac{4 \div 2}{10 \div 2} = \frac{2}{5}$ ✓

Example 2: Write 3 minutes 20 seconds as a fraction of an hour.

Remember: The quantities must be in the same units before we can successfully work out the calculations.

This is written as $\frac{3 \text{ min } 20 \text{ sec}}{1 \text{ hour}}$

Convert everything to seconds $= \frac{(3 \times 60) \text{ sec} + 20 \text{ sec}}{60 \times 60}$

$= \frac{(180 + 20) \text{ sec}}{3600 \text{ sec}}$

$= \frac{200 \text{ sec}}{3600 \text{ sec}}$

Remember:

60 seconds = 1 minute

60 minutes = 1 hour

1 hour = 60 × 60
= 3600 seconds

In its lowest term, first, divide both numbers by 100

$= \frac{200 \div 100}{3600 \div 100}$

$= \frac{2}{36}$ (then divide both numbers by 2)

$= \frac{1}{18}$ ✓

EXERCISE 3C

1) Express 5 as a fraction of 7

2) Express 4 as a fraction of 20. Leave your answer in its lowest term.

3) What fraction of 5 cm is 2 m? Leave your answer in its lowest term.

4) What fraction of 30 cm is 3 m? Leave your answer in its lowest term.

5) Write the first number as a fraction of the second number. Leave your answers in their lowest term.

a) 12 weeks, one year

b) 30 minutes, 2 hours

c) 7 mm, 21 cm

d) $34, $100

e) 6 min 15 seconds, 1 hour

f) 16 cm, 10 m

g) 20m, 3 km

3.6 WRITING ONE QUANTITY AS A PERCENTAGE OF ANOTHER

Two steps to follow:

1) Write the first number as a fraction of the second number (See section 3.5).

2) Multiply the fraction or decimal formed by 100

Example 1: Write 6 as a percentage of 24

Write 6 as a fraction of 24: $\frac{6}{24}$

Multiply by 100 to convert to a percentage

$$= \frac{6}{24} \times 100$$

To make it easier, reduce $\frac{6}{24}$ to its lowest term which is $\frac{6 \div 6}{24 \div 6} = \frac{1}{4}$

Therefore, $\frac{1}{4} \times 100 = \frac{1 \times 100}{4} = \frac{100}{4} = \mathbf{25\%}$ ✓

Alternatively, $\frac{6}{24} \times 100 = \frac{6 \times 100}{24} = \frac{600}{24} = \mathbf{25\%}$

Example 2: Elizabeth scored 20 out of 25 in a maths test.
Work out Elizabeth's percentage mark.

As a fraction, Elizabeth's mark $= \frac{20}{25}$

To convert to percentage, multiply by 100.

$$= \frac{20}{25} \times 100$$

A simpler way is to divide 100 by 25 and multiply by 20.

$$= 100 \div 25 = 4$$

$$4 \times 20 = \mathbf{80\%}$$

Example 3: A kettle is in a sale and reduced to ₦4000. What is the percentage reduction?

Find the reduction:
₦5 000 – ₦4 000 = ₦1 000

The reduction as a fraction is $\frac{1000}{5000}$

To its lowest term the fraction is $\frac{1}{5}$

Percentage reduction $= \frac{1}{5} \times 100$
$= \mathbf{20\%}$

EXERCISE 3D

1) Write

a) 20 as a percentage of 50

b) 5 as a percentage of 20

c) £100 as a percentage of £500

d) 12 as a percentage of 36

2) Express these times as a percentage of an hour.

a) 6 minutes

b) 30 minutes

c) 15 minutes

d) 60 minutes

3) Express 2 hours as a percentage of 5 days.

4) Express 300 ml as a percentage of 2 litres

5) 

A polo shirt is reduced from ₦50 to ₦35 in a sale. What is the percentage reduction?

6) The Nigerian Football Association has 80 members as shown in the table below.

Female	Male
30	50

a) What percentages of the members are male?

b) What percentages of the members are female?

7) The results of a science test for three students are shown below.

Names	**Out of 40**
Abdul	8
Uduak	28
Amina	20

a) What is Abdul's percentage score?

b) What is Amina's percentage score?

c) Okonkwo says "Uduak scored 71%."
Is Okonkwo correct?
Explain fully.

Chapter 3 Review Section

Assessment

1) 73% of students walk to school. What percentage of students goes to school by other means?

…………….. **1 mark**

2) Copy and shade in

a) 25%

b) 50%

c) 75%

………….. **3 marks**

3) Write 4cm as a fraction of 2m

…………… **1 mark**

4) Write the following numbers as a percentage of the second.

a) 15 weeks, 1 year

b) 500 g, 2 kg

c) ₦4 000, ₦8 000

…………...**6 marks**

5) Express these fractions as percentages.

a) $\frac{1}{2}$ b) $\frac{15}{25}$ c) $\frac{7}{10}$ d) $\frac{3}{20}$

………….. **8 marks**

6) In summer examinations, Rebecca scored these marks:

Maths: 17 out of 25
Economics: 2 out of 40
English: 30 out of 50.
Physics: 7 out of 20

a) Work out Rebecca's percentage for **each** subject.

.... **8 marks**

b) Arrange the percentages in question 6a above in order of size, smallest first.

.... **1 mark**

7) Copy and complete the table below.

Fractions	Percentages	Decimals
$\frac{2}{5}$		
		0.35
	60%	
$\frac{6}{25}$		

.... **8 marks**

8) Arrange in order of size, highest first.

$\frac{1}{2}$, 0.65, $\frac{3}{5}$, 51%

.... **2 marks**

9) Which is bigger, 30% or $\frac{8}{25}$?
Explain your decision.

.... **2 marks**

10) There are 45 women and 15 men in a theatre.

a) What percentage are women? .... **2 marks**

b) What percentage are men? .... **2 marks**

4 Percentages 2

This section covers the following topics:

- Percentage of a quantity
- Proportions using percentages
- Find percentage change
- Simple Interest
- Reverse Percentages
- Compound interest

LEARNING OBJECTIVES

By the end of this unit, you should be able to:

a) Find the percentage of a quantity
b) Compare proportions using percentages
c) Find a percentage increase
d) Find a percentage decrease
e) Find percentage change
f) Find Simple Interest
g) Find reverse percentages
h) Find compound interest

KEYWORDS

- Percentage Change
- Increase and Decrease
- Quantity
- Proportion
- Simple Interest
- Compound Interest

4.1 PERCENTAGE OF A QUANTITY

Finding the percentage of a quantity is the same as finding the fraction of a quantity. Remember, 10% means $\frac{10}{100} = \frac{1}{10}$, therefore, finding 10% of a number simply means **dividing the number by 10**.

Example 1: Find 10% of 20.
Since 10% means $\frac{1}{10}$, we have to find $\frac{1}{10}$ of 20 which means $\frac{20}{10} = \mathbf{2}$

Example 2: Find 10% of 70.........................It means 70 ÷ 10 = **7**
Example 3: Find 10% of 635......................It means 635 ÷ 10 = **63.5**

From 10%, a lot of percentages can be obtained.

- To find 20%, find 10% and multiply by 2
- To get 30%, find 10% and multiply by 3
- To get 70%, find 10% and multiply by 7, and so on.
- To get 5%, find 10% and divide by 2.
- To get 50%, find 10% and multiply by 5 **or** simply divide the number by 2.

As you get familiar with percentages, there are other ways you could work out any percentage of an amount. For example: to get 70%, you could work out 50% and then add it on to 20%.

Also remember, 1% is the same as $\frac{1}{100}$. Therefore, finding 1% of a number simply means **dividing the number by 100.**

Example 3: Find 1% of 34........................It means $\frac{34}{100} = \mathbf{0.34}$
Example 4: Find 1% of 436......................It means $\frac{436}{100} = \mathbf{4.36}$

Mixed examples

Example 5: Find 20% of 70
Find 10% and multiply by 2. 10% of 70 = 7 7 × 2 = **14**

Example 6: Find 40% of ₦120 10% of 120 = 12 12 × 4 = **₦48**

Example 7: Find 5% of 30 10% of 30 = 3 3 ÷ 2 = **1.5** or $\mathbf{1\frac{1}{2}}$
Example 8: Find 15% of 80
10% of 80 = 8 5% of 80 = 10% ÷ 2 = 8 ÷ 2 = 4 **15%** of 80 = 8 + 4 = **12**

EXERCISE 4A

Work out without a calculator. Show all working out.

1) a) 10% of 30	b) 10% of 4	c) 10% of 60	d) 10% of 8
e) 10% of 16	f) 10% of 40	g) 10% of ₦80	h) 10% of ₦300
i) 20% of 30	j) 20% of 4	k) 20% of 60	l) 20% of 8
m) 40% of 40	n) 50% of 36	o) 80% of 20	p) 100% of 50

2) a) 5% of 20	b) 5% of 140	c) 60% of 100	d) 1% of 200
e) 5% of 38	f) 1% of 24	g) 50% of 2440	h) 35% of 80kg

Example 9: Find 13% of 30

Method 1

10% of 30 = 30 ÷ 10 = 3

\+ 1% of 30 = 30 ÷ 100 = 0.3

1% of 30 = 30 ÷ 100 = 0.3

1% of 30 = 30 ÷ 100 = 0.3

13% of 30 = **3.9**

Method 2

13% means $\frac{13}{100}$

Therefore, $\frac{13}{100} \times 30$

$= 0.13 \times 30 = \mathbf{3.9}$

Example 10: Find 7% of $150

1% of 150 = $\frac{1}{100} \times 150 = \frac{150}{100} = 1.5$

$1.5 \times 7 =$ **$10.50**

Example 11: Find 11% of ₦200

10% of ₦200 = 200 ÷ 10 = 20

\+ 1% of ₦200 = 200 ÷ 100 = 2

11% of ₦200 = **₦22**

Method 2

11% means $\frac{11}{100} = 0.11$

$0.11 \times 200 = 22$

Therefore, 11% of 200 = **₦22**

Example 12: Work out 2.3% of 19.35

2.3% means $\frac{2.3}{100} = 0.023$

$0.023 \times 19.35 = \mathbf{0.44505}$

Example 13: Work out $4\frac{1}{2}$% of 90

$4\frac{1}{2}$% means $\frac{4.5}{100} = 0.045$

$0.045 \times 90 = \mathbf{4.05}$

EXERCISE 4B

Work out the following without using a calculator.

1) a) 1% of 130 b) 3% of 35 c) 20% of 80 d) 80% of 30
e) 21% of 33 f) 14% of 55 g) 4% of 130 h) 7% of ₦300
i) 55% of £60 j) 90% of 750 k) 15% of 8 l) 35% of 80
m) 61% of 40 kg n) 7% of 890 o) 23% of ₦2000 p) 4% of 9

2) a) 2.7% of £70 b) 3.2% of 56 c) 2.8% of 200 d) 17.5% of 460
e) $2\frac{1}{2}$% of 30 f) $7\frac{1}{2}$% of 120 g) 3.5% of 68kg h) $6\frac{1}{2}$% of ₦5000

4.2 COMPARING PROPORTIONS USING PERCENTAGES

Okechukwu sat tests in Physics, Chemistry, Biology and Mathematics. His results were:

Physics: $\frac{20}{30}$, Chemistry: $\frac{77}{90}$, Biology: $\frac{16}{25}$, Mathematics: $\frac{18}{20}$
Which subject did he do best?

Solution: Before making any decision(s), we can write each fraction as a percentage. (Refer to section 6.4).

Physics: $\frac{20}{30} \times 100 = 66.7\%$ Biology: $\frac{16}{25} \times 100 = 64\%$
Chemistry: $\frac{77}{90} \times 100 = 85.6\%$ Mathematics: $\frac{18}{20} \times 100 = 90\%$

From the above calculations, Okechukwu did best in Mathematics test.

4.3: PERCENTAGE INCREASE AND DECREASE

A quantity can be increased or decreased by a percentage.

Example 1: Increase 100 by 10%

First, find 10% of 100 which is $\frac{100}{10} = 10$.

Therefore, increasing 100 by 10 means

100 + 10 = **110**

Example 2: Decrease ₦510 by 20%
First, find 20% of ₦510
Find 10% and multiply by 2

10% of ₦510 = $\frac{510}{10}$ = ₦51
20% of 510 = 51 × 2 = ₦102
Decreasing ₦510 by 20% means 510 – 102
= **₦408**

4.4 FINDING A PERCENTAGE CHANGE

Percentage change is calculated by using the formula below.

$$\text{Percentage change} = \frac{\text{Change (increase or decrease)}}{\text{Original amount}} \times 100$$

Example 1: The price of a TV set was increased from £80 to £120. Work out the percentage increase.

Solution: First work out the increase.

$120 - 80 = 40$

Percentage increase $= \frac{40}{80} \times 100 = \mathbf{50\%}$

Example 2: The original price of a carpet was ₦2 400. The price is now ₦2 000. Find the percentage decrease.

Solution: Decrease = 2400 – 2000 = ₦400

Percentage decrease $= \frac{400}{2400} \times 100 = \mathbf{16.7\%}$

EXERCISE 4C

1) Increase the following amounts by 10%.

a) 20 c) ₦200 e) 840
b) 80 d) ₦340 f) 3000

2) Increase the amounts in question 1 by 15%.

3) Decrease the following quantities by 20%.

a) 40 kg c) £123 e) 4000 g
b) 92 litres d) $183.52 f) 245 cm

4) Nonso earns ₦25 000 for selling CD's a day. How much **extra** will he get if he has a 10% salary increase?

5) The price of a Renault car is £4 200. There is a discount of 25% on all Renault cars. How much will Tochukwu pay for a Renault car?

6) Which of the following statements represent a 40% change?

a) 35 to 40 b) 50 to 70 c) 70 to 80.

7) *Shoprite* is having a 35% sale on all men's shirts. The original prices for three shirts are £15, £32 and £45 respectively. Work out the sale price for each shirt.

8) A dog weighs 2 kg and while running away from a fox, it loses one of its legs. As a result, the weight was reduced by 3%. What is the weight of the dog now?

9) Emma buys a car for ₦540 000 and sold it two weeks later for ₦720 000. Work out the percentage profit.

10) Increase ₦5500 by 2%
11) Decrease ₦6000 by 13%
12) Increase 72 kg by 15%
13) Decrease £780 by 1%

4.5 SIMPLE INTEREST

If you deposit money in a building society or bank, they will pay you **money (interest)** on this money you deposited.

Also, if you borrow money from a bank, you also have to pay interest on the borrowed money. Simple interest is normally calculated on a yearly basis (annual) and depends on the **interest rate**.

Example 1: Andy saves £500 in a bank for a year. If the interest rate is 5% per annum, a) Calculate the interest on the amount. b) The total amount Andy will have after a year.

Solution

a) Interest = 5% of £500

$= \frac{5}{100} \times 500 = \textbf{£25}$

b) After one year, Andy will

£25 + £500 = **£525**.

Interest paid out this way is known as **simple interest**.

2) Henry's bank account pays interest at the rate of 4.5% per annum. If he deposits ₦20 000 into his account, how much simple interest will he receive after a) 1 year b) 6 years

Solutions

a) $\frac{4.5}{100} \times$ ₦20 000 = **₦900**

b) ₦900 × 6 = **₦5 400**

Generally, the formula for simple interest is $\mathbf{I} = \frac{\mathbf{P\ R\ T}}{\mathbf{100}}$, where I is the simple interest, P = amount invested, R = rate and T = time invested.

EXERCISE 4D

1) Work out the simple interest on the following:

a) £700 at 3% per annum for 1 year
b) £8 000 at 4% per annum for 1 year
c) ₦30 000 at 2% per annum for 4 years
d) ₦25000 at 5% per annum for 2 years
e) ₦2000 at 7.5% per annum for 6 years
f) $5500 at $2\frac{1}{2}$% per annum for 3 years
g) $7000 at 3.5% per annum for 2 years
h) ₦45 000 at 5% per annum for 4 years

2) Chibogu saves ₦5000 in a bank for a year. If the interest rate is 2.5% per annum, calculate
a) the interest on the amount.
b) the total amount Chibogu will have after a year.

3) Sarah applied for a £35 000 car loan and was successful. If the interest is 5% per annum, how much simple interest will she pay after a) 1 year b) 3 years?

4)

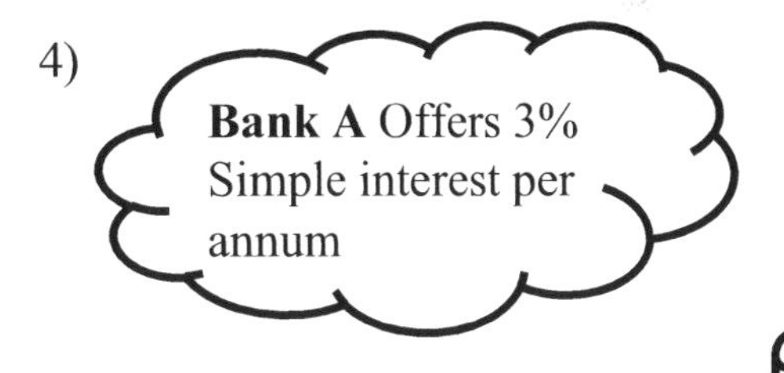

Bank **B** offers 5% interest for the first year and then 2% simple interest per annum for the following years.

Nicole wants to invest £10 000 for 5 years. Which bank will she deposit her money to earn more interest?

5) Ajayi borrowed ₦3 500 000 from Bank X for a second-hand car. The agreement with Bank X was for three years at a simple interest of 8% per annum.

a) What is the simple interest on the borrowed money at 8% per annum for three years?
b) What is the total amount Ajayi would pay back at the end of 3 years?

c) If Ajayi was to pay the total amount due in monthly instalments over three years, how much would that be each month?

VALUE ADDED TAX CALCULATIONS (VAT)

VAT is a form of tax that everybody must pay when buying goods and services. The rates vary from country to country, and in the UK, it is currently 20%.

Example 1: How much will a customer pay for the microwave below?

Solution: VAT = 20% of £30
10% of £30 = £3.
Therefore, 20% = 2 × £3 = £6
Amount to pay = £30 + £6 = **£36**

EXERCISE 4E

1) A bicycle is on sale at £80 + 20% VAT.
a) Work out the amount of VAT to be paid.
b) Work out the total cost of the bike.

2) A bill for garden clearance is £75 + 20% VAT. How much is the total bill?

3) Calculate the total price of the following items:
a) Dinning table: £350 + 20% VAT b) Lexus Car: £12 000 + 20% VAT
c) 40-inch TV: £900 + 20% VAT

4)

a) How much would the government receive if VAT rate is 20%?

b) How much does the customer pay for the laptop?

PROBLEM-SOLVING

1) A dining table originally costs £200 but a $3\frac{1}{2}$% discount is given for cash payments.

a) What is the cash price for the dining table?

b) Mark wants to buy two dining tables. He has up to £1000 on his credit card and £195 cash.

i) How much will Mark pay for the two dining sets if he must use the cash and credit card?

ii) How much change will he receive in cash?

2) Bradley wants to buy the Lexus car but decided to buy on credit. He deposited 40% of the cash value and agreed on £500 monthly plan for 12 months.

a) How much did Bradley deposit?

b) How much is the total of the monthly payments?

c) How much will it cost Bradley to buy on credit?

3) Iconic Concepts Limited paid £240 electricity bill **inclusive** of VAT for 6 months usage at the rate of 20%. How much will the government receive as VAT?

4)

a) How much is the total monthly instalments?

b) If not paying by cash, how much is the deposit?

c) How much more is buying on credit than paying with cash?

d) How much is the total amount of option 2?

4.6 COMPOUND INTEREST

When interest is given on the interest, we talk about **compound interest**.

Example 1: Jude invests £300 in a building society which pays 3% interest each year. What is the value of the investment after 3 years?
Solution:
1st year: 3% of £300 = **£9**

2nd year: £300 + £9 = £309
3% of £309 = **£9.27**

3rd year: £9.27 + £309 = £318.27
3% of £318.27 = £9.55
9.55 + 318.27 = **£327.82** ✓

However, the value after 3 years could have been worked out in one calculation.

> 3% + 100% = 103% = 1.03
> 1.03 is now the multiplier.
>
> £300 × 1.03^3 ← Number of years
>
> = **£327.82** ✓

You may use the formula for compound interest which is

$$A_n = \left[1 + \frac{r}{100}\right]^n A_0$$

Where A_n = Total amount after *n* years
r = Interest rate (%)
A_0 = Initial amount invested

Example 2: Sandra's father invested ₦100 000 in a bank for her at 20% compound interest. **a)** Find the total amount due after 15 years. **b)** Find the total interest on the investment for the 15 years.

Solution:
a) 100% + 20% = 120% = 1.2
₦100 000 × $(1.2)^{15}$ = **₦1 540 702.16**

b) Interest = ₦1 540 702.16 – 100 000
= **₦1 440 702.16** ✓

Example3: How much do you need to invest now to receive £20 000 in 8 years at 7% interest rate?

Solution: This is working backwards.

100% + 7% = 107% = 1.07 (multiplier)
Working backwards:

$$\text{Present value} = \frac{20\ 000}{(1.07)^8} = \frac{20\ 000}{1.71818618}$$

= **£11 640.18** ✓

DEPRECIATION

Depreciation is a reduction in the value of an asset due to wear and tear or other factors.

Example 4: Tom bought a car for £8500, and its value depreciates (decreases) by 7% each year. Work out the value of the car after 5 years.

Depreciation: 100% – 7% = 93% = 0.93
After 5 years: £8500 × 0.93^5
= **£5 913.35** ✓

EXERCISE 4F

1) Callum invests £2800 in a bank which earns 3% compound interest per annum. Find the value of Callum's investment after:
a) 2 years b) 3 years c) 7 years.

2) Find the compound interest on £3000 invested for 3 years at 8%.

3) £13500 is invested at $7\frac{1}{2}\%$ per annum compound interest for 6 years. Find
a) the interest at the end of 6 years.
b) the total amount payable at the end of 6 years.

4) The price of a car depreciates at the rate of 15% per annum. Find the value of the car after 5 years if the price of the car is £12 000.

5) The population of Nigeria is 200 million and growing at 1.5% per annum.

a) In how many years' time will Nigerian population exceed 210 million?
b) What size will the population be in 5 years' time?

6) Leanne, head of mathematics department in a UK school, **borrows** £8500 for 5 years at 4% simple interest. Niamh, Leanne's second in charge borrows the same amount at 4% per annum compound interest.

a) Who pays more interest?
b) How much more will the person pay back?

4.7 REVERSE PERCENTAGES

Reverse percentages simply mean working backwards to find the **original** amount after an increase or a decrease.

Example 1: A laptop sells for £350 after a 40% increase in the cost price. What is the cost of the laptop before the increase?

Solution: 100% + 40% = 140% = 1.4

Cost price = $\frac{350}{1.4}$ = **£250**

Example 2: John was offered 15% discount when buying a new sofa. The discounted price is £595. What is the full price of the sofa?

Solution: Full price means the price before the discount or the original/cost price.

Because it is discount, we subtract the percentages: 100% - 15% = 85% = 0.85.

Full price = $\frac{595}{0.85}$ = **£700**

Example 3: Dylan invests some funds in a bank at 5% interest per annum. After 5 years, the value of Dylan's investment is £25000. Calculate the amount Dylan invested.

100% + 5% = 105% = 1.05

Amount invested = $\frac{25\,000}{(1.05)^5}$
= **£19 588.15**

EXERCISE 4G

All questions to 2 decimal places where possible.

1) After an increase of 2%, the price of an air ticket is £725. What was the price before the increase?

2) The volume of a metal increased by 5% to 258.9 cm^3 after heating. What is the volume of the metal before heating?

3) The price of a dining table is £600 including VAT at 20%. What is the price of the dining table without VAT?

4) The sale prices of two shirts are shown below in a sale of 25% off.

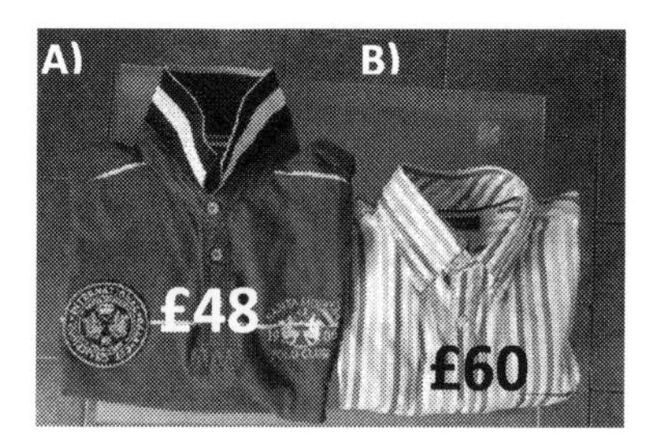

Find the original prices of the shirts.

5) Kola visited London and bought some items inclusive of VAT at 20%.

Laptop	£115.20
Electric kettle	£45
Samsung mobile phone	£420

a) Find the total cost of the items Kola bought without VAT.
b) How much can Kola reclaim in VAT?

Chapter 4 Review Section

Assessment

1) Work out without a calculator.

a) 10% of 130 b) 1% of 34 c) 30% of 600 d) 21% of ₦6000

…………………. **6 marks**

2) Decrease ₦4500 by 20% …………………. **2 marks**

3) Increase 45kg by 50% …………………. **2 marks**

4) Work out the percentage change for each of the following.

a) Increasing ₦200 by ₦50
b) Decreasing ₦5 000 by ₦200
c) Increasing 72 kg by 3 kg

…………………. **6 marks**

5) A man bought a Ferrari car. The value depreciates from £200 000 when new to £180 000 two years later. Calculate the percentage depreciation. ………… **2 marks**

6) A price of ₦960 is increased by 5%, and then three weeks later, it increased by a further 6%. Find the final price. …………………. **3 marks**

7) The price of a bicycle is reduced by 30% in a sale. The original price was ₦35 000.

a) What is the sale price?

b) What would 10 bicycles cost before the sale?

c) Nneamaka bought 3 bicycles while on sale, what would be her loss if she had bought them before the sale?

………………….**6 marks**

8) A bicycle is on sale at £130 + 20% VAT.

a) Work out the amount of VAT to be paid.

b) Work out the total cost of the bike.

……………………...**4 marks**

9) A bill for garden clearance is £95 + 20% VAT. How much is the total bill?

…………………….. **2 marks**

10) Calculate the total price of the following items:

a) Dining table: £300 + 20% VAT

b) Volvo Car: £15 000 + 20% VAT

c) 50-inch TV: £1100 + 20% VAT

……………………..**6 marks**

11) £13500 is invested at $7\frac{1}{2}$% per annum compound interest for 6 years. Find:
a) The interest at the end of 6 years. b) The total amount payable at the end of 6 years.

…………………….**4 marks**

12) William invests some funds in a bank at 6% interest per annum. After 3 years, the value of William's investment is £35000. Calculate the amount William invested.

…………………….**3 marks**

5 Algebra 1

This section covers the following topics:

- Order of operations
- Substitution
- Identifying equations
- Solving equations
- Forming and solving equations
- Additive inverse
- Inequalities

LEARNING OBJECTIVES

By the end of this unit, you should be able to:

a) Remove brackets from basic algebraic expressions
b) Expand/multiply algebraic terms
c) Divide algebraic terms
d) Work out calculations using BIDMAS
e) Identify and solve equations with brackets
f) Understand additive inverse
g) Solving inequalities

KEYWORDS

- Brackets
- Algebraic terms
- Multiply
- BIDMAS
- Algebraic expressions
- Equations

5.1 ORDER OF OPERATIONS

When there are several operations in one calculation, the correct order must be followed to get the calculation right.

$4 + 2 \times 3$ will have two answers; 18 and 10 if the correct order is not followed. To make sure that everybody is doing the same thing, we use BIDMAS or BODMAS to help us.

↓
Brackets
Other things **or Indices** like **(3^2), (2^3), √, ……**
Division and Multiplication
Addition and Subtraction

In a calculation, do the brackets first, followed by other things like powers (3^2) and roots (√), then division and multiplication come next – ***do them in the order they appear***, and finally, addition and subtraction - **do them in the order they appear.**

Also, you may use **bracket(s)** to indicate the calculation you want to perform first.

Example 1: Work out the values of

a) $4 + 2 \times 3$
   Multiplication (×) comes first before addition (+). $4 + (2 \times 3) = 4 + 6 =$ **10**
   Notice that 4 retained the position on the left even though we had to multiply first.

b) $4 - 2 \times 3 = 4 - (2 \times 3) = 4 - 6 =$ **-2**

c) $6 + 3 \times 5 - 4 = 6 + (3 \times 5) - 4 = 6 + 15 - 4 =$ **17**

d) $6 - 4 + 3$ …Since the operations are addition and subtraction, perform the calculation in the order they appear in the question. $6 - 4 = 2$ and $2 + 3 =$ **5**

e) $20 - 6 + 3 \times 4 = 20 - 6 + (3 \times 4) = 20 - 6 + 12 =$ **26**

f) $7 \times 10 \div 2$ …Since the operations are multiplication and division, perform the calculation in the order they appear in the question. $7 \times 10 = 70$ and $70 \div 2 =$ **35**

g) $3x + 8x \div 4 = 3x + (8x \div 4) = 3x + 2x =$ **5x**

h) $24 \div 2 + 5 - 3 \times 4 = (24 \div 2) + 5 - 3 \times 4$
$= 12 + 5 - 3 \times 4 = 12 + 5 - (3 \times 4) = 12 + 5 - 12 =$ **5**

EXERCISE 5A

1) Calculate

a) $4 \times 3 + 10$
b) $3 + 4 \times 2$
c) $5 \times 10 - 3$
d) $5 + 18 \div 6$
e) $3 + 2 - 1$
f) $3 - 2 + 4$
g) $10 \div 5 \times 9$
h) $3 \times 4 + 5 \times 2$
i) $5 \times 8 - 3 + 7$
j) $4 + 4^2 \div 2 - 3$
k) $9 + 3 \times (17 - 7)$
l) $54 \div (20 - 5 + 3)$
m) $27 - (3^2 \div 1)$
n) $7 - 4 \div 2 + 17$
o) $35 - (7 \times 3)$
p) $4 \times 5 - 5 - 10 \div 2$
q) $5^2 + (8 - 2 + 4) - 20$
r) $2 + 3 \times (8 - 5)$
s) $6 - 4 \times 2$
t) $100 + (34 - 14) \times 10$

2) Copy the calculations below and put in **brackets** where necessary to make the answer correct.

a) $5 + 6 \div 2 = 8$
b) $21 \div 3 + 4 = 3$
c) $2 \times 7 - 4 = 6$
d) $8 - 4 \div 4 = 7$
e) $9 + 3 + 3 \times 2 = 21$
f) $40 - 8 \times 7 = 224$

3) Simplify

a) $4w \times 3 + 10$
b) $3 + 4 \times 2y$
c) $5 \times 10c - 3c$
d) $5 + 18x \div 6$
e) $3n + 2n - 1$
f) $3w - 2 + 4w$
g) $10s \div 5s \times 9$
h) $3 \times 4m + 5 \times 2$
i) $5p \times 8 - 3 + 7p$
j) $4w + 4^2 \div 2 - 3$
k) $9 + 3 \times 17x - 7$
l) $200x \div 20 - 5 + 3$
m) $27p - 3^2 \div 1$
n) $7 - 4n \div 2 + 17$
o) $35k - 7 \times 3$
p) $4 \times 5 - 5 - 10k \div 2k$
q) $8b \times 8 - 5 \times 4b$
r) $2 + 3g \times 8g - 5$
s) $6x - 4x \times 2$
t) $4v \times 3 + 4 \times 6v - 4v \times 2$

5.2 SUBSTITUTION AND FORMULAE

Substitution simply means putting numbers in place of letters in expressions and formulae and performing the calculation(s).

For example, $6x + 3y$ is an algebraic expression. $6x$ and $3y$ cannot be added together because they are not like terms. However, if the values of the letters x and y are known, we can then substitute (replace) them for the letters in the expression and find its value.

Remember: $p \times p \times p = p^3$, $p + p + p = 3p$, $2y = 2 \times y$, $abc = a \times b \times c$, $2x^2 = 2 \times x \times x$

Example 1: If $x = 10$ and $y = 2$, the value of the expression $6x + 3y$ will be

$6 \times \mathbf{10} + 3 \times \mathbf{2} = 60 + 6 = \mathbf{66}$

Example 2: If $n = 3$, $y = ½$ and $t = 9$, find the values of

a) $t - n \longrightarrow 9 - 3 = \mathbf{6}$
b) $n^2 \longrightarrow 3 \times 3 = \mathbf{9}$
c) $4t \longrightarrow 4 \times 9 = \mathbf{36}$
d) $4(n + 9) \longrightarrow 4 \times (3 + 9) = 4 \times 12 = \mathbf{48}$
e) $8y \longrightarrow 8 \times ½ = \mathbf{4}$

EXERCISE 5B

1) If m = 3, n = 10 and y = ½, work out the value of

a) $2n$
b) $10m$
c) $m + n$
d) $n - m$
e) $20y$
f) $4y + m$
g) n^2
h) $6n + \frac{1}{2}y$
i) y^2
j) $2m^2$
k) $13n$
l) $15 - n$

2) If q = 7 and c = 0, work out the value of

a) $3q$
b) $40 – q$
c) $145c + 37$
d) q^2
e) $q - q$
f) $\frac{1}{3}c + 10q$
g) $5q + c + 7$
h) $8 – 2q$
i) $3(q + c)$
j) $16 + 10c - q$
k) $91 – 6q$
l) $20 - \frac{3}{4}c$

3) Work out the perimeter of the shapes below if
i) $x = 3$ cm ii) $x = 1.5$ m

a)
3x
x - 1

b)
x
x
Square
Square

c)
$4x + 1$
$5x + 2$
$2x$

d)
x
x
$x + 2$

4) If c = 4 and d = 10, find the value of

a) $3(d - c)$
b) $4c + 10d$
c) $40 \div c$
d) $10(d + c)$
e) $\frac{42}{c + d}$
f) $c\ (c - d)$
g) $\frac{cd}{20}$
h) $\frac{100}{d}$

5) If x = 2 and y = 3, calculate

a) $(y + x)^2$
b) $\sqrt{2x + 5}$
c) $x^2 + y^3$
d) $(3x)^2$
e) $y^0 + x^0$
f) $10(x – y)$

6) If s = 10, t = 3 and u = 0, calculate the value of:

a) $\frac{st}{3}$ b) t^3s c) $3st – t^3$ d) $(s + t + u)^2$

7) If $c = \frac{1}{2}$, d = 0.25 and e = -6, work out
The value of:

a) $20c$ b) $8c + d$ c) c^3 d) $d + e$

5.3 SUBSTITUTION INVOLVING NEGATIVE NUMBERS

In algebra, it is important to understand the operations with negative numbers. A negative number multiplied by a positive number always gives a **negative result**. A negative number multiplied by another negative number always gives a **positive result**.

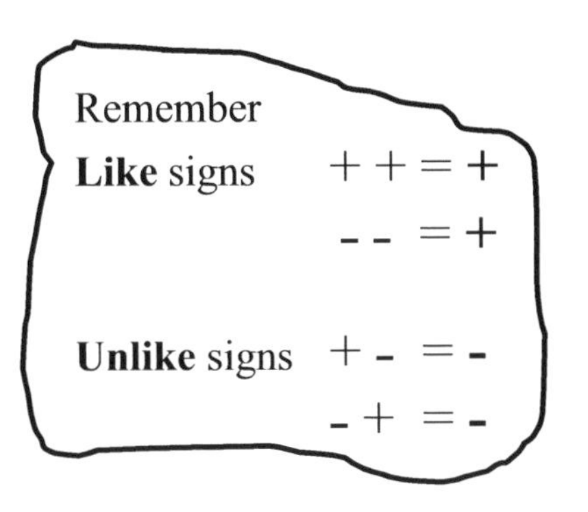

$2 \times 4 = \mathbf{8}$
$-2 \times (-4) = \mathbf{8}$

$2 \times (-4) = \mathbf{-8}$
$-2 \times 4 = \mathbf{-8}$

Example 1: If $x = 10$ and y = -5, find the value of

a) $x + y$ — $10 + (-5) = 10 - 5 = \mathbf{5}$
b) $2x - y$ — $2 \times \mathbf{10} - (-5) = 20 + 5 = \mathbf{25}$
c) y^2 — $(-5) \times (-5) = \mathbf{25}$
d) x^2 — $10 \times 10 = \mathbf{100}$
e) $4x^2$ — $4 \times x^2 = 4 \times x \times x = 4 \times 10 \times 10 = \mathbf{400}$

EXERCISE 5C

1) If p = -3, find the value of

a) 3p
b) 4p + 6
c) p + 1
d) 10 - p
e) 2p - 4
f) 18 - 2p
g) p - 100
h) p + p + p
i) $p^2 + 12p$
j) p - p
k) p ÷ p
l) 16p - 10

2) If $x = 3$ and y = -5 find the value of

a) $x + y$
b) $3x + 4y$
c) $6x - 2y$
d) $2x + y$
e) $x + 1 + y$
f) x - y
g) 70 - 5y
h) $3x + 10y + 20$
i) 9y - 3
j) y + 3x
k) $x^2 + y^2$
l) 19 + y - y

3) If a = -12 and b = -6, work out the value of

a) a + b
b) b + a
c) a + a + b
d) b – a
e) a – b
f) 2a
g) 3b + 5
h) 3a – 2b
i) 3 – b
j) a ÷ b
k) a ÷ 2b
l) 40 + b

5.4 SUBSTITUTION INTO A FORMULA

A **formula** is a rule to work out a value.

Example 1: In the formula V= IR, you may work out the value of V if the values of I and R are known. Likewise, you may work out the values of I or R if the remaining values are known.
When I = 3 and R = 5, the value of V in the above formula will be $3 \times 5 =$ **15**.

Example 2: The formula for area of a circle is given as $\mathbf{A = \pi r^2}$, where r is the radius.
If r = 5 cm and
$\pi = 3.142$,
$A = 3.412 \times 5^2$
$= 3.142 \times 25$
$= \mathbf{78.55\ cm^2}$

Example 3: Find the value of P is r = 3 from the formula P = 40r - 6
$P = (40 \times \mathbf{3}) - 6 =$ **114**

EXERCISE 5D

1) $S = \frac{P}{6} - 8$. Find the value of S when
a) P = 18 b) P = 48

2) $T = \frac{3u-2}{5}$. Find T when u = 4

3) S = 2t + 30. Find the value S when
a) t = 10 b) t = -30

4) V = u + at. Find the value of V when u = 7, a = 3 and t = 4.

5) $C = \frac{T}{4} - 3$. Find C when T = 28.

6) S = 200 + 3y. Find the value of S when a) $y = \frac{1}{3}$ b) y = -10

7) Volume of an object is given by the formula $V = r^2h$. Find
a) V when r = 5.7 and h = 2.5
b) h when V = 294 and r = 7

8) Copy the diagram below and find the value of each expression using y = 3 in the middle.

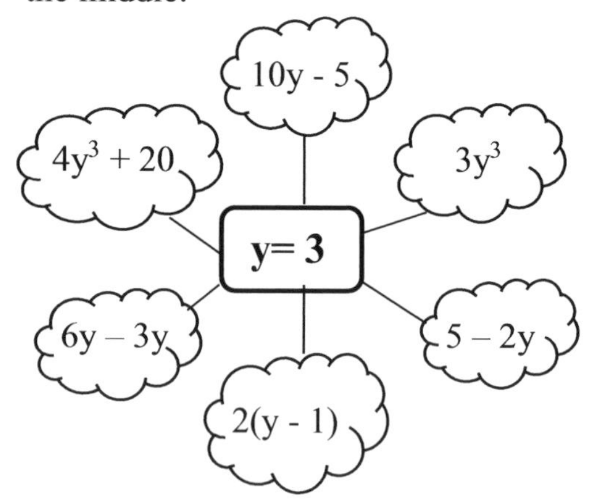

9) $f = \frac{1}{u} + \frac{1}{v}$. Find the value of F when
a) u = 2 and v = 3
b) u = 0.6 and v = 0.3

10) The equation of a straight line is in the form **y = mx + c**, where m is the gradient and c the intercept on the y – axis. Find

a) y when m = 3, x = 2 and c = -5
b) y when m = -2, x = 4 and c = 0.4

5.5 SOLVING EQUATIONS

In expressions, there are **no equal** signs. Examples $5x$, $2x + 5$, $n + 6y$….
In equations, there are equal signs and letters stand for a particular number - the solution of the equation.

Linear equations have only the first power of the unknown quantity.
Examples of linear equations are: $2x = 10$, $x + 5 = 20$…

RULES FOR SOLVING EQUATIONS

Do the same thing to both sides. You may

- Add the same number to both sides
- Subtract the same number from both sides
- Multiply both sides by the same number
- Divide both sides by the same number

Example 1: Solve $x + 5 = 9$

To solve means to find the value of the unknown. The left side of the equation MUST equal the right side. $x + 5$ must equal 9. To make the left side equal to 9, find the opposite of **+ 5**, which is -5. Subtract 5 from both sides as explained below.

So in $x + 5 = 9$
(-5) (-5) …………………. Subtract 5 from both sides
$x + 5 - \mathbf{5} = 9 - \mathbf{5}$
$x = 4$ ✓

Check that $x = 4$ is the right answer by replacing x with 4 in the original equation. If it equals 9, then 4 is the solution. From the original equation
$x + 5 = 9$
$\mathbf{4} + 5 = 9$
$9 = 9$…. this is the confirmation needed. The left side is now equal to the right side.

Example 2: Solve $2x = 10$
This means 2 multiply by a number gives 10. We must find that number.
Find the opposite of times 2 (×2). This is (÷ 2). We divide both sides by 2.

$$\frac{\overset{1}{\cancel{2}}x}{\underset{1}{\cancel{2}}} = \frac{\overset{5}{\cancel{10}}}{\underset{1}{\cancel{2}}}$$

$x = 5$ ✓

Or $\mathbf{2x = 10}$
(÷2) (÷2)
$\mathbf{x = 5}$

Check your answer by multiplying 2 by 5 in the original equation. $2 \times \mathbf{5} = 10$

Example 3: Solve $\frac{x}{5} = 7$

Multiply both sides by 5 to make a whole number.

$\frac{x}{\cancel{5}} \times \cancel{5} = 7 \times \mathbf{5}$ Therefore, $x = 35$ ✓

Check: From the original equation, **35** ÷ 5 = 7

Example 4: Solve $x - 3 = -7$
Add 3 to both sides. $x - 3 + \mathbf{3} = -7 + \mathbf{3}$ $x = -4$ ✓
Check: - **4** - 3 = -7

Example 5: solve 40 = 8y
Divide both sides by 8 because 8y means 8 × y. 5 = y which implies that y = 5 ✓
Check: 8 × **5** = 40. Left side equals right side.

Example 6: Solve $7 = \frac{n}{3}$

A number divides 3 and the answer is 7. To solve this, find the opposite of divide by 3. It is multiply (×) by 3. So we multiply both sides by 3.

$7 \times \mathbf{3} = \frac{n}{\cancel{3}} \times \cancel{3}$ 21 = n Therefore, n = 21 ✓

Check: **21** ÷ 3 = 7

Example 7: Solve 2x = -8

Divide both sides by **2** $\frac{\cancel{2}x}{\cancel{2}} = \frac{\overset{-4}{\cancel{-8}}}{\cancel{2}}$ $x = -4$ ✓

Check: 2 × - **4** = - 8

Equations could be thought of as a set of balanced scales. The middle section represents the equal sign.

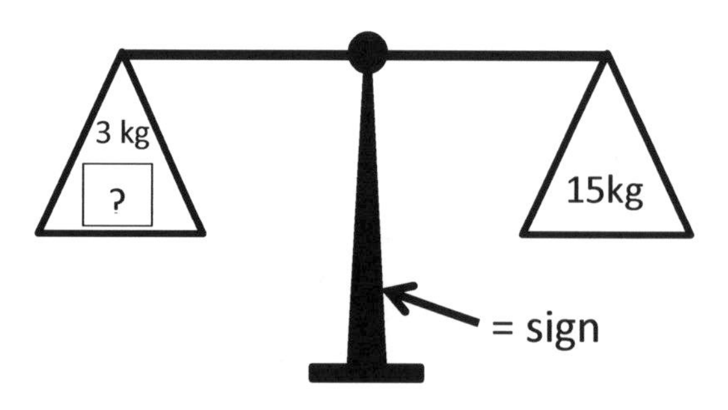

The left side must equal 15kg.
Therefore, 15 - 3 = 12
The missing weight is **12 kg**.
Check: 3 + 12 = 15

EXERCISE 5E

1) Solve the equation below.

a) $c + 5 = 7$
b) $c - 4 = 9$
c) $c + 5 = 20$
d) $9 - x = 8$
e) $x + 11 = 22$
f) $x - 90 = 1$
g) $x + 19 = 37$
h) $w + 6 = 16$
i) $w - 6 = 16$
j) $w + 7 = -2$
k) $u + 5 = -8$
l) $6 + x = -10$
m) $u + 0 = -7$
n) $r - 49 = 4$
o) $g - 3 = -8$
p) $w - 7 = 16$
q) $13 + y = -3$
r) $56 = r - 8$
s) $4 + y = 34$
t) $n - 7 = 15$
u) $9 + e = 2$
v) $20 = x - 4$
w) $c - 20 = -10$
x) $x + 6.8 = 13.6$
y) $x - 5.5 = 2.3$
z) $u + 256 = 600$

EXERCISE 5F

1) Solve

a) $8x = 16$
b) $7x = 21$
c) $42 = 7x$
d) $63 = 9x$
e) $5x = 55$
f) $13w = 39$
g) $3w = 27$
h) $-2c = 4$
i) $-7c = 14$
j) $20 = -5w$
k) $8x = 64$
l) $-7x = 70$
m) $-x = -7$
n) $-3n = -9$
o) $\frac{n}{7} = 6$
p) $\frac{x}{3} = 10$
q) $\frac{w}{9} = -2$
r) $\frac{3}{n} = 1$
s) $-9x = -63$
t) $-n = 9$
u) $18 = -9n$
v) $-20m = -60$
w) $\frac{w}{1.5} = 3$
x) $9z = 3$
y) $67n = 670$
z) $\frac{y}{2} = -47$

5.6 SOLVING EQUATIONS IN TWO-STEPS

Example 1: Solve $2x + 5 = 11$
Solution: Get rid of (+5) by doing the exact opposite. Therefore, subtract 5 from both sides.

$2x + 5 - \mathbf{5} = 11 - \mathbf{5}$
$2x + 0 = 6$ Therefore, $2x = 6$
Divide both sides by 2 to give $\boldsymbol{x = 3}$ ✓

Check: Put $x = 3$ in the original equation
$2 \times \mathbf{3} + 5 = 11$ ……………..this gives $6 + 5 = 11$
11 = 11 (*Left hand side is equal to right hand side***)**

So, the solution to the above equation is $x = 3$

Example 2: Solve $7x - 1 = -8$
Add 1 to both sides
$7x - 1 + 1 = -8 + 1$
$7x = -7$
Divide both sides by 7
$\mathbf{x = -1}$ ✓

Example 3: Solve $\frac{2x}{5} = 10$
Multiply both sides by 5 to get
$2x = 50$
Divide both sides by 2
$\boldsymbol{x = 25}$ ✓
Check: $2 \times 25 = 50$, $50 \div 5 = 10$

EXERCISE 5G

1) Solve the equations

a) $2c + 5 = 13$
b) $5c + 10 = 20$
c) $5x - 10 = 10$
d) $4w + 1 = 9$
e) $5 + 3x = 8$
f) $8 + 2x = 6$
g) $7x + 5 = 26$
h) $20w - 7 = 13$
i) $8y + 3 = 51$
j) $6x - 16 = -4$
k) $3u - 7 = -10$
l) $10x - 10 = -20$
m) $4y + 6 = 14$
n) $16y - 5 = 155$
o) $\frac{4n}{3} = 8$
p) $\frac{6v}{9} = 2$
q) $7 = 6x - 5$
r) $19 = 5x + 4$
s) $8c + 9 = 49$
t) $40w - 4 = 76$
u) $3x + 20 = 44$
v) $2n - 5 = 9$
w) $8w + 2 = 66$
x) $8 + 4x = 20$
y) $13c - 6 = 46$
z) $4 + 17x = -13$

5.7 FRACTIONAL EQUATIONS

Example 1: Solve $\frac{x}{3} + 7 = 5$
Subtract 7 from both sides

$\frac{x}{3} + 7 - \mathbf{7} = 5 - \mathbf{7}$

$\frac{x}{3} = -2$

Multiply both sides by 3
$\boldsymbol{x = -6}$ ✓

It could also be seen as a function machine

x → [÷3] → $\frac{x}{3}$ → [+7] → $\frac{x}{3} + 7$

Like all function machines, the inverse of the above operation is subtract 7 then multiply by 3.

$\frac{x}{3} + 7 = 5$
(-7) (-7)…………Inverse of +7 is -7

$\frac{x}{3} = -2$

$\frac{x}{3} \times \mathbf{3} = -2 \times \mathbf{3}$ ……...Inverse of ÷3 is ×3

$x = -6$

Example 2: Solve $\frac{2x}{5} = 10$
Multiply both sides by 5 to get
$2x = 50$
Divide both sides by 2
$\boldsymbol{x = 25}$ ✓

Check: $2 \times 25 = 50$, $50 \div 5 = 10$

Example 3: Solve $\frac{c}{2} = \frac{1}{4}$

Multiply both sides by 2 and 4 to make the numbers horizontal.

$\frac{c}{2} \times \mathbf{2} \times \mathbf{4} = \frac{1}{4} \times \mathbf{2} \times \mathbf{4}$

$\frac{8c}{2} = \frac{8}{4}$ ….cancelling down gives

$4c = 2$….divide both sides by 4

$c = \frac{2}{4}$

$c = \frac{\mathbf{1}}{\mathbf{2}}$ ✓

EXERCISE 5H

1) Solve the equations below

a) $\frac{n}{5} + 1 = 4$

b) $\frac{n}{4} - 3 = 7$

c) $\frac{n}{2} + 5 = 10$

d) $8 = \frac{40}{n}$

e) $\frac{n}{-3} = 3$

f) $\frac{n}{10} - 8 = 22$

g) $\frac{n}{-7} = 9$

h) $\frac{3n}{5} = 3$

i) $6 + \frac{n}{7} = 8$

j) g) $\frac{n}{8} + 8 = 23$

k) $\frac{4n}{3} = 4$

l) $\frac{n}{6} - 2 = 11$

m) $8 = \frac{1}{4}n$

n) $\frac{n}{6} = \frac{2}{3}$

o) $\frac{n}{5} = \frac{8}{20}$

p) $\frac{3n}{6} = \frac{2}{5}$

q) $\frac{n}{-10} = \frac{2}{5}$

r) $\frac{n}{6} + 5 = 2$

s) $15 + \frac{n}{5} = 16$

t) $\frac{20}{n} = 5$

u) $\frac{35}{n} = 7$

v) $3 - \frac{3}{n} = 0$

w) $\frac{1}{2} = \frac{4n}{8}$

x) $\frac{n}{6} = \frac{2}{3}$

y) $20 = 6 + \frac{n}{2}$

z) $\frac{n}{4} - 1 = -3$

EXTENSION QUESTIONS

1) The rectangle below has a perimeter of 26 cm. Work out the length of the longest side of the rectangle.

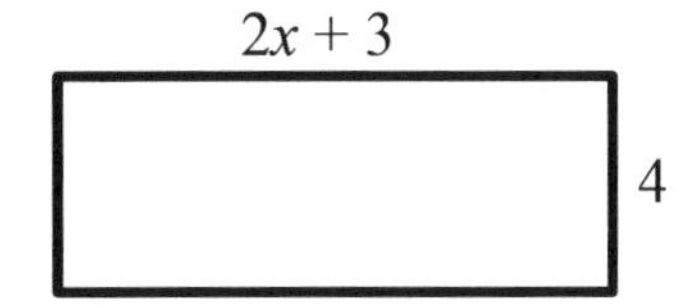

2) Work out the area of the rectangle in question 1 above.

3) Solve the equations below.

a) $\frac{d}{5} - 3 = -6$

b) $\frac{4}{5}w = -8$

c) $\frac{7-n}{3} = 2$

5.8 EQUATIONS WITH BRACKETS

Example 1: Solve $3(2x + 2) = 30$

Expanding Method

$3(2x + 2) = 30$
$(3 \times 2x) + (3 \times 2) = 30$
$6x + 6 = 30$
(Now subtract 6 from both sides)
$6x = 24$
Divide both sides by 6
$x = \frac{24}{6} = \mathbf{4}$

Dividing by multiplier method

$\frac{\not{3}(2x+2)}{\not{3}} = \frac{30}{3}$

$2x + 2 = 10$
(-2) (-2)
$2x = 8$
$x = \frac{8}{2} = \mathbf{4}$

Check: Put $x = 4$ in the original equation. Left hand side (LHS) = 30 and right-hand side (RHS) = 30.

Example 2: Solve $8 = 4(5x – 1)$

Using the expanding method,

$8 = 20x - 4$
(+4) (+4)
$12 = 20x$
(÷20) (÷20)
$\frac{12}{20} = x$
$\frac{3}{5} = x$ or $x = \frac{3}{6}$ or **0.6**

Example 3: Solve $3(x+2) + 4(x-2) = 12$

$3x + 6 + 4x – 8 = 12$
$7x - 2 = 12$
$7x = 12 + 2$
$7x = 14$
$x = \frac{14}{7} = \mathbf{2}$

Check: Put $x = 2$ in the original equation.

$3(\mathbf{2}+2) + 4(\mathbf{2}-2) = 12$
$3 \times 4 \quad + 4 \times 0 \quad = 12$
$12 \qquad\qquad = 12$
LHS = RHS = 12

EXERCISE 5I

Solve the equations

1) $2(3b – 7) = 4$
2) $3(2t + 5) = 21$
3) $3(x – 4) = 3$
4) $2(12 – x) = 0$
5) $3(5 – c) = 6$
6) $21 = 3(2p + 5)$
7) $4(2x + 5) = 19$
8) $6(x – 3) = 32$
9) $6 = 2(4 – x)$
10) $4(x – 1) = 16$
11) $11(4y–2) = 43$
12) $1 + 3(d -1) = 4$
13) $3(4 – x) = 6$
14) $3(x – 2) = -3$
15) $5(x + 2) = 60$
16) $7(x + 3) = -7$

Solve the equations

17) $3(2y + 1) + 2(y – 1) = 25$
18) $4(2 – 4x) – 2(3 + 5x) = 28$
19) $6x – (3 – x) = 0$
20) $0.2(3x – 4) + 0.3(5x + 2) = 30$

5.9 EQUATIONS WITH UNKNOWN ON BOTH SIDES

Sometimes, the unknown terms are on both sides of the equality sign. It is best solved by collecting the unknown terms either to the left or right of the equality sign.

Example 1: Solve $3n - 5 = 2n + 8$

A faster approach would be to take 5 over to the right side and take 2n to the left of the equality sign. However, we must follow the rules of moving terms and numbers. *If it is a positive term or number, subtract on both sides. If it is a negative term or number, add on both sides.*

$$3n - 5 = 2n + 8$$

To take 5 to the right side, add 5 on both sides.

$$3n - 5 + \mathbf{5} = 2n + 8 + \mathbf{5}$$
$$3n = 2n + 13$$

Now subtract 2n on both sides.

$$3n - 2n = 13$$
$$\mathbf{n = 13} \checkmark$$

Point to note: You have a choice of getting rid of any of the terms or numbers first, but it must be systematic (One after another). You will still have the same answer. For example, you may decide to get rid of 2n first.

$$3n - 5 = 2n + 8$$
$$3n - 5 - \mathbf{2n} = 2n + 8 - \mathbf{2n}$$
$$n - 5 = 8$$

Now add 5 on both sides

$$n - 5 + \mathbf{5} = 8 + \mathbf{5}$$
$$\mathbf{n = 13}$$

Example 2: Solve $3(x + 5) = 2(4 - x)$

Expand first to give $3x + 15 = 8 - 2x$

Now, add 2x to both sides

$$3x + 15 + \mathbf{2x} = 8 - 2x + \mathbf{2x}$$
$$5x + 15 = 8$$

Take 15 away from both sides

$$5x + 15 - \mathbf{15} = 8 - \mathbf{15}$$
$$5x = -7$$

Dive both sides by 5

$$5x \div \mathbf{5} = -7 \div \mathbf{5}$$
$$\mathbf{x} = \frac{-7}{5} = \mathbf{-1\tfrac{2}{5}} \checkmark$$

Exercise 3: Solve $\frac{3}{4}y - 5 = 9 - y$

Solution: When fractions are involved in an equation, remove the denominator. To remove the denominator, multiply both sides by the denominator, 4.

$$\mathbf{4} \times (\tfrac{3}{4}y - 5) = \mathbf{4} \times (9 - y)$$
$$3y - 20 = 36 - 4y$$

Now, add 20 to both sides

$$3y - 20 + \mathbf{20} = 36 - 4y + \mathbf{20}$$
$$3y = 56 - 4y$$

Add 4y to both sides

$$3y + \mathbf{4y} = 56 - 4y + \mathbf{4y}$$
$$7y = 56$$

Divide both sides by 7

$$\frac{7y}{7} = \frac{56}{7}$$
$$\mathbf{y = 8} \checkmark$$

As usual, you may check your answer by replacing y in the original equation with 8. LHS = RHS = 1. Therefore, y = 8 is the correct solution.

Example 4: Solve $\frac{2x-1}{3} = \frac{3x+1}{4}$

If we have a fraction, the first step is to remove the denominator(s). Since there are two denominators, multiply both sides by the LCM of the two numbers. The LCM of 3 and 4 is 12.

$$\frac{\overset{4}{\cancel{12}}(2x-1)}{\cancel{3}} = \frac{\overset{3}{\cancel{12}}(3x+1)}{4}$$

$4(2x - 1) = 3(3x + 1)$

Expand

$8x - 4 = 9x + 3$

Add 4 to both sides

$8x - 4 + \mathbf{4} = 9x + 3 + \mathbf{4}$

$8x = 9x + 7$

Subtract 9x from both sides

$8x - \mathbf{9x} = 9x + 7 - \mathbf{9x}$

$-x = 7$

Divide both sides by -1

Therefore, $\mathbf{x = -7}$ ✓

EXERCISE 5J

Solve the equations below.

1) 5y – 8 = 3y
2) 2a – 7 = 8 – 3a
3) 2x + 15 = 8 – 4x
4) 18 – 5x = 3x + 2
5) 11x – 5 = x + 25
6) y + 5 = 3y + 9
7) 4 – x = 8x + 20
8) 3w – 15 = 8 – 4w
9) y + 9 = 3y + 16
10) 7c – 11 = 2c + 4
11) d + 3 = 14 – 3d
12) 3t = 2t + 7
13) 1 + 7x = 4 - x
14) 15x = 10 – 5x
15) y – 3 = 3y + 7
16) 4x = 3x + 6
17) 7 – 3u = 5–2u
18) 6x – 2 = 1-3x
19) f–16 = 16– 2f
20) 14 –3d = d +3
21) 4x + 5 = x +5
22) k – 3 = 3k - 2
23) 8c – 1 = c - 5
24) 6w + 8 = w+1

EXERCISE 5K

Find the solution to

1) 5(x + 2) = 2(x + 6)
2) 8(y - 3) = 2y
3) 7(3a – 5) = 2(5a – 1)
4) 4(3x + 1) = 32 – 2x
5) x(3 + 5) = 6 – 4x
6) 3(2x – 1) = 8x + 1
7) 2(c – 3) – (c – 2) = 5
8) 5(g + 1) = 2g + 3 + g
9) 4(1 – 2x) = 3(2 – x)
10) 5y – 3(y – 1) = 39

11) $\frac{2f-1}{3} = \frac{f}{2}$

12) $\frac{t-1}{5} - \frac{t-1}{3} = -2$

13) $\frac{d}{2} - 3 = \frac{d}{5}$

14) $\frac{3}{x+3} = \frac{9}{x+5}$

15) $\frac{3}{2x-1} = \frac{4}{3x+1}$

FORMING AND SOLVING EQUATIONS

Example 1: I subtract 3 from a number and then multiply the result by 10, the answer is 100. a) Write an **equation** and b) **solve** it to find the original number.

Solution: Let the number be **y**.
If I subtract 3, it will give y – 3.
If I multiply (y - 3) by 10, it will give 10(y – 3). **a)** Since the answer is 100, the equation is **10(y – 3) = 100.** ✓

b) Expand the bracket to give
10y – 30 = 100
(+30) (+30)
10y = 130
(÷10) (÷10)
y = 13. The original number is **13.** ✓

Example 2: Conrad doubles a number, added 12 and then divides the result by 5. He got the same answer when he multiplied the number by 1.
a) Write an equation for Conrad's number. **b)** Solve the equation to find Conrad's number.
Solution
Let Conrad's number be **n**. He doubles the number to give 2n. He added 12 to give 2n + 12. He then divides 2n + 12 by 5 to give $\frac{2n + 12}{5}$. The result is the same as 1 × n = n.
a) So, the equation is $\frac{2n + 12}{5} = n$ ✓
b) Multiply both sides by 5 to give
2n + 12 = 5n
(Subtract 2n from both sides)
12 = 3n. Therefore, **n = 4**
Conrad's number is **4** ✓

EXERCISE 5L

In questions 1 – 5, Helen is thinking of a number. Write an equation and hence, solve it to find Helen's number.

1) She doubles the number and adds 10. The answer is 44.

2) She adds 3 to the number and multiplies the result by 7. The answer is 70.

3) She multiplies the number by 3 and adds 6. She then divides the result by 3. She got the same number by multiplying the number by 2.

4) She is thinking of a fraction. She subtracts 6 from the number and divides the result by 4. She got the same answer when she subtracts 4 from the number.

5) Helen trebles the number, adds 2 and multiplies the result by 2. She got the same answer when she multiplied the number by 2 and add 12.

6) The sum of three consecutive numbers is 66. Find the numbers.

7) Philip is t years old. His mother is 25 years older.

a) How old will Philip be in 10 years' time?
b) How old will Philip's mother be in 10 years' time?
c) How old was Philip 5 years ago?
d) If the mother is 50 years old, how old is Philip?

8) The perimeter of the rectangle below is 22 cm.

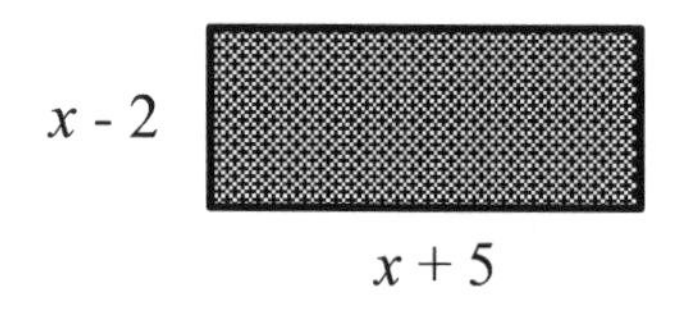

a) Write an equation for the perimeter.
b) Hence, solve the equation to find the value of x.
c) Work out the length and width of the rectangle.
d) Calculate the area of the rectangle.

9) Dr Brown has £x. Jude, his eldest child, has $\frac{2}{3}$ of his money. Leanne, his only daughter, has $\frac{1}{5}$ of his money.
If the total amount of money Jude and Leanne have is £13 000, work out how much their father have in Pounds.

10) I add 38 to a number. I then divide the result by 2. The overall result is the same as ten times the number. Work out the original number by first forming an equation.

11) I am thinking of a number. I added 6 to the number and multiplied the result by 7. I then subtract 40, and the result is 16.
a) Write an equation to find my number.
b) Hence, solve the equation to find the number.

12) Joe is thinking of a number. He subtracts 9 from the number and then doubles it. He got the same answer as dividing the number by 5. Write an equation and solve it to find Joe's number.

13) For each diagram, write down an equation involving w and solve it.

a)

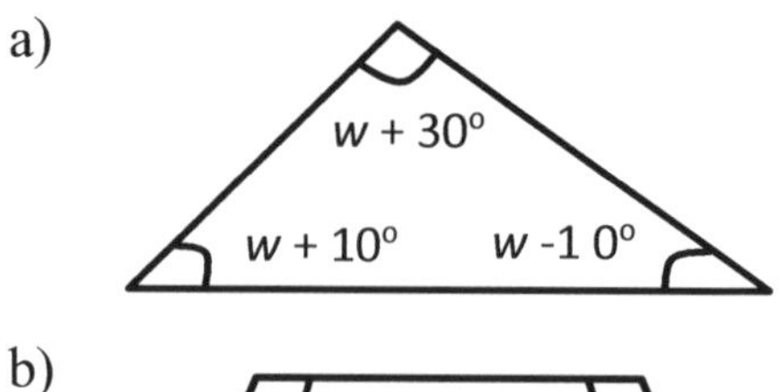

b)

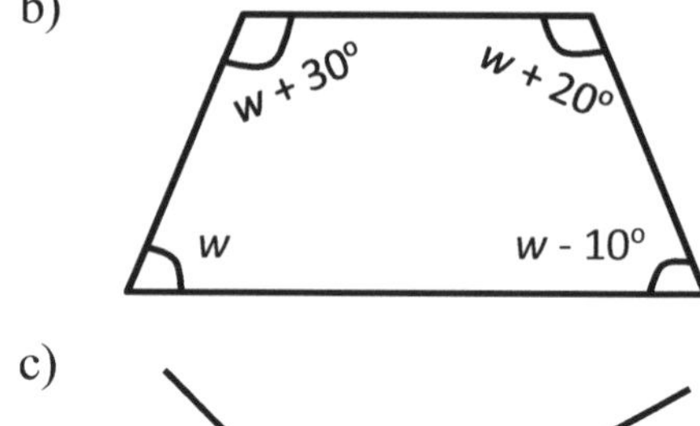

c)

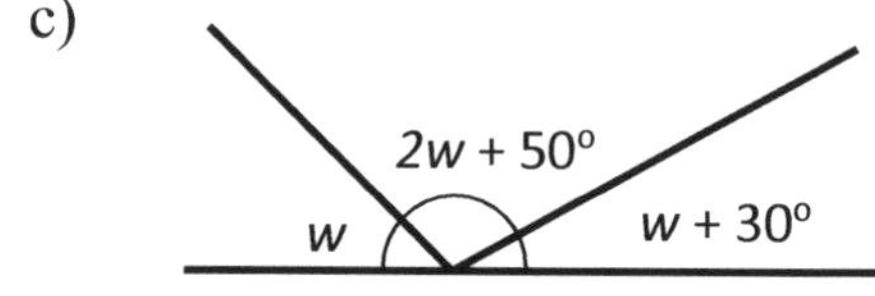

14) Tony added four consecutive numbers and the total was 450.
a) Write down an equation. 4x + 6= 450
b) Solve your equation to find the consecutive numbers. 111,112,113,114

15) The triangle below has a perimeter of 49 cm.

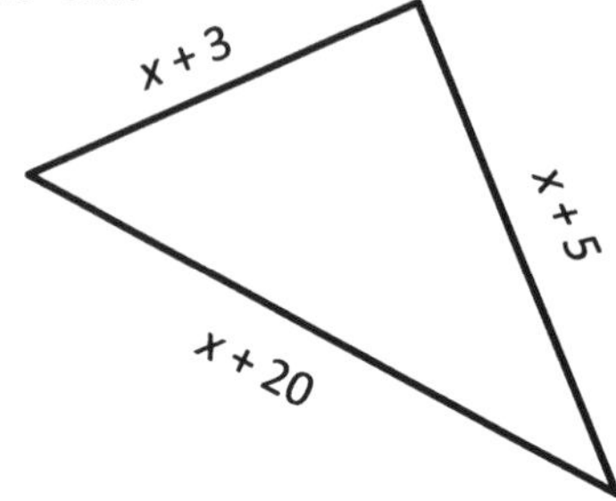

a) Write an expression for the perimeter of the triangle. 3x + 28
b) Form an equation. 3x + 28 = 49
c) Solve your equation. x = 7
d) Work out the lengths of all the sides. 10 cm, 12cm, 27cm

5.10 INEQUALITIES

The symbols for inequalities are:

$<$	Less than
$\leq$	Less than or equal to
$>$	Greater than
$\geq$	Greater than or equal to

$x < 3$ means that the value of x is less than 3, but **not equal** to 3. The values that satisfy this inequality could be 2, 1, 0, -2,.....but 3 is **not** included.

$x \leq 3$ means that the value of x **must** be 3 or less. For example, 3, 2, 1, 0,......

$x > 3$ means that the value of x **must** be greater than 3 and **not equal** to 3. Examples are 4, 5, 6, ...

$x \geq 3$ means that the value of x is greater than or equal to 3. The values that satisfy this inequality must be 3 itself and above. For example, 3, 4, 5....

We encounter inequalities in everyday life.

- "The number of students in class A is more than 10." We can write as $\mathbf{s > 10}$ where s is the number of students.
- "The penalty for stealing must be at least 6 months in prison." We can write $\mathbf{p \geq 6}$ months.
- "The speed limit in Harlow is 30 m.p.h." We can write $\mathbf{s \leq 30}$, where s is the speed limit.

INEQUALITIES ON A NUMBER LINE

To successfully represent inequalities on a number line, the following concepts should be followed.

1) **Open circle** (O) means that the number is **not included**. For example,

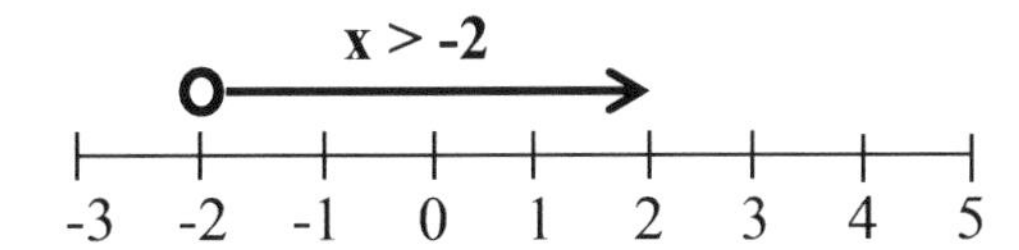

The direction of the arrow represents the values covered by the inequality. $x > -2$, means that -2 is not included because of the open circle. The integer values that satisfy the inequality $x > -2$ are -1, 0, 1, 2, 3.....

2) Closed circle (●) means that the number **is included**. For example,

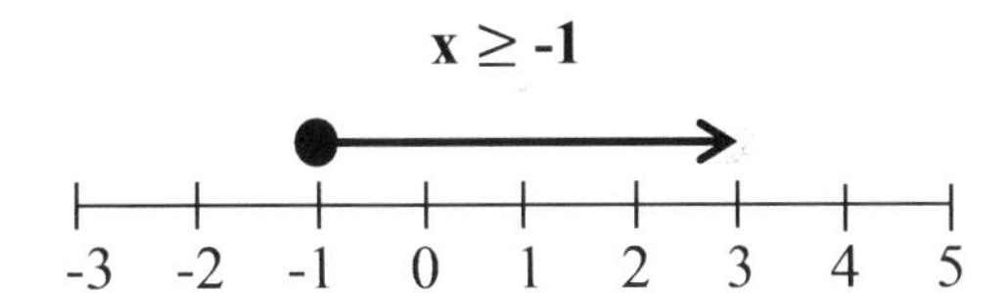

The integer values that satisfy the inequality $x \geq -1$ are -1, 0, 1, 2, 3......

3) $-3 < x \leq 2$ represents all the numbers between -3 and 2 but -3 is not included. The integer numbers that satisfy this inequality are -2, -1, 0, 1 and 2. Notice that -3 is not included. On a number line, $-3 < x \leq 2$ is

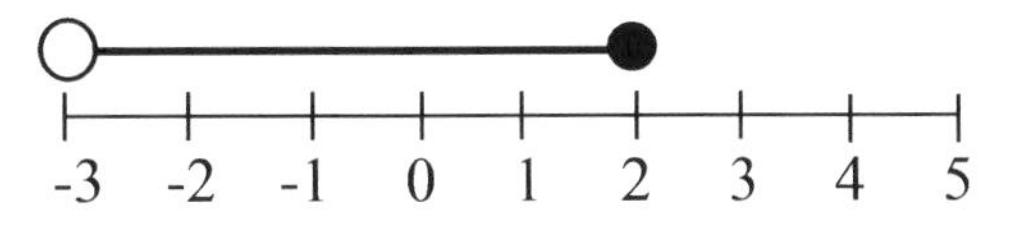

EXERCISE 5M

Using *x* for the variables, write down the inequalities shown.

1)

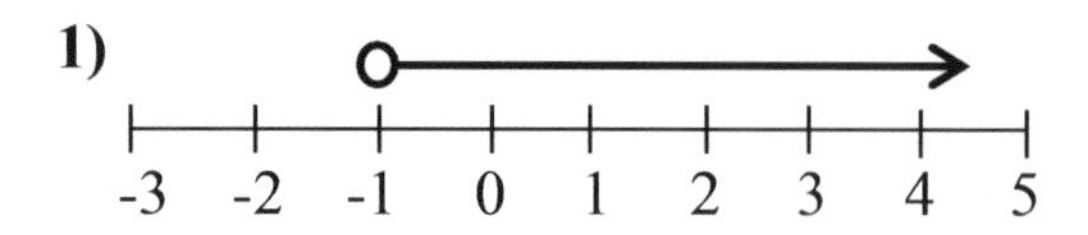

2)

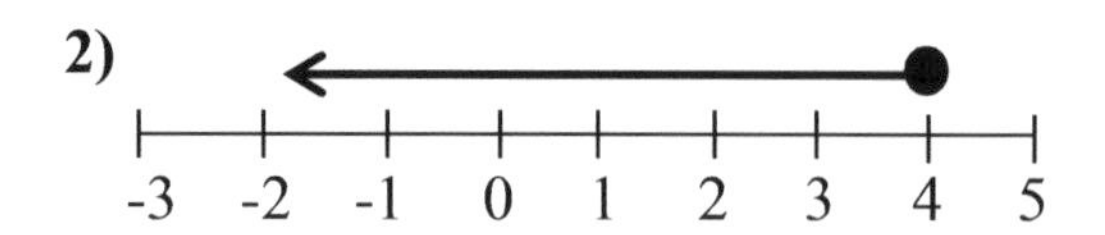

3)

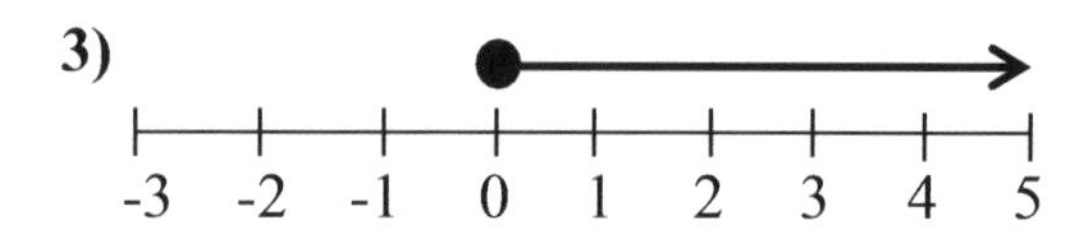

4)

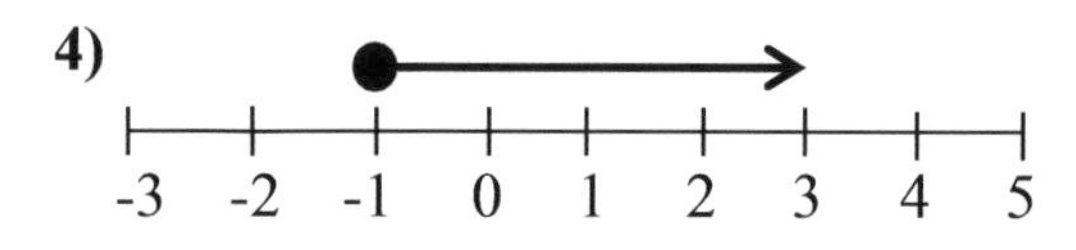

5)

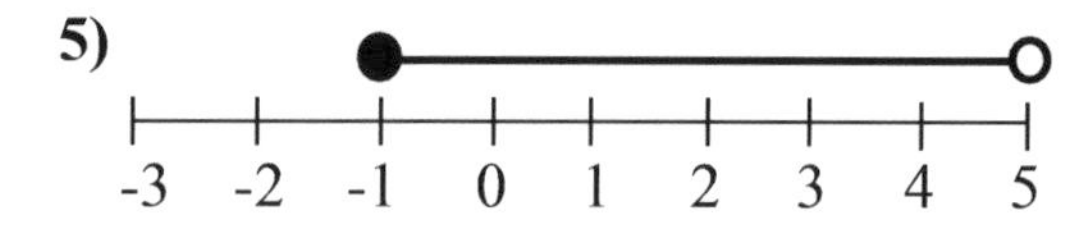

6)

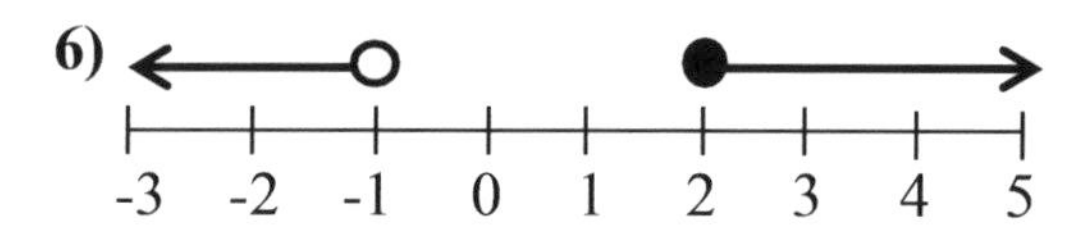

7)

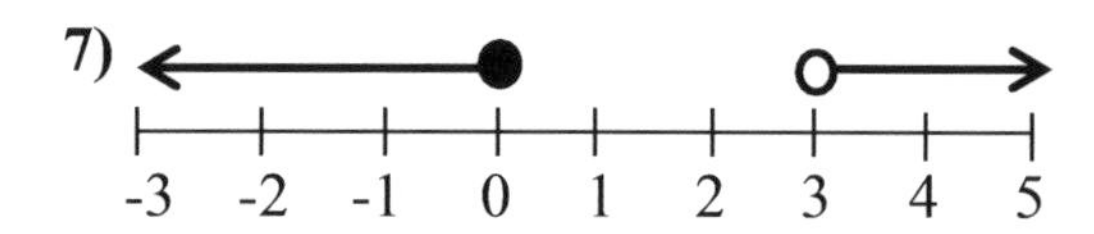

Unlike equality sign where $x = 5$ is the same as $5 = x$, inequalities are different. $x < 5$ is **not** the same as $5 < x$.
$x < 5$ is **the same** as $5 > x$. The inequality sign is reversed if we want to start with the number.

The reason for this is simple; for $x < 5$, let's assume that x is 4. $4 < 5$ is correct. If we want to start from 5 and write $5 < 4$. This is not correct as 5 is not less than 4. Therefore, we reverse the sign to give **$5 > 4$,** which is now correct.

EXERCISE 5N

1) Using inequality symbols, rewrite the following statements.

a) x is less than 5
b) x is less than 3
c) x is greater than 1
d) x is less than or equal to -2
e) y is greater than or equal to 1

2) By drawing a number line, display the inequalities below.

a) $x < -3$	d) $x \geq 5$
b) $x < 3$	e) $w > 6$
c) $x \leq -4$	f) $w < 5$

3) Write **true** or **false** for each statement.

a) $5 < 6$	e) $0 < 7$
b) $4 < 3$	f) $4 \geq 4$
c) $5 \leq 7$	g) $4^3 < -20$
d) $10 \geq 10$	h) $-40 > -10$

4) Write down the **smallest** integer that satisfies the following inequalities.
a) $x > 1$ b) $x > -2$ c) $x \geq 5$ d) $x \geq -9$

5) Write down the **largest** integer that satisfies each of these inequalities.
a) $x < 2$ b) $x \leq 1\frac{1}{2}$ c) $x < -0.5$
d) $x \leq -3\frac{3}{4}$

SOLVING INEQUALITIES

If we solve normal equations like $2x = 6$, the solution is **one** value which is 3 in this case. However, when solving inequalities, we find **a range of values** while following most of the rules of solving linear equations.

RULES FOR SOLVING INEQUALITIES

1) Add or subtract the same thing to both sides.

2) Multiply or divide both sides by the same positive number while retaining the inequality sign.

3) If we must multiply or divide by a **negative** number, then the inequality sign must be **reversed**.

Observe the inequality $3 < 4$. If we multiply both sides by a negative number say (-2), we have -6 and -8.

$$3 < 4$$

$$\times(-2) \qquad -6 \; > \; -8$$

↑ Sign is reversed

If we retain the inequality sign, <, the statement would be incorrect. -6 is not less than -8. It is greater than -8. Therefore, we reverse the sign to greater than (>). **$-6 > -8$.**

Same principle for dividing by a negative number! ***Reverse the sign.***

Example 1: Solve the inequalities

a) $x + 2 > 9$ d) $-2x \geq 9$

b) $x - 1 \leq 3$ e) $4 < x + 2 \leq 7$

c) $3x > -9$

Solutions:

a) $x + 2 > 9$

Subtract 2 to both sides, $\mathbf{x > 7}$

b) $x - 1 \leq 3$

Add 1 to both sides, $\mathbf{x \leq 4}$

c) $3x > -9$

Divide both sides by 3, $\mathbf{x > -3}$

d) $-2x \geq 9$

Divide both sides by -2, $\mathbf{x \leq -4.5}$

Remember, the inequality sign is reversed because we divided by a negative number.

e) $4 < x + 2 \leq 8$

Subtract 2 from all sides (to have only x in the middle)

$4 - \mathbf{2} < x + 2 - \mathbf{2} \leq 8 - 2$

This gives $\mathbf{2 < x \leq 6}$

We can represent the inequality on a number line like

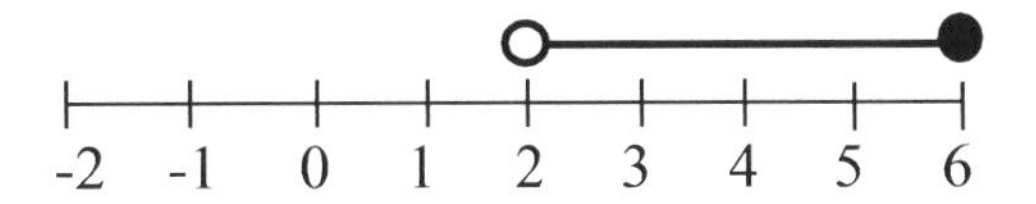

Example 2: Write down the integer values that satisfy the inequality

$-4 \leq x - 3 < 2$

Add 3 to all sides to give $-1 \leq x < 5$.

The integer values are **-1, 0, 1, 2, 3, 4**

Notice that 5 is not included.

EXERCISE 5O

Solve the inequalities below.

1) $x + 2 < 7$
2) $x - 3 < 2$
3) $x + 2 \leq -4$
4) $x - 4 \geq 8$
5) $6x + 3 > 9$
6) $3x - 3 > -2$
7) $4 + x < 2$
8) $6 - x \leq 8$
9) $-4 - x > 9$
10) $9 - 2x < 4$
11) $-7n \geq -28$
12) $-2w + 3 < 2$

Solve the inequalities and represent on a number line.

13) $-5n \leq 10$
14) $\frac{n}{4} \geq -2$
15) $\frac{n}{3} < 1$
16) $\frac{n}{10} < \frac{3}{5}$
17) $2 < x + 1 < 3$
18) $3x + 3 > 4$
19) $\frac{x+5}{4} > -2$
20) $-15 < x - 2 \leq 3$
21) $-6 \leq 3x < 3$
22) $-2 < \frac{1}{2}x < 3$

Where **possible**, find the range of values of ***n*** for which both inequalities are true.

23) $n > -2$ and $n < 1$
24) $n > 5$ and $n \leq 10$
25) $n < 2$ and $n > 4$

Two pairs of inequalities are given below. Solve them and write down integer values that satisfy both inequalities.

26) $x + 5 < 10$ and $2x + 3 \geq 1$

27) $3x - 1 > 5$ and $4x < 20$

28) $-4x < 12$ and $x + 3 < 9$

INEQUALITIES AND REGIONS

When two or more variables are involved, the inequality can be represented by a **region** on a graph. The region is an area where **all the points** obey a given rule.

RULES FOR THE REGIONS

1) We use a **broken** line for inequalities of < or >. The points on the broken line are **not included** in the region.

2) We use a **solid** line for inequalities of ≤ or ≥. The points on the solid line are **included** in the region.

Example 1: Represent the regions marked by the inequalities below.

a) $x < 1$ b) $y \geq 1$ c) $x + y \geq 1$

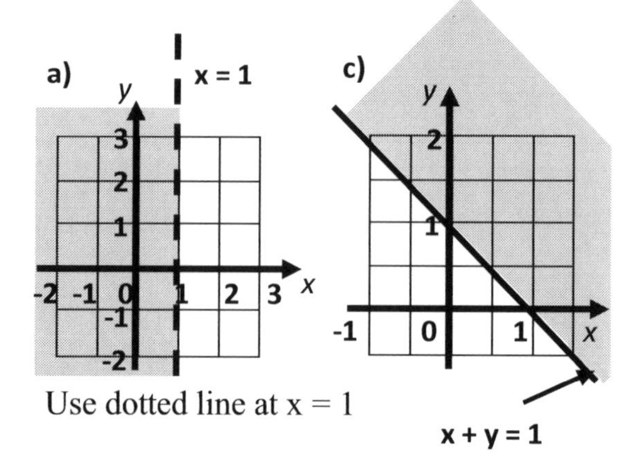

Use dotted line at x = 1

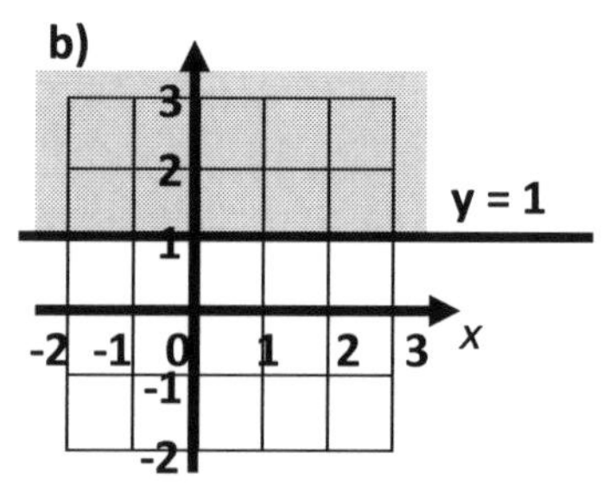

Use solid line at y = 1

Example 2: Shade the region which satisfies the inequalities $x > 1$, $y \geq 1$ and $x + y < 4$.

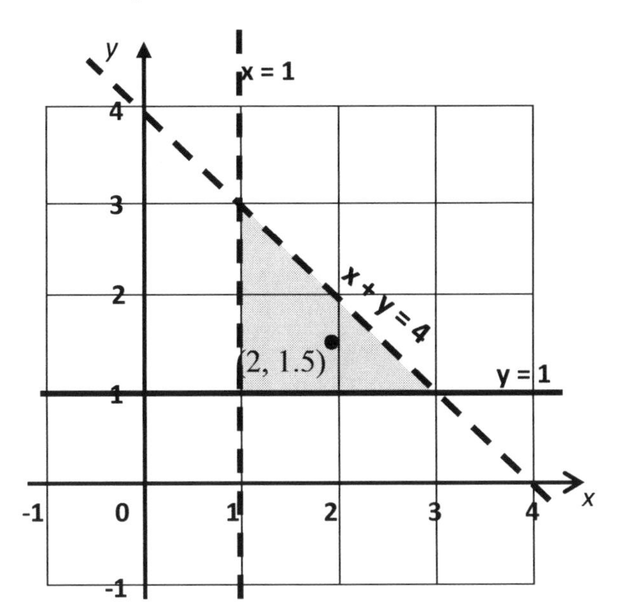

To **check** if the shaded region is correct, choose a coordinate point **inside** the shaded area and **not** on any of the lines. Let us choose (2, 1.5), see diagram. For

$x > 1$, the x coordinate is greater than 1. $y \geq 1$, the y coordinate is greater than or equal to 1 and $x + y < 4$, x coordinate (2) + y coordinate (1.5) is 3.5 which is less than 4.

Therefore, the shaded area is correct.

EXERCISE 5P

Draw diagrams to show the regions which satisfy the inequalities below.

1) $x > 3$
2) $x \leq 5$
3) $x > -1$
4) $y > 5$
5) $y > x + 2$
6) $x + y \leq 3$
7) $x + y \geq -2$
8) $y > 3 - 2x$
9) $x > 1$, $y \leq 3$, $y \geq x - 3$

Find the inequalities which describe the **unshaded** regions.

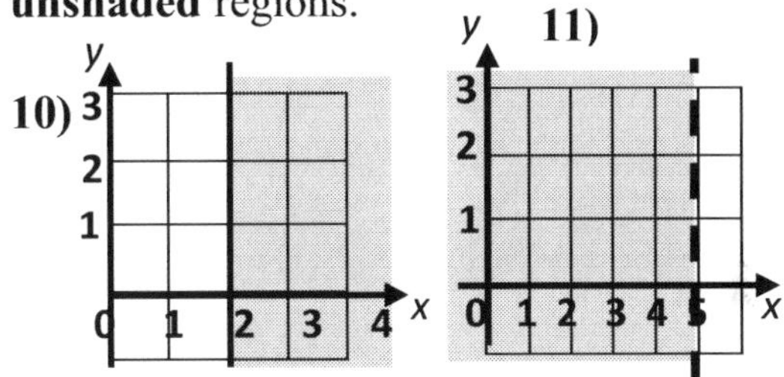

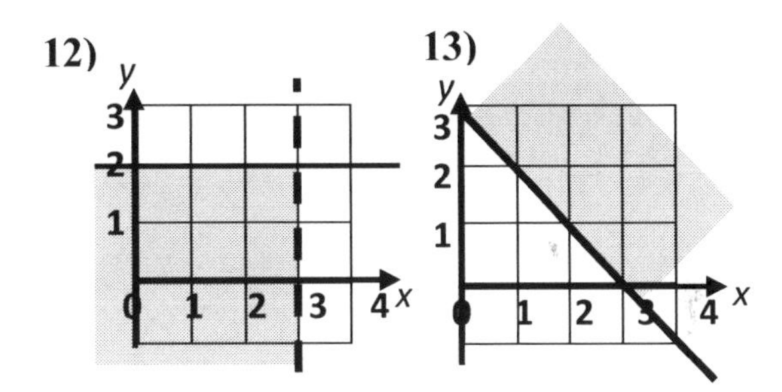

14) Arinze's supermarket employs people as security men or cleaners. Every evening, a maximum of 8 people are employed, with at least 3 people as security and one person as a cleaner.

a) If s = the number of security men and c = the number of cleaners, form inequalities to represent the information.
b) Draw a diagram to represent the region which satisfies the inequalities.

Chapter 5 Review Section

Assessment

1) Simplify by removing brackets

a) $8 + (13 - 5)$ b) $10 - (6 + 7)$ c) $(4 + 2) + (13 - 4) - 8$

……………………...**3 marks**

2) Write without brackets and simplify where possible.

a) $(c - d) + (e - f)$ b) $9w + 5n + (n + 4w)$ …………………… **2 marks**

3) Simplify

a) $(5f - b) - (4f - 5b)$ b) $7a - (3d - 2e)$

c) $8 \times w^2$ d) $30a^3b^2 \div 10ab$ ……………………...**4 marks**

4) Work out the values of

a) $6 + 2 \times 3$ b) $5 \times 8 - 3 + 7$ c) $4^2 + (5 - 2 + 4)$ …………………… **3 marks**

5) If $x = 5$ and $y = -2$, work out the value of

a) $x + y$ b) $x - y$ c) y^2 d) $7 - y$ …………………… **4 marks**

6) Solve the equations

a) $n + 8 = 10$

b) $2 + n = 13$

c) $n - 7 = 20$

d) $6x = 24$

e) $x + 4 = -5$

f) $3x = -3$

g) $17 - y = 3$

h) $x - 14 = -4$

…………………… **8 marks**

7) Solve the equations

a) $\frac{n}{5} = 3$

b) $\frac{x}{7} = 10$

c) $\frac{2x}{7} = 2$

d) $\frac{x}{9} = -5$

e) $\frac{x}{20} = -3$

f) $13 + \frac{x}{3} = 19$

g) $11 - \frac{y}{6} = 3$

h) $5x - 4 = 11$ ………………… **16 marks**

8) Solve the equations

a) $3(2y + 1) + 2(y - 1) = 25$

b) $y + 9 = 3y + 17$

c) $7c - 11 = 2c + 4$

d) $4(2 - 4x) - 2(3 + 5x) = 80$

e) $7 - 5d = d + 3$

f) $8x + 5 = x + 4$

g) $\frac{4}{x + 4} = \frac{8}{x + 6}$

h) $\frac{1}{4x - 1} = \frac{3}{3x + 1}$

……………**24 marks**

9) Dorothy is thinking of a fraction. She subtracts 6 from the number and divides the result by 4. She got the same answer when she subtracts 4 from the number. Find Dorothy's fraction.

……………..**3 marks**

10) The perimeter of the rectangle below is 30 cm.

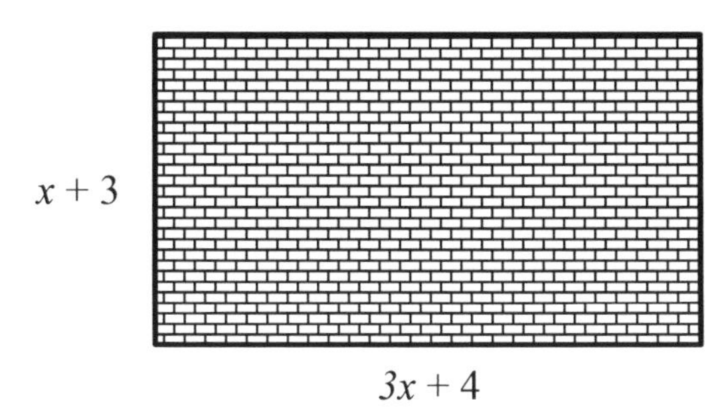

a) Write an equation for the perimeter. ………………**2 marks**

b) Hence, solve the equation to find the value of x. ………………**3 marks**

c) Work out the length and width of the rectangle. ………………**2 marks**

d) Calculate the area of the rectangle. ……………....**2 marks**

11) Complete the following inequalities by inserting $>$ or $<$ in each box.

a) 4 □ 5 b) 8 □ 5 c) 9 □ -3 d) -5 □ -7 e) -10 □ -9

…………5 marks

12) Show the following inequalities on a number line.

a) $x < -5$ b) $x \geq 3$ c) $x < -7$ d) $-2 < x < -1$ e) $0 < x \leq 5$

…………5 marks

13) Solve the inequalities.

a) $6x - 2 > 5$ …………2 marks

b) $x + 7 \leq -3$ …………2 marks

c) $-7w \leq 49$ …………2 marks

d) $\frac{a - 5}{3} < -10$ …………2 marks

e) $-3 < x + 5 < 7$ …………2 marks

14) List all the integer (whole number) values that satisfy the following inequalities.

a) $3 < x < 9$ …………2 marks

b) $-4 \leq n \leq 2$ …………2 marks

c) $-2 < 2(n - 3) < 7$ …………2 marks

d) $-8 \leq 4x < 4$ …………2 marks

15) Shade the region that satisfies the inequality $y \geq 2x - 3$ …………3 marks

16) Write down the inequalities that describe the **shaded** regions.

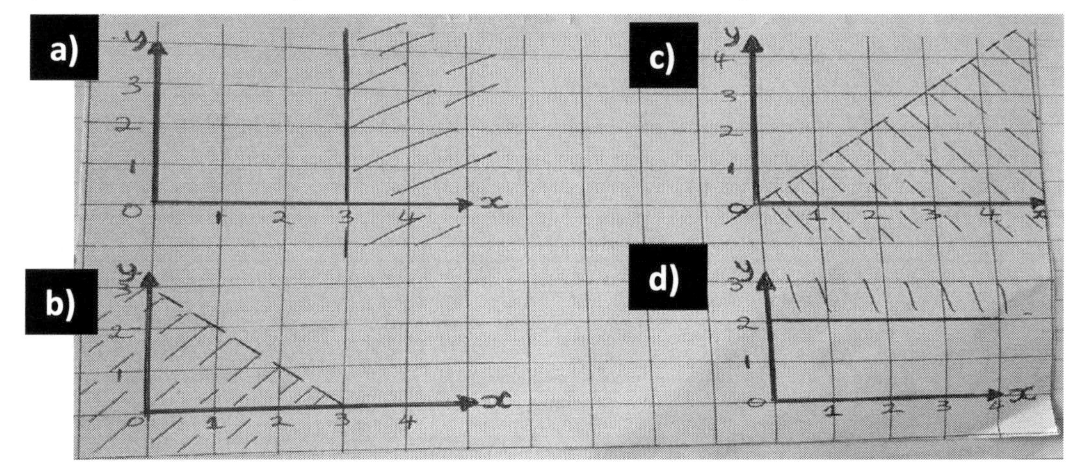

…………8 marks

6 Polygons and circles

This section covers the following topics:

- Geometrical properties of 2D shapes
- Angles in 2D shapes
- Polygons
- Circles

LEARNING OBJECTIVES

the end of this unit, you should be able to:

a) Understand and use properties of lines and angles
b) Calculate angles on a straight line
c) Calculate angles at a point
d) Work out vertically opposite angles
e) Understand geometrical properties of triangles, quadrilaterals and polygons
f) Calculate angles in parallel lines
g) Calculate angles in triangles and quadrilaterals
h) Understand corresponding and alternate angles
i) Understand the properties of circles

KEYWORDS

- Angles
- Parallel line
- Corresponding and Alternate angles
- Polygons
- Circles
- Quadrilaterals

6.1 LINES AND ANGLES

Angles on one side of a straight line always add to 180 degrees.

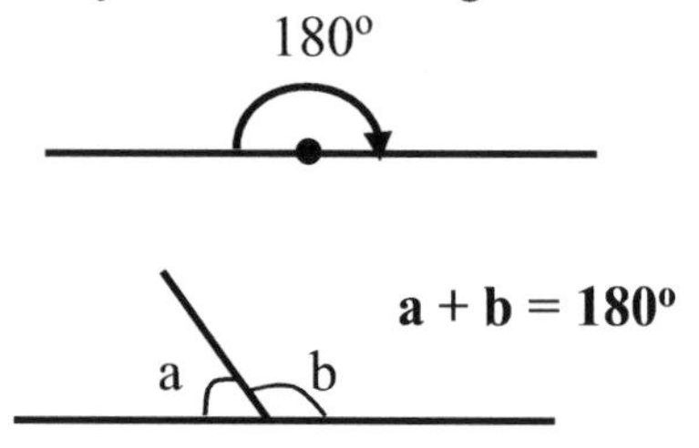

Example 1: Calculate the value of the missing angle.

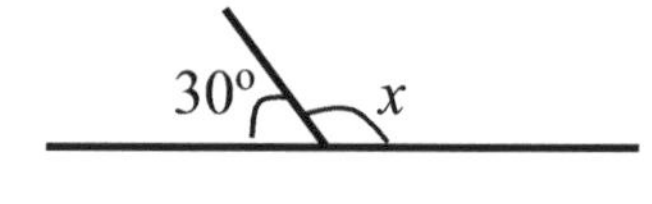

$30 + x = 180$……… $x = 180 - 30 = \mathbf{150°}$

Check: 30° + 150° = 180 …angles on a straight line

Example 2: work out the missing angle.

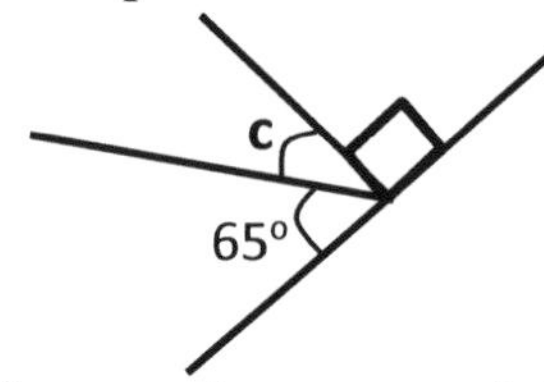

Since angles on a straight line add to 180°,

$65° + c + 90° = 180°$

$155° + c = 180°$

$c = 180 - 155 = \mathbf{25°}$

Check: $65° + 25° + 90° = 180°$

Example 3: Work out the value of x.

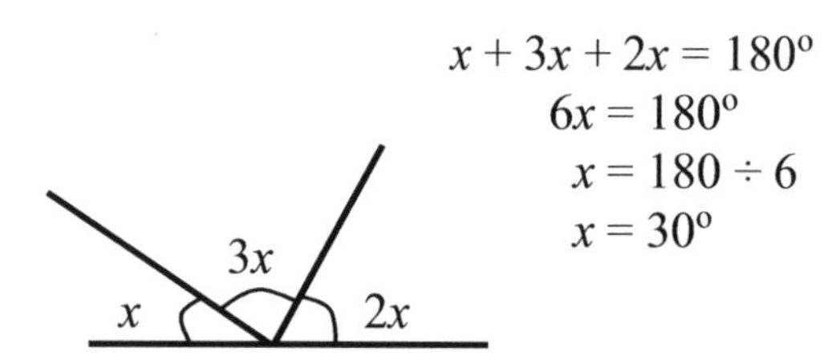

$x + 3x + 2x = 180°$

$6x = 180°$

$x = 180 \div 6$

$x = 30°$

6.2 ANGLES AT A POINT

The sum of angles at a point is 360°.

Note: There are 360° in a ***full*** turn.

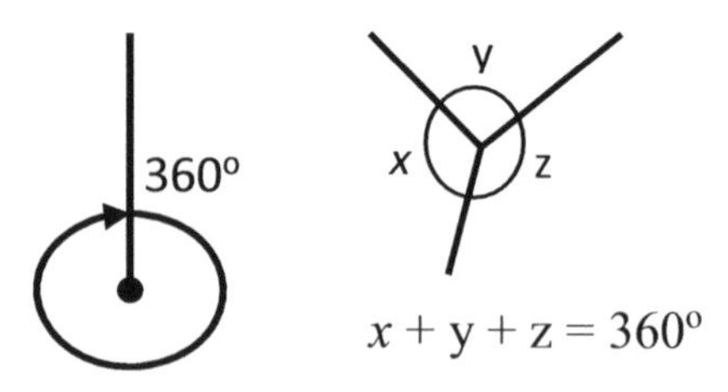

$x + y + z = 360°$

Example 1: Work out the missing angle

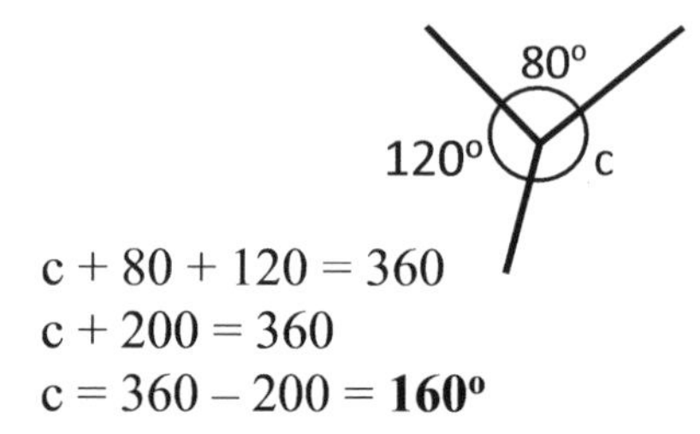

$c + 80 + 120 = 360$

$c + 200 = 360$

$c = 360 - 200 = \mathbf{160°}$

Example 2:

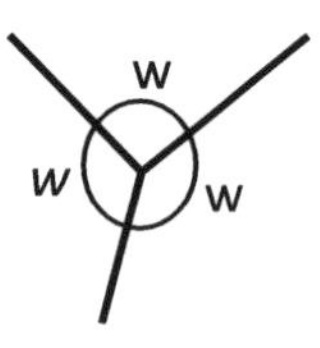

$w + w + w = 360°$

$3w = 360$

$w = 360 \div 3$

$w = 120°$

6.3 VERTICALLY OPPOSITE ANGLES

Vertically opposite angles are equal.

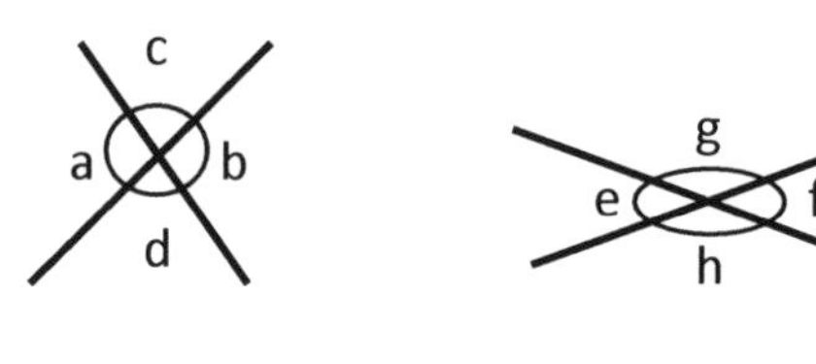

a = b, c = d

e = f and g = h

Example 1: Work out the missing angle

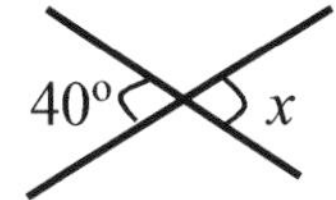

x = **40°**….vertically opposite angles

Example 2 Work out the missing angles

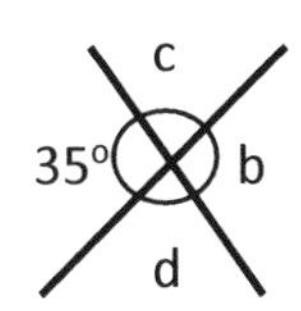

b = **35°** (vertically opposite angles)
c = 180 – 35 = **145°**
(Angle on a straight line)
d must be **145°** (Vertically opposite to angle c)

Example 3 Work out the missing angles

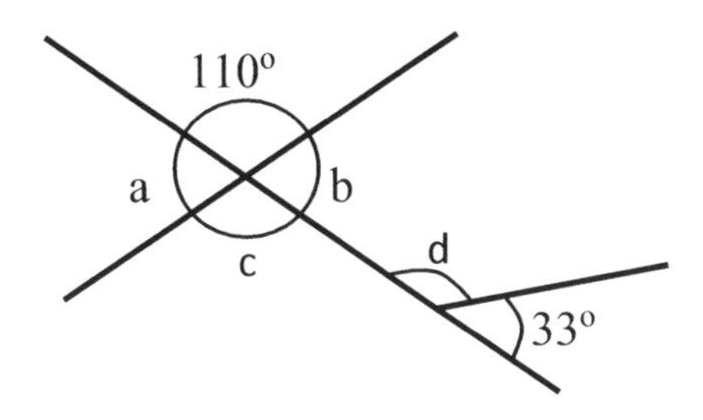

c = 110°……Vertically opposite angles
b = 180 – 110
= 70°……..angle on a straight line
a = b = 70°….vertically opposite angles
d = 180 – 33 = 147°
……..angle on a straight line

EXERCISE 6A

1) Calculate the value of the angles marked with letters in each diagram.

a)

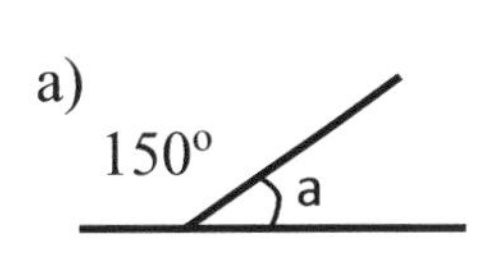

b)

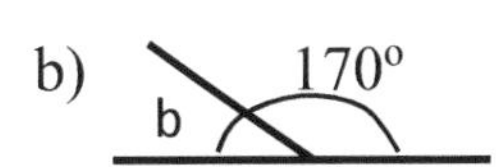

c)

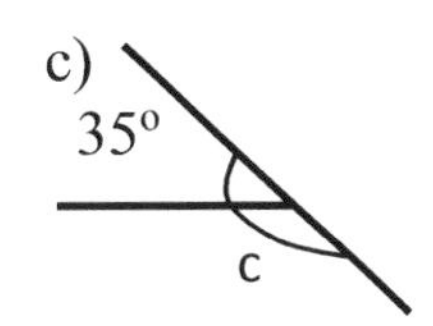

d)

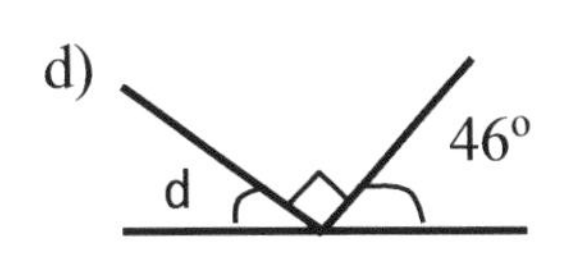

e)

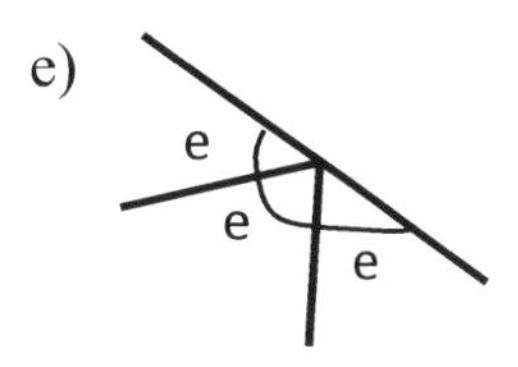

f)

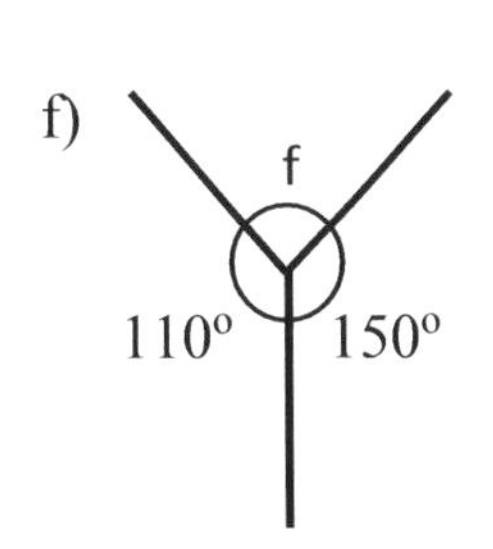

g)

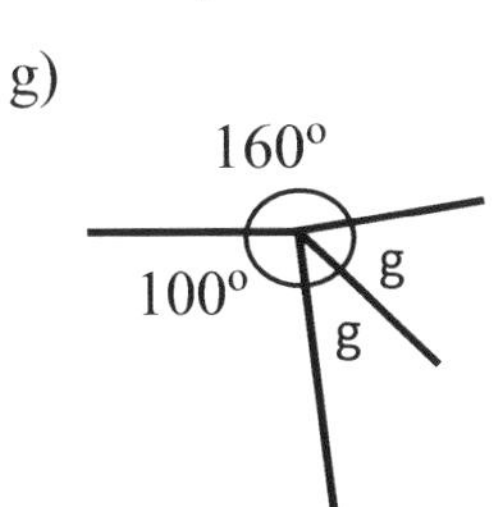

h)

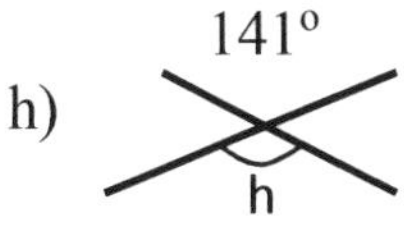

i)

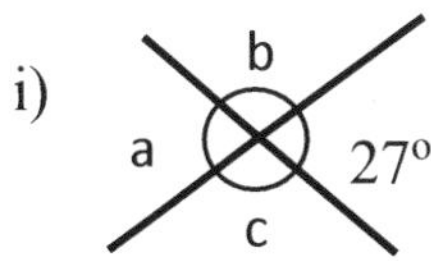

j)

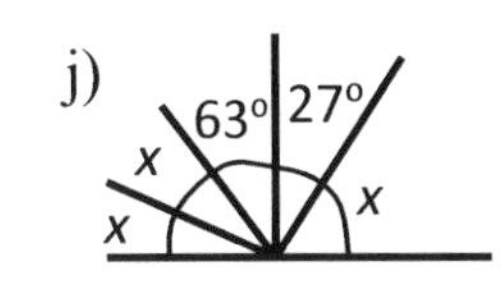

k)

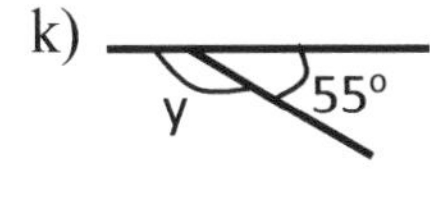

l)

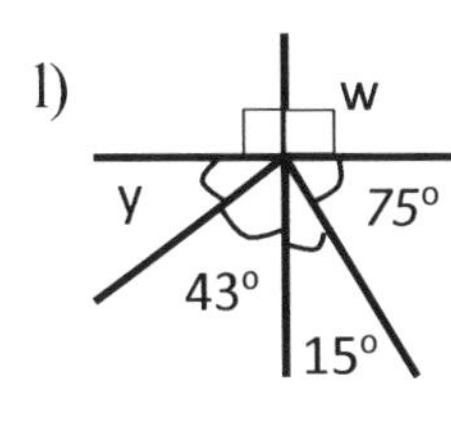

m)

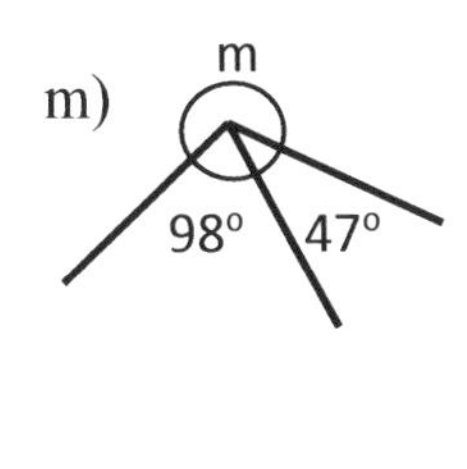

n) x 50° n

6.4 TRIANGLES AND ITS PROPERTIES

A triangle is a two-dimensional shape made up of three sides and three angles. All the interior (inside) angles in a triangle add to **180°**.

There are basically four types of triangles.

1) Right-angled triangle
2) Isosceles triangle
3) Equilateral triangle
4) Scalene triangle

RIGHT-ANGLED TRIANGLE

A right-angled triangle has one angle that is 90°. It is usually marked as a square.

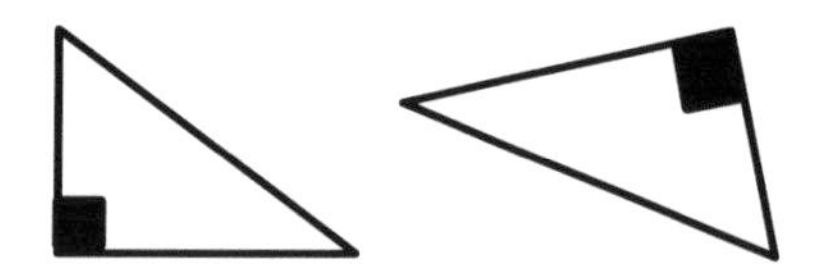

A right-angled triangle has **no line of symmetry** but has rotational symmetry of **order 1**.

However, if the right-angled triangle is also isosceles, there is only **one** line of symmetry.

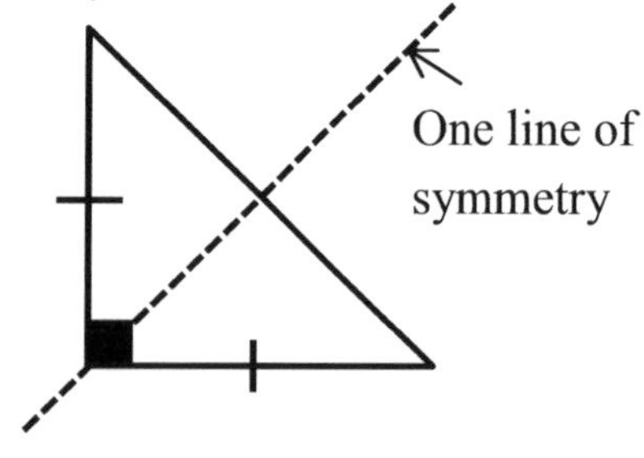

Right-angled isosceles triangle

ISOSCELES TRIANGLE

Isosceles triangle has two equal sides and two equal *base* angles. It has **one line of symmetry** and a rotational symmetry of **order 1**.

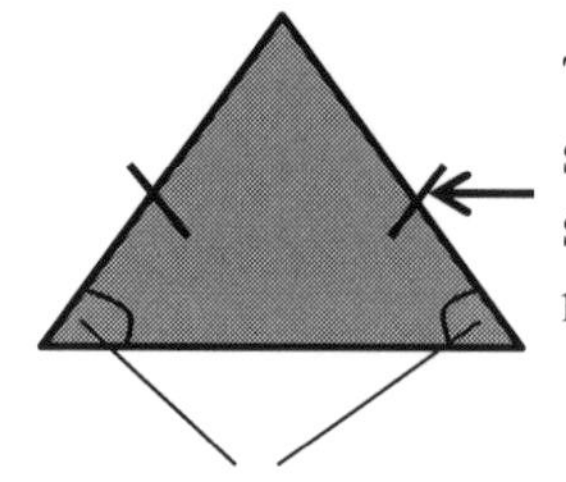

EQUILATERAL TRIANGLE

Equilateral triangle has three equal sides and three equal angles. All the three angles are 60° each.

It has **three lines of symmetry** and rotational symmetry of **order 3**.

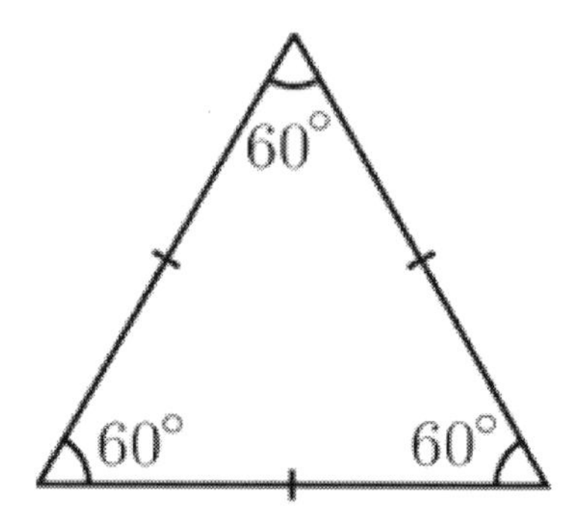

Most times, three marks are used on the three sides indicating three equal sides.

SCALENE TRIANGLE

A scalene triangle has three unequal sides and angles

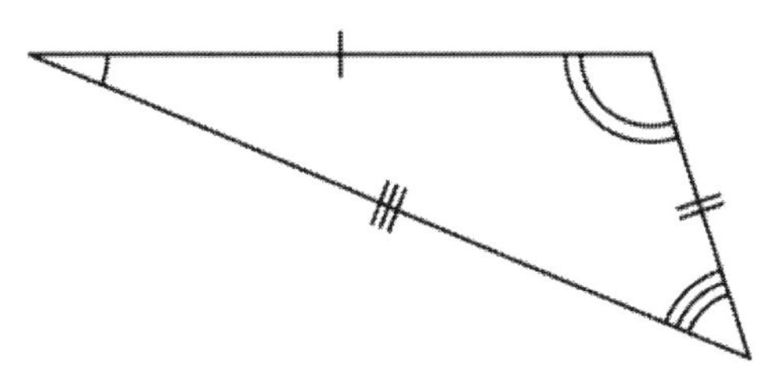

It has **no line of symmetry** and rotational symmetry of **order 1**.

Remember: Rotational symmetry is the number of times a shape fits exactly onto itself in a complete turn (360°).

CALCULATING ANGLES IN TRIANGLES

The interior (inside) angles in any triangle add up to 180°.

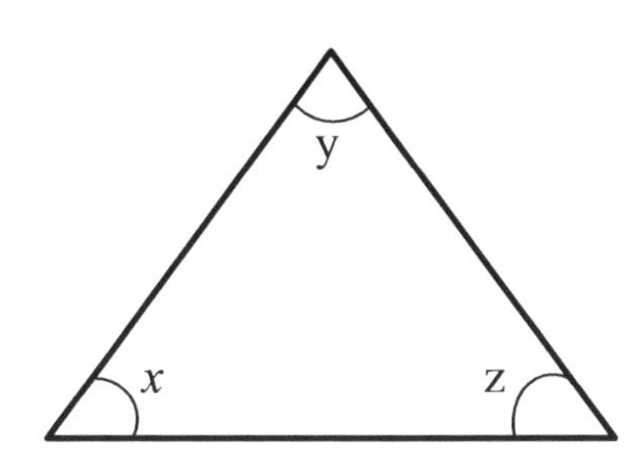

$x + y + z = 180°$

Once two angles are known, the third angle can be calculated.

Example 1: Calculate the missing angle

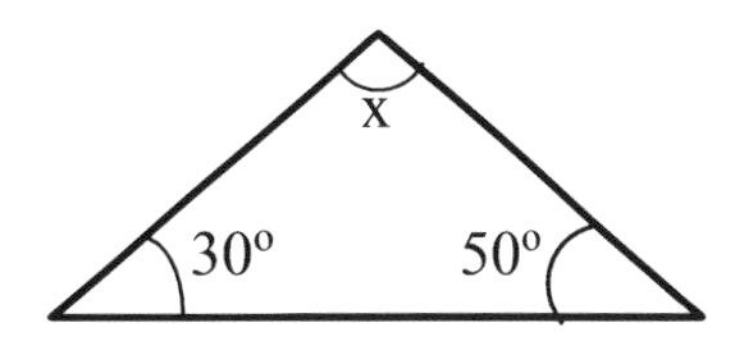

$x + 30 + 50 = 180°$
$x + 80 = 180°$
$x = 180 - 80 =$ **100°**

Check: 100 + 30 + 50 = 180

Example 2: Calculate the missing angles $A\hat{C}B$ and $C\hat{A}B$

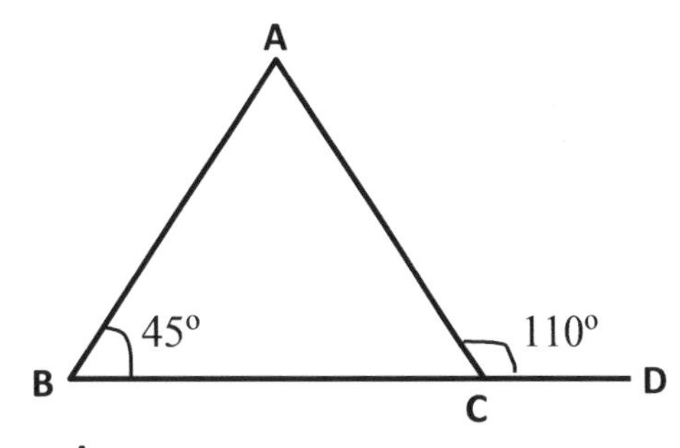

$A\hat{C}B = 180 - 110 =$ **70°**
(Angles on a straight line)

$C\hat{A}B = 180 - (45 + 70) =$ **65°**
(Angles in triangles add to 180°)

Example 3: Calculate the missing angles.

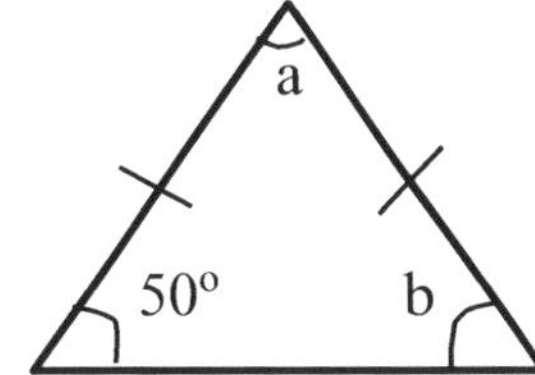

Triangle is isosceles; therefore, the base angles must be the same. b = **50°**

Angles in a triangle add to 180°, so angle a = 180 – (50 + 50) = **80°**

EXERCISE 6B

All diagrams are not accurately drawn.
Calculate the size of angles marked by letters in the diagrams below.

1)

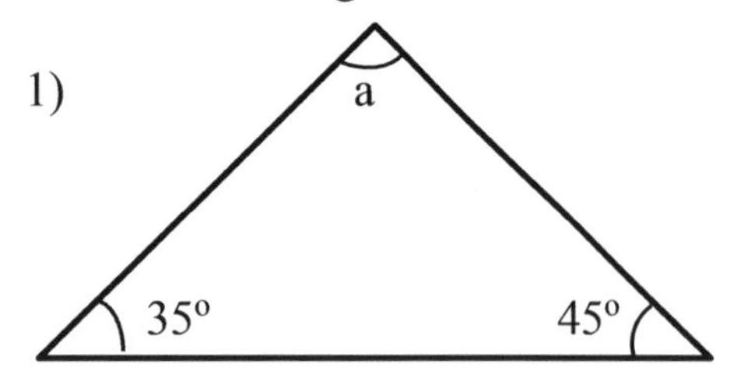

2)

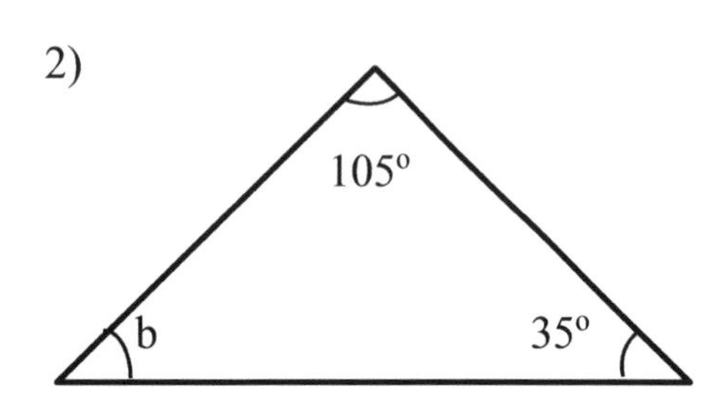

3)

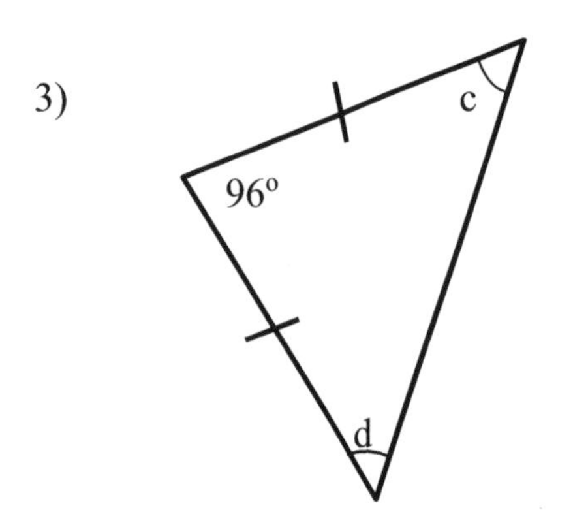

4)

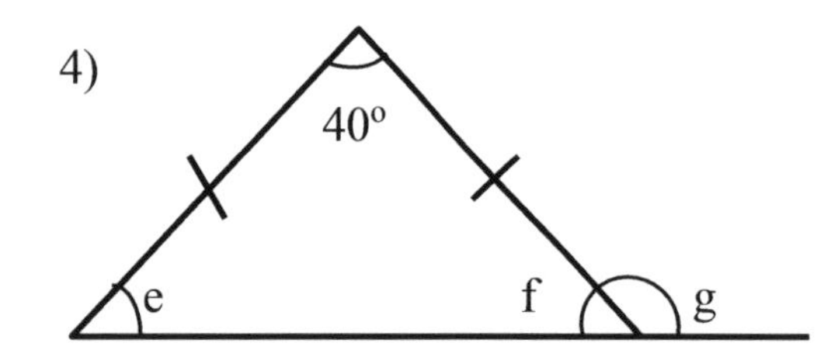

5)

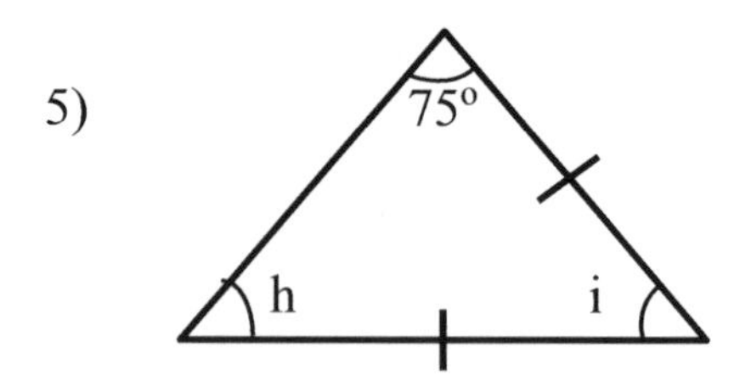

6)

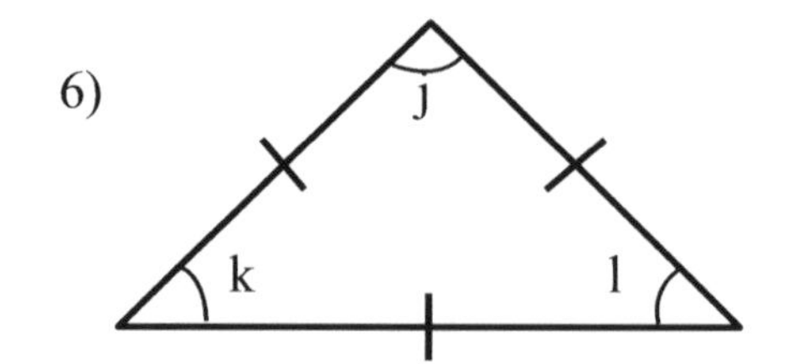

7)

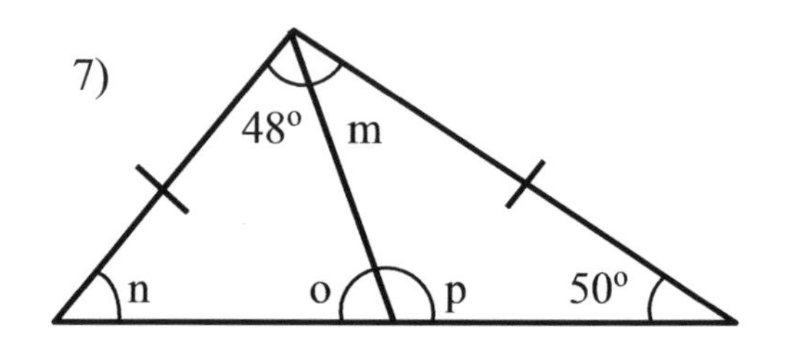

8)

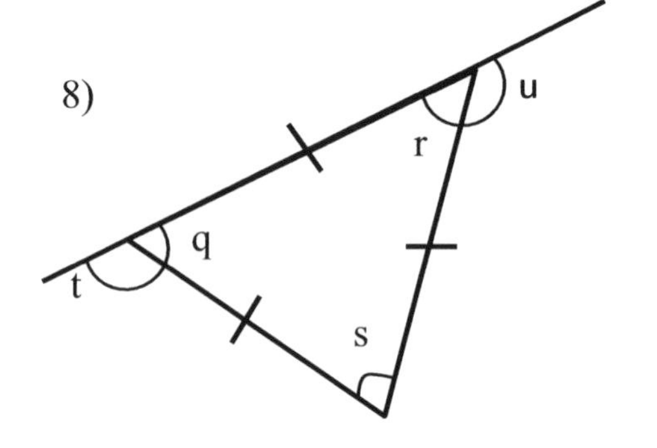

9)

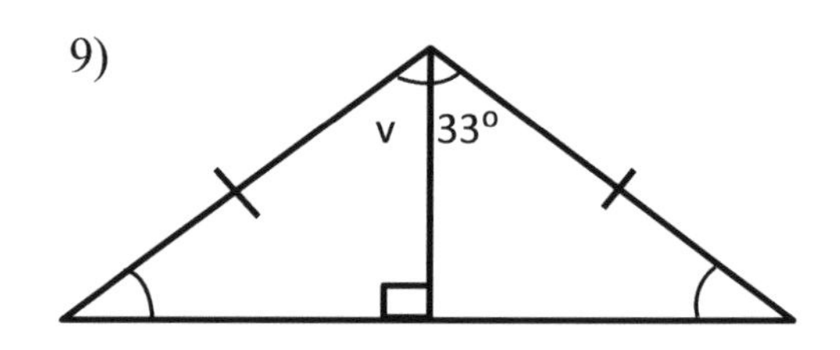

10) Indicate whether the triangles below are isosceles, right-angled, equilateral or scalene. *Give a reason for your answer.*

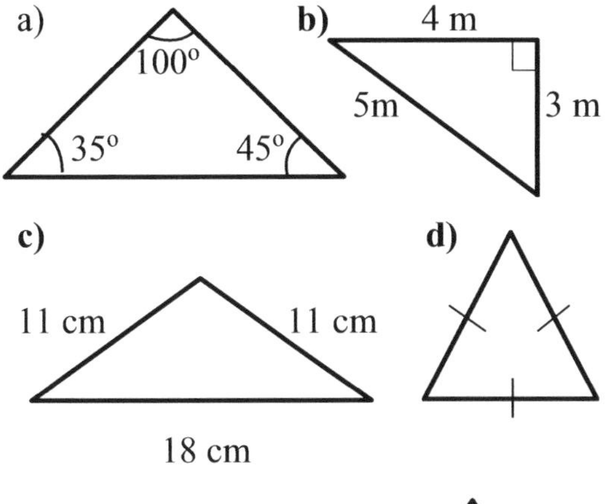

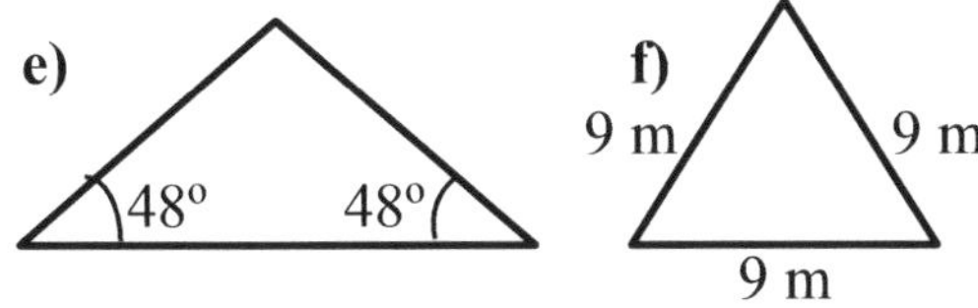

11)

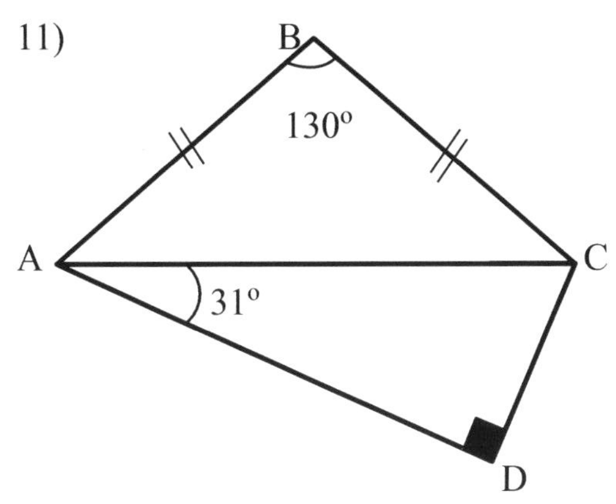

Work out the following angles.

a) $B\hat{A}C$

b) $A\hat{C}B$

c) $A\hat{C}D$

d) $B\hat{A}D$

e) $B\hat{C}D$

12) Obiora says "Angles 54°, 85° and 42° will form a triangle."
Is Obiora correct? Explain fully.

13) If two angles of a triangle are given below, calculate the third angle.

a) 30° and 56°
b) 130° and 20°
c) 23° and 76°
d) 114° and 56°
e) 100° and 40°
f) 60° and 60°
g) 45° and 45°
h) 20° and 79°
i) 56° and 85°
j) 55° and 70°

14) Look at the triangle below.

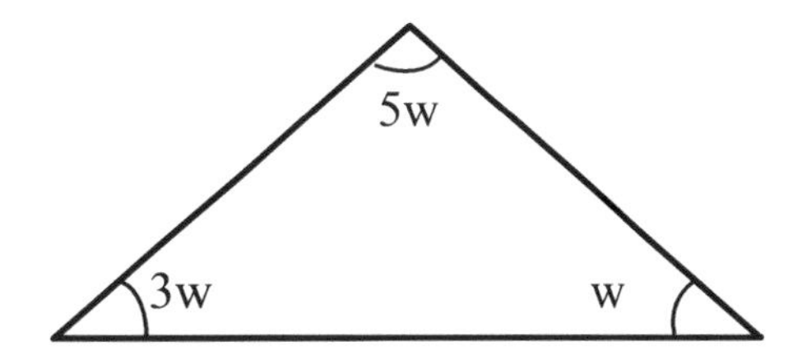

a) Form an equation in w
b) Sole the equation to find w
c) Work out the biggest angle.
d) Work out the range of the angles

15)

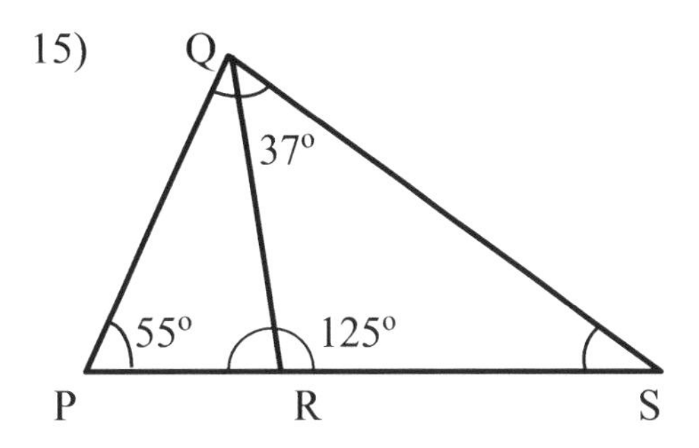

a) Calculate angles

i) $P\hat{R}Q$ ii) $P\hat{Q}R$ iii) $Q\hat{S}R$

b) What type of triangle is

i) $P\hat{Q}R$ ii) $Q\hat{R}S$?
Give a reason for each answer.

6.5 QUADRILATERALS

A quadrilateral is a two-dimensional shape with four straight sides and four angles.

A diagonal is a line joining two opposite corners. By using a diagonal, a quadrilateral may be divided into two triangles. The interior angles of a quadrilateral will always **add to 360º**.

Examples of quadrilaterals are square, rectangle, parallelogram, rhombus, trapezium and kite.

SQUARE

A square has four right angles. All the lengths are equal and the angles add up to 360 degrees.

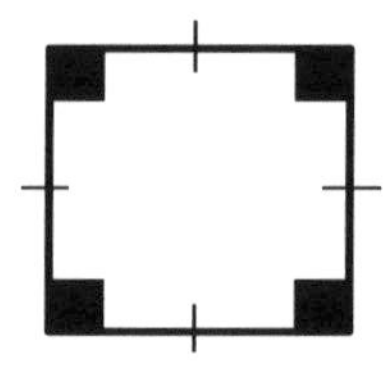

Also, opposite sides are parallel as shown with the arrows.

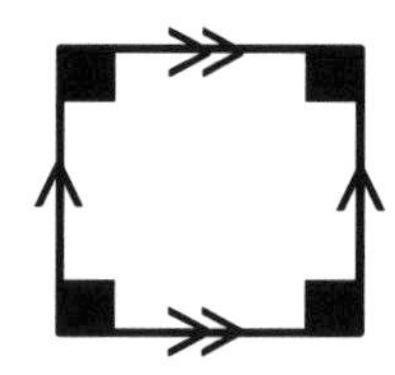

The diagonals are equal in length and bisect each other at 90 degrees.

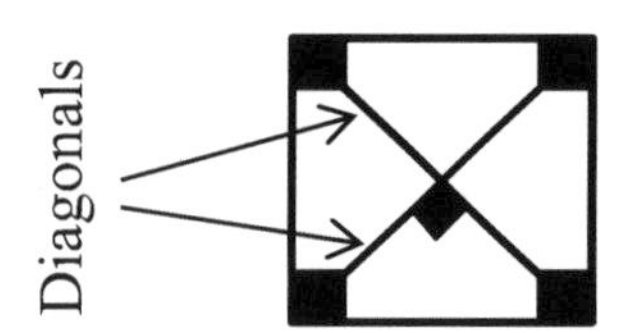

A square has four lines of symmetry and rotational symmetry of order 4.

RHOMBUS

A rhombus is a flat shape with four equal straight sides. None of the angles is 90º. They are like diamonds.

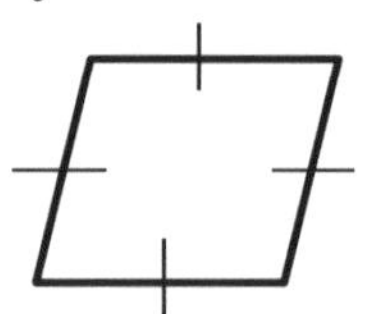

PROPERTIES

Four equal sides
Opposite angles are equal but not 90º
Opposite sides are parallel
Diagonals bisect each other at 90º
All angles add up to 360º
Two lines of symmetry
Rotational symmetry of order 2

RECTANGLE

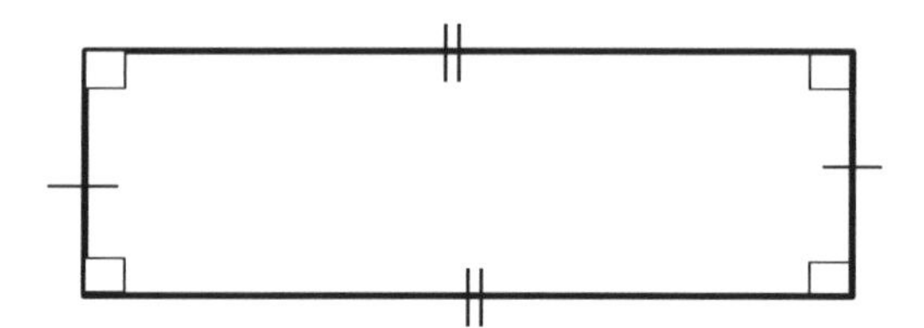

PROPERTIES

Opposite sides are parallel
Opposite sides are equal in length
All angles are 90º each
Diagonals bisect each other and are equal in length
All angles add up to 360º
2 lines of symmetry
Rotational symmetry of order 2

Remember: A square is a special type of rectangle with all the sides equal in length.

PARALLELOGRAM

A parallelogram is like a rectangle pushed out of shape. The angles **are not** 90° each.

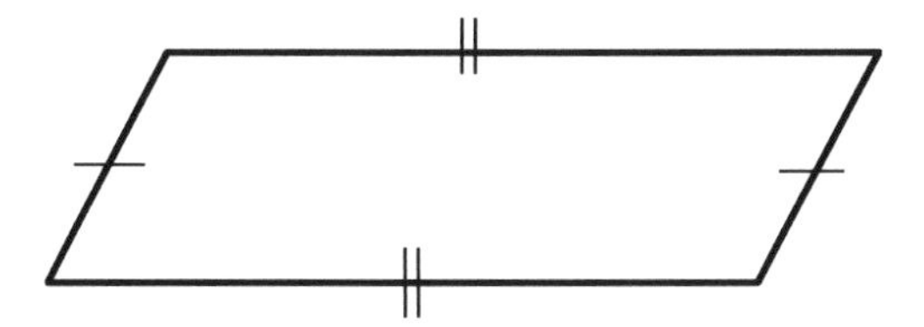

PROPERTIES

Opposite angles are equal
Opposite sides are equal in length
Opposite sides are parallel
Diagonals bisect each other
All interior angles add to 360°
Rotational symmetry of order 2
A general parallelogram like the shape above has **no line of symmetry**.

However, special parallelograms like rhombus, square and rectangle have lines of symmetry.

TRAPEZIUM

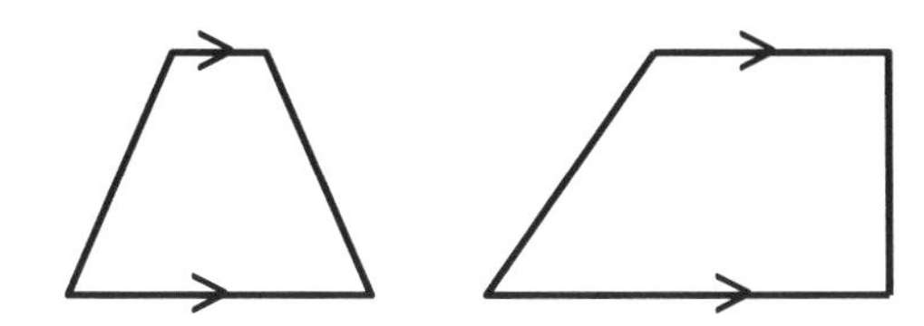

A trapezium has **one pair** of parallel sides.

PROPERTIES

One pair of parallel sides
No line of symmetry (unless it is an isosceles trapezium)
Rotational symmetry of order 1

A quadrilateral may also be called **isosceles trapezium** if the non-parallel sides are equal in length.

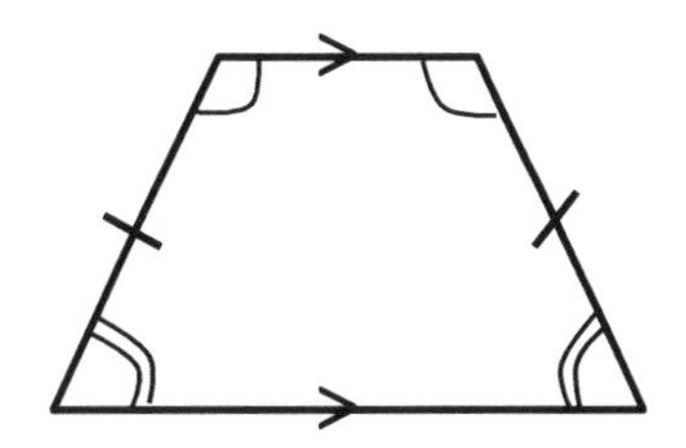

PROPERTIES

Two sets of equal angles
A set of equal sides
One line of symmetry
Rotational symmetry of order 1

KITE

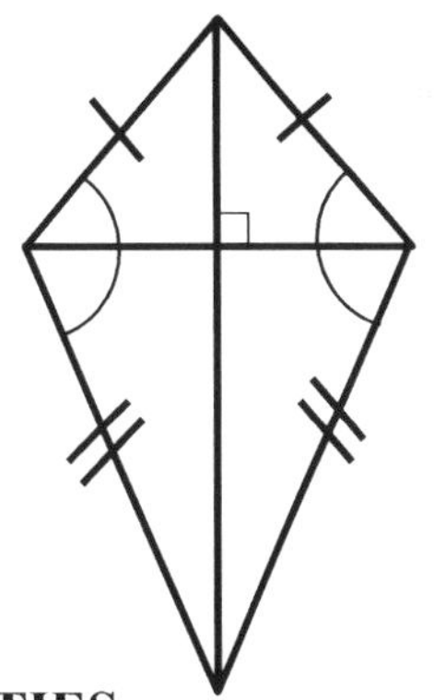

PROPERTIES

A pair of opposite angles is equal
Two pairs of adjacent sides are equal
Diagonal intersect at 90°
Note: A kite is made up of two isosceles triangles with a common base.

EXERCISE 6C

1) Write down the mathematical name for each shape below.

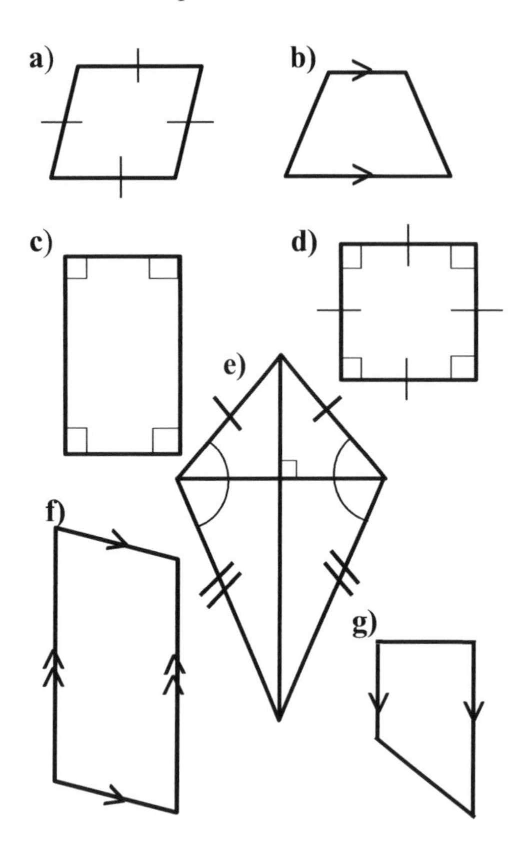

2) Show (by drawing) how two isosceles triangles can be joined together to form a kite.

3) Show by drawing how two of these shapes can be joined to make a rectangle.

4) Mention one difference between a square and a rhombus.

5) Is shape B a quadrilateral? Explain.

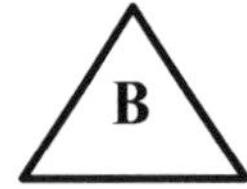

6) From the list of quadrilaterals below, copy and complete each statement.

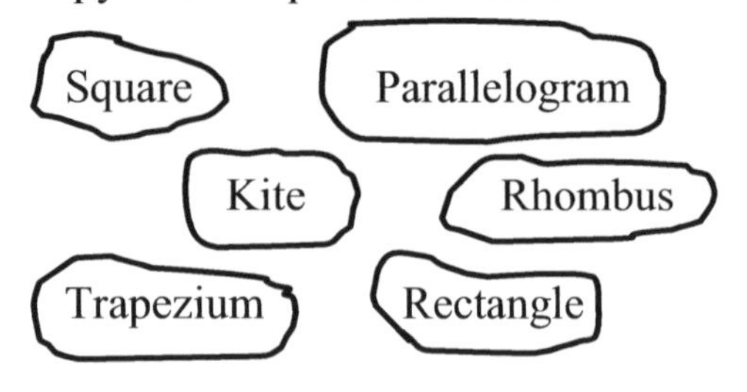

a) I have one pair of parallel sides. Therefore, I am a ……………………..

b) My diagonals bisect each other at right angles. All my sides are equal. All my angles are **not** 90°.
I am a …………………………………

c) I have one pair of opposite angles equal. I am made up of two isosceles triangles. My name is a ………………

d) I am a ……………………...... because all my sides are equal, and all my angles are 90°.

e) My opposite sides are parallel and equal in length. All my angles are 90° each. I am a …………………. because I also have 2 lines of symmetry?

f) My opposite angles are equal. Also, my opposite sides are parallel and equal in length. However, my angles are not 90° and I have **no line** of symmetry. My name is a ……………………

7) Kunle says "The shape below is a trapezium because it has two sets of parallel sides." Is Kunle correct? Explain fully.

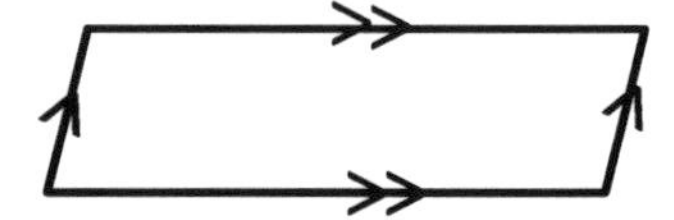

6.6 ANGLES IN QUADRILATERALS

The sum of the interior angles of any quadrilateral is 360 degrees.

Example 1: Calculate the size of the missing angle.

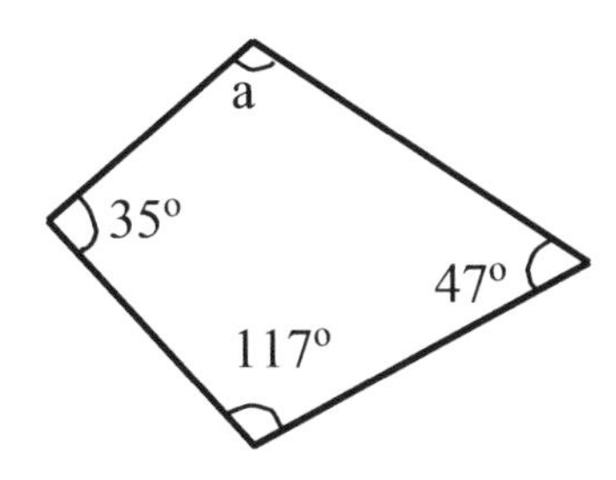

a + 35° + 117° + 47° = 360°
a + 199° = 360°
a = 360° – 199° = **161°** ✓

Check: 161 + 35 + 117 + 47 = 360

Example 2: Calculate the missing angles.

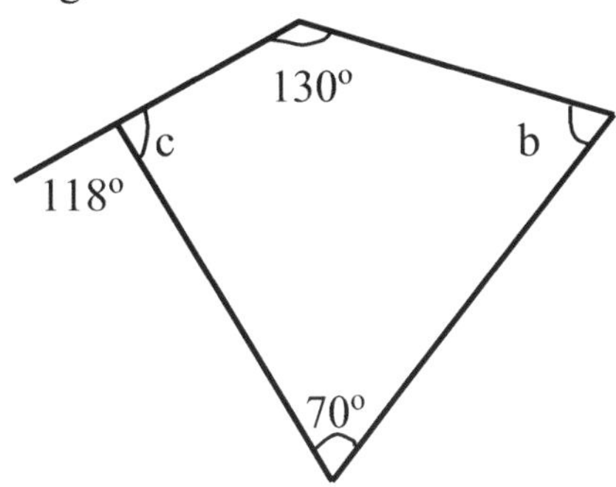

c = 180 – 118 = **62°** ✓
.....angle on a straight line add to 180°

62 + 130 + 70 + b = 360°
...angles in a quadrilateral add to 360°

262 + b = 360
b = 360 – 262 = **98°** ✓

Example 3
a) Form an equation in x.
b) Calculate the value of x.
c) Work out the value of $B\hat{C}D$

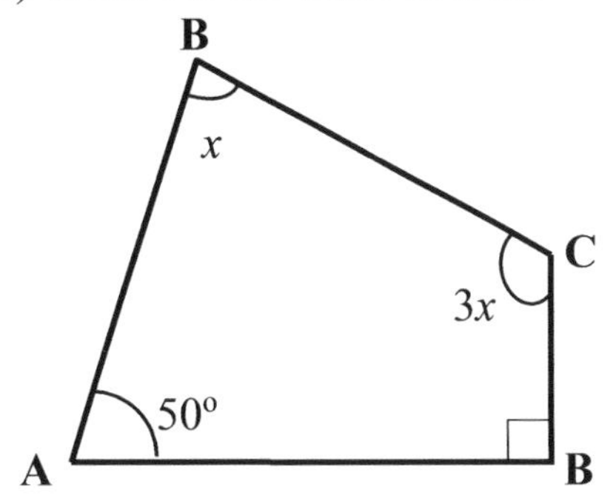

a) Form an equation in x.
$x + 3x + 90 + 50 = 360$
$4x + 140 = 360$
$4x = 360 - 140$
$\mathbf{4x = 220}$ ✓

b) $4x = 220$
$x = \frac{220}{4} = \mathbf{55^o}$ ✓

c) $B\hat{C}D = 3x = 3 \times 55 = \mathbf{165^o}$ ✓

Example 4 Calculate angles w, c and y.

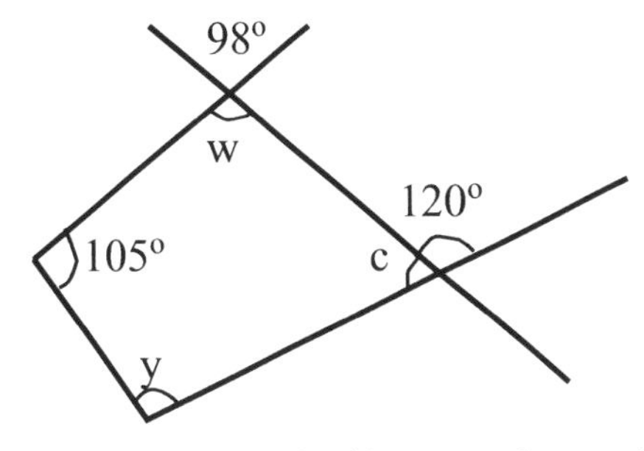

w = **98°**......vertically opposite angles
c = 180 – 120 = **60°**....... straight line
y = 360 – (105 + 98 + 60)
= 360 – 263 = **97°**

EXERCISE 6D

1) Work out the size of the unknown angles in each diagram.

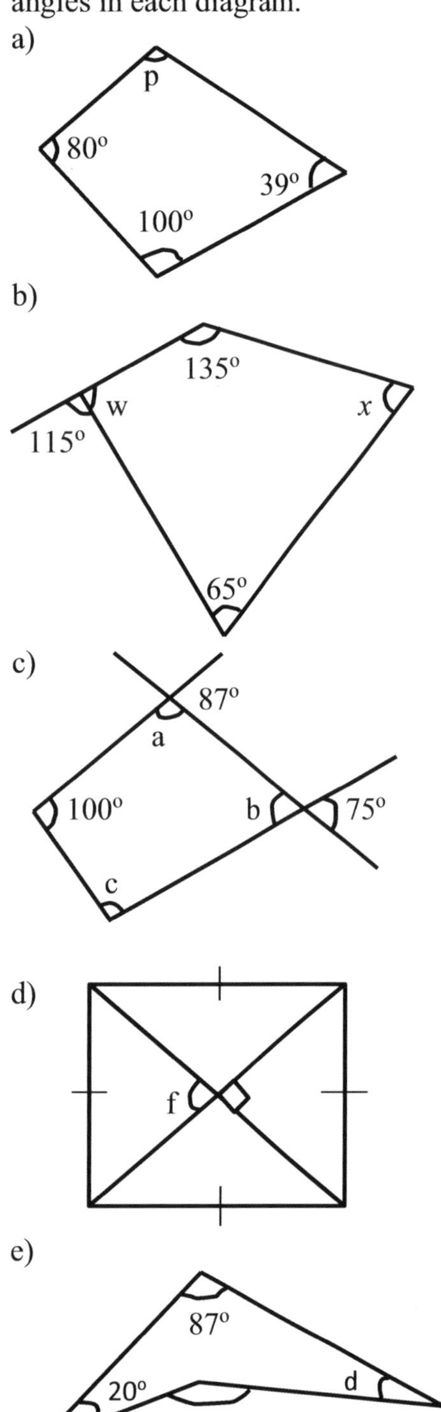

2) The smallest angle is 40°. The opposite angle to the smallest angle is 60° more. The third angle is twice the smallest angle. Calculate the value of the remaining angle.

3) For the quadrilaterals drawn below, work out the missing angles by forming an equation first.

a)

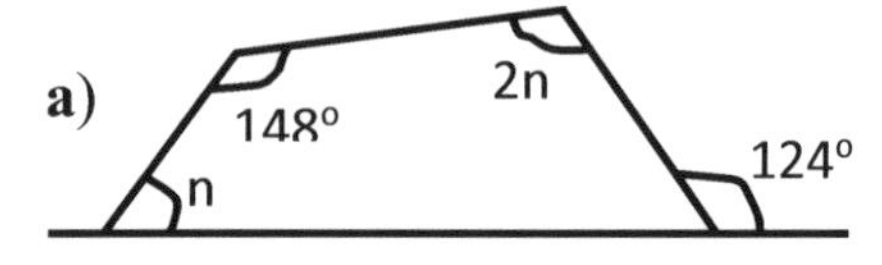

b)

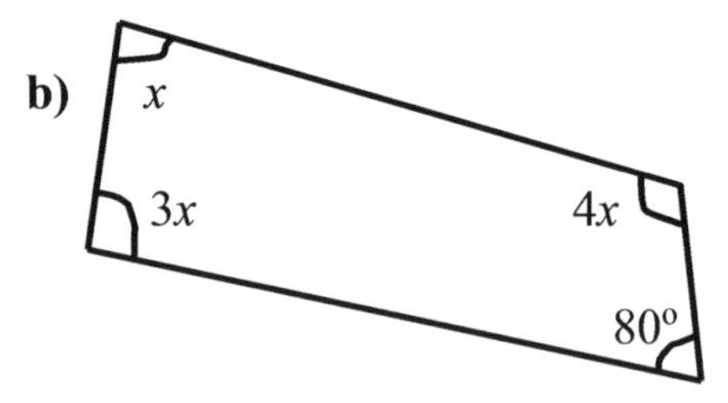

c)

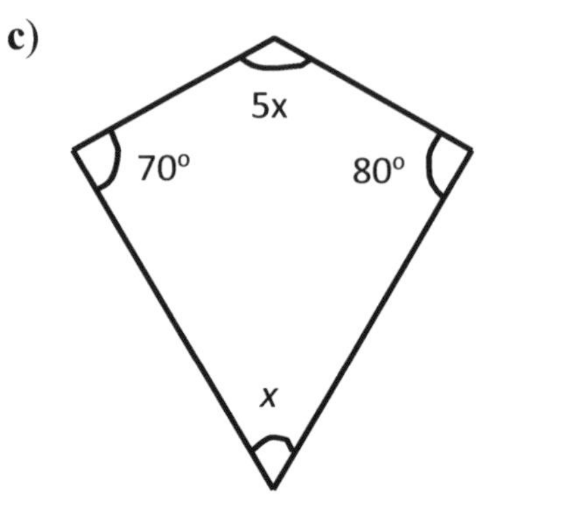

d)

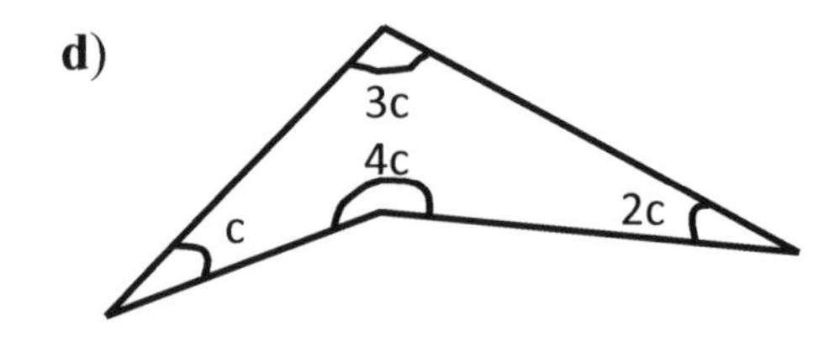

6.7 POLYGONS

Any close two-dimensional shapes with three or more straight sides are called **polygons**.

POLYGON	NUMBER OF SIDES	SUM OF INTERIOR ANGLES
Triangle	3	180°
Quadrilateral	4	360°
Pentagon	5	540°
Hexagon	6	720°
Heptagon	7	900°
Octagon	8	1080°
Nonagon	9	1260°
Decagon	10	1440°

Note: The number of interior angles goes up by 180° each time.

REGULAR POLYGON

A regular polygon has all its angles the same and all the lengths equal. Therefore, squares, equilateral triangles, regular pentagons etc. are all examples of regular polygons.

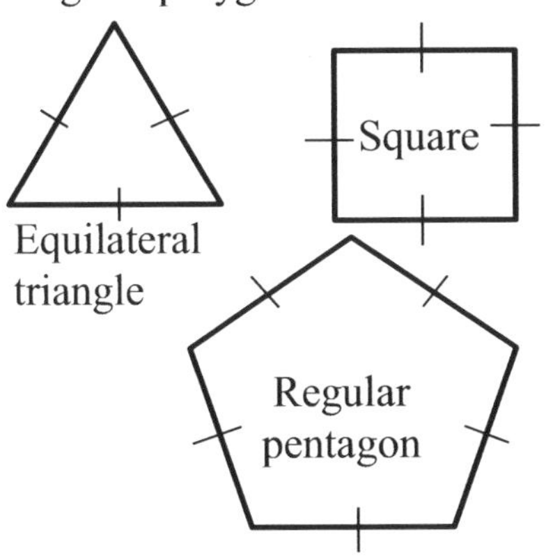

6.8 CIRCLES

We need a pair of compasses to draw a circle accurately.

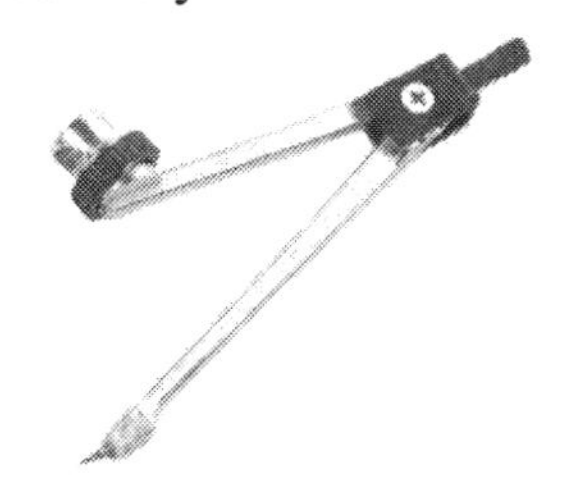

PARTS OF A CIRCLE

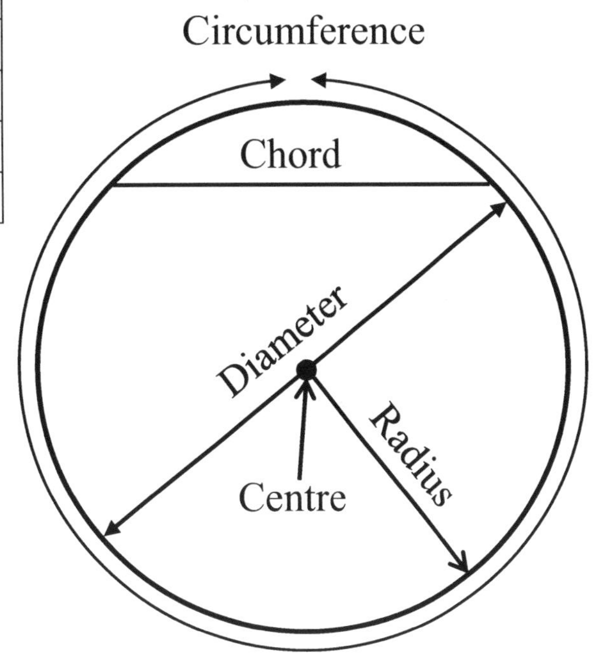

Diameter is a straight line passing through the **centre** and joining two points on the circumference

Radius is a straight line from the centre to the edge of the circle. Twice the radius will equal a diameter.

Circumference is the distance around the circle. It is similar to the perimeter.

Chord is a straight line joining two points on the circumference. If the chord passes through the centre, then it is a diameter.

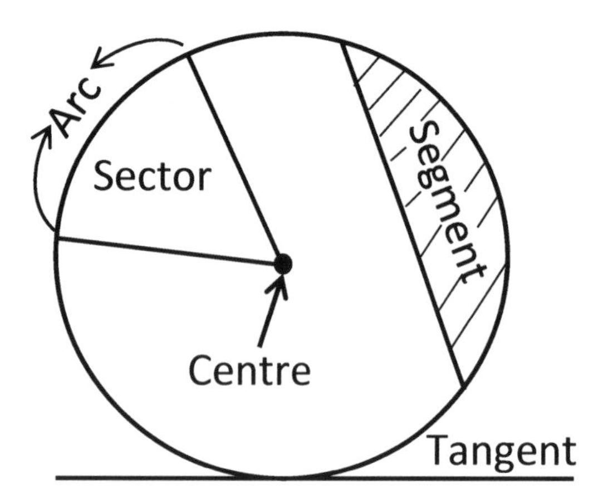

Arc is a part of the circumference of the circle.

The **sector** is the part of the circle lying between two radii and an arc.

The **segment** is the part of the circle between a chord and the circumference.

Tangent is a straight line that touches the outside of a circle at one point only.

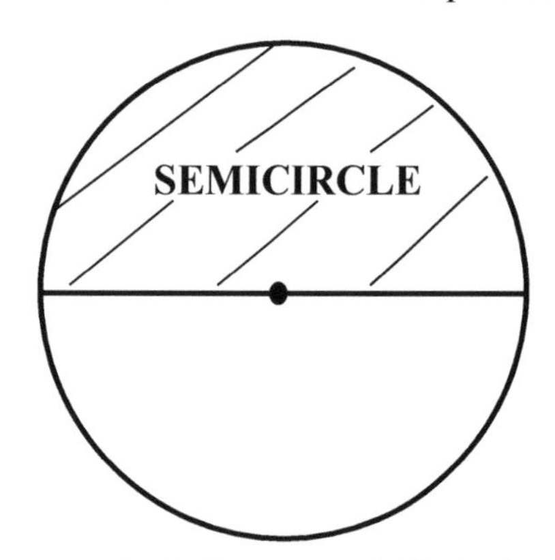

Two semi-circles = one full circle.

EXERCISE 6E

1) Is a circle a polygon? Explain fully

2) Name any four polygons together with the sum of their interior angles.

3) What would a shape with 12 sides be called? What would the sum of their interior angles add up to?

4) By using a pair of compasses, draw a circle with
a) radius of 4 cm
b) radius of 2.5 cm
c) diameter of 5 cm.

5) a) Draw a circle of diameter 12 cm.
   b) Draw a chord and label it P.
   c) Draw a tangent and label it Q.
   d) Mark a point R on the circumference.

6) a) Draw a circle of radius 5.5 cm.
   b) Draw a diameter of the circle.
   c) Shade the area between the diameter and the circumference.
   d) What is the name of the shaded part?

7) a) Draw a circle of radius 6 cm.
   b) Identify and name any 6 parts of the circle.

8) Okoro says "A circle with a diameter of 20 cm must have a radius of 40 cm." Is Okoro correct? Explain fully

6.9 ANGLES AND PARALLEL LINES

Parallel lines are lines that will **never** meet, no matter how far they are extended.

Parallel lines exist everywhere in real life including rail tracks and the sides of a piece of paper. Also, in a square or a rectangle, the two opposite sides are parallel.

Small arrows are used to identify two or more parallel lines.

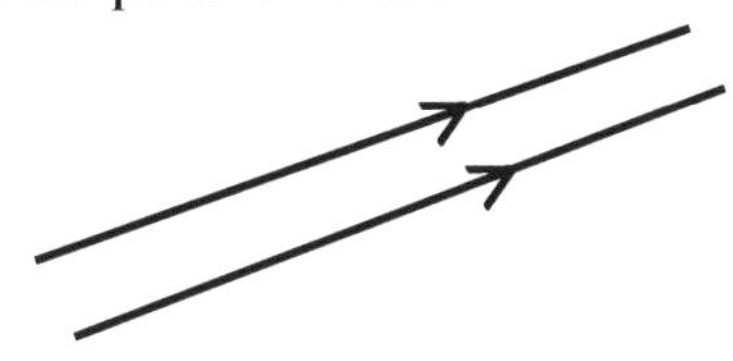

A line that cuts a pair of parallel lines is called a **transversal**.

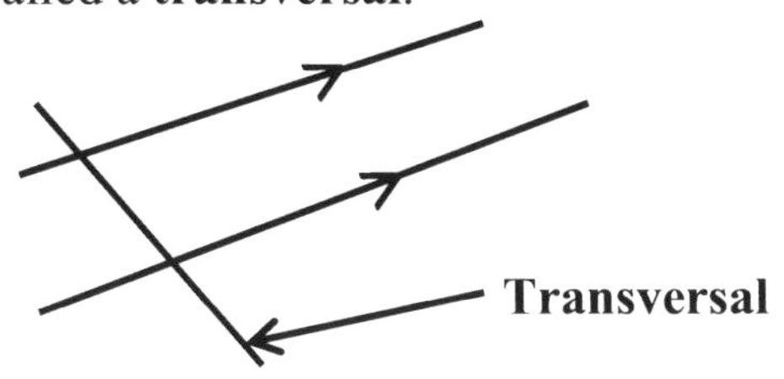

Eight angles are formed when a line cuts through a pair of parallel lines.

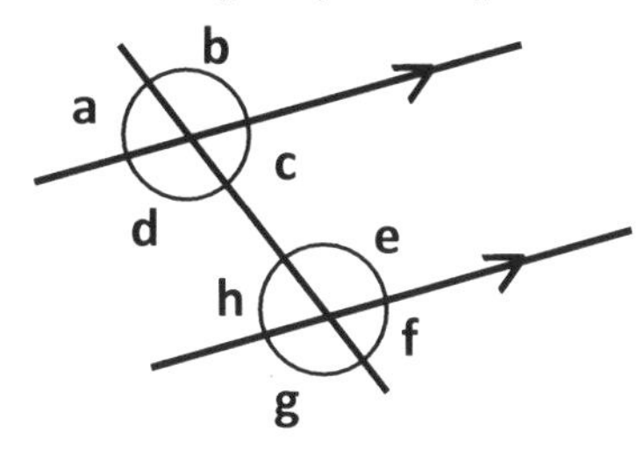

a = c = h = f and b = d = e = g
If the transversal is not perpendicular (at right angles) to the parallel lines, 4 acute and 4 obtuse angles are formed.

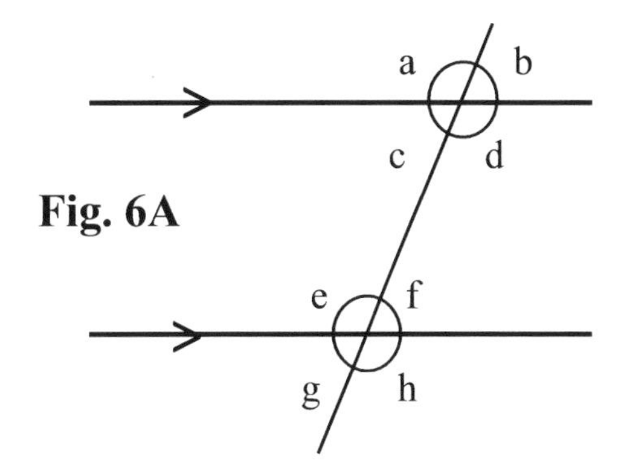

ALTERNATE ANGLES

Alternate angles are equal. They are angles on the opposite sides of the transversal. They form a **Z - SHAPE**.

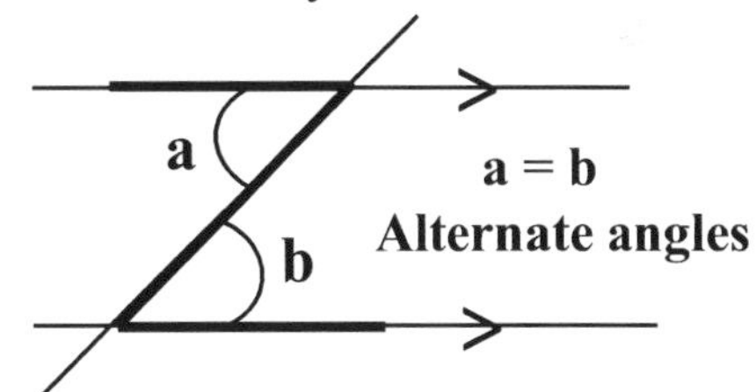

In Fig 8A above, angles **c** and **f** are alternate angles and are equal.
Also, angles **d** and **e** are alternate angles and are equal too.

Example 1: Write down the value of angles *x* and *y*. Give a reason for your answer.

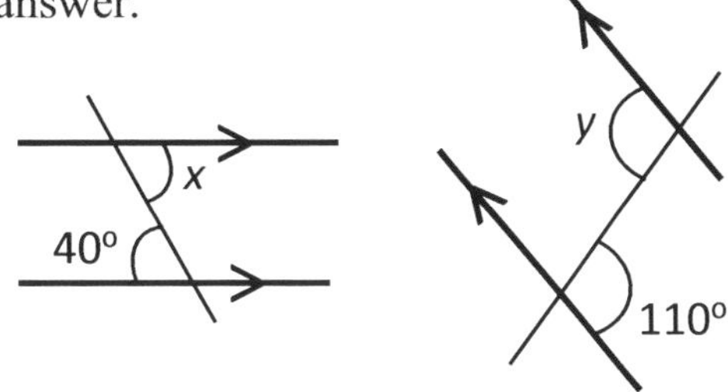

x = **40º** because they are alternate angles. y = **110º** because they are alternate angles.

CORRESPONDING ANGLES

The angles in **matching corners** when a line crosses two parallel lines are called **corresponding angles**.

Also, corresponding angles are **equal**. They also form an F- SHAPE.

In figure 8A above, angles **b** = **f**, **a** = **e**, **d** = **h** and **c** = **g** and all are corresponding angles.

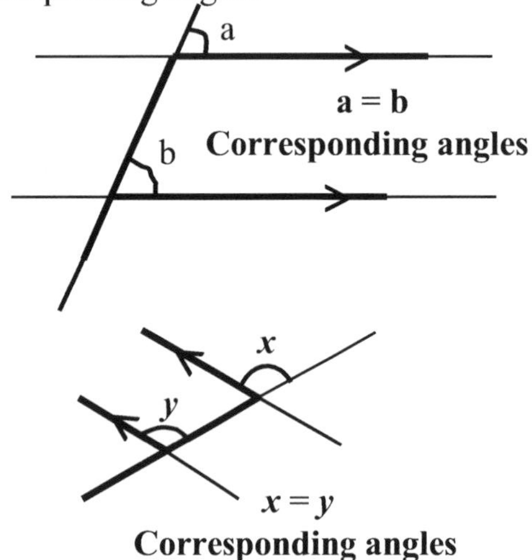

In the above examples, if **a** = 30°, **b** will also be 30° because they are corresponding angles.

Also, if x = 120°, y will be 120° because they are corresponding angles.

Note: Alternate and corresponding angles are always equal when the lines are parallel. If the lines are not parallel, **do not** assume that the angles formed are corresponding or alternate. They may not be equal.

EXERCISE 6F

1) Calculate the value of the angles marked by letters. Give a reason for your answer.

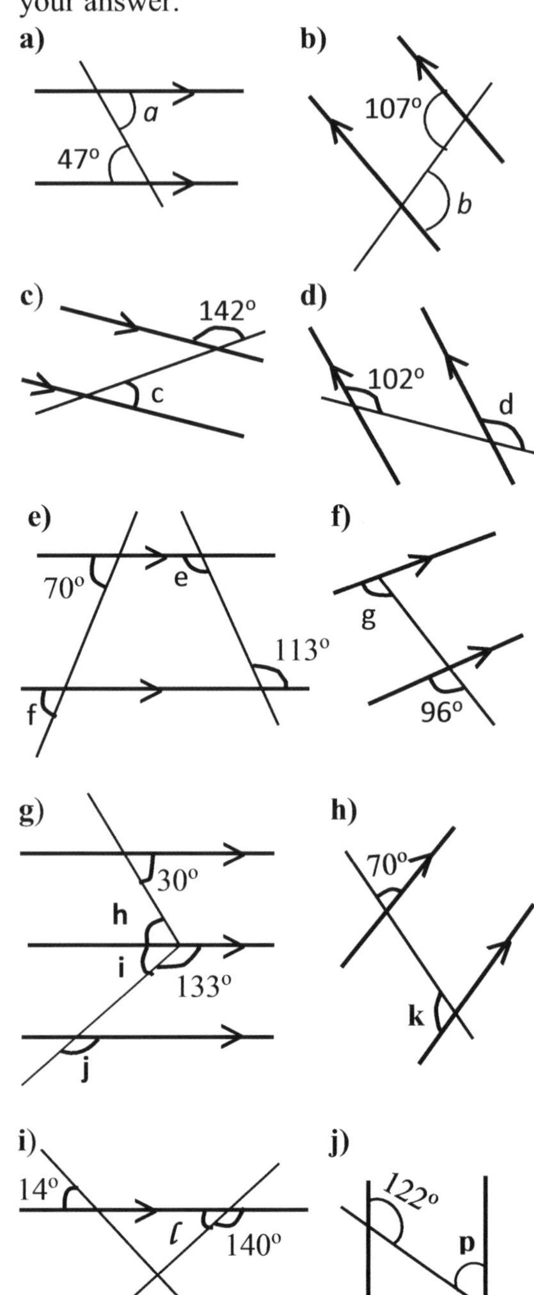

Chapter 6 Review Section

Assessment

1) Work out the missing angles

a)

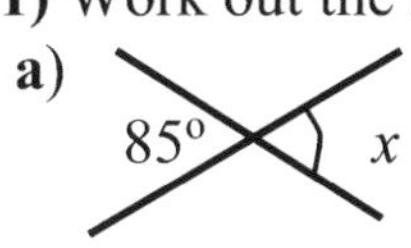

b)

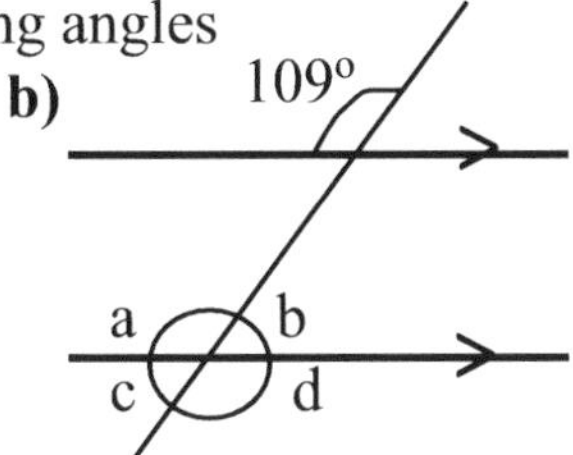

c)

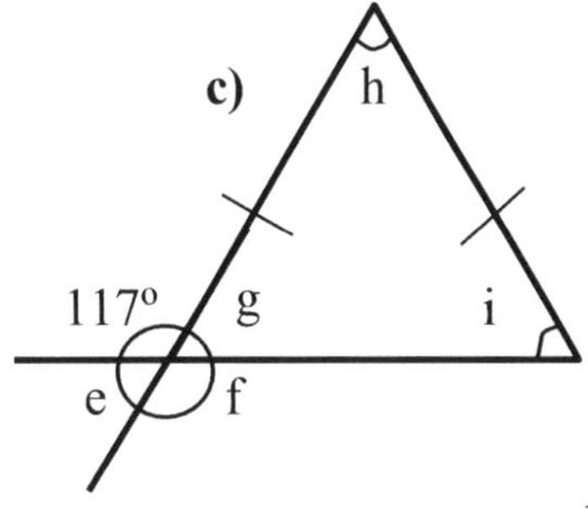

……………………**10 marks**

2) Work out the missing angles.

a)

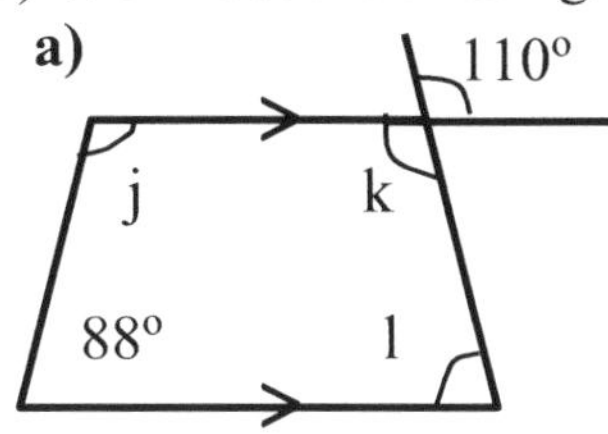

b)

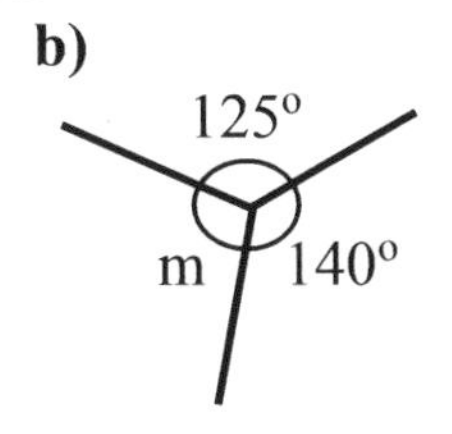

c)

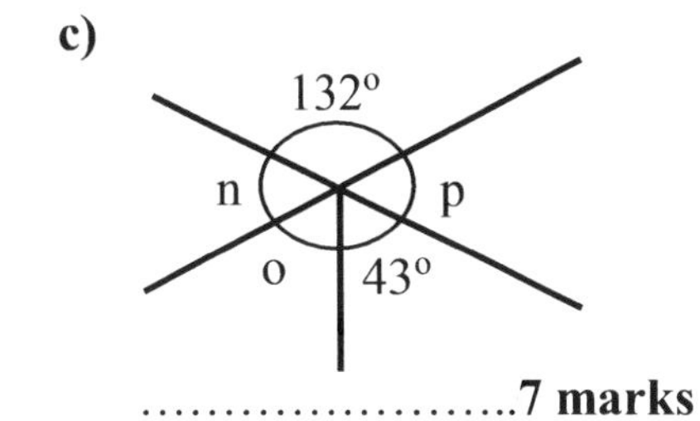

……………………**7 marks**

3) **A** **B** **F** **C** **E** **D**

a) How many sets of parallel lines are there in the shape shown?

b) Name the line which is parallel to AF.

c) Name the line which is parallel to ED.

d) What is the mathematical name for shape ABCDEF?

e) How many lines of symmetry does shape ABCDEF have?

f) What type of angle is EDC?

……………….. **6 marks**

7 Circle, area and perimeter

This section covers the following topics:

- Area of circles
- Perimeter/circumference of circles and 2-d shapes
- Area and perimeter of compound shapes

LEARNING OBJECTIVES

By the end of this unit, you should be able to:

a) Calculate the area and perimeter of 2-d shapes
b) Calculate the area and circumference circular objects
c) Calculate the area and perimeter of compound shapes
d) Calculate the area of shaded parts

KEYWORDS

- Circle
- Circumference
- Pi
- Area, perimeter
- Compound shape

7.1 CIRCLES AND PERIMETER

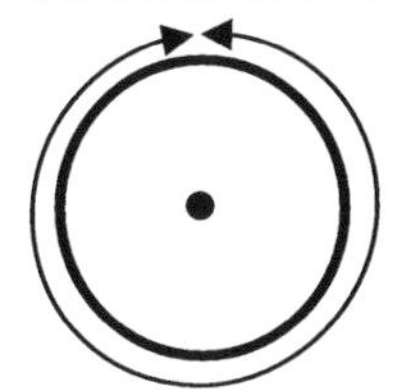

The total distance around the circle is called the circumference. Therefore, the perimeter of a circle is known as its circumference.

Since the outer surface of a circle is circular, it is difficult to use a ruler for measuring the circumference.

A more simplistic way of finding the circumference of a circle is by using a thread.

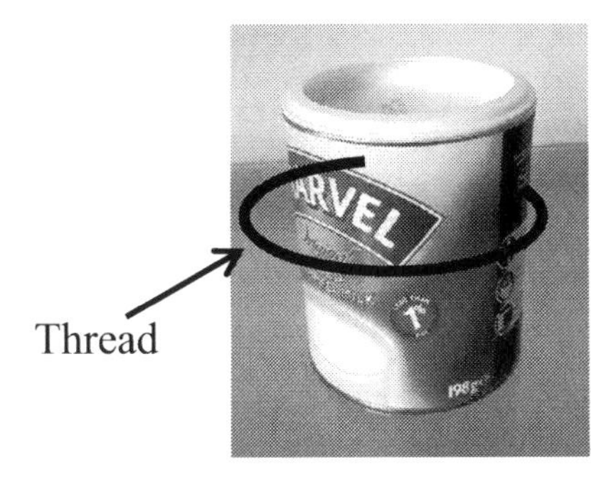

To find the circumference of the circular base of the cylindrical container, wind a piece of thread around the container once. Mark the thread where they cross each other. Pull the thread straight and measure against a ruler.

The length of the thread is the circumference of the circle.

Known fact: If you measure the distance around a circle (circumference) and then divide it by the distance across the circle through the centre (diameter), your answer will always come close to a particular value. This depends on the accuracy of the measuring instrument(s).

That particular value is approximately 3.141592653…….

In mathematics, we use the Greek letter π (pronounced as ***pi***) to represent this number. In some textbooks, you might see π as 3.14, 3.141, $\frac{22}{7}$ or$3\frac{1}{7}$. All these numbers are not even accurate as it is impossible to express the number as an exact fraction or a decimal number.

Therefore,

Circumference (c) = π × diameter (d)

C = πd

We also know that twice the radius is the diameter of a circle. Therefore, the above formula could also be written as

C = 2π*r*
where *r* is the radius of the circle.

Example 1
Calculate the circumference of a circle with diameter of 14 cm. Use π as $\pi\frac{22}{7}$.

Solution: Using the formula, C = πd
$C = \frac{22}{7} \times 14$ cm
C = 22 × 2 = **44 cm**

Example 2: Calculate the circumference of the circle below. Use π as 3.14.

Solution: $C = 2\pi r$
$C = 2 \times 3.14 \times 3$
$C =$ **18.84 cm**

Example 3:

The front wheel of the bicycle has a radius of 35 cm.

a) What is the circumference of the front wheel?
b) How many complete revolutions does the wheel make when the bicycle travels 500 **metres**?

a) $C = \pi d$ or $C = 2 \pi r = 2 \times \frac{22}{7} \times 35$
$C =$ **220 cm**

b) 100 cm = 1 m and
220 cm = 2.2 m

Complete revolution will be
$500 \div 2.2 =$ **227**

Example 4: Calculate the perimeter of the semi-circle below. Use π as 3.14.

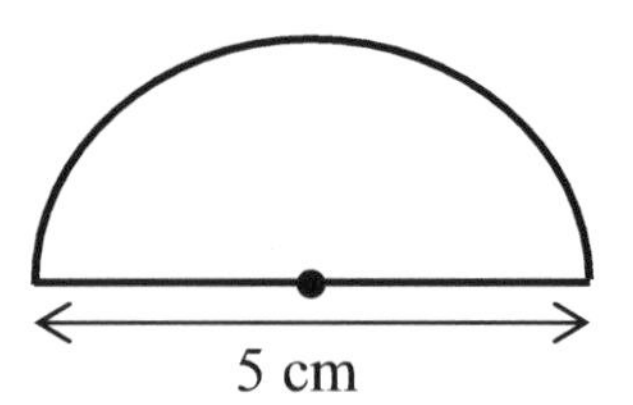

The perimeter of the semi-circle is the length of the circular face (circumference) plus the diameter.

Length of the circular face is the circumference ÷ 2.

$\frac{\pi \times d}{2} = \frac{3.14 \times 5}{2} = \frac{15.7}{2} = 7.85$ cm
Perimeter of the semi-circle
$= 7.85 + 5 =$ **12.85 cm**

Example 5: Calculate the perimeter of the object below. Use π as 3.14.

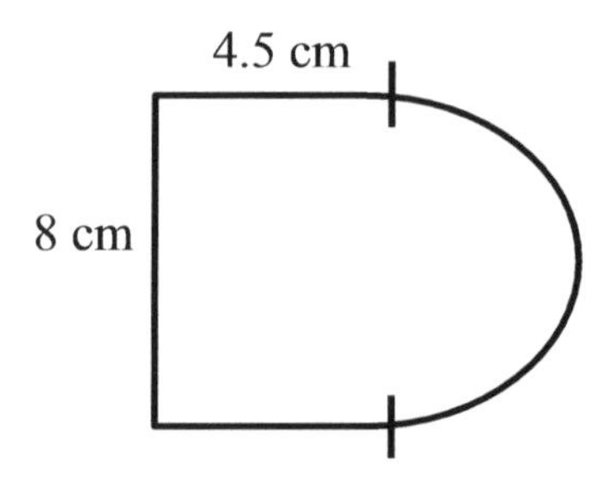

The total perimeter will be 8 cm + 4.5 cm + 4.5 cm + length of the circular end (semi-circle).

Length of the semi-circle
= circumference ÷ 2
$= (\pi \times 8) \div 2 = (3.14 \times 8) \div 2$
= 12.56 cm

Perimeter = 8 + 4.5 + 4.5 + 12.56
= **29.56 cm**

EXERCISE 7A

1) **Group work:** Look around and find three circular objects. Measure the circular ends and record your answers. Remember to include the unit(s) you may have used.

In this exercise, you require a measuring instrument like a thread or a tape.

2) By measurements, work out the perimeter of the shapes below.

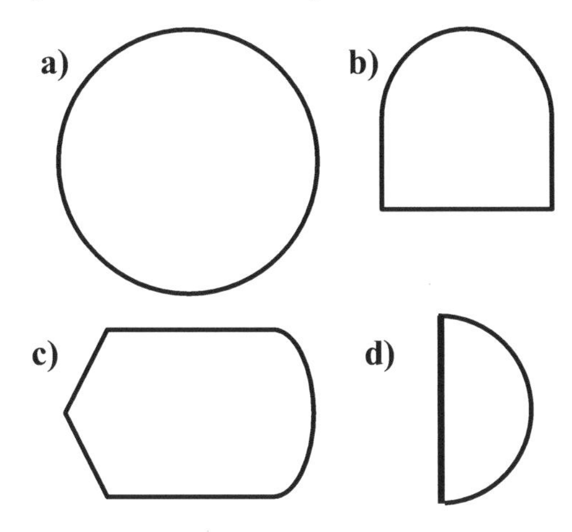

3) Calculate the circumference of the circles below. Use π as $\frac{22}{7}$.

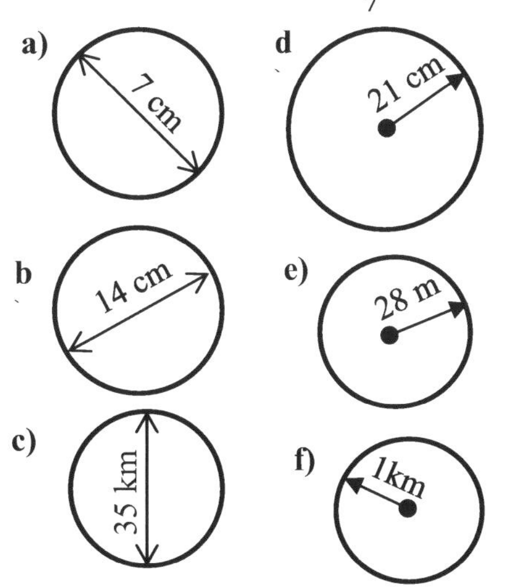

4) Calculate the perimeter of these shapes to one decimal place where possible and use π as $3\frac{1}{7}$.

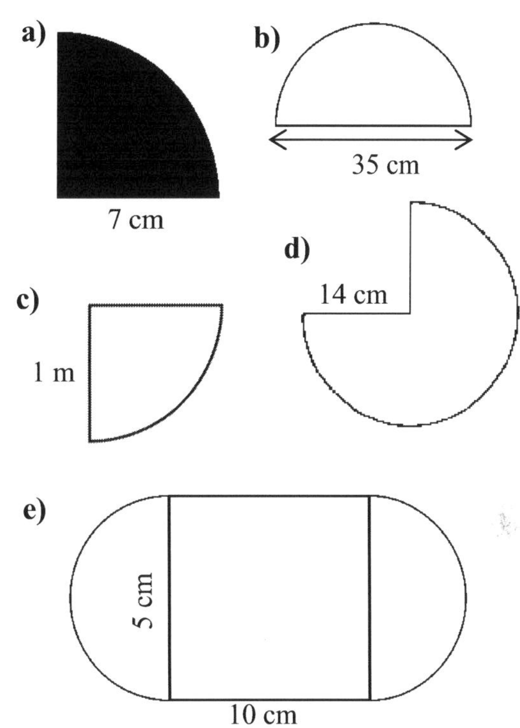

5) A circular piece of string has a radius of 21 cm. What is the circumference of the string? Use π as 3.14.

6) A duct tape is wound 20 times round a cylinder of diameter 8 cm. How long is the duct tape? Use π as 3.14

7) Calculate the diameter of a circle with a circumference of 7 cm. Take π as $\frac{22}{7}$.

8) A bicycle wheel has a diameter of 40 cm. How many complete revolutions does the wheel make when the bicycle travels 200 metres? Take π as 3.14.

7.2 AREA OF PLANE SHAPES

In simple terms, the **area** of a shape is the amount of **space** inside it.

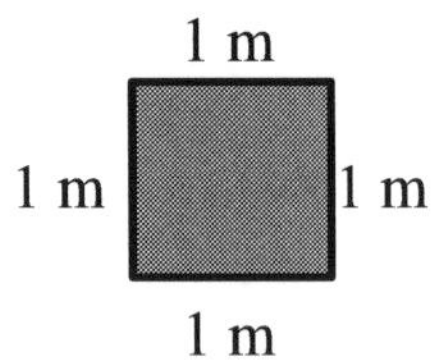

The square above has a length of 1metre and covers an area of $1 \times 1 =$ **1 m²**.

Conventionally, the square is used as the shape for the unit of area. Similarly, squares will have units depending on the unit of its length. If a length is in centimetres, the unit of the area will be **cm²**.

AREA BY COUNTING SQUARES

When shapes are drawn on a centimetre square grid, the area can be worked out by simply **counting** the squares.

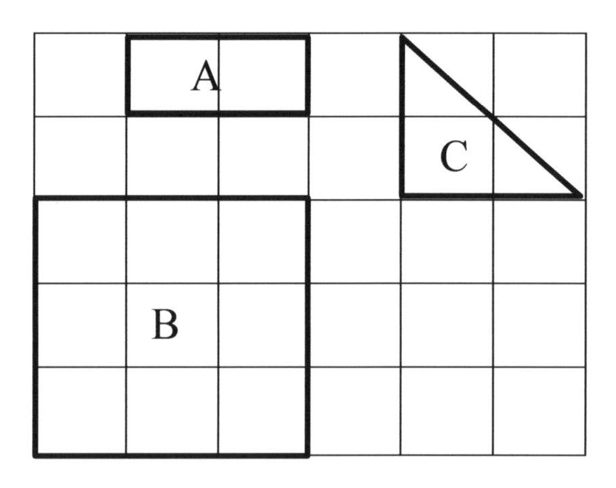

The area by counting the squares will be A = 2 cm², B = 9 cm² and C = 2 cm².

AREA BY ESTIMATE

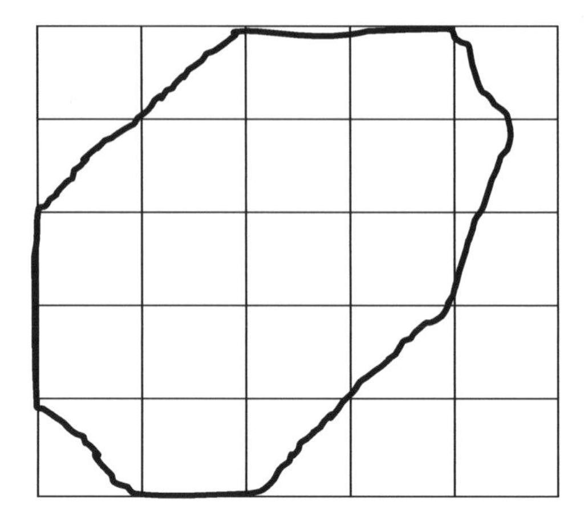

It will be almost impossible to find the exact area of the above shape. We estimate by counting the full squares and add up smaller parts to make up.

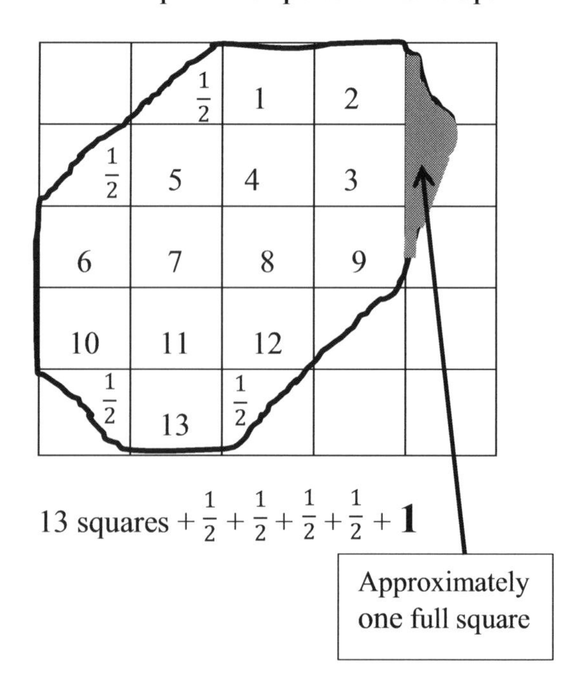

13 squares $+ \frac{1}{2} + \frac{1}{2} + \frac{1}{2} + \frac{1}{2} + \mathbf{1}$

The approximate area of the irregular shape above is **16 square units**.

AREA BY CALCULATIONS

AREA OF A RECTANGLE

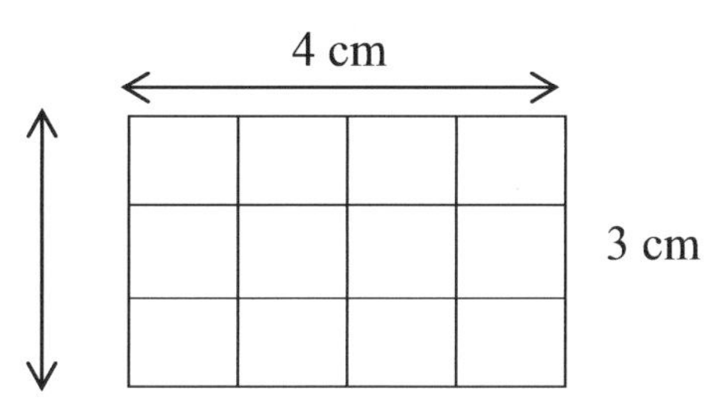

Area = length × breadth (width)
$= 4 \times 3 = \mathbf{12\ cm^2}$

AREA OF A SQUARE

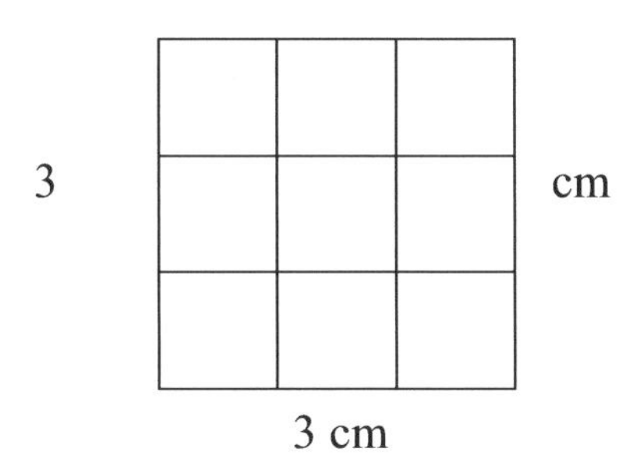

Area of a square = Length × length
$= (\text{Length})^2$
$= 3 \times 3 = 9\ cm^2$

Example 1: The area of a rectangle is 42 cm^2. What is the width of the rectangle is the length is 7 cm?

Area = Length × breadth (width)
$42 = 7 \times \text{width}$
Width $= 42 \div 7 =$ **6 cm**

Example 2: Work out the area of a square with length 2 cm.
Area of a square = length × length
$= 2 \times 2$
$= \mathbf{4\ cm^2}$

Example 3: Work out the area of the shaded part.

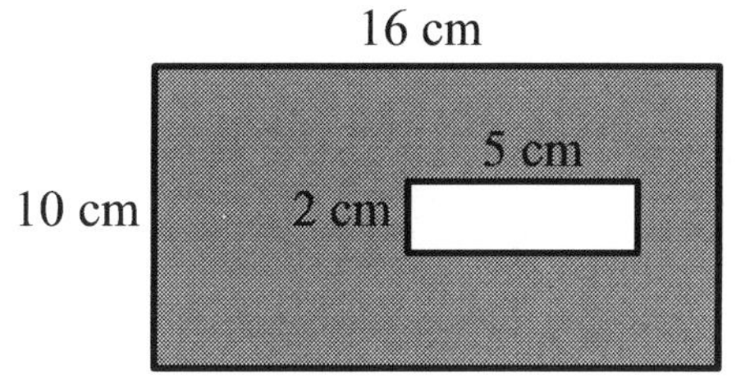

Area of big rectangle $= 16 \times 10$
$= 160\ cm^2$
Area of small rectangle $= 2 \times 5$
$= 10\ cm^2$
Area of shaded part $= 160 - 10$
$= \mathbf{150\ cm^2}$

Example 4: Calculate the length of a side of a square with an area of 49 km^2.

The length $= \sqrt{49} =$ **7 km**

Check: $7 \times 7 = 49$

Example 5: Calculate the area of the shape below.

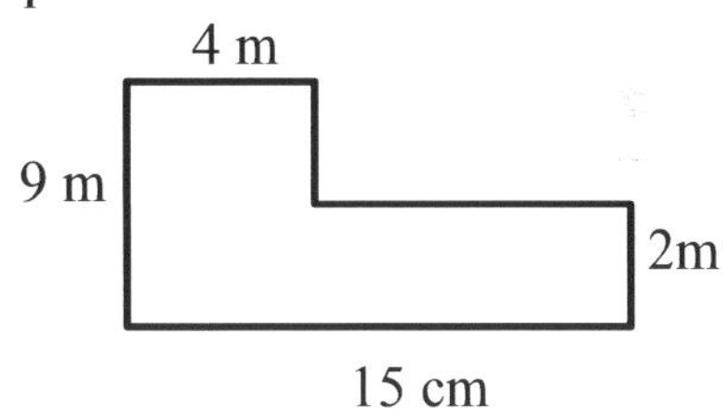

Solution
Split the shape into two rectangles.

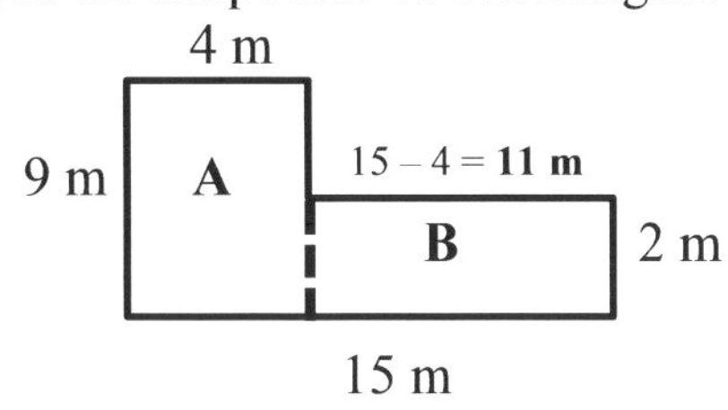

Area of A $= 4 \times 9 = 36\ m^2$
Area of B $= 11 \times 2 = 22\ m^2$
Total area $= 36\ m^2 + 22\ m^2 = \mathbf{58\ m^2}$

EXERCISE 7B

1) Calculate the area of the rectangles and square below. All lengths in cm.

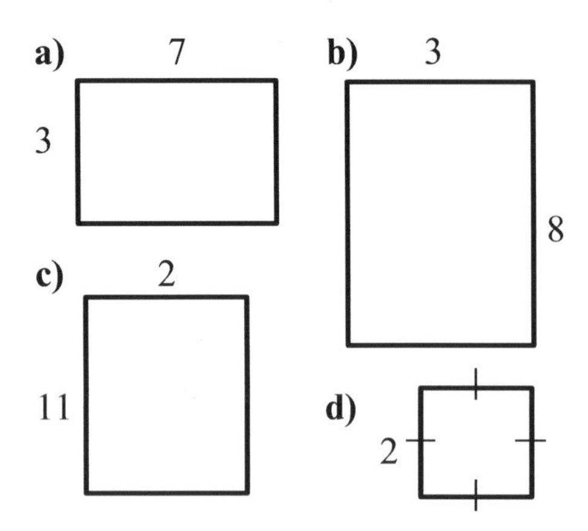

2) Below is a centimetre square grid. Work out the area of the shapes.

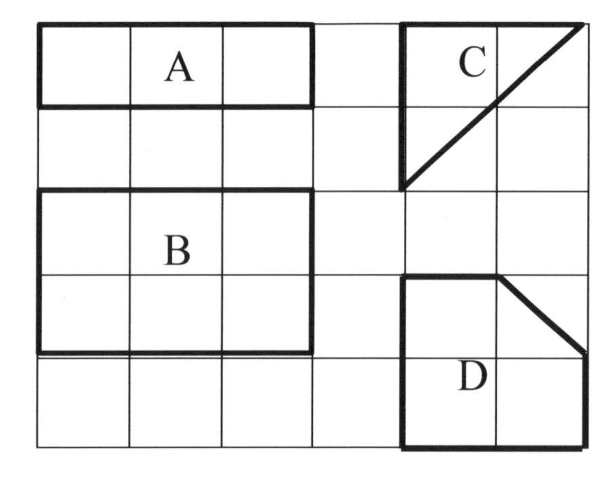

3) Estimate the area of the shapes drawn on a centimetre square grid below.

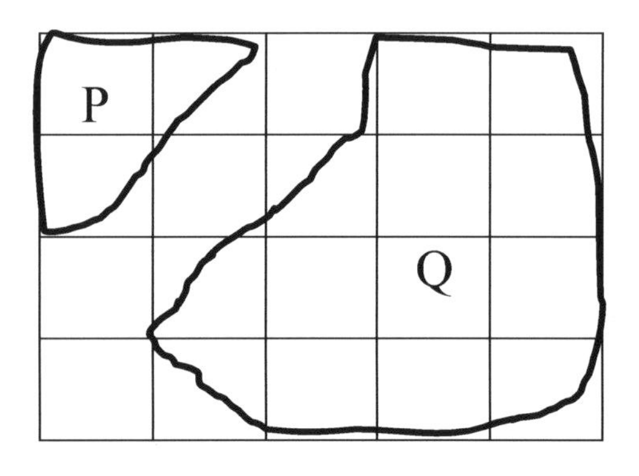

4) The area of a rectangle is 8 cm^2. What could be the length and breadth (width) of the rectangle?

5) A square has an area of 36 m^2. A length is 6 m, calculate the size of the other length.

6) Work out the area of the shaded region of the two rectangles.

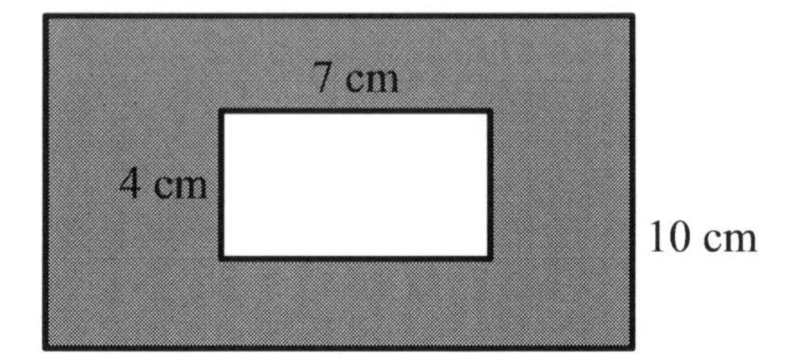

7) The floor below is to be carpeted.

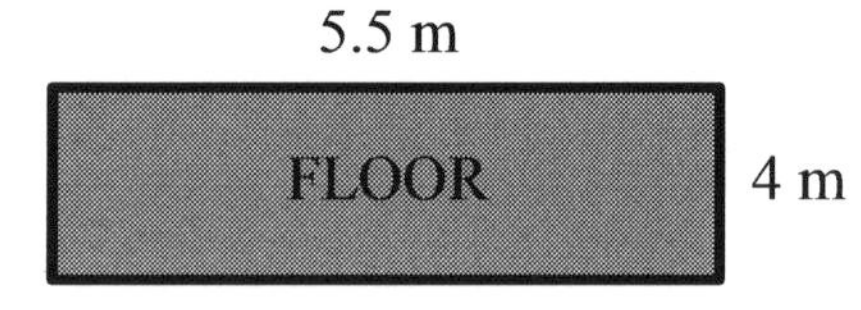

a) Work out the area of the floor.
b) A vinyl carpet costs ₦1 500 per square metre. The cost of labour to lay the carpet is ₦4 300.
What is the total cost to successfully lay the carpet on the floor?

8) Complete the table of squares below.

Area	Length of side
1 cm^2	
9 m^2	
	8 cm
	4.3 cm
81 cm^2	

AREA OF A PARALLELOGRAM

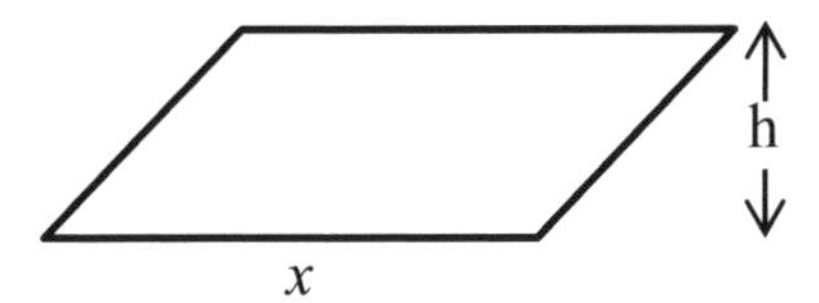

The area of a parallelogram is **base × perpendicular height**

Perpendicular means at 90°.

Example 1: Calculate the area of the parallelogram below.

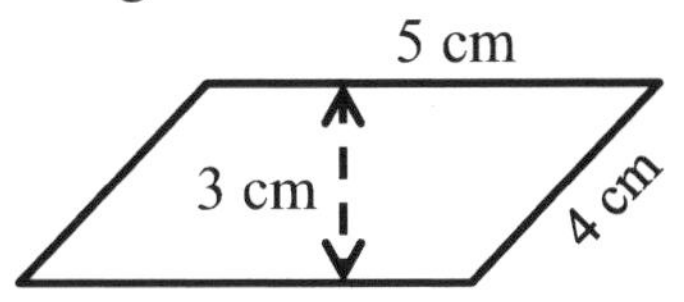

Area = 5 × 3 = **15 cm²**

Notice that 4 cm was not used. It is not perpendicular (at 90°) with the base.

Do **not use** the slant height for calculating the area of a parallelogram.

Example 2: Calculate the area of a parallelogram with a base of 10 cm and perpendicular height of 5 cm.

Area = 10 × 5 = **50 cm²**

Example 3: Calculate the height of the parallelogram below.

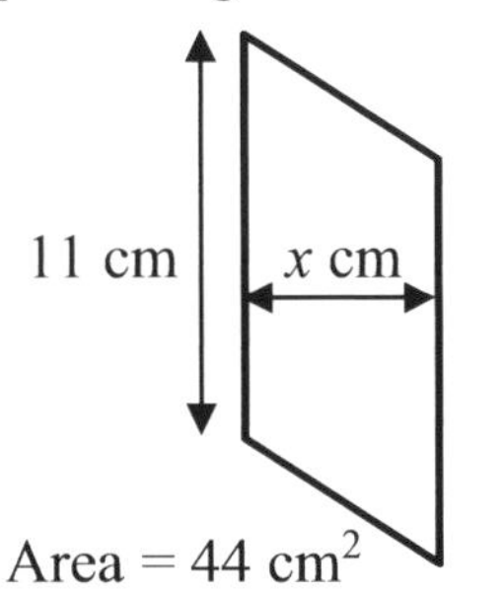

Area = $11 \times x$
$44 = 11 \times x$
$x = 44 \div 11$
$= 4$

Therefore, the height = **4 cm**

Useful formula:

Height of a Parallelogram = area ÷ its base

Base of a Parallelogram = area ÷ perpendicular height

AREA OF A TRIANGLE

Two identical triangles will always join to produce a parallelogram.

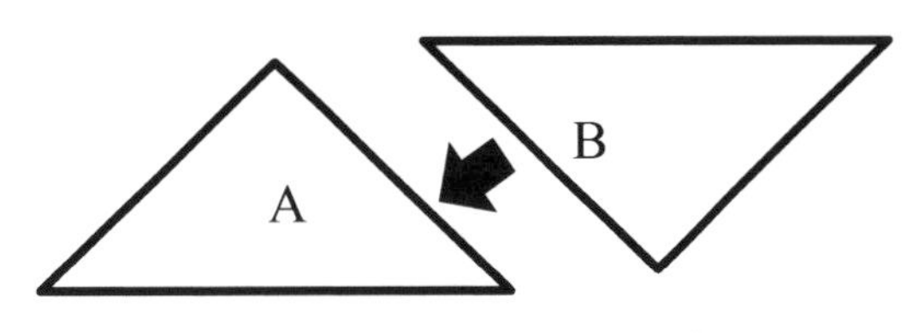

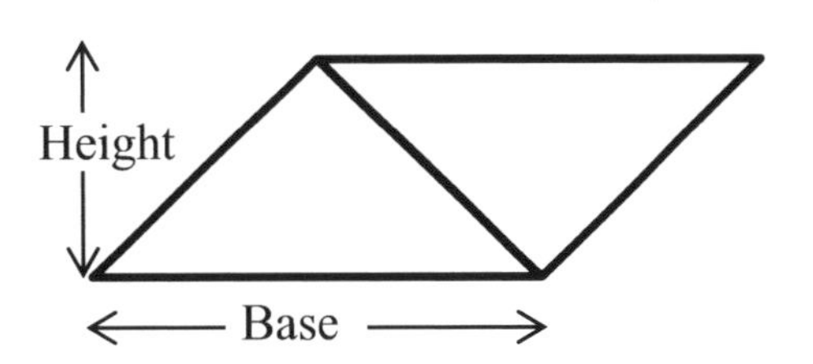

It goes to say that two triangles make up one parallelogram.

The area of a parallelogram is base × perpendicular height; therefore, the area of a triangle must be the area of a parallelogram ÷ 2.

Area of a triangle $= \frac{1}{2}$ **base × perpendicular**

Example 4: Calculate the area of the area of the triangle below.

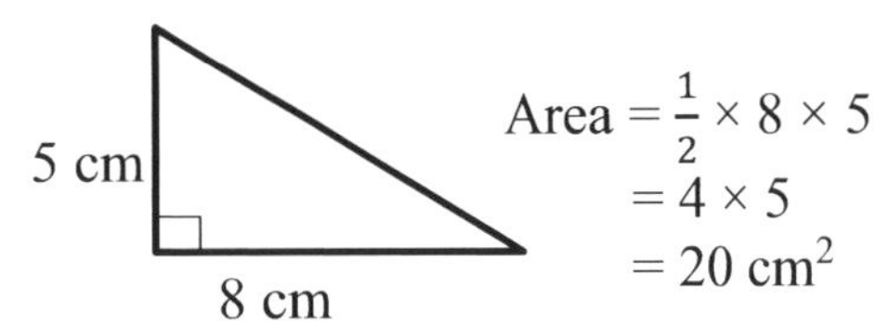

Area $= \frac{1}{2} \times 8 \times 5$
$= 4 \times 5$
$= 20 \text{ cm}^2$

Example 5: Calculate the area of triangle PQR.

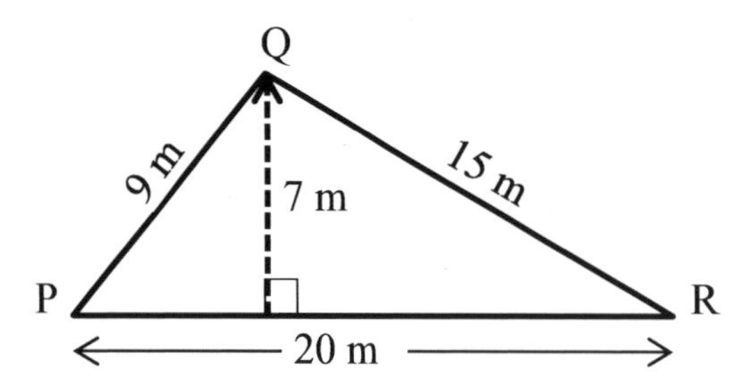

Area of PQR $= \frac{1}{2} \times 20 \times 7$
$= \mathbf{70\ m^2}$

Notice that 9 m and 15 m were not used. Only the lengths that are perpendicular to each other are used for calculating the area of a triangle.

In this case, 7 m height is perpendicular to the base of 20 m.

Example 6: Work out the area of triangle STU

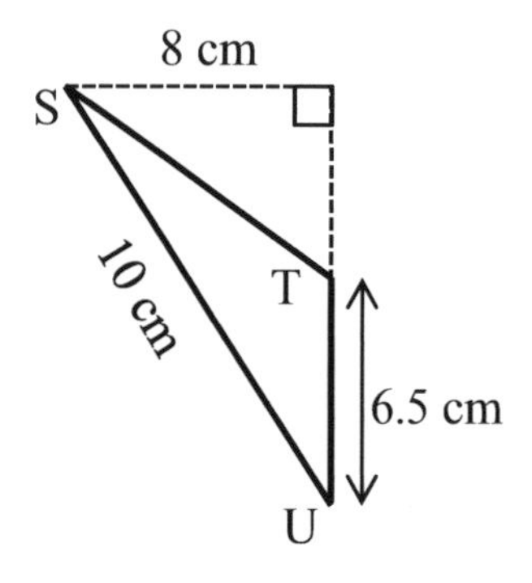

Area $= \frac{1}{2} \times 6.5 \times 8$
$= \mathbf{26\ cm^2}$

EXERCISE 7C

1) All lengths are in cm. Work out the area of the triangles below.

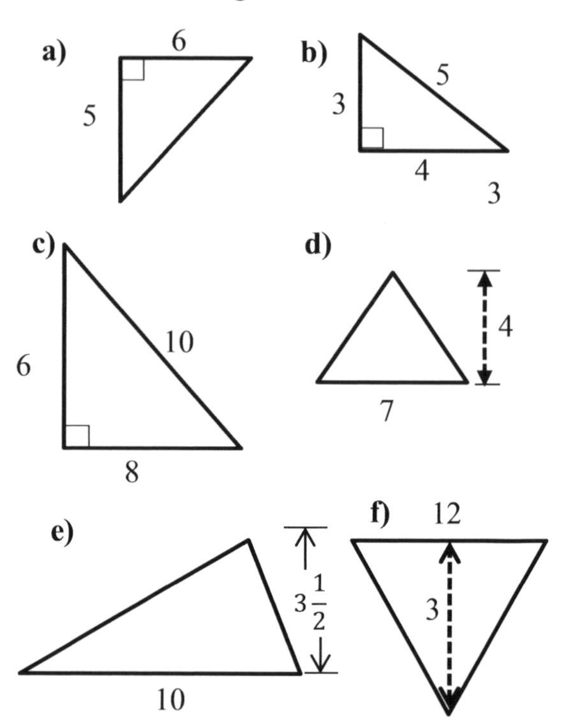

2) Calculate the total area of the quadrilateral ABCD. All lengths are in metres.

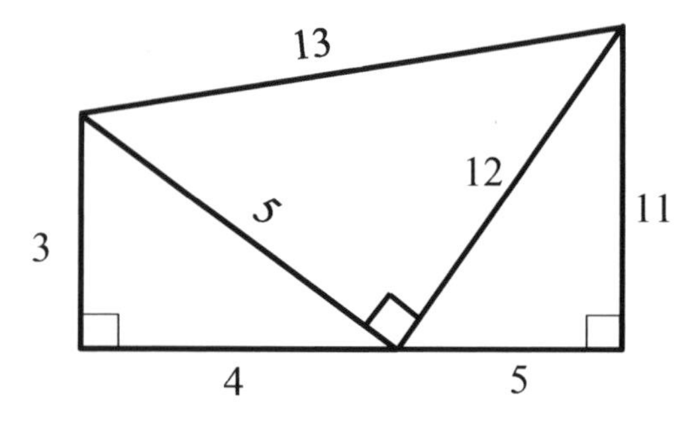

3) A triangle has an area of 30 cm^2 with a base length of 10 cm. Calculate the perpendicular height of the triangle.

4) List a possible base length and height of a triangle with area 28 cm^2.

5) All shapes below are parallelograms. Calculate the area of each shape. All lengths are in cm.

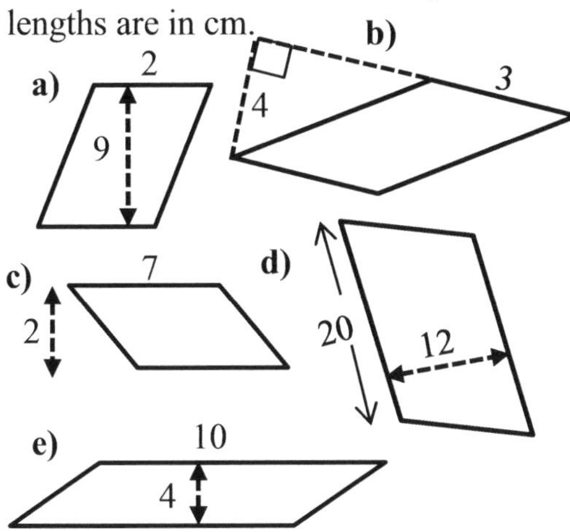

6) All lengths are in cm. Work out the area of the parallelograms and triangles below.

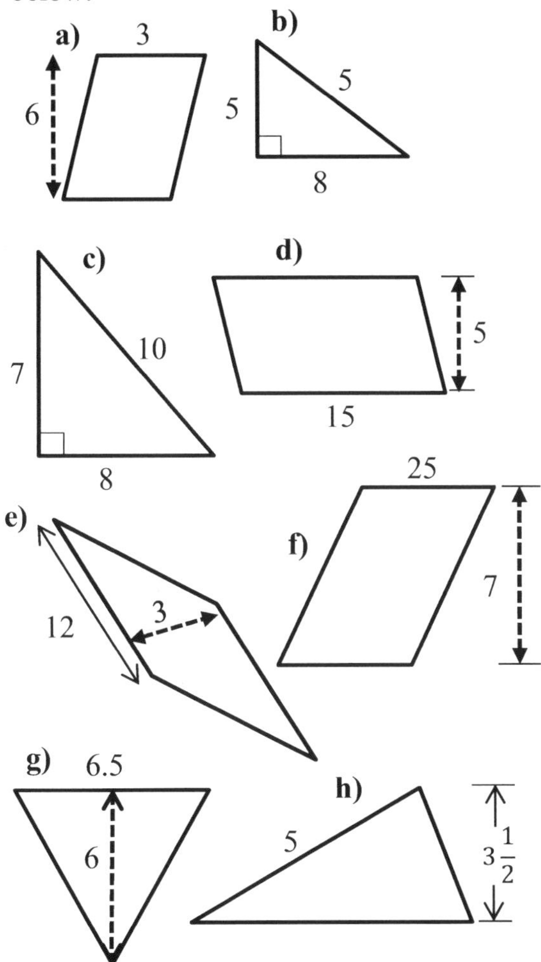

7) In questions 7a - d, calculate the base, *y*.

6

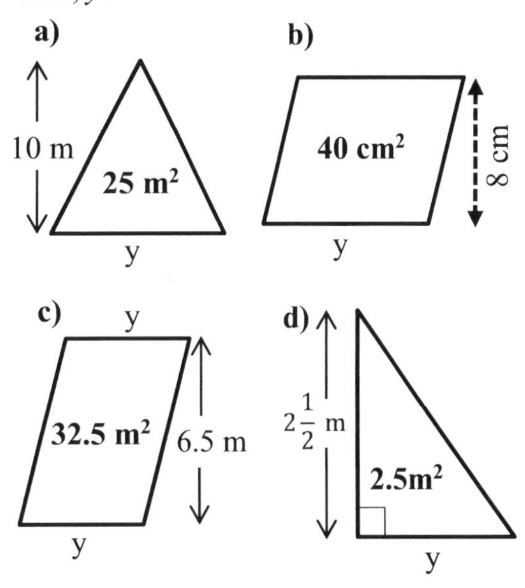

8) Calculate the area of the plane shapes below. All lengths are in cm.

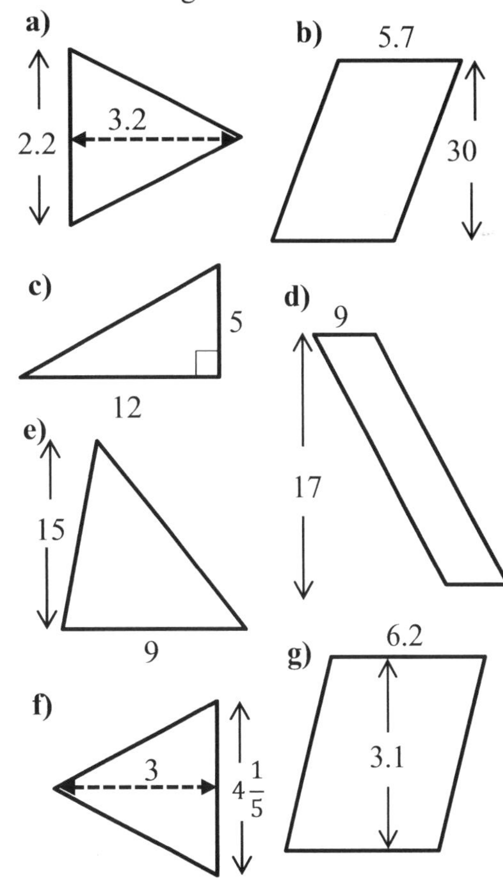

AREA OF A TRAPEZIUM

A trapezium is a four-sided straight shape with **one pair** of parallel sides.

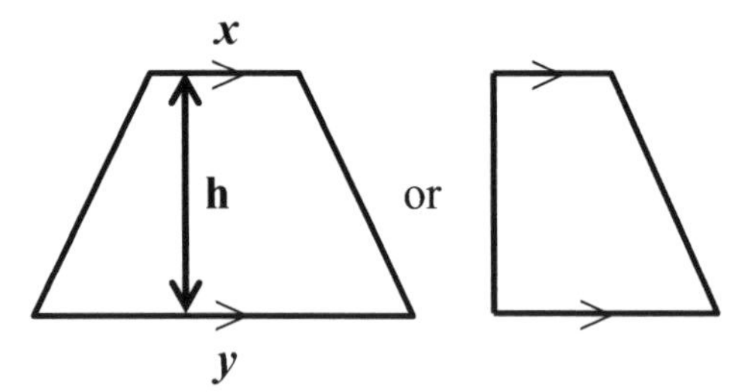

If two identical trapezia are put together, they form a parallelogram.

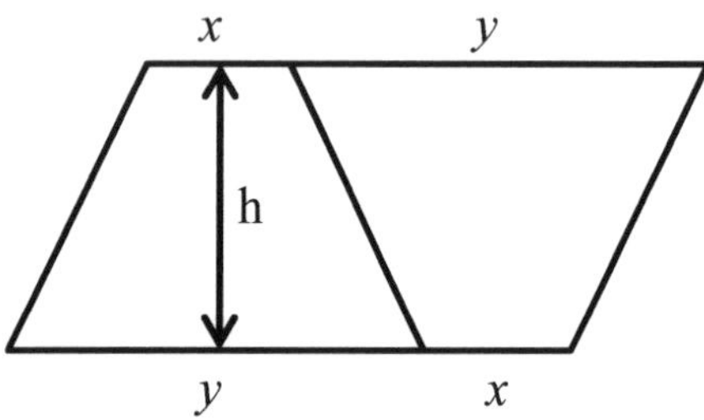

Remember:
Area of a parallelogram = base × perpendicular height $(x + y) \times h$

Since two trapezia make up one parallelogram, the area of a trapezium
$= \frac{1}{2} \times$ sum of the parallel sides × height
$= \frac{1}{2} \times (x + y) \times h$

Example 1: Calculate the area of the trapezium below.

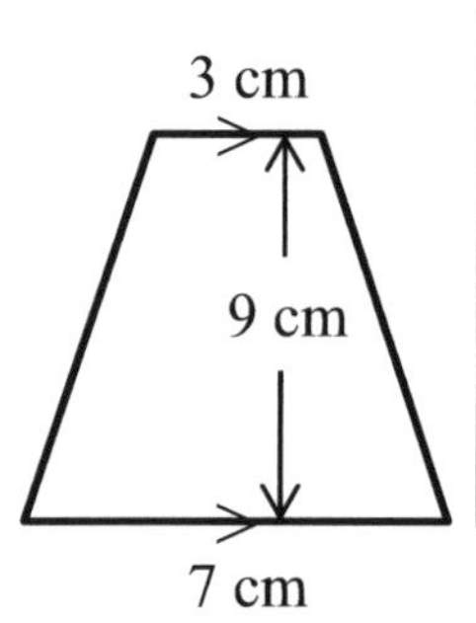

Add the parallel sides 3 + 7 = 10
Divide by 2
10 ÷ 2 = 5
Multiply by perpendicular height
5 × 9 = **45 cm²**

Example 2: Calculate the area of the trapezium below.

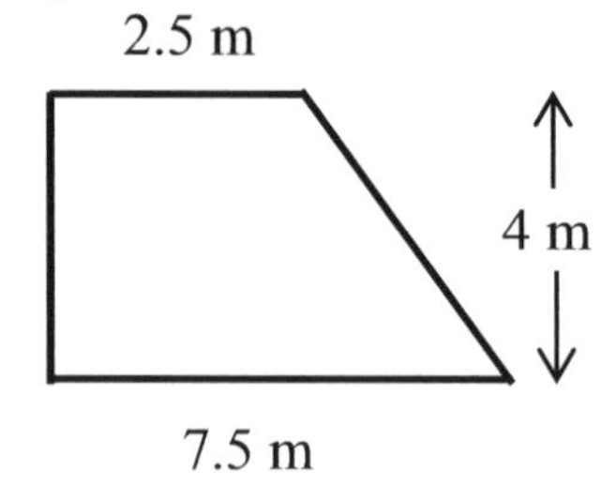

Area $= \frac{1}{2} \times (2.5 + 7.5) \times 4$
$= \frac{1}{2} \times 10 \times 4$
$= \mathbf{20\ m^2}$

EXERCISE 7D

1) All the lengths are in cm. Work out the area of the trapezia below.

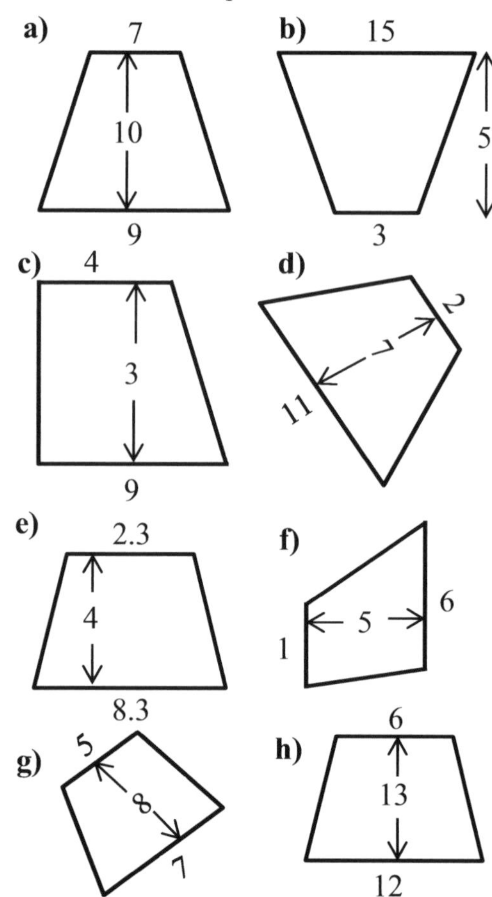

2) Calculate the length of the missing sides in the trapezia below. All the lengths are in *m*.

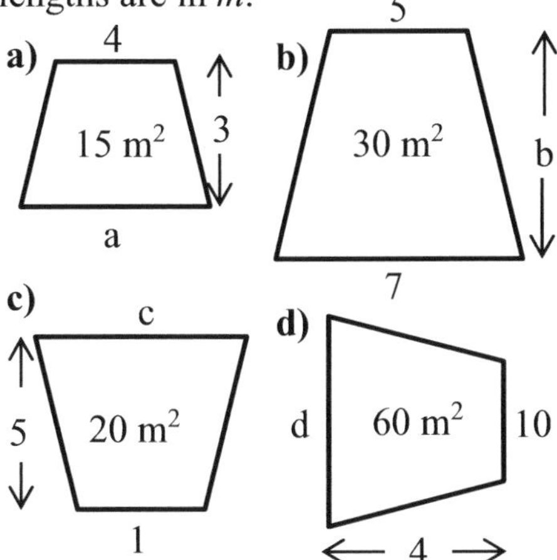

7.3 COMPOUND SHAPES

Any shape made up of more than one primary figure is called a **compound** or **composite** shape.

Areas of compound shapes are calculated by splitting them into its individual shapes and then add together.

Example 1: Calculate the area of the compound shape below.

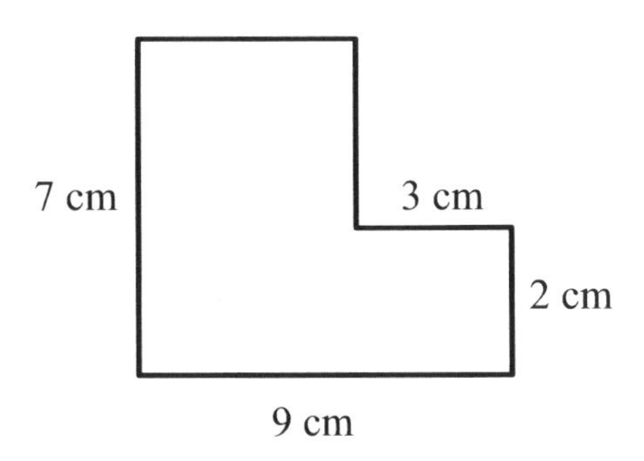

Solution: Split the shape into two rectangles.

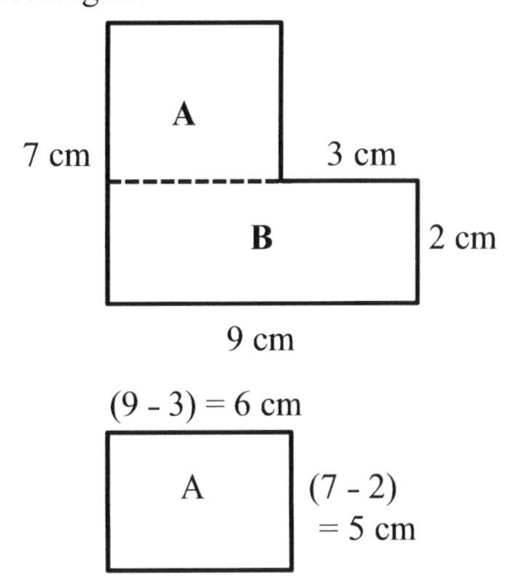

Area of A $= 6 \times 5 = 30$ cm^2
Area of B $= 9 \times 2 = 18$ cm^2
Total area $= 30 + 18 =$ **48 cm^2**

Example 2: Work out the area of the shape.

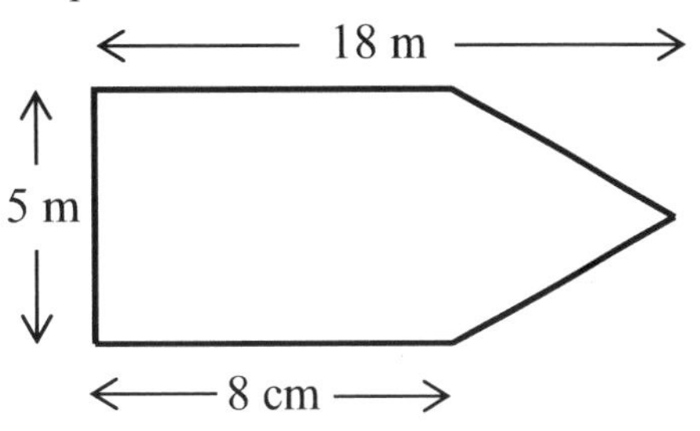

Solution: Split the shape into a rectangle and triangle.

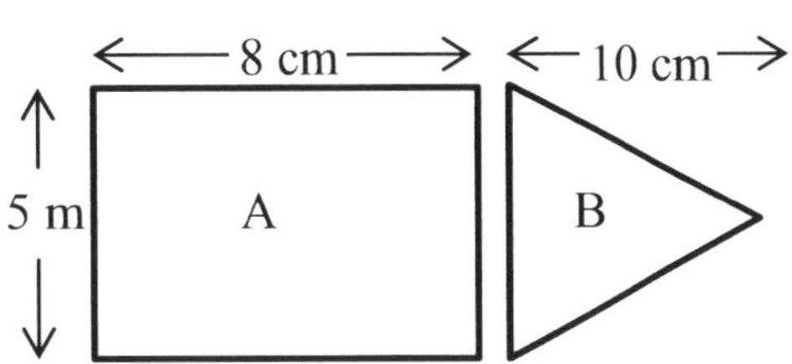

Area of A $= 8 \times 5 = 40$ cm^2
Area of B $= \frac{1}{2} \times 5 \times 10 = 25$ cm^2
Total area $= 40 + 25 =$ **65 cm^2**

EXERCISE 7E

1) Calculate the area of the compound shapes. All lengths are in cm.

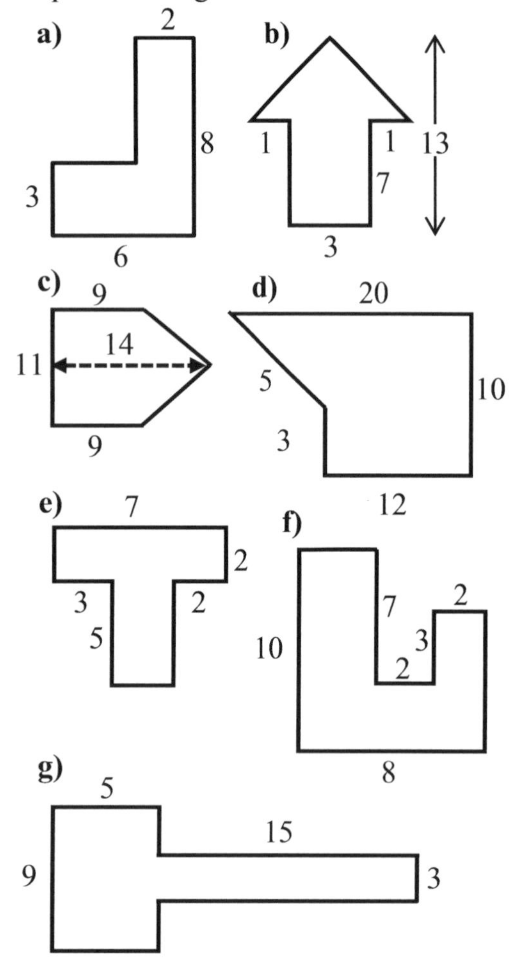

2) The shape below contains two identical triangles at both ends and a rectangle. Calculate the area of the shape. All lengths are in centimetres.

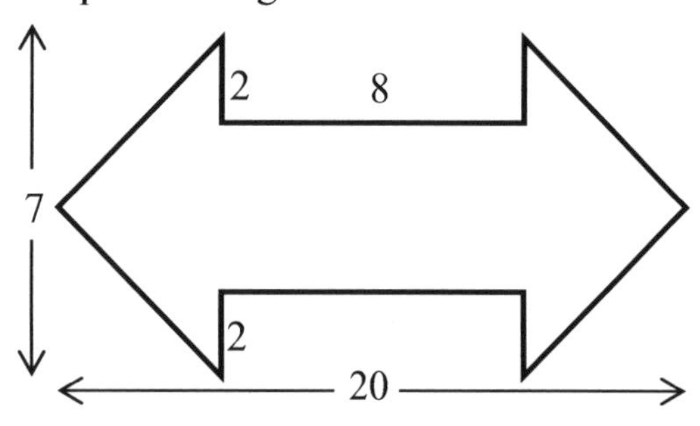

3) Obiora showed his working out for the area of this shape.

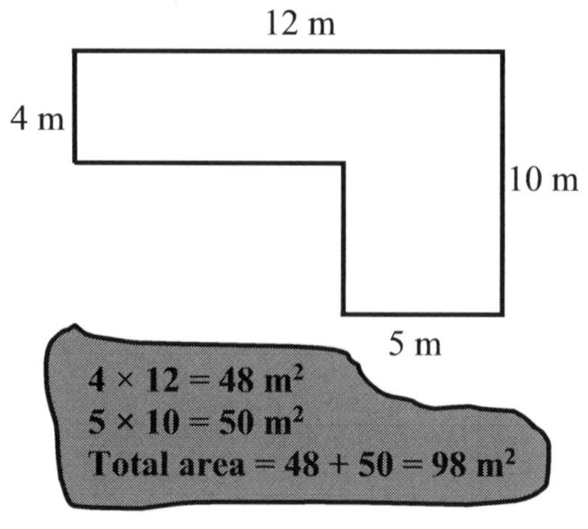

Show that Obiora is wrong.

4) Calculate the area of the shapes below. All lengths are in metres.

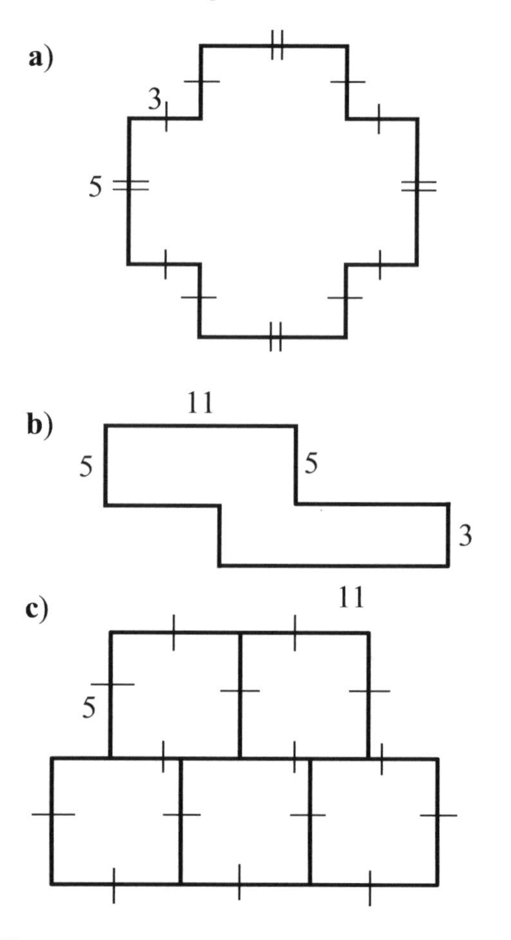

7.4 SHADED AREA

Example 1: Calculate the shaded area.

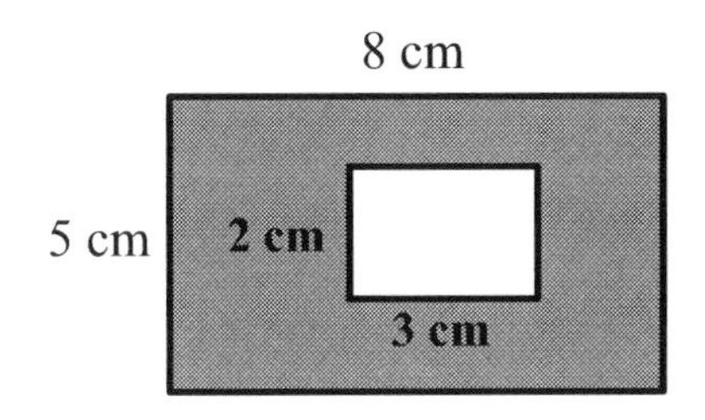

Area of big rectangle = $8 \times 5 = 40$ cm^2
Area of small rectangle = $3 \times 2 = 6$ cm^2
So, are of shaded part = 40 - 6 = 34 **cm²**

Example 2: Calculate the shaded area

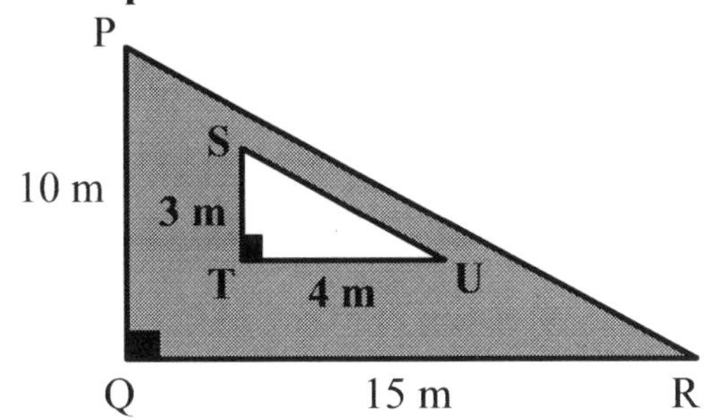

Area of ΔPQR = $\frac{1}{2} \times 10 \times 15 = 75$ m^2
Area of ΔSTU = $\frac{1}{2} \times 4 \times 3 = 6$ m^2
Shaded area = 75 - 6 = **69 m²**

Example 3: Work out the shaded area.

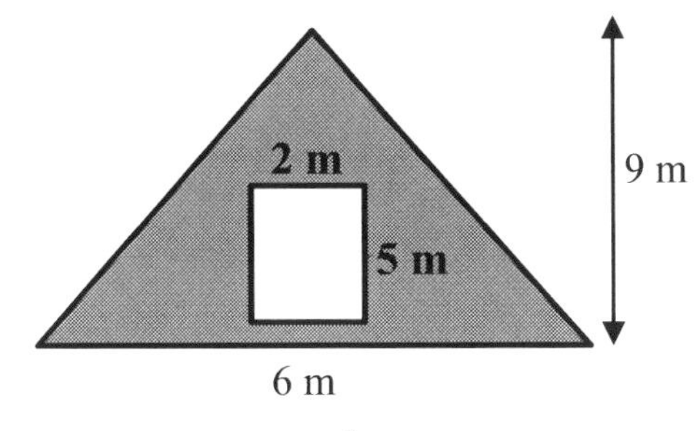

Area of triangle = $\frac{1}{2} \times 6 \times 9 = 27$ m^2
Area of rectangle = $2 \times 5 = 10$ m^2
Shaded area = 27 - 10 = **17 m²**

EXERCISE 7F

1) Calculate the shaded area of the shapes below. All lengths are in metres.

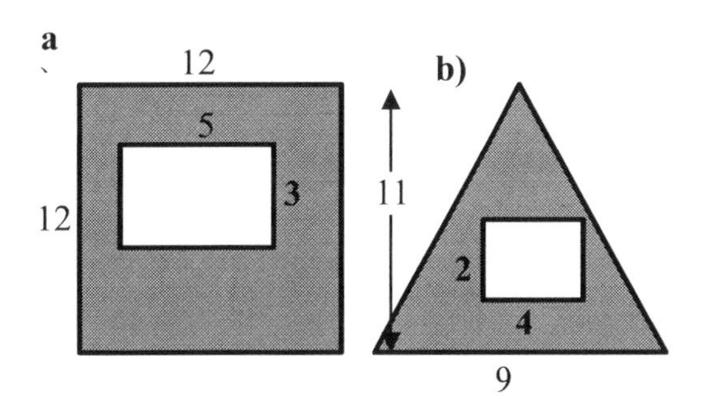

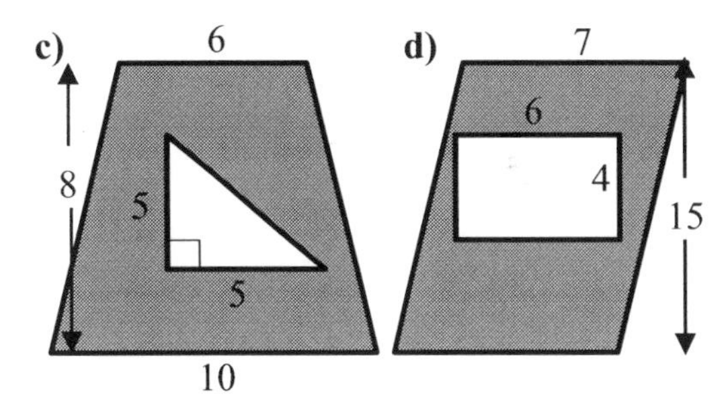

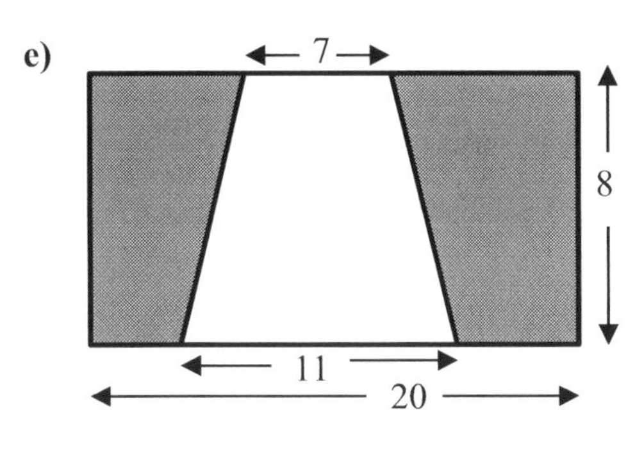

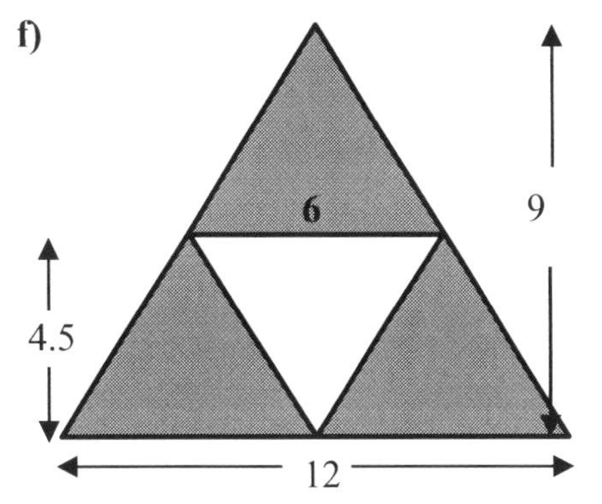

7.5 AREA OF A CIRCLE

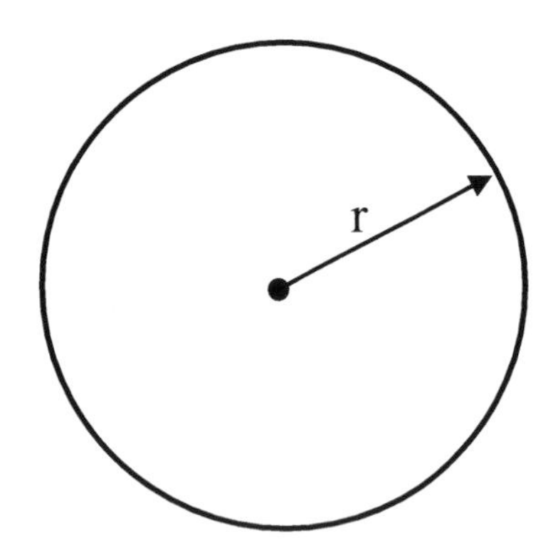

Divide the circle above into different sectors, cut them out and rearrange to form a rectangle. The length of the rectangle will be half the circumference, that is $\frac{1}{2} \times \pi \times$ diameter.
However, diameter = 2 × radius(r)

Length $= \frac{1}{2} \times \pi \times 2 \times r = \pi r$

The shape formed is **close to** a rectangle as shown below.

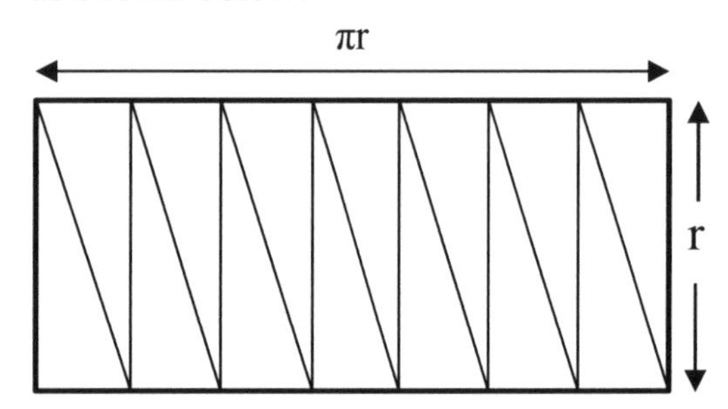

Therefore, the **area of the circle** which is also the area of the rectangle will equal $\pi \times r \times r = \pi r^2$

Area of circle = $\boldsymbol{\pi r^2}$
Always remember to use the radius instead of the diameter when working out the area of a circle.

Example 1: A circle has a radius of 7 cm. Calculate the area of the circle. Use π as $\frac{22}{7}$.

Area $= \pi r^2 = \frac{22}{7} \times 7^2$
$= \frac{22}{7} \times 7 \times 7 = \mathbf{154\ cm^2}$

Example 2: Calculate the area of the circle. Use π as $\frac{22}{7}$ and round to two decimal places.

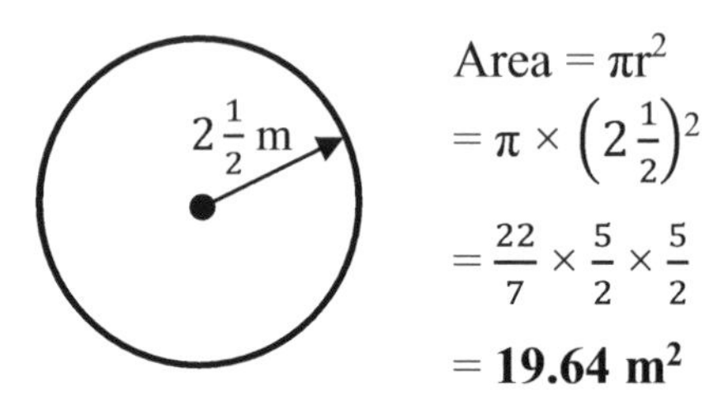

Area $= \pi r^2$
$= \pi \times \left(2\frac{1}{2}\right)^2$
$= \frac{22}{7} \times \frac{5}{2} \times \frac{5}{2}$
$= \mathbf{19.64\ m^2}$

Example 3: Work out the area of the circle. Use π as 3.14 and round to one decimal place.

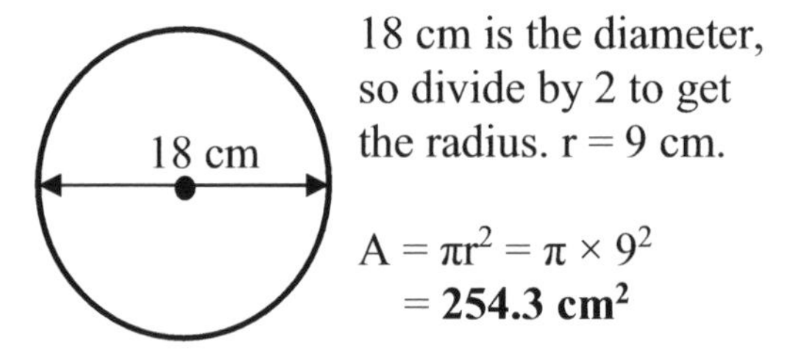

18 cm is the diameter, so divide by 2 to get the radius. r = 9 cm.

$A = \pi r^2 = \pi \times 9^2$
$= \mathbf{254.3\ cm^2}$

Example 4: Calculate the area of the semicircle. Use π as $\frac{22}{7}$.

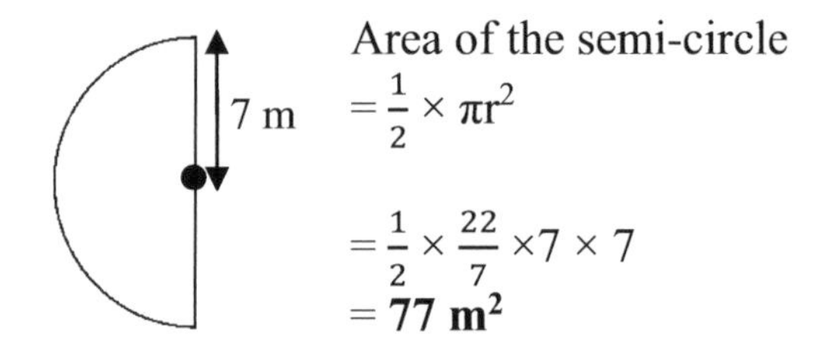

Area of the semi-circle
$= \frac{1}{2} \times \pi r^2$

$= \frac{1}{2} \times \frac{22}{7} \times 7 \times 7$
$= \mathbf{77\ m^2}$

EXERCISE 7G

1) Calculate the area of the shapes below. Use π as $\frac{22}{7}$ and round to one decimal place where possible.

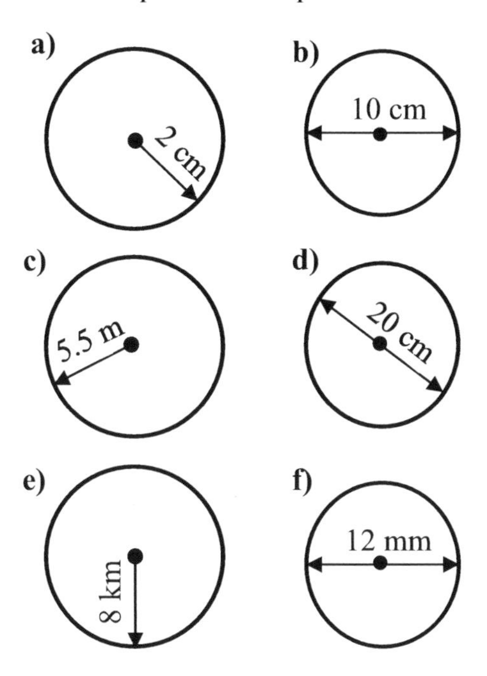

2) A circular track has a radius of 70 m.
a) Calculate the diameter of the track.
b) Calculate the area of the track.
Use π as 3.14.

3) Copy and complete the table below.
Use π as $\frac{22}{7}$.

Diameter(m)	Radius(m)	Area(m^2)
14		
42		
	21	
	14	

4) Calculate the area of the semicircles and quadrants. Give your answers to one decimal place where possible. All lengths are in metres. Use π as 3.14.

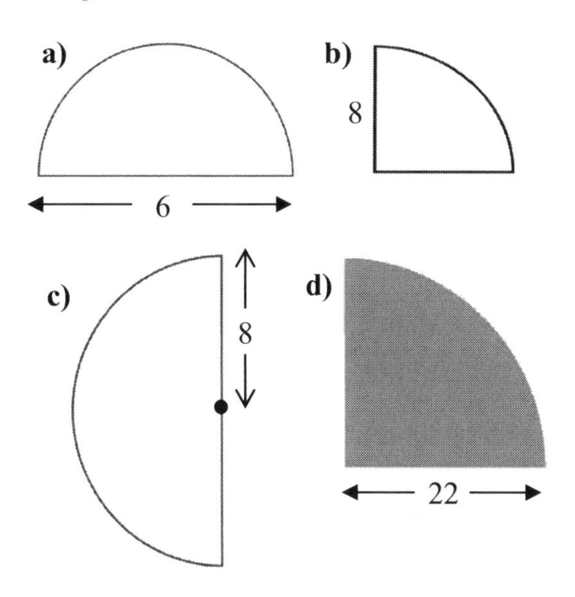

5) Calculate the area of the compound shapes made up of semicircles and rectangles. Use π as 3.14.

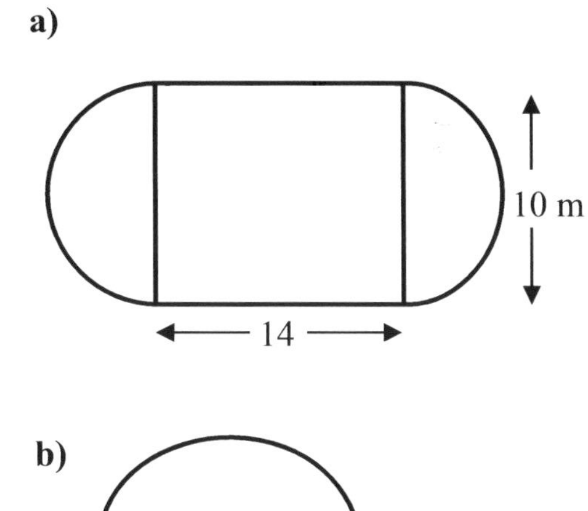

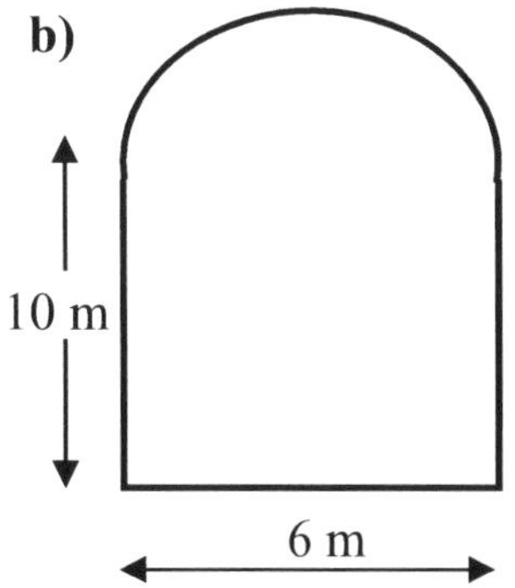

Example 5: Calculate the area of the shaded part. Use π as 3.14.

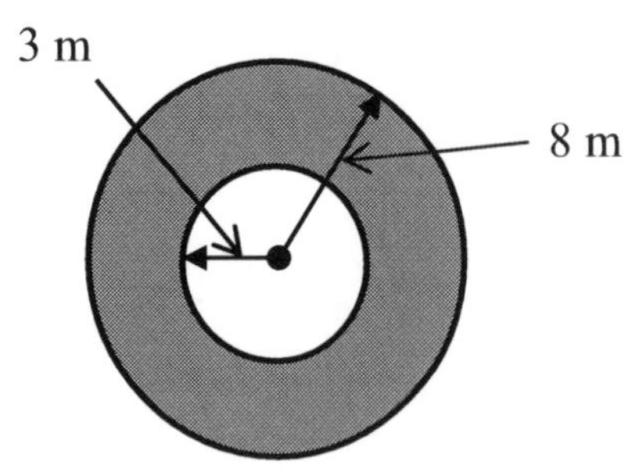

Area of the big circle = πr^2
$= 3.14 \times 8^2$
$= 200.96 \text{ m}^2$

Area of the small circle = πr^2
$= 3.14 \times 3^2$
$= 28.26 \text{ m}^2$

Area of shaded part = 200.96 - 28.26
= **172.7 m²**

Example 6: A rectangle is inscribed in a circle shown below. Calculate the shaded area to one decimal place. Take π as $\frac{22}{7}$.

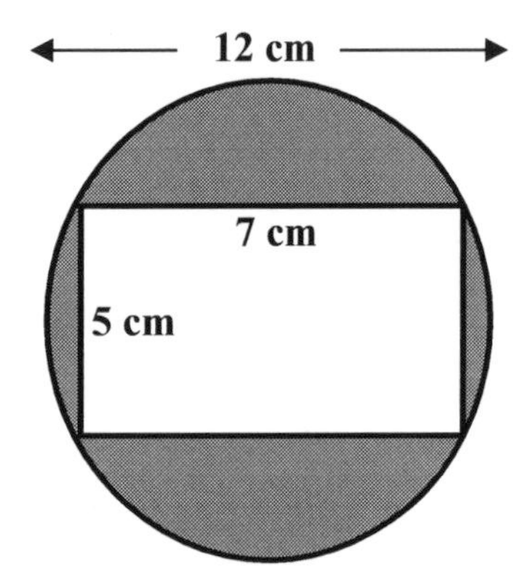

Area of rectangle = $7 \times 5 = 35 \text{ cm}^2$

Area of circle = πr^2
...But diameter = 12 cm, therefore radius = 12 ÷ 2 = 6 cm.

Area = $\frac{22}{7} \times 6^2 = \frac{22}{7} \times 6 \times 6$
$= 113.1428571 \text{ cm}^2$

Shaded area = 113.1428571 - 35
= **78.1 cm²**

EXERCISE 7H

1) Calculate the area of the shaded part of each of the diagrams below. Take π as 3.14 and round your answers to one decimal place where possible.

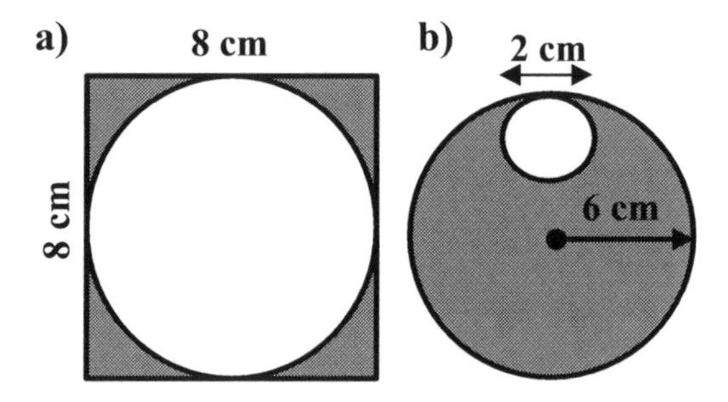

2) Three identical circles are placed in a rectangular box as shown below.

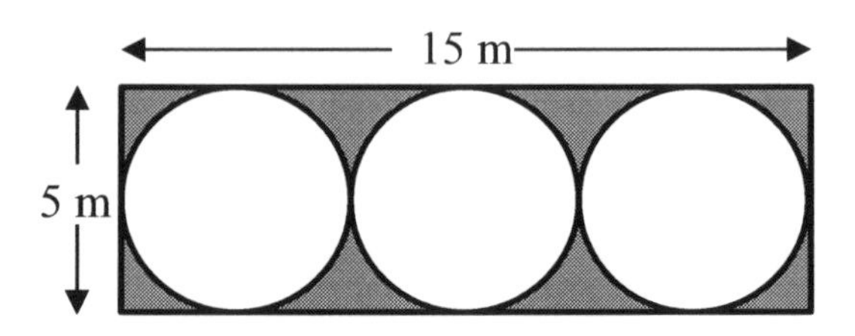

a) Work out the area of a circle.
b) Work out the total area of the three circles.

c) Work out the shaded area.
(Take π as 3.14 and round to two decimal places where possible)

3) Calculate the percentage of the shaded area. Take π as 3.14.

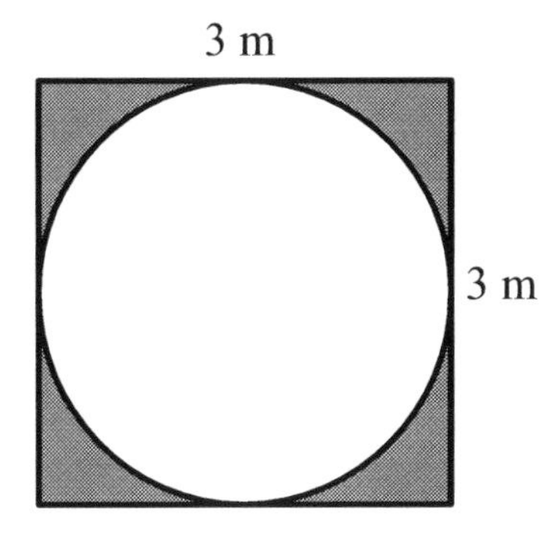

4) A circle of diameter 20 cm fits inside a semi-circle. Calculate the shaded area. Take π as 3.14 and round to one decimal place.

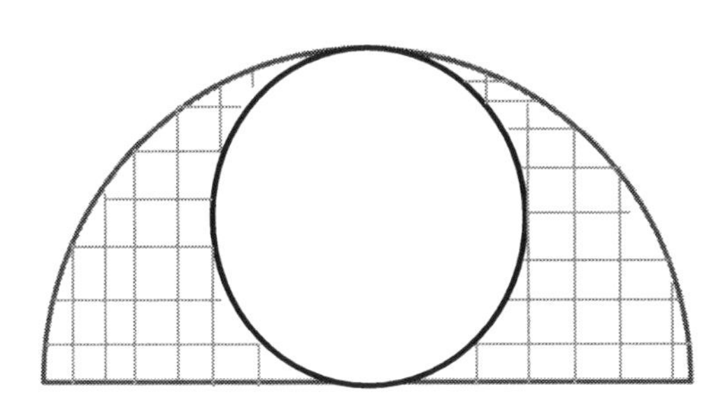

5) Three identical circles of diameter 10 cm each are placed inside a large circle of diameter 30 cm.

a) Calculate the area of the shaded part.
b) Work out the percentage of the shaded part. Use π as 3.14.

6) Calculate the area of the shaded part. Use π as 3.142.

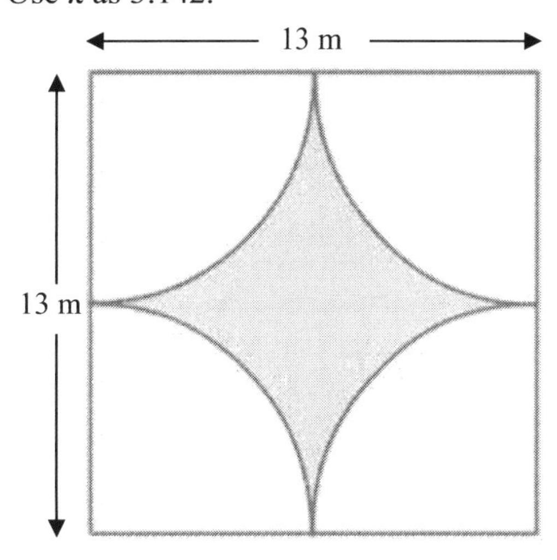

7) You have three semi-circles. The diameter of the large semicircle is 30 m. Work out the area of the shaded part and round your answer to one decimal place.
Use π as 3.142.

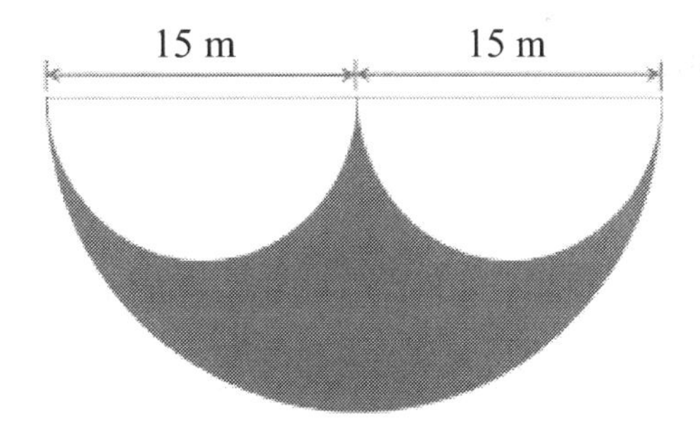

Chapter 7 Review Section

Assessment

1) What number belongs in each box?

a)

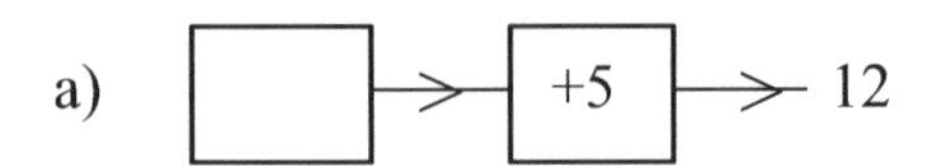

b)

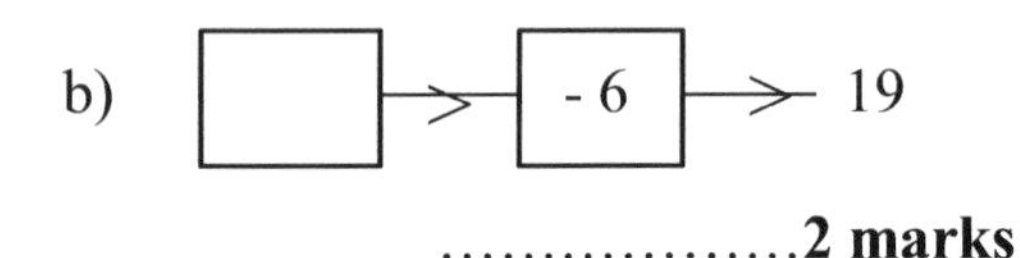

………………**2 marks**

2) What rule goes in the box?

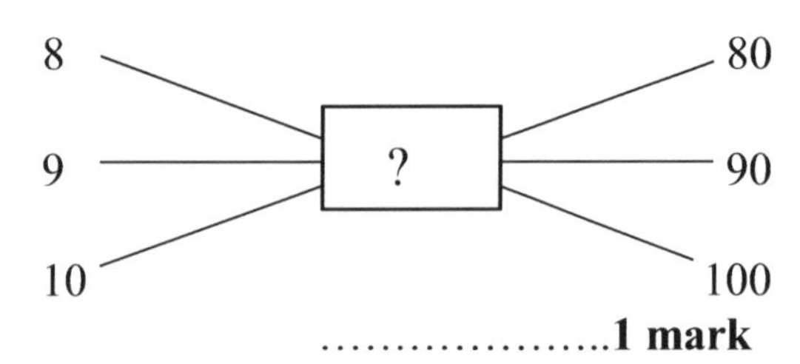

…………………**1 mark**

3) Copy and complete the table for the function machine.

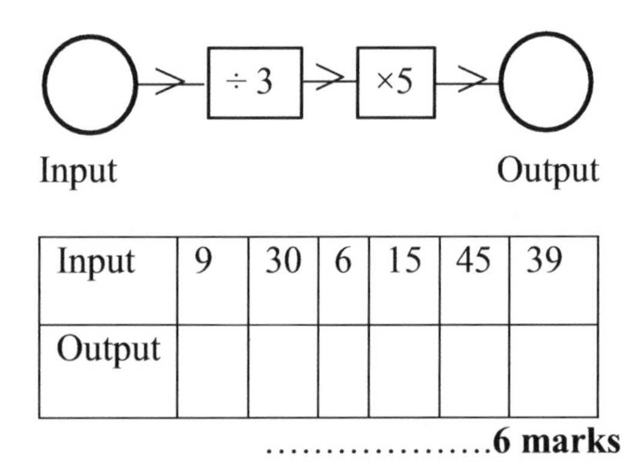

Input	9	30	6	15	45	39
Output						

………………**6 marks**

4) Complete the table for the magic squares below.

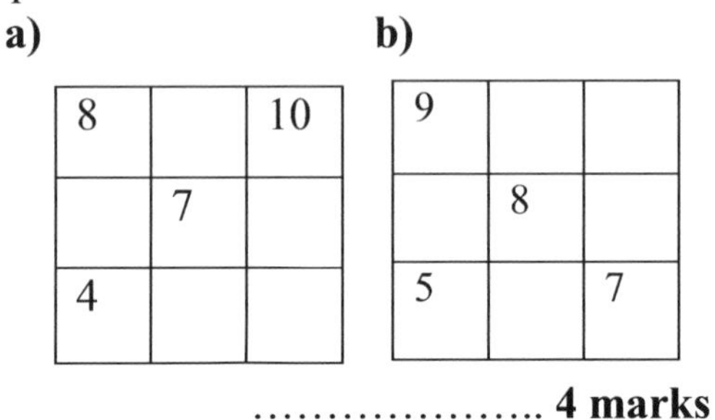

a)

8		10
	7	
4		

b)

9		
	8	
5		7

……………….. **4 marks**

5) On the centimetre grid below, calculate the perimeter of the shapes.

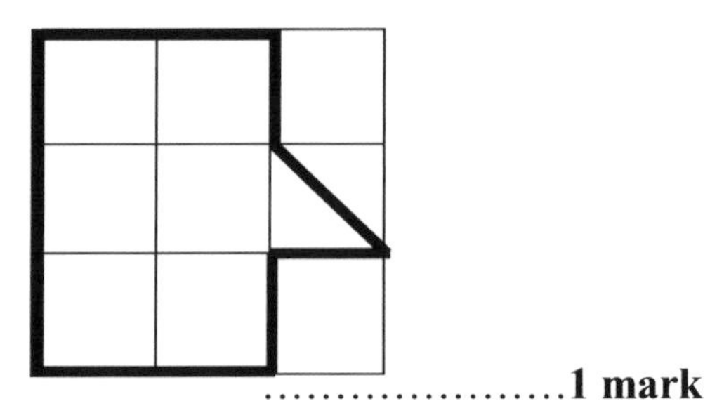

…………………**1 mark**

6) Calculate the perimeter of the object below. Use π as 3.14.

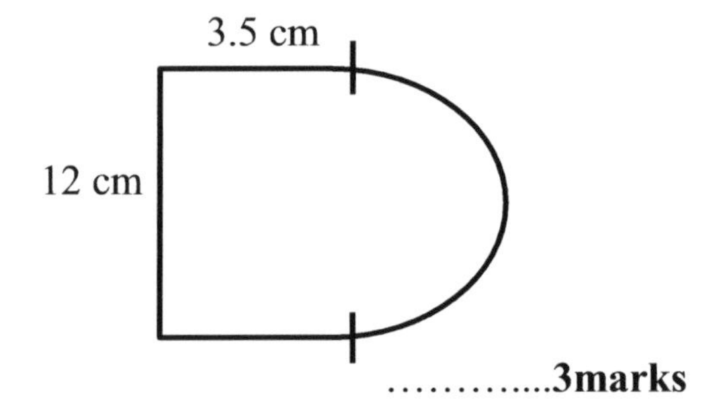

………....**3marks**

7) Calculate the circumference of the circles below. Use π as $\frac{22}{7}$.

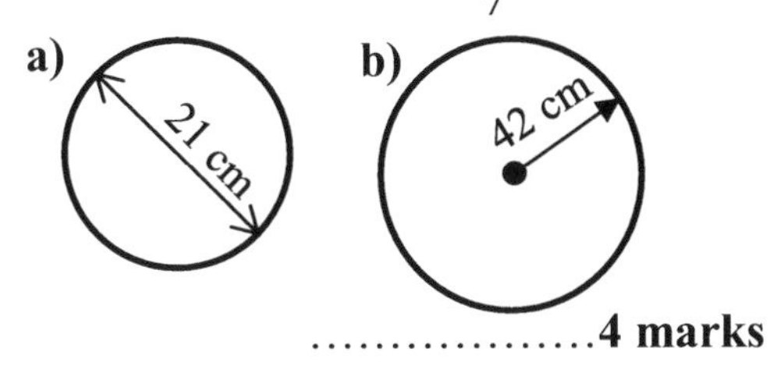

..................**4 marks**

8) Work out the area of the shaded part.

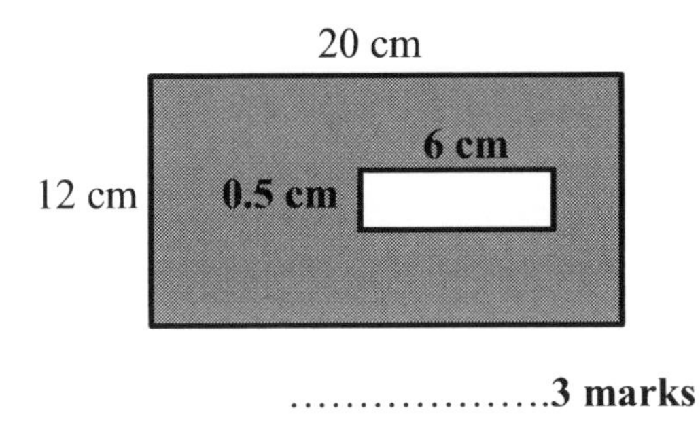

..................**3 marks**

9) All lengths are in cm. Work out the area of the triangles below.

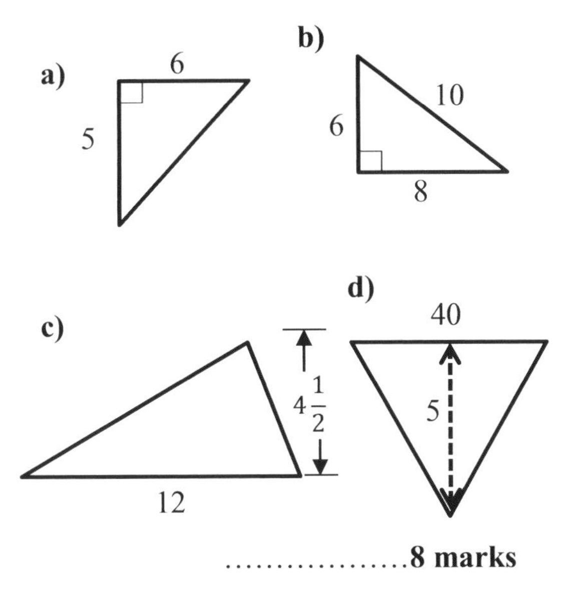

..................**8 marks**

10) A triangle has an area of 21 cm^2 with a base length of 7 cm. Calculate the perpendicular height of the triangle.

..................**2 marks**

11) All lengths are in cm. Work out the area of the parallelograms and triangles below.

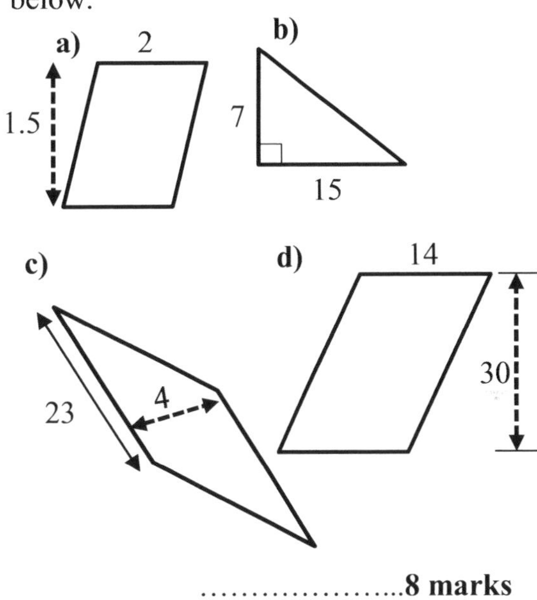

.....................**8 marks**

12) Calculate the length of the missing sides in the trapezia below.

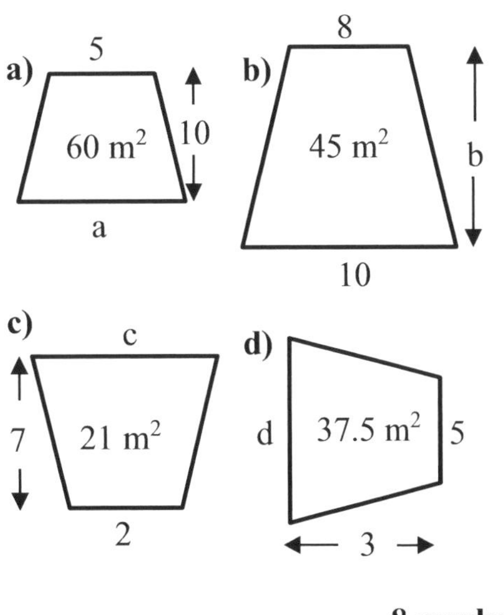

..................**8 marks**

13) Calculate the area of the compound shapes. All lengths are in cm.

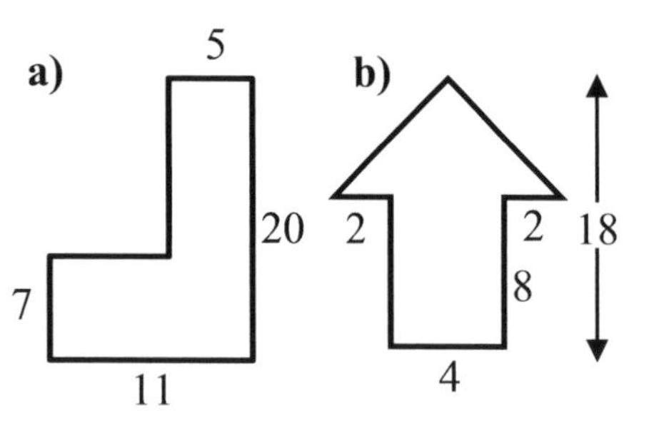

....................**4 marks**

14) Calculate the area of the shapes below. Use π as $\frac{22}{7}$ and round to one decimal place where possible.

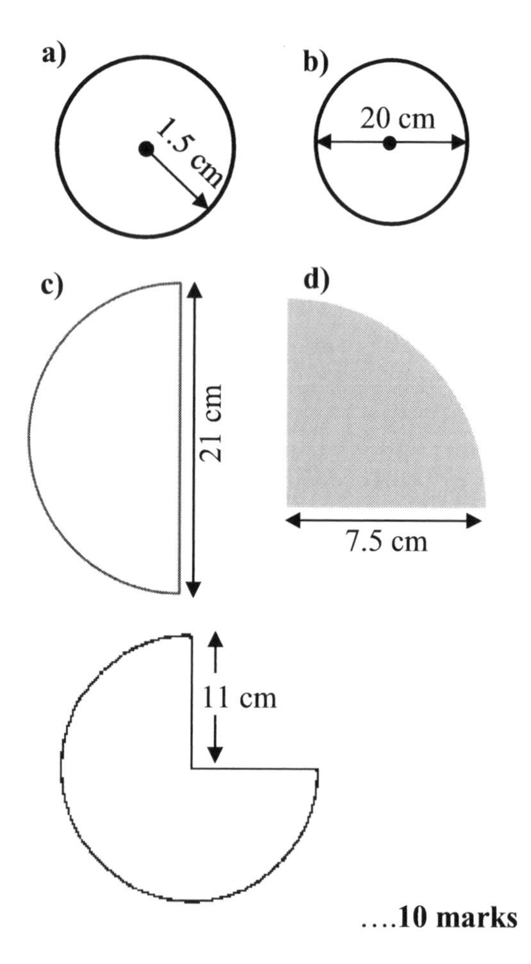

....**10 marks**

15) Four identical white squares of length 7 cm each are placed inside a large circle of diameter 25 cm.

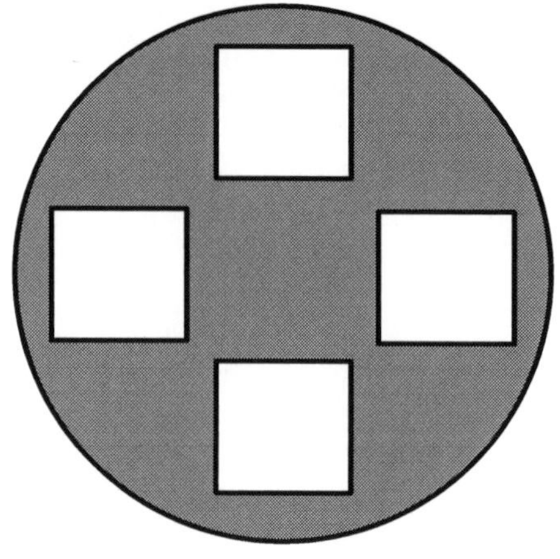

a) Calculate the area of the shaded part.

.... **3 marks**

b) Work out the percentage of the shaded part. Use π as 3.14.

...................**2 marks**

16) An arts theatre has a circular floor as the base. The diameter is 40 metres. A tile specialist charges ₦2 500 per square metre to tile the whole floor. Use π as $\frac{22}{7}$ and work out the cost of tiling the entire floor of the art theatre.

.... **4 marks**

17) A school track is made up of two straights and two semi-circular ends. Calculate the enclosed area within the track. Take π as 3.14 and round to 2 decimal places.

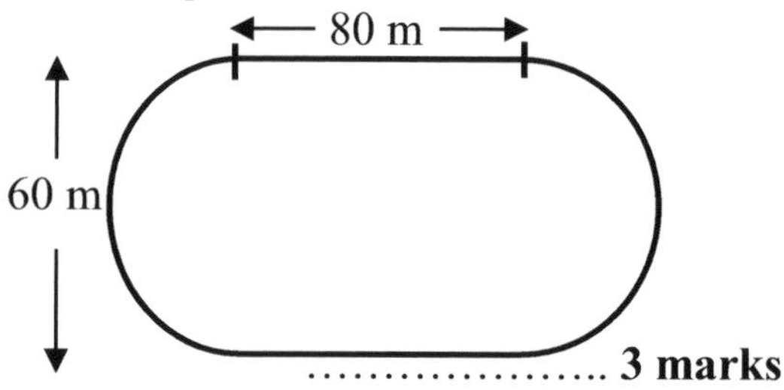

.... **3 marks**

8 Sequences

This section covers the following topics:

- Sequence of numbers
- Finding terms from the nth term of a sequence
- Graph of a linear sequence
- Finding the nth term of a number sequence
- Quadratic sequences

LEARNING OBJECTIVES

By the end of this unit, you should be able to:

a) Produce a sequence of numbers
b) Understand terms and nth terms
c) Find a formula for the nth term of a linear sequence
a) Draw graphs to show linear sequences
b) Find quadratic terms and nth terms

KEYWORDS

- Sequence
- Terms
- Nth terms
- Patterns
- Rule
- Quadratic

8.1 LINEAR SEQUENCE

A number **sequence** is a list of numbers that follow a particular rule (pattern). It is linear if it can be represented by a straight line graph. It increases or decreases in equal-sized steps.

Each number in a sequence is called a **term**.

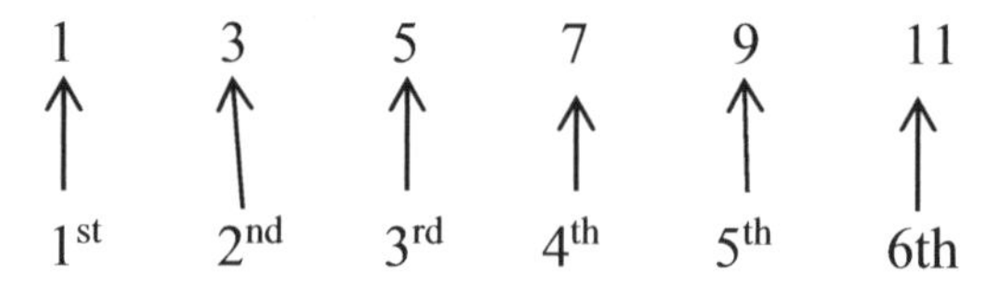

Consecutive terms are terms that are next to each other. They are usually separated by commas. In the example above, 1 and 3, 3 and 5, 5 and 7, 7 and 9, 9 and 11 are consecutive odd numbers.

There are lots of different number patterns. A sequence can continue forever (infinite). For example, 2, 4, 6, 8, 10…

Sequences follow a **rule**.

The sequence 4, 8, 12, 16, 20 will continue if we keep on adding 4.

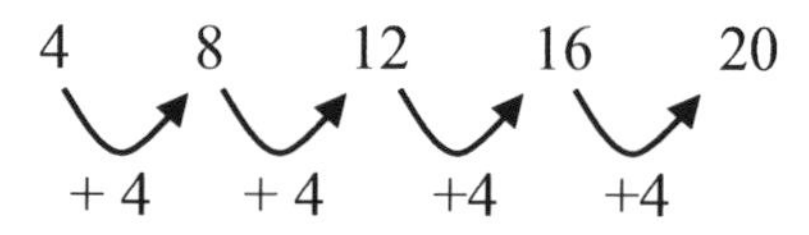

The rule for this sequence is to **add 4** each time. It is often called **term-to-term** rule.
Therefore, 4, 8, 12, 16, 20 is a linear sequence because it has a common difference of 4.

Some common sequences to remember are:

1, 3, 5, 7, 9…odd numbers
1, 4, 9, 16, 25… square numbers
2, 4, 6, 8, 10 … even numbers
10, 100, 1000, 10000… powers of ten
1, 3, 6, 10, 15… triangular numbers
1, 1, 2, 3, 5, 8… Fibonacci sequence

TRIANGULAR NUMBERS

Triangular numbers are formed by using consecutive numbers. The first triangular number is 1.

1
1 + 2 = **3**
1 + 2 + 3 = **6**
1 + 2 + 3 + 4 = **10**
1 + 2 + 3 + 4 + 5 = **15**
1 + 2 + 3 + 4 + 5 + 6 = **21**

FIBONACCI NUMBERS (SEQUENCE)

Fibonacci was an Italian mathematician regarded as the "Greatest European mathematician of the middle ages." His full name was Leonardo Pisano.

Source: Wikipedia

Fibonacci numbers are:1,1,2,3,5,8….

Each term in the Fibonacci sequence is the sum of the previous two terms.

1
$1 + 0 = \mathbf{1}$
$1 + 1 = \mathbf{2}$
$2 + 1 = \mathbf{3}$
$3 + 2 = \mathbf{5}$
$5 + 3 = \mathbf{8}$
$8 + 5 = \mathbf{13}$

Example 1: For each of the sequence, work out the missing terms.

a) 4, 7, 10, 13, …, …, …
b) 1, …, 13, …, …, 31, 37

Answers:
a) The rule is to add 3 each time, so the missing numbers are 16, 19 and 22.

b) Look at the 6th and 7th terms. The difference between 31 and 37 is 6. The rule is + 6 each time. The missing numbers are: 7, 19, 25

EXERCISE 8A

1) Find the next two terms in each sequence.
a) 4, 5, 6, 7, ___ , ___
b) 7, 9, 11, 13, ___, ___
c) 1, 6, 11, 16, ___, ___
d) 4, 7, 13, 25, ___, ___
e) 3, 7, 11, 15, ___, ___
f) 22, 16, 10, 4, ___, ___
g) 5, 6, 8, 11, ___, ___
h) 7, 6.9, 6.8, 6.7, ___, ___
i) 2, 4, 8, 16, ___, ___
j) 1, 4, 8, 13, ___, ___
k) 3, 9, 27, ___, ___
l) $\frac{1}{2}, \frac{2}{6}, \frac{3}{18}, \frac{4}{54},$ ___, ___

2) Write the term-to-term rule for questions 1a, b and c.

3) The term-to-term rule is +7. Write two different sequences that fit the rule.

4) For the sequence 4, 9, 14, 19, write down
a) the term-to-term rule
b) the first term
c) the 10thterm

5) You are given the first term and the rule of different sequences. Write down the first four terms of each sequence.

First term	Rule
4	Add 6
47	Add 15
2	Subtract 7
3	Triple
200	Divide by 2

6) Copy and complete the sequences.

$3 \times 88 = 264$
$4 \times \square = 352$
$\square \times 88 = 440$
$6 \times \square = 528$
$7 \times 88 = \square$
. . .
. . .
$19 \times 88 = \square$

7) Find i) the rule ii) the missing numbers for each sequence

a) 7, 15, 31, 63, ___
b) 225, 180, 135, ___ , ___
c) ___, 80, 77, 74, 71
d) 0.3, ___, 30, 300, ___
e) 8, 9, 11, 14, ___

8) Look at the sequence below.

$5^2 =$	25
$55^2 =$	3025
$555^2 =$	308025
$5555^2 =$	30858025

What is the value of 55555^2?

TERM NUMBER FROM A FORMULA

From the sequence, 4, 7, 10, 13, 16, a formula can be written as $3 \times n + 1$ or $3n + 1$ where n is the number of terms. Using the formula $3n + 1$, you could find any number of terms. For example, the 2000th term will be $3 \times 2000 + 1 =$ **6001**

Example 2: Write down the first three terms of the formula (nth term) $8n + 5$.

First term is when $n = 1$,
$(8 \times 1) + 5 =$ **13**.
Second term is when $n = 2$,
$(8 \times 2) + 5 =$ **21**.
Third term is when $n = 3$,
$(8 \times 3) + 5 =$ **29**
Therefore, the first three terms are:
13, 21 and 29

Example 3: Find the first three terms of the formula 5n.
First term: $5 \times 1 =$ **5**, second term: $5 \times 2 =$ **10**, third term: $5 \times 3 =$ **15**

Example 4: Find the 20th term of the nth term $2n - 3$.
20th term $= (2 \times 20) - 3 =$ **37**

Example 5: Write down the first five terms of the nth term n - 5

First term: $1 - 5 =$ **- 4**
Second term: $2 - 5 =$ **-3**
Third term: $3 - 5 =$ **-2**
Fourth term: $4 - 5 =$ **-1**
Fifth term: $5 - 5 =$ **0**
The first 5 terms are: -4, -3, -2, -1, 0

Example 6: Find the 10th term of the nth term formula, $3n^2$.
Note: $3n^2 = 3 \times n^2$. Therefore, the 10th term will be $3 \times 10^2 = 3 \times 100 =$ **300**

Example 7: Write down the first three terms of the nth term $4n^2 + 1$.

1st term: $4 \times (1)^2 + 1 = 4 \times 1 + 1 =$ **5**
2nd term: $4 \times (2)^2 + 1 = 4 \times 4 + 1 =$ **17**
3rd term: $4 \times (3)^2 + 1 = 4 \times 9 + 1 =$ **37**

A common mistake would be to multiply 4 by n and then square it. It is wrong.
Example: To find the 1st term of $4n^2 + 1$, a common mistake will be $(4 \times 1)^2 + 1 = 4^2 + 1 = 17$. ✖
This is wrong.

Example 8: For the sequence 11, 14, 17, 20, write down a) $T_{(1)}$, b) $T_{(2)}$, c) $T_{(10)}$

Solution: $T_{(1)}$ means the 1st term, $T_{(2)}$ means the 2nd term and $T_{(10)}$ means the 10th term.

$T_{(1)} = 11$
$T_{(2)} = 14$
$T_{(10)} = 38$.......from continuing the sequence.

EXERCISE 8B

1) Write down the first **three** terms of the sequences produced by these nth term formulae.

a) $2n$	g) $3 - n$	m) $20n - 3$
b) $14n$	h) $4 + 2n$	n) $2n^2$
c) $2n + 2$	i) $10 - 5n$	o) $3n^2 - 1$
d) $3n - 1$	j) $8 + 3n$	p) $4n^2 + 10$
e) $4n + 1$	k) $n^2 + 7$	q) $13 - 2n$
f) n^2	l) $9 + n^2$	r) $\frac{1}{n^2}$

2) Write down the first **five** terms of the sequences below.

a) First number is 40, the rule is add 4
b) First number is 7, the rule is add 5
c) First number is 13, the rule is minus 2
d) First number is 2, rule is subtract 3
e) First number is -7, the rule is add 9

3) The third term of a sequence is 16, fourth term, 22 and fifth term, 28. Find the first and second terms.

4) For the sequence -14, -9, -4, 1…
a) Find the term-to-term rule
b) Find the next term in the sequence
c) Find the 100th term in the sequence.

5) For the sequence below
7, 11, 15, 19, 23…,
Write down
a) $T_{(1)}$ b) $T_{(2)}$ c) $T_{(97)}$

6) Each number in this sequence is -3 times the previous number.
-4, 12, -36, …
Write down
a) the 4th term b) the 6th term

7) Fill in the missing numbers

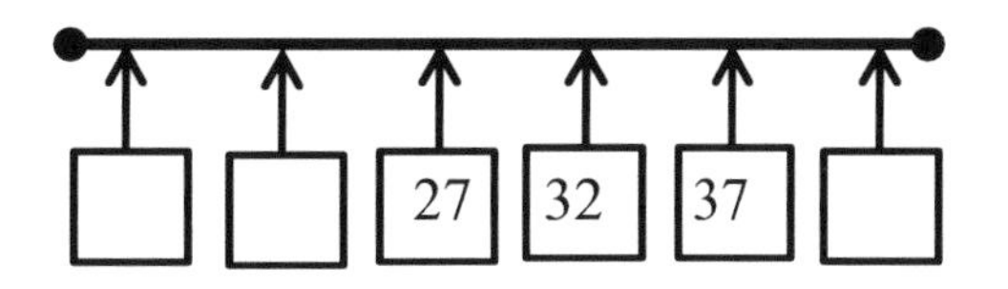

8) Write down the first three terms of the sequences where $T_{(n)}$ is:

a) $2n + 11$ c) n^2
b) $n - 3$ d) $3n^2 + 3n$

SEQUENCES AND GRAPHS

Number patterns can be drawn on a graph. For example, the number pattern 3, 6, 9, 12, 15 can be shown on a graph as follows.

Term	1	2	3	4	5
Magnitude	3	6	9	12	15

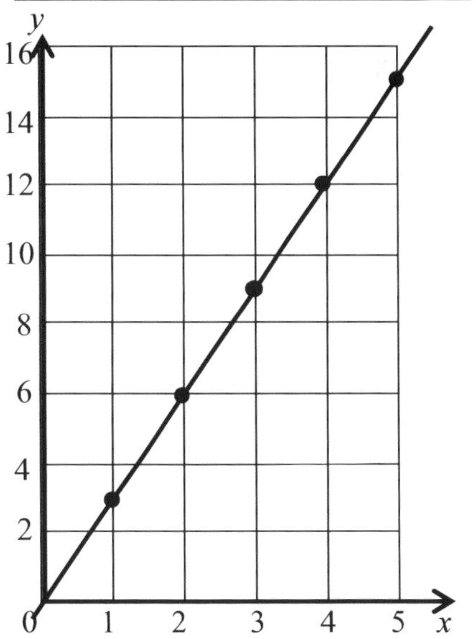

The graph of the sequence is a straight line indicating linear sequence.

8.2 NTH TERM FORMULA

For most mathematical sequences, there are formulae (nth terms). The nth term is very important in predicting the number of terms in a sequence or the term number.

A general term in the sequence is the **nth term**, where *n* stands for any number of terms.

For a linear sequence, the difference is constant. The sequence 4, 6, 8, 10 …have a constant difference of 2, all through. The sequence is linear. To find the nth term of a linear sequence, try to memorise **DNO**.

D = difference between the terms
N = Number of terms
O = **Zero** term
(Term before the 1st term)

Example 1: Find the nth term for the sequence 9, 14, 19, 24, 29…

Solution:

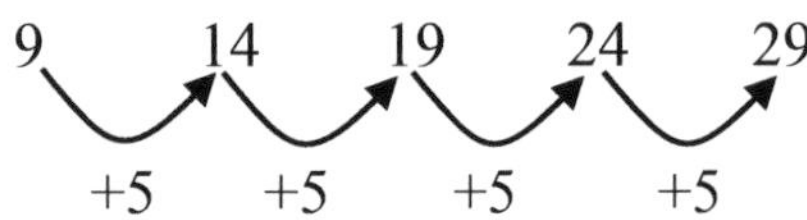

Difference between the terms is 5.
N, which is the number of terms, is written beside the difference to give 5n.
0, which is the zero term is the number before 9.
So we work backwards to get the zero term. 9 – 5 = **4**
Our sequence would look like this:

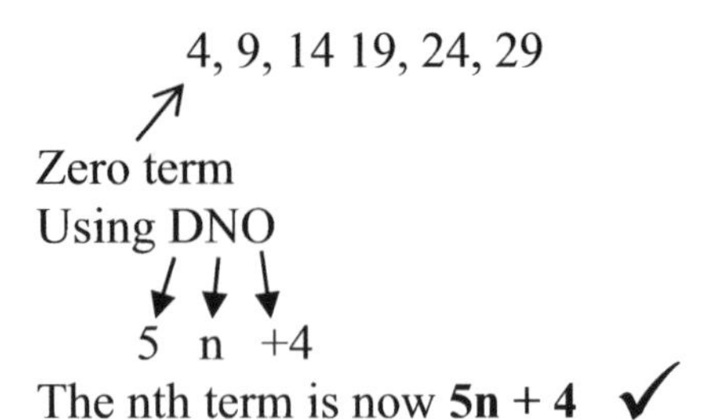

The nth term is now **5n + 4** ✓

CHECK YOUR ANSWER
Using the nth term 5n + 4 obtained, find the first 2 or three terms. If they equal the original sequence, then the nth term is correct.
1st term: $5 \times \mathbf{1} + 4 = 9$
2nd term: $5 \times \mathbf{2} + 4 = 14$
3rd term: $5 \times \mathbf{3} + 4 = 19$

Since the numbers obtained are the original numbers in the sequence, the nth term 5n + 4 is correct.

Example 2: Find the nth term formula for the sequence 2, 5, 8, 11, 14,..
Solution
Difference = 3, which implies a 3n term
Zero term = 2 – 3 = -1.
Nth term using DNO = **3n - 1** ✓

Example 3: Find the nth term formula for the sequence 20, 18, 16, 14, 12…
Solution
Difference = 18 – 20 = -2
which implies -2n term
Zero term = 20 – (-2) = 22
Nth term = **-2n + 22** ✓

Example 4: Find the nth term of the sequence $\frac{5}{7}, \frac{6}{9}, \frac{7}{11}, \frac{8}{13}, \ldots$
For numerator, nth term = n + 4
For denominator, nth term = 2n + 5
Nth term = $\frac{\mathbf{n+4}}{\mathbf{2n+5}}$ ✓

EXERCISE 8C

1) Find the nth term of each of the sequences below.

a) 5, 7, 9, 11, 13…
b) 11, 14, 17, 20, 23…
c) 4, 5, 6, 7, 8, 9 …
d) 30, 25, 20, 15, 10 …
e) 1, 5, 9, 13, 17 …
f) 9, 12, 15, 18, 21 …
g) 7, 14, 21, 28, 35 …
h) 0, 4, 8, 12, 16 …
i) -1, 1, 3, 5, 7 …
j) 17, 14, 11, 8, 5 …
k) 500, 550, 600, 650, 700 …

2) Copy and complete the mapping diagram below.

Term number (n)	4n	Term
1	4	7
2	8	11
3		
4		19
70		

3) Write down each sequence and select the correct expression for the nth term.

a) 4, 6, 8, 10, 12…
b) 33, 30, 27, 24…
c) 2, 4, 6, 8…
d) 2, 5, 8, 11, 14…
e) 3, 4, 5, 6…
f) $1^2, 2^2, 3^2, 4^2$…

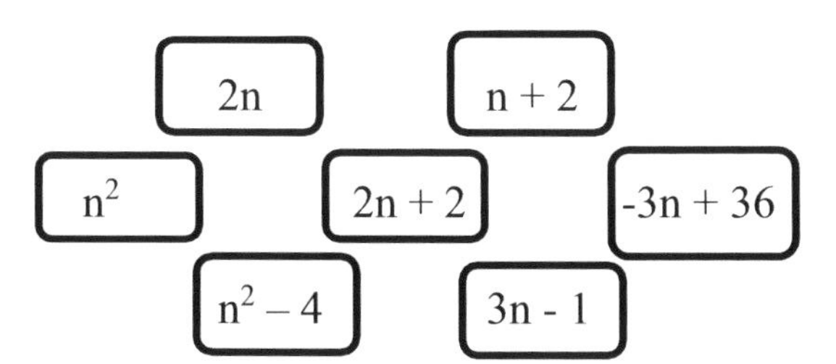

4) This sequence of patterns is made from regular pentagons with sides of 1 centimetre.

Pattern 1 Pattern 2 Pattern 3

a) What is the perimeter of pattern number 3?
b) Draw pattern numbers 4 and 5
c) Copy and complete the table below.

Pattern number	1	2	3	4	5	6
Perimeter (cm)						

d) Find a formula for the perimeter of the nth pattern.
e) Find the perimeter of the 100th pattern.
f) Which pattern number would have a perimeter of 3150?

5) Rods are placed to form a pattern.

Pattern 1 Pattern 2 Pattern 3

a) How many rods will there be in pattern number 5?
b) How many rods are needed for the nth pattern?
c) What pattern number will have 59 rods?
d) Copy and complete the table below.

Pattern number	1	2	3	50
Number of rods				

8.3 QUADRATIC SEQUENCES

In a linear sequence, the first difference is constant whereas, in a **quadratic sequence**, the second difference is constant (the same).

Nth Term formula for a quadratic sequence

Example 1: Find the nth term for 1, 4, 9, 16, 25

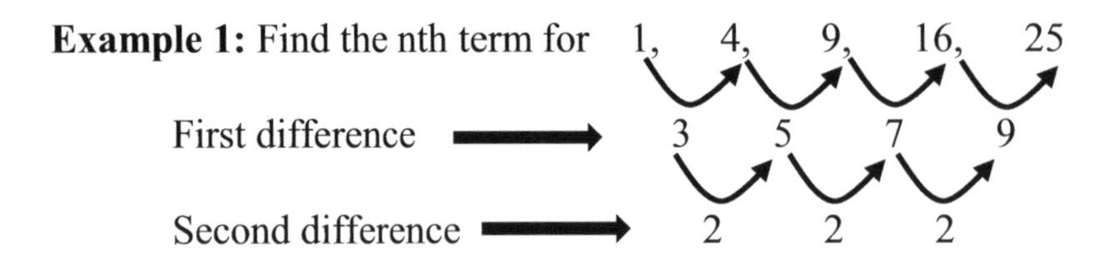

The sequence 1, 4, 9, 16, 25 is **quadratic** because the second difference is the same.

Once the second difference is constant, we know that the nth term formula must contain $\mathbf{n^2}$. To know the coefficient of n^2 (the number before n^2), we divide the constant by **2**. $2 \div 2 = 1$. It follows that the nth term formula for the quadratic sequence must start with $1n^2$. However, in algebra, it is not the convention to write the number, 1, in front of a letter. Therefore, the nth term of 1, 4, 9, 16, 25 must start with $\mathbf{n^2}$.

Now, make a table

n	n^2	Sequence	Difference
1	1	1	0
2	4	4	0
3	9	9	0
4	16	16	0
5	25	25	0

Since the difference from the table is the same, the nth term for 1, 4, 9, 16, 25 = $\mathbf{n^2}$ ✓

Example 2: Find the nth term for 5, 8, 13, 20, 29..

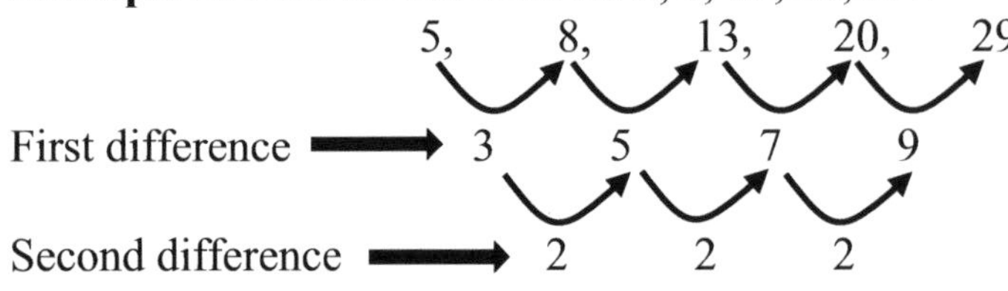

$2 \div 2 = 1$, therefore the nth term must start with n^2.

n	n^2	Sequence	Difference
1	1	5	+4
2	4	8	+4
3	9	13	+4
4	16	20	+4
5	25	29	+4

Since the difference from the table is the same, the nth term of 5, 8, 13, 20, 29 = $\mathbf{n^2 + 4}$ ✓

Example 3: Find the nth term of the sequence 15, 25, 39, 57,....

First difference = 10, 14, 18
Second difference = 4, 4, 4

$4 \div 2 = 2$, therefore, the nth term must start with $2n^2$. Next, draw the table. Remember, $2n^2 = \mathbf{2 \times n^2}$.
Find n^2 first and then multiply by 2.

n	$2n^2$	Sequence	Difference
1	2	15	13
2	8	25	17
3	18	39	21
4	32	57	25

As you can see, the difference is **not** the same from the table. We, therefore, find the linear sequence of 13, 17, 21, 25 because the difference is constant. Refer to section 10.2 above. The nth term for 13, 17, 21, 25 is $4n + 9$.
Therefore, the nth term for 15, 25, 39, 57,.. is $\mathbf{2n^2 + 4n + 9}$ ✓

Check the result: $2n^2 + 4n + 9$
1st term = $2(1)^2 + 4(1) + 9 = 15$
2nd term = $2(2)^2 + 4(2) + 9 = 25$
3rd term = $2(3)^2 + 4(3) + 9 = 39$
4th term = $2(4)^2 + 4(4) + 9 = 57$
Since the terms are the same with the original sequence, the nth term is correct.

Points to remember:
1) If the second difference is 2, the nth term must start with n^2.
2) If the second difference is 4, the nth term must start with $2n^2$.
3) If the second difference is 6, the nth term must start with $3n^2$.

Example 4: Find the nth term of the sequence 4, 22, 52, 94

Fist difference = 18, 30, 42
Second difference = 12, 12, 12
$12 \div 2 = 6$, therefore, the nth term must start with $6n^2$.

N	$6n^2$	Sequence	Difference
1	6	4	-2
2	24	22	-2
3	54	52	-2
4	96	94	-2

Since the difference from the table is constant (-2), the nth term of 4, 22, 52, 94 is $\mathbf{6n^2 - 2}$ ✓

EXERCISE 8 D

1) Look at the sequences below:
a) 1, 3, 5, 7,....
b) 11, 9, 7, 5, ...
c) 7, 17, 31, 49, ...
d) 15, 25, 39, 57, ...
e) 20, 30, 40, 50, ...
f) 33, 47, 63, 81, ...

i) Which of these sequences are linear?
ii) Which of these sequences is quadratic?
iii) Write down the next two terms of the quadratic sequences.
iv) Find the nth terms of the quadratic sequences above.

2) Find the nth terms of each sequence.
a) 2,5,10,17,26
b) 6,9,14,21,30
c) 3,9,19,33,51
d) 0,6,16,30,48
e) 6,15,30,51,78
f) 4,11,22,37,56
g) 4,13,26,43
h) 3,13,29,51
i) 3,10,21,36
j) 6,24,54,96

3) A quadratic sequence begins with 6, 9, 14, 21,….. Find the 100th term.

4) The fourth, fifth and sixth terms of a quadratic sequence are 30, 41 and 54. Find the first, second and third terms of the sequence.

5) Squares are arranged to form a sequence of rectangular patterns as shown below.

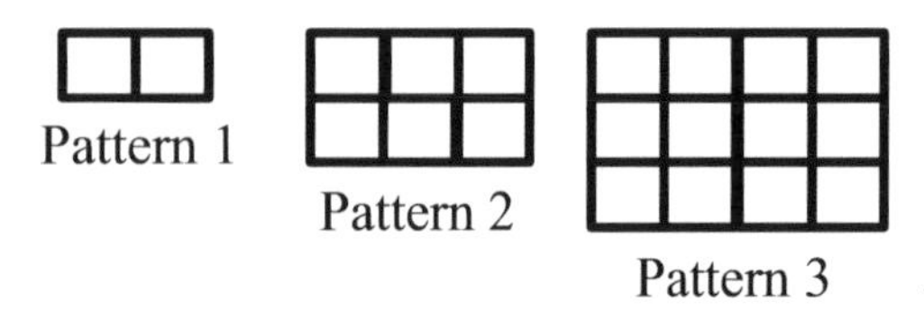

a) How many squares will be needed in pattern 4?

b) Write an expression in terms of n for the number of squares required to form the nth pattern.

c) How many squares will there be in pattern number 300?

6) Triangular numbers are used to make the patterns below.

a) Draw the next pattern.

b) If $\otimes$ represents a circle, find the number of circles in the nth term of the sequence.

c) How many circles will be in the 100th pattern?

TERMS FROM QUADRATIC NTH TERMS

Example 1: Write down the first three terms of the nth term $n^2 + 1$.

1st term is when n = 1: $(1)^2 + 1 =$ **2**
2nd term is when n = 2: $(2)^2 + 1 =$ **5**
3rd term is when n = 3: $(3)^2 + 1 =$ **10**
The first three terms are: 2, 5 and 10.

Example 2: Write down the first five terms of $3n^2 - 5$

This is $(3 \times n^2) - 5$
1st term $= (3 \times (1)^2) - 5$
$= (3 \times 1) - 5 = 3 - 5 =$ **-2**
2nd term $= 3 \times (2)^2 - 5 = 3 \times 4 - 5 =$ **7**
3rd term $= 3 \times (3)^2 - 5 = 3 \times 9 - 5 =$ **22**
4th term $= 3 \times (4)^2 - 5 = 3 \times 16 - 5 =$ **43**
5th term $= 3 \times (5)^2 - 5 = 3 \times 25 - 5 =$ **70**

Example 3: Find the 50th term of the nth term $2n^2 + n$.

Replace n with 50 and work out the value of the expression.
$2 \times (50)^2 + 50 = 2 \times 2500 + 50 =$ **5050**

EXERCISE 8E

1) Write down the first three terms of

a) $n^2 + 3$
b) $n^2 - 1$
c) $n^2 + 10$
d) $4 + n^2$
e) $3n^2 + 1$
f) $3n^3 + 2n$
g) $2n^2 + 2$
h) $2n^2 - 6$
i) $2n^2 + n + 2$
j) $3n^2 + 3n - 3$

2) Write down the 70th term for

a) $2n^2 + 9$
b) $3n^2 - 3n + 2$

Chapter 8 Review Section

Assessment

1a) What is the next number in the sequence 4, 7, 10, 13, 16..? **1mark**

b) One number in the sequence is k.

i) Write in terms of k, the next number in the sequence...........................**1mark**

ii) Write in terms of k, the number in the sequence before k. ....**1 mark**

2) A sequence begins 6, 15, 30, 51, 78,…

a) Write an expression, in terms of n, for the nth term of this sequence..........**3 marks**

b) Work out the 200th term.. **2 marks**

3) A sequence begins with 4, 13,……. The next term in the sequence is found by using the rule: ***multiply the previous number by 3 and add 1.*** Use this rule to find the next four numbers in the sequence..**2 marks**

4) Write down i) the nth term ii) the 20th term of the sequences below.

a) 5, 8, 11, 14, 17, ...**2 marks**

b) 70, 60, 50, 40, 30, ...**2 marks**

c) 0, 6, 16, 30, 48, ...**3 marks**

d) 31, 34, 39, 46, ...**3 marks**

e) $\frac{3}{5}, \frac{4}{7}, \frac{5}{9}, \frac{6}{11},$ ....**3 marks**

f) $(1 \times 3), (2 \times 4), (3 \times 5), (4 \times 6),$ ....**2 marks**

5) Look at patterns 1 and 2. Find the values of ***c*** and ***d***.

Pattern 1: 4 9 16 25 **d**

Pattern 2: 5 7 9 **c** ....**2 marks**

6) The term of a particular sequence is given by $\mu n = 3n^2 - 3$.

a) Write down the first five terms. ....**5 marks**

b) Work out the 25th term of the sequence. ....**2 marks**

9 Ratio and proportion

This section covers the following topics:

- Simplifying ratios
- Dividing/sharing ratios
- Simple proportion
- Maps, scales and ratios

LEARNING OBJECTIVES

By the end of this unit, you should be able to:

a) Simplify ratios
b) Divide a quantity in a given ratio
c) Solve word problems involving ratio and proportion
d) Understand direct and indirect proportions
e) Interpret and solve mathematical problems involving maps

KEYWORDS

- Ratio
- Share/divide
- Simplify
- Proportion
- Rate

9.1 RATIOS

Ratio is used to compare quantities.

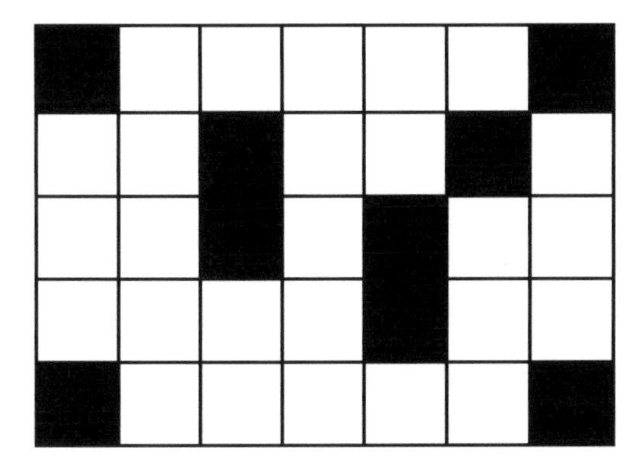

There are 9 shaded squares and 26 plain squares. The ratio of shaded squares to plain squares can be written as **9:26**. Ratios compare the sizes of parts to each other.

Example 1: What is the ratio of shaded triangles to plain triangles in the diagram below?

Shaded: plain
5: 4
Notice that the order is 5 and 4 and **not** 4 and 5.

EQUIVALENT RATIOS

Equivalent ratios can be obtained by multiplying or dividing numbers in a ratio by the same number.

4:7 is equivalent to 8:14 because 4 and 7 are multiplied by 2.
Also, 15:10 is equivalent to 3:2 because 15 and 10 are divided by 5.

SIMPLIFYING RATIOS

The ratios 4:5 and 3:4 are in their simplest forms or terms. This is because there is no whole number which will divide into both sides **exactly,** without any remainder(s).

The easiest way to simplify a ratio is by dividing each part by the highest common factor.

Example 2: Simplify the following ratios to their lowest terms.
a) 4:10 b) 24:16 c) 15:5:35

a)

4 : 10
÷2 ÷2
2 : 5

b)

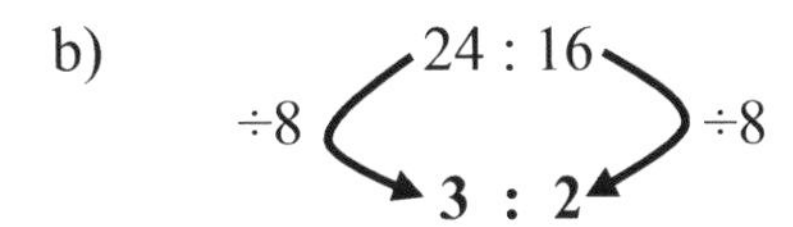

c)

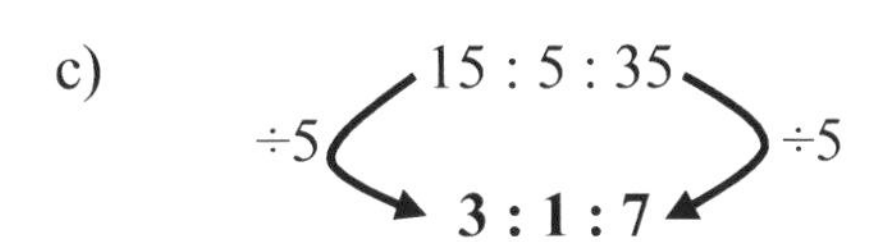

Note: Ratios **must** be simplified in the **same unit** even if they are expressed in different units.

Example 3: Simplify 40p: £1
The units are not the same, so change £1 to 100p since £1 = 100 pence.

Example 4: Simplify to its lowest form
4000 g: 2 kg: 1 tonne

The ratios have different units, so we must convert to the same unit before simplifying.

Convert all to kg
1000g = 1 kg, therefore 4000 g = *4 kg*
1 tonne = *1000 kg*

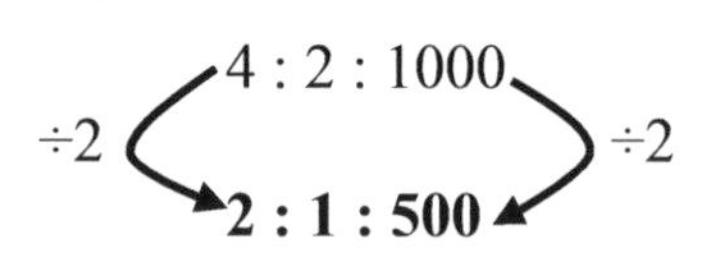

Example 5: Express 35:105 as a fraction in its lowest term.
$35{:}105 = \frac{35}{105} = \frac{\mathbf{1}}{\mathbf{3}}$

Example 6: Simplify 0.6:3
Change the decimal to a whole number to make the calculations easier. To achieve this, multiply both sides by 10.

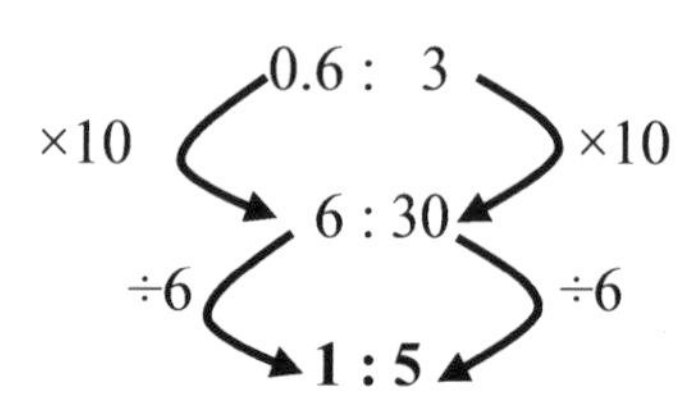

Example 7: Simplify $\frac{3}{5} : 3$
Change the fraction $\frac{3}{5}$ to a whole number by multiplying both sides by 5.

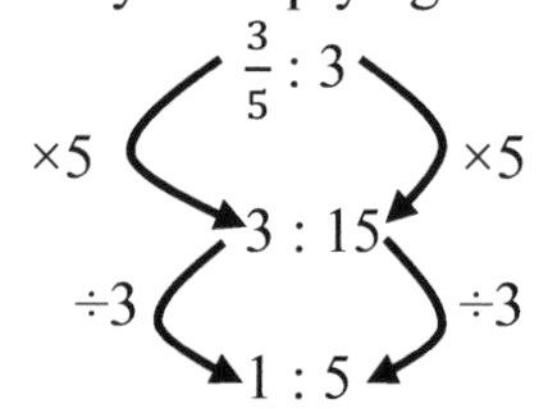

Therefore, $\frac{3}{5}$:3 in its simplest form is **1:5**.

Example 8: In a studio, the ratio of women to men is 4:7. There are 12 women, how many men are in the studio?

Solution: Women: Men
4 : 7
12: **?**
Using the knowledge of equivalent ratios,

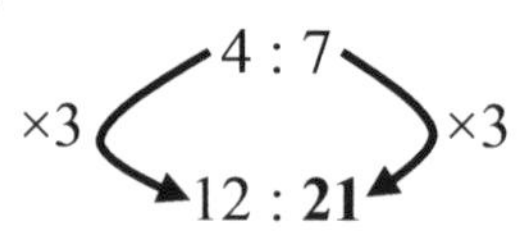

There are 21 men in the studio.

EXERCISE 9A

1) Write down the ratio of shaded to unshaded shapes in each diagram below. Simplify where possible.

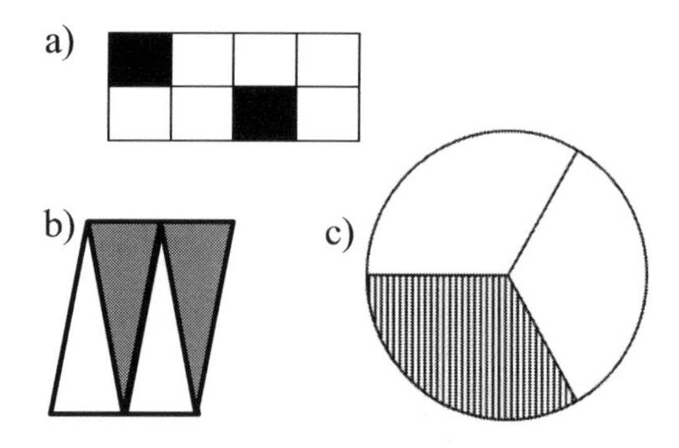

2) Express each of the ratios below in its simplest form.

a) 4:8
b) 3:6
c) 8:4
d) 6:3
e) 26:13
f) 5:40
g) 38:60
h) 12:36
i) 9:6
j) 12:18
k) 14:22
l) 40:48
m) 2:8:20
n) 100:75:25
o) 7:14:49
p) 3:9:3
q) 4:6:12
r) 60:20:30

3) Write each ratio in its simplest form.

a) 1.2 : 1.8 e) 2.7 : 10.8

b) 1.2 : 2 f) $\frac{1}{2}$: 5

c) 4.9 : 9.8 g) $\frac{3}{4}$: 3

d) 0.5 : 1.5 h) $1\frac{1}{2}$: 3

4) In a shop, there are 84 chairs and 14 tables. Find the ratio of chairs to tables in their lowest term.

5) In a class of 50 students, there are 14 boys. What is the ratio of girl to boys in their lowest form?

6) In a box, the ratio of blue pens to red pens to black pens is 4:3:5. If there are seventy black pens, how many red pens and blue pens are there in the box?

7) In a class of 33 pupils, 17 are girls. What fraction are boys?

8) Write each ratio in its simplest form.

a) 4 kg : 2000 g c) 3 kg : 6 tonnes

b) 140 cm : 7 m d) £8 : 40p

9) Fill in the missing numbers.

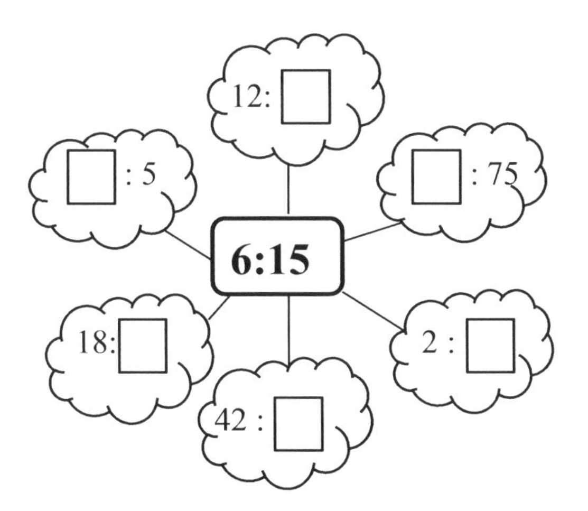

DIVIDING/SHARING RATIOS

Example 1: Divide £20 in the ratio 2:3.

We could write the ratio as
2 parts**:** 3parts since ratios are made up of parts.
Total number of parts = 2 + 3 = 5 parts.
Each part = 20 ÷ 5= 4 parts.

Therefore, 2 parts = 2 × 4 = **£8** ✓
3 parts = 3 × 4 = **£12** ✓

Always check that the parts add up to the total amount. £8 + £12 = £20

Example 2: 75 oranges are shared between Isabel, Edward and Angela in the ratio 3:5:7. How many oranges will each person get?

Solution:
Add up all the ratios: 3 + 5 + 7 = 15
Each share = 75 ÷ 15 = **5** oranges

Isabel gets 3 × 5 = **15 oranges**
Edward gets 5 × 5 = **25 oranges** and
Angela gets 7 × 5 = **35 oranges**

Check: 15 + 25 + 35 = 75

Also, remember that the order of the ratios matters a lot. Three shares/parts **must** be for Isabel because Isabel's name was mentioned first, and the first ratio is 3. Likewise, five shares/parts are for Edward and seven parts for Angela.

EXERCISE 9B

1) Share £36 in the ratio
a) 1:2 b) 2:7 c) 5:7

2) Divide £420 in the ratio
a) 1:6 b) 2:3 c) 4:3

3) Joseph and Mark shared £12000 in the ratio 7:3.
a) How much did Mark receive?
b) How much more is Joseph's share than Marks share?
c) What is Joseph's percentage share?

4) The amount £880 is divided in the ratio 4:5:2. What is the difference between the largest and the smallest share?

5) A quadrilateral has angles which are in the ratio 5:7:2:4

a) Find all the angles in the quadrilateral
b) What is the difference between the largest angle and the smallest angle?

6) Two equilateral triangles are shown below with their perimeters, P. What is the ratio of the lengths of their sides?

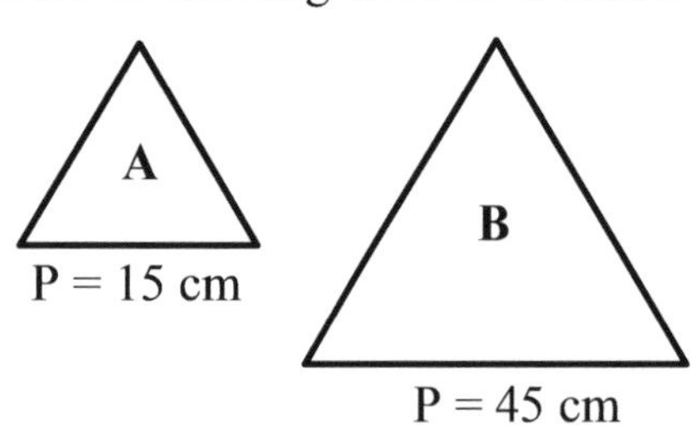

7) x and y are two numbers and their sum is 35. However, x is 9 less than y. Find the ratio of x:y.

8) Look at the spider diagram below. Divide the amount given in the ratios provided.

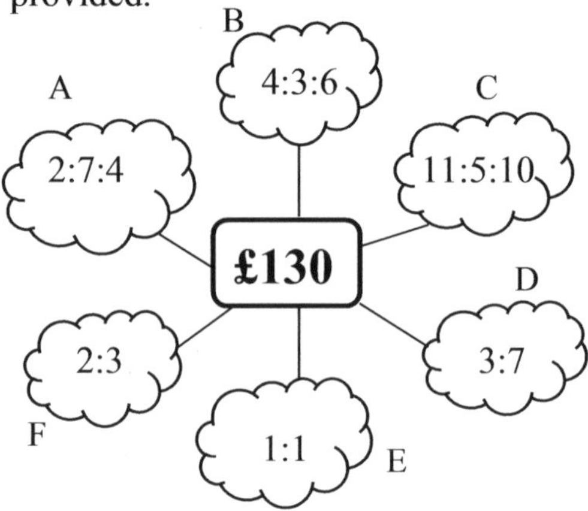

9.2 PROPORTION

Proportion is a bit different from the ratio. Proportion compares the size of a part to the size of a whole.

Example 1: What proportion of these squares is shaded?

Answer

$\frac{4}{16} = \frac{1}{4}$ or 0.25

Proportion can be expressed as fraction, decimal or percentage.

DIRECT PROPORTION

If two quantities increase or decrease at the same rate, they are **directly proportional**. A typical example is petrol consumption in a car. The more petrol, the more the distance travelled. Also, the less petrol in a car, the less distance travelled.

Example 1: The cost of 7 books is £24.50. Find the cost of a book.

÷7 (7 books = £24.50) ÷7

1 book = **£3.50**

The cost of a book is £3.50

Example 2: The cost of 5 oranges is £2. Find the cost of 15 oranges.

×3 (5 oranges = £2) ×3

15 oranges = **£6**

The cost of 15 oranges is £6.

Example 3: If three paintings cost £300, find the cost of 11 paintings.

Method 1: Unitary method

The unitary method requires finding the cost of **one item** and then multiplying by the quantity required.

÷3 (3 paintings = £300) ÷3

1 painting = **£100**

×11 ( ) ×11

11 paintings = **£1 100**

Therefore, 11 paintings will cost **£1100**.

Method 2: Using fractions.

3 paintings cost £300

11 paintings will cost $\frac{11}{3}$ × £300

= £1 100

EXERCISE 9C

1) If ten computers cost £4 500,

a) find the cost of one computer.

b) find the price of 12 computers.

2) Four books cost £20. Find the cost of 9 books.

3) 5 red pens cost ₦220. Find the cost of 3 red pens.

4) The car boot of a car contains one tyre, two jacks, five spanners and two buckets. Find the proportion of

a) Jacks

b) Spanners

c) Buckets

d) Equipment that are not spanners.

5) Yvonne has £5000. She spends $\frac{2}{5}$ of her money and saves the rest.

a) What proportion did he save?

b) How much did he spend?

6) Seven second-hand dining tables cost £350. How much will eight of such dining tables cost?

7) At a seminar, there are 69 delegates. There are twice as many women as men.

a) What proportion are men?

b) What proportion are women?

c) How many men are in the seminar?

d) How many women are there?

INVERSE PROPORTION

Two quantities are inversely proportional if an increase or decrease in one quantity produces a decrease or increase in the second quantity.

Example 1:

It took ten days for three men to complete the building of a conservatory. How long would it take 15 men to complete it?

Solution:
One man will take $3 \times 10 = 30$ days to complete the conservatory.
Therefore,
15 men will take $(30 \div 15) = $ **2 days**.

EXERCISE 9D

1) 5 women can clean a shop in 7 hours. How many hours would it take seven women to clean the shop?

2) 20 cows can feed in a field for six days. How long would it take to feed a) 3 cows b) w cows?

3) 12 people take 4 hours to cut grasses in a field. How long will it take eight people to do the work?

4) 60% of biology books in a cabinet is red. The rest are blue.
a) What is the ratio of red books to blue books?
b) If the numbers of blue books are 18, how many red books are in the cabinet?

RATE

A ratio which involves two quantities that are different is known as **rate**. Assuming a parent gives a child £500 every week as pocket money; this is an example of rate. It could be written as £500 per week. This tells us the amount of money spent every week by the child. Other examples of rate are metre per second (m/s), km per litre (km/l) and so on.

Example 1: Joe ran 120m in 15 seconds.
a) Find the rate in metre per second.
b) Find the rate in metre per minute.

a) Rate $= \frac{120\ m}{15\ s} =$ **8 m/s**

b) Rate $= \frac{120\ metre}{15\ sec}$

....but 60 seconds = 1 minute
15 seconds = 0.25 or $\frac{1}{4}$ minute

Therefore, rate $= \frac{120\ m}{0.25\ m} =$ **480 m/min**

EXERCISE 9E

1) Emily gets paid £150 for 10 hours of work. What is the rate of pay per hour?

2) Joseph ran 200 m in 20 seconds. Find the rate in a) metres per second, b) metres per minute.

3) Jude pays £9600 per annum to rent a shop. What would be the monthly rate?

9.3 MAP, SCALES AND RATIO

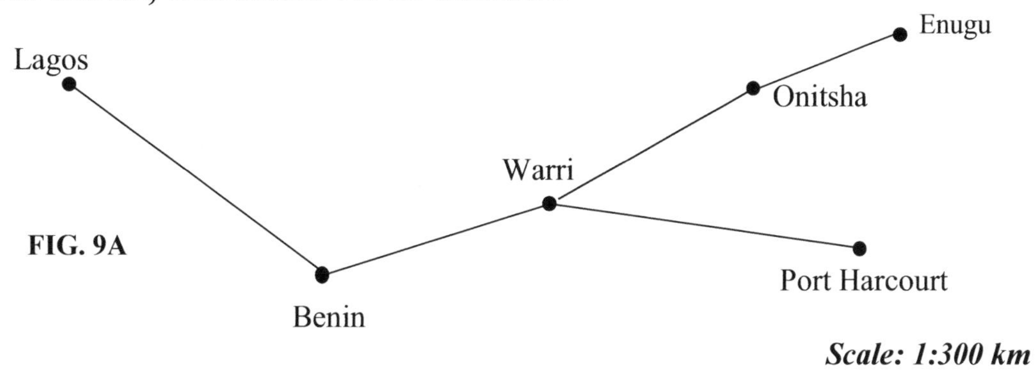

Example 1: Work out the actual distance between Lagos and Benin.

1 cm on the map = 300 km on land.

By measurement, the distance from Lagos to Benin is 4 cm.

Using the scale, 4 cm = 4×300 = **1200 km**

EXERCISE 9F

Use Fig. 9A for questions 1 and 4.

1) Benin and Port Harcourt appear 7cm apart. Find the actual distance between the two towns.

2) Find the actual distance between Benin and Enugu which appears 8 cm apart. Write your answer **in metres**.

3) Find the actual distance between Warri and Port Harcourt.

4) Find the actual distance between Lagos and Enugu.

5) A rectangular museum has a length of 5 cm and a width of 3 cm. If the scale on a map is 1: 2000, calculate a) the real-life dimensions of the museum b) the area of the museum in real life.

6) 3.5 km is the distance between two points. How far apart will the two points be on **the map** with a scale of 1:50 000?

10 Brackets & factorisation

This section covers the following topics:

- Expanding brackets
- Factorisation
- Substitution
- Algebraic Fractions

LEARNING OBJECTIVES

By the end of this unit, you should be able to:

a) Expand single and double brackets
b) Factorise expressions
c) Solve equations involving brackets and fractions
d) Form and solve equations
e) Understand additive inverse
f) Find HCF AND LCM of algebraic fractions
g) Simplify algebraic fractions

KEYWORDS

- Expand
- Factorise
- Substitute
- Algebraic fractions
- Algebraic expression
- Terms

10.1 EXPANDING BRACKETS

Removing or expanding a bracket means to **multiply** each term in the bracket by the term outside the bracket.

Example 1: Simplify/Expand $2(3 + 5)$

Multiply 2 by 3 and then 2 by 5 and add them together.
$2 \times 3 = 6$ and $2 \times 5 = 10$, therefore, $6 + 10 =$ **16**

Alternatively, using BIDMAS perform the operation in the bracket first and multiply by 2. $3 + 5 = 8$. $8 \times 2 =$ **16**

Example 2: Expand the following:
a) $5(n + 7)$, b) $3(n - 10)$, c) $(4 + n)3$
d) $7(5p - 3)$, e) $w(w + 2)$, f) $3c(4c - 10)$
g) $5(x + y - z)$
Solutions:
a) $5(n + 7)$
↑
(Invisible multiplication sign)

There is an invisible multiplication (x) sign between 5 and the bracket. This means $5 \times (n + 7)$
5 is multiplying **everything** in the bracket, not just *n*.

$5(n + 7)$

$= (5 \times n)$ and $5 \times (+7)$
⬇ ⬇
$= 5n$ $+35$
$=$ $\mathbf{5n + 35}$

b) $3(n - 10)$
$= (3 \times n)$ and (3×-10)
⬇ ⬇
$3n$ $- 30$
$=$ $\mathbf{3n - 30}$

c) $(4 + n)3$
This is the same as $(4 + n) \times 3$ or $3 \times (4 + n)$
$= 3 \times 4$ and $3 \times (+n)$
⬇ ⬇
12 $+3n$
$=$ $\mathbf{12 + 3n}$

d) $7(5p - 3)$
$= 7 \times 5p$ and $7 \times (-3)$
$=$ $\mathbf{35p - 21}$
e) $w(w + 2)$
$= w \times w$ and $w \times (+2)$
$=$ $\mathbf{w^2 + 2w}$

f) $3c(4c - 10)$
$= 3c \times 4c$ and $3c \times (-10)$
$=$ $\mathbf{12c^2 - 30c}$

g) $5(x + y - z)$
$= 5 \times x$ and $5 \times y$ and $5 \times (-z)$
$=$ $\mathbf{5x + 5y - 5z}$

EXERCISE 10A

1) Expand the following:

a) $2(x + 3)$
b) $3(x + 4)$
c) $4(x + 2)$
d) $5(x - 3)$
e) $7(x - 4)$
f) $10(4 - x)$
g) $8(3 + x)$
h) $12(x + 5)$
i) $8(x - 10)$
j) $99(x + 10)$
k) $12(5 - x)$
l) $a(a + 3)$
m) $w(w + 7)$
n) $3x(x + 5)$
o) $m(m + 9)$
p) $6(7 - 3c)$
q) $4d(d + 2)$
r) $a(7 - a)$
s) $2(x + 2c - 9)$
t) $0.5x(x + 4)$
u) $\frac{1}{4}(12x - 36)$
v) $0.9(0.2c + 3)$

EXPANDING WITH NEGATIVE NUMBERS AND TERMS

Always be careful when multiplying a negative number or term outside a bracket.

Recollect:	Rule	
	+ + = + - - = +	**Like** signs will give positive
	+ - = - - + = -	**Unlike** signs will give negative

Example: Expand the brackets.
a) -3(n + 5) **b)** -6(n – 3) **c)** -4(5n + 3)
d) $-2x\,(-3x - 5y)$

a) -3(n + 5)
= -3 × n and -3 × (+5)
⬇ ⬇
-3n -15
= **-3n - 15**

b) -6(n – 3)
= -6 × n and -6 × (-3)
= **-6n + 18**

c) -4(5n + 3)
= -4 × 5n and -4 × (+3)
= **-20n - 12**

d) $-2x\,(-3x - 5y)$
= $-2x \times -3x$ and $-2x \times -5y$
⬇ ⬇
= $6x^2$ and $(+10xy)$

= $\mathbf{6x^2 + 10xy}$

EXERCISE 10B

1) Expand each bracket.

a) -2(x + 2)
b) -8(x + 3)
c) -3(x + 10)
d) -5(x + 2)
e) -9(x + 4)
f) –(x + 2)
g) –4(x + 2)
h) -7(x + 7)
i) -10(x + 6)
j) -20(x + 5)
k) -6(7 + x)
l) -3(x – 4)
m) -9(x – 1)
n) -4(x – 2)
o) -8(x – 5)
p) -7(5 – x)
q) –a(a + 3)
r) –w(w + 10)
s) $-6w(6w^2 + 5w)$
t) –v(3v – 2)
u) $-4w^2(7w + 6)$
v) –a(a + 4n - 2)

EXPANDING AND SIMPLIFYING BRACKETS

To expand and simplify simply means to multiply (remove) bracket and collecting like terms. Each bracket is treated separately and then combined by collecting like terms.

Example: Remove the brackets and simplify.

a) 3(a + 2) + 4(a + 1)
b) 4(a + 5) + 3(a - 2)
c) 6y - 4(y + 2) - 3
d) 5(y - 3) - (2y - 7)
e) -2(a + 7) + 3(2a - 5)
f) 13(2k - 2) - 4k

a) 3(a + 2) + 4(a + 1)
= 3 × a + 3 × 2 + 4 × a + 4 × 1
= 3a + 6 + 4a + 4
Collecting like terms gives **7a + 10**

b) 4(a + 5) + 3(a – 2)
= 4a + 20 + 3a - 6
= **7a + 14**

c) $6y - 4(y + 2) - 3$
$= 6y - 4y - 8 - 3$
$= \mathbf{2y - 11}$

d) $5(y - 3) - (2y - 7)$
$= 5y - 15 \quad - 2y + 7$
$= \mathbf{3y - 8}$

e) $-2(a + 7) + 3(2a - 5)$
$= -2a - 14 \quad + 6a - 15$
$= \mathbf{4a - 29}$

f) $13(2k - 2) - 4k$
$= 26k - 26 - 4k$
(Do not multiply 13 and 4k)
$= \mathbf{22k - 26}$

EXERCISE 10C

1) Expand and simplify.

a) $2(x + 3) + 3(x + 1)$
b) $3(x + 5) + 6(x + 2)$
c) $4(x + 7) + 8(x + 3)$
d) $(x + 4) + (x + 4)$
e) $5(x + 3) - 2(x + 2)$
f) $2(x + 4) - 3(x + 5)$
g) $3(x - 2) - 2(x - 5)$
h) $2(x + 3) + 3x$
i) $8(2y - 2) - 4y$
j) $7y - y(3 - y)$
k) $5(6t + 2) - 8$
l) $3(x + 3) + 3(x + 3)$
m) $4(4d - 2) + 6(4d - 3)$
n) $5d + 6(d - 1) + 7$
o) $x(x - 1) + x(x - 2)$
p) $-3d(d + 4) - d(d - 3)$
q) $n(n - 7) + 4n(n - 2)$
r) $-5(7w + 2) - 9(7w - 4)$
s) $-2s(3s - 6) - (7s - 9) + 50$

WORDED PROBLEMS

Example 1: Write an expression for the perimeter of the rectangle below.

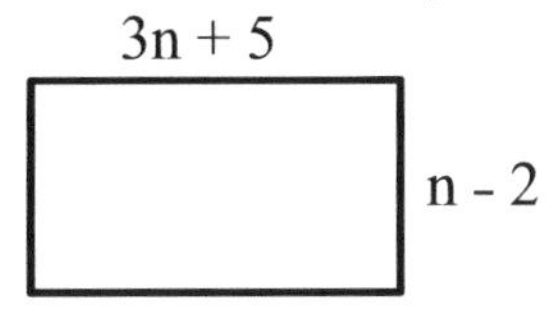

Add all the sides.
$3n + 5 + 3n + 5 + n - 2 + n - 2$
Collect like terms
$= 3n + 3n + n + n + 5 + 5 - 2 - 2$
$= \mathbf{8n + 6}$

EXERCISE 10D

1) A rectangular swimming pool has length $(5x - 3)$ m and width $(x + 7)$ m. Write an expression for the pool's perimeter.

2) Three shapes W, X and Y have lengths c, c + 7 and c + 11 respectively.

In the diagram below, express the length, **a** in terms of **c**. Simplify your answers where possible.

a)
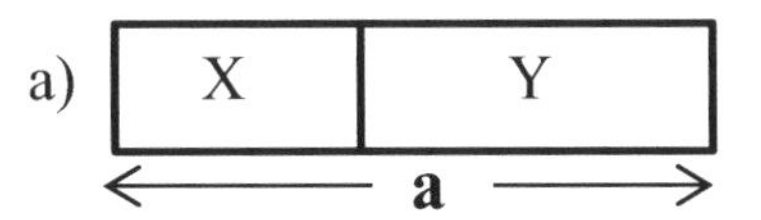

b)
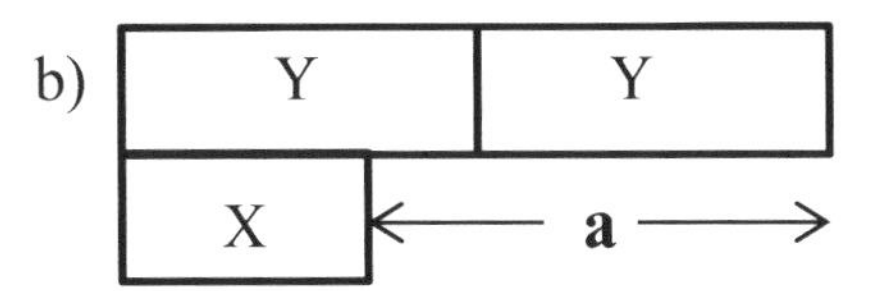

c)
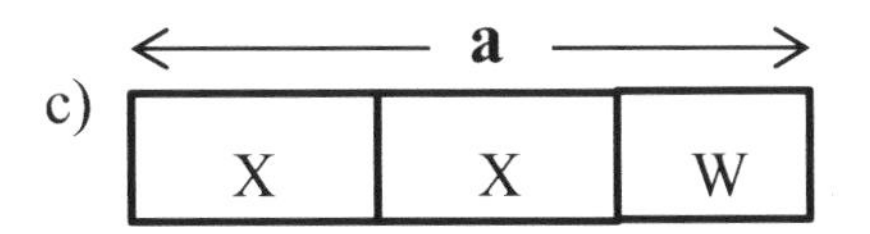

3) A cuboid measures **f** cm by **f** cm by (**f** – 2) cm as shown.

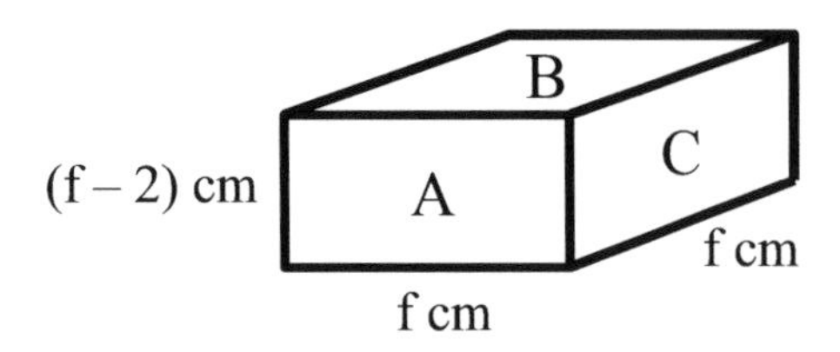

Write an expression for the area of
a) face A
b) face B.
c) Write an expression for the total surface area of the cuboid.
d) Write an expression for the volume of the cuboid.

4) The top of a luxurious dining table is rectangular which consists of 30 small rectangular pieces as shown below.

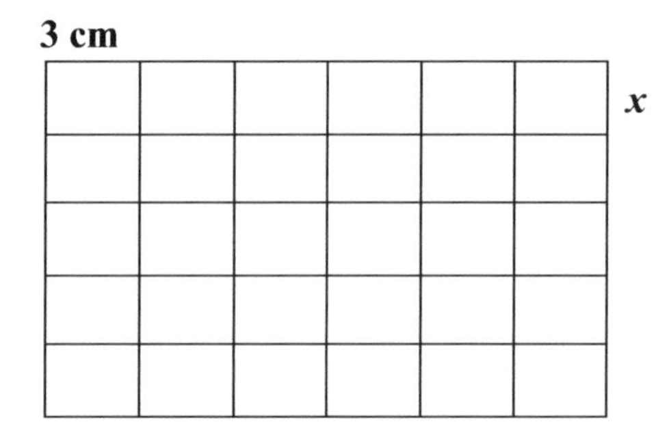

a) Write an expression for the area of a small rectangular piece.
b) Write an expression for the perimeter of a small rectangular piece.
c) Write an expression for the perimeter of the tabletop.
d) Write an expression for the area of the table top.
e) What is the difference between the perimeter and area of the table top?
f) If 7 rectangular pieces are removed from the table top, what is the area of the remaining table top?

10.2 EXPANDING DOUBLE BRACKETS

Example 1: Expand (n + 1) (n + 2)

Method 1: Grid method

	n	2
n	$n \times n =$ $\mathbf{n^2}$	$2 \times n =$ **2n**
1	$n \times 1 =$ **n**	$2 \times 1 =$ **2**

Therefore, the area of the big rectangle $= n^2 + n + 2n + 2 = \mathbf{n^2 + 3n + 2}$

Method 2: FOIL Method

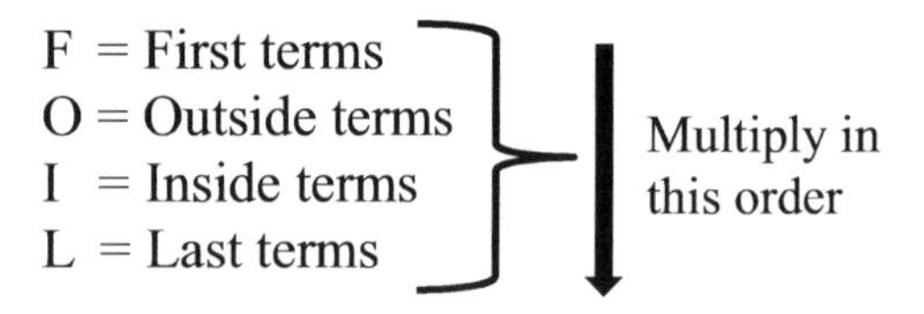

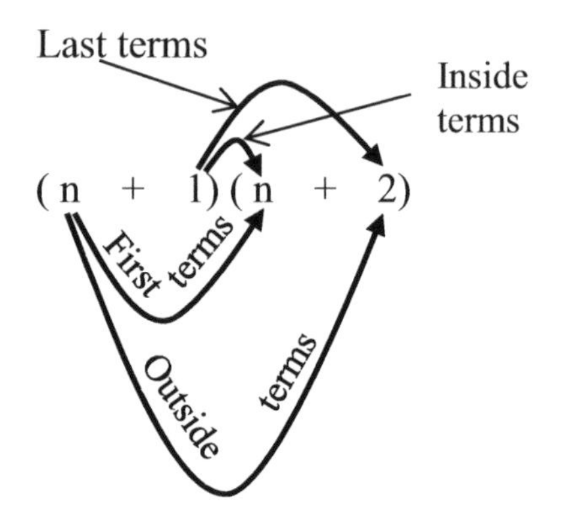

(n + 1)(n + 2)
$n \times n = \mathbf{n^2}$, $1 \times n = \mathbf{n}$,
$n \times 2 = \mathbf{2n}$ $1 \times 2 = \mathbf{2}$
$n^2 + 2n + n + 2 = n^2 + 3n + 2$
Therefore, $(n+1)(n+2) = \mathbf{n^2 + 3n + 2}$

Example 2: Expand $(x - 2)(x + 4)$
You may use any method. Using the FOIL method,

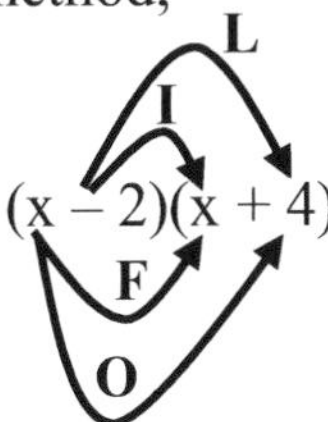

First: $x \times x = x^2$
Outside: $x \times 4 = 4x$
Inside: $-2 \times x = -2x$
Last: $-2 \times (+4) = -8$

Therefore,
$x^2 + 4x - 2x - 8$
$= \mathbf{x^2 + 2x - 8}$

Example 3: Expand $(n - 4)(n - 3)$

F: $n \times n = n^2$
O: $n \times (-3) = -3n$
I: $-4 \times n = -4n$
L: $-4 \times (-3) = 12$

Therefore,
$n^2 - 3n - 4n + 12$
$= \mathbf{n^2 - 7n + 12}$

Example 4: Expand $(2x - 5)^2$

$(2x - 5)^2 = (2x - 5)(2x - 5)$

Using the foil method,

F: $2x \times 2x = 4x^2$
O: $2x \times (-5) = -10x$
I: $-5 \times 2x = -10x$
L: $-5 \times (-5) = 25$

Therefore,
$4x^2 - 10x - 10x + 25$
$= \mathbf{4x^2 - 20x + 25}$

Example 5: Work out the area of the rectangle in terms of w.

2w + 1
3 - w

Area $= (2w + 1) \times (3 - w)$
$= (2w + 1)(3 - w)$
Using Foil method
$= 6w - 2w^2 + 3 - w$
$= \mathbf{-2w^2 + 5w + 3}$

EXERCISE 10 E

1) Expand and simplify.

a) $(x + 1)(x + 3)$
b) $(x + 2)(x + 5)$
c) $(x + 3)(x + 7)$
d) $(n + 4)(n + 4)$
e) $(n + 6)(n + 9)$
f) $(x + 2)(x - 1)$
g) $(x + 3)(x - 7)$
h) $(w - 5)(w - 1)$
i) $(w - 10)(w + 1)$
j) $(c + 9)(c - 12)$

2) Remove the brackets and simplify fully.

a) $(6m + 3)(2m + 5)$
b) $2(3x + y)(x + 3y)$
c) $(x - 8)^2$
d) $(a + b)^2$
e) $(7c - 5)(4c - 3)$
f) $w(w + 2)(3w - 8)$

3) In the figure below, slabs A, B, C and D are identical and surround a pond. Find an expression for the:

a) The perimeter of PQRS
b) Area of PQRS
c) Area of the pond.

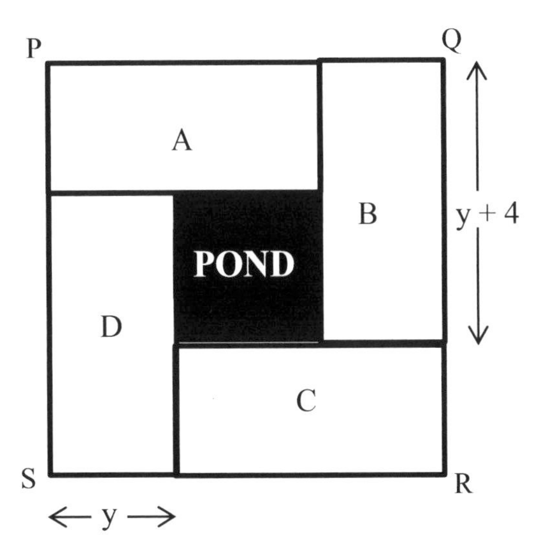

10.3 FACTORISATION

Factorisation is the opposite of expanding brackets. Consider the expression 3y + 15. The terms 3y and 15 have a common factor. The common factor is 3. We may write the expression using a bracket.

Example 1: Factorise 3y + 15

The highest common factor (HCF) of 3y and 15 is 3. Write 3 and then put a bracket like 3(   ).
Now, divide each term by 3.

$$3\left[\frac{\not{3}y}{\not{3}} + \frac{\overset{5}{\not{15}}}{\not{3}}\right] = \mathbf{3(y + 5)}$$ ✓

Check that you have factorised properly by expanding the bracket. If the original expression is obtained, then it is correct. $(3 \times y) + (3 \times 5) = 3y + 15$. This is the same as the original expression.

Example 2: Factorise $5w$ - 10
HCF = 5 and it comes outside the bracket.

Therefore, $5\left[\frac{\not{5}w}{\not{5}} - \frac{\overset{2}{\not{10}}}{\not{5}}\right] = \mathbf{5(w - 2)}$ ✓

Example 3: Factorise $12m + 18$
HCF = 6.

Therefore, $6\left[\frac{\overset{2}{\not{12}}m}{\not{6}} + \frac{\overset{3}{\not{18}}}{\not{6}}\right] = \mathbf{6(2m + 3)}$ ✓

Example 4: Factorise $2x^2 + 4xy$
HCF = $2x$.
Therefore, $\mathbf{2x(x + 2y)}$ ✓

Example 5:
Factorise $20x^2y + 8xy^2 + 6xy$

Solution
HCF of 20, 8 and 6 = 2
HCF of x^2, x and $x = x$
HCF of y, y^2 and y = y

Therefore, the overall HCF = $2xy$

$$2xy\left[\frac{20x^2y}{2xy} + \frac{8xy^2}{2xy} + \frac{6xy}{2xy}\right]$$

= **2xy (10x + 4y + 3)** ✓

EXERCISE 10 F

1) Factorise the following expressions.

a) $4x + 16$
b) $6x + 9$
c) $12x + 16$
d) $5a + 10$
e) $9x + 90$
f) $6x - 3$
g) $9x - 18$
h) $5x - 30$
i) $30x - 45$
j) $7m - 49$
k) $28n + 21$
l) $7x - 14b + 21c$
m) $4n - 6ny$
n) $12x + 14$
o) $16y - 16$
p) $abcd - bcd$
q) $5m + 20n - 5$
r) $75k - 450$

EXERCISE 10 G

1) Factorise the expressions below.

a) $x^2 + 5x$
b) $4t^2 - 16t$
c) $bc^2 - dc$
d) $7p^2 + 28p$
e) $c^2d + cd^2$
f) $f^3 - 2f$
g) $ps + qs - prs$
h) $11x^2 + x$
i) $d^2f^2 - mnd^2f^2 + df$
j) $3x^2 - 9x + 27x^3$
k) $14y - 7y^2 + 28y^3$
l) $cd + ce + fd + fe$
m) $x(b + c) + 3(b + c)$
n) $-30y^2 - 9$

10.4 ALGEBRAIC FACTORS

Examples of algebraic terms are 4x, 3xy, $12x^2y$.....

The factors of 8 are 1, 2, 4 and 8. The factors of 20 are 1, 2, 4, 5, 10 and 20. The number, 1, has one factor, 1.

Other than the number 1, all whole numbers have two or more factors. Numbers that have only two factors are called prime numbers.

Example 1: Write the factors of 10x.
10x: 1, 2, 5, 10, x, 2x, 5x, 10x

Example 2: Write down the factors of $4cd^2$.
$4cd^2$: 1, 2, 4, c, 2c, 4c, d, 2d, 4d, d^2, $2d^2$, $4d^2$, cd, 2cd, 4cd, cd^2, $2cd^2$, $4cd^2$

EXERCISE 10 H

1) Write down all the factors of

a) 4c
b) 12t
c) ac
d) 7xy
e) y^2
f) 6xy
g) $10x^2$
h) 35ef

10.5 HCF OF ALGEBRAIC EXPRESSIONS

First, work out the prime factors (using factor tree) of the numerical terms and then find the factors of the letters.

Example 1: Find the HCF of 18xy and $40x^2y$.

Prime factors of 18 from factor tree are $2 \times 3 \times 3$.
Therefore, $18xy = 2 \times 3 \times 3 \times x \times y$

Prime factors of 40 from factor tree are $2 \times 2 \times 2 \times 5$.
Therefore,
$40x^2y = 2 \times 2 \times 2 \times 5 \times x \times x \times y$

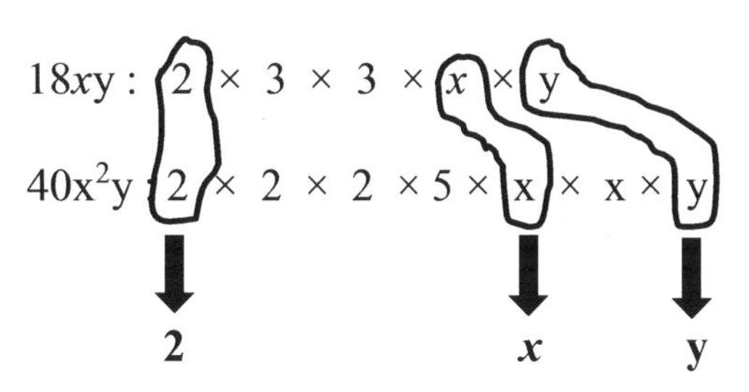

The HCF of 18xy and $40x^2y$ is $2 \times x \times y$ = **2xy** ✓

Example 2: Find the HCF of abc and bcd.

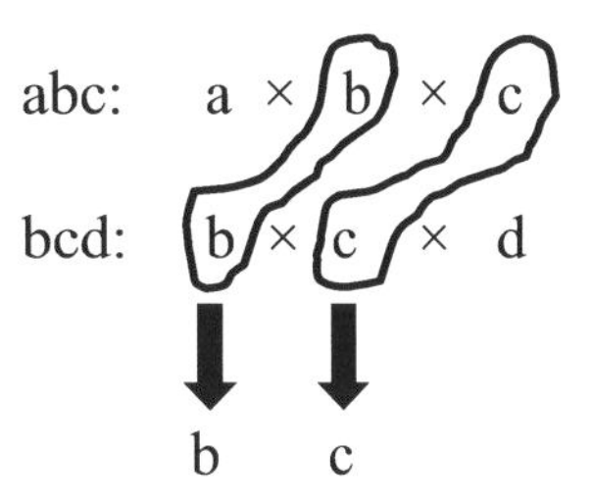

HCF of abc and bcd = b × c = **bc** ✓

EXERCISE 10 I

1) Find the HCF of the following algebraic expressions.

a) pqr and qrs	f) $25de^2$ and 5cf
b) 3mn and 7gn	g) 6jk and 6jm
c) $10c^2$ and $25c^2$	h) n^2 and 7n
d) $30m^2n$ and 60mn	i) 5ab and $15a^2b$
e) ab and cb	j) $6n^2y$ and $24ny^2$

10.6 LCM OF ALGEBRAIC EXPRESSIONS

First, work out the HCF using the method described in section 12.5 and multiply the HCF together with the leftovers to give the LCM.

Example 1: Find the LCM of 18xy and $40x^2y$.

Prime factors of 18 from factor tree are $2 \times 3 \times 3$.
Therefore, $18xy = 2 \times 3 \times 3 \times x \times y$

Prime factors of 40 from factor tree are $2 \times 2 \times 2 \times 5$.
Therefore,
$40x^2y = 2 \times 2 \times 2 \times 5 \times x \times x \times y$

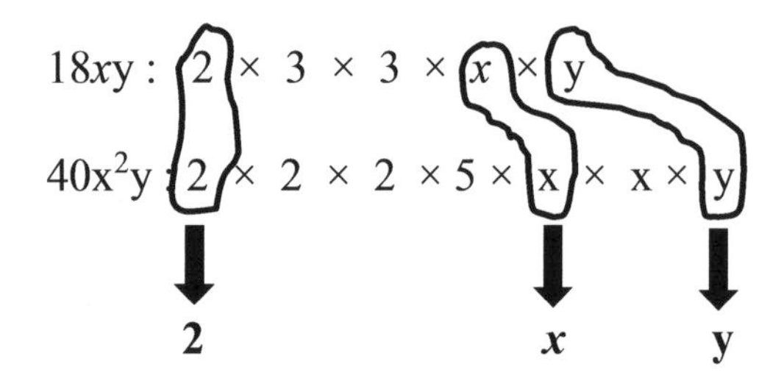

The HCF $= 2 \times x \times y = 2xy$

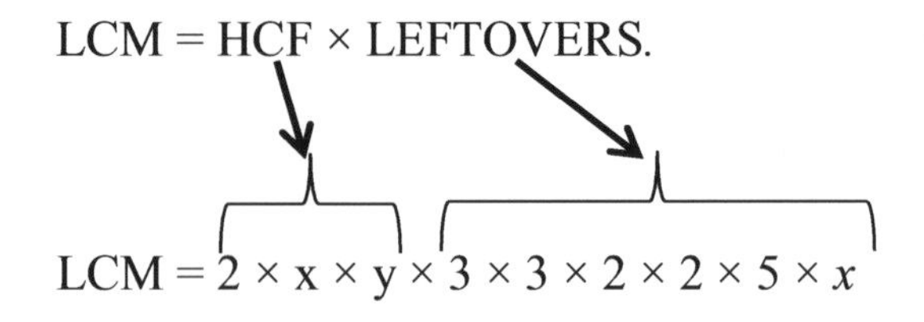

LCM = $\mathbf{360x^2y}$ ✓

Example 2: Find the LCM of abc and bcd.

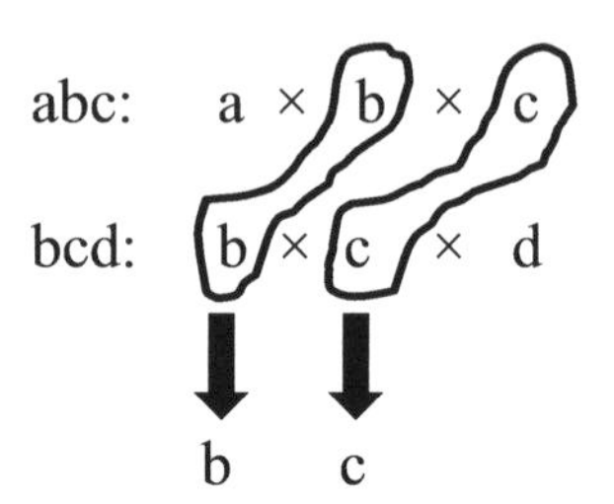

HCF of abc and bcd = b × c = **bc**

Therefore,
LCM = HCF × LEFTOVERS
= b × c × a × d = **abcd** ✓

USING VENN DIAGRAM

Refer to Example 1 above.

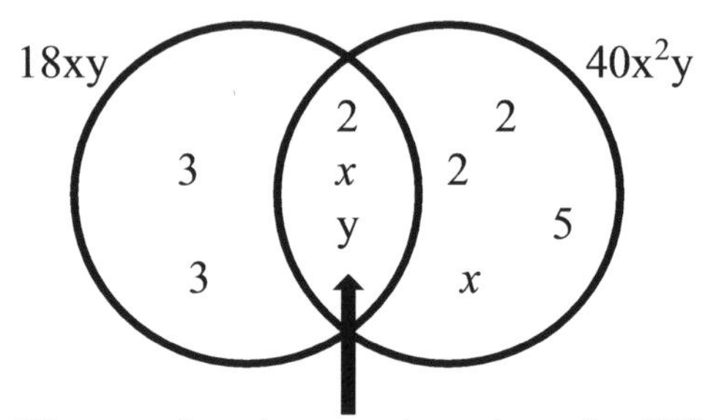

The overlapping section gives the HCF which is $2 \times x \times y = 2xy$.

LCM = HCF × LEFTOVERS
$= 2 \times x \times y \times 3 \times 3 \times 2 \times 2 \times 5 \times x$
$= \mathbf{360x^2y}$ ✓

EXERCISE 10 J

1) Work out the LCM of
a) 4p and 6p
b) c and d
c) 3t and 5u
d) n^2 and 7n
e) 8cd and $4cd^2$
f) $5p^2q$ and $30pq^2$
g) abc and $10b^2$
h) mn and an
i) 15mn and $7n^2$

10.7 ALGEBRAIC FRACTIONS

Algebraic fractions are worked out the same way normal fractions are calculated. They follow the same principle(s).

MULTIPLYING ALGEBRAIC FRACTIONS

Remember: $\frac{1}{8} \times \frac{2}{3} = \frac{1 \times 2}{8 \times 3} = \frac{2}{24} = \frac{1}{12}$

Example 1: Simplify $\frac{f}{g} \times \frac{h}{t}$

$= \frac{f \times h}{g \times t} = \frac{fh}{gt}$ ✓

Example 2: Simplify $\frac{20x}{5xy}$

$\frac{20x}{5xy} = \frac{\overset{4}{\cancel{20}} \times \cancel{x}}{\cancel{5} \times \cancel{x} \times y}$

$= \frac{4}{y}$ ✓

Example 3: Simplify $\frac{8c}{3g} \times \frac{d}{3h} \times \frac{f}{k}$

Numerator: $8c \times d \times f = 8cdf$
Denominator: $3g \times 3h \times k = 9ghk$

Therefore: $\frac{8cdf}{9ghk}$ ✓

Example 4: Simplify $\frac{16w}{4} \times \frac{7}{w^2}$

$= \frac{\overset{4}{\cancel{16}} \times \cancel{w}}{\cancel{4}} \times \frac{7}{\cancel{w} \times w}$

$= \frac{4 \times 7}{w} = \frac{28}{w}$ ✓

DIVIDING ALGEBRAIC FRACTIONS

Remember: Keep, Change, Flip

$\frac{3}{4} \div \frac{1}{2} = \frac{3}{4} \times \frac{2}{1} = \frac{3 \times 2}{4 \times 1} = \frac{6}{4} = 1\frac{1}{2}$

Example 5: Simplify $\frac{11}{x} \div \frac{2}{x}$

$\frac{11}{x} \times \frac{x}{2} = \frac{11 \times \cancel{x}}{\cancel{x} \times 2} = \frac{11}{2} = \mathbf{5\frac{1}{2}}$ ✓

Example 6: Simplify $\frac{y^2 + y}{y^2 - y}$

Factorise numerator and denominator.

$\frac{y^2 + y}{y^2 - y} = \frac{\cancel{y}(y + 1)}{\cancel{y}(y - 1)} = \frac{y + 1}{y - 1}$ ✓

EXERCISE 10 K

Simplify the following.

1) $\frac{10t}{5}$

2) $\frac{8x}{4y}$

3) $\frac{2x}{7x}$

4) $\frac{2ac}{10a}$

5) $\frac{26abc}{13a}$

6) $\frac{7ab}{4} \times \frac{2}{b}$

7) $\frac{16xy}{3} \times \frac{9}{2x}$

8) $\frac{d}{e} \times \frac{f}{g}$

9) $\frac{7pqr}{14qr}$

10) $\frac{5p}{50pq} \times \frac{75q}{ps}$

Simplify and following

11) $\frac{7x^2}{x}$

12) $\frac{5pq^2}{25pq}$

13) $\frac{3ab^2}{27a^2b}$

14) $\frac{10st}{3s^2t}$

15) $\frac{x^2}{3x}$

16) $\frac{42c^2}{7c^2}$

17) $\frac{3w^2}{4} \times \frac{2}{3y}$

18) $\frac{c}{8} \times \frac{d^2}{c^2}$

19) $\frac{6p}{a^2} \times \frac{a}{3p^2}$

20) $\frac{cd^2}{de^2} \div \frac{de}{4d}$

21) $\frac{18y^2}{8w^2} \div \frac{y^3}{w^3}$

22) $\frac{g^2 + g}{g^2 - g}$

23) $\frac{y^2}{y + 2y} \div \frac{y}{y + 2}$

24) $\frac{(3x)^2}{3x}$

ADDING / SUBTRACTING ALGEBRAIC FRACTIONS

Adding and subtracting fractions follow the same rule as arithmetic fractions.

Points to remember

1) Before adding or subtracting fractions, the denominators must be the same.

2) Do to the numerator what you did to the denominator.

3) Add or subtract the fraction.

Example 1: Simplify $\frac{x}{3} + \frac{y}{4}$

Solution: Make the denominators the same.

$$\frac{x \times 4}{3 \times 4} + \frac{y \times 3}{4 \times 3} = \frac{4x}{12} + \frac{3y}{12} = \frac{\mathbf{4x + 3y}}{\mathbf{12}} \checkmark$$

*Note: Since 4x and 3y are not like terms, **do not** add them.*

Example 2: Simplify $\frac{2c}{3} - \frac{c}{5}$

$$\frac{2c \times 5}{3 \times 5} - \frac{c \times 3}{5 \times 3} = \frac{10c}{15} - \frac{3c}{15} = \frac{\mathbf{7c}}{\mathbf{15}} \checkmark$$

Example 3: Simplify $\frac{2n - 4}{8} + \frac{n - 2}{4}$

$$\frac{(2n - 4) \times 1}{8 \times 1} + \frac{(n - 2) \times 2}{4 \times 2}$$

$$= \frac{2n - 4}{8} + \frac{2n - 4}{8} = \frac{2n - 4 + 2n - 4}{8}$$

$$= \frac{4n - 8}{8} \text{ (Now factorise the numerator)}$$

$$= \frac{4(n - 2)}{8} = \frac{\cancel{4}(n - 2)}{\cancel{8}_{2}} = \frac{\mathbf{n - 2}}{\mathbf{2}} \checkmark$$

Example 4: Simplify $\frac{3}{y} + \frac{4}{2y} + \frac{5}{3y}$

Make the denominators the same.

$$\frac{3 \times \mathbf{6}}{y \times \mathbf{6}} + \frac{4 \times 3}{2y \times 3} + \frac{5 \times 2}{3y \times 2}$$

$$= \frac{18}{6y} + \frac{12}{6y} + \frac{10}{6y} = \frac{18 + 12 + 10}{6y}$$

$$= \frac{40}{6y} = \frac{\mathbf{20}}{\mathbf{3y}}$$ ✓

EXERCISE 10 L

Simplify the following:

1) $\frac{3x}{7} + \frac{x}{7}$

2) $\frac{5n}{6} + \frac{2n}{6}$

3) $\frac{7x}{7} - \frac{4x}{7}$

4) $\frac{1}{y} + \frac{n}{y}$

5) $\frac{2xy}{4w} + \frac{3xy}{4w}$

6) $\frac{2c}{3} - \frac{d}{6}$

7) $\frac{3x}{u} + \frac{t}{4}$

8) $\frac{5}{y} + \frac{7}{2y} + \frac{9}{3y}$

9) $\frac{x}{10} + \frac{3}{c}$

10) $\frac{5y}{7} - \frac{x}{3}$

11) $\frac{c-1}{5} + \frac{c+1}{2}$

12) $\frac{b}{5} + \frac{b-3}{3}$

13) 4d - $\frac{5f}{7g}$

14) 8u + $\frac{7u+v}{9}$

15) $\frac{k}{3} - \frac{2k}{9}$

16) $\frac{2x-4}{8} + \frac{x-2}{4}$

17) $\frac{(4x+7)}{7} - \frac{(x-4)}{7}$

Chapters 9 & 10 Review Sections

Assessment

1) Write each ratio in its simplest form.

a) 2.2 : 2 c) $\frac{1}{2}$: 7

b) 0.5 : 3.5 d) $1\frac{1}{2}$: 6

……..**4 marks**

2) In a class of 70 students, there are 14 boys. What is the ratio of girl to boys in their lowest form? …………..**2 marks**

3) Fill in the missing numbers.

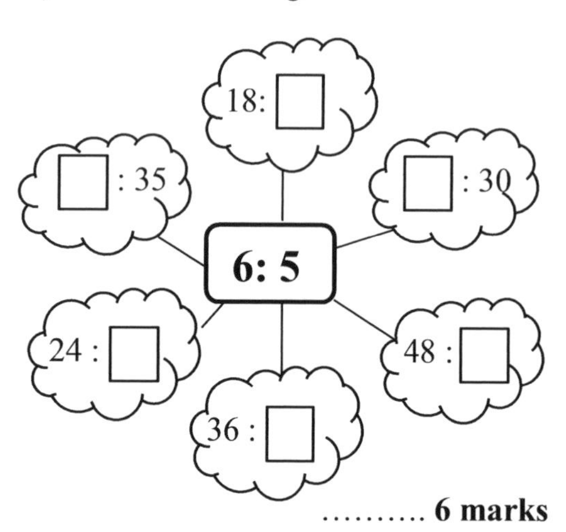

………. **6 marks**

4) 75 oranges are shared between Isabel, Edward and Angela in the ratio 3:5:7. How many oranges will each person get?

………...**3 marks**

5) Four books cost £20. Find the cost of 9 books. ………**2 marks**

6) Expand each bracket.

a) -4(x + 5) b) -3(x – 4)

c) -3s(3s – 4) – (7s – 9) + 7

……....6 **Marks**

7) Factorise the expressions below.

a) $w^2d + wd^2$ c) $16ry^2 + 8r^2y - 4ry$

b) a(b + c) + 3(b + c) ………**6 marks**

8) Find the HCF and LCM of the following algebraic expressions.

a) $15w^2x$ and $30wx$

b) $5n^2y$ and $25ny^2$ ……………**3 marks**

9) Simplify the following.

a) $\frac{54abc}{9a}$ b) $\frac{15p}{30pq} \times \frac{q}{ps}$

………..**4 marks**

10) Simplify the following:

a) $\frac{5x}{9} + \frac{x}{9}$

b) $\frac{3n}{3} + \frac{2n}{3}$

c) $\frac{3c}{3} - \frac{d}{7}$

……… **6 marks**

11 Reciprocal and inverse

This section covers the following topics:

- Reciprocals
- Inverse
- Identity

LEARNING OBJECTIVES

By the end of this unit, you should be able to:

a) Find reciprocal of a number
b) Understand and use identities for addition and multiplication

KEYWORDS

- Reciprocal
- Multiplicative inverse
- Identity
- Chance

RECIPROCALS

When two numbers are multiplied, and the result (answer) is **1**, we say that each number is the **reciprocal or multiplicative inverse** of the other. 1 is the **identity** for multiplication.

Examples

1) $\frac{1}{7} \times 7 = 1$, we say that 7 is the reciprocal of $\frac{1}{7}$ and $\frac{1}{7}$ is also the reciprocal of 7.

2) $\frac{2}{3} \times \frac{3}{2} = 1$, we say that $\frac{3}{2}$ is the reciprocal or multiplicative inverse of $\frac{2}{3}$

3) Write down the reciprocals of

a) $\frac{5}{7}$ b) 9 c) $\frac{1}{8}$ d) 0.7 e) $4\frac{2}{5}$ f) -3

Answers

a) $\frac{5}{7}$: $1 \div \frac{5}{7} = \mathbf{\frac{7}{5}}$

b) 9: $1 \div 9 = \mathbf{\frac{1}{9}}$

c) $\frac{1}{8}$: $1 \div \frac{1}{8} = \mathbf{8}$

d) 0.7: $1 \div 0.7 = 1 \div \frac{7}{10} = \mathbf{\frac{10}{7}}$

e) $4\frac{2}{5}$:

(First, change to improper fraction)

$\frac{(5 \times 4)+2}{5} = \frac{22}{5}$

Therefore, $1 \div \frac{22}{5} = \mathbf{\frac{5}{22}}$

f) -3: $1 \div -3 = \mathbf{-\frac{1}{3}}$

Therefore, swapping/interchanging the denominator and the numerator of a fraction gives the reciprocal of that fraction.

EXERCISE 11

1) Write down the reciprocal of the numbers below.

a) 2 f) $-\frac{5}{7}$ k) -5

b) 4 g) -11 l) $-\frac{1}{12}$

c) $\frac{1}{7}$ h) $\frac{9}{14}$ m) $\frac{2}{11}$

d) $\frac{5}{9}$ i) $\frac{4}{7}$ n) - 4

e) -2 j) 12 o) $\frac{3}{17}$

2) Write down the multiplicative inverse of the following:

a) - 0.9 f) $3\frac{5}{7}$

b) - 1.5 g - $2\frac{1}{2}$

c) - 1.25 h) - 4.5

d) -100 i) $\frac{1}{w}$

e) $5\frac{1}{7}$ j) $\frac{n}{7}$

3) Andrew says "The reciprocal of 6 is greater than the reciprocal of $\frac{1}{4}$. Is Andrew correct? Explain fully.

4) a) Write $5\frac{2}{3}$ as an improper fraction.

b) What is the reciprocal of $5\frac{2}{3}$?

c) Find the reciprocal of - $5\frac{2}{3}$

d) Multiply your answers from part a and part b.

12 Graphs and gradients

This section covers the following topics:

- Straight line graphs
- Gradient
- Intercept
- Quadratic Graphs
- Mid-point of a line segment

LEARNING OBJECTIVES

By the end of this unit, you should be able to:

a) Draw straight line graphs
b) Find gradients and y-intercepts
c) Calculate mid-point of a line segment
d) Find the equations of straight lines using $y = mx + c$
e) Understand parallel and perpendicular lines
f) Draw quadratic graphs

KEYWORDS

- Straight line
- Gradients
- Y-intercept
- Parallel and perpendicular lines
- Quadratic Graphs

12.1 STRAIGHT LINE GRAPHS

By using the knowledge of coordinates, you can find points and draw a graph for any relationship. Before plotting a graph, we need to find the coordinate points (x, y) to locate a point. The coordinate pairs are better found using a **table of values.** It is a table that contains the coordinate pairs needed for drawing straight line graphs.

Example 1: Draw the graph of $y = x + 1$. Use x values from 0 to 5.

Solution: Draw a table of values.

Table of values →

x	0	1	2	3	4	5
y = x + 1	1	②	3	4	5	⑥
Coordinates	(0,1)	(1,2)	(2,3)	(3,4)	(4,5)	(5,6)

The number 2 in circle, was obtained by substituting (replacing) the value of x (1) in the equation original equation $y = x + 1$. From $y = x + 1$, $y = 1 + 1 = 2$. Also, the number 6 in circle, was obtained by substituting the value of 5 in the equation $y = x + 1$. From $y = x + 1$, $y = 5 + 1 = 6$. Do same to all the y values.

Therefore, from the table of values, we proceed to draw the graph.

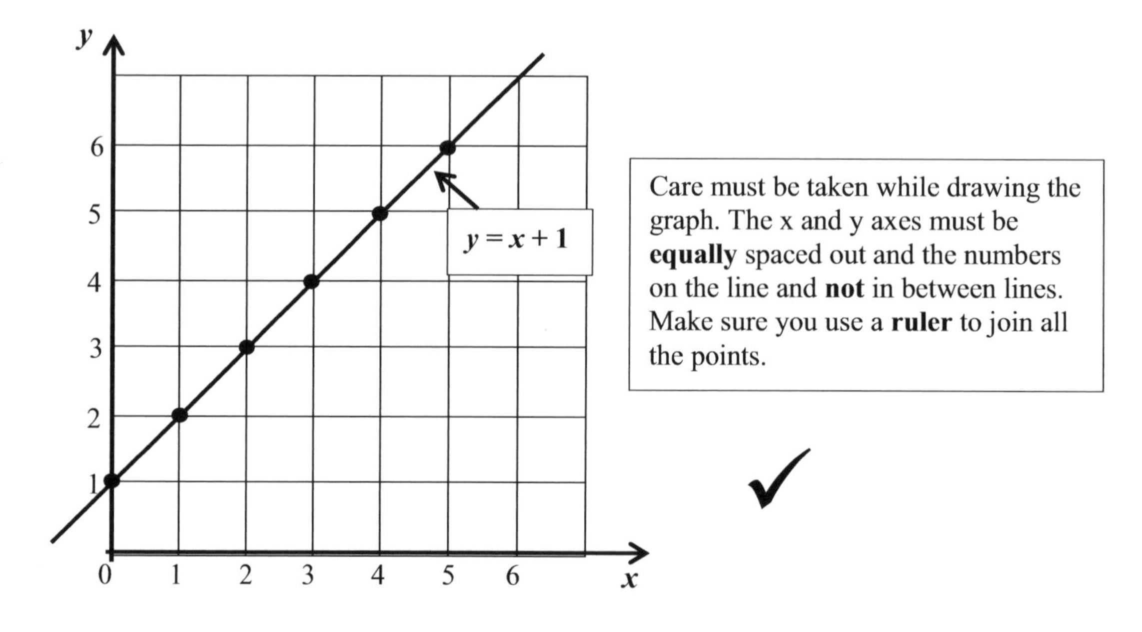

Example 2 Daw the graph of y = 2x - 2, x values from -2 to 3.

First, draw the table of values to get your coordinate pairs.

x	-2	-1	0	1	2	3
y = 2x – 2	**-6**	-4	-2	0	**2**	4
Coordinates	(-2, -6)	(-1, -4)	(0, -2)	(1, 0)	(2, 2)	(3, 4)

The number -6 in circle was obtained by substituting the value of x (-2) in the original equation, $y = 2x - 2$. $y = (2 \times -2) - 2 = -4 - 2 = -6$. Also, the value of 2 in circle was obtained by substituting the value of x (2) in the original equation y = 2x - 2. $y = (2 \times 2) - 2 = 4 - 2 = 2$.

By using the coordinates from the table of values above, plot the graph. Remember: Since there are positive and negative numbers in the x and y coordinates, there must be four quadrants.

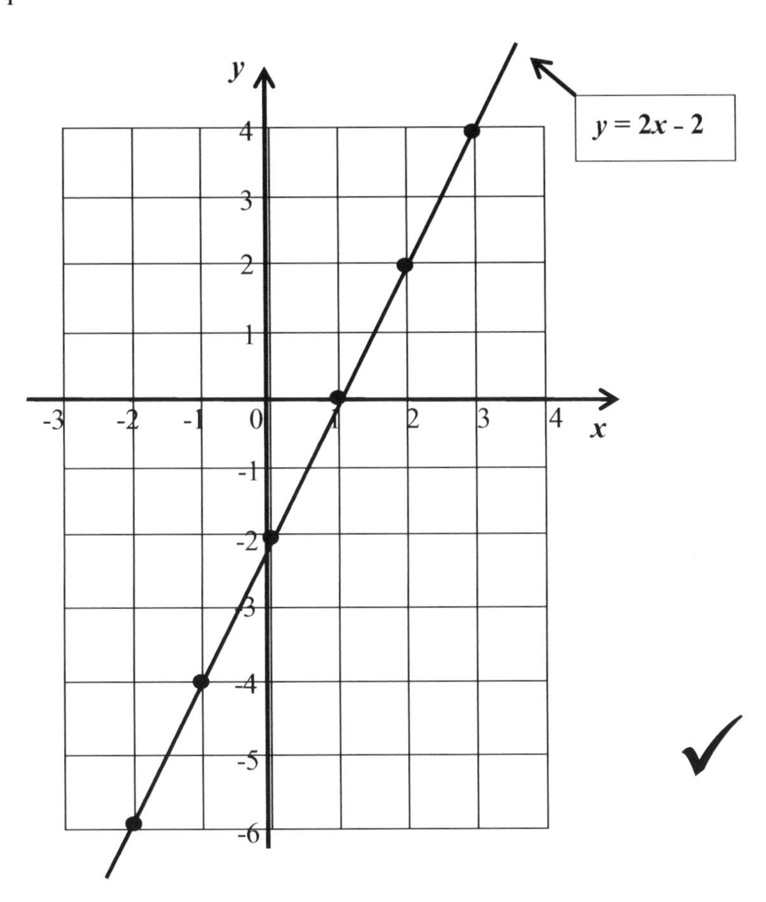

EXERCISE 12 A

1a) Copy and complete the table of values below for equation y = 3x + 1.

x	1	2	3	4	5
y = 3x + 1			10		
Coordinates			(3,10)		

b) Draw the graph of y = 3x + 1.
c) What is the coordinate of the point where the line touches the y-axes?

2a) Copy and complete the coordinates below using equation y = x + 3.

(-1, ?), (0, ?), (2, ?), (6, ?)

b) Draw a set of axes with x-values from -1 to 7 and y values from 1 to 10.

c) Plot the points with coordinates found in question 2a.

d) What is coordinate of the point where the line crosses the y-axis?

3) Draw the graph of the following equations. Use x values from -3 to 3.

a) y = x + 4
b) y = 2x
c) y = 5 – x
d) y = 3x - 3
e) $y = \frac{1}{2}x + 1$
f) $y = \frac{1}{2}x - 5$

4) A physicist uses the relationship C = 3d to convert the diameter (d) of a circle to its circumference (c).
a) Copy and complete the table below.

d	0	5	13	20
C				

b) Using d on the horizontal axis and c on the vertical, plot the points from 4a.

c) Copy the following table and use the graph to fill in the missing entries.

Diameter	Circumference
4	
16	
90	
	3

5) Two points P and Q are shown.

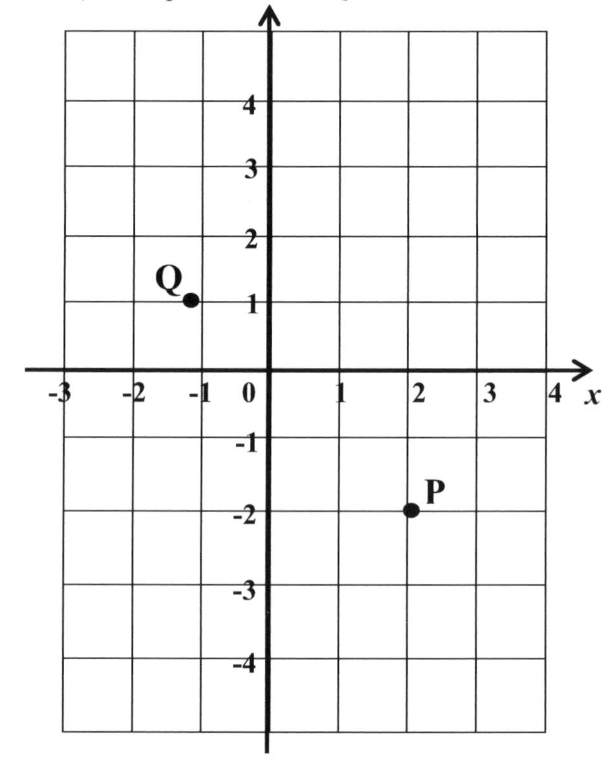

a) Write down the coordinates of P and Q.
b) Copy the graph above and join line PQ with a ruler.
c) Draw the graph of y = 3x – 4 on the same graph in 5b above.
d) Write down the coordinates of the point where line PQ cuts the graph of y = 3x - 4.

12.2 HORIZONTAL LINES

Horizontal lines have equations of the type y = 2, y = - 2 and so on as shown below. It starts with y because it touches the y-axes.

The horizontal line on top of the **x-axes** and parallel to it has the equation **y = 0.**

12.3 VERTICAL LINES

Vertical lines have equations of the type $x = 1$, $x = -2$ and so on. It starts with x because it touches the x-axes.

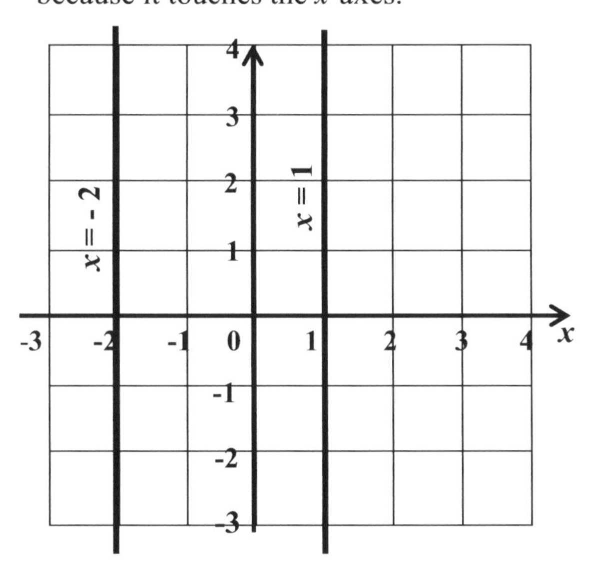

The vertical line on top of the **y-axes** and parallel to it has the equation **x = 0.**

EXERCISE 12B

1) Write down the equation of each line in the diagram below.

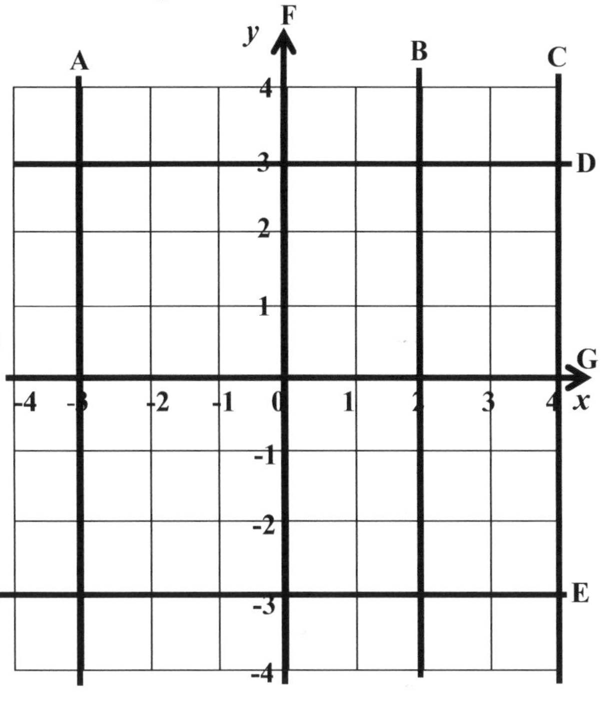

2a) Draw a rectangle with coordinates (1, 4), (-3, 4), (-3, -2) and (1, -2)

2b) Write down the equation of the lines that form the sides of the rectangle.

3a) Draw the lines y = 3, x = 3 and y = x on the same graph.
3b) Write down the coordinates of the triangle formed.
3c) Work out the area of the triangle formed if the lengths are in cm.

4) The table below shows the conversion rate from Pounds Sterling (£) to Euro (€).

£	0	5	10	15	20
€	0	10	20	30	40

a) Draw the conversion graph.

b) Use the graph to convert
i) £6 ii) £14 iii) £17 into Euros.

c) Using the graph, convert these amounts into Pounds Sterling:
i) €5 ii) €22 iii) €37

12.4 EQUATIONS OF STRAIGHT LINES

Equations of straight lines can be obtained by finding the **gradient** or **slope** of the line and the **y-intercept**.

The intercept of a graph is the value of **y** at the point where the graph crosses the y-axes.

Example 1

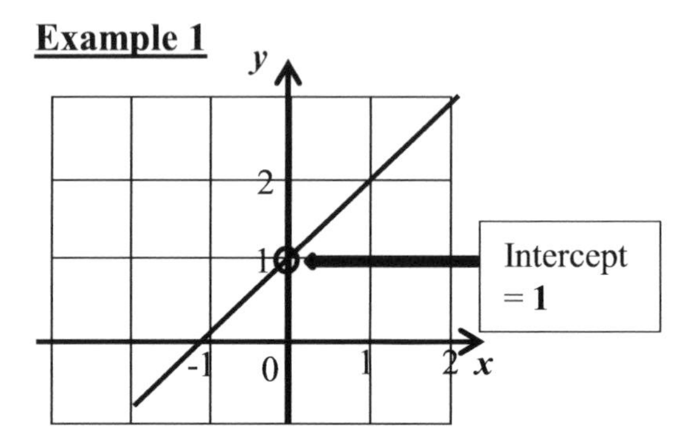

Example 2

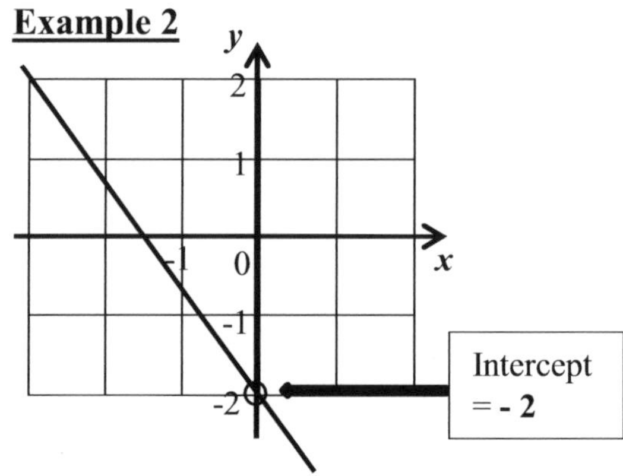

Example 3

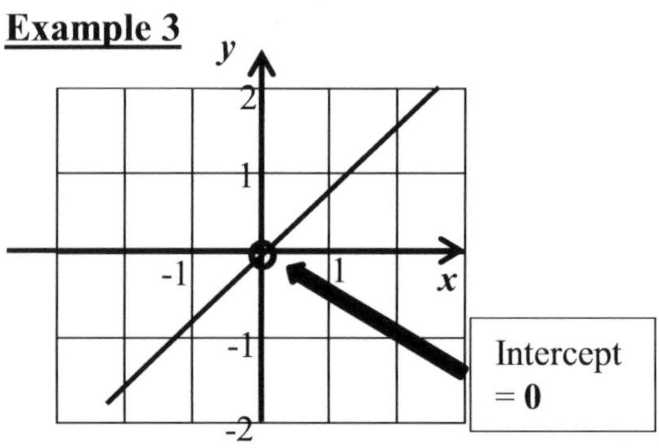

GRADIENT/SLOPE

The **gradient** of a line describes how steep the line is. The gradient can be positive, negative or zero and the direction of the line helps in determining the type of gradient.

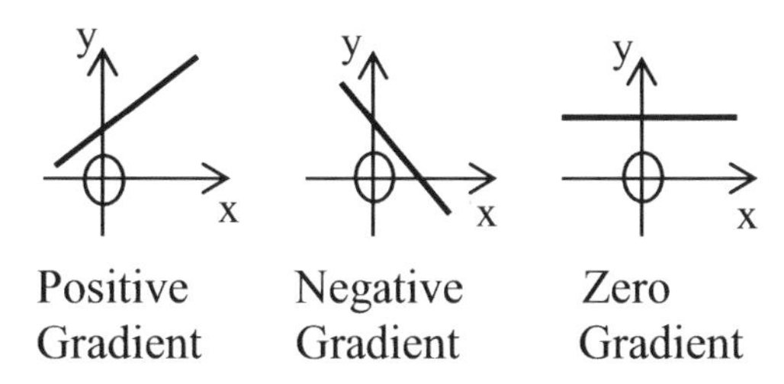

Positive Gradient   Negative Gradient   Zero Gradient

In other words, the gradient of a graph is the mathematical way of measuring its **steepness** or rate of change.

Note: If a line is vertical, the gradient **cannot** be specified.

The gradient can be measured between any two convenient points on the graph.

$$\text{Gradient} = \frac{\text{Change in } \mathbf{y}}{\text{Change in } \mathbf{x}}$$

Example 1:
Find the gradient of the line joining the points A (1, 1) and B (4, 3).

Solution: You may work this out with drawing a graph. Identify the *x* and *y* coordinates and apply the formula above.
For points A (1, 1) and B (4, 3)
x, y *x1, y2*

Change in y = 3 – 1 = 2
Change in x = 4 – 1 = 3

Therefore, gradient of AB = $\frac{2}{3}$

To use the graph to find the gradient of line AB, join the point with a ruler. Draw a perpendicular line from B to meet the horizontal line at C.

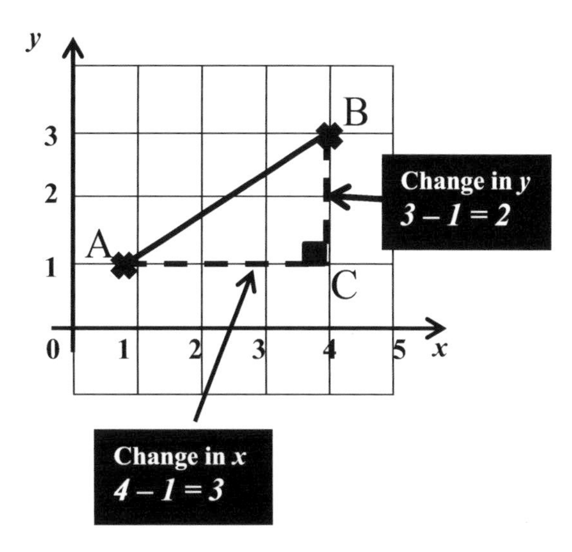

The gradient of AB = $\frac{2}{3}$ and it is **positive**.

Example 2: Find the gradient of the line joining the points P and Q.

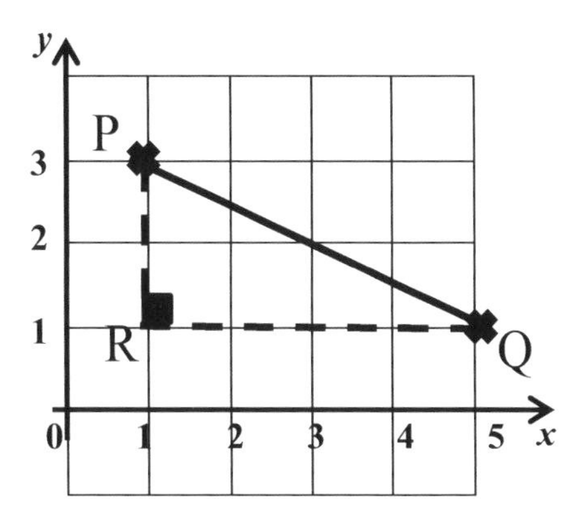

$$\text{Gradient} = \frac{3-1}{1-5} = \frac{2}{-4} = -\frac{1}{2}$$

The gradient is negative as the line slopes downwards.

Example 3: Without drawing the graph, work out the gradient of the line joining points A (-3, 1) and B (-1, 5).

Gradient $= \frac{Change\ in\ y}{Change\ in\ x} = \frac{5-1}{-1-(-3)}$
$= \frac{4}{2} = \mathbf{2}$

Example 3: Find the gradient of the line joining the points (0, 6) and (2, 1).

Gradient $= \frac{Change\ in\ y}{Change\ in\ x} = \frac{1-6}{2-0}$
$= \frac{-5}{2} = \mathbf{-2.5}$

The negative gradient indicates that the graph slopes downwards.

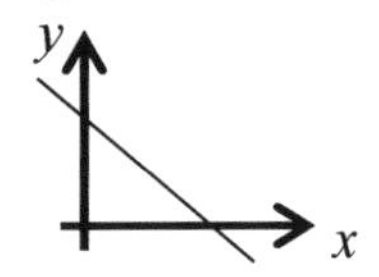

Example 4: Find the gradient of the line

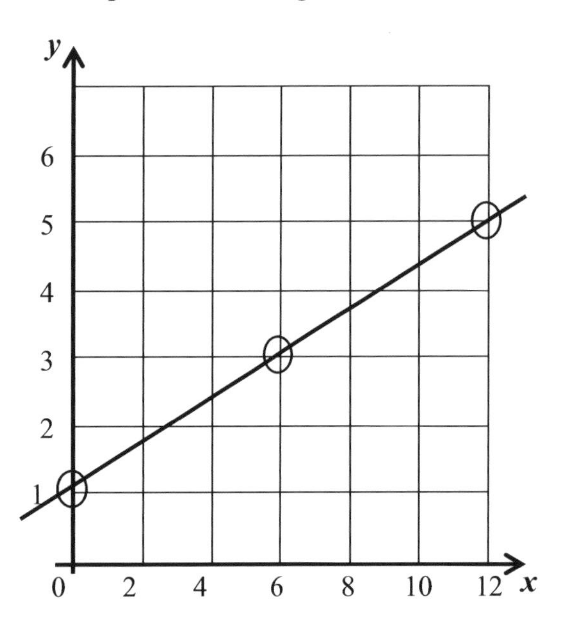

Three **convenient** points are circled.

No other point(s) is convenient as we must know the exact coordinates (integers).

Now, choose **any** of the two convenient points and draw a right-angle triangle.

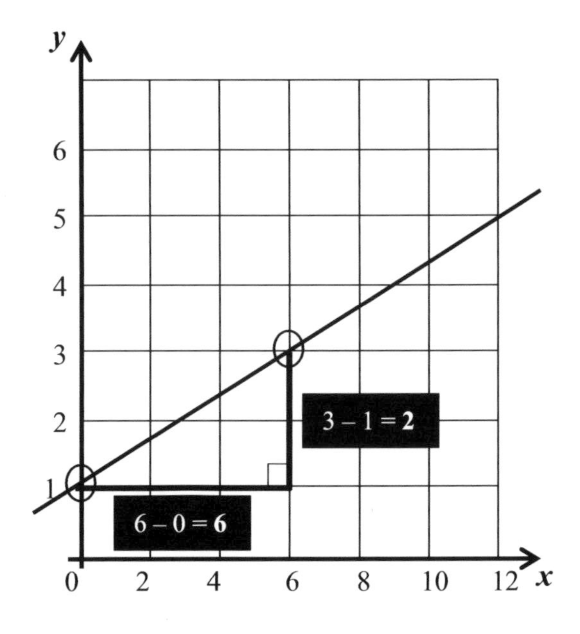

Gradient $= \frac{Change\ in\ y}{Change\ in\ x} = \frac{2}{6} = \mathbf{\frac{1}{3}}$

EXERCISE 12 C

1) Calculate the gradients of the lines joining each of these pairs of coordinates.

a) (1,1) and (2,2)
b) (4,4) and (7,7)
c) (2,1) and (3,7)
d) (1,2) and (4,8)
e) (3,1) and (5,4)
f) (3,0) and (4,3)
g) (0,4) and (2,-6)
h) (-7, 10) and (0,0)
i) (2,3) and (3,7)
j) (-1,4) and (-2,5)
k) (1,4) and (-1,2)
l) (-2,3) and (6,1)
m) (-1,8) and (-2,4)
n) (-3,5) and (-1,1)
o) (6,11) and (-2,-5)
p) (-5, -3) and (1,9)

2) Find the gradient of each line.

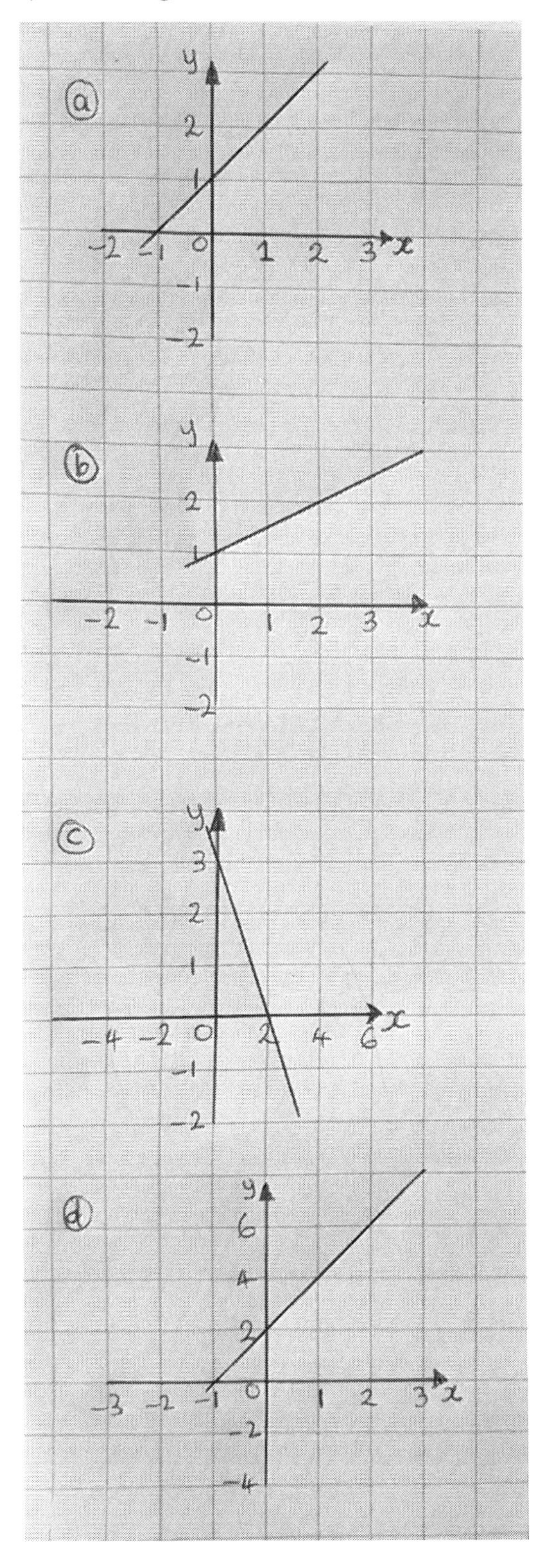

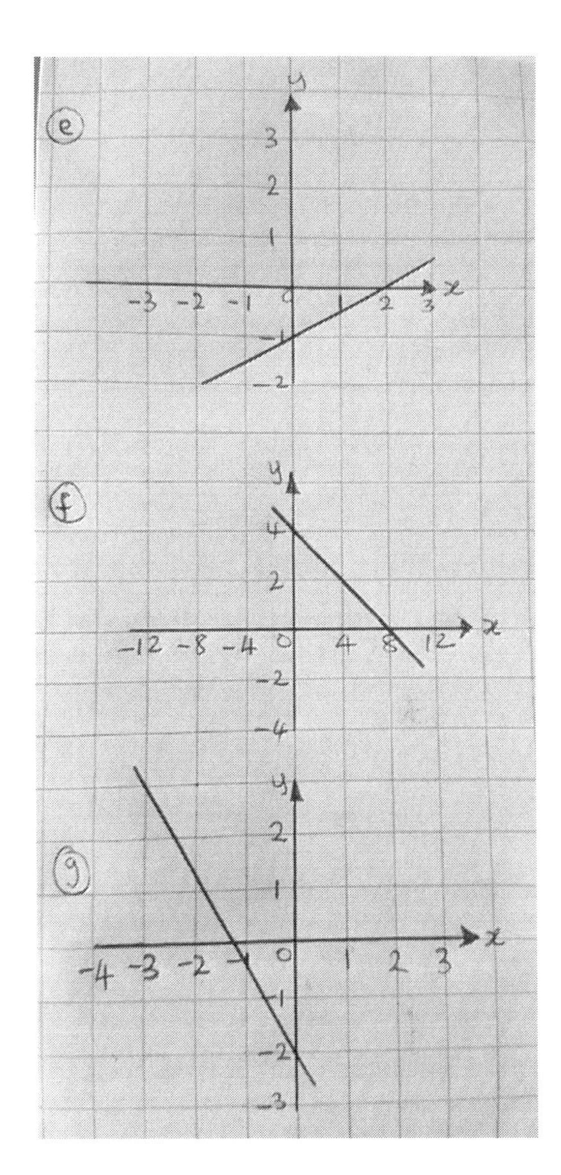

3) A quadrilateral ABCD has coordinates A (2, 9) B (1, 5) C) (4, 1) and D (10, 6). Find the gradient of each side of the quadrilateral.

4) For each pair of coordinate values, work out the slope of the line that joins them.

a) (b, b) and (0, 0)
b) (-p, 4p) and (2p, - 2p)
c) (0, 0) and (7q, 7q)
d) (g, 3g) and (-4g, - 4g)

12.5 MIDPOINT OF A LINE SEGMENT

The middle of a line segment is its **midpoint.** Remember that a **line segment** has a beginning and an end.

Example 1: Find the coordinates of the midpoint of line segment joining P(2, 5) and Q(7, 10).

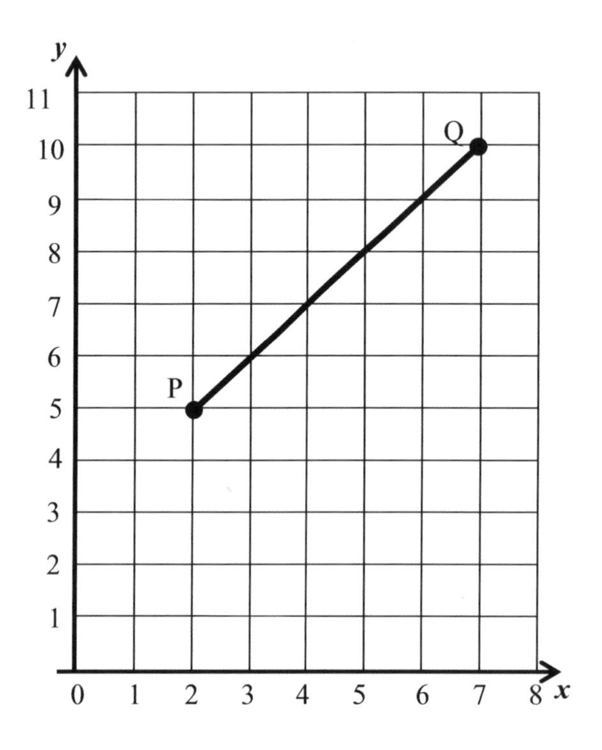

Mid-point (x) = $\frac{2+7}{2} = \frac{9}{2} = 4.5$

Mid-point (y) = $\frac{5+10}{2} = \frac{15}{2} = 7.5$

Therefore, the coordinates of the midpoint of PQ = **(4.5, 7.5)**

EXERCISE 12D

1) Calculate the coordinates of the mid-point of the line segment joining the pairs of points below.

a) (1, 6) and (7, 6) d) (0, 5) and (4, 1)
b) (1, 6) and (1, 2) e) (-2, -5) and (1, 6)
c) (0, 2) and (7, 6) f) (-3, -4) and (5, 0)

2) Work out the coordinates of the midpoints of the given line segments.

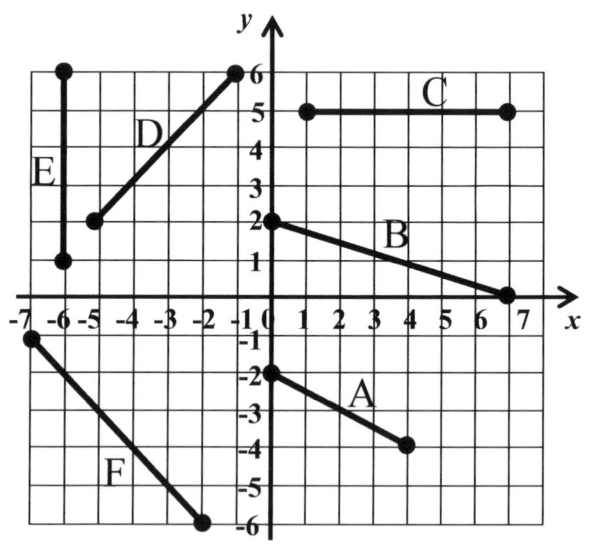

3) The coordinates of the midpoint of line AB are (-2, 2.5). If A is at (0, 0), work out the coordinates of B.

4) The coordinates of the vertex of a triangle are A(1,2), B (1, 6) and C(5, 2).
a) Work out the coordinates of the midpoint of AB, BC and AC.

b) Work out the area of the triangle if the length of each side of the square is 1cm.

12.6 Y = MX + C

The equation of a straight line is usually written in the form **y = mx + c,** where **m** is the gradient of the line and **c** the intercept on the y-axis.

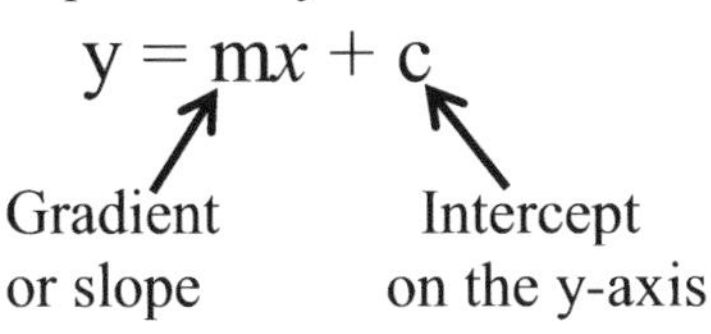

Example 1: Write down the gradient and intercept of the following:

a) $y = 3x + 5$ d) $y = 3 - 2x$

b) $y = 2x - 3$ e) $y = 7 + \frac{1}{2}x$

c) $y = -4x$ f) $y = -\frac{3}{4}x - 5$

Solutions:

a) $y = 3x$ (+5)

Gradient = **3** Intercept = 5

b) $y = 2x$ (-3)

Gradient = 2 Intercept = **-3**

c) $y =$ (-4) x

Gradient = **- 4** Intercept = **0**

d) $y = 3$ (-2) x

Intercept = **3** Gradient = **-2**

e) $y = 7$ $\left(+\frac{1}{2}\right)$ x

Intercept = **7** Gradient = $\frac{1}{2}$

f) $y =$ $\left(-\frac{3}{4}\right)$ x (-5)

Gradient = $-\frac{3}{4}$ Intercept = **-5**

Example 2: Write down the equation of the graph given that the gradient is 2 and intercept is -3.

Solution: Remember the form of a straight-line graph: $y = mx + c$.
Substitute the values of m and c.
The equation is $\mathbf{y = 2x - 3}$

Example 3: Write down the equation of the graph with a gradient of $-\frac{1}{4}$ and 0 intercepts.
The equation is $\mathbf{y = -\frac{1}{4}x}$ or $y = -\frac{x}{4}$

Example 4: Find the equation of line A.

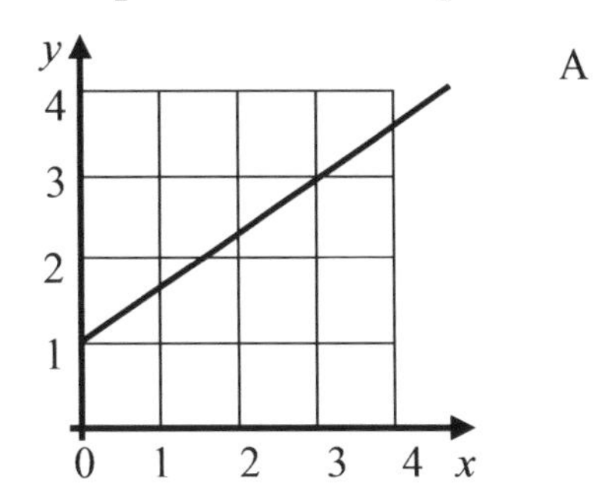

We need to find the gradient and the y-intercept. For the gradient, choose two convenient points and make a triangle.

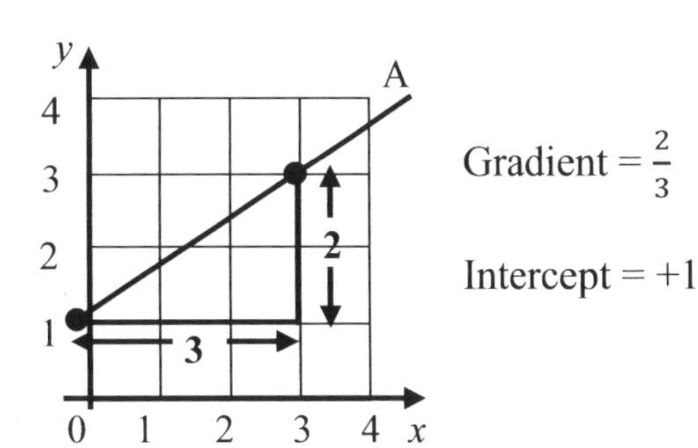

Therefore, the equation of line A using $y = mx + c$

$y = \frac{2}{3}x + 1$

Example 5: Find the equation of the straight line which passes through points (0, 2) and (6, 8).

Solution: You may decide to draw a graph and then find the gradient by drawing a triangle as shown earlier.

However, you may work out the gradient without drawing a graph as follows:

$$\text{Gradient (m)} = \frac{\text{Change in y}}{\text{Change in } x}$$

$$m = \frac{8-2}{6-0} = \frac{6}{6} = \mathbf{1}$$

Since one of the coordinate pairs is (0, 2), it means that the line must pass through y-axis at 2. Therefore, c = 2

Using the y = mx + c, y = 1x + 2.
The equation of the line is **y = *x* + 2**

Example 6: Sketch the line y = 2x + 1.
Solution: We know that the graph must pass through +1 on the y-axis (intercept). Since the gradient is positive, the line will slope upwards from left to right.

Also, the gradient is 2 from the equation which means that the steepness of the line must be 1 across and 2 upwards.

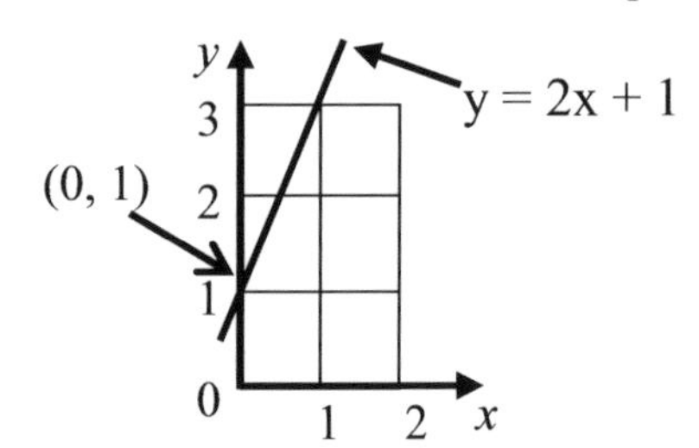

EXERCISE 12 E

1) Write down the equations of the straight lines with
a) gradient of 4 and through point (0,4)
b) gradient of -5 and through point (0,5)
c) gradient of 4 and y-intercept of -6
d) gradient of $\frac{2}{3}$ and y-intercept of 4
e) gradient of -4 and y-intercept of -5
f) gradient of $-\frac{5}{6}$ and y-intercept of -4

2) Find the equation of the line which passes through
a) (0,4) and (3,2) b) (5, 2) and (0, -2)
c) (0, -1) at a gradient of 3

3) A shop has a conversion table below.

Pounds (£)	5	10	15	20	25
Naira (₦)	8	21	34	47	60

a) Draw a graph for the information.
b) Find the equation of the line connecting £ and ₦.

4) Find the equation of each line.

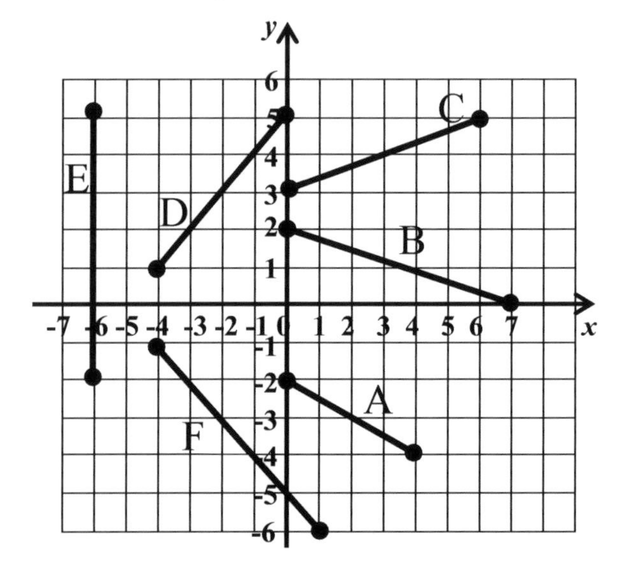

12.7 PARALLEL LINES

Parallel lines can never meet. They have the same distance apart and are always seen with arrows.

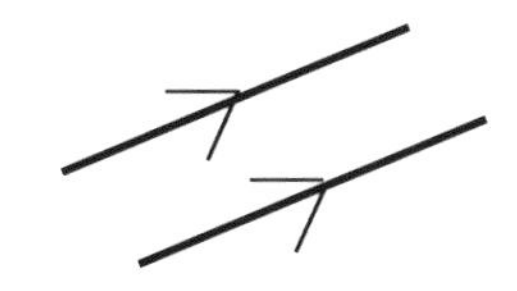

Lines with the **same gradients** are parallel.

Example 1: These lines are parallel because they have the same gradient.

$y = \mathbf{3}x$ $\quad$ $y = \mathbf{3}x - 1$ $\quad$ $y = 9 + \mathbf{3}x$

gradient $\quad$ gradient $\quad$ gradient

All have 3 as their gradient(s).

Example 2: Is the line $2y - 6x = 8$ parallel to the line $y = 3x - 1$?

Solution: First put $2y - 6x = 8$ in the form $y = mx + c$ by making y the subject of the formula.

$2y - 6x = 8$

(+6x) $\quad$ **(+ 6x)**

$2y = 6x + 8$

$(\div 2)$ $\quad$ $(\div 2)$

$y = 3x + 4$

The gradient is 3 likewise the gradient for $y = 3x - 1$.

Therefore, the two lines are parallel.

12.8 PERPENDICULAR LINES

Two straight lines are perpendicular (at right angles) if the **product** of their gradients is **-1**.

$$m \times m_1 = -1$$

If a line has gradient, **m**, then a line perpendicular to it must have a gradient of $-\frac{1}{m}$.

The two lines below are perpendicular. Hence, the product of their gradients $1 \times -1 = -1$.

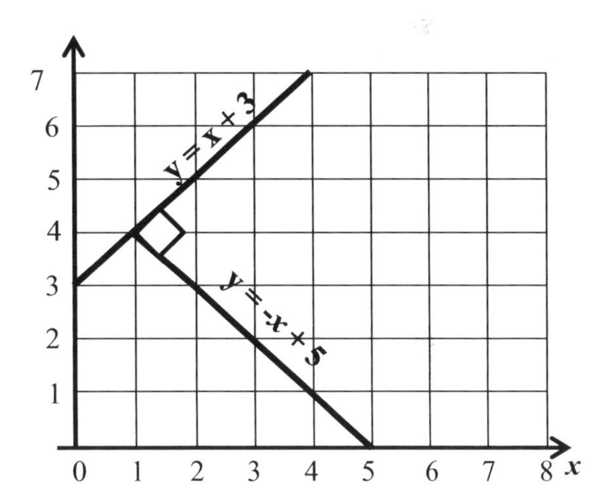

Example 3: Show that the two lines $y = 4x + 5$ and $y = -\frac{1}{4}x + 3$ are perpendicular.

$y = 4x + 5$ $\qquad$ $y = -\frac{1}{4}x + 3$

Gradient = 4 $\qquad$ Gradient = $-\frac{1}{4}$

$4 \times (-\frac{1}{4}) = -1$

Since the product of the two gradients is **-1**, the two lines are perpendicular.

Example 4: A line joins (1, 2) and (5, -6). Find the gradient of a line perpendicular to this line.

Solution: Find the gradient of the line with points (1, 2) and (5, -6).

Gradient (m) = $\frac{-6-2}{5-1} = \frac{-8}{4} = -2$

Therefore $-2 \times m_l = -1$

$m_l = \frac{-1}{-2} = \frac{1}{2}$ ✓

....where m_l is the gradient of the line perpendicular to the original line

Check: $-2 \times \frac{1}{2} = \mathbf{-1}$

EXERCISE 12 F

1) Look at the pairs of equations below:

P) $y = 2x - 2$ and $y = -\frac{1}{2}x + 5$

Q) $y = 4x + 3$ and $y = 5 + 4x$

R) $y = 5x - \frac{1}{2}$ and $y = -\frac{1}{5}x + \frac{1}{2}$

S) $y = 4 + 9x$ and $y = 9x - 3$

T) $y = 8x - 7$ and $y = 7x - 7$

State which of the lines are

a) parallel to each other

b) perpendicular to each other

c) neither perpendicular nor parallel to each other.

2) Draw the line $y = 2x + 1$. State its **gradient** and draw a line perpendicular to the line. What is the gradient of the line perpendicular to $y = 2x + 1$?

3) Line $y = 5x + c$ passes through the point (3, 20). What is the value of c?

4) A straight line parallel to $y = x + 1$ passes through the point (0,4) What is the equation of the line?

12.9 QUADRATIC GRAPH

Quadratic functions will always have an **x^2 term** with no other higher powers. The graph of the quadratic function is a curve known as a **parabola**.

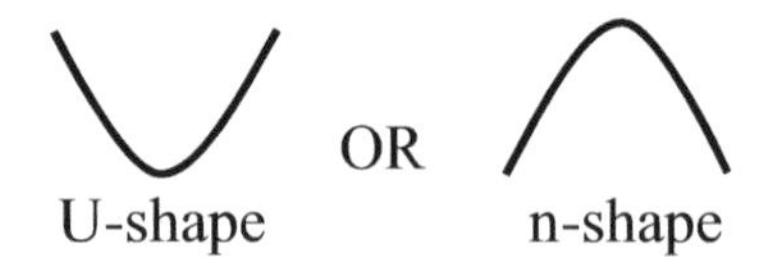

Drawing a quadratic graph is like drawing a linear graph by substituting the values of x to find the corresponding y values. The difference is that the graph of a quadratic function will always be a U or n-shape while that of a linear function will always be a straight line.

Example 1: Draw the graph of $y = x^2 + 1$, x values from -2 to 2

Solution: First draw a table of values.

x	**-2**	**-1**	**0**	**1**	**2**
x^2	4	1	0	1	5
$\mathbf{y = x^2 + 1}$	**5**	**2**	**1**	**2**	**5**

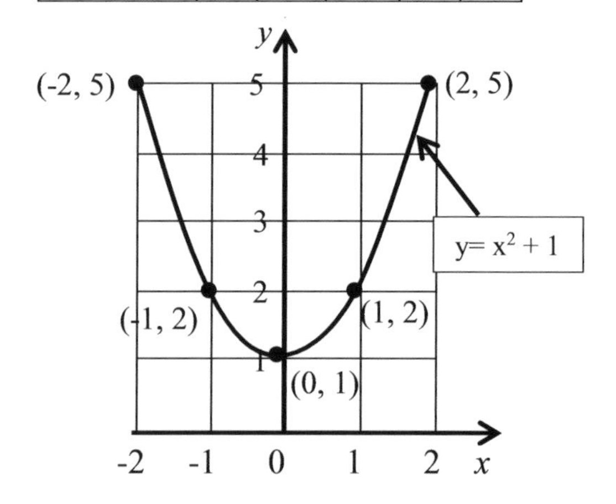

Note that the points are joined by a smooth curve.

Example 2

a) Draw the graph of $y = x^2 - 3$, x values from -3 to 3.

b) Use your answer from part a to estimate the value of y when

i) x = 0.5 ii) x = -2.5

x	-3	-2	-1	0	1	2	3
x^2	9	4	1	0	1	4	9
$y = x^2 - 3$	6	1	-2	-3	-2	1	6

a)

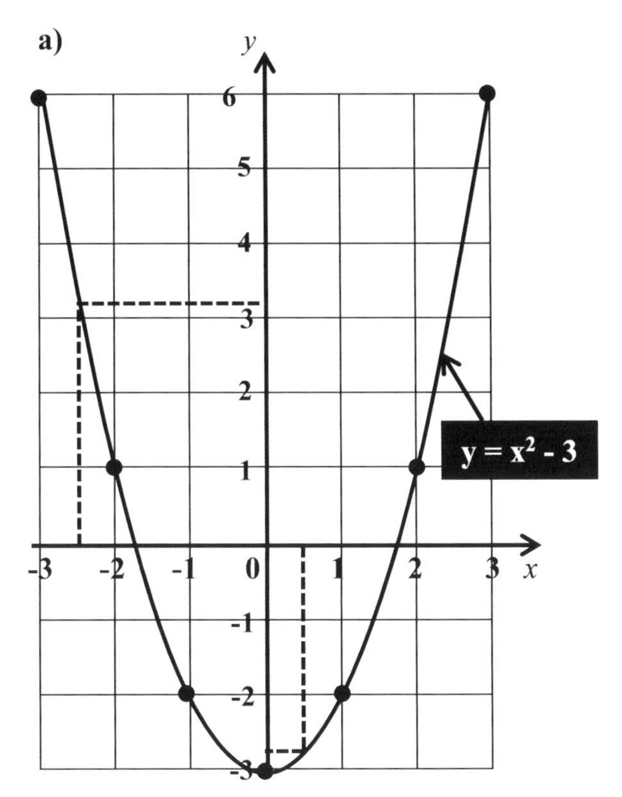

From the graph,

b) i) when x = 0.5, y = - 2.75
ii) when x = - 2.5, y = 3.25

EXERCISE 12 G

1a) Copy and complete the tables below to find the value of $y = x^2 + 4$.

x	-3	-2	-1	0	1	2	3
x^2	9					4	
$y = x^2 + 4$			5				

1b) Draw the graph of $y = x^2 + 4$.

2) For x values from -3 to 3, plot the graph of

a) $y = x^2 - 5$ c) $x^2 + x$

b) $y = x^2 - 1$ d) $5 + x^2$

3a) Draw the graph of $y = x^2 - 4$, x values from -5 to 5.

b) Use your answer from part **a** to estimate the value of *y* when

i) x = 1.5 ii) x = -2.5

EXTENSION WORK

4) Draw the graph of the following equations, x values from -3 to 3.

a) $y = x^2 + 3x + 1$

b) $y = x^2 + 2x - 1$

c) $y = -x^2 - x - 2$

5a) Draw the graph of $y = x^2 + 1$, x values from -3, to 3.

b) Draw the graph of y = 5 on the same axes as in part **a** above.

c) Use your graph to solve the equation $x^2 + 1 = 5$

Chapters 11 &12 review Sections

Assessment

1) a) Write $3\frac{2}{3}$ as an improper fraction. b) What is the reciprocal of $3\frac{2}{3}$?
c) Find the reciprocal of $-3\frac{2}{3}$ d) Multiply your answers from part a and b. …**4 marks**

2a) Copy and complete the table of values below for equation $y = 5x - 1$.

x	1	2	3	4	5
y = 5x - 1			14		
Coordinates			(3,14)		

………. **3marks**

b) What is the coordinate of the point where the line touches the y-axes? ………1**mark**

3) A quadrilateral ABCD has coordinates A (2, 5) B (5, 9) C) (9, 5) and D (5, 1).

Find a) the gradient of line CD ………...**2 marks**

b) the equation of line CD ………...**2 marks**

4) Calculate the coordinates of the mid-point of the line segment joining the pairs of points. a) (1, 6) and (1, 2) b) (-2, -5) and (1, 6) ………….. **4 marks**

5) Find the equation of the line which passes through
a) (0,7) and (3,2) b) (0, -1) at a gradient of 4 ………….. 4 **marks**

6) Show that the two lines $y = 4x + 5$ and $y = -\frac{1}{4}x + 3$ are perpendicular……… **3 marks**

13 Averages and range

This section covers the following topics:

- Mean
- Median
- Mode
- Range

LEARNING OBJECTIVES

By the end of this unit, you should be able to:

a) Calculate the mean of a set of numbers
b) Find the median of a set of numbers
c) Identify the mode from a given set of data
d) Write down mode from a bar chart
e) Find the range from a bar chart
f) Work out the mean from a frequency table
g) Find the mode from a frequency table
h) Find the median from a frequency table
i) Find the averages and range from grouped data

KEYWORDS

- Averages
- Mean
- Median
- Mode
- Range
- Grouped data

13.1 AVERAGES

An **average** is a central number that is representative of all the numbers in a set. The **mean**, **median** and **mode** are the three most commonly used averages.

There is also a concept known as the ***range,*** which **is not** an average, but the difference between the highest and lowest numbers.

Example 1:
Joe completed seven mathematics homework and his marks out of 20 recorded as shown below.

9 11 7 8 15 14 6
A set of values like the one above is called a **distribution**.

MEAN

The mean of a probability distribution is obtained by adding all the numbers in the distribution and then divide by the number of items.

The mean for Joe's marks would be

$= \frac{9 + 11 + 7 + 8 + 15 + 14 + 6}{7}$

$= \frac{70}{7} = \mathbf{10}$ ✓

Example 2: Calculate the mean of 6, 7, 3, 4, 1, 9.

Mean $= \frac{6 + 7 + 3 + 4 + 1 + 9}{6} = \frac{30}{6} = \mathbf{5}$ ✓

Example 3:
Calculate the mean of 4, 7, 4, 2, 7.

Mean $= \frac{4 + 7 + 4 + 2 + 7}{5} = \frac{24}{5} = \mathbf{4.8}$ ✓

EXERCISE 13 A

1) Calculate the mean of each distribution.
a) 3, 7, 2
b) 10, 5, 12, 6, 12
c) 2, 4, 6, 6, 8, 10
d) 7, 9, 2, 4, 2, 1, 3
e) 5, 8, 4, 9, 10, 12, 23, 35

2) In a sale, Eric bought: three books at ₦250 each, two books at ₦450 each, six books at ₦125 each and four books at ₦300 each.
a) How much did Eric pay for all the books he bought?
b) What was the mean cost of the books?

3)

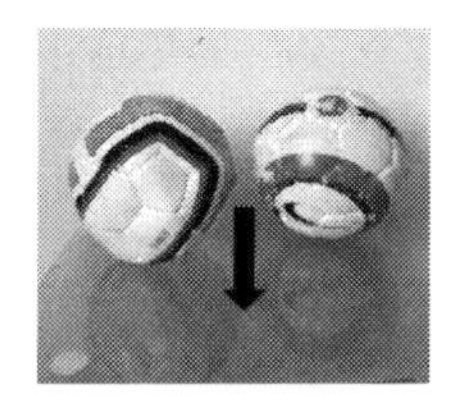

410 g each

450 g

a) Calculate the overall mass of the three balls.
b) Calculate the mean mass of the three footballs.

4)

Two baby pandas weigh 225 g and 350g respectively. What is the average weight of the two pandas?

5) Work out the mean of the following distributions. Leave some answers as ***mixed numbers*** in their simplest form.

a) 7, 5, 9, 9, 5, 6, 4, 3
b) 10, 7, 12, 7, 24, 20
c) 9, 7, 4, 3
d) 12, 3, 4, 5, 2, 5, 4
e) 7 m, 2 m, 10 m, 3 m, 8m
f) 3.75, 0.5, 0.8, 0.95

6) Five students took a geography examination and scored the following marks: 12%, 33%, 50%, 30%, and 60%. Calculate the mean percentage mark.

7) The mean of three numbers is 9.
a) What is the sum of the three numbers?
b) If two of the numbers are 15 and 4, what is the third number?

8) Ifeoma covered a number from the set of numbers below and said: "The mean of the numbers is 6.34."

3.5 4 9.2  8

What is the value of the covered number?

9) To one decimal place, work out the mean of these numbers.
a) 8, 11, 3, 4, 3, 2
b) 10, 3.3, 0.7, 6.5, 9.5, 3, 6, 7
c) 6, 5, 7, 4, 3, 2, 1, 6, 7, 8, 9
d) $\frac{1}{2}$, $2\frac{1}{4}$, $5\frac{1}{5}$

10) Enugu Rangers football team's goal average was 2.3 after 30 matches. How many goals did Enugu Rangers football team score?

MEDIAN

The **median** is the middle value in the list of numbers when arranged in **order of size**. The order could be ascending or descending.

However, if there are two middle numbers, find the average of the two numbers (add them and divide by 2). That would give the median of the numbers.

Example 1: Ebuka sometimes cycles to school. He recorded the number of times he cycled to school in the last seven weeks, as shown below.

1, 2, 5, 0, 3, 2, 4
To find the median, first, arrange in order of size.

It becomes: 0, 1, 2, (2,) 3, 4, 5

There are seven numbers in the list, the 4th number ($\frac{7+1}{2}$) is the median. So, the 4th number which is also the middle value is 2; therefore the median number of times Ebuka cycled to school is **2**. ✓

Notice that you would still get the same answer if you had arranged from highest to lowest. 5, 4, 3, (2,) 2, 1, 0

Example 2:
Find the median of 5, 4, 7, 2, 3, 1

First arrange in order: 1, 2, (3, 4) 5, 7

Notice that 3 and 4 are the middle numbers. (3 + 4) ÷ 2 = 7 ÷ 2 = 3.5
Therefore, **3.5** is the median. ✓

EXERCISE 13 B

1) Write down the median value of each set of numbers.

a) 3, 6, 5, 2, 9, 8, 7
b) 9, 7, 1, 0, 3, 2
c) 12, 8, 4, 3, 1, 7, 6, 4, 2
d) 8, 1, 6, 4
e) 10, 20, 80, 40, 50
f) 4, 3, 2, 8, 7, 6, 5, 4

2) Two students, Chuba and Tochukwu, had five maths tests over a term. The tests were marked out of 30, and the results are shown below:

Chuba: 19, 23, 17, 16, 20
Tochukwu: 6, 21, 27, 3, 20

a) Find the median mark for Chuba.
b) Find the median mark for Tochukwu.
c) Comment on the marks of both students.

3) The morning temperatures (°C) in Lagos for a week in January were:

27, 19, 30, 16, 21, 15, 20
Find the median temperature in Lagos.

4) The show sizes of eight sailors are:
12, $10\frac{1}{2}$, 10, 11, 8, 10, 9, 7
Find the median shoe size for the data.

5) The height of some tables in metres was recorded as 0.9, 1, 1.2, 0.8, 0.85
Work out the median of the heights.

6) Find the median value of
a) 8, 9, 12.5, 4.5, 3, 2.5, 7, 10.5, 4, 3
b) 60, 70, 45, 35, 75

MODE

The most occurring number or value in a distribution is called the **mode** or **modal value**.

You may also have two modes which are explained in Example 2 below. The distribution with two modes is sometimes said to be **bimodal**. Other distributions with more than two modes are said to be **multimodal**. The mode can also represent qualitative data.

Example 1:
Find the mode of the set of numbers.
5, 3, 4, 5, 6, 5, 5, 7, 2

It is always advisable to arrange the distribution in order of size (Not a must). It helps in identifying the mode or modes quicker.

In order of size, the above distribution is 2, 3, 4, 5, 5, 5, 5, 6, 7

It is very clear from the numbers above that the most occurring number is 5, as it appeared four times and no other number or numbers appeared four times. Therefore, **5** is the mode. ✓

Example 2:
Find the mode of the numbers below.
6, 9, 10, 15, 6, 3, 10, 7

In order of size: 3, **6, 6**, 7, 9, **10, 10**, 15
Two numbers appeared twice.
Therefore, the modes are **6** and **10**. ✓

Example 3: Work out the mode for 1, 4, 3, 5. There is **no mode.** ✓

EXERCISE 13 C

1) From the list of numbers below, work out the mode.

a) 2, 5, 4, 1, 2, 1, 1, 7,
b) 3, 5, 8, 9, 3,
c) 4, 2, 12, 7, 12, 7, 5
d) 1, 9, 8, 15, 9, 3, 9, 4, 9
e) 20, 30, 40, 45, 58

2) Kolade asked six people what size of shoes they wear, and he recorded their sizes. 7, $8\frac{1}{2}$, 9, 8, $10\frac{1}{2}$, 8. What is the mode of their shoe sizes?

3) In a school survey, six students said their favourite colours. The results are given below.
Blue Red Blue Yellow Pink Blue
Work out the mode of their colours.

4) A pyramid was designed with different colours as shown. What is the mode of the colours?

5) Write down the mode for each set of data below.

a) 13, 13, 43, 34, 53, 13, 50, 34
b) 7, -8, 6, -3, -8, 2, 3, 8
c) 3.4, 3.5, 1.2, 3.5, 6.7, 2.8
d) $\frac{2}{5}, \frac{1}{5}, \frac{2}{5}, \frac{3}{5}, \frac{3}{6}, \frac{2}{5}, \frac{2}{4}, \frac{2}{5}, \frac{2}{4}$

6) Write down the modal type for each set of data below.
a) Blue, white, yellow, black, blue, red,
b) π, £, √, ₦, π, Θ, π, Θ
c) Dog, cat, rabbit, cat, mouse, dog

RANGE

The **range** is not an average, but the difference between the highest and lowest values in the **distribution**.

Example 1: Work out the range of the set of numbers: 4, 7, 2, 9, 13, 4, 2

Solution:
You may decide to first arrange the numbers in order of size before looking for the highest and lowest numbers.
In order of size: 2, 2, 4, 4, 7, 9, 13

Highest number = 13,
Lowest number = 2
Therefore, the range is 13 - 2 = **11** ✓

EXERCISE 13 D

1) Work out the range of the set of numbers below.

a) 6, 11, 8, 2, 10
b) 4, 5, 3, 12, 15
c) 19, 23, 35, 14, 8, 5
d) 90, 40, 30, 85, 20, 10, 60

2) The number of goals scored in seven football matches in one state is shown.
2 1 3 1 2 2 1
Work out the range.

3) In a small class of five pupils, their heights are 140, 156, 134, 170, and 165. Work out the range of their heights in centimetres.

4) In exercise 20 C, work out the range of the data in question 5.

13.2 AVERAGES AND RANGE FROM FREQUENCY TABLES AND DIAGRAMS

AVERAGES AND RANGE FROM A FREQUENCY TABLE

Example 1: The frequency table below shows the marks obtained by a class in a history test.

Marks	Frequency
5	1
8	2
12	5
15	3

The **modal mark** is the mark with the highest frequency. The highest frequency is 5, therefore the modal mark is **12.**

The **range** is
(Highest mark - the lowest mark) $15 - 5 =$ **10**

For the **median mark**, first, add up the frequencies. $1 + 2 + 5 + 3 = 11$.
The median is $\left(\frac{n+1}{2}\right)th$ value. $(11 + 1) \div 2 = 6$.

It means that the median lies on the **6th value.** Go back to the frequency and start from the top, keep adding the numbers until you get to 6 or more and read off the corresponding ***Mark's*** value.

1 + 2= 3……..But 3 is not up to 6, so consider adding the next number which is 5.
3 + 5 = 8 this is more than 6 but still within limits. Read across to give 12. Therefore, the **median mark** is **12.**

To calculate the mean, extend the column as shown below.

Mark	Frequency	Mark × Frequency
5	1	$5 \times 1 = 5$
8	2	$8 \times 2 = 16$
12	5	$12 \times 5 = 60$
15	3	$15 \times 3 = 45$
Total	11	126

Mean $= \frac{\text{total of (mark × frequency)}}{\text{total number of frquency}} = \frac{126}{11} =$ **11.45** to 2 decimal places.

Note: A common mistake is to divide by 4. Always divide by the total of all the frequencies and in the above example, 11.

EXERCISE 13 E

1) The table below shows the weights of the apples in a bag.

Weight (g)	Frequency
100	5
110	7
115	10

Work out the
a) modal weight b) median weight
c) range d) mean weight.

2) A six-sided dice is rolled in an experiment and the results shown in the frequency table below.

Score	Frequency
1	10
2	20
3	6
4	25
5	12
6	15

Work out the following:
a) the modal score b) the range
c) the median score d) the mean score.

3) The table below shows the number of goals scored in each football match by Enyimba FC for a season in Nigeria.

Goals	Frequency
0	8
1	19
2	11
3	2

a) Work out the mean number of goal scored per match.
b) How many games did Enyimba FC play in the season?

4) In a primary school, pupils are awarded merits for 90% attendance or more. The table shows the number of merits awarded to pupils in two different year groups.

Number of merits	Year 3 frequency	Year 4 frequency
0	5	3
1	15	27
2	20	12
3	14	8
4	7	6

a) How many students are in Year 3?
b) How many students are in Year 4?
c) How many merits were awarded in total to pupils in Year 4?
d) Work out the mean number of merits given per pupil for pupils in year 3?
e) How many more pupils received three merits in Year 3 than in year 4?

5) The table below shows the number of off days taken by some employees at a company in August 2017.

Number of off days	0	1	2	3	4	5
Frequency	5	4	1	0	3	2

a) Calculate the modal number of off days.
b) Calculate the mean number of off days.
c) Calculate the range
d) Work out the median number of off days.

AVERAGES AND RANGE FROM DIAGRAMS

Example 1: Some adults were asked about their favourite hair colour, and the bar chart shows the results.

Hair Colour

a) What is the modal hair colour?
The longest bar is red. Therefore, ***red*** is the modal colour.

b) How many adults took part in the survey?
2+6+4+5+8 = 25
25 adults took part in the survey.

c) How many adults have blonde hair?
6 adults

2) Study the bar chart shown below.

Shoe sizes of surveyed students

Frequency

0 1 2 3 4 5 6 7 8

5 6 $7\frac{1}{2}$ 8 $8\frac{1}{2}$

Shoe size

a) How many students were surveyed?

1 + 4 + 5 + 7 + 5 = **22 students**

b) What is the modal shoe size? The longest bar has shoe size of 8. Therefore, **8** is the modal shoe size.

c) Complete the frequency table below.

Shoe size	5	6	$7\frac{1}{2}$	8	$8\frac{1}{2}$
Frequency	1	4	5	7	5

d) Work out the median shoe size.
1 + 4 + 5 + 7 + 5 = 22…….(22 + 1) ÷ 2 =11.5. The median lies on the 11.5th value which is 8 (Refer to example 1 above). Therefore, the median shoe size is **8**.

e) How many more students wear size eight shoes than size 6?
7 – 4 = **3 more students**

EXERCISE 13 F

1) Michael stood in front of his house and counted the number of children in each of the cars that passed for 20 minutes. He drew a bar chart to represent his findings as shown below.

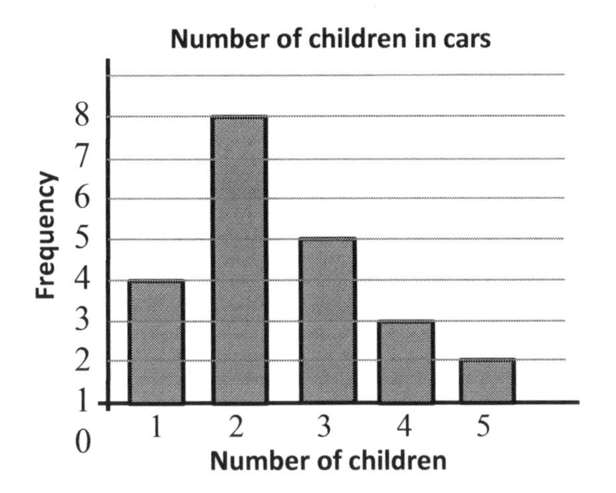

a) How many cars passed Michael's house in total?

b) Find the modal number of children in the car.

c) Work out the range of the data.

d) Calculate the median number of children in the car.

2) The pie chart shows types of pets owned by some students.

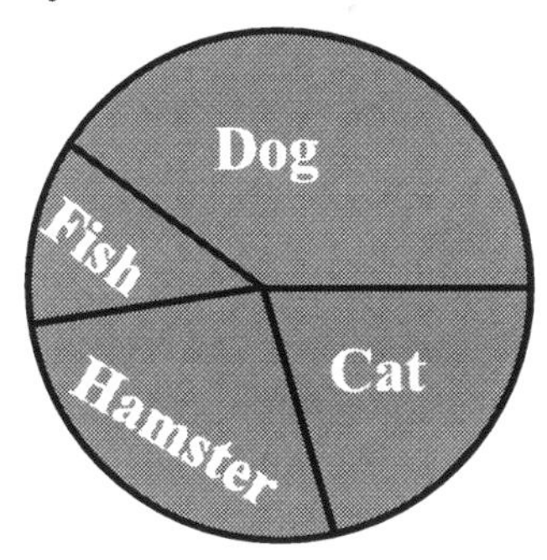

Write the mode of this data.

3) The pie chart below shows information about mock grades of some SSS3 students in a school.

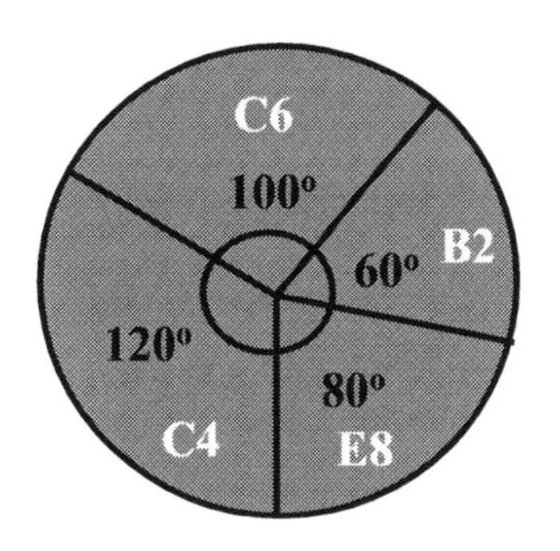

a) What fraction of the students achieved grade E8? Simplify your answer to its lowest form.

b) What grade was the mode?

c) If 12 students achieved grade B2,

i) how many students took part in the survey?

ii) how many students achieved grades C6 and C4?

4) The pictogram represents the number of apples sold each day by Tunji.

Key: ⊕ Represent 8 apples

Monday	⊕ ⊕
Tuesday	⊕ ⊕ ⊕ ◖
Wednesday	⊕ ⊕ ◔
Thursday	⊕ ⊕ ⊕ ⊕
Friday	⊕ ⊕ ⊕ ⊕ ⊕ ◖

a) Work out the range.

b) How many apples were sold on Tuesday?

c) On which day were the most apples sold?

d) How many apples were sold altogether?

5) The bar chart represents the number of horror books owned by four boys.

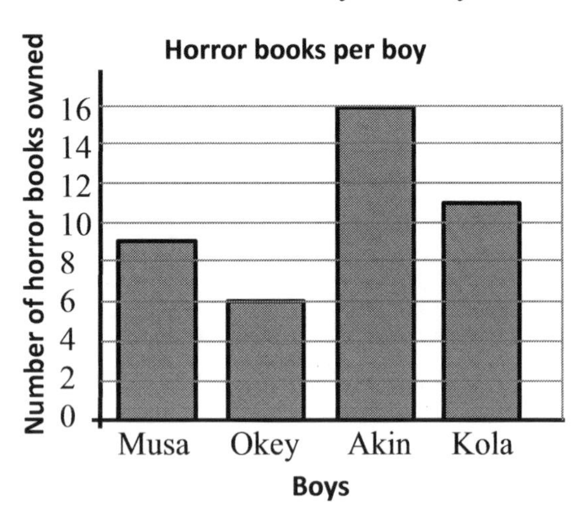

a) Who owned the most horror books?
b) Work out the range of the number of horror books owned by the boys.
c) Work out the mean number of horror books.
d) Musa and Kola owned how many horror books?

6) An advert in the Sun Newspaper reads:

Profession	**Wages per month(₦)**
Carpenter	40 000
Electrician	65 000
Handyperson	35 000
Cleaner	45 000
Plumber	55 000

a) What is the median wage?
b) Work out the mode of these wages.
c) Work out the mean wage.

The advert further says "Carpenter needed, average salary more than ₦53 000 per month."

d) Show why the advert is deceptive.

7) Name three things that are wrong with this bar chart.

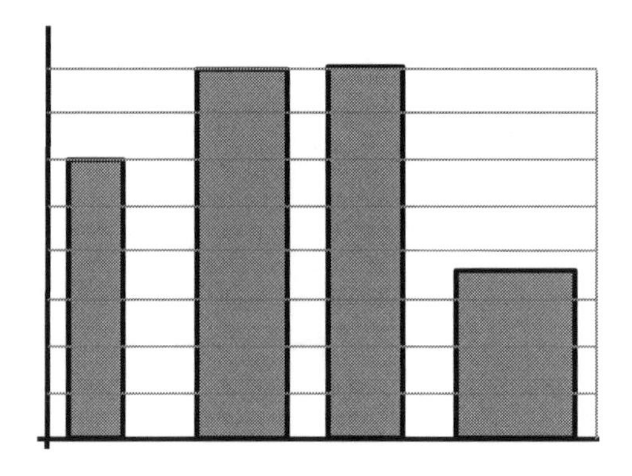

8) The numbers of students living in 50 households in **two** different cities in Nigeria are shown in the tables below.

Lagos

No. of Students	Frequency
0	7
1	15
2	20
3	5
4	2
5	1

Enugu

No. of students	Frequency
0	12
1	16
2	14
3	6
4	2
5	0

a) Draw bar charts for the two cities.
b) Work out the mean, median, mode and range for Lagos and Enugu.
c) Comment on the two cities.

13.3 AVERAGES FROM GROUPED DATA

When data is **grouped**, we can only estimate the mean using the midpoint of each group. For the median, we can only say which group it lies in and not the exact numerical value. You will get to know the **modal group** instead of mode since the data is in a group. For grouped data, class intervals are equal.

Example 1: The physics test results for 20 people are shown below.

5	10	12	4	10	6	18	20	22	25
7	19	5	22	14	18	3	9	19	20

a) Draw a grouped frequency table to represent the information.
b) Work out the modal group.
c) Which group contains the median?
d) Find an estimate for the mean.
e) Why is the answer to part d above only an estimate?

a)

Score	Tally	Frequency
0 – 5	////	4
6 – 10	~~////~~	5
11 – 15	//	2
16 – 20	~~////~~ /	6
21 - 25	///	3

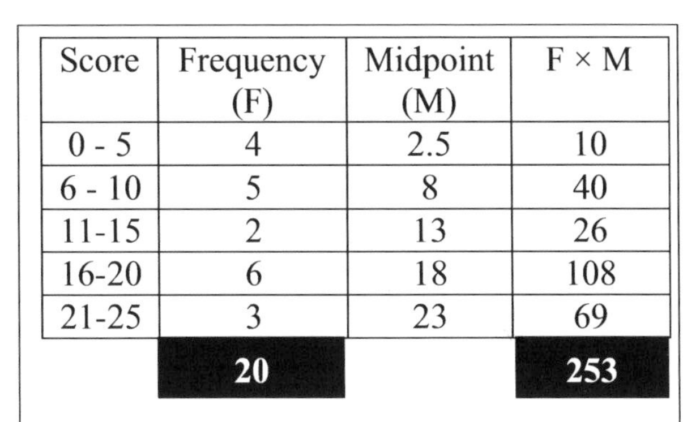

Score	Frequency (F)	Midpoint (M)	F × M
0 - 5	4	2.5	10
6 - 10	5	8	40
11-15	2	13	26
16-20	6	18	108
21-25	3	23	69
	20		**253**

$$\text{Mean} = \frac{253}{20} = \mathbf{12.65}$$

e) The mean is only an **estimate** since we used the midpoints instead of the actual values.

b) The modal group is the group with the highest frequency. It is **16 - 20**

c) There are 20 values (add up all the frequencies). Therefore, the median lies halfway between the 10th and 11th values. Starting from the top of the frequency, start adding until you get to 10.5 or more. Then, read of the group across. This means that the median is in the group of **11 – 15.**

d) To work out the mean, the **midpoint** of each interval must be used. Then multiply the midpoint and the frequency and add them together. For 0 - 5, midpoint = **2.5.**

EXERCISE 13G

1) The table below shows information about the weights of some oranges.

Weight (g)	Frequency
0 – 5	2
6 – 10	4
11 – 15	7
16 - 20	10
21 - 25	12

a) Work out the modal class of the weight of the oranges.
b) Which group contains the median?
c) Find an estimate for the mean.

2) The weights of some pupils were recorded, and the results are shown below.

Weight (kg)	Frequency
$40 \leq w < 45$	2
$45 \leq w < 50$	4
$50 \leq w < 55$	7
$55 \leq w < 60$	10
$60 \leq w < 65$	12

a) Estimate the median weight.
b) What is the modal class?
c) Find an estimate for the mean.
d) Why is the answer to part *c* only an estimate?

3) The lengths of a specific type of grass are given below, in millimetre.

30 31 36 39 40 43 47 52 55 60 61 65
70 71 73 75 78 79 81 83 85 87 88 89

a) Work out the mean length.
b) In class intervals of 30 – 39, calculate the mean length.
c) What are your findings from parts **a** and **b**?

4) The table shows the heights of some sailors.

Height (m)	2.5 – 2.6	2.7 – 2.8	2.9 – 3.0	3.1 – 3.2
Frequency	3	5	2	1

a) Find the modal class.
b) Estimate the median.

14 Pythagoras' theorem

This section covers the following topics:

- Pythagoras' theorem
- Length of a line segment

LEARNING OBJECTIVES

By the end of this unit, you should be able to:

a) Calculate the missing lengths in a right-angled triangle
b) Solve problems using Pythagoras' theorem
c) Work out the length of a line segment

KEYWORDS

- Pythagoras' theorem
- Hypotenuse
- Right-angled triangle
- Square

PYTHAGORAS' THEOREM

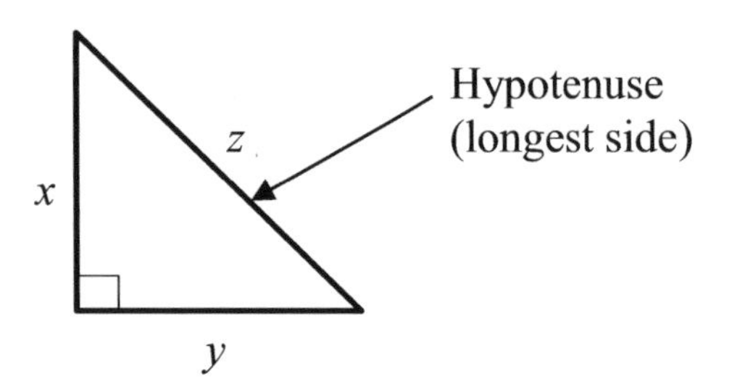

x and y are the shorter sides.

In a right-angled triangle, the sum of the squares on the other two sides is equal to the square of the hypotenuse. This is Pythagoras' theorem.

$$z^2 = x^2 + y^2$$

Note:
* Pythagoras' theorem can only be applied to a right-angled triangle.
* Identify the longest side before applying Pythagoras' theorem.

Example 1: Identify the longest sides (hypotenuse) of these right-angled triangles.

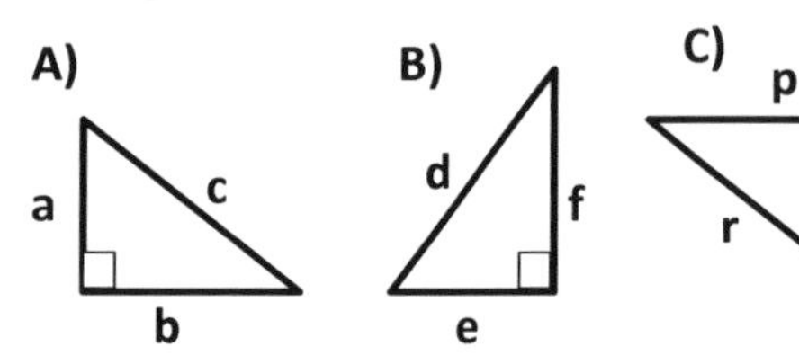

Solutions: A = c, B = d, C = r

Example 2: Calculate the missing length.

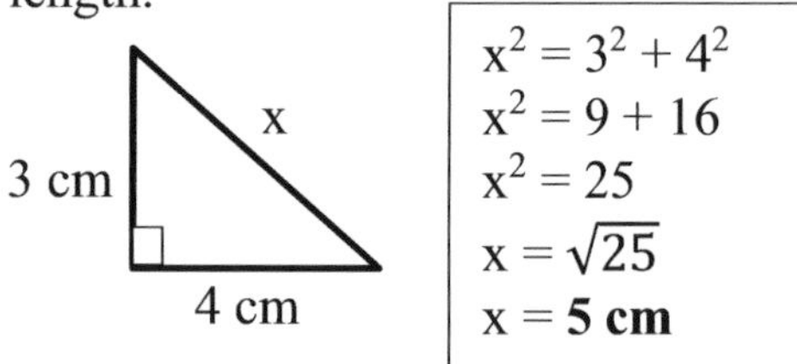

Example 3: Calculate the missing length.

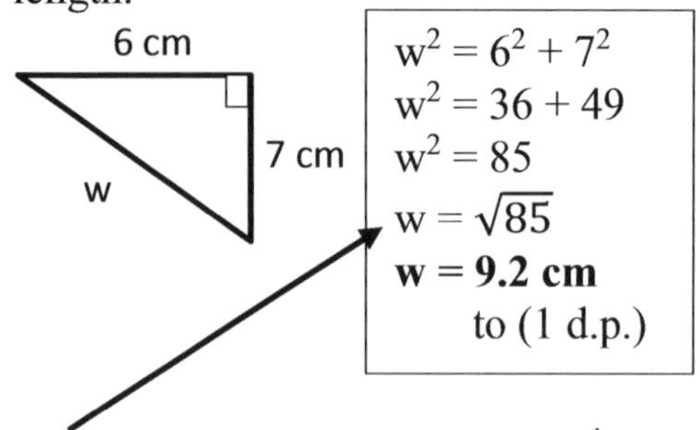

$\sqrt{85}$ cm is the **exact** answer. Working out the decimal part is only an approximation. Always check the question for the format.

EXERCISE 14A

1) Calculate the side marked ***w***. All lengths are in cm and give your answers correct to two decimal places where applicable.

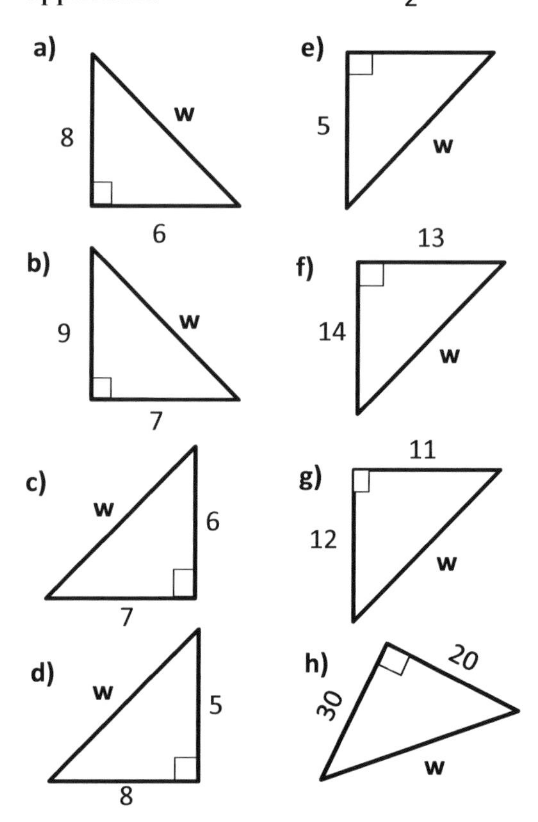

WHEN FINDING A SHORTER SIDE

Example 4: Calculate the missing length.

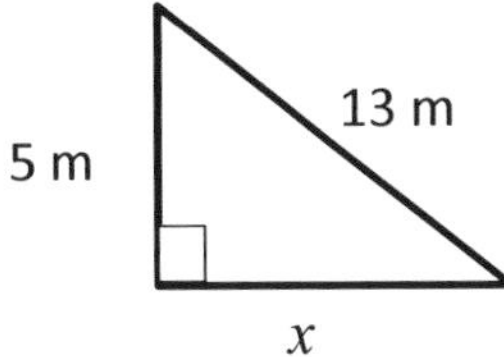

It is obvious we are not looking for the longest side. We still apply Pythagoras' theorem. Identify the hypotenuse, which is 13 m.

$x^2 + 5^2 = 13^2$
$x^2 + 25 = 169$
$x^2 = 169 - 25$
$x^2 = 144$
$x \sqrt{144}$
$\mathbf{x = 12\ m}$

Note: When looking for any of the shorter sides, **take away**.
$x^2 = 13^2 - 5^2$
$x^2 = 169 – 25 = 144$
$x = \sqrt{144}$
$\mathbf{x = 12\ m}$

EXERCISE 14 B

1)

Calculate the side marked ***x***. All lengths are in **m** and give your answers correct to two decimal places where applicable.

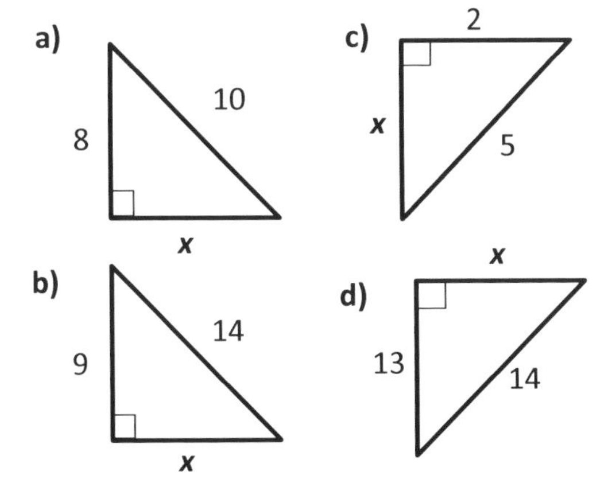

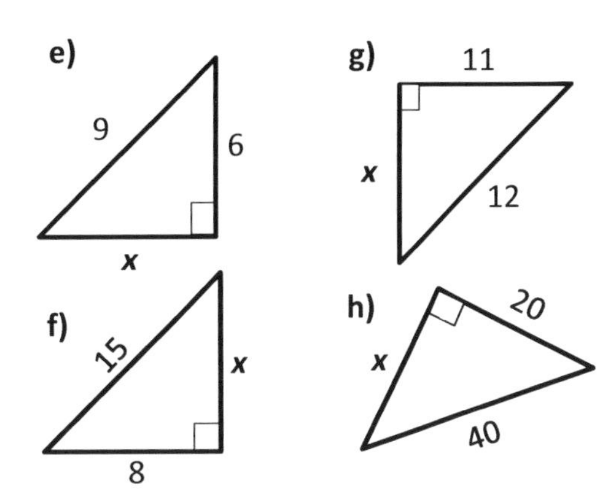

MIXED PROBLEMS

Example 5: Calculate the diagonal of the rectangle below.

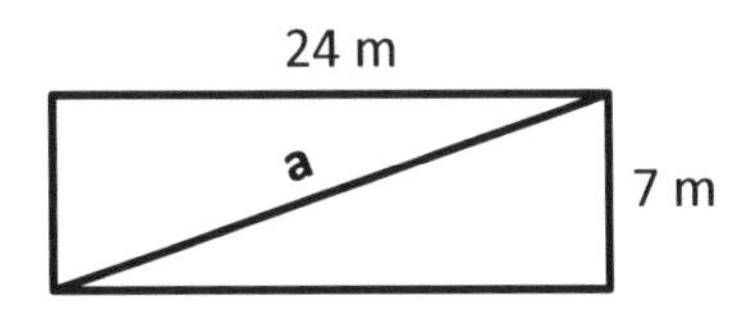

Half of the rectangle is a right-angled triangle. Therefore, we can apply Pythagoras' theorem.

$a^2 = 7^2 + 24^2$
$a^2 = 49 + 576$
$a^2 = 625$
$a = \sqrt{625} = \mathbf{25\ m}$

Example 6: A cone is shown below. Work out the radius to 1 decimal place.

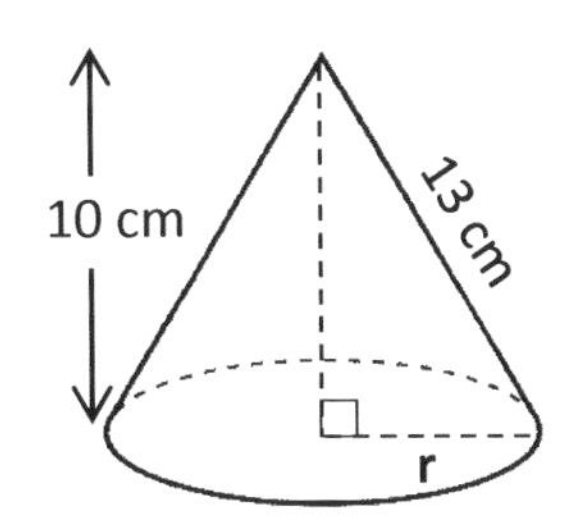

$r^2 = 13^2 - 10^2$
$r^2 = 169 - 100$
$r^2 = 69$
$r = \sqrt{69}$
$\mathbf{r = 8.3\ cm}$

EXERCISE 14 C

1) The front of a school field looks like this. Work out the missing side, to one decimal place.

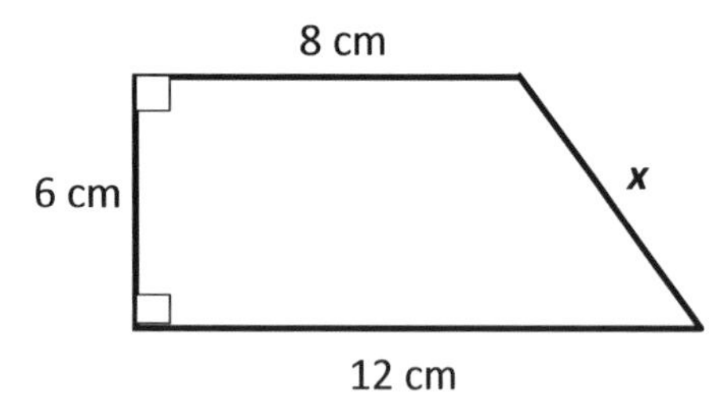

2) Calculate the length of the side marked ***m*** in the diagram below.

a)

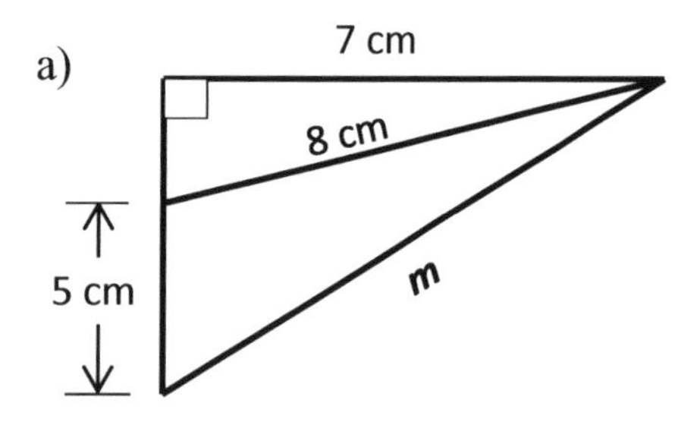

3) A square EFGH is drawn inside a bigger square PQRS, as shown below. Work out the length of the bigger square.

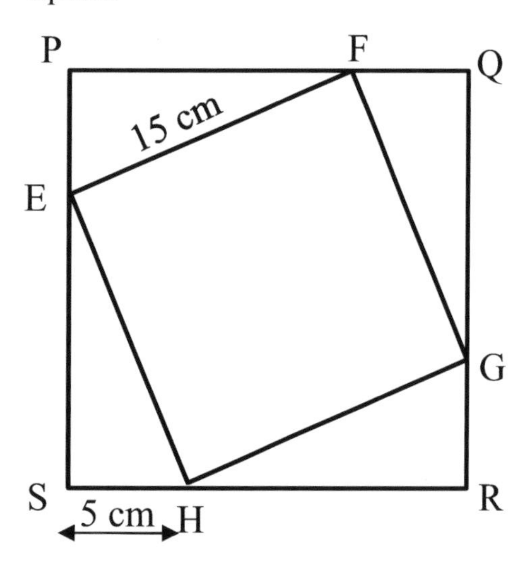

4) Work out the missing length w, to 2 decimal places.

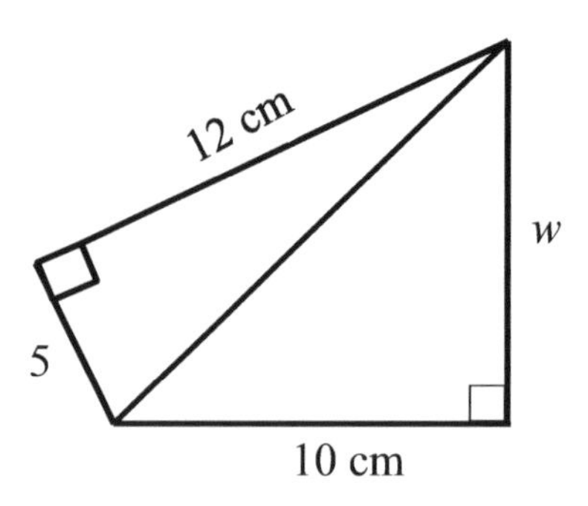

5) The length of a ladder is 15 m, and it rests against a vertical wall. The foot of the ladder is 5.6 m from the wall. How far up the wall does the ladder reach?

6) Find the length of the sides marked ***y*** in the diagrams below.

a)

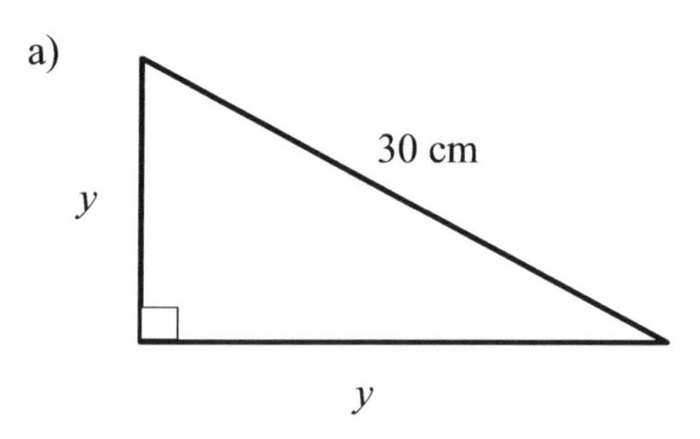

b)

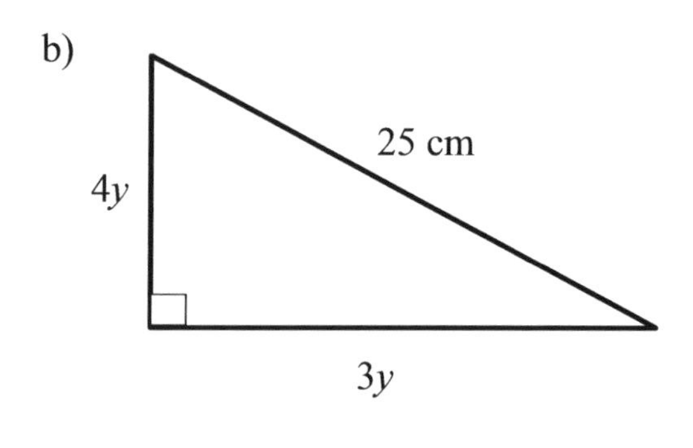

LENGTH OF A LINE SEGMENT

A line that has a beginning and an end is called a **line segment**.

Example 1: Calculate the length PQ.

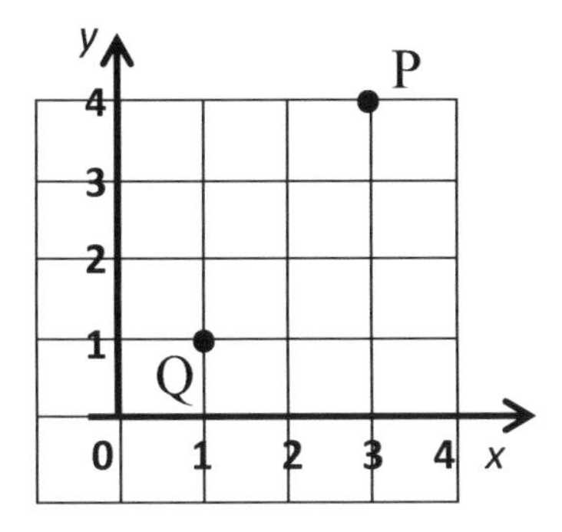

Solution: Join the points P and Q with a straight line and draw a triangle with a right angle at R as shown below.

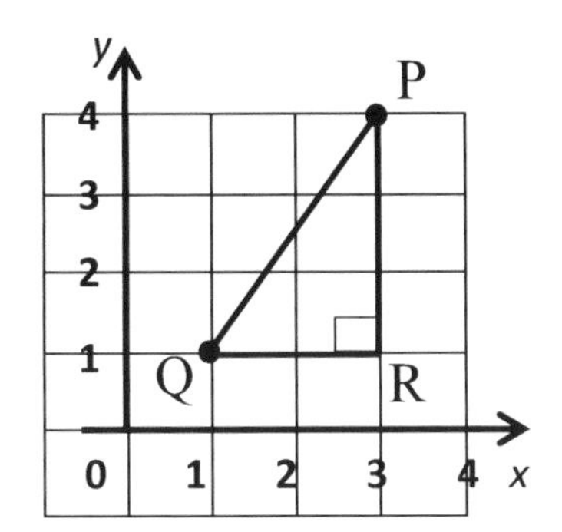

PR = 4 – 1 = 3
QR = 3 – 1 = 2

Using Pythagoras' theorem,
$PQ^2 = PR^2 + QR^2$
$PQ^2 = 3^2 + 2^2$
$PQ^2 = 9 + 4 = 13$
PQ = $\sqrt{\mathbf{13}}$ units (Exact length) ✓

PQ = **3.61** to 2 decimal places

EXERCISE 14D

1) Work out the length of the line segments shown to two decimal places where applicable.

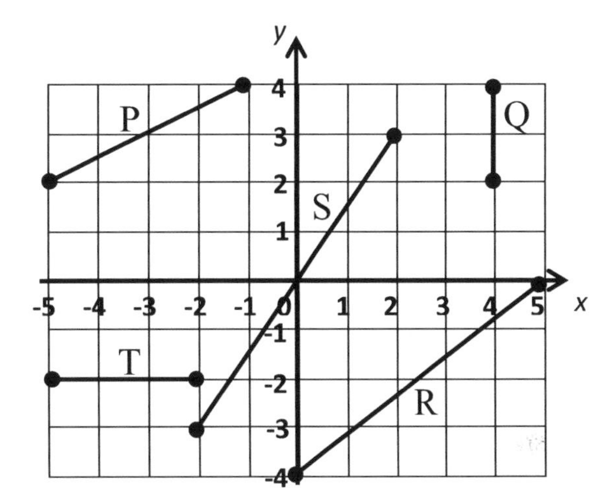

2) Work out the **exact** lengths of these line segments. Do not draw a sketch/graph.

a) A (1, 1) and B (5, 4)
b) P (-5, 1) and Q (-1, 4)
c) D (0, -2) and E (-5, 0)
d) S (3, 4) and T (0, 7)
e) L (-3, -2) and M (3, 3)

15 Bearings

This section covers the following topics:

- Bearings

LEARNING OBJECTIVES

By the end of this unit, you should be able to:
a) Understand bearings
b) Measure bearings
c) Calculate bearings of return journeys

KEYWORDS

- Bearings
- North
- Clockwise direction
- Angle

BEARINGS

Bearing are angles measured **clockwise** from the **North**. Bearings are always written with three digits. 7° is 007°, and 35° is 035°.

COMPASS BEARINGS

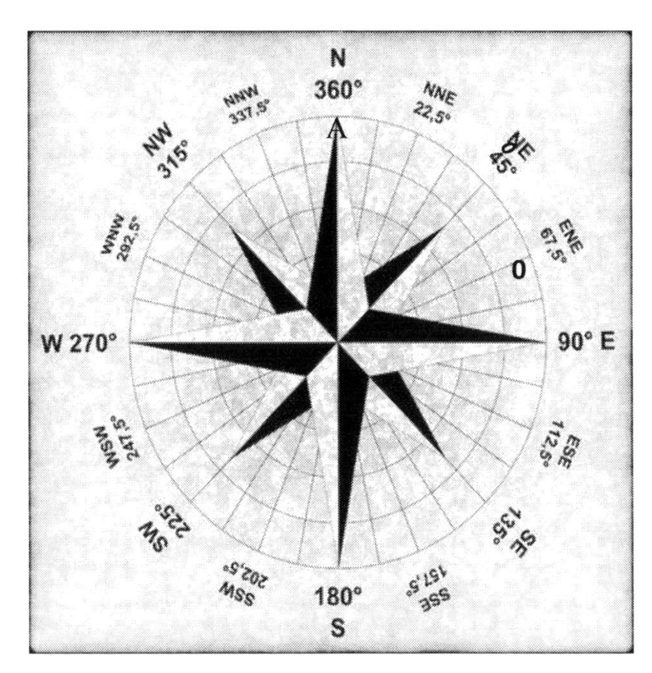

Points of a compass are useful in understanding bearing and angles around it. Bearings are used by sailors and other navigational purposes.

Example 1: Measure the bearing of B from A.

Solution: "From A" is a keyword. Start from **A** and draw a North line.

Place a protractor from the north and measure the angle.

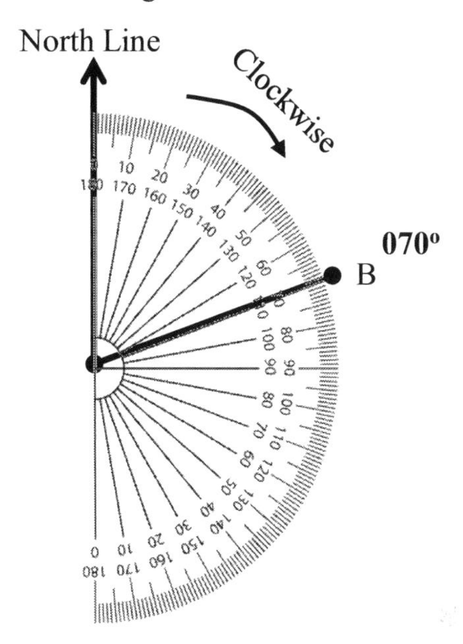

The bearing of B from A is **070°**

Example 2: Work out the bearing of i) P from Q ii) Q from P

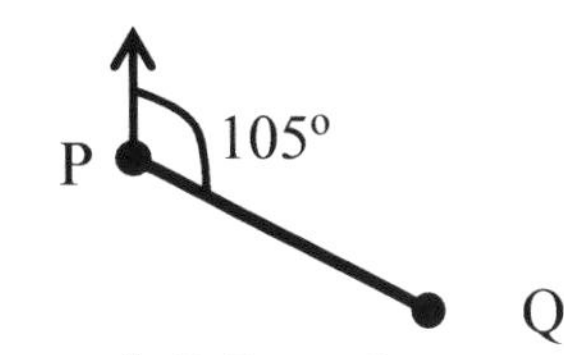

Solution: i) P from Q.
Start from Q and draw a North line.

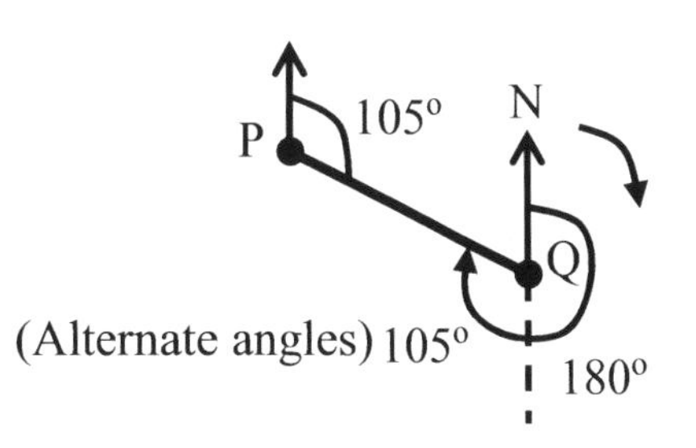

Therefore, the bearing of P from Q is 180° + 105° = **285°**

ii) Bearing of Q from P is **105°** as we go clockwise from the North to the line.

EXERCISE 15A

1) Measure the bearing of
i) B from A ii) A from B

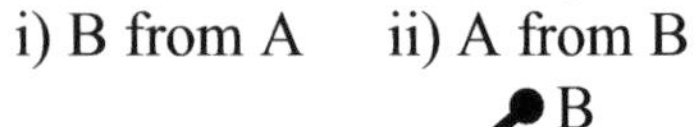

2) Work out the bearing of
i) P from Q ii) Q from P

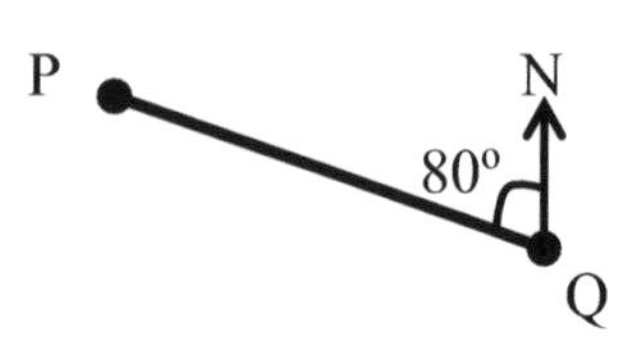

3) Below is a map of four towns.

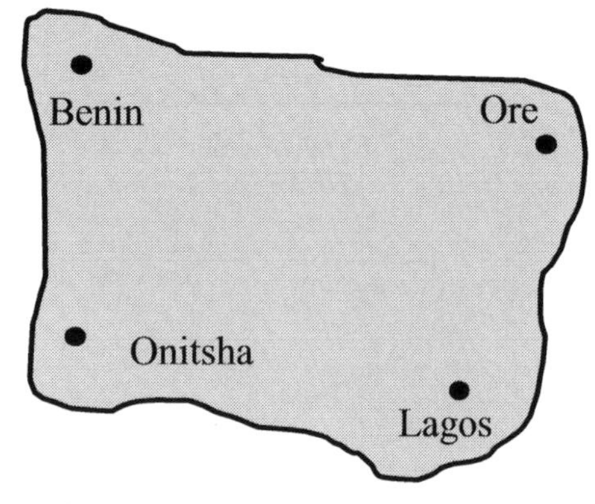

Find the bearings of
a) Benin from Ore
b) Benin from Lagos
c) Onitsha from Ore

4) Work out the bearing of
i) P from Q ii) Q from P

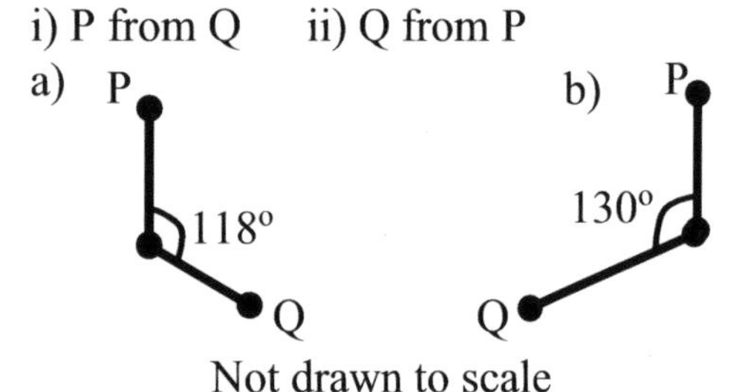

Not drawn to scale

5)

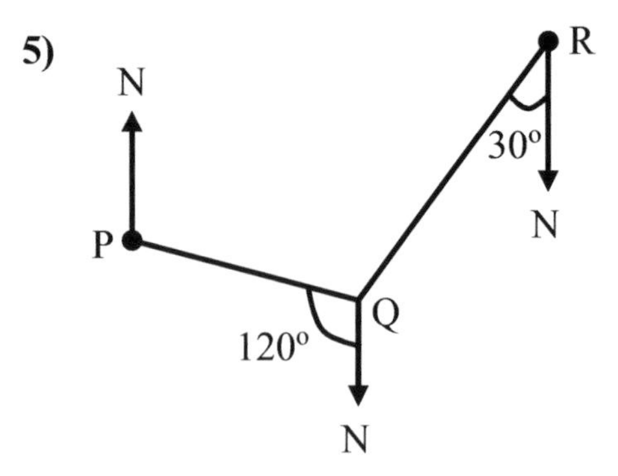

a) Write down the bearing of Q from P.
b) Write down the bearing of P from Q.
c) Write down the bearing of R from Q.
d) Write down the bearing of Q from R

6) Birmingham is 200 km on a bearing of 055° from London. Milton Keynes is 80 km on a bearing of 300° from London.
By using a scale of 1 cm = 40 km, draw accurate positions of London, Birmingham and Milton Keynes.

7) Mrs Ginger is a pilot and flies 400 km from Manchester to Scotland, on a bearing of 060°. She then flies a further 650 km to Wales on a bearing of 150°.

a) Calculate the direct return **distance** from Wales to Manchester to two decimal places.

b) By using a scale of 1 cm to 50 km, draw an accurate diagram for Mrs Ginger's flight.

c) **Measure** the return distance from Wales to Manchester from your diagram in part b.

d) Compare the result from part c to that of part a. What do you notice?

16 Simultaneous equations

This section covers the following topic:

- Simultaneous Linear Equations

LEARNING OBJECTIVES

By the end of this unit, you should be able to:
- Solve a pair of simultaneous equations

KEYWORDS

- Solve
- Simultaneous equation

SOLVING SIMULTANEOUS EQUATIONS

These are equations with two unknowns. They also have two solutions which are true at the same time.

Example 1: Solve the pair of simultaneous equations

$x + 2y = 15$
$x + y = 8$

First label the equations

$x + 2y = 15$ ....(1)
$x + y = 8$ ....(2)
$y = 7$

Think of what letter to eliminate first. Since the x's have the same coefficients, eliminate x first by subtracting.

Next, find the value of x by replacing (substituting) $y = 7$ into equation 1 or 2. Put x $y = 7$ into equation 2. $x + 7 = 8$ implying that $x = 1$.
Therefore, the solution to the simultaneous equations is $\mathbf{x = 1}$ and $\mathbf{y = 7}$

Check: put $x = 1$ and $y = 7$ into (1) or (2). Using equation (2), $1 + 7 = 8$

Example 2: Solve the pair of simultaneous equations

$4c + 3d = 49$ ....(1)
$c - 2d = 4$ .... (2)

Solution: It is easier to eliminate ***c***. Therefore, multiply equation (2) by 4.

$$\begin{array}{r} 4c + 3d = 49 \\ - \quad 4c - 8d = 16 \\ \hline 11d = 33 \end{array}$$

$\mathbf{d = 3}$

To get the value of c, substitute the value of $d = 3$ into equation (1).

$4c + 3(3) = 49$
$4c + 9 = 49$
$4c = 49 - 9$
$4c = 40$ Therefore, $\mathbf{c = 10}$

The solutions to the simultaneous equations is $c = 10$ and $d = 3$.

Check: Replace $c = 10$ and $d = 3$ in equation (1)

$4(10) + 3(3) = 49$
$40 + 9 = 49$

Since Left Hand Side is equal to Right Hand Side, the solutions are correct.

Example 3: The difference between the ages of Paul and Sasha is 12 years. The sum of their ages is 62 years.

a) Form an equation to find their ages.
b) Solve the equation to find the ages of Paul and Sasha.
c) What is the product of Paula and Sasha's ages?

Solution:

a) Let Paul's age = x
   Let Sasha's age = y
Difference between their ages is $x - y = 12$
Sum of their ages is $x + y = 62$
The equations are $\mathbf{x - y = 12}$ and $\mathbf{x + y = 62}$

b) Solve simultaneously

$$\begin{array}{rl} & x - y = 12 \quad \ldots\ldots(1) \\ - & x + y = 62 \quad \ldots\ldots(2) \\ \hline & -2y = -50 \\ & y = 25 \end{array}$$

Replace y = 25 in equation (2)

$x + 25 = 62$
$x = 62 - 25$
$x = 37$

Paul is 37 years and **Sasha, 25 years**.

c) The product of their ages is **925 years**.

EXERCISE 16

Solve the simultaneous equations for Questions 1 to 6.

1) $x + y = 7$
   $x - y = 3$

2) $x + y = 13$
   $x - y = 3$

3) $x - y = 15$
   $x + y = 25$

4) $2x + y = 13$
   $3x - y = 22$

5) $4x - 2y = 10$
   $3x + y = 10$

6) $2x - 3y = 4$
   $5x - 2y = 21$

7) Try and solve the simultaneous equations. Why do you think you were not able to solve the equations?

$y - 2x = 4$
$y - 2x = 5.$

8) The difference between the ages of Tom (t) and Harry (h) is 37 years. The sum of their ages is 67 years.
a) Form an equation to find their ages.
b) Solve the equation to find the ages of Tom and Harry.
c) What is the product of Tom and Harry's ages?

17 Algebra 2

This section covers the following topics:

- Multiplying more than two binomials
- Factorising quadratic expressions
- Solving quadratic equations
- Completing the square
- Harder simultaneous equations
- Changing the subject of a formula

LEARNING OBJECTIVES

By the end of this unit, you should be able to:

a) Expand three brackets

b) Factorise quadratic expressions involving the difference of two squares, factorise expressions in the form $a^2 + bx + c$ and $ax^2 + bx + c$

c) Solve quadratic equations by factorisation, completing the square and Quadratic formula

d) Find turning points and to sketch a quadratic graph

e) Solving harder quadratic and linear inequalities

f) Change the subject of a formula involving squares, square roots and factorisation

17.1 MORE THAN TWO BINOMIALS

Expressions that contain two terms are called **binomials**.

The expression **a + 5** contains two terms, a and +5, hence it is called a binomial.

In this section, we shall consider multiplying more than two binomials.

Example 1: Expand and simplify (x + 2) (x + 3) (x + 4)

There are lots of ways we can multiply these binomials. However, only two methods shall be considered - the "FOIL" method and the "GRID" method.

FOIL method

F – first
O – outside
I – inside
L – last

This means to multiply the first terms in each bracket followed by the outside terms, then the inside terms and finally, the last terms.

Multiply the first two brackets (x + 2) (x + 3) using the foil method.

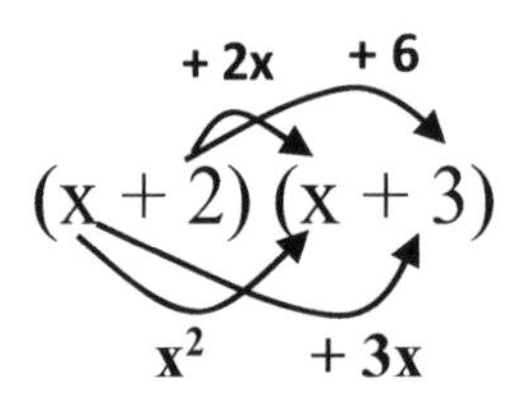

$x \times x = \mathbf{x^2}$, $x \times 3 = \mathbf{3x}$, $2 \times x = \mathbf{2x}$ and $2 \times 3 = \mathbf{6}$

Add all the terms will give $x^2 + 3x + 2x + 6$
Collecting like terms: $x^2 + 5x + 6$

Now, multiply by the remaining bracket (x + 4).

$(x^2 + 5x + 6)\ (x + 4)$

$x^2 \times x = \mathbf{x^3}$, $x^2 \times 4 = \mathbf{4x^2}$
$5x \times x = \mathbf{5x^2}$, $5x \times 4 = \mathbf{20x}$
$6 \times x = \mathbf{6x}$, and $6 \times 4 = \mathbf{24}$

Collecting like terms:
$x^3 + 4x^2 + 5x^2 + 20x + 6x + 24$
$= \mathbf{x^3 + 9x^2 + 26x + 24}$ ✓

GRID method

First multiply (x + 2) and (x + 3)

×	**x**	**2**
x	x^2	2x
3	3x	6

$x^2 + 3x + 2x + 6 = x^2 + 5x + 6$

Then, multiply by (x + 4)

x	x^2	5x	6
x	x^3	$5x^2$	6x
4	$4x^2$	20x	24

$x^3 + 4x^2 + 5x^2 + 20x + 6x + 24$
$= \mathbf{x^3 + 9x^2 + 26x + 24}$ ✓

Example 2

Expand and simplify $(3x - 2)(x - 4)^2$

Solution: Expand $(x - 4)^2$ and then multiply by $(3x - 2)$

$(x - 4)^2 = (x - 4)(x - 4)$

Using the grid method:

x	**x**	**-4**
x	x^2	$-4x$
-4	$-4x$	16

$= x^2 - 4x - 4x + 16$
$= x^2 - 8x + 16$

Next, multiply by $(3x - 2)$

x	$\mathbf{x^2}$	**-8x**	**16**
3x	$3x^3$	$-24x^2$	$48x$
-2	$-2x^2$	$+ 16x$	-32

$= 3x^3 - 2x^2 - 24x^2 + 16x + 48x - 32$
$= \mathbf{3x^3 - 26x^2 + 64x - 32}$ ✓

EXERCISE 17A

1) Expand and simplify.

a) $(x + 1)(x + 2)(x + 4)$
b) $(x + 2)(x + 3)(x + 1)$
c) $(x - 2)^2 (x - 5)$
d) $(b - c)(c + 3)(d - 2)$
e) $(w^2 - 3w + 5)(w - 7)$

2) Expand and simplify.

a) $(2x + 6)(3x + 2)(5x + 1)$
b) $(5x - 4)^3$

3) a) Work out the expression for the volume of the cuboid below.

b) Work out the expression for the surface area of the cuboid.

All lengths in cm.

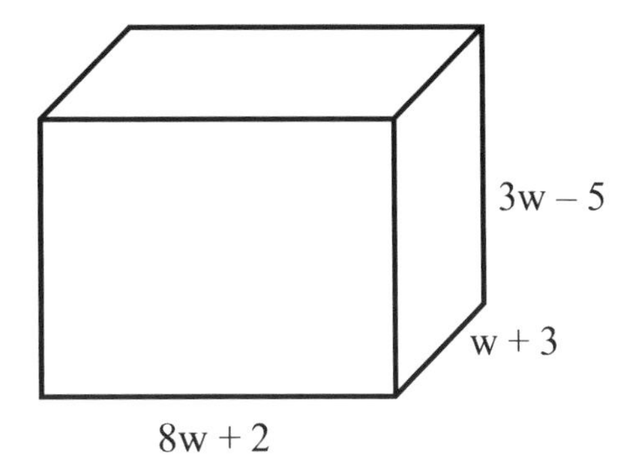

c) If the volume is 3360 cm³, work out the dimensions of the cuboid.

4) $(2c + 1)(c - 1)(3c + 2) = gc^3 + c^2 - 5c - 2$.

Find the value of g and c.

17.2 FACTORISING QUADRATIC EXPRESSIONS

If the highest power of *x* is *2*, then the algebraic expression is called a quadratic expression. Quadratic expressions are usually in the form $\mathbf{ax^2 + bx + c}$, where a is not equal to zero (0) and a, b and c are numbers.

Examples of quadratic expressions are:

$x^2 + 5x$, $2w^2 + 3w - 2$, $4x^2 - 36$

In all the above examples, the highest power of the letters is 2, making them quadratic expressions.

17.3 DIFFERENCE OF TWO SQUARES

If an expression is in the form $a^2 - b^2$, where c and d are algebraic terms or just numbers, it is called the difference of two squares.

Check: $(a - b)(a + b) = a^2 - b^2$

Note: $a^2 + b^2 \neq (a - b)(a + b)$

Example 1: Factorise $a^2 - 16$

Since 16 is a square number, split the letter ***a*** into two brackets together with each value for the square root.

$\sqrt{16} = +4$ or -4
Therefore, $a^2 - 16 = \mathbf{(a - 4)(a + 4)}$ ✓

Example 2: Factorise $25w^2 - 36$

$\sqrt{25} = +5$ or -5.
$\sqrt{36} = +6 \text{ or } -6$

Therefore,

$25w^2 - 36 = \mathbf{(5w - 6)(5w + 6)}$ ✓

Always check your answer by multiplying (expanding) the two brackets. It will give back the original expression.

Example 3:

From a list of numbers in Ebuka class, he thinks of a number. He squares it and then subtracts 1.

a) Write an expression to show this information.
b) Factorise the expression.

Solution:

a) Let his number be ***x***.
He squares his number to get $\mathbf{x^2}$.
Subtracts 1 gives $\mathbf{x^2 - 1}$. ✓

b) $\sqrt{1} = +1 \text{ or } -1$
Therefore, factorising $x^2 - 1$ gives:

$\mathbf{(x - 1)(x + 1)}$ ✓

EXERCISE 17B

1) Factorise the quadratic expressions.

a) $a^2 - 4$
b) $a^2 - 9$
c) $x^2 - 100$
d) $x^2 - 144$
e) $x^2 - 225$
f) $r^2 - 1$
g) $k^2 - 81$
h) $w^2 - 169$
i) $w^2 - 324$
j) $m^2 - 729$

2) Factorise the expressions below.
a) $16m^2 - 1$
b) $25x^2 - 49$
c) $100t^2 - 9$
d) $400w^2 - 64$
e) $100x^2 - 25$
f) $20x^2 - 45$

3) Emily thinks of a number.
She squares it and subtracts 81.

a) Form an algebraic expression for the information.

b) Factorise your expression.

4) Mark says
"When $x^2 + 9$ is factorised, the answer is $(x - 3)(x + 3)$."
Is Mark correct?
Explain fully.

5) Edward squared a number and multiplied the outcome by 25. He then subtracts 400 from it.

a) Form an algebraic expression for the information.

b) Factorise the expression in part **a**.

17.4 FACTORISING QUADRATIC EXPRESSION OF THE FORM $x^2 + bx + c$

The knowledge of expanding double brackets comes in handy when factorising three-part quadratic expressions.

Expansion

$$(x + 1)(x + 3) = x^2 + 4x + 3$$

Factorisation

To factorise expressions in the form $x^2 + bx + c$, take your time to find two numbers that multiply to give the ***c*** part, but the same numbers will add up to give the ***b*** part (the coefficient of x).

Example 1: Factorise $x^2 + 7x + 10$

<u>Solution</u>: 2 and 5 will multiply to give 10. Also, 2 and 5 will add to give 7. Therefore, the two numbers are 2 and 5. Split the brackets to accommodate the 2 and 5. $(x + 2)(x + 5)$

$x^2 + 7x + 10$ will factorise to:
$(x + 2)(x + 5)$ ✓

Example 2: Factorise $x^2 + x - 6$

-2 and 3 will multiply to give -6.
-2 and 3 will add to give 1 (the coefficient of x).

Therefore, $x^2 + x - 6$ will factorise to:
$(x - 2)(x + 3)$ ✓

Example 3: Factorise $w^2 - 11w + 24$

-3 and -8 will multiply to give 24.
-3 and -8 will add to give – 11.

Therefore,
$w^2 - 11w + 24$ will factorise to:
(w – 3) (w – 8) 

EXERCISE 17C

1) Factorise the quadratic expressions.

a) $x^2 + 4x + 4$
b) $x^2 + 6x + 5$
c) $x^2 + 7x + 12$
d) $x^2 + 9x + 14$
e) $x^2 + 14x + 48$

2) Factorise the quadratic expressions.

a) $x^2 - 4x - 5$
b) $x^2 + x - 12$
c) $x^2 + 5x - 14$
d) $x^2 + 4x - 21$
e) $x^2 - 2x - 48$

3) Factorise

a) $w^2 - 6w - 55$
b) $w^2 - 8w - 48$
c) $w^2 + 4x - 60$
d) $w^2 - 16x + 64$
e) $w^2 - 13x + 42$

4) Explain why $x^2 + 3x + 10$ cannot factorise.

5) Factorise $w^{12} - 597w^6 - 1800$

17.5 FACTORISING QUADRATIC EXPRESSSION OF THE FORM $ax^2 + bx + c$

We shall consider two methods for factoring quadratics in the form $ax^2 + bx + c$, where $a \neq 0$ and a, b and c are numbers.

Example 1: Factorise $3x^2 + 16x + 5$

Longer but assured method

The coefficient of x^2 is not 1, but 3. Therefore, multiply 3 by the last number, 5.
$3 \times 5 =$ **15.**

Next, find two numbers that multiply to give 15, but add to give 16.

$1 \times 15 = 15$ and $1 + 15 = 16$.

The numbers are **1** and **15**.

Using the original expression $3x^2 + 16x + 5$, replace 16x with **x** and **15x** (since $x + 15x = 16x$).

The replaced expression is
$3x^2 + x + 15x + 5$

Note: It doesn't matter which expression you write first.

Factorise in pairs. $(3x^2 + x) + (15x + 5)$
$x(3x + 1) + 5(3x + 1)$

At this point, the terms in the brackets **must** be the same. Almost done!

x $(3x + 1)$ **+ 5**$(3x + 1)$

Now, collect the outside terms, $(x + 5)$ with one of the brackets, $(3x + 1)$.

The answer is **(x + 5) (3x + 1)** 

Check your answer by multiplying the two brackets. You will get back the original expression $3x^2 + 16x + 5$.

Alternative method

Factorise $3x^2 + 16x + 5$

Both signs are positive, SO FORM THE BRACKET WITH POSITIVE SIGNS.

(□x + □) (□x + □)

Next, the coefficient of x^2, which is 3, has only two factors, 3 and 1. The brackets must have (3x + □) (x + □)

Also, the factors of 5 are 5 and 1 only. Find pairs of factors of 5 that will combine with 3 to give 16. The pairs are 3 x 5 = 15 and 1 x 1 = 1. 15 + 1 = 16.

Therefore, the correct combination is

(3x + 1) (x + 5), same as with the previous method.

EXERCISE 17D

1) Factorise the expressions below.

a) $5x^2 + 16x + 3$
b) $2x^2 + 15x + 28$
c) $10x^2 + 23x + 12$
d) $4x^2 + 41x + 10$
e) $6x^2 + 31x + 5$

2) Factorise

a) $8x^2 - 10x - 12$
b) $6x^2 - 19x + 15$
c) $4x^2 - 13x + 10$
d) $42x^2 + 17x - 15$
e) $30x^2 + 20x - 10$

3) If $(6x - 5)$ is a factor of $18x^2 - 3x - 10$, work out the other factor.

4) The shape below is a rectangle with length **b** and width **w**.

$(30w^2 + 9w - 12)$ cm^2 *a*

b

a) Work out possible expressions for length, *b* and width, a.

b) Work out a simplified expression for the perimeter of the rectangle.

c) If w = 17, calculate the perimeter of the rectangle.

17.6 SOLVING QUADRATIC EQUATIONS

By now, you must be familiar with factoring quadratics. Some expressions will factorise while some will not. In this section, we shall consider solving quadratic equations by factorisation method, quadratic formula and by completing the square.

BY FACTORISATION

Example 1: Solve $x^2 - 4 = 0$
Factorise the left side.
$(x - 2)(x + 2) = 0$
Either $(x - 2) = 0$ or $(x + 2) = 0$
When $x - 2 = 0$, $x = 2$
When $x + 2 = 0$, $x = -2$.

Therefore, the solutions are
$\mathbf{x = 2}$ and $\mathbf{x = -2}$

Example 2: Solve $5x^2 - 45 = 0$

Add 45 to both sides
$5x^2 = 45$
Divide both sides by 5
$x^2 = 9$
$x = \sqrt{9}$
$\mathbf{x = \pm 3}$

Example 3:
Solve $x^2 + 12x + 35 = 0$

Factorise the left side
$(x + 5)(x + 7) = 0$
From $x + 5 = 0$, $x = -5$
From $x + 7 = 0$, $x = -7$
The solutions are $\mathbf{x = -5}$ and $\mathbf{x = -7}$.

Example 4: Solve $5x^2 + 16x + 3 = 0$

Factorise the left-hand side
$(x + 3)(5x + 1) = 0$

From $x + 3 = 0$, $x = -3$
From $5x + 1 = 0$, $x = -\frac{1}{5}$
The solutions are $\mathbf{x = -3}$ and $\mathbf{x = -\frac{1}{5}}$

BY QUADRATIC FORMULA

The quadratic formula is used where a quadratic expression cannot be factorised.

The quadratic equation is in the form $ax^2 + bx + c = 0$ where $a \neq 0$. The quadratic formula

$$x = \frac{-b \pm \sqrt{b^2 - 4ac}}{2a}$$

gives the solutions (roots) of the quadratic equation.

Example 1:
Solve the equation $x^2 + 5x + 5 = 0$

Solution: The expression cannot be factorised, so we use the quadratic formula.

Comparing $\mathbf{1}x^2 + \mathbf{5}x + \mathbf{5} = 0$ with

$$\mathbf{a}x^2 + \mathbf{b}x + \mathbf{c}$$

gives $a = 1$, $b = 5$ and $c = 5$

Next is to substitute the values of a, b and c into the quadratic formula

$$x = \frac{-b \pm \sqrt{b^2 - 4ac}}{2a}$$

$$x = \frac{-(5) \pm \sqrt{(5)^2 - 4 \times 1 \times 5}}{2 \times 1}$$

$$x = \frac{-5 \pm \sqrt{25 - 20}}{2}$$

$$x = \frac{-5 \pm \sqrt{5}}{2}$$

$x = \frac{-5 + \sqrt{5}}{2}$ or $x = \frac{-5 - \sqrt{5}}{2}$

x = - 1.38 or **x = -3.62**

General curve of a quadratic equation.

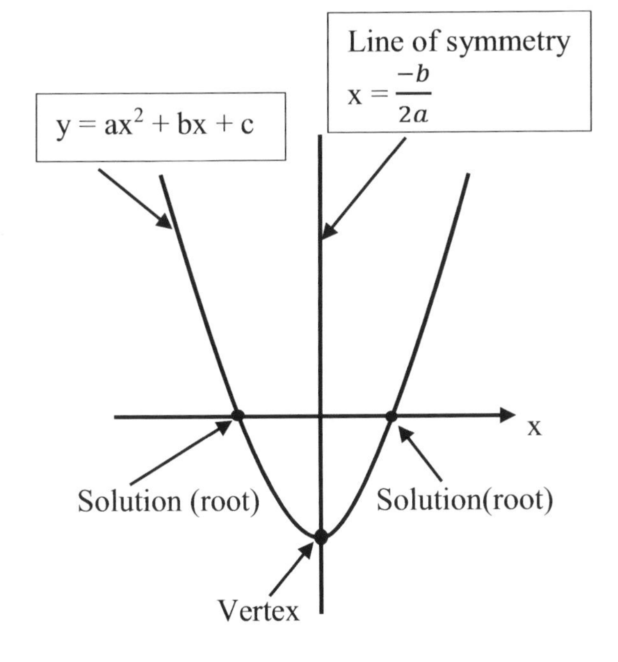

Example 2: Solve the quadratic equation $2x^2 - 2x - 5 = 0$ and round to 2 decimal places.

Solution: a = 2, b = -2 and c = -5
Using the quadratic equation

$$x = \frac{-b \pm \sqrt{b^2 - 4ac}}{2a}$$

$$x = \frac{-(-2) \pm \sqrt{(-2)^2 - 4 \times 2 \times -5}}{2 \times 2}$$

$$x = \frac{2 \pm \sqrt{4 + 40}}{4}$$

$$x = \frac{2 \pm \sqrt{44}}{4}$$

$x = \frac{2 + \sqrt{44}}{4}$ or $\frac{2 - \sqrt{44}}{4}$

x = 2.16 or **x = -1.16**

Example 3:
Sole the equation $x(5 + x) = 7$ and round to 2 decimal places.

Solution: Expand the brackets.
$5x + x^2 = 7$
Subtract 7 on both sides
$5x + x^2 - 7 = 0$
Rearrange in the form $ax^2 + bx + c$
$x^2 + 5x - 7 = 0$
a = 1, b = 5 and **c = -7**
Using the quadratic formula

$$x = \frac{-5 \pm \sqrt{25 - (4 \times 1 \times -7)}}{2 \times 1}$$

$$x = \frac{-5 \pm \sqrt{25 + 28}}{2} = \frac{-5 \pm \sqrt{53}}{2}$$

$x = \frac{-5 + \sqrt{53}}{2}$ or $x = \frac{-5 - \sqrt{53}}{2}$

x = 1.14 or **x = -6.14**

Example 4: Solve $3x(x-2) = (x+2)^2 - 5$

Solution: Expand the brackets

$3x^2 - 6x = x^2 + 4x + 4 - 5$

$3x^2 - 6x - x^2 - 4x - 4 + 5 = 0$

$2x^2 - 10x + 1 = 0$

Using the quadratic formula when $a = 2$, $b = -10$ and $c = 1$,

$$x = \frac{-b \pm \sqrt{b^2 - 4ac}}{2a}$$

$$x = \frac{-(-10) \pm \sqrt{(-10)^2 - 4 \times 2 \times 1}}{2 \times 2}$$

$$x = \frac{10 \pm \sqrt{100 - 8}}{4}$$

$$x = \frac{10 \pm \sqrt{92}}{4}$$

$$x = \frac{10 + \sqrt{92}}{4} \quad \text{or} \quad x = \frac{10 - \sqrt{92}}{4}$$

$\mathbf{x = 4.90}$ or $\mathbf{x = 0.10}$

Please note: In surd form, the roots are

$$x = \frac{10 + \sqrt{92}}{4} \quad \text{or} \quad x = \frac{10 - \sqrt{92}}{4}$$

EXERCISE 17E

1) Solve the quadratic equations by factorisation.

a) $5x^2 + 16x + 3 = 0$
b) $2x^2 + 15x + 28 = 0$
c) $10x^2 + 23x + 12 = 0$
d) $4x^2 + 41x + 10 = 0$
e) $6x^2 + 31x + 5 = 0$

2) Solve the quadratic equations.

a) $8x^2 - 10x - 12 = 0$
b) $6x^2 - 19x + 15 = 0$
c) $4x^2 - 13x + 10 = 0$
d) $42x^2 + 17x - 15 = 0$
e) $30x^2 + 20x - 10 = 0$

3) Solve the quadratic equations.

a) $a^2 - 4 = 0$
b) $a^2 - 9 = 0$
c) $x^2 - 100 = 0$
d) $x^2 - 144 = 0$
e) $x^2 - 225 = 0$
f) $r^2 - 1 = 0$
g) $k^2 - 81 = 0$
h) $w^2 - 169 = 0$
i) $w^2 - 324 = 0$
j) $m^2 - 729 = 0$

4) Solve the equations below.

a) $w^2 - 6w - 55 = 0$
b) $w^2 - 8w - 48 = 0$
c) $w^2 + 4x - 60 = 0$
d) $w^2 - 16x + 64 = 0$
e) $w^2 - 13x + 42 = 0$

EXERCISE 17F

Solve the quadratic equations using the quadratic formula. Leave your answers in surd form where possible.

1) $x^2 + 9x + 2 = 0$
2) $x^2 + 4x + 1 = 0$
3) $x^2 + 7x - 6 = 0$
4) $x^2 + 5x - 3 = 0$
5) $x^2 - 3x - 2 = 0$
6) $2x^2 + 7x + 3 = 0$
7) $3x^2 + 8x - 2 = 0$
8) $6x^2 - 5x - 3 = 0$
9) $4x^2 + 5x - 3 = 0$
10) $x^2 - x - 7 = 0$

EXERCISE 17G

1) Round your answers from exercise 17F question 3 to 3 significant figures.

2) Use the quadratic formula to solve the equations below.

a) $x^2 = x + 7$ d) $y(y - 5) = 3$
b) $2x^2 = 11 - x$ e) $5x(x - 3)=(x + 3)^2 - 7$
c) $x(x + 7) = -9$ f) $(y - 9)^2 = 11$

3) Th difference between two numbers is 7. When multiplied together, the answer is 5. Work out the two numbers and leave your answers to 2 decimal places.

4) Find two consecutive whole numbers whose product is 306.

5) The base of a right-angled triangle is 10m longer than the perpendicular height. The area of the triangle is 37.5 m^2. Work out the perpendicular height.

17.7 THE DISCRIMINANT

In the quadratic formula

$$x = \frac{-b \pm \sqrt{b^2 - 4ac}}{2a},$$

the expression under the square root sign $\mathbf{b^2 - 4ac}$ is called the **discriminant**.

Knowledge of the Discriminant

1) $b^2 - 4ac > 0$

The quadratic equation has two real roots (solutions) if $b^2 - 4ac$ is greater than zero (0). The parabola which is the shape of the quadratic curve cuts the x-axis at two distinct places. Consider the parabola represented by the equation $x^2 + 5x - 7 = 0$.

The discriminant $b^2 - 4ac$
$= 5^2 - 4 \times 1 \times -7 = 25 + 28 = \mathbf{53}$

This is greater than zero. It means that there are two real roots (solutions) and the curve cuts the x-axis at two points.

To show this, solve the quadratic equation. $a = 1$, $b = 5$ and $c = -7$.

$$x = \frac{-5 \pm \sqrt{5^2 - (4 \times 1 \times -7)}}{2 \times 1}$$

$\mathbf{x = 1.14}$ or $x = \mathbf{-6.14}$

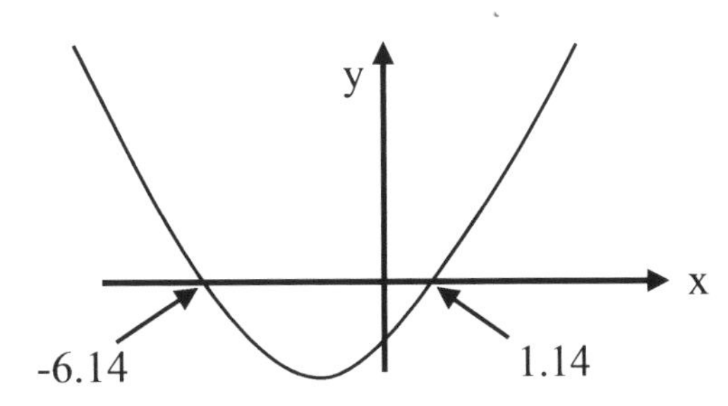

2) $b^2 - 4ac < 0$

The quadratic equation has **no real roots** or solutions if the discriminant $b^2 - 4ac$ is less than zero. This is because the square root of a negative number does not exist (will give a math error on a calculator) and therefore, the parabola **does not** cut the x-axis.

Consider the quadratic equation $2x^2 - 3x + 5 = 0$, where a = 2, b = -3 and c = 5. The discriminant $b^2 - 4ac$ $= (-3^2) - 4 \times 2 \times 5 = 9 - 40 = \mathbf{-31}$

When used in the quadratic equation, it becomes $\sqrt{-31}$. The square root of a negative number is undefined and does not exist. Therefore, the curve of the quadratic equation **will not** cut the x-axis and will have no solution.

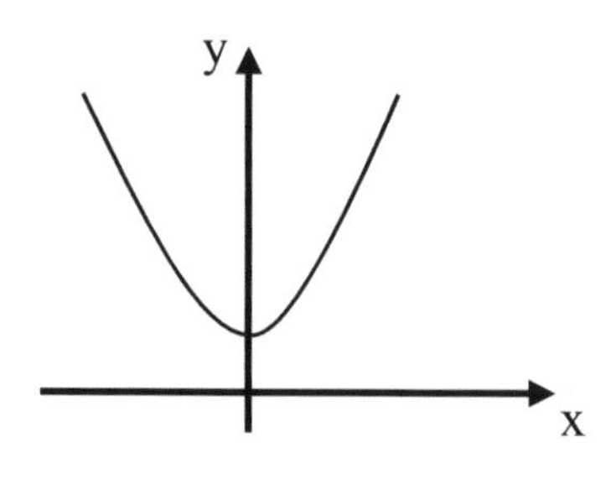

3) $b^2 - 4ac = 0$

When the discriminant $b^2 - 4ac = 0$, the x-axis is the horizontal tangent to the parabola (curve). The quadratic equation has only **one root** (solution). Consider the quadratic equation $3x^2 + 6x + 3 = 0$ where a = 3, b = 6 and c = 3.

Using the discriminant $b^2 - 4ac$,
$6^2 - (4 \times 3 \times 3) = 36 - 36 = \mathbf{0}$
Applying this in the quadratic formula gives:

$$x = \frac{-b \pm \sqrt{0}}{2a} = \frac{\mathbf{-b}}{\mathbf{2a}}$$

It is now evident that when the discriminant = 0, the part of the quadratic equation that contains the square root sign becomes non-existent as the square root of zero is 0. Therefore, we have only one root or solution. If we continue solving the quadratic equation above, $x = \frac{-6}{6} = -1$, showing only one root at (-1, 0).

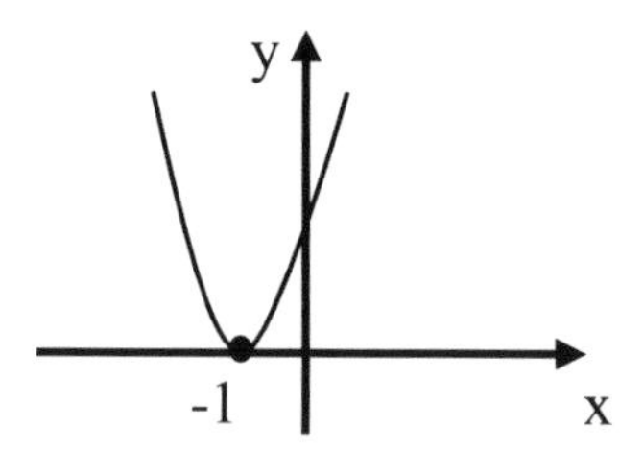

Example 1: Using only the discriminant, decide whether the quadratic equation $2x^2 + 3x - 1 = 0$ has two, one or no solutions.

Solution: a = 2, b = 3 and c = -1. Using the discriminant $b^2 - 4ac$,
$= 3^2 - (4 \times 2 \times -1) = 9 + 8 = 17$
Since **17 > 0**, there are two solutions.

Example 2:
Decide whether $5x^2 - 2x + 3 = 0$ has one, two or no solutions.
Solution: $b^2 - 4ac = (-2)^2 - (4 \times 5 \times 3)$ $= 4 - 60 = -56$.
Since **-56 < 0**, there are no solutions.

EXERCISE 17 H

For questions **1 – 8**, show whether the quadratic equations below have one, two or no solutions. Sketch the graphs.

1) $x^2 + 3x + 6 = 0$

2) $3x^2 + 5x - 9 = 0$

3) $2x^2 - 2x - 7 = 0$

4) $3x^2 + 6x + 3 = 0$

5) $5x^2 + 10x + 5 = 0$

6) $x^2 - 4x + 5 = 0$

7) $x^2 + 3x + 2 = 0$

8) $6x^2 + 9x - 3 = 0$

9) **a** Work out the coordinates of P and Q
b Write down the equation of the line of symmetry.

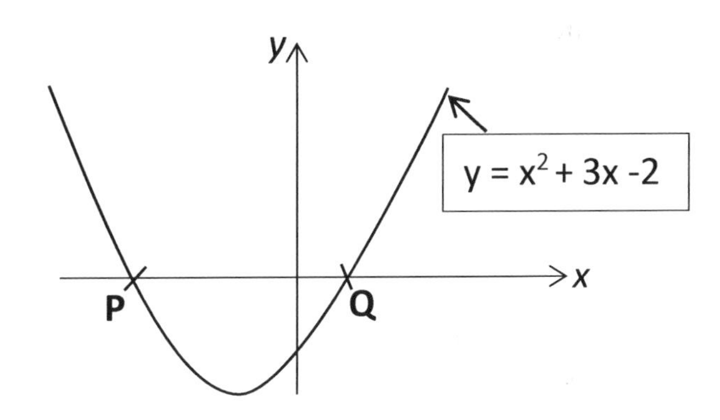

17.8 COMPLETING THE SQUARE

Completing the square is an important technique for solving quadratic equations.
The maximum and minimum **turning points** of the parabola can be found using the method of completing the squares.

Consider the function $y = x^2 + 7x - 3$
From graphs of a quadratic function, where the coefficient of x^2 is positive, any of the graphs below is possible.

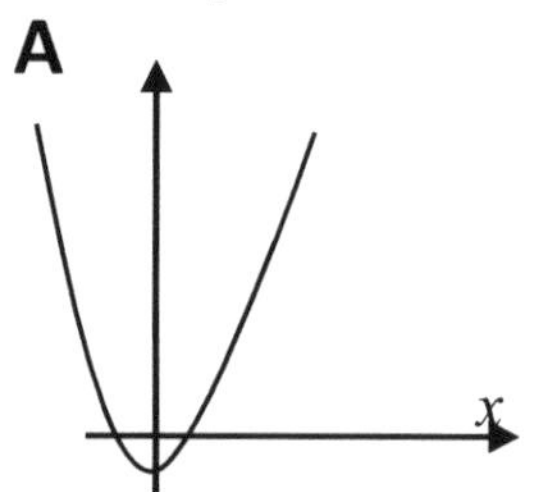

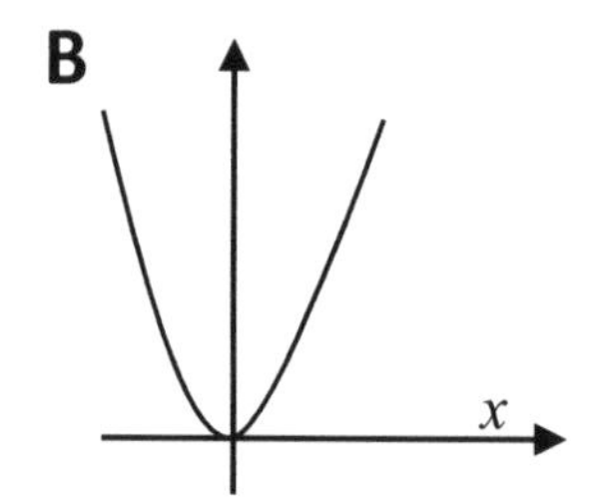

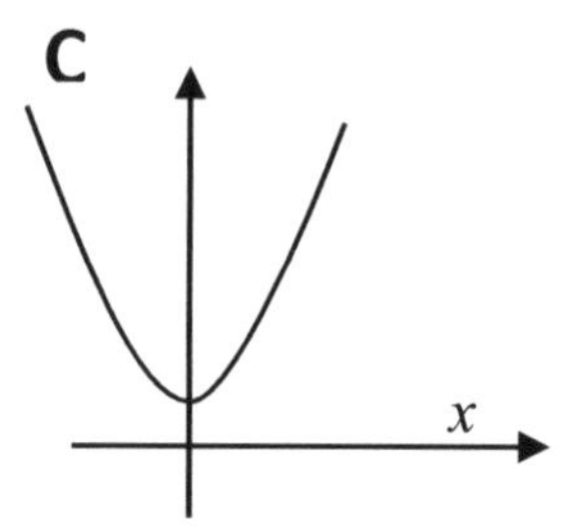

We may be interested in finding the coordinates of the minimum point. The parabola will have a minimum (turning) point on the graph and we may be interested in finding the $\boldsymbol{x}$ and $\boldsymbol{y}$ values at this point. Is the turning point below or above the x- axis?
We may use completing the square technique to answer the question.

What is the meaning of complete or perfect square?

An expression of the form $(x + a)^2$ is called a **complete** or **perfect** square.
Expanding will give $(x + a)^2 = (x + a)(x + a) = x^2 + 2ax + a^2$

Also, $(x - a)^2$ is a complete square. Expanding the bracket will give
$(x - a)^2 = (x - a)(x - a) = x^2 - 2ax + a^2$

Consider the expression $(x + 3)^2$

$(x + 3)^2 = (x + 3)(x + 3)$

$\quad = x^2 + 6x + 9$

$\mathbf{x^2 + 6x + 9}$ is a perfect square

Methods of completing the square

When the coefficient of x^2 is 1

Before we solve quadratic equations using completing the square method, we must first write the expression in completed square form. It must be written in a form which has a square term and a constant.

Method 1: General method

Example 1: Write $x^2 + 6x + 3$ in the form $(x + p)^2 + q$
Consider the expressions $\mathbf{x^2 + 6x}$.

Halve the coefficient of x in the expression 6x. This is 3.
As a perfect square, $(x + 3)^2 = x^2 + 6x + 9$

To get back $x^2 + 6x$, we subtract 9
$x^2 + 6x + 9 - \mathbf{9} + 3$
but $(x + 3)^2 = x^2 + 6x + 9$

Therefore, $(x + 3)^2 - 9 + 3$
$= (x + 3)^2 - 6$

Finally, $x^2 + 6x + 3$ in the form $(x + p)^2 + q = \mathbf{(x + 3)^2 - 6}$

We shall look at turning points at a later stage and its implications on the quadratic graph.

Also, we can solve quadratic equations using the completed square form. All these will be dealt with at a later stage in this chapter.

Method 2: The co-efficient method

This method of completing the square compares the coefficients of the original quadratic to those of the multiplied out completed form.

Example 1: Write $x^2 + 6x + 3$ in the form $(x + p)^2 + q$

Expanding gives ➡ $(x^2 + 2px + p^2) + q$

This compares to:

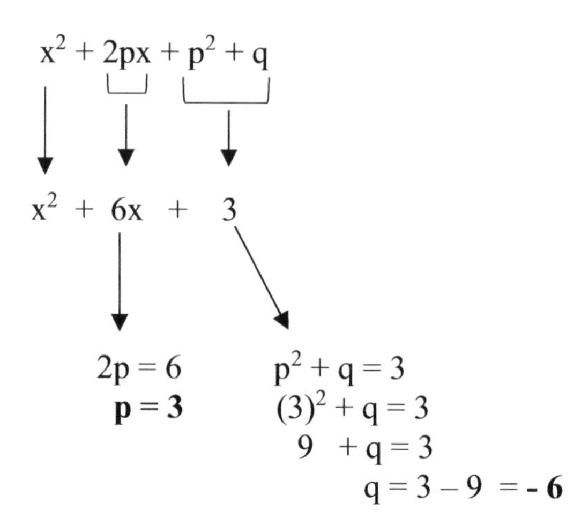

Putting the values of p and q into the form $(x + p)^2 + q$ will give: $\mathbf{(x + 3)^2 - 6}$ 

EXERCISE 17 I

Write in completed square form the expressions below using any method.

1) $x^2 + 8x + 2$
2) $x^2 + 10x + 7$
3) $x^2 + 6x - 9$
4) $x^2 - 6x - 1$
5) $x^2 - 16x + 5$
6) $x^2 + 2x - 5$
7) $x^2 + 5x + 5$
8) $x^2 - 7x - 9$

Write the following expressions in the form $(a + b)^2 + c$. Write down the values of b and c.

9) $x^2 + 20x + 7$
10) $x^2 - 7x + 5$
11) $x^2 - 3x - 6$
12) $x^2 + 14x + 8$
13) $x^2 + 9x + 2$
14) $x^2 - 4x - 4$
15) $x^2 + 2x + 5$
16) $x^2 + 6x + 13$

17) Work out the values of **p** and **q** such that $x^2 + 8x - 8 = (x + p)^2 - q$

17. 9 SOLVING QUADRATIC EQUATIONS BY THE METHOD OF COMPLETING THE SQUARE

Example 1: Solve the equation $x^2 - 6x + 3 = 0$ by completing the square.

$(x - 3)^2 = x^2 - 6x + 9$
$(x^2 - 6x + 9) - 9 + 3 = 0$
$(x - 3)^2 - 6 = 0$
Add 6 to both sides
$(x - 3)^2 = 6$
$x - 3 = \sqrt{6}$
$x - 3 = \pm\sqrt{6}$
Add 3 to both sides
$x = 3 \pm \sqrt{6}$ .... this is the answer in surd form. It is the exact answer.

However, we may proceed to evaluate the whole answer.
$x = 3 + 2.449....$ $x = 5.449...$
or
$x = 3 - 2.449...$ $x = 0.5505...$

By using completing the square method, the solutions to the quadratic equation $x^2 - 6x + 3 = 0$ is **x = 5.449...** or **x = 0.5505...**

Example 2:
Solve the equation $n^2 + 5n - 2 = 0$ using the completing the square method.
Leave your answer in surd form.

Solution:
$5 \div 2 = 2.5$
$(n + 2.5)^2 = n^2 + 5n + 6.25$
$(n^2 + 5n + 6.25) - 6.25 - 2 = 0$
$(n + 2.5)^2 - 8.25 = 0$

Add 8.25 to both sides
$(n + 2.5)^2 = 8.25$
$n + 2.5 = \pm\sqrt{8.25}$

Subtract 2.5 from both sides
$n = \pm\sqrt{8.25} - 2.5$

$\mathbf{n = \sqrt{8.25} - 2.5}$ or $\mathbf{n = -\sqrt{8.25} - 2.5}$

EXERCISE 17 J

1) Solve the equations by completing the square. Leave your answers in surd form where appropriate.

a) $m^2 + 8m - 3 = 0$
b) $x^2 + 8x - 5 = 0$
c) $x^2 - 4x - 1 = 0$
d) $x^2 + 10x + 6 = 0$
e) $x^2 - 6x - 2 = 0$
f) $x^2 + 12x + 9 = 0$

2) Solve the following equations by completing the square. Leave your answers to 2 decimal places where possible.

a) $c^2 + 2c - 3 = 0$
b) $x^2 + 8x - 5 = 0$
c) $x^2 - 4x - 1 = 0$
d) $x^2 + 10x + 6 = 0$
e) $x^2 - 6x - 2 = 0$
f) $y^2 + 12y + 9 = 0$

17.10 When the coefficient of x^2 is not 1

When the coefficient of x^2 is not 1, take out the coefficient as a **factor** and complete the squares as usual.

Example 1: Write $5x^2 + 10x + 2$ in the form $\mathbf{a(x + p)^2 + q}$.

The coefficient of x^2 is not 1. It is 5.

Take out **5** as a factor.

$5x^2 + 10x + 2 \quad = 5[x^2 + 2x + 2/5]$

$= 5[(x + 1)^2 - 3/5]$

Multiply out the brackets

$= 5(x + 1)^2 - 3$

Therefore, $a = 5$, $p = 1$ and $q = -3$

Example 2: Write $\frac{1}{4}x^2 - 5x + 4$ in the form $a(x + p)^2 + q$

Take out ¼ as a factor.

$$\tfrac{1}{4}x^2 - 5x + 4 = \tfrac{1}{4}\left(x^2 - 20x + 16\right)$$

$$= \tfrac{1}{4}\left[(x - 10)^2 - 84\right]$$

Multiply out the brackets

$$= \tfrac{1}{4}(x - 10)^2 - 21$$

EXERCISE 17K

Write the expressions in the form $a(x + p)^2 + q$

1) $3x^2 + 6x + 9$
2) $5x^2 + 15x + 10$
3) $4x^2 - 6x + 1$
4) $7x^2 - 14x + 6$
5) $\frac{1}{2}n^2 + 5n + 6$
6) $\frac{3}{4}k^2 - 3k + 6$

Example 3: Solve the quadratic equation $5x^2 + 10x + 2 = 0$ by completing the square.

Look back to example 1. In completed square form, $5x^2 + 10x + 2 =$ $\mathbf{5(x + 1)^2 - 3}$
Therefore, $5(x + 1)^2 - 3 = 0$

add 3 to both sides

$5(x + 1)^2 = 3$

divide both sides by 5

$(x + 1)^2 = 3/5 = 0.6$

$x + 1 = \pm\sqrt{0.6}$

subtract 1 from both sides

$x = \sqrt{0.6} - 1$ or $x = -\sqrt{0.6} - 1$

$x = -0.23$ to 2 dp or $x = -1.77$ to 2dp.

EXERCISE 17L

Solve the equations using completing the square method to 2 decimal places where appropriate.

1) $5x^2 + 15x + 10 = 0$
2) $4x^2 - 6x + 1 = 0$
3) $7x^2 - 14x + 6 = 0$
4) $\frac{1}{2}n^2 + 5n + 6 = 0$

17.11 TURNING POINT

The vertex of a parabola (quadratic curve) is the place where it **turns**. It is therefore called the turning point. The perpendicular line that passes through the vertex is also known as the axis of symmetry of the parabola.

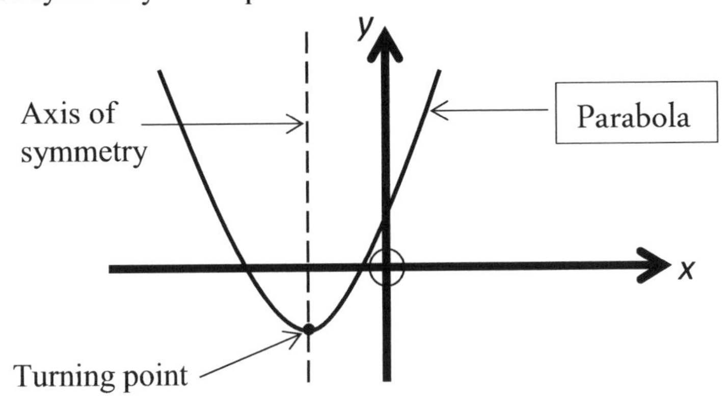

When equations are written in the form $\mathbf{y = a(x + p)^2 + q}$, the quadratic graph formed will have axis of symmetry x = - p and turning point (-p, q).

Example 1: **a)** Write $3x^2 + 6x + 7$ in the form $\mathbf{a(x + p)^2 + q}$
b) Find the turning point of the graph
c) Write the equation of the line of symmetry

Solution:

a) Factor out 3 in $3x^2 + 6x + 7$ $= 3\left[(x^2 + 2x + 7/3)\right]$

$= 3\left[(x + 1)^2 + 4/3\right]$

$= \mathbf{3\,(x + 1)^2 + 4}$

b) The minimum value is obtained when the expression in the bracket is equal to zero. This happens when x = -1. Therefore, the turning point occurs at (-1, 4)

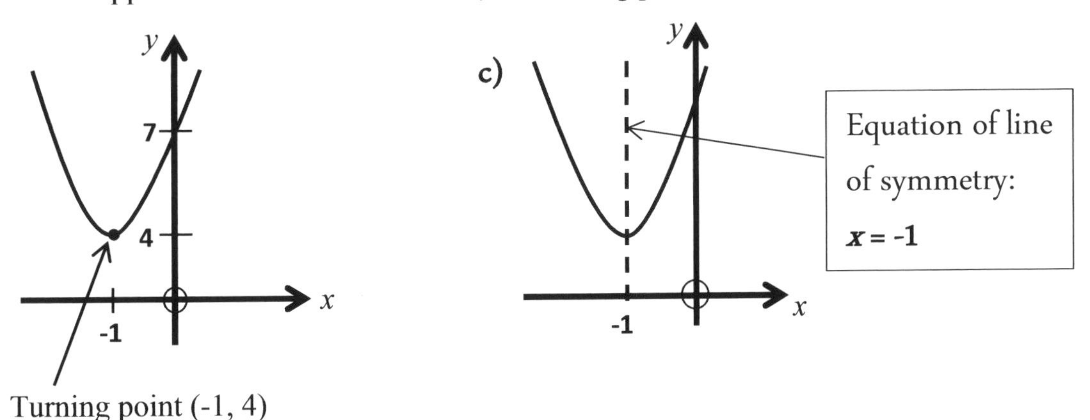

17.12 SKETCHING THE GRAPH OF A QUADRATIC EQUATION IN COMPLETED SQUARE FORM

$5x^2 + 10x + 2 = \mathbf{5(x + 1)^2 - 3}$ in completed square form

1) The shape of the graph is U- shaped, since it has a positive gradient (5).
2) The turning point is **(-1, -3)** and the axis of symmetry is the line x = -1.
3) The minimum value is y = -3
4) The y – intercept occurs when x = 0. In this case, it is **y = 2**
5) The solutions of the quadratic equation are where the graph cuts the x – axis. This occurs when y = 0. We need to solve the equation to find out the two points.

Therefore, $5(x + 1)^2 - 3 = 0$
add 3 to both sides
$5(x + 1)^2 = 3$
divide both sides by 5
$(x + 1)^2 = 3/5 = 0.6$
$x + 1 = \pm\sqrt{0.6}$
subtract 1 from both sides
$x = \sqrt{0.6} - 1$ or $x = -\sqrt{0.6} - 1$
x = - 0.23 to 2 dp or **x = -1.77** to 2dp.

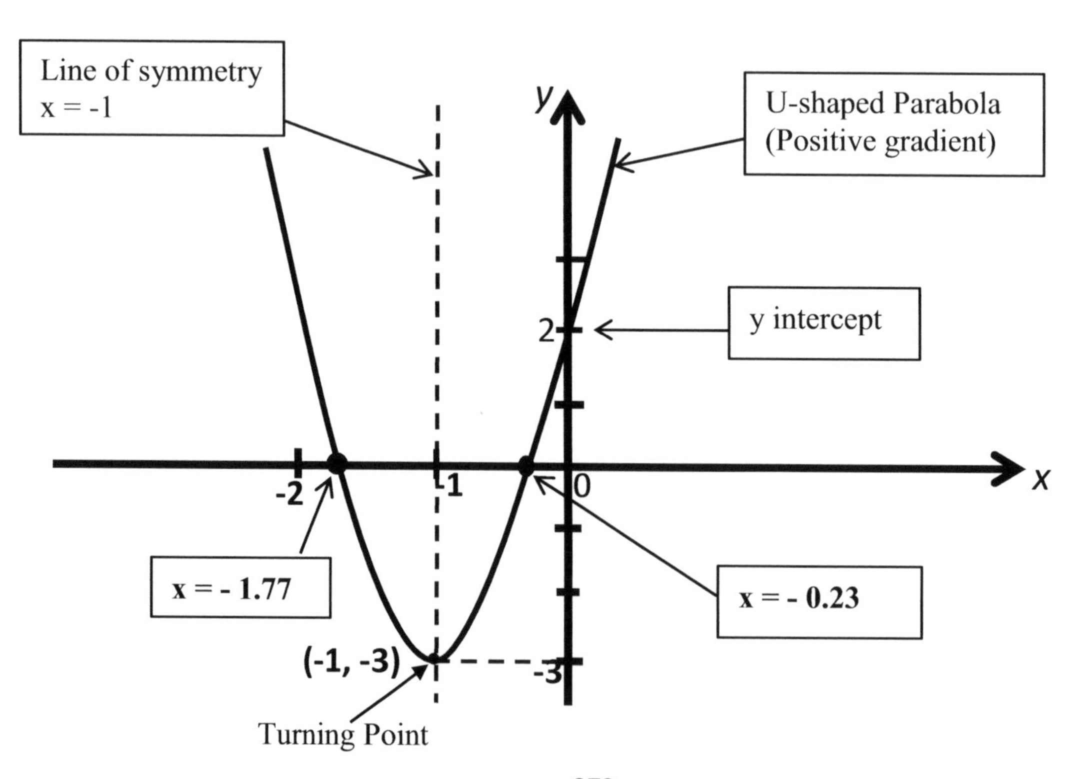

EXERCISE 17M

Write the quadratic equations in the form $a(x + p)^2 + q$

1) $2x^2 + 6x + 7$
2) $3x^2 - 9x + 1$
3) $4b^2 + 4b + 3$
4) $7c^2 - 21c + 8$
5) $5x^2 - 10x - 5$
6) $9x^2 + 9x - 5$

7) In questions 1 – 6 above, work out the values of a, p and q.

8) Sketch the graphs of (i) $2x^2 + 6x + 7$
(ii) $3x^2 - 9x + 1$

9) Write down (i) the turning points in questions 1 - 6
(ii) the equation of the line of symmetry in questions 1 - 6

10) a Form a quadratic equation using the lengths of the triangle below
b Find the solutions of this equation by completing the squares.
c Write down the lengths of the triangles in centimetres to 1 decimal place.

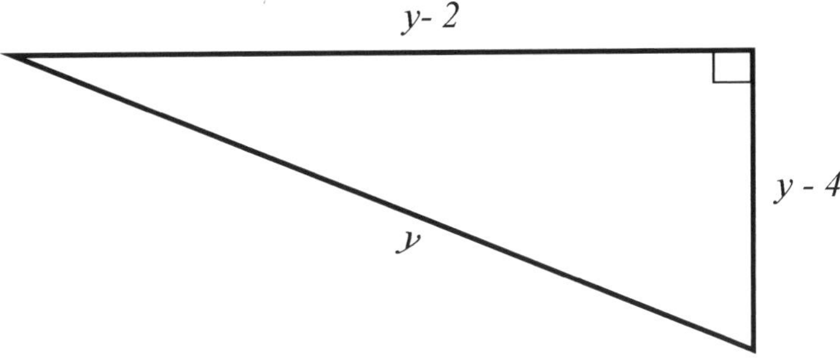

11) Solve the quadratic equations below.
a) $2x^2 + 6x + 7 = 0$
b) $3x^2 - 9x + 1 = 0$
c) $4b^2 + 4b + 3 = 0$
d) $7c^2 - 21c + 8 = 0$
e) $5x^2 - 10x - 5 = 0$
f) $9x^2 + 9x - 5 = 0$

12) a Find ***c*** and ***d*** such that $3 + 6x - 3x^2 = d - (x - c)^2$
b Solve the quadratic equation $3 + 6x - 3x^2 = 0$
c Write down the turning point on the graph.
d Equation of line of symmetry

17.13 HARDER SIMULTANEOUS EQUATIONS

In this section, we shall consider solving linear and non-linear simultaneous equations by graphical and algebraic methods.

GRAPHICAL METHOD

Example 1:
Solve the pair of simultaneous equations $y = x^2 + 1$ and $y = 2.5$

Solution: First draw a table of values for the coordinate pairs and then draw the graph of $y = x^2 + 1$.
Choose your x values and find the corresponding y values.

x	-3	-2	-1	0	1	2	3
y	10	5	2	1	2	5	10

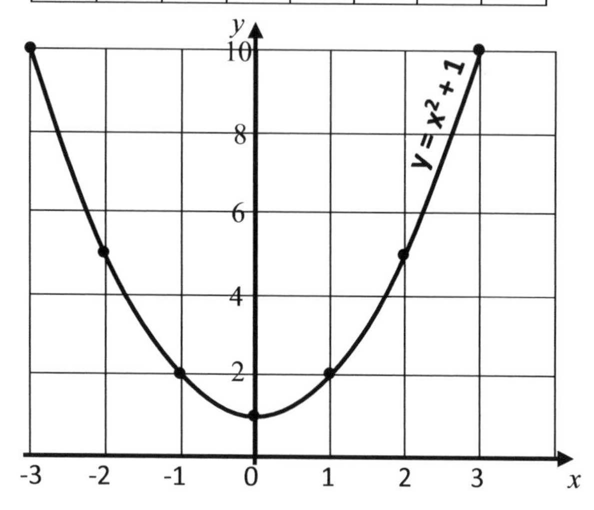

Next step is to draw the graph of $y = 2.5$ on the same graph.

Now, find where the two lines meet and record the coordinates.

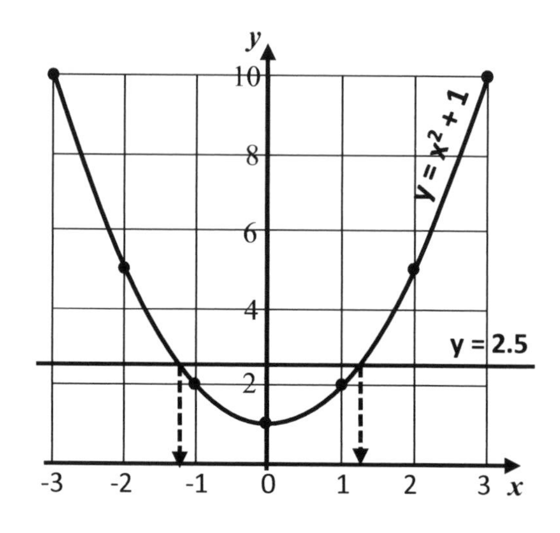

There are two points of intersection: **(-1.22, 2.5)** and **(1.22, 2.5).** ✓

They are also the solutions to the simultaneous equations $y = x^2 + 1$ and $y = 2.5$.

ALGEBRAIC METHOD

Example 2:
Solve $y = x^2 + 5x + 5$ and $y = 10$

Solution: Since **y** is the subject of both equations, it means that $x^2 + 5x + 5 = 10$

Take 10 away from both sides

$x^2 + 5x + 5 - 10 = 0$
$x^2 + 5x - 5 = 0$

The left-hand side cannot factorise, so we use the quadratic formula:

$$x = \frac{-b \pm \sqrt{b^2 - 4ac}}{2a}$$

a = 1, b = 5 and c = -5

$$x = \frac{-5 \pm \sqrt{5^2 - (4 \times 1 \times -5)}}{2 \times 1}$$

$$x = \frac{-5 \pm \sqrt{25 + 20}}{2}$$

$$x = \frac{-5 \pm \sqrt{45}}{2}$$

$x = \dfrac{-5 + \sqrt{45}}{2}$ or $x = \dfrac{-5 - \sqrt{45}}{2}$

x = 0.854 or x = -5.854

When x = 0.854, y = 10
When x = -5.854, y = 10 ✓

Same answers as with the graphical method.

Example 3:
Solve the simultaneous equations.
$y = x^2 - 6$ and $y = 2x - 3$.

Solution: **y** is the subject of both equations. Logically, $x^2 - 6 = 2x - 3$

$x^2 - 6 - 2x + 3 = 0$
$x^2 - 2x - 3 = 0$

Factorising gives (x - 3) (x + 1)

Therefore, (x – 3) (x + 1) = 0

Either x – 3 = 0, which means that **x = 3**
or, x + 1 = 0, which means that **x = -1**.

Substitute in any of the original equations of your choice.
From y = 2x – 3: when **x = 3, y = 3**
From y = 2x – 3: when **x = -1, y = -5**.

The two points of intersection are:
(3, 3) and (-1, -5). ✓

Example 4: Find the coordinates of the points of intersection of the simultaneous equations.

$y^2 + 4^2 = 8x$ and $y + 3 = 2x$

Solution:
Rearrange the equations to
$y^2 = 8x - 16$ and $y = 2x - 3$

Since the subject of the formulae are **not** the same (y^2 and y), we substitute $y = 2x - 3$ into $y^2 = 8x - 16$ to give

$(2x - 3)^2 = 8x - 16$

$(2x - 3)(2x - 3) = 8x - 16$

$4x^2 - 12x + 9 = 8x - 16$

$4x^2 - 12x + 9 - 8x + 16 = 0$

$4x^2 - 20x + 25 = 0$

FACTORISE

$(2x - 5)(2x - 5) = 0$

$2x - 5 = 0$, $\mathbf{x = 2.5}$.

Put $x = 2.5$, into $y = 2x - 3$
to give $\mathbf{y = 2}$.

The coordinates of the points of intersection are **(2.5, 2)**. ✓

Example 5:
a) Find the coordinates of the point of intersection of the equations
$x + y = 5$ and $x^2 + y^2 = 17$

b) Sketch the diagram/graph.

Solution:
From $x + y = 5$, $x = 5 - y$

Substitute $x = 5 - y$ in $x^2 + y^2 = 17$

$(5 - y)^2 + y^2 = 17$

Expanding $(5 - y)^2$ gives $25 - 10y + y^2$

$25 - 10y + y^2 + y^2 = 17$

$25 - 10y + 2y^2 - 17 = 0$

$2y^2 - 10y + 8 = 0$

Factorising gives

$(2y - 8)(y - 1) = 0$

Either $2y - 8 = 0$, so $y = 4$
Or $y - 1 = 0$, so $y = 1$

When $y = 4$, $x = 1$and
When $y = 1$, $x = 4$

Therefore, the points of intersection are

(1, 4) and **(4, 1)** ✓

b)

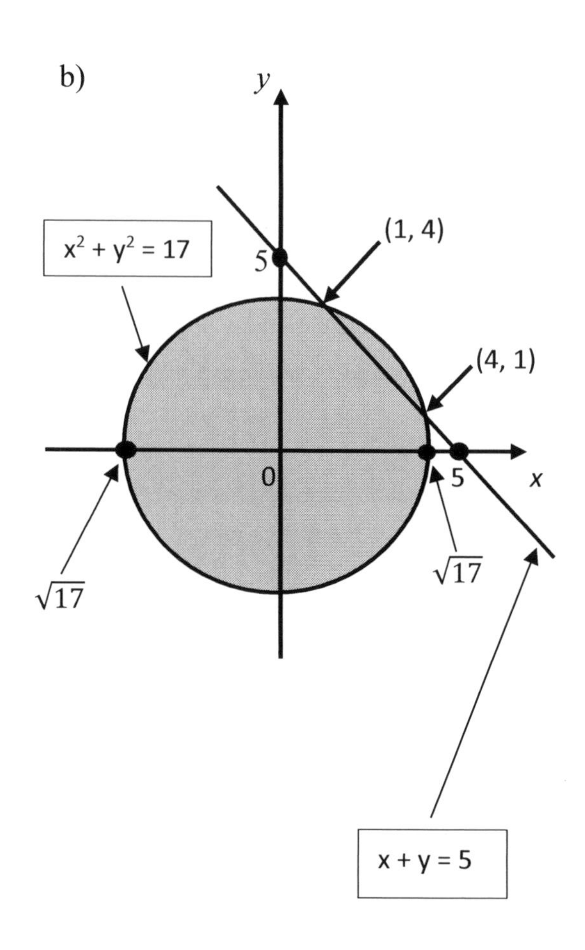

EXERCISE 17N

Solve these pairs of simultaneous equations.

1) $y = x^2$
$y = 1$

2) $y = x^2$
$y = 4$

3) $y = x^2$
$y = 2x + 3$

4) $y = x^2$
$y = -x + 12$

5) $y = x^2$
$y = x + 12$

6) $y = 2x^2 + 3x - 5$
$y = 7x + 25$

7) $y = 2x^2 + 3x - 5$
$y = -3x + 3$

8) $x^2 + y^2 = 25$
$y = \frac{4}{3}x$

9) $x^2 + y^2 = 25$
$y = \frac{1}{2}x + 5$

10) Find the points of intersection of the simultaneous equations below. Round to 2 decimal places.

$x^2 + y^2 = 36$
$y = 2x - 4$

17.14 CHANGE OF SUBJECT

In this section, we shall look at rearranging a formula to make a variable, the subject. The letter always on its own on one side of the equation is known as the **subject of the formula**.

Tips on changing the subject of formula are the same procedure when solving equations.

Example 1: Make w the subject of the formula $y = 3w + 7$.

Solution: Take 7 away from both sides.
$y - 7 = 3w$

Finally, divide both sides by 3.

$\frac{y-7}{3} = w$ ✓

Example 2: Make c the subject of the formula $p = 4c$.

Solution: Divide both sides by 4.

$c = \frac{P}{4}$ ✓

Example 3: Make t the subject of the formula $k = 3w + 5t$

Solution:

Subtract 3w from both sides.
$k - 3w = 5t$

Divide both sides by 5
$\frac{k-3w}{5}$ ✓

Example 4: Make c the subject of the formula $b = 4c^2 - 10$.

Solution: Add 10 to both sides.
$b + 10 = 4c^2$

Divide both sides by 4.
$\frac{b+10}{4} = c^2$

Square root the left side to get rid of the square on c.

$c = \sqrt{\frac{b+10}{4}}$ ✓

Example 5: Make w the subject of the formula $d = \sqrt{4w + k}$

Solution: To remove the square root sign, square d. $d^2 = 4w + k$

Subtract k from both sides.
$d^2 - k = 4w$

Divide both sides by 4.

$\frac{d^2 - k}{4} = w$ ✓

Example 6: Make b the subject of the formula $c = \frac{1}{4}b + 20$

Solution: Subtract 20 from both sides.
$c - 20 = \frac{1}{4}b$

Multiply both sides by 4.
$4(c - 20) = b$

$4c - 80 = b$ ✓

Example 7: Make w the subject of the formula $t = \frac{w^2}{p} - y$

Solution: Add y to both sides.
$t + y = \frac{w^2}{p}$
Multiply both sides by p.
$p\,(t + y) = w^2$
$pt + py = w^2$

Square root the left side to remove the square on w.

$w = \sqrt{pt + py}$ ✓

Example 8: Make w the subject of the formula $5(3w - c) = 8(w + d)$

Solution: Expand the brackets.
$15w - 5c = 8w + 8d$
Add 5c to both sides
$15w = 8w + 8d + 5c$

Subtract 8w from both sides.
$15w - 8w = 8d + 5c$
$7w = 8d + 5c$
Divide both sides by 7

$w = \frac{8d+5c}{7}$ ✓

Example 9: Make k the subject of the formula $p(qk - r) = e(k + t)$

Solution: Expand both sides
$pqk - pr = ek + et$
$pqk - ek = et + pr$
Factorise
$k(pq - e) = et + pr$
Divide both sides by $(pq - e)$
$k = \frac{et+pr}{pq-e}$ ✓

EXERCISE 17O

1) Make **w** the subject of the formula.

a) $c = w + 7$
b) $y = 6w - 2$
c) $y = aw + t$
d) $b = \frac{2}{w}$

e) $4t + 5f = t + 6w$
f) $k = 7w - 7f$
g) $k = w^2 + 18$
h) $m = \frac{1}{5}w^2 + y$

i) $a = \sqrt{7w + k}$
j) $b = cw^2 - 10.$
k) $t = \frac{w^2}{9} - y$
l) $a(3w - c) = 7(w + d)$

2) Make **c** the subject of the formula.

a) $t = g - \frac{y}{5c+w}$

b) $\frac{m+c}{m-c} = r$

c) $g = 5n\sqrt{\frac{c}{w}}$

d) $hc^2 - 5r = e + pc^2$

e) $\frac{y}{n} = \frac{c}{b(a-c)}$

f) In question 2e, find the value of c when $a = 8$, $b = 4$, $y = 7$, $n = 4$

18 Iterations

This section covers the following topics:

- Process of iteration
- Solutions using iterations

LEARNING OBJECTIVES

By the end of this unit, you should be able to:

a) Find approximate solutions to equations numerically using iterations

b) Understand the process of iteration.

18.1 Iterative process

An **iterative method** is a mathematical procedure that generates a sequence of improving approximate solutions for a class of problems.

Iteration is an act of repeating a process with the aim of approaching a desired goal, target or results. Each repetition of the process is also called ***iteration.***

An iterative method is called ***convergent*** if the corresponding sequence converges for given initial approximations.

In the process of finding the root of an equation (or a solution of a system of equations), an iterative method uses an initial guess to generate successive approximations to a solution.

In contrast, direct methods attempt to solve the problem by finite sequence of operations.

Example 1: Solve the equation $x^3 - 3x + 5 = 0$ using iteration method, correct to 2 dp.

Steps to iteration

Step1: The equation must be ***re-arranged*** to form an iterative formula.

$x^3 - 3x + 5 = 0$

add 3x to both sides

$x^3 + 5 = 3x$

subtract 5 from both sides

$x^3 = 3x - 5$

$x = \sqrt[3]{3x - 5}$

The iterative formula is

$$x_{n+1} = \sqrt[3]{3x_n - 5}$$

Step 2: Guess or choose the starting value. This could be given or not.
Let the **starting** value be 5. Therefore, $\boldsymbol{x_1} = 5$

Step 3: Substitute the value of x_1 into the iterative formula

$$x_{n+1} = \sqrt[3]{3x_n - 5}$$

Use $x_1 = 5$ in the iterative formula above to find x_2

$$x_2 = \sqrt[3]{(3 \times \mathbf{5}) - 5} \qquad = 2.1544..........$$

$$x_3 = \sqrt[3]{(3 \times 2.154.....) - 5} = 1.135...........$$

Continue with the iteration process until the values tend to converge at a particular number. That is the required degree of accuracy.

$x_4 = -1.168.......$
$x_5 = -2.041.......$
$x_6 = -2.232.......$
$x_7 = -2.269.......$
$x_8 = -2.277.......$
$x_9 = -2.278.......$
$x_{10} = -2.278.......$
$x_{11} = -2.279.......$
$x_{12} = -2.279.......$

There is no point continuing the iteration process at this point as the values converge to -2.28 to 2 decimal places. Therefore, the solution is $\boldsymbol{x = -2.28}$ to 2 decimal places.
Note: This could be quite tedious due to time constraints. A quicker method of doing this is by using a scientific calculator with the ANS sign.

Type in the first approximation **(x_1)** into your calculator and press the = sign button. This sets it up for use with the iteration formula while entering 'ANS' and pressing the '= sign' button each time to get x_3, x_4, x_5..........

Example 2: Work out the first five iterations using the iterative formula $x_{n+1} = 4x_n - 3$ with $x_1 = 2$.
Solution: $x_2 = 4 \times 2 - 3 = \mathbf{5}$, $x_3 = 4 \times 5 - 3 = \mathbf{17}$, $x_4 = \mathbf{65}$, $x_5 = \mathbf{257}$ and $x_6 = \mathbf{1025}$.

Facts about iteration

If the x value does not converge to a particular number during iteration, re-arrange the equation in another way.

The equation $x^3 - 7x + 3 = 0$ can be re-arranged in the following ways:

1) $x^3 - 7x + 3 = 0$

Make this x the subject of the formula
add (**7*x***) to both sides
$x^3 + 3 = 7x$

divide by **7** on both sides

$$x = \frac{x^3 + 3}{7}$$

Using the starting number as **4**, the solution will not tend to a limit. It is **diverging** as the values are not tending to a particular number. In this case, we must re-arrange using the other unknown in the other expression of the same equation (though same letter, x).

2) $x^3 - 7x + 3 = 0$

Make this x the subject of the formula.
add (7x) to both sides
$x^3 + 3 = 7x$

Subtract 3 from both sides
$x^3 = 7x - 3$

$$x = \sqrt[3]{7x - 3}$$

Using the value of $\mathbf{x_1} = \mathbf{4}$ as the starting number, x equals 2.40 to 2 decimal places after several iterations. It tends to **converge** at x = 2.40.

Finally, if the starting number is not given, you must estimate its value and use it in the iteration formula created. Consult your teacher for assistance in using the scientific calculator as you will continue the process until the required degree of accuracy is obtained.

Care must also be taken to enter the correct **starting number** and other forms of calculator manipulations.

Using the scientific calculator for subsequent iterations is vital to avoid unnecessary time-wasting which could lead to frustrations and mistakes.

EXERCISE 18A

1) Work out the first five iterations of the iterative formula $x_{n+1} = 2x_n + 3$.
Use the starting number as 3.

2) (a) Show how to rearrange the equation $x^3 + 7x - 3 = 0$ to give $\mathbf{x} = \frac{3}{x^2 + 7}$

(b) If $x_1 = 5$, find a solution to the equation using iteration correct to 2 decimal places.

(c) Show that there is another way of rearranging the equation $x^3 + 7x - 3 = 0$ to give $x = \sqrt[3]{3 - 7x}$. Show all working out.

3) A field is in the form of a rectangle. The width is 3m less than the length and the area of the field is $64m^2$.

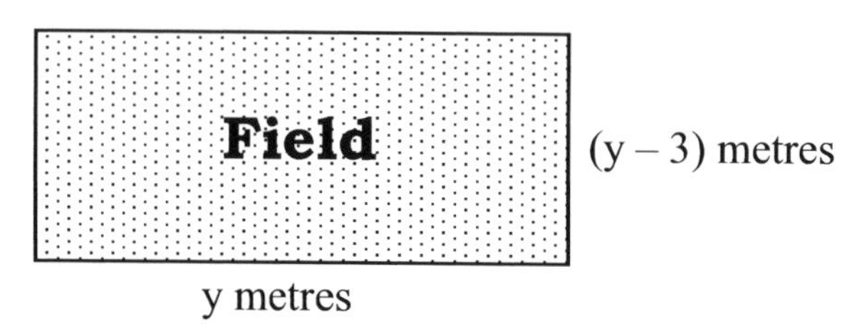

(a) Write an expression for the area of the field.
(b) Write a quadratic equation for the area of the field
(c) Write an iterative formula and (d) Find the length of the field ***y***, using the iterative formula to 2 decimal places. Let the starting number = 2.

4) Look at this equation: $x^3 - 8x + 7 = 0$

(a) Show that the equation can be rearranged to give: $x = \sqrt[3]{8x - 7}$
(b) Write the iterative formula.
(c) Using **4** as the starting number find a solution using the iterative formula, correct to 2 decimal places.

5) By using a reasonable starting number, find a solution to the quadratic equation $x^2 - x - 7 = 0$ using iteration. Show all working out and correct to 3 significant figures

6) $x_{n+1} = \frac{x_n^2 + 7}{3}$ is the iterative formula used to find a solution to the quadratic equation $x^2 - 3x + 7 = 0$.

If $x_1 = 2$ as the starting no, what happens with the iteration?

19 Quadratic Inequalities

This section covers the following topics:

- Representing quadratic inequalities
- Solving quadratic inequalities

LEARNING OBJECTIVES

By the end of this unit, you should be able to:

a) Understand and represent quadratic inequalities on a number line

b) Solve quadratic inequalities

19.1 INEQUALITIES ON A NUMBER LINE

Recap: The number line can be used to show solutions to linear inequalities. The following conventions show how to represent inequalities.

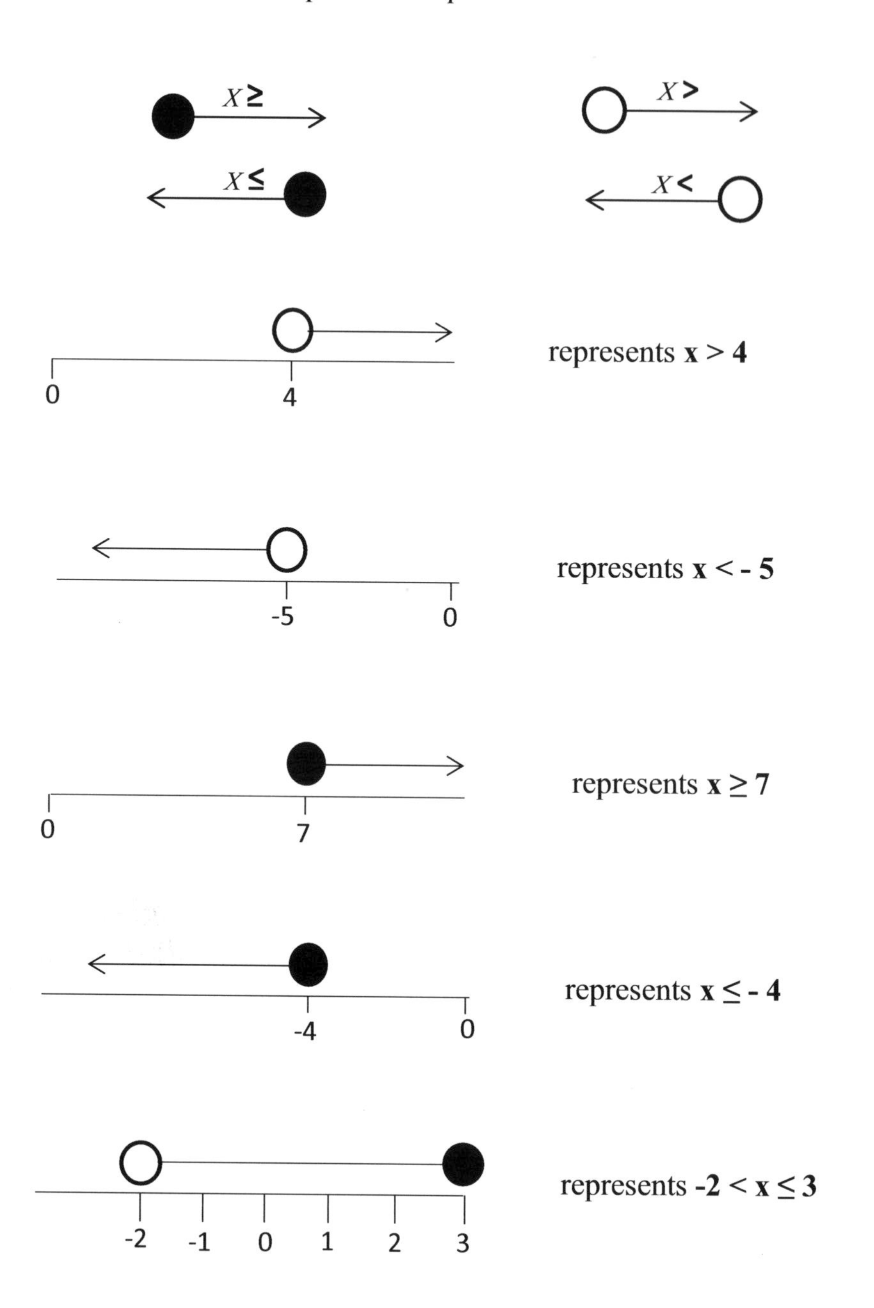

19.2 SOLVING QUADRATIC INEQUALITIES

When solving inequalities, imagine the inequality sign to be an **equal sign** and solve the equation. However, remember to put back the inequality sign when completing the equation. The inequality sign may change to the opposite sign if: you are dividing or multiplying by a negative number.

Example 1

Solve this quadratic inequality $x^2 < 25$ and represent the solution on a number line.

Solution:
Imagine the inequality sign ($<$) to be an equal sign.
$x^2 = 25$
$x = \pm\sqrt{25}$

The solutions are $x = +5$ and $x = -5$

Now, change the equality sign to the inequality sign ($<$).

The solution $x = 5$ would suggest the condition $x < 5$.

Therefore, $\mathbf{x < 5}$ satisfies the inequality $x^2 < 25$.

From the solution $x = -5$, the condition to be obtained cannot be $x < -5$ as this **does not** satisfy the inequality $x^2 < 25$. For example, if x is -6 which is less than -5, $(-6)^2$ is 36 and 36 is not less than 25.

We must now reverse the sign to $\mathbf{x > -5}$ as it satisfies the inequality $x^2 < 25$.
The solutions to the quadratic inequality $x^2 < 25$ are $\underline{\mathbf{x < 5}}$ and $\underline{\mathbf{x > -5}}$.

To represent the solutions on a number line, you must remember that $x > -5$ can be written as $-5 < x$.

Therefore, the solution to $x^2 < 25$ could be written as $\mathbf{-5 < x < 5}$ and will be represented on a number line as

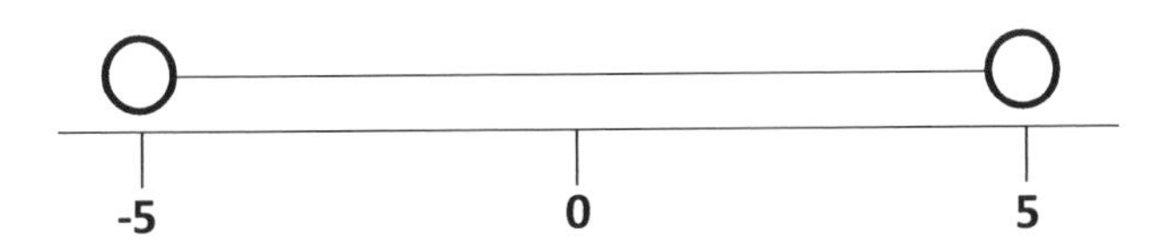

Example 2
Solve the quadratic inequality $\mathbf{x^2 \geq 9}$ and show the solution on a number line.

Solution:
Change the inequality sign to the equality sign, (=)
$x^2 = 9 \Longrightarrow x = \pm\sqrt{9}$

Therefore, x = 3 and x = -3

Looking at the positive value of x would suggest the condition $x \geq 3$.

Therefore, $\mathbf{x \geq 3}$ satisfies the inequality $x^2 \geq 9$.

From the solution x = -3, the condition obtained cannot be $x \geq -3$ as this **does not** satisfy the inequality $x^2 \geq 9$. The inequality sign must now be reversed to $\mathbf{x \leq -3}$ and this satisfies the inequality $x^2 \geq 9$.

The solutions to the quadratic inequality $x^2 \geq 9$ is $\mathbf{x \geq 3}$ and $\mathbf{x \leq -3}$.
On a number line, the solution will be represented as

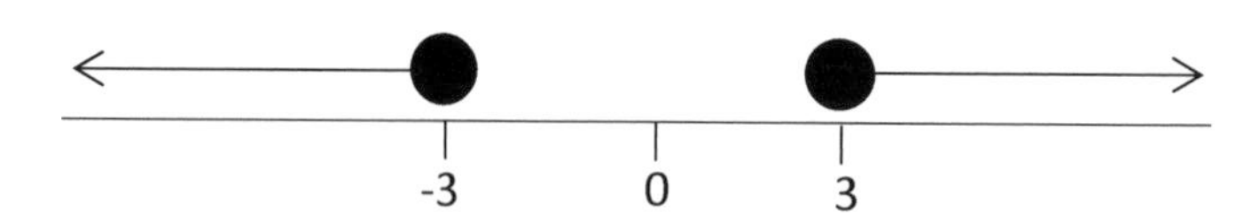

Example 3 Solve $x^2 - 3x - 10 < 0$

Factorise the left side to $(x + 2)(x - 5)$ ⟶

Find two numbers that will multiply to give -10 but will add to give -3. The numbers are -5 and 2

If $(x - 5)(x + 2) = 0$,
Either $x - 5 = 0 \Longrightarrow x = 5$
or $x + 2 = 0 \Longrightarrow x = -2$

From the solution x = 5, the condition would suggest the solution $\mathbf{x < 5}$ which really satisfies the inequality $x^2 - 3x - 10 < 0$.

From the solution x = - 2, the condition obtained cannot be $x < -2$ as this **does not** satisfy the inequality $x^2 - 3x - 10 < 0$. The sign is then reversed to $\mathbf{x > -2}$ which will now satisfy the inequality given.

Therefore, the solution to the quadratic inequality $x^2 - 3x - 10 < 0$ is $\mathbf{-2 < x < 5}$

EXERCISE 19A

Solve the quadratic inequalities below **and** represent their solutions on a number line.

1) $x^2 < 36$
2) $x^2 < 4$
3) $x^2 < 100$
4) $x^2 > 36$
5) $x^2 > 4$
6) $x^2 > 100$
7) $x^2 \leq 64$
8) $x^2 \leq 49$
9) $x^2 \leq 100$
10) $x^2 \geq 64$
11) $x^2 \geq 1$
12) $x^2 \geq 100$

EXERCISE 19B

Solve the inequalities **and** show their solutions on a number line.

1) $x^2 + 9 < 25$
2) $x^2 - 10 < 26$
3) $x^2 - 10 > 54$
4) $x^2 + 7 \leq 56$
5) $x^2 + 20 < 101$
6) $x^2 - 3 \leq 13$
7) $2x^2 - 4 > 28$
8) $2x^2 + 7 \leq 207$
9) $x^2 + 2x - 35 < 0$
10) $x^2 + 5x - 6 \geq 0$
11) $x^2 - 7x - 30 \leq 0$
12) $x^2 + x - 20 > 0$
13) $x^2 + 4x < 32$

20 Surds

This section covers the following topics:

- Simplifying surds
- Solving problems involving surds

LEARNING OBJECTIVES

By the end of this unit, you should be able to:

a) Simplify surds
b) Multiply surds
c) Add and subtract surds
d) Rationalise the denominator
e) Expand brackets involving surds
f) Solve problems involving surds

Simply put, surds are roots of rational numbers. **Rational numbers** are numbers that can be put in the form $\frac{a}{b}$.

Examples of surds are $\sqrt{2}, \sqrt{3}, \sqrt{11}, \sqrt{19}$ etc. Numbers in surd form are the **exact values.**

20.1 SIMPLIFYING SURDS

Rule 1: $\sqrt{x} \times \sqrt{y} = \sqrt{xy}$

$\sqrt{2} \times \sqrt{3} = \mathbf{\sqrt{6}}$
$\sqrt{3} \times \sqrt{3} = \sqrt{9} = \mathbf{3}$

Therefore,
$\sqrt{7} \times \sqrt{7} = \mathbf{7}$
$\sqrt{2} \times \sqrt{2} \times \sqrt{2} = \mathbf{2\sqrt{2}}$

Rule 2: $p\sqrt{x} \times q\sqrt{y} = pq\sqrt{xy}$

$4\sqrt{3} \times 2\sqrt{5} = 4 \times 2 \times \sqrt{3} \times \sqrt{5}$
$= \mathbf{8\sqrt{15}}$

Rule 3: $\sqrt{x} \div \sqrt{y} = \sqrt{\frac{x}{y}}$

$\sqrt{8} \div \sqrt{2} = \sqrt{\frac{8}{2}} = \sqrt{4} = \mathbf{2}$

$\sqrt{100} \div \sqrt{5} = \sqrt{\frac{100}{5}} = \sqrt{20} = \mathbf{2\sqrt{5}}$

Rule 4: $p\sqrt{x} \div q\sqrt{y} = \frac{p}{q} \times \sqrt{\frac{x}{y}}$

$50\sqrt{8} \div 5\sqrt{2} = \frac{50}{5} \times \sqrt{\frac{8}{2}}$
$= 10\sqrt{4} = 10 \times 2 = \mathbf{20}$

Example 1: Simplify $\sqrt{20}$

To simplify a surd, find factors which are **square numbers**.

$\sqrt{20} = \sqrt{4 \times 5} = \sqrt{4} \times \sqrt{5} = \mathbf{2\sqrt{5}}$

Example 2: Simplify $\sqrt{48}$

$\sqrt{48} = \sqrt{16 \times 3} = \sqrt{16} \times \sqrt{3} = \mathbf{4\sqrt{3}}$

Example 3: Simplify $\sqrt{3} \times \sqrt{5}$

$= \sqrt{3 \times 5} = \mathbf{\sqrt{15}}$

Example 4: Simplify $5\sqrt{27} \times 2\sqrt{3}$

$= 5 \times 2 \times \sqrt{27} \times \sqrt{3}$
$= 10 \times \sqrt{81}$
$= 10 \times 9$
$= \mathbf{90}$

EXERCISE 20A

1) Simplify each expression and leave your answers in surd form where necessary.

a) $\sqrt{3} \times \sqrt{7}$
b) $\sqrt{2} \times \sqrt{3} \times \sqrt{5}$
c) $\sqrt{5} \times \sqrt{5}$
d) $\sqrt{7} \times \sqrt{7} \times \sqrt{7}$
e) $\sqrt{9} \times \sqrt{4}$
f) $\sqrt{11} \times \sqrt{12}$
g) $\sqrt{2} \times \sqrt{8} \times \sqrt{20}$
h) $12\sqrt{4} \times 3\sqrt{7}$
i) $\sqrt{100} \div \sqrt{2}$
j) $\sqrt{50} \div \sqrt{2}$

k) $5\sqrt{10} \div 20\sqrt{2}$
l) $2\sqrt{63} \times 3\sqrt{3}$
m) $2\sqrt{6} \times 4\sqrt{3}$
n) $\sqrt{20} \times \frac{2\sqrt{15}}{4\sqrt{10}}$

o) $\left[\frac{3}{\sqrt{5}}\right]^2$

2) Simplify each of these surds in the form $x\sqrt{y}$.

a) $\sqrt{8}$
b) $\sqrt{18}$
c) $\sqrt{40}$
d) $\sqrt{75}$
e) $\sqrt{112}$
f) $\sqrt{28}$

g) $\sqrt{320}$

3) Simplify each expression.

a) $x\sqrt{y} \times a\sqrt{b}$
b) $p\sqrt{q} \times c\sqrt{d}$
c) $a\sqrt{b} \div p\sqrt{q}$

4) Find the value of c that makes the equations correct.

a) $\sqrt{4} \times c = \sqrt{36}$

b) $2\sqrt{5} \times c = 6\sqrt{35}$

c) $\frac{20\sqrt{15}}{c\sqrt{3}} = 5\sqrt{5}$

20.2 ADDITION AND SUBTRACTION OF SURDS

Before surds can be added or subtracted, the number under the square root signs **must** be the same for all the surds. If they are not the same, simplify the surds.

Example 1: $3\sqrt{5} + 7\sqrt{5}$

Solution: the numbers under the square root sign is the same for both surds. Therefore, they can be added together.

$3\sqrt{5} + 7\sqrt{5} = (3 + 7)\sqrt{5} = \mathbf{10\sqrt{5}}$

Example 2: $4\sqrt{3} - \sqrt{3}$
$= (4 - 1)\sqrt{3} = \mathbf{3\sqrt{3}}$

Example 3: $4\sqrt{8} + \sqrt{50}$

Solution: the numbers under the square root sign are not the same. Therefore, simplify the surds until they have a common number under the square root sign.

$4\sqrt{8} + \sqrt{50}$

↓ ↓

$4 \times 2\sqrt{2} + 5\sqrt{2}$

$8\sqrt{2} + 5\sqrt{2}$

Note:
$\sqrt{8} = 2\sqrt{2}$
and
$\sqrt{50} = 5\sqrt{2}$

$\sqrt{2}$ is now common to both surds.

$= (8 + 5)\sqrt{2}$
$= \mathbf{13\sqrt{2}}$

EXERCISE 20B

Simplify each of these expressions as much as possible.

1) $2\sqrt{6} + 3\sqrt{6}$
2) $5\sqrt{5} + \sqrt{5}$
3) $3\sqrt{10} - \sqrt{10}$
4) $4\sqrt{11} + 7\sqrt{11}$
5) $3\sqrt{8} + \sqrt{18}$
6) $\sqrt{63} + \sqrt{175}$
7) $\sqrt{27} + 5\sqrt{3}$
8) $\sqrt{20} + \sqrt{45}$
9) $\sqrt{405} - \sqrt{80}$
10) $3\sqrt{6} + 9\sqrt{24} - \sqrt{6}$

20.3 RATIONALISING THE DENOMINATOR

To rationalise the denominator means that the denominator **should not** include surds.

Steps to rationalising the denominator include multiplying the numerator and denominator by a suitable square root.

Example 1: Rationalise the denominator of the expression $\frac{1}{\sqrt{5}}$.

Solution: Multiply both numerator and denominator by $\sqrt{5}$.

$$\frac{1}{\sqrt{5}} \times \frac{\sqrt{5}}{\sqrt{5}} = \frac{1 \times \sqrt{5}}{\sqrt{5} \times \sqrt{5}} = \frac{\sqrt{5}}{5} \checkmark$$

Example 2:

Rationalise the denominator $\frac{3}{2\sqrt{3}}$

Solution: Multiply the numerator and denominator by the surd only.

$$\frac{3}{2\sqrt{3}} \times \frac{\sqrt{3}}{\sqrt{3}}$$

$$= \frac{3 \times \sqrt{3}}{2\sqrt{3} \times \sqrt{3}} = \frac{3\sqrt{3}}{2 \times 3} = \frac{3\sqrt{3}}{6} = \frac{\sqrt{3}}{2} \checkmark$$

Example 3:

Rationalise the denominator $\frac{4}{3 + \sqrt{5}}$.

Solution: When the denominator contains a number with a plus or minus sign and a surd, **multiply** the numerator and denominator by the denominator with the reverse sign.

The denominator is $3 + \sqrt{5}$, therefore, multiply the numerator and denominator by **$3 - \sqrt{5}$.**

$$\frac{4}{3 + \sqrt{5}} \times \frac{3 - \sqrt{5}}{3 - \sqrt{5}}$$

$$= \frac{4 \times (3 - \sqrt{5})}{(3 + \sqrt{5})(3 - \sqrt{5})}$$

$$= \frac{12 - 4\sqrt{5}}{9 - 3\sqrt{5} + 3\sqrt{5} - 5)}$$

$$= \frac{12 - 4\sqrt{5}}{9 - 5} = \frac{12 - 4\sqrt{5}}{4} = \mathbf{3 - \sqrt{5}} \checkmark$$

EXERCISE 20C

Simplify by rationalising the denominator.

1) $\frac{1}{\sqrt{6}}$

2) $\frac{2}{\sqrt{5}}$

3) $\frac{7}{\sqrt{11}}$

4) $\frac{2}{4\sqrt{5}}$

5) $\frac{5\sqrt{8}}{10\sqrt{20}}$

6) $\frac{2}{5+\sqrt{2}}$

7) $\frac{3+\sqrt{5}}{\sqrt{5}}$

8) $\frac{7+6\sqrt{3}}{\sqrt{3}}$

9) $\frac{11-3\sqrt{5}}{\sqrt{5}}$

10) $\frac{2+3\sqrt{5}}{3+\sqrt{5}}$

11) $\frac{5+\sqrt{3}}{3+\sqrt{2}}$

12) $\frac{2+\sqrt{7}}{2-\sqrt{2}}$

20.4 BRACKETS AND SURDS

Example 1:
Expand and simplify $\sqrt{3}\,(\sqrt{2}+5)$

Solution: $\sqrt{3} \times \sqrt{2} + \sqrt{3} \times 5$
$= \boldsymbol{\sqrt{6} + 5\sqrt{3}}$

Example 2:
Expand and simplify $2\sqrt{5}\,(\sqrt{5} - 3\sqrt{7})$

Solution: $2\sqrt{5} \times \sqrt{5} - 2\sqrt{5} \times 3\sqrt{7}$
$= \mathbf{10 - 6\sqrt{35}}$

Example 3:
Expand and simplify
$(\sqrt{5}+\sqrt{3})\,(\sqrt{7}-\sqrt{2})$

Solution: Using the FOIL method,

$\sqrt{5} \times \sqrt{7} - \sqrt{5} \times \sqrt{2} + \sqrt{3} \times \sqrt{7} - \sqrt{3} \times \sqrt{2}$

$= \boldsymbol{\sqrt{35} - \sqrt{10} + \sqrt{21} - \sqrt{6}}$

Example 4: Expand and simplify
$(2-\sqrt{5})(\sqrt{5}-1)$

Solution: Using the FOIL method $2 \times \sqrt{5} - 2 - 5 + \sqrt{5}$

$= 2\sqrt{5} - 2 - 5 + \sqrt{5}$
$=2\sqrt{5} + \sqrt{5} - 7$

$= \boldsymbol{3\sqrt{5} - 7}$

EXERCISE 20D

Expand and simplify:

1) $\sqrt{2}(4+\sqrt{3})$

2) $\sqrt{3}(\sqrt{3}-4)$

3) $2\sqrt{5}(3\sqrt{3}+5)$

4) $3\sqrt{3}\,(4+\sqrt{3})$

5) $(\sqrt{2}+\sqrt{5})(\sqrt{5}-\sqrt{3})$

6) $(\sqrt{2}+\sqrt{7})^2$

7) $(3-\sqrt{5})(\sqrt{5}-2)$

8) $(5\sqrt{7}-\sqrt{2})(\sqrt{7}+2\sqrt{3})$

EXERCISE 20E

EXTENDED PROBLEMS

1) Work out the mean of the three numbers $\sqrt{18}+\sqrt{8}+\sqrt{32}$. Leave your answer in the form $a\sqrt{b}$. You must show all working out.

2) Simplify the following:

a) $\sqrt{45}$
b) $\sqrt{128}$
c) $\frac{4\sqrt{6}\times 2\sqrt{5}}{\sqrt{6}}$

d) $\frac{-5\sqrt{50}}{\sqrt{8}}$

3) The shape below is a rectangle.

$(\sqrt{3}+5)\ cm$

$\sqrt{5}\ cm$

a) Work out the exact area.
b) Work out the exact perimeter.

4) The circle below has a diameter of $\mathbf{6(\sqrt{20}-2)}\ cm$.

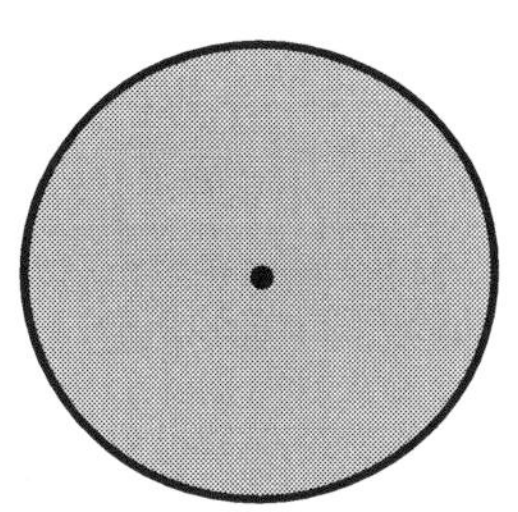

Work out
a) the exact area of the circle
b) the exact circumference of the circle.

5) Shape A has an area of 300 m^2.

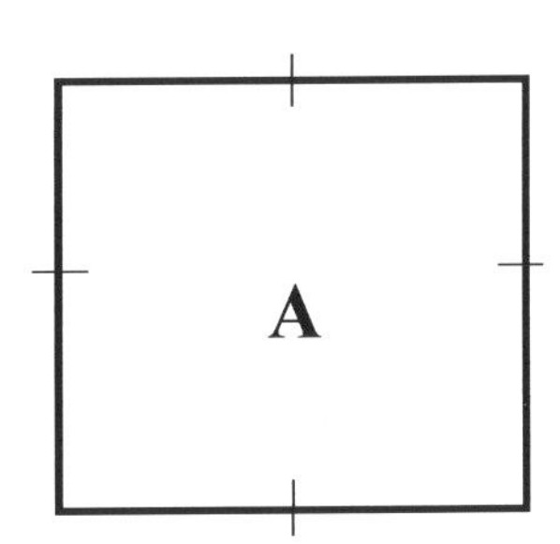

Work out the exact value of the perimeter of shape A.

21 Vectors

This section covers the following topics:

- Describing vectors
- Resultant vectors
- Vector geometry
- Parallel vectors

LEARNING OBJECTIVES

By the end of this unit, you should be able to:

a) Represent as column vectors
b) Find the resultant vector
c) Add and subtract vectors
d) Understand vector geometry
e) Find parallel vectors
f) Understand the conditions for vectors to be collinear

21.1 REPRESENTING VECTORS

Vector quantities have both **magnitude** and **direction**. Examples of vector quantities are velocity, force, acceleration, weight, momentum and displacement.

Velocity will only change when their magnitude and or direction changes.

Vectors can be represented in different ways:

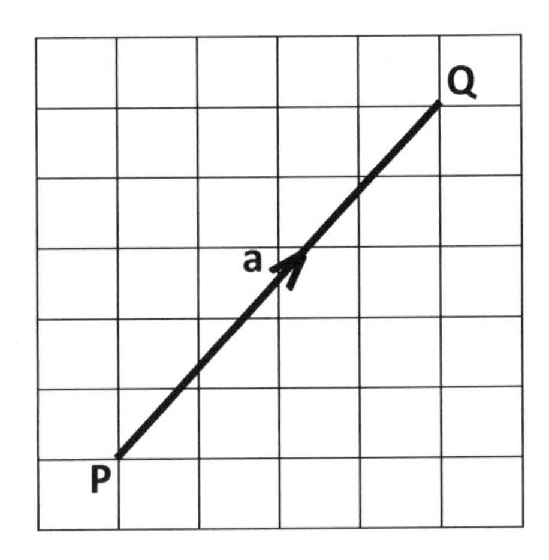

1) **Bold letter:** Vectors can be denoted by a bold letter.
In the diagram, **a** is the vectors. Some mathematicians would underline the bold letter if handwritten, like $\underline{\mathbf{a}}$.

2) **Arrow**: From the diagram, the vector could be represented as $\overrightarrow{PQ}$. The arrow indicates the direction from P to Q.

3) **Column vector** $\begin{pmatrix} x \\ y \end{pmatrix}$ $\mathbf{a} = \begin{pmatrix} 4 \\ 5 \end{pmatrix}$

Going in the opposite direction from Q to P, the vector would be represented as:

1) **-a**

2) $\overrightarrow{QP}$

3) $\begin{pmatrix} -4 \\ -5 \end{pmatrix}$

Vectors can be shown on a coordinate grid in **bold letter(s)** and as column pair(s).

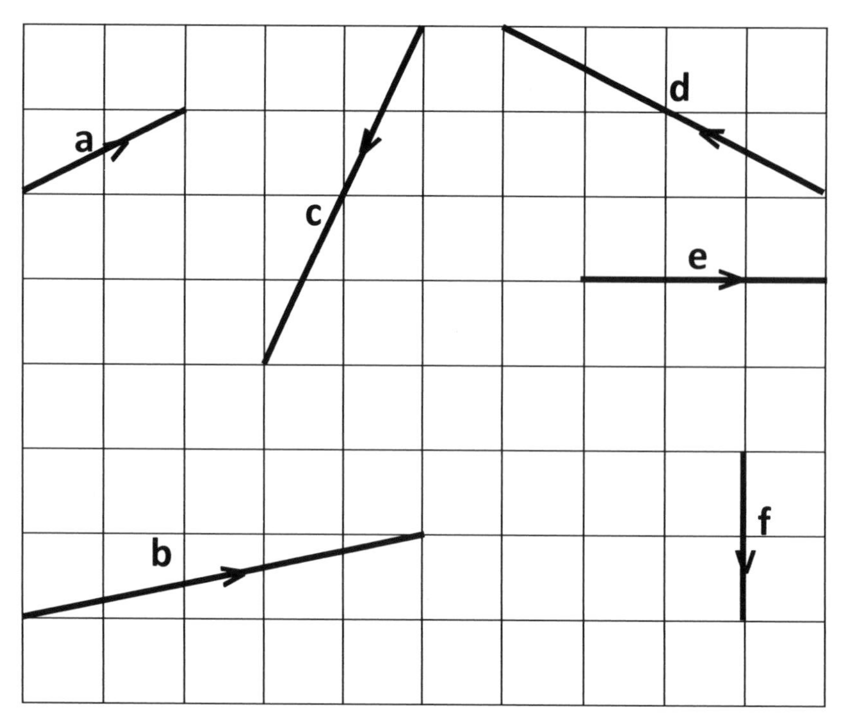

Fig. 21

The x and y elements give the displacement as measured from **tail to nose** using the direction of the arrow head.

From fig.21, the horizontal displacement is like the **x-coordinate.** If moving from left to right, the value is positive. If moving from right to left, the value is negative.

The vertical displacement is like the **y-coordinate**. If moving up, the value is positive, and when moving down, the value is negative.

The column vectors from fig.21 are:

$$\mathbf{a} = \begin{pmatrix} 2 \\ 1 \end{pmatrix} \quad \mathbf{b} = \begin{pmatrix} 5 \\ 1 \end{pmatrix} \quad \mathbf{c} = \begin{pmatrix} -2 \\ -4 \end{pmatrix} \quad \mathbf{d} = \begin{pmatrix} -4 \\ 2 \end{pmatrix} \quad \mathbf{e} = \begin{pmatrix} 3 \\ 0 \end{pmatrix} \quad \mathbf{f} = \begin{pmatrix} 0 \\ -2 \end{pmatrix}$$

21.2 MAGNITUDE AND DIRECTION OF A VECTOR QUANTITY

The magnitude (size) of a vector quantity can be found by using Pythagoras' theorem.

Example 1: Work out a) the magnitude and b) the direction of $\mathbf{a} = \begin{pmatrix} 5 \\ 3 \end{pmatrix}$

Solution:

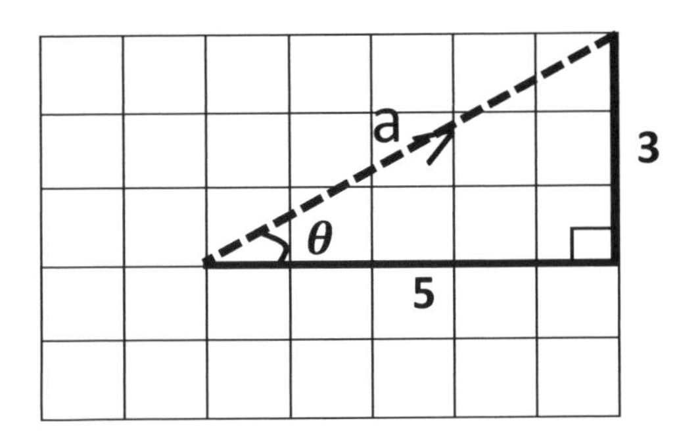

a) $\mathbf{a}^2 = 5^2 + 3^2$

$a^2 = 25 + 9$

$a^2 = 34$

$a = \sqrt{\mathbf{34}}$ or $a = 5.83$ to 2 d.p

b) The direction is calculated using trigonometry. Tan $\theta = \frac{3}{5} = 0.6$

$\theta = tan^{-1}\, 0.6$

$\theta = 30.96 \approx 31^0$

Example 2: Work out the magnitude of the vector $\mathbf{w} = \begin{pmatrix} -3 \\ -4 \end{pmatrix}$

Solution: $w^2 = (-3)^2 + (-4)^2$

$w^2 = 9 + 16$

$w^2 = 25$

$w = \sqrt{25} = \mathbf{5}$

Example 3: Work out the size and direction of the vector $\mathbf{b} = \begin{pmatrix} 0 \\ 7 \end{pmatrix}$

Solution:

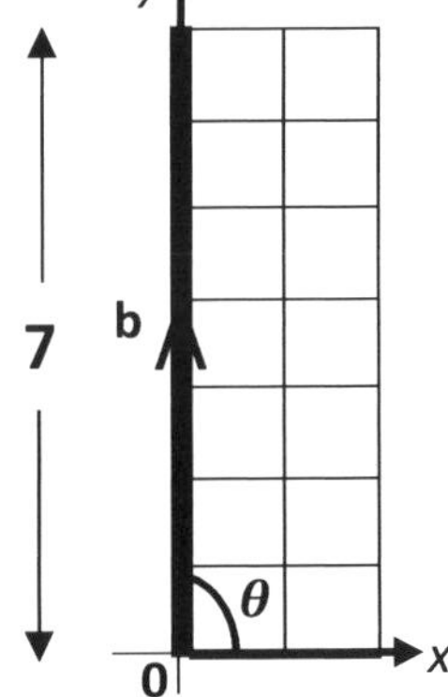
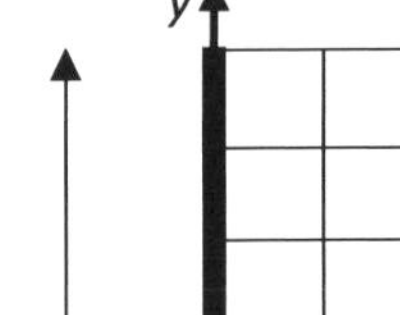

a) Size, which is the same as magnitude, is **7**.
There is **no** horizontal movement; only vertical.

b) The direction is upwards (vertical) which is 90°.

21.3 ADDITION AND SUBTRACTION OF VECTORS

Column vectors can be added together or subtracted. Also, vector representations by line segments can be added or subtracted. The result of addition or subtraction of vectors is known as the **resultant vector**.

Example 1: If $\mathbf{a} = \begin{pmatrix} 2 \\ 5 \end{pmatrix}$ and $\mathbf{b} = \begin{pmatrix} 3 \\ 2 \end{pmatrix}$, work out a) $\mathbf{a} + \mathbf{b}$ b) $\mathbf{b} - \mathbf{a}$.

Confirm by representing the vectors in a grid.

Solution:

a) $\mathbf{a} + \mathbf{b} = \begin{pmatrix} 2 \\ 5 \end{pmatrix} + \begin{pmatrix} 3 \\ 2 \end{pmatrix} = \begin{pmatrix} 2+3 \\ 5+2 \end{pmatrix} = \begin{pmatrix} \mathbf{5} \\ \mathbf{7} \end{pmatrix}$

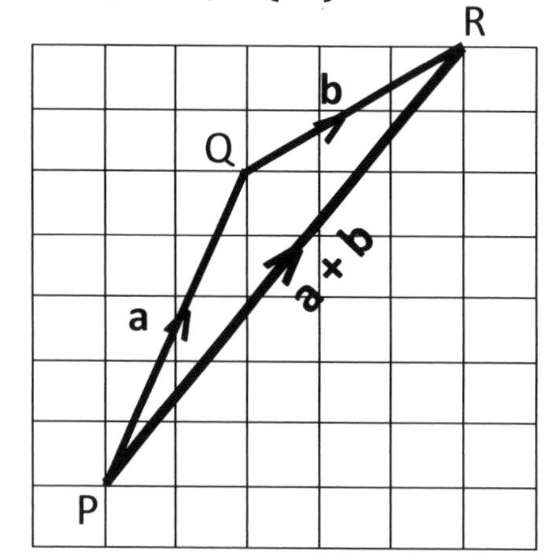

a + **b** is the resultant vector. This is moving **directly** from P to R. It has the same effect as moving from P to Q and then to R.

b) $\mathbf{b} - \mathbf{a} = \begin{pmatrix} 3 \\ 2 \end{pmatrix} - \begin{pmatrix} 2 \\ 5 \end{pmatrix} = \begin{pmatrix} 3-2 \\ 2-5 \end{pmatrix} = \begin{pmatrix} \mathbf{1} \\ \mathbf{-3} \end{pmatrix}$

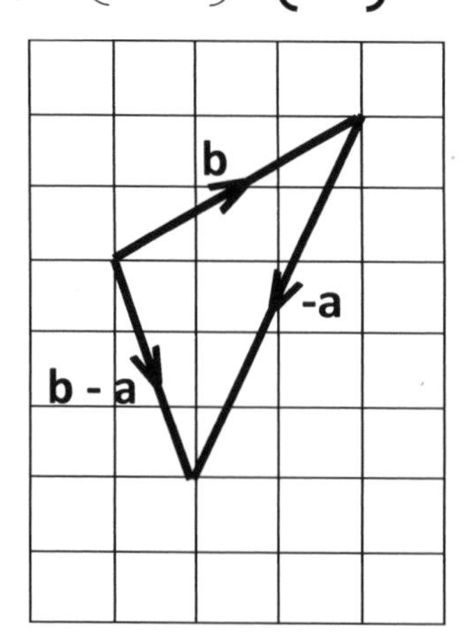

Example 2: Work out the resultant vector, AC.

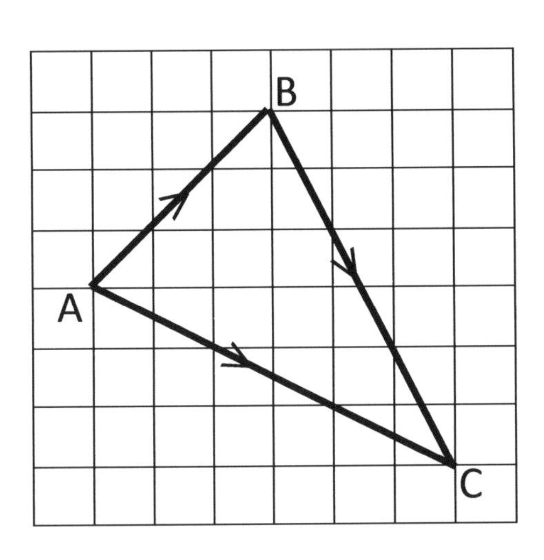

The combination of the two displacements AB and BC will give the resultant displacement, AC. In vector notations,

$\overrightarrow{AB} + \overrightarrow{BC}$ will give $\overrightarrow{AC}$

$\overrightarrow{AB} = \begin{pmatrix} 3 \\ 3 \end{pmatrix}$, $\overrightarrow{BC} = \begin{pmatrix} 3 \\ -6 \end{pmatrix}$

$\overrightarrow{AB} + \overrightarrow{BC} = \begin{pmatrix} 3+3 \\ 3+-6 \end{pmatrix} = \begin{pmatrix} 6 \\ -3 \end{pmatrix}$

The resultant vector $\overrightarrow{AC} = \begin{pmatrix} \mathbf{6} \\ \mathbf{-3} \end{pmatrix}$

Example 3: Write column vectors for the vectors shown in the diagram.

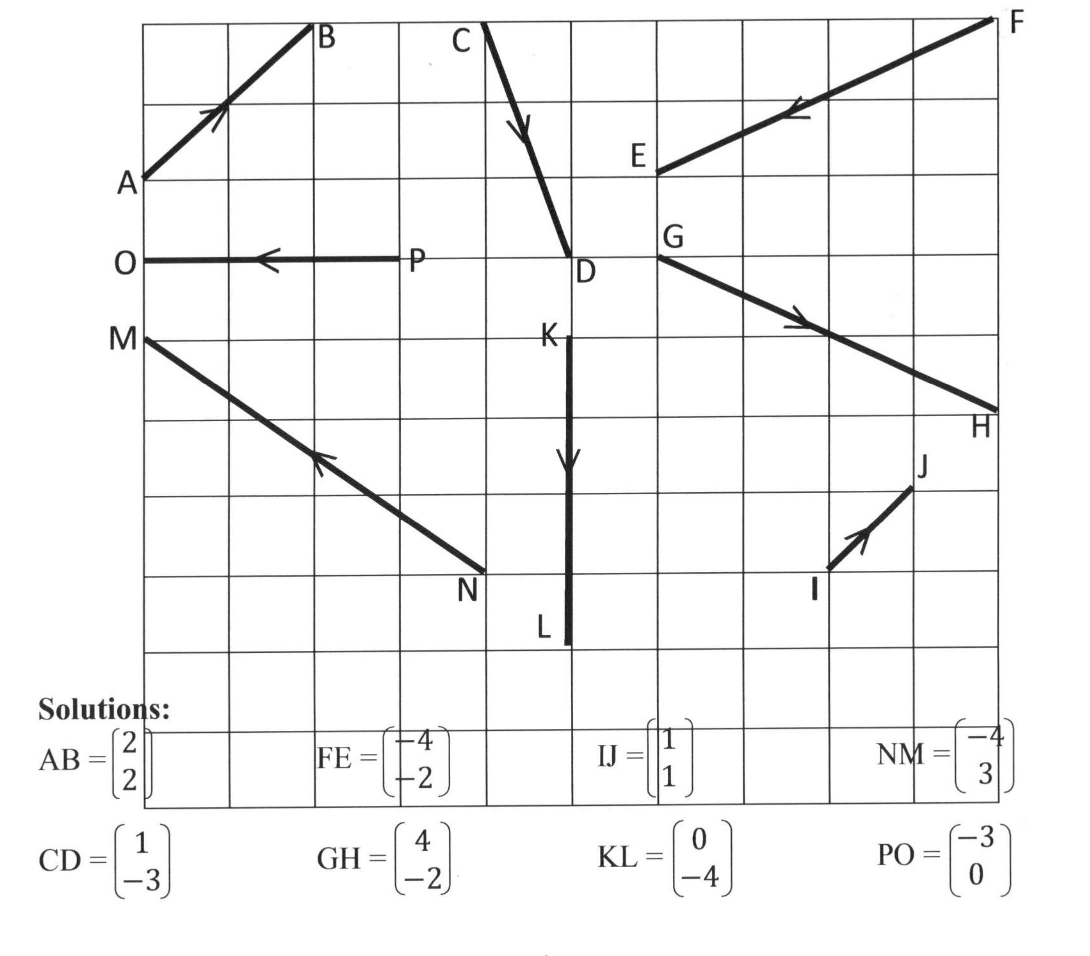

Solutions:

$AB = \begin{pmatrix} 2 \\ 2 \end{pmatrix}$ $FE = \begin{pmatrix} -4 \\ -2 \end{pmatrix}$ $IJ = \begin{pmatrix} 1 \\ 1 \end{pmatrix}$ $NM = \begin{pmatrix} -4 \\ 3 \end{pmatrix}$

$CD = \begin{pmatrix} 1 \\ -3 \end{pmatrix}$ $GH = \begin{pmatrix} 4 \\ -2 \end{pmatrix}$ $KL = \begin{pmatrix} 0 \\ -4 \end{pmatrix}$ $PO = \begin{pmatrix} -3 \\ 0 \end{pmatrix}$

21.4 EQUAL VECTORS

Vectors are equal if:

- they have the same length(magnitude) **and** are in the same direction.

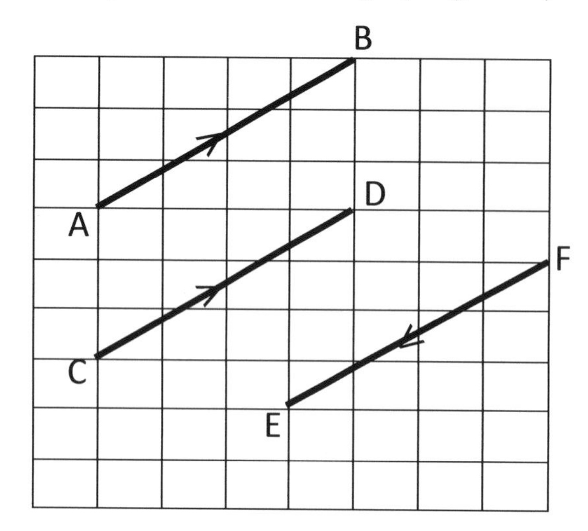

$\overrightarrow{AB} = \overrightarrow{CD} = \begin{pmatrix} 4 \\ 3 \end{pmatrix}$

Vectors $\overrightarrow{AB}$ and $\overrightarrow{CD}$ are equal and in the same direction (arrows).

Vectors $\overrightarrow{AB}$ or $\overrightarrow{CD}$ **and** $\overrightarrow{FE}$ are equal in magnitude **but not** in the same direction.

$\overrightarrow{FE} = \begin{pmatrix} -4 \\ -3 \end{pmatrix}$

Therefore, $\overrightarrow{AB}$ or $\overrightarrow{CD} = -\overrightarrow{FE}$

21.5 MULTIPLYING VECTORS

Example1: If $\mathbf{a} = \begin{pmatrix} -1 \\ 4 \end{pmatrix}$, $\mathbf{b} = \begin{pmatrix} -3 \\ -5 \end{pmatrix}$ and $\mathbf{c} = \begin{pmatrix} 2 \\ -3 \end{pmatrix}$, work out a) $3\mathbf{a}$ b) $4\mathbf{c}$ c) $2\mathbf{a} + 4\mathbf{b}$.

Solutions: a) $3\mathbf{a} = 3 \times \begin{pmatrix} -1 \\ 4 \end{pmatrix} = \begin{pmatrix} 3 \times -1 \\ 3 \times 4 \end{pmatrix} = \begin{pmatrix} \mathbf{-3} \\ \mathbf{12} \end{pmatrix}$

b) $4\mathbf{c} = 4 \times \begin{pmatrix} 2 \\ -3 \end{pmatrix} = \begin{pmatrix} 4 \times 2 \\ 4 \times -3 \end{pmatrix} = \begin{pmatrix} \mathbf{8} \\ \mathbf{-12} \end{pmatrix}$

c) $2\mathbf{a} + 4\mathbf{b} = 2 \times \begin{pmatrix} -1 \\ 4 \end{pmatrix} + 4 \times \begin{pmatrix} -3 \\ -5 \end{pmatrix}$

$= \begin{pmatrix} -2 \\ 8 \end{pmatrix} + \begin{pmatrix} -12 \\ -20 \end{pmatrix} = \begin{pmatrix} -2 + -12 \\ 8 + -20 \end{pmatrix} = \begin{pmatrix} \mathbf{-14} \\ \mathbf{-12} \end{pmatrix}$

Note: When a vector is multiplied by a negative sign, it reverses the direction in the plane (x and y components are reversed).

If $\mathbf{a} = \begin{pmatrix} 2 \\ 3 \end{pmatrix}$, $\mathbf{-a} = -\begin{pmatrix} 2 \\ 3 \end{pmatrix} = \begin{pmatrix} \mathbf{-2} \\ \mathbf{-3} \end{pmatrix}$

21.6 PARALLEL VECTORS

Vectors are parallel if one is a multiple of the other. It relates to multiplying a vector by a scalar quantity as in section 21.5.

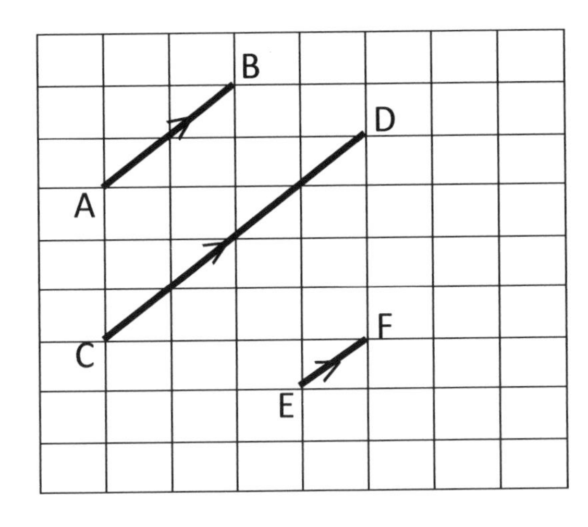

$\overrightarrow{AB} = \begin{pmatrix} \mathbf{2} \\ \mathbf{2} \end{pmatrix}$ and

$\overrightarrow{CD} = 2AB = 2\begin{pmatrix} 2 \\ 2 \end{pmatrix} = \begin{pmatrix} \mathbf{4} \\ \mathbf{4} \end{pmatrix}$

Since $\overrightarrow{CD} = 2AB$, lines AB and CD are parallel. CD is a multiple of AB.

Also, vector $\overrightarrow{EF}$ is parallel to $\overrightarrow{AB}$ and $\overrightarrow{CD}$ because they are multiples of $\overrightarrow{EF}$.

EXERCISE 21 A

1) Write the column vectors for each of the diagrams labelled **a** to **j**.

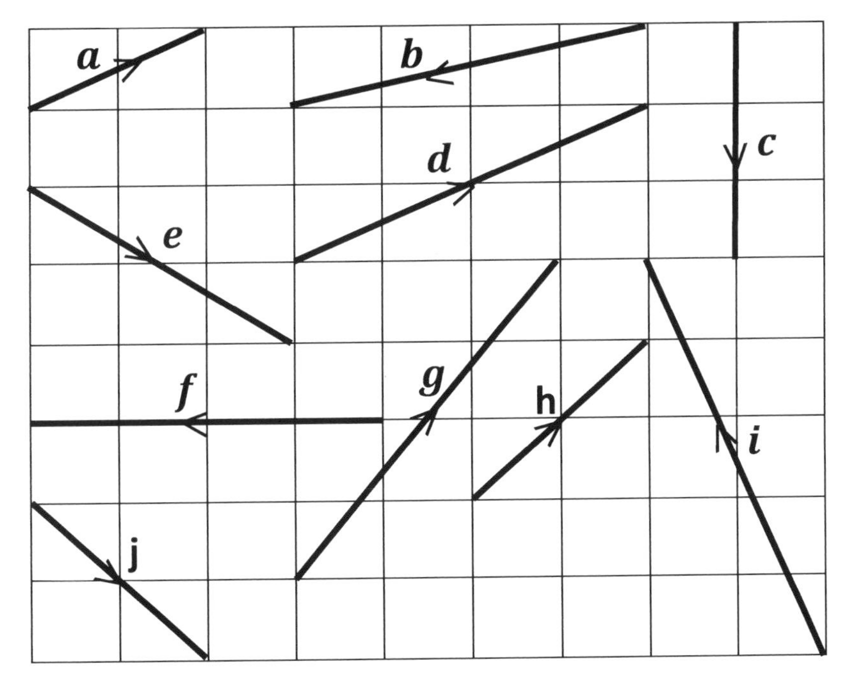

2) Which line are parallel?

3) $\mathbf{a} = \begin{pmatrix} 2 \\ 5 \end{pmatrix}$ $\mathbf{b} = \begin{pmatrix} 4 \\ -2 \end{pmatrix}$ $\mathbf{c} = \begin{pmatrix} -2 \\ -3 \end{pmatrix}$ $\mathbf{d} = \begin{pmatrix} -5 \\ 4 \end{pmatrix}$

On a square grid, draw the suitable diagrams to show these vectors.

a) $\mathbf{a} + \mathbf{b}$ b) $\mathbf{b} + \mathbf{d}$ c) $\mathbf{c} - \mathbf{b}$
d) $2\mathbf{a} - \mathbf{d}$ e) $\mathbf{b} + \mathbf{c}$

4) Work out the **magnitude** and **direction** of the angle made with the positive x-axis of each vector below.

$\mathbf{a}\begin{pmatrix} 2 \\ 7 \end{pmatrix}$ $\mathbf{b}\begin{pmatrix} 0 \\ 3 \end{pmatrix}$ $\mathbf{c}\begin{pmatrix} 5 \\ -3 \end{pmatrix}$

5) Point **T** has coordinates (4, -5 and point **U** has coordinates (-3, 9).

a) Write down the column vector that describes the movement from T to U.

b) Write down the column vector that describes the movement from U to T.

c) Work out the coordinates of the midpoint of TU.

6)

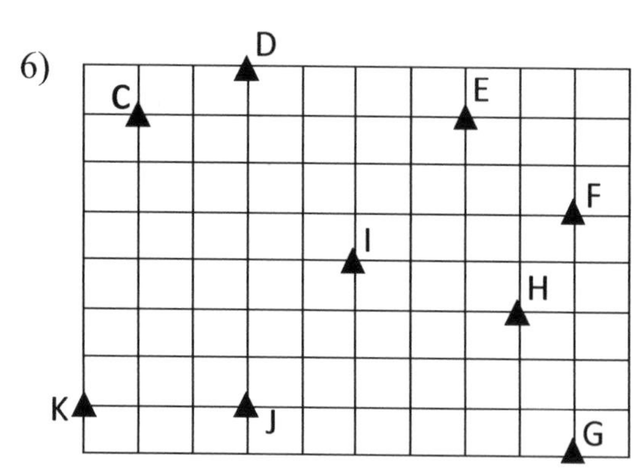

a) Use the diagram above to write down the column vectors for:

i) $\overrightarrow{CD}$ ii) $\overrightarrow{JI}$ iii) $\overrightarrow{HD}$ iv) $\overrightarrow{GK}$

v) $\overrightarrow{KF}$ vi) $\overrightarrow{DE}$ vii) $\overrightarrow{HE}$ viii) $\overrightarrow{KE}$

7) Work out the values of **p**, ***q*** and ***r*** in each of the calculations below.

a) $\begin{pmatrix} 3 \\ 5 \end{pmatrix} + \begin{pmatrix} p \\ 3 \end{pmatrix} = \begin{pmatrix} 7 \\ q \end{pmatrix}$

b) $\begin{pmatrix} -4 \\ 7 \end{pmatrix} + \begin{pmatrix} r \\ p \end{pmatrix} = \begin{pmatrix} -2 \\ 10 \end{pmatrix}$

c) $-3\begin{pmatrix} p \\ 10 \end{pmatrix} = \begin{pmatrix} -12 \\ r \end{pmatrix}$

d) $\mathrm{p}\begin{pmatrix} -6 \\ q \end{pmatrix} = \begin{pmatrix} 18 \\ 21 \end{pmatrix}$

e) $\mathrm{p}\begin{pmatrix} -6 \\ 5 \end{pmatrix} + \mathrm{q}\begin{pmatrix} 3 \\ -4 \end{pmatrix} = \begin{pmatrix} 39 \\ -46 \end{pmatrix}$

8) $\overrightarrow{AB} = \begin{pmatrix} 2 \\ 2 \end{pmatrix}$, $\overrightarrow{AD} = \begin{pmatrix} 4 \\ 0 \end{pmatrix}$

$\overrightarrow{BC} = \begin{pmatrix} 4 \\ 0 \end{pmatrix}$ and $\overrightarrow{DC} = \begin{pmatrix} 2 \\ 2 \end{pmatrix}$

If the vertices of a 2-d shape are A, B, C and D, the vectors above describe how to navigate from one vertex to another.

a) What kind of shape is ABCD?

b) Draw the shape using the vectors given.

c) What can you say about the sides of this 2-d shape?

21.7 VECTOR GEOMETRY

The direction of the arrows is IMPORTANT in solving vector geometry.

The rule is: the vectors with the arrows in the same direction will add to give the vectors with the arrow(s) in the opposite direction. The diagram could be a triangle or any polygon. You must take a route of known vectors irrespective of the shape.

Example 1:

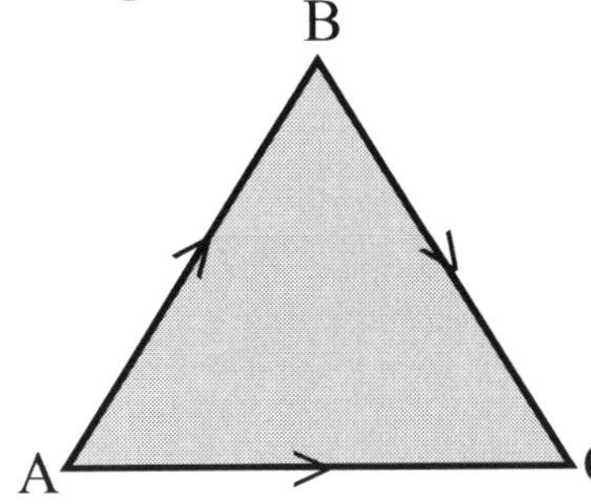

Looking at the arrows, $\overrightarrow{AB}$ and $\overrightarrow{BC}$ are in the same direction (clock-wise) while $\overrightarrow{AC}$ is opposing them (anti-clockwise).

Therefore, $\overrightarrow{\mathbf{AB}} + \overrightarrow{\mathbf{BC}} = \overrightarrow{\mathbf{AC}}$

$\overrightarrow{AC}$ is also the **resultant vector**.

Example 2: Find the vector $\overrightarrow{BA}$ in terms of **p** and **q.**

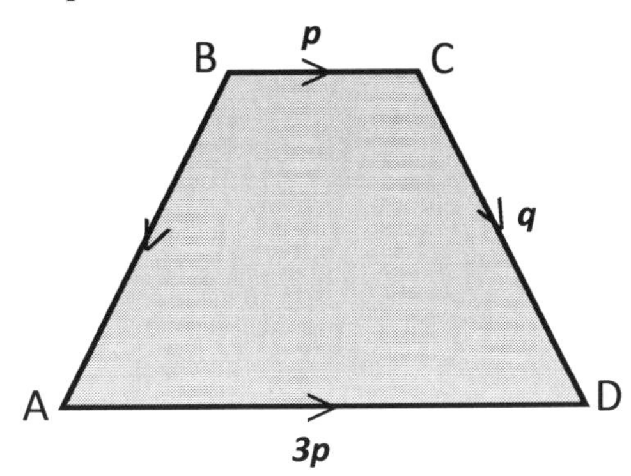

$\overrightarrow{BC} = \mathbf{p}$ $\quad$ $\overrightarrow{AD} = \mathbf{3p}$ $\quad$ $\overrightarrow{CD} = \mathbf{q}$

$\overrightarrow{CB} = \mathbf{-p}$ $\quad$ $\overrightarrow{DA} = \mathbf{-3q}$ $\quad$ $\overrightarrow{DC} = \mathbf{-q}$

Looking at the arrows again, $\overrightarrow{BC} + \overrightarrow{CD}$ (clockwise) = $\overrightarrow{BA} + \overrightarrow{AD}$ (anti-clockwise).

$$p + q = \overrightarrow{BA} + 3p$$

Solving the equation: $\quad p + q - 3p = \overrightarrow{BA}$

$$-2p + q = \overrightarrow{BA}$$

$$\text{or } q - 2p = \overrightarrow{BA}$$

Example 3: The diagram is a triangle PQR.

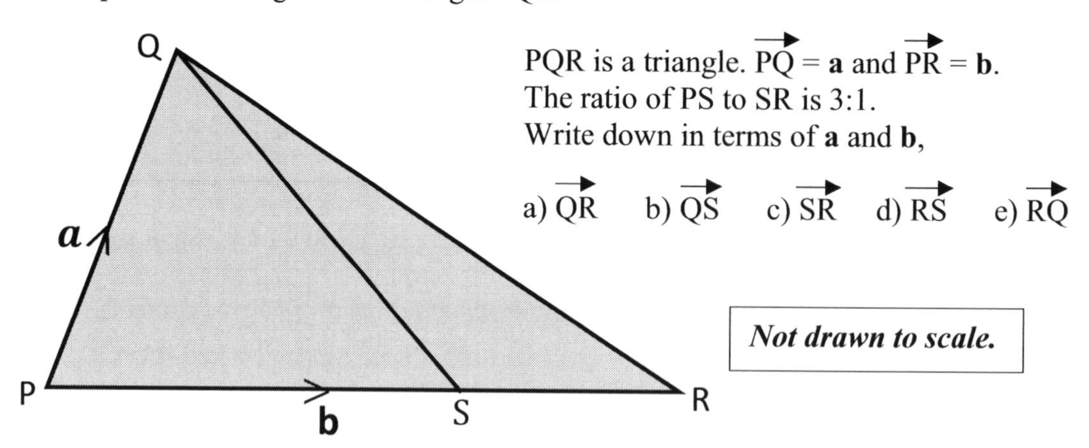

PQR is a triangle. $\overrightarrow{PQ} = \mathbf{a}$ and $\overrightarrow{PR} = \mathbf{b}$.
The ratio of PS to SR is 3:1.
Write down in terms of **a** and **b**,

a) $\overrightarrow{QR}$ b) $\overrightarrow{QS}$ c) $\overrightarrow{SR}$ d) $\overrightarrow{RS}$ e) $\overrightarrow{RQ}$

Solution: a) $\overrightarrow{PQ} + \overrightarrow{QR} = \overrightarrow{PR}$

$$\mathbf{a} + \overrightarrow{QR} = \mathbf{b}$$

$$\overrightarrow{QR} = \mathbf{b} - \mathbf{a}$$

b) To find $\overrightarrow{QS}$, $\overrightarrow{PS}$ or $\overrightarrow{SR}$ must be worked out.
From the question, PS:SR is 3:1.

$PS = \frac{3}{4} PR.$

$= \frac{3}{4}\mathbf{b}$

Therefore, from triangle PQS,

$\mathbf{PQ} + \overrightarrow{QS} = \overrightarrow{PS}$

$a + \overrightarrow{QS} = \frac{3}{4}\mathbf{b}$

$\overrightarrow{QS} = \frac{3}{4}\mathbf{b} - \mathbf{a}$

c) $\overrightarrow{SR} = \frac{1}{4}\overrightarrow{PR}$

$\overrightarrow{SR} = \frac{1}{4}\mathbf{b}$

d) $\overrightarrow{RS} = -\frac{1}{4}\mathbf{b}$

e) $\overrightarrow{RQ} = -(\overrightarrow{QR})$

$= -(\mathbf{b} - \mathbf{a})$

$= -\mathbf{b} + \mathbf{a}$ or $\mathbf{a} - \mathbf{b}$

Example 4:

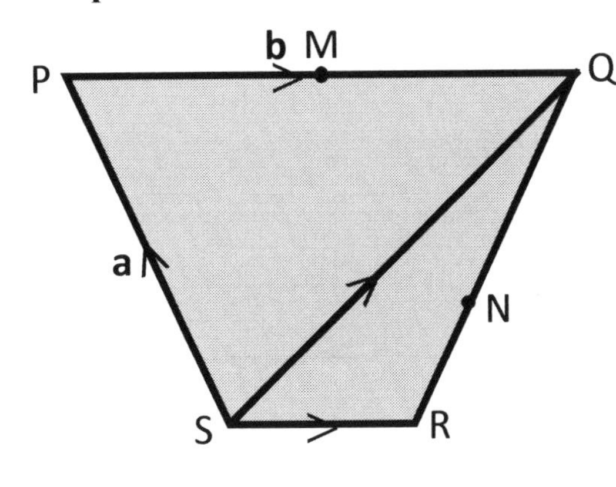

PQRS is a quadrilateral. $\overrightarrow{SP} = \mathbf{a}$, $\overrightarrow{PQ} = \mathbf{b}$ and $\overrightarrow{SR} = \frac{2}{3}$ of PQ. M is the mid-point of PQ and N is a point on QR such that QN:NR = 4:3. Work out the expressions for the vectors in terms of **a** and **b** for

a) $\overrightarrow{SQ}$ b) $\overrightarrow{SR}$ c) $\overrightarrow{QR}$ d) $\overrightarrow{SM}$ e) $\overrightarrow{NM}$ f) $\overrightarrow{SN}$

Solutions:

a) From triangle PQS, $\overrightarrow{SP} + \overrightarrow{PQ} = \overrightarrow{SQ}$

$$\mathbf{a} + \mathbf{b} = \overrightarrow{SQ}$$

b) From the question, $\overrightarrow{SR} = \frac{2}{3}\overrightarrow{PQ}$

$$= \frac{2}{3}\mathbf{b}$$

c) To find QR, you either consider the whole quadrilateral or from triangle SQR.

From triangle SQR, $\overrightarrow{SQ} + \overrightarrow{QR} = \overrightarrow{SR}$.

$$\mathbf{a} + \mathbf{b} + \overrightarrow{QR} = \frac{2}{3}\mathbf{b}$$

$$\overrightarrow{QR} = \frac{2}{3}\mathbf{b} - \mathbf{a} - \mathbf{b}$$

$$= -\frac{1}{3}\mathbf{b} - \mathbf{a}$$

d) $\overrightarrow{SP} + \overrightarrow{PM} = \overrightarrow{SM}$

$$\mathbf{a} + \frac{1}{2}\mathbf{b} = \overrightarrow{SM}$$

e) To work out $\overrightarrow{NM}$, work out $\overrightarrow{QN}$.

From the question, QN:NR = 4:3.

$$\overrightarrow{QN} = \frac{4}{7}\overrightarrow{QR}$$

$$= \frac{4}{7}(-\frac{1}{3}\mathbf{b} - \mathbf{a}) = -\frac{4}{21}\mathbf{b} - \frac{4}{7}\mathbf{a}$$

From triangle NMQ,

$$\overrightarrow{MQ} + \overrightarrow{QN} = \overrightarrow{MN}$$

$$\frac{1}{2}\mathbf{b} + (-\frac{4}{21}\mathbf{b} - \frac{4}{7}\mathbf{a}) = \overrightarrow{MN}$$

$$\frac{1}{2}\mathbf{b} - \frac{4}{21}\mathbf{b} - \frac{4}{7}\mathbf{a} = \overrightarrow{MN}$$

$$\frac{13}{42}\mathbf{b} - \frac{4}{7}\mathbf{a} = \overrightarrow{MN}$$

.....the question says $\overrightarrow{NM}$, therefore,

$$\overrightarrow{NM} = -(\overrightarrow{MN})$$

$$= -(\frac{13}{42}\mathbf{b} - \frac{4}{7}\mathbf{a})$$

$$= -\frac{13}{42}\mathbf{b} + \frac{4}{7}\mathbf{a}$$

f) From triangle SNR,

$$\overrightarrow{SN} + \overrightarrow{NR} = \overrightarrow{SR}$$

$$\overrightarrow{SN} + \frac{3}{7}\overrightarrow{QR} = \frac{2}{3}\mathbf{b}$$

$$\overrightarrow{SN} = \frac{2}{3}\mathbf{b} - \frac{3}{7}(-\frac{1}{3}\mathbf{b} - \mathbf{a})$$

$$= \frac{17}{21}\mathbf{b} + \frac{3}{7}\mathbf{a}$$

Example 5:

$\overrightarrow{BC} = \mathbf{a}$ and $\overrightarrow{CA} = \mathbf{b}$. N is a point on BA such that $\overrightarrow{BN} = 2\overrightarrow{NA}$. K is a point on CA such that $\overrightarrow{CK} = 2\overrightarrow{KA}$.

1) Work out the expressions in terms of **a** and **b** for a) $\overrightarrow{BA}$ b) $\overrightarrow{BN}$ c) $\overrightarrow{NA}$

2) Show that $\overrightarrow{NK}$ is parallel to $\overrightarrow{BC}$.

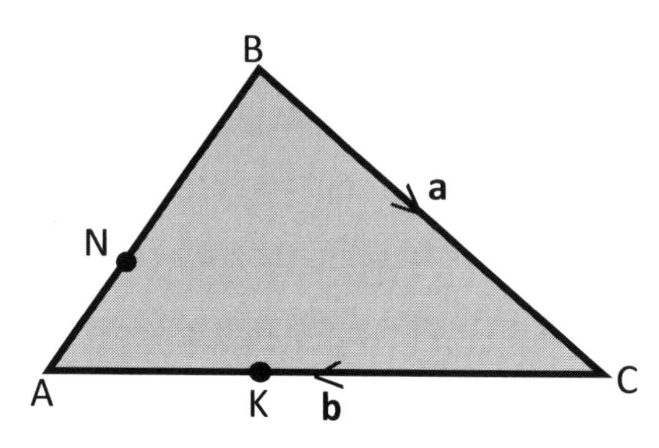

Solutions:

1a) $\overrightarrow{BC} + \overrightarrow{CA} = \overrightarrow{BA}$

$\mathbf{a} + \mathbf{b} = \overrightarrow{BA}$

b) $\overrightarrow{BN} = \frac{2}{3}\overrightarrow{BA} = \frac{2}{3}(\mathbf{a} + \mathbf{b}) = \frac{2}{3}\mathbf{a} + \frac{2}{3}\mathbf{b}$

c) $\overrightarrow{NA} = \frac{1}{3}\overrightarrow{BA} = \frac{1}{3}(\mathbf{a} + \mathbf{b}) = \frac{1}{3}\mathbf{a} + \frac{1}{3}\mathbf{b}$

2) $\overrightarrow{NK} + \overrightarrow{KA} = \overrightarrow{NA}$

$\overrightarrow{NK} + \frac{1}{3}\mathbf{b} = \frac{1}{3}\mathbf{a} + \frac{1}{3}\mathbf{b}$

$\overrightarrow{NK} = \frac{1}{3}\mathbf{a} + \frac{1}{3}\mathbf{b} - \frac{1}{3}\mathbf{b}$

$\overrightarrow{NK} = \frac{1}{3}\mathbf{a} = \frac{1}{3}\mathbf{BC}$

Conclusion: BC is a multiple of NK. 3NK = BC and ST is a third of length BC. Therefore, NK is parallel to BC. ✓

Example 6: ABCD is a parallelogram. U is the midpoint of CB and T is the midpoint of CD. $\overrightarrow{DA} = \mathbf{p}$ and $\overrightarrow{CD} = \mathbf{q}$. Show that $\overrightarrow{UT}$ is parallel to $\overrightarrow{BD}$.

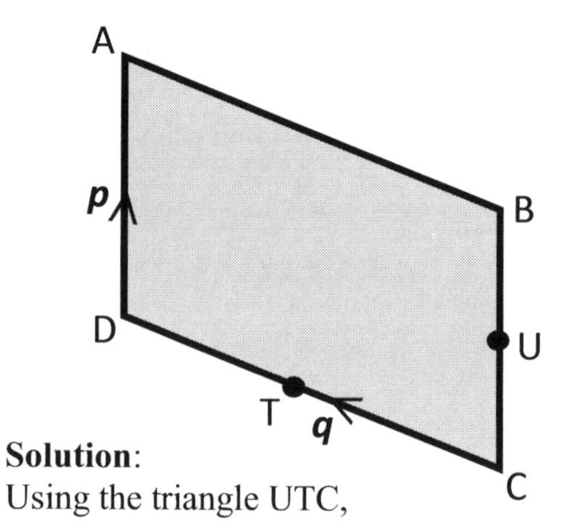

Solution:
Using the triangle UTC,

$\overrightarrow{CU} + \overrightarrow{UT} = \overrightarrow{CT}$

$\frac{1}{2}\mathbf{p} + \overrightarrow{UT} = \frac{1}{2}\mathbf{q}.$

Therefore, $\overrightarrow{UT} = \frac{1}{2}\mathbf{q} - \frac{1}{2}\mathbf{p} = \frac{1}{2}(\mathbf{q} - \mathbf{p}).$

Using triangle ABD, $\overrightarrow{BD} + \overrightarrow{DA} = \overrightarrow{BA}$
$\overrightarrow{BD} = \overrightarrow{BA} - \overrightarrow{DA} = \mathbf{q} - \mathbf{p}.$
Since UT is a multiple of BD, they are parallel.

21.8 COLLINEAR

Three or more point are said to be **collinear** if they lie on a single straight line.

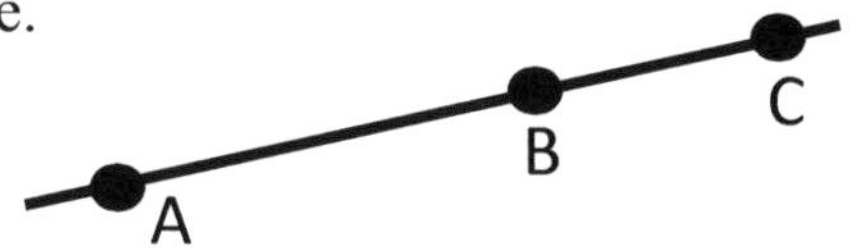

To prove that points are lying on the same straight line using vectors, you must prove that two lines are **parallel** and have a **common** point.
You must prove that line AB is parallel to line BC. Since they have a common point B, the points are collinear.

Example 7:

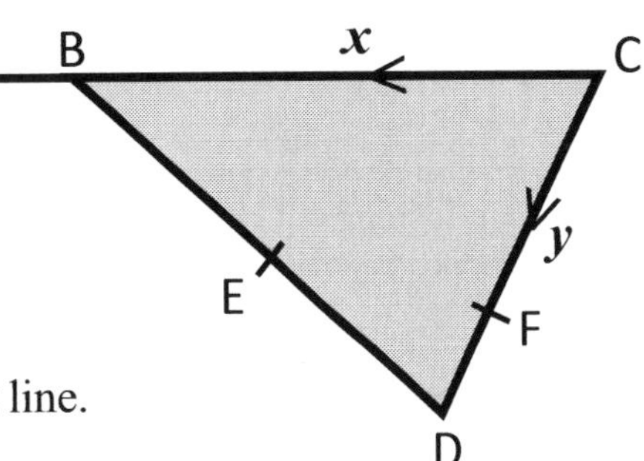

$\overrightarrow{CB} = \boldsymbol{x}$ and $\overrightarrow{CD} = \boldsymbol{y}$.
The ratio of CF: FD = 2:1.
B is the mid-point of $\overrightarrow{CA}$ and
E is the mid-point of $\overrightarrow{BD}$.
Show (prove) that points F, E and A lie on a straight line.

Solution: We must prove that $\overrightarrow{FE}$ is parallel to $\overrightarrow{EA}$ and that they have a common point at E.

$$\overrightarrow{CB} + \overrightarrow{BD} = \overrightarrow{CD}$$
$$\mathbf{x} + \overrightarrow{BD} = \mathbf{y}$$
$$\overrightarrow{BD} = \mathbf{y} - \mathbf{x}$$

$$\overrightarrow{FD} = \frac{1}{3}\overrightarrow{CD} = \frac{\mathbf{1}}{\mathbf{3}}\mathbf{y}$$

$$\overrightarrow{ED} = \frac{1}{2}\overrightarrow{BD} = \frac{1}{2}(\mathbf{y} - \mathbf{x}) = \frac{1}{2}\mathbf{y} - \frac{1}{2}\mathbf{x}$$

From triangle DEF,

$$\overrightarrow{FE} + \overrightarrow{ED} = \overrightarrow{FD}$$
$$\overrightarrow{FE} + \frac{1}{2}\mathbf{y} - \frac{1}{2}\mathbf{x} = \frac{\mathbf{1}}{\mathbf{3}}\mathbf{y}$$

$$\overrightarrow{FE} = \frac{1}{3}\mathbf{y} - \frac{1}{2}\mathbf{y} + \frac{1}{2}\mathbf{x}$$

$$\overrightarrow{FE} = -\frac{1}{6}\mathbf{y} + \frac{1}{2}\mathbf{x} = \boxed{\frac{\mathbf{1}}{\mathbf{2}}\mathbf{x} - \frac{\mathbf{1}}{\mathbf{6}}\mathbf{y}}$$ ✓

From triangle EAB,

$$\overrightarrow{BE} + \overrightarrow{EA} = \overrightarrow{BA}$$

$$\frac{1}{2}\mathbf{y} - \frac{1}{2}\mathbf{x} + \overrightarrow{EA} = \mathrm{x}$$

$$\overrightarrow{EA} = \mathrm{x} - \left(\frac{1}{2}\mathbf{y} - \frac{1}{2}\mathbf{x}\right) = \mathrm{x} - \frac{1}{2}\mathbf{y} + \frac{1}{2}\mathbf{x}$$

$$\overrightarrow{EA} = \boxed{\frac{3}{2}\mathbf{x} - \frac{1}{2}\mathbf{y}}$$ ✓

Conclusion

From the calculations, $\overrightarrow{FE}$ is parallel to $\overrightarrow{EA}$ (multiple of FE). **E** is a common point; therefore, points F, E and A lie on a straight line.

EXERCISE 21B

1) ABCD are the vertices of a square.

$\overrightarrow{AB} = \mathbf{a}$ and $\overrightarrow{AD} = \mathbf{b}$.

a) Sketch the square.
b) Find the following vectors in terms of **a** and **b**.
i) $\overrightarrow{DC}$ ii) $\overrightarrow{BC}$ iii) $\overrightarrow{AC}$, iv) $\overrightarrow{DB}$ v) $\overrightarrow{BA}$

2) The diagram is a triangle PQR.

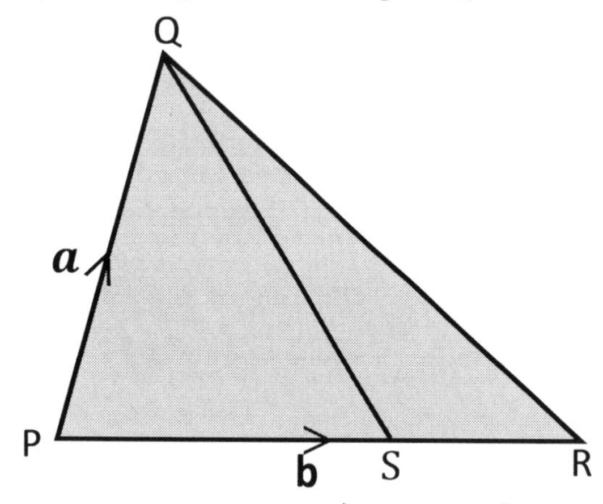

PQR is a triangle. $\overrightarrow{PQ} = \mathbf{a}$ and $\overrightarrow{PR} = \mathbf{b}$.
The ratio of PS to SR is 5:3.
Write down in terms of **a** and **b**,

a) $\overrightarrow{QR}$ b) $\overrightarrow{QS}$ c) $\overrightarrow{SR}$ d) $\overrightarrow{RS}$

e) $\overrightarrow{RQ}$

3) The diagram shows a regular hexagon. Write each vector in terms of **a** and **b**.

a) $\overrightarrow{AC}$
b) $\overrightarrow{FD}$
c) $\overrightarrow{FC}$
d) $\overrightarrow{AD}$

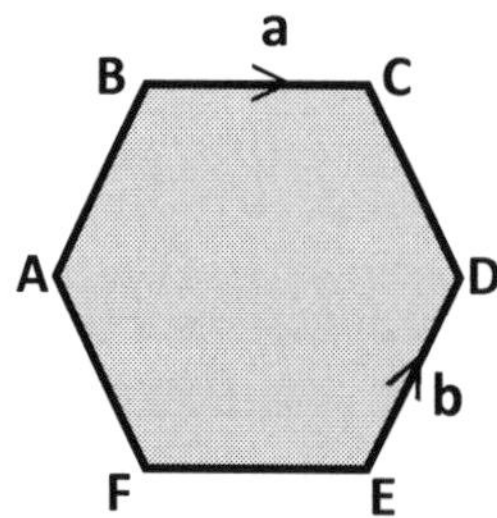

4) The ratio of BM:MC is 2:3.
$\overrightarrow{AC} = \mathbf{a}$ and $\overrightarrow{AB} = \mathbf{b}$.

Find a) BC b) AM

5) $\overrightarrow{CB} = \mathbf{a}$ and $\overrightarrow{CD} = \mathbf{b}$. The ratio of CF: FD = 5:3. **B** is the mid-point of CA and **E** is the mid-point of BD. Are points F, E and A in a straight line? Show your working out.

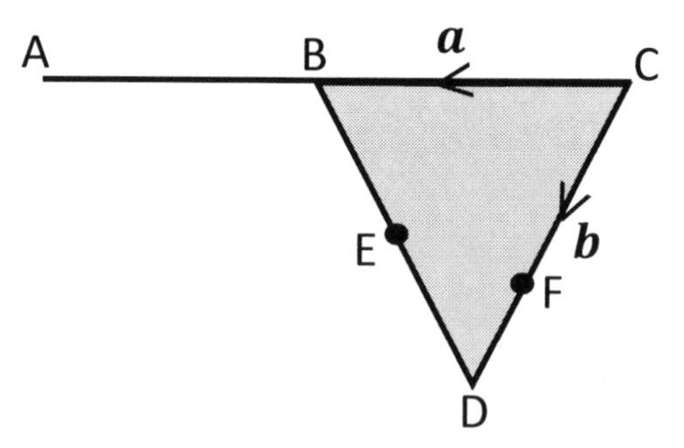

6)

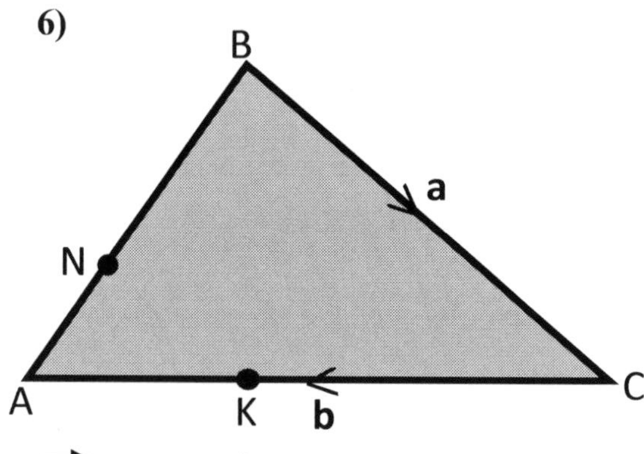

$\overrightarrow{BC} = \mathbf{a}$ and $\overrightarrow{CA} = \mathbf{b}$. BN = 3NA.
CK = 3KA.
a) Work out the expressions for
i) $\overrightarrow{BA}$ ii) $\overrightarrow{BN}$ iii) $\overrightarrow{NA}$.
b) Show that $\overrightarrow{NK}$ is parallel to $\overrightarrow{BC}$.

7)

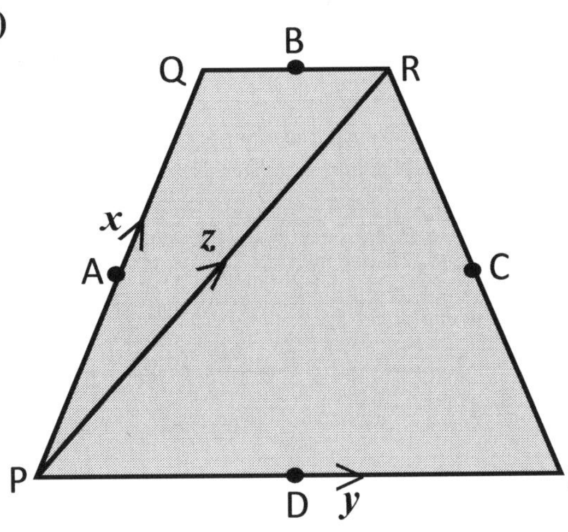

PQRS is a trapezium. $\overrightarrow{PQ} = \mathbf{x}$ and $\overrightarrow{PS} = \mathbf{y}$ and $\overrightarrow{PR} = \mathbf{z}$.
A, B, C and D are midpoints of PQ, QR, RS and PS respectively.

a) Find these vectors in terms of x, y and z.

i) $\overrightarrow{QR}$ ii) $\overrightarrow{RS}$ iii) $\overrightarrow{PC}$ iv) $\overrightarrow{CB}$ v) $\overrightarrow{AS}$

b) Is $\overrightarrow{AD}$ parallel to $\overrightarrow{BC}$? Explain fully.

c) What type of quadrilateral is ABCD?

8) PQRS is a trapezium with PQ parallel to SR.

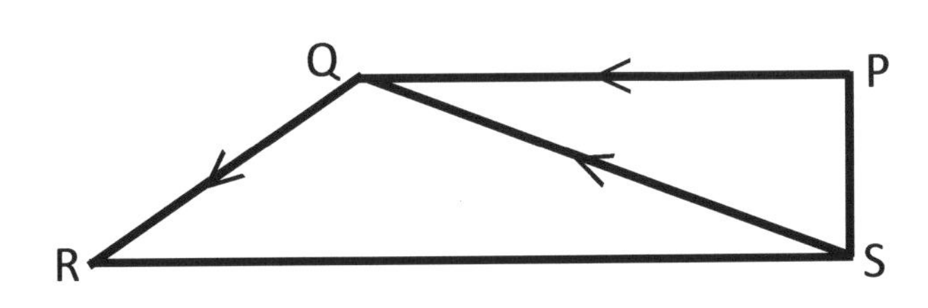

$\overrightarrow{PQ} = 5\mathbf{x} + \alpha\mathbf{y}$, $\overrightarrow{QR} = 3\mathbf{x} + 6\mathbf{y}$ and

$\overrightarrow{SQ} = 5\mathbf{x} + 4\mathbf{y}$.

a) Find $\overrightarrow{SR}$ in terms of **x** and **y**.

b) Work out the value of α.

9) Three triangles are joined to form a tessellating diagram. The triangles are congruent.

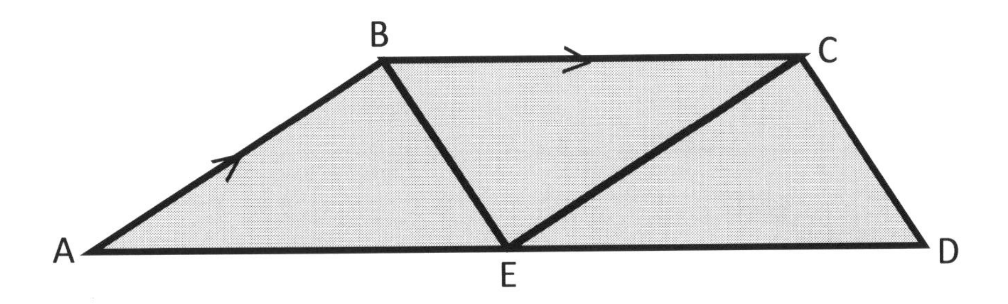

$\overrightarrow{AB} = \mathbf{b}$ and $\overrightarrow{BC} = \mathbf{a}$. Find these vectors in terms of a and b, simplifying your answers.

a) $\overrightarrow{EB}$ b) $\overrightarrow{AE}$ c) $\overrightarrow{AC}$ d) $\overrightarrow{DB}$ e) $\overrightarrow{EC}$ f) $\overrightarrow{DC}$

22 Functions

This section covers the following topics:

- Basic functions
- Inverse functions
- Composite functions

LEARNING OBJECTIVES

By the end of this unit, you should be able to:

a) Find the input and output of a function
b) Find the inverse function
c) Find the composite of two functions

KEYWORDS

- Input
- Output
- Inverse function
- Composite function

22.1 FUNCTIONS – INPUT AND OUTPUT

A rule for changing one number into another is referred to as a **function**.

$y = 5x + 1$ is a function. The value of x is the variable and can take any value(s).

Function notation could be used in an equation. $f(x) = 5x + 1$.

The (f) identifies the expression as a function, and the (x) is for the **input**. We may also use a **function machine** to find the **output** when an input is used in the expression or equation.

Example 1: A function machine for the function notation $f(x) = 5x + 1$ could be represented as:

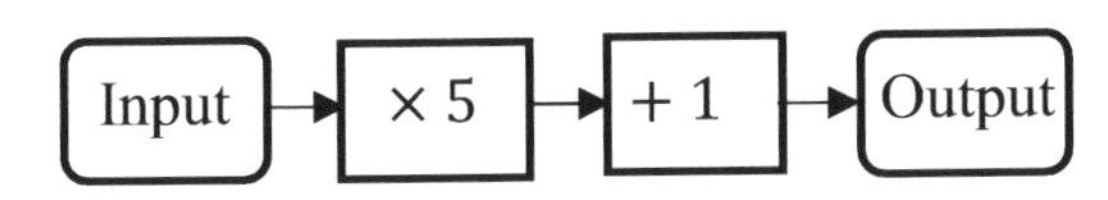

If x = 2, it means the input is 2.
$2 \times 5 + 1 = 11$.
It means the output is 11.

Example 2: For the function $f(x) = 3x^2 + 5x + 1$, find
a) f(4) b) f(-5) c) $f(\sqrt{36})$

Solutions:
a) $f(4) = 3(4)^2 + 5(4) + 1$
$= 48 + 20 + 1 = \mathbf{69}$

b) $f(-5) = 3(-5)^2 + 5(-5) + 1$
$= 75 - 25 + 1 = \mathbf{51}$

c) $f(\sqrt{36}) = 3(\sqrt{36})^2 + 5(\sqrt{36}) + 1$
$f(6) = 108 + 30 + 1 = \mathbf{139}$

Example 3: The function f(x) is defined as $f(x) = 2x^2 - 4$. Find the value of each of the following: a) f(-3) b) $f(\sqrt{5})$.

Solutions:
a) $f(-3) = 2(-3)^2 - 4 = 18 - 4 = \mathbf{14}$
b) $f(\sqrt{5}) = 2(\sqrt{5})^2 - 4 = 10 - 4 = \mathbf{6}$

EXERCISE 22A

1) For the function $f(x) = 3 + 7x$, find the values of:
a) f(3) b) f(-2) c) $f(\frac{1}{7})$ d) $f(\sqrt{4})$

2) The function machine is as shown below:

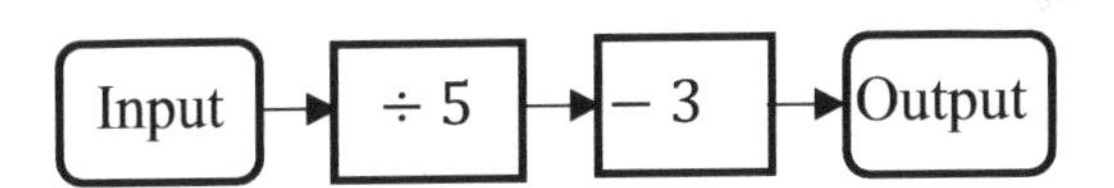

Find the output when the input is
a) 35 b) $\sqrt{100}$ c) 6 d) $\frac{1}{4}$

3) From the function machine in question 2, work out the **input** if the output is
a) 52 b) -43 c) $\sqrt{225}$ d) $\frac{3}{4}$.

4) For the function $f(x) = 2x^2 - 5$, find the values of: a) f (4) b) f(-10) c) $f(\sqrt[3]{8})$ d) $f(\sqrt{3})$ e) When f(x) = 27, work out **both** values of x.

5) If $g(x) = 5g^3 - 3g - 3$, find the values of: a) g(3) b) g(-7) c) $g(\sqrt[4]{81})$ d) g(0).

6) Four consecutive numbers were inserted into the function $f(x) = 6x + 2$ and the sequence created was 68, 74, 80 and 86. Find the four numbers.

22.2 INVERSE FUNCTION

If a function performs the opposite process of the initial or original function, it is said to be an **inverse function.** If the original function is multiplying and the function performs division, it is an inverse function.

The notation that will be used for inverse function is $\mathbf{f^{-1}(x)}$. However, other letters may be used.

Example 1: Find the inverse of the function $f(x) = x + 5$.

RULES FOR INVERSE FUNCTION

a) Write the function f(x) as y.

b) Make x the subject of the formula or function.

c) Finally, replace x with $f^{-1}(x)$ and then replace **y** with **x**.

Solution: From $f(x) = x + 5$,

Write f(x) as y.

$y = x + 5$.

Make x the subject. $y - 5 = x$

Finally, replace x with $f^{-1}(x)$ and change

y to x. This will give:

$\mathbf{f^{-1}(x) = x - 5}$ ✓

Example 2: Find the inverse function

$f(x) = 6x - 3$.

Solution: Write f(x) as y ⟶ $y = 6x - 3$.

Make x the subject of the formula.

$y + 3 = 6x$ ⟶ $x = \frac{y+3}{6}$

$\mathbf{f^{-1}(x) = \frac{x+3}{6}}$ ✓

Example 3: Find the inverse function

$f(x) = x^2 - 7$.

Solution: $y = x^2 - 7$

$y + 7 = x^2$ ⟶ $x = \sqrt{y+7}$

Finally, $\mathbf{f^{-1}(x) = \sqrt{x+7}}$ ✓

Example 4: Find the inverse function

$f(x) = \frac{x+5}{2x-3}$.

Solution: $y = \frac{x+5}{2x-3}$

To make x the subject, multiply both sides by $2x - 3$.

$y(2x - 3) = x + 5$

$2xy - 3y = x + 5$

$2xy - x = 5 + 3y$

$x(2y - 1) = 5 + 3y$

$x = \frac{5+3y}{2y-1}$

Finally, $\mathbf{f^{-1}(x) = \frac{5+3x}{2x-1}}$ ✓

EXERCISE 22B

1) Find an expression for the inverse function $f^{-1}(x)$ for:

a) $f(x) = x - 7$

b) $f(x) = 2x + 9$

c) $f(x) = x^2 - 1$

d) $f(x) = \frac{x+2}{3}$

e) $f(x) = \frac{3}{x+5}$

f) $f(x) = x^3 - 10$

2) From questions 1a - f above, find the value of $f^{-1}(8)$.

3) If $f(x) = \frac{5x-7}{2x+9}$, find an expression for $f^{-1}(x)$.

4) Find the inverse function of the following:

a) $f(x) = \frac{4}{x} + 3$

b) $f(x) = \frac{3x+3}{17}$

c) $f(x) = \frac{13}{x}$

d) $f(x) = \frac{px+q}{rx-p}$

5) Given that $f(x) = \frac{8x^2 - 22x + 5}{4x^2 - 25}$,

Find a) $f^{-1}(x)$

b) $f^{-1}(5)$

22.3 COMPOSITE FUNCTIONS

A **composite function** is when a third function is created by combining two functions.

If **f(x)** and **g(x)** are two functions, the function formed by substituting f(x) into g(x) is known as **gf(x)**. Also, the function formed by substituting g(x) into f(x) is called **fg(x).**

Example 1: If $f(x) = 3x + 2$ and $g(x) = x - 1$, find a) fg(x) b) gf(x).

Solution: **a)** fg(x) means substitute g(x) into f(x). $f(x) = 3x + 2$

$= 3(\mathbf{x - 1}) + 2$

$= 3x - 3 + 2 = \mathbf{3x - 1}$

b) gf(x) means substitute f(x) into g(x).

$g(x) = x - 1 = (3x + 2) - 1$

$= 3x + 2 - 1 = \mathbf{3x + 1}$

Example 2: $f(x) = \frac{2}{5}x + 5$ and $g(x) = 2x - 1$. Find i) fg(3) and ii)gf(-2).

Solution: **a)** $g(3) = 2(3) - 1 = 5$

$fg(3) = f(5) = \frac{2}{5}(5) + 5 = \mathbf{7}$

b) $f(-2) = \frac{2}{5}(-2) + 5 = 4.2$

$gf(-2) = g(4.2) = 2(4.2) - 1 = \mathbf{7.4}$

Example 3: Given that $f(x) = 6x + 3$ and $g(x) = 4x$, find a) $fg(x)$ b) $ff(x)$.

Solution: **a)** $fg(x) = 6(4x) + 3 = \mathbf{24x + 3}$

b) $ff(x) = 6(6x + 3) + 3$

$= 36x + 18 + 3 = \mathbf{36x + 21}$

Example 4:

Given that $g(x) = 5x^2 + 6x - 4$ and

$f(x) = x + 2$, find $g(fx)$.

Solution: Put $f(x)$ into $g(x)$.

$gf(x) = 5(x + 2)^2 + 6(x + 2) - 4$

$= 5(x^2 + 4x + 4) + 6x + 12 - 4$

$= 5x^2 + 20x + 20 + 6x + 8$

$= \mathbf{5x^2 + 26x + 28}$

Example 5: $f(x) = x^3 - 3$ and

$g(x) = 2(x - 5)$. Find simplified expressions for **a)** $gf(x)$ **b)** $gg(x)$.

Solution:

a) $gf(x)$ means substitute $f(x)$ into $g(x)$. Expand $g(x) = 2(x - 5)$ to $2x - 10$.

$gf(x) = 2(x^3 - 3) - 10$

$= 2x^3 - 6 - 10 = \mathbf{2x^3 - 16}$

b) $gg(x)$ means substitute $g(x)$ into $g(x)$.

$gg(x) = 2(2x - 10) - 10$

$= 4x - 20 - 10$

$= \mathbf{4x - 30}$

EXERCISE 22C

1) Given the functions

$f(x) = 5x - 1$, $g(x) = x^2 - 3$ and

$h(x) = 9 - x^2$, determine the simplified expressions for each of the following.

a) $fg(x)$

b) $gf(x)$

c) $fh(x)$

d) $ff(x)$

2) Given the functions $f(x) = 4 - 2x$ and $g(x) = \frac{x+2}{3}$, find the value of:

a) $fg(2)$

b) $fg(-5)$

c) $gf(2)$

d) $gg(-10)$

3) If $f(x) = 6 - x$, $g(x) = 7(x - 3)$, find expressions for each of the following.

a) $gg(x)$ b) $fg(x)$ c) $gf(x)$

Using the functions above, find the value of each of the following:

d) $fg(1.2)$

e) $gf(-3)$

4) Given $f(x) = 2x^2 - 3x + 5$ and

$g(x) = (x - 7)$, find $gf(x)$.

23 Circles and Equation

This section covers the following topics:

- Equation of a circle
- Tangents to a circle

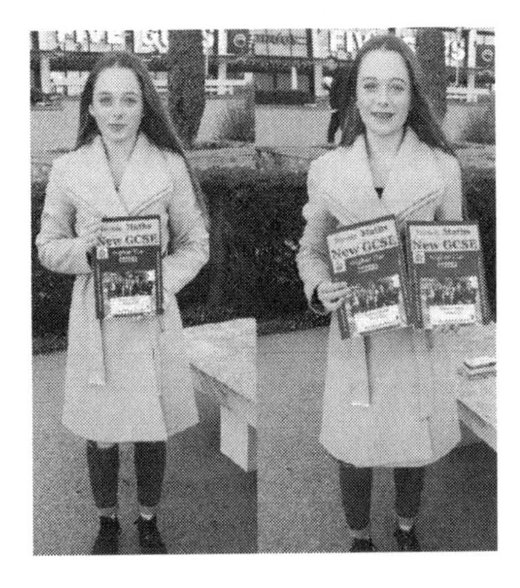

LEARNING OBJECTIVES

By the end of this unit, you should be able to:

a) Find the radius and diameter of a circle
b) Find the equation of a tangent to a circle
c) Form the equation of a circle

KEYWORDS

- Circle
- Tangent
- Equation

23.1 EQUATION OF A CIRCLE

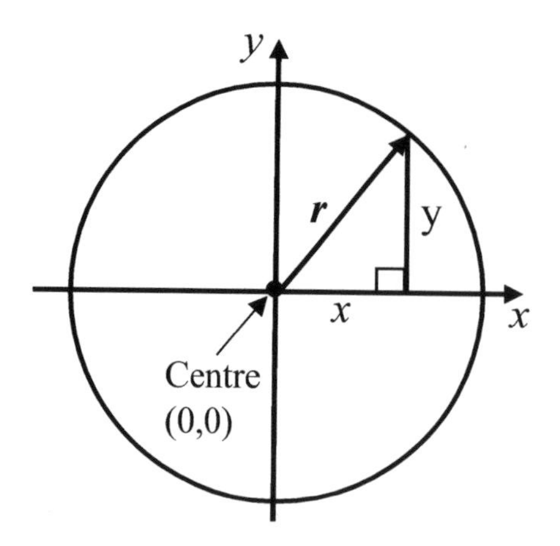

Every point on the circumference is the same distance from the centre of the circle. It is the radius.

When centre is (0,0) and radius is r, the circle has equation $\mathbf{x^2 + y^2 = r^2}$.

Example 1: The graph of a circle is given by the equation $x^2 + y^2 = 4$.

a) Write down the radius of the circle.

b) Draw the graph of the circle.

c) What are the coordinates of the points where the graph cuts the x and y-axes?

d) What is the diameter of the circle.

Solutions:

a) $r^2 = 16$, therefore, $r = \sqrt{4} = 2$. Radius is **2** units.

b)

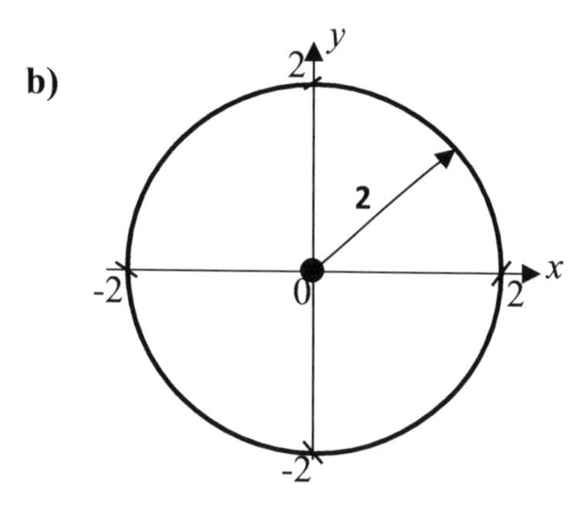

c) x – axes (-2, 0) and (2, 0)

y – axes (0, -2) and (0, 2)

d) Diameter $= 2 \times radius = 2 \times 2 = \mathbf{4\ units}$.

Example 2: A circle has equation $x^2 + y^2 = 25$.

Determine if the points a) (2, 5), b) (1, 2) and c) (-3,4) lie ***outside*** the circle, ***inside*** the circle or on the ***circumference*** of the circle.

Solutions: Pythagoras' theorem will be used to test the points.

a) For point (2, 5):

$2^2 + 5^2 = 4 + 25 = 29$.
29 is greater than 25, therefore the point will lie **outside** the circumference.

b) For point (1, 2):

$1^2 + 2^2 = 1 + 4 = 5$.
5 is less than 25, therefore the point will lie **inside** the circle.

c) For point (-3, 4):

$(-3)^2 + 4^2 = 9 + 16 = 25$.
25 is equal to 25, therefore the point will lie on the **circumference** of the circle.

Example 3: A circle has equation $x^2 + y^2 = 25$. Point P (3, 4) is on the circumference of the circle. Work out the equation of the tangent of the circle at point P.

Solution: Sketch the circle.

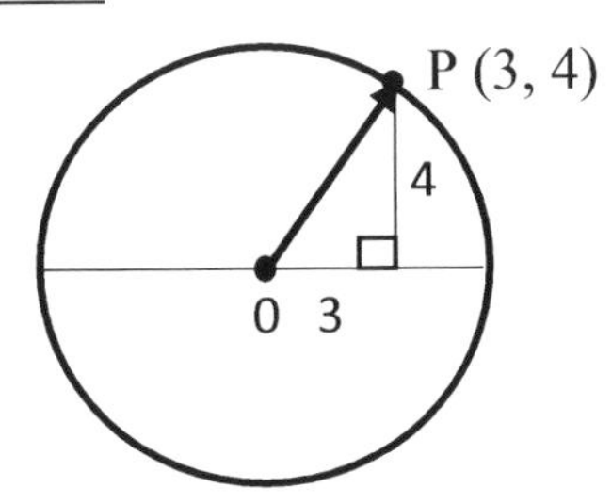

Work out the gradient of the radius.

Gradient of the radius (m_1) $= \frac{4}{3}$

Next is to find the gradient of the tangent (perpendicular) at (3, 4).

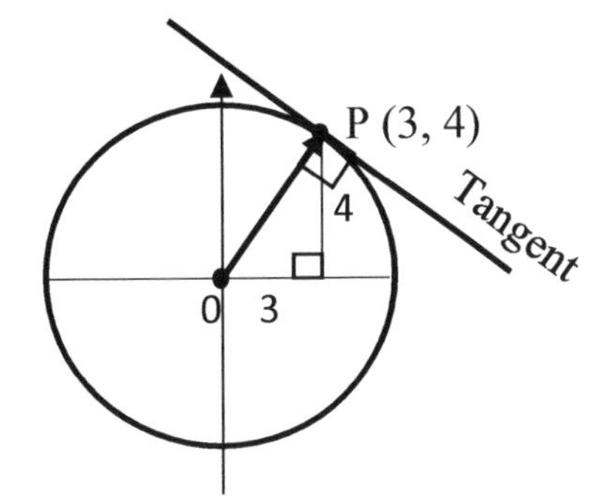

Remember: Gradient of a perpendicular line (m_2) × gradient of the radius (m_1) must equal **-1**.

Therefore, $m_2 = -\frac{3}{4}$

From $y = mx + c$, and using point (3, 4),

$4 = -\frac{3}{4} \times 3 + c$

$4 = -\frac{9}{4} + c$.

Add $\frac{9}{4}$ *to both sides*

$c = \frac{25}{4} = 6.25$

Replacing $c = 6.25$ in the equation,

$y = mx + c$,

$y = -\frac{3}{4}x + 6.25$

Therefore, the equation of the tangent at point P is:

$\mathbf{y = -\frac{3}{4}x + 6.25}$ ✓

EXERCISE 23

1a) Draw x and y-axes from -4 to 4.

b) Draw the graph of $x^2 + y^2 = 9$.

c) What is the diameter of the circle?

d) Write down the coordinates of the points where the circle touches the x and y-axes.

2a) Draw the graph of $x = 3$ using the axes for x and y marked from -6 to 6.

b) On the same graph, draw a circle with a diameter of 10 units, centre (0,0).

c) Work out the equation of the circle.

d) Write down the coordinates of the points where the circle touches the x and y-axes.

e) Write down the coordinates of the points where line $x=3$ touches the circle.

3a) Draw the graph of the circle $x^2 + y^2 - 64 = 0$.

b) On the same axes, draw the graph of $y = 3x + 2$.

c) Find the x - coordinates of the points where the graphs intersect.

d) What is the gradient of the graph $y = 3x + 2$?

e) Draw another line parallel to $y = 3x + 2$ and passing through the point, (0,6). Label it E.

f) Find the x - coordinates of the points where line E touches the circle.

4) A circle has equation $x^2 + y^2 = 80$. Point C (-4, 8) lies on the circumference of the circle.

a) Find the gradient of the tangent to the circle at point C.

b) Work out the equation of the tangent to the circle at point C.

5) The circle $x^2 + y^2 - 121$ has tangents at points (0,11) and (11,0). Work out the equations of the tangents.

6) Form the equation of a circle with centre at the origin (0,0) when the radius is

a) 3 b) 13 c) $\sqrt{5}$ d) $3\sqrt{3}$ e) 3.9

7) A circle has equation $x^2 + y^2 = 625$.

Find the equation of the tangent to the circle at these points.

a) (7, 24) b) (-7, 24)

8) Point P (0,12) is the point where two tangents meet. The angle between the two tangents is 64°. Work out the equation of the circle.

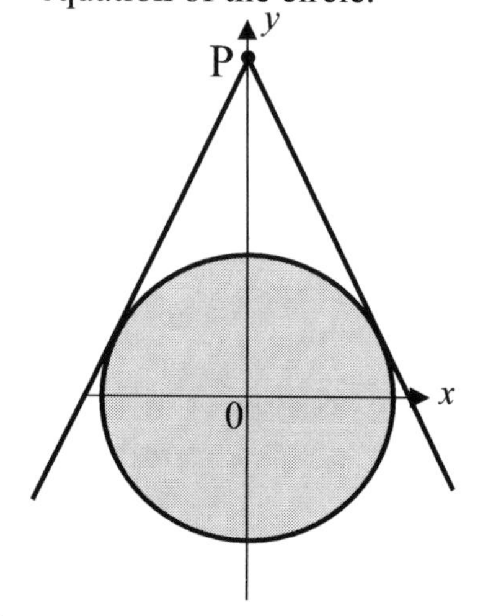

9)

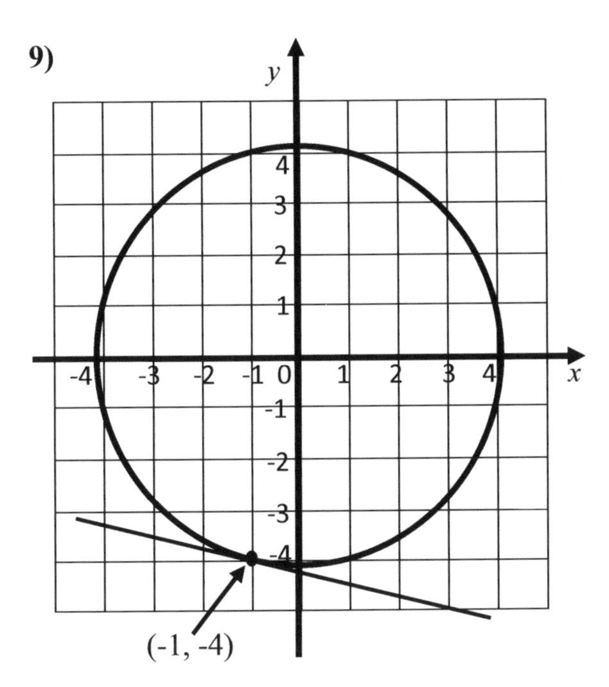

The circle above has equation
$x^2 + y^2 = 17$.

a) Work out the exact value of the diameter of the circle.

b) Find the equation of the tangent at point (-1, -4).

c) Point T $(-3, \sqrt{8})$ is on the circumference. Find the equation of the tangent at point T.

10) The equation of a circle is

$x^2 + y^2 = 3721$.

Find the coordinates of the points where
a) $x = 0$

b) $x = 11$

c) $y = 60$.

24 Velocity-time graphs

This section covers the following topics:

- Velocity-time graphs
- Distance from a velocity-time graph
- Work out the speed and acceleration
- The area under a curve
- Rates of change

LEARNING OBJECTIVES

By the end of this unit, you should be able to:

a) Read information from a velocity-time graph
b) Find the distance from a velocity-time graph
c) Work out/estimate area under the graph
d) Draw a tangent to estimate acceleration
e) Estimate area under a curve

KEYWORDS

- Velocity-time graph
- Acceleration
- Deceleration
- Area under the graph
- Zero gradient

24.1 VELOCITY-TIME GRAPH

Fig.24 fully explains the basic knowledge and mathematics behind the velocity-time graph.

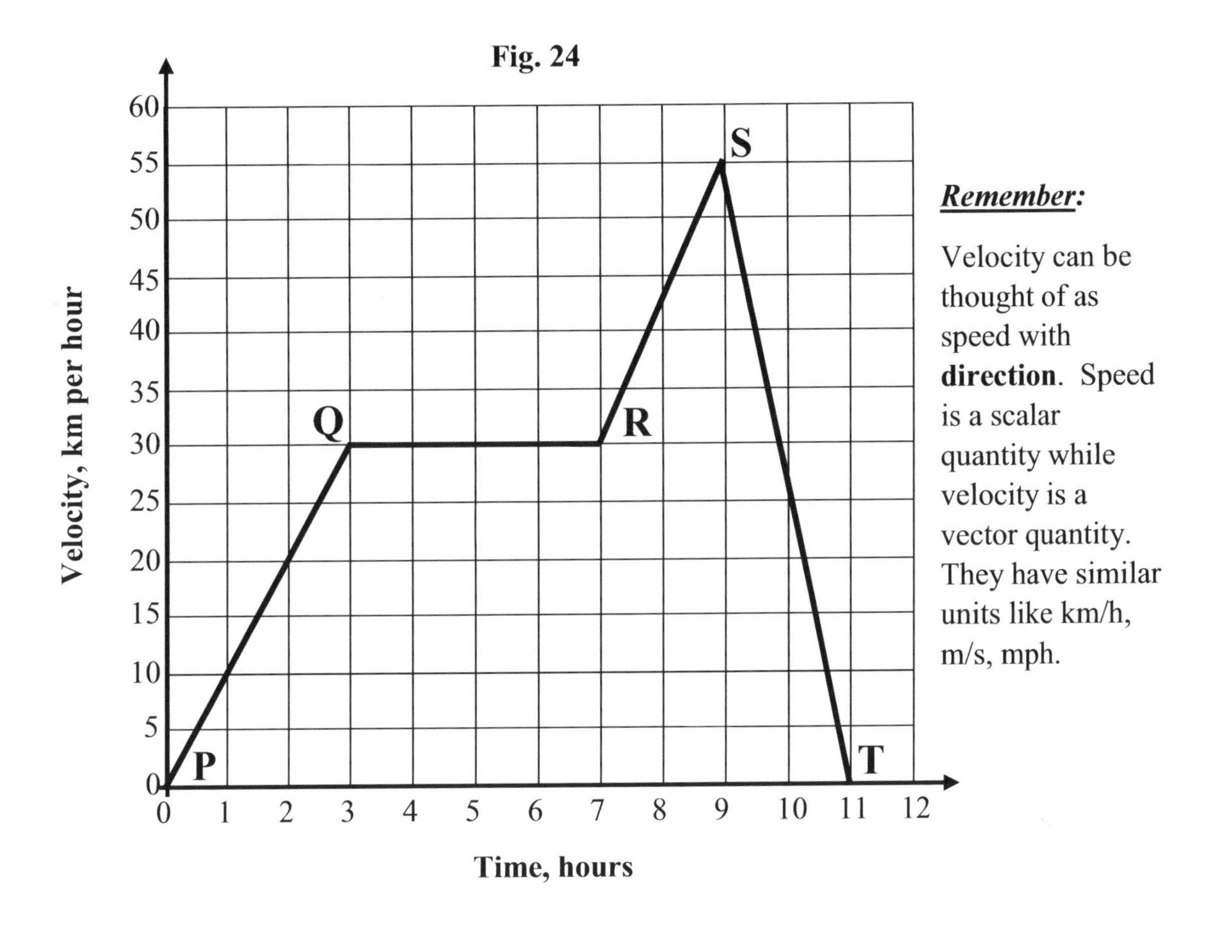

Positive gradients mean the velocity is increasing. Sections PQ and RS have positive gradients. P to Q takes 3 hours and the speed increases from 0 km/h to 30 km/h.
R to S takes 2 hours (9 hrs – 7 hrs) and the speed increases from 30 km/h to 55 km/h.

Section QR is horizontal signifying that the speed is **constant** at 30 km/h for 4 hours (7 hrs – 3 hrs).

Section ST takes 2 hours as the speed decreases from 55 km/h to 0 km/h (deceleration).

For a velocity-time graph, the **distance** covered is the **area** under the graph. The total area under the graph can be calculated by splitting the graph into different sections of 2-d shapes to make it manageable. The area under the graph will be covered at a later stage in this chapter.

Example 1: From the diagram below, calculate

a) the distance covered in the first 3 hours
b) the acceleration in the first 3 hours
c) the initial velocity
d) the total distance covered
e) the average speed for the whole journey.

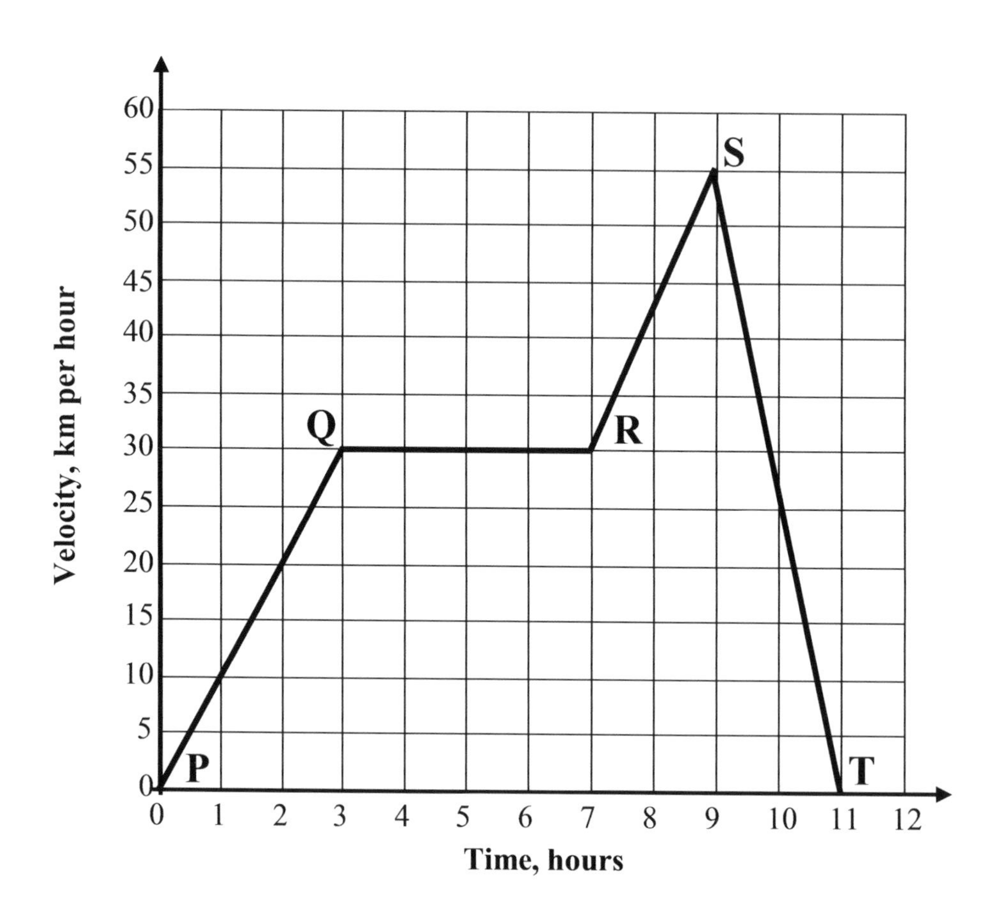

Solutions:

a) The distance in the first 3 hours = area of the triangle. $\frac{1}{2} \times 3 \times 30 = \mathbf{45}\ \boldsymbol{km}$.

b) Acceleration in the first 3 hours $= \frac{change\ in\ veleocity}{change\ in\ time} = \frac{30-0}{3-0} = \frac{30}{3} =$ **10 km/s²**.

c) Initial velocity is the velocity at the start. This is **0 km/hour**.

d) Total distance covered is the area under the graph. Split the graph into triangles and rectangles as shown with sections A, B, C, D, E, F and G.

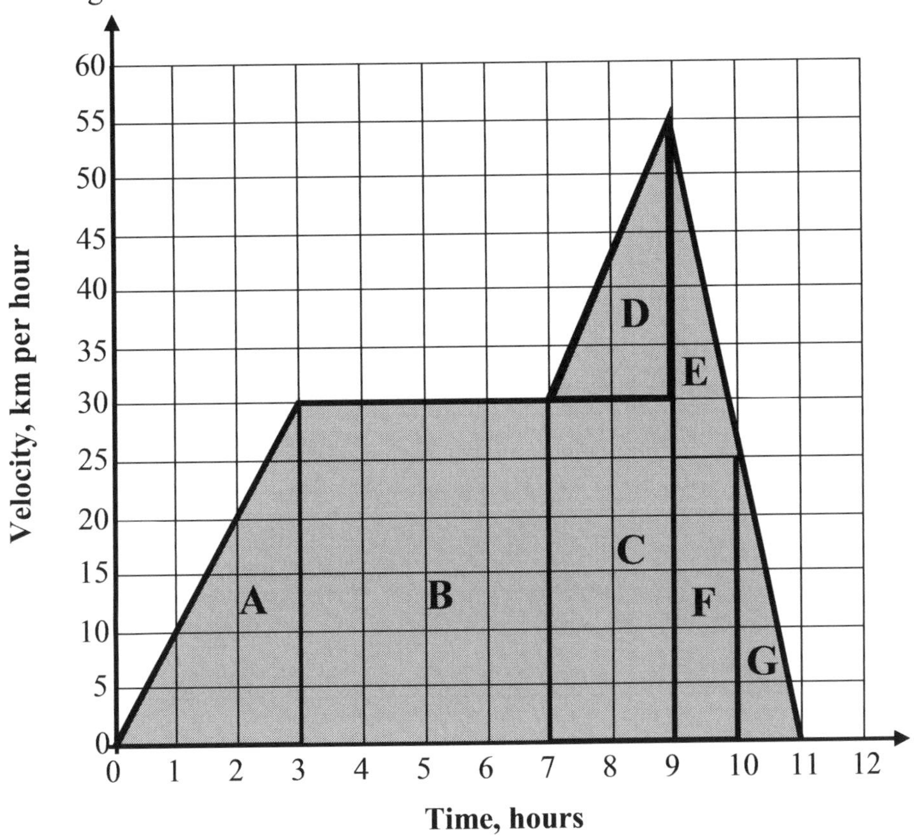

Area of triangle A = $\frac{1}{2} \times 3 \times 30 = 45$ Km^2

Area of rectangle B = $4 \times 30 = 120$ km^2

Area of rectangle C = $2 \times 30 = 60$ km^2

Area of triangle D = $\frac{1}{2} \times 2 \times 25 = 25$ km^2

Area of triangle E = $\frac{1}{2} \times 1 \times 30 = 15$ km^2

Area of rectangle F = $1 \times 25 = 25$ km^2

Area of triangle G = $\frac{1}{2} \times 1 \times 25 = 12.5$ km^2

Total area under the graph
= 45 + 120 + 60 + 25 + 15 + 25 + 12.5
= 302.3 km^2

However, area under the graph is the distance travelled.

Therefore, the total distance covered = **302.3 km**.

e) Average speed for the whole journey is total distance travelled ÷ total time taken.

Average speed = $\frac{302.3}{11}$ = **27.48 km/hour**.

EXERCISE 24A

1) The velocity-time graph below represents a bus journey between two bus stops.

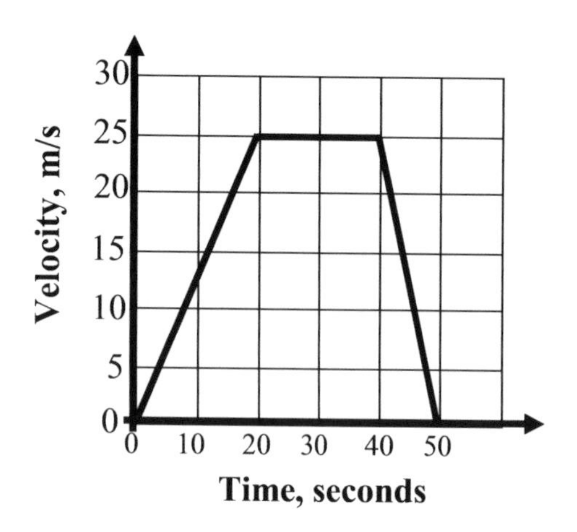

a) What is the velocity at 10 seconds?

b) What is the steady speed of the bus?

c) Find the distance between the two bus stops.

d) While decelerating, what was the distance covered by the bus?

e) Work out the average speed for the whole journey.

2)

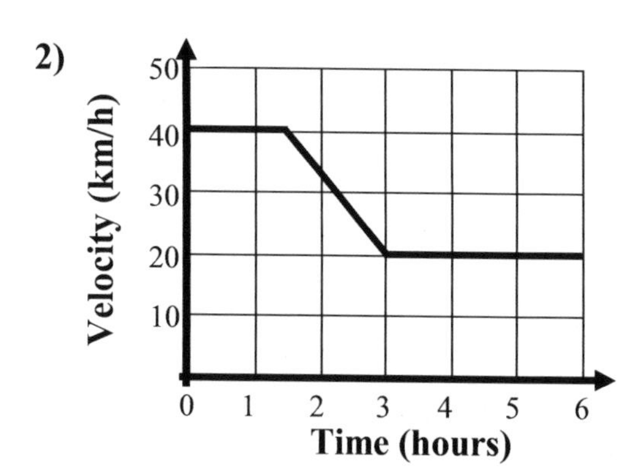

Work out the distance travelled in the first 5 hours.

3) This is a velocity-time graph.

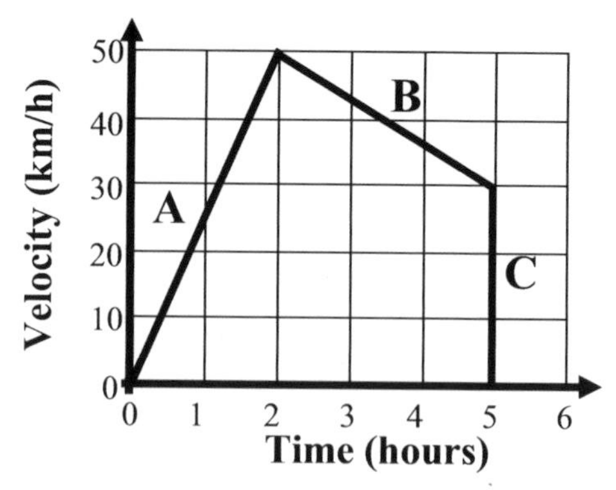

a) Work out the average speed of the entire journey.
b) Work out i) the acceleration for sections A and ii) the deceleration for B.

4) The graph below shows the velocity of a sports bike.

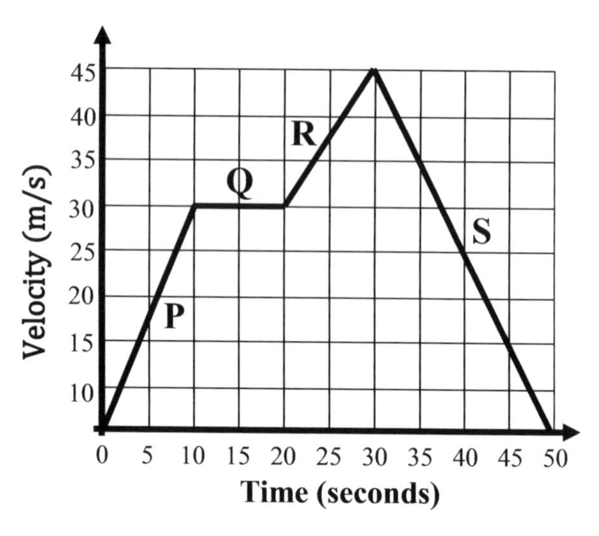

a) What is the velocity of the bike after 25 seconds?

b) Describe the movement(s) of this sports bike during P, Q, R, S sections.

c) Describe what is happening after 30 seconds.

d) At what rate is the bike decelerating?

24.2 AREA UNDER A CURVE

To work out the area under a curve accurately is beyond GCSE and would be dealt with in Advanced level.

However, we estimate the area under a curve by splitting (dividing) into smaller shapes of rectangles, triangles or trapezium.

Also, it could be **under-estimating** or **over-estimating** depending on the divisions made on the curve.

Example 1: Calculate the estimate of the total distance travelled.

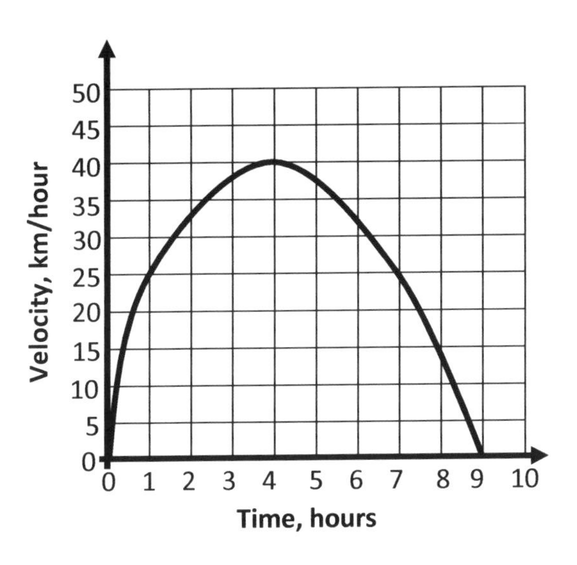

Solution: Split the shape into A, B, C and D as shown in the diagram. This is just a guide. You may split into manageable parts to help in your calculations.

Remember: Area under the graph is the **distance** travelled. Therefore, the unit will be in **km**, to represent the distance.

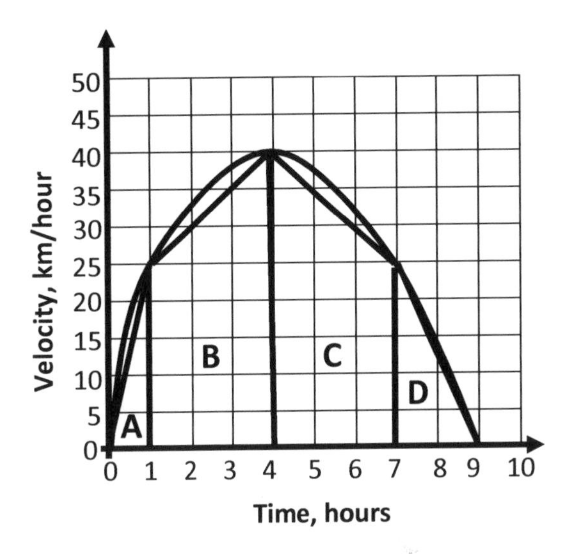

Area of triangle A $= \frac{1}{2} \times 1 \times 25$

$= 12.5$ km

Area of trapezium B $= \frac{1}{2}(25 + 40) \times 3$

$= 97.5$ km

Area of trapezium C $= \frac{1}{2}(40 + 25) \times 3$

$= 97.5$ km

Area of triangle D $= \frac{1}{2} \times 2 \times 25$

$= 25$ km

Total area $= 12.5 + 97.5 + 97.5 + 25$

$= 232.5$ km

Therefore, the total distance travelled is **232.5 km**. Since all the split shapes are inside the graph (curved area), this will represent a slight **under-estimation** of the real distance travelled.

Example 2: Estimate the distance covered.

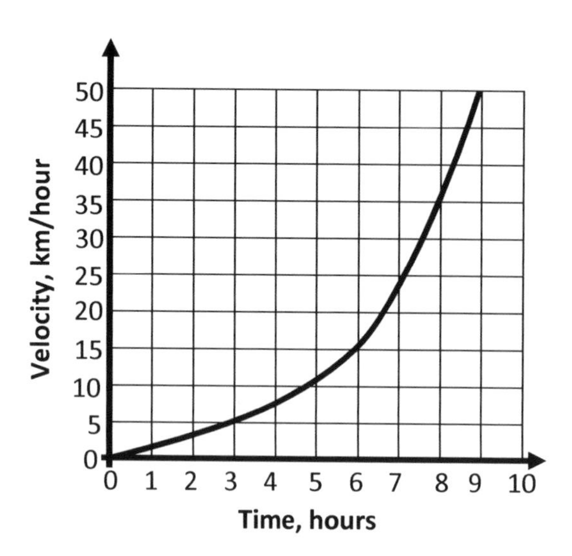

Solution: Split the shape into A and B.

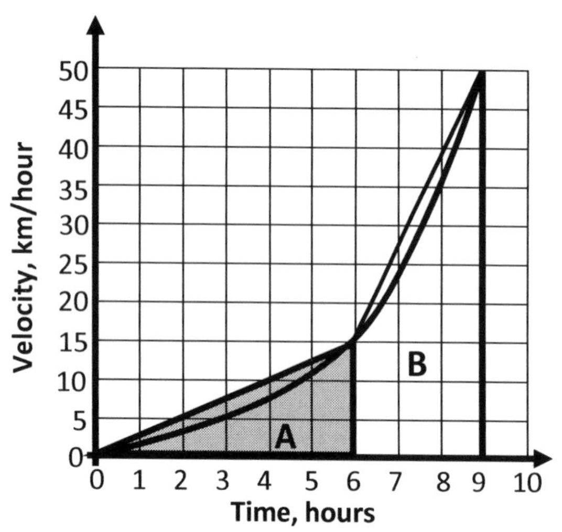

Area of triangle A $= \frac{1}{2} \times 6 \times 15 = 45$ km

Area of trapezium B $= \frac{1}{2}(15 + 50) \times 3$
$= 97.5$ km

Total distance travelled $= 45 + 97.5$
$=$ **142.5 km**

The areas calculated are greater than the area under the graph. Therefore, it is an **over-estimation**.

EXERCISE 24 B

1) For the velocity-time graph below, estimate the distance travelled. Also, state if it was under or over-estimation.

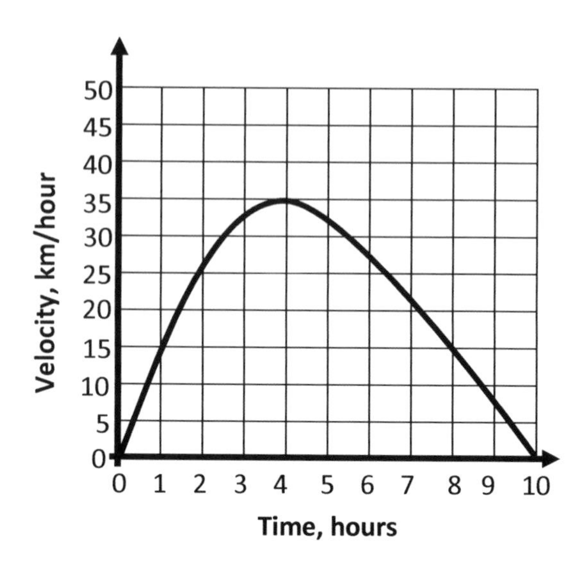

2) This is a velocity-time graph.

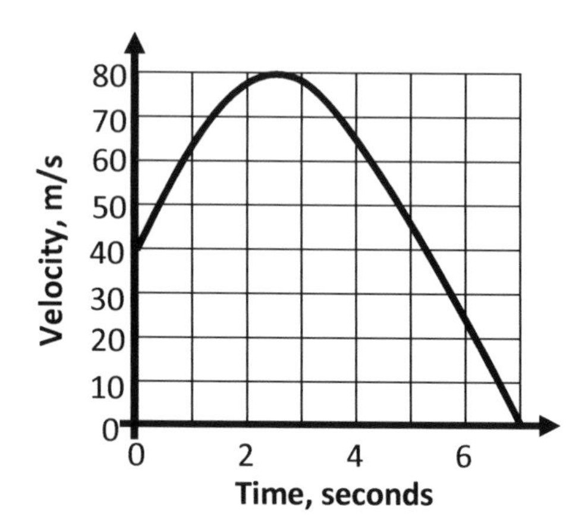

a) Find the initial velocity.

b) Find the maximum velocity.

c) Work out an estimate of the total distance travelled.

3) a) Estimate the total distance covered. b) State if it was under or over-estimation.

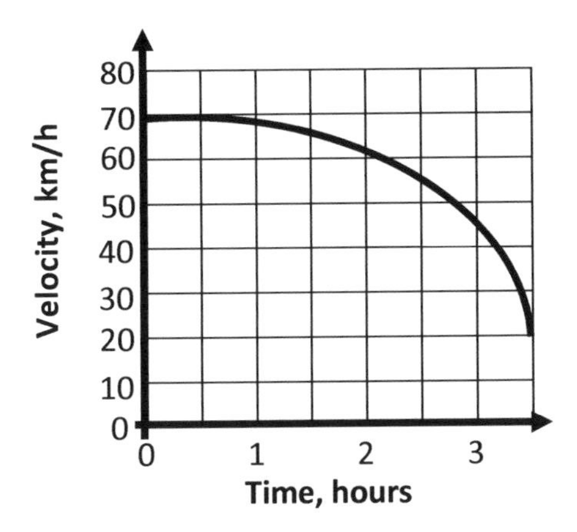

4) a) Estimate the distance covered.
b) State whether it was under or over-estimation.

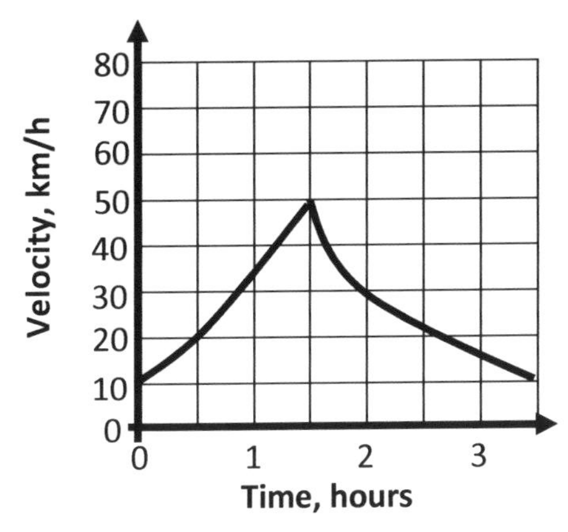

5) Estimate the total distance travelled.

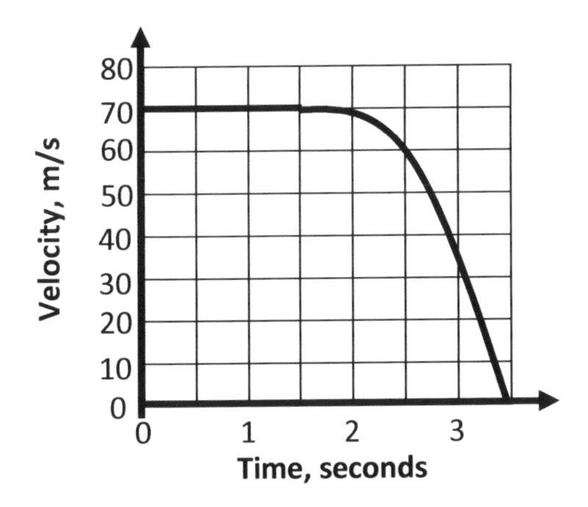

24.3 TANGENTS, GRADIENTS AND RATE OF CHANGE

This section reinforces the use of the gradient of a straight line. We shall use this knowledge to calculate acceleration at a time in a velocity-time graph.

Recall: A tangent is a straight line that touches the outside of a circle or a curve at a point only.

To accomplish this, we shall draw a tangent at the point of interest in the curve and calculate the gradient of the tangent. Draw a right-angled triangle from the tangent and calculate the $\frac{change\ in\ y}{change\ in\ x}$. That will give the acceleration at that point in time.

DISTANCE-TIME GRAPH

In a distance-time graph, the gradient at any point gives the **velocity** at that point on the graph. ***NOTE***: Velocity (vector quantity) and Speed (scalar) have the same unit(s).

Example 1: Estimate the velocity at 1 hour.

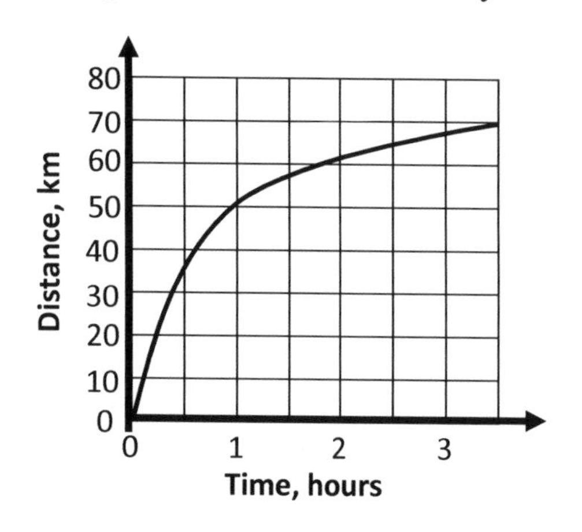

Solution: Draw a tangent at 1 hour

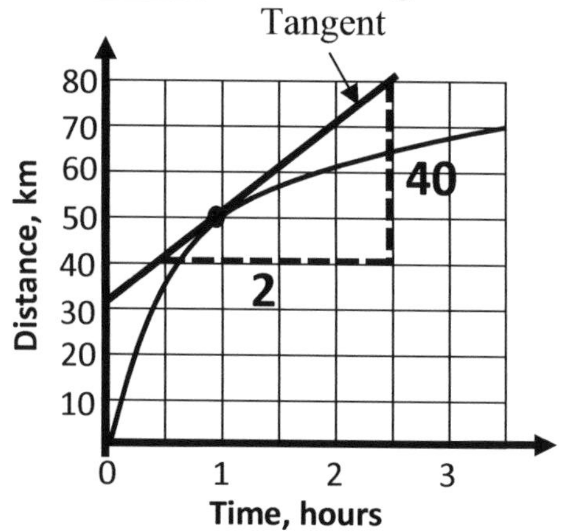

At 1 hour, velocity $= \frac{40}{2} =$ **20 km/h**

Average speed/velocity between two times can also be calculated by working out the gradient of the chord (line joining two points on the curve).

VELOCITY-TIME GRAPH

In a velocity-time graph, the gradient at any point gives the **acceleration** at that point on the curve.

Example 2: Estimate the acceleration at 2 hours.

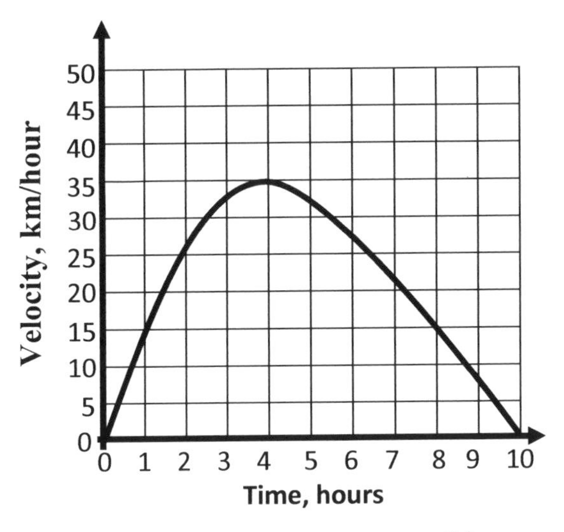

Solution: Draw a tangent at 2 hours.

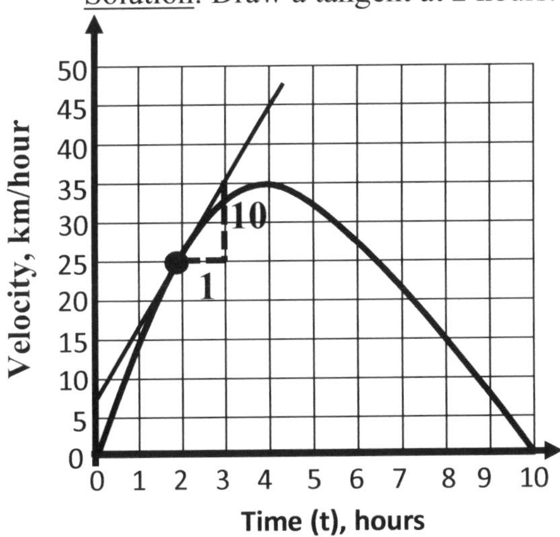

The acceleration at 2 hours is

$\frac{10}{1}$ = **10 km/h²**

Notice the unit of acceleration. In this case km/h².

From example 2, acceleration **is zero** at 4 hours. This is the highest or maximum point on the graph before deceleration starts. The gradient is zero at that point.

EXERCISE 24 C

1) A distance -time graph is shown below.

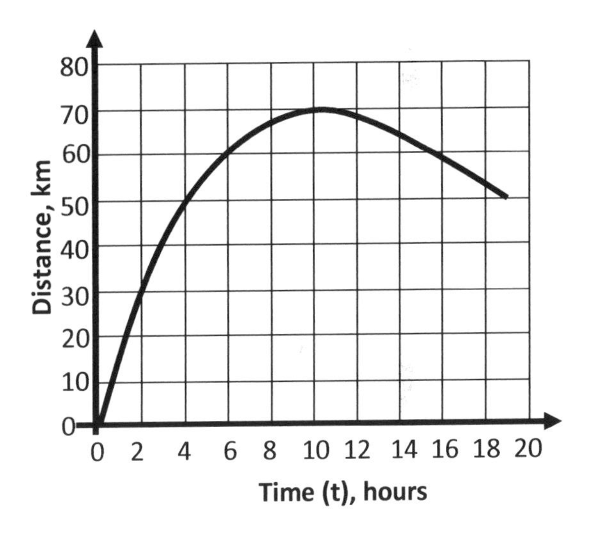

a) At what time is the velocity zero? Explain fully.

b) Estimate the velocity when:

i) t = 2 hours   ii) t = 6 hours

c) Estimate the average velocity from:

i) t = 0 hours to t = 4 hours

2) The graph below is a velocity-time graph.

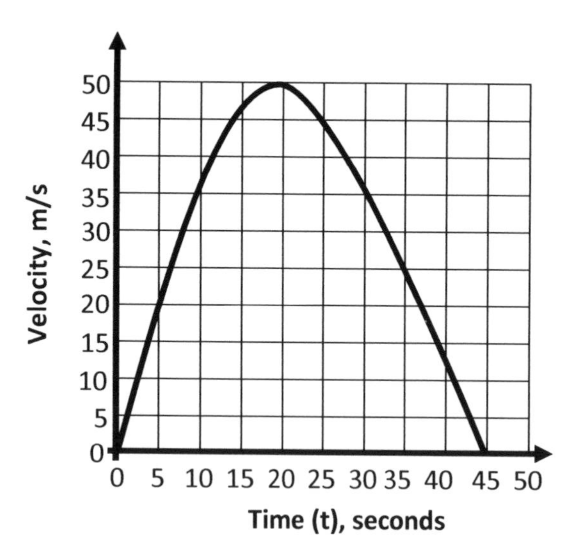

a) Estimate the acceleration when t = 10 s.

b) Estimate the deceleration when t = 30 s.

c) At what time is the acceleration zero? Explain fully.

3)

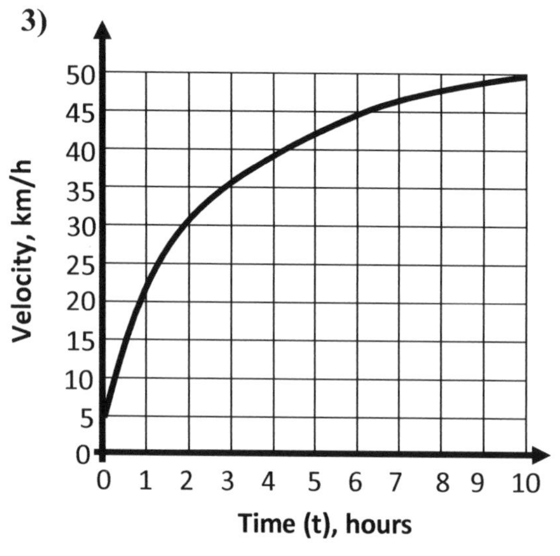

a) Estimate the acceleration when t = 2 hrs.

b) Estimate the total distance travelled.

4)

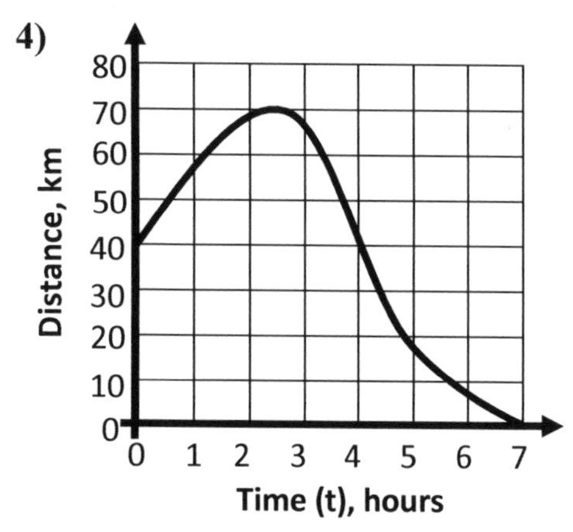

From the distance-time graph, estimate the velocity when:

a) t = 1.5 hrs

b) t = 2.5 hrs.

c) Estimate the average velocity from

t = 4 hours to t = 7 hours.

5) Consider the function $\mathbf{d = 6t - t^2}$ which represents the distance, in metres, travelled by a football with respect to time of landing, t in seconds.

a) Draw the graph of the function for **t** values from 0 to 6.

b) At what time is the gradient zero?

c) What is the maximum speed?

d) At point (4,8), is the gradient positive or negative?

6) The diagram below represents a velocity-time graph of a bus moving between two counties.

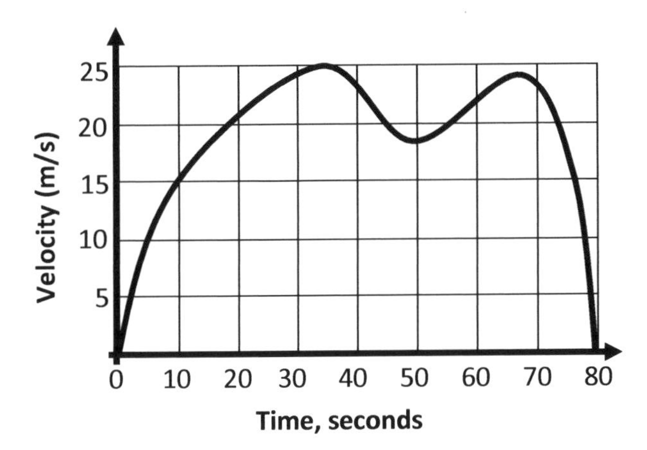

a) Estimate the acceleration of the bus when the time is equal to 10 seconds.

b) What is the acceleration when the time is equal to 35 seconds.

c) estimate the total distance travelled by the bus.

7) Consider the function $\mathbf{d = 8t - t^2}$ which represents the distance travelled in metres by a football with respect to time of landing in seconds, t.

a) Draw the graph of the function for t values from 0 to 8.

b) Calculate the maximum speed.

c) Work out the speed when: i) $t = 2$ ii) $t = 4$.

<ins>POINTS TO NOTE:</ins>

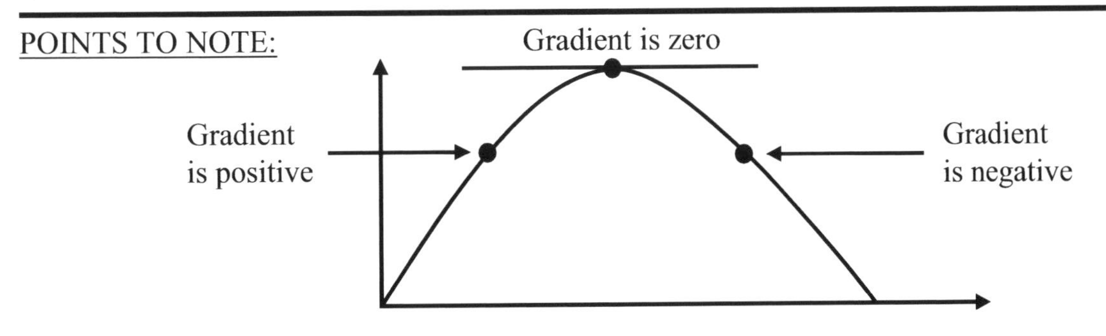

25 Trigonometry

This section covers the following topics:

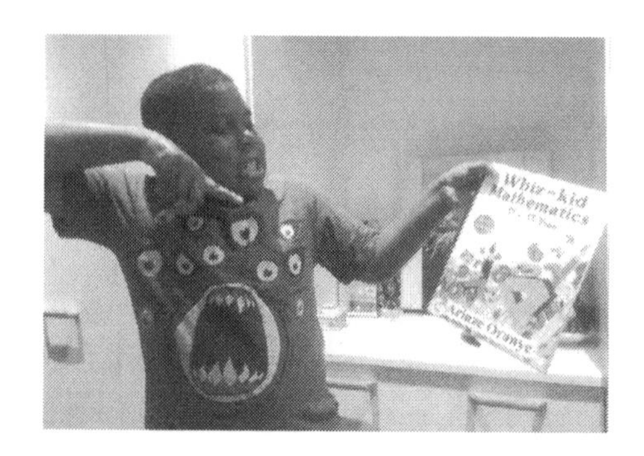

- Trigonometric ratios
- Sine, cosine and tangent functions
- Sine rule
- Cosine rule
- Area of a triangle using trigonometry
- Trigonometry in three-dimensions

LEARNING OBJECTIVES

By the end of this unit, you should be able to:

a) Understand Trigonometry
b) Use trigonometric ratios to find lengths and angles in right-angled triangles
c) Find and memorise exact values of important trigonometric ratios
d) Use sine and cosine rule to find lengths and angles in any triangle
e) Find the area of a triangle using trigonometry
f) Solve trigonometry problems in three-dimensional figures

KEYWORDS

- Trigonometry
- Right-angled triangle
- Sine, cosine and tangent
- Sine rule
- Cosine rule
- Area

Trigonometry simply means **triangle measurements.** It is used in the calculation of sides and angles in triangles.

Trigonometry is helpful when solving problems that Pythagoras' theorem cannot solve. Two lengths must be given before using Pythagoras theorem in a right-angled triangle.

However, when only a length and an angle is given, trigonometry comes in handy.

25.1 TRIGONOMETRIC RATIOS (Sin, Cos, Tan)

In a right-angled triangle, the longest side is known as the **hypotenuse**, the side opposite the angle which universally is known as **theta** (θ)is called the **opposite** while the side next to a marked angle and the right angle is known as the **adjacent**.

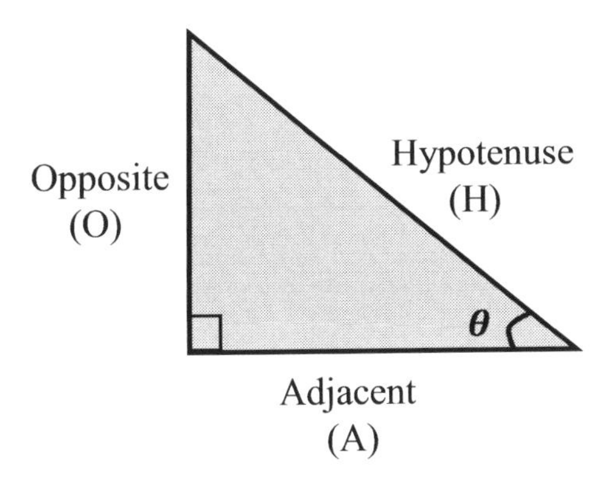

It is vital that the system of naming the sides is mastered. It is fundamental to working with trigonometry.

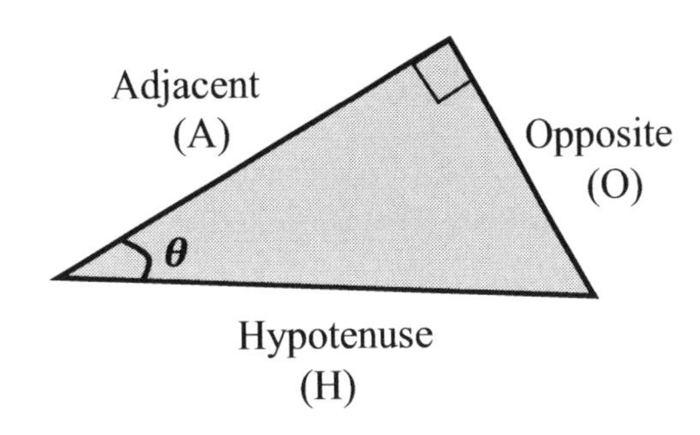

Similar triangles could be used to find trigonometric ratios.

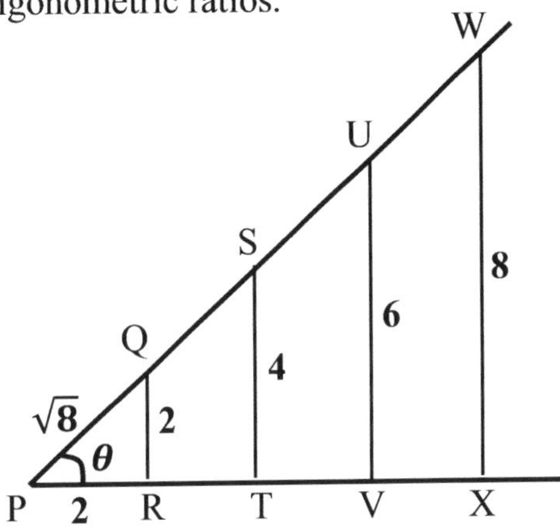

The diagram above shows four similar triangles, PQR, PST, PUV and PWX.

The ratio of $\frac{opposite}{Hypotenuse}$ is $\frac{2}{\sqrt{8}}$ or 0.7071

for all the similar triangles.

The ratio of $\frac{opposite}{adjacent} = 1$ for all the similar triangles.

Finally, the ratio of $\frac{adjacent}{Hypotenuse} = \frac{2}{\sqrt{8}}$ for all the similar triangles.

The ratios are called **trigonometric ratios.**

The **sine**, **cosine** and **tangent** ratios for theta (θ) will be defined as:

$$\text{Sin } \theta = \frac{Opposite}{Hypotenuse} = \frac{O}{H}$$

$$\text{Cos}\theta = \frac{Adjacent}{Hypotenuse} = \frac{A}{H}$$

$$\text{Tan } \theta = \frac{Opposite}{Adjacent} = \frac{O}{A}$$

Note: $Sin\ \theta, Cos\ \theta$ *and* $Tan\ \theta$ will always have the same value for any particular angle, irrespective of the size of the triangle.

Example 1: From the right-angled triangle below, find

a) Sin θ b) Cos θ c) Tan θ

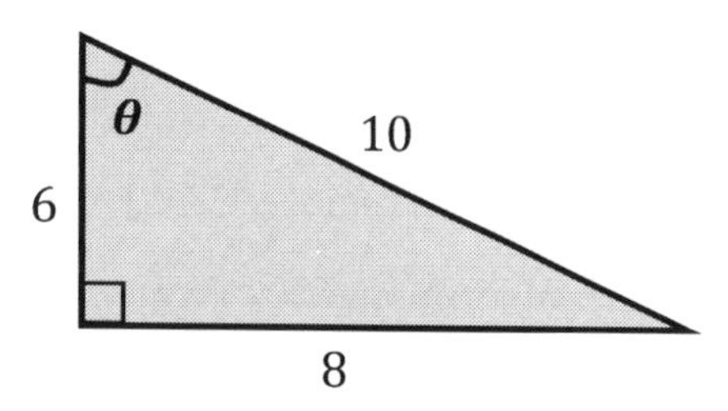

Solutions: Identify the opposite, hypotenuse and adjacent with respect to the angle, θ.

a) $\text{Sin } \theta = \frac{Opposite}{Hypotenuse} = \frac{8}{10} = \frac{4}{5}$

b) $\text{Cos } \theta = \frac{Adjacent}{Hypotenuse} = \frac{6}{10} = \frac{3}{5}$

c) $\text{Tan } \theta = \frac{Opposite}{Adjacent} = \frac{8}{6} = \frac{4}{3}$

Example 2: State the opposite, adjacent and hypotenuse for the triangle below in relation to the angle, θ.

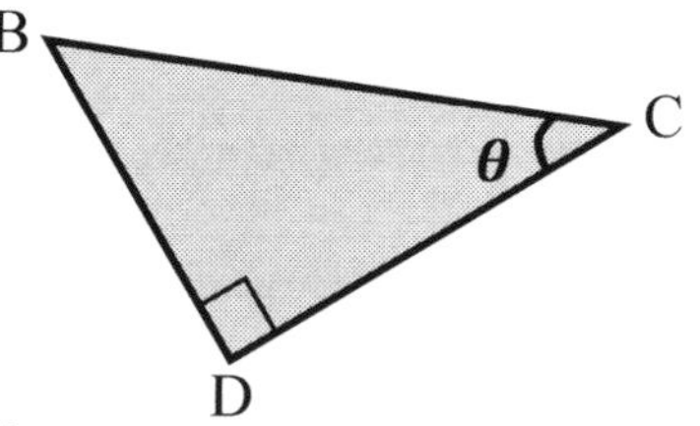

Solution:
The opposite is the side facing angle $\boldsymbol{\theta}$ This is BD.

Adjacent is the side next to angle $\boldsymbol{\theta}$ and the right angle. This is CD.

Hypotenuse is the longest side of the triangle. This is also the side opposite the right-angle sign. This is BC.

Example 3: Use your calculator to find the values of the following.
a) sin 30° b) cos 43° c) tan 84°

Solutions:
a) PRESS the **sin** button and type 30. Press the equal button. Sin 30° = **0.5**.

b) For cos 43°, press the **cos** button and type 43. Then press the equal sign (=) button to give **0.7314** to 4 decimal places.

c) For tan 84°, press the **tan** button and type 84. Then press the equal sign button (=) to give **9.5144** to 4 decimal places.

Example 4: Work out the values of

a) 3 sin 67° b) 11 cos 123° c) $\frac{8}{\sin 79^{\circ}}$

and round your answers to 3 significant figures.

Solutions: a) 3 sin 67° = 3 × sin 67°

= 3 × 0.920504853 = **2.76** to 3 s.f.

b) 11 cos 123° = 11 × cos 123°

= 11 × -0.544639035 = **-5.99** to 3 s.f.

c) $\frac{8}{\sin 79} = \frac{8}{0.981627183}$ = **8.15** to 3 s.f

Example 5: What angle has a cos of $\frac{2}{8}$?

Solution: $\frac{2}{8}$ = 0.25

We must find the inverse of cos 0.25.

This is written as $\cos^{-1}$ 0.25, and you should press the SHIFT button and then the cos button before typing 0.25.

Therefore, $\cos^{-1}$ 0.25 = **75.5°** to 3 s.f.

NOTE: Use the SHIFT button from the calculator before any of the trig ratios (sin, cos, tan) when finding angles.

USING MNEMONIC

Trigonometric ratios are easily remembered using the formula triangle and mnemonic SOH CAH TOA.

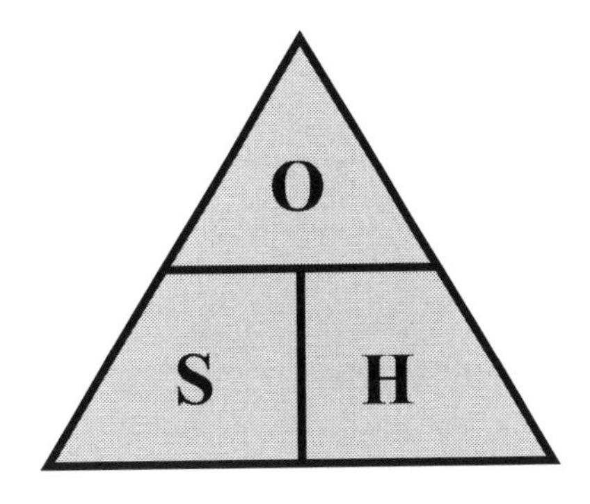

SOH: $\sin\theta = \frac{Opposite\ (O)}{Hypotenuse\ (H)}$

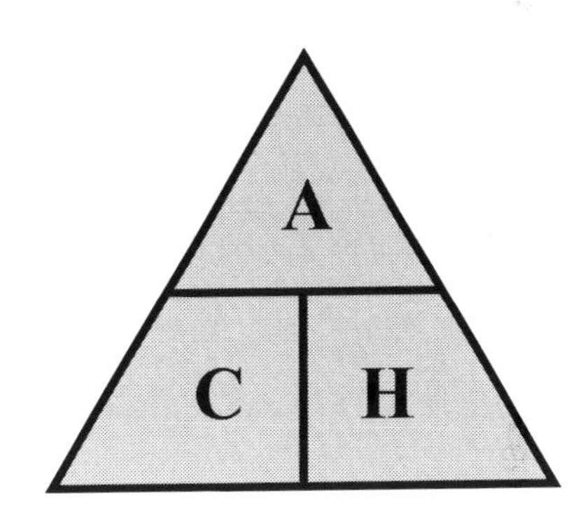

CAH: $\cos\theta = \frac{Adjacent\ (A)}{Hypotenuse\ (H)}$

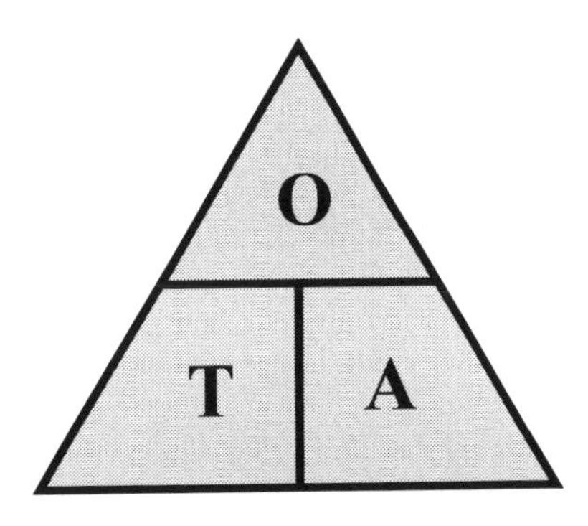

TOA: $\tan\theta = \frac{Opposite\ (O)}{Adjacent\ (A)}$

EXERCISE 25 A

1) Use a calculator to find the following and give your answers correct to 2 decimal places.

a) sin 15° d) cos 19°

b) sin 34° e) tan 34.5°

c) tan 76° f) cos 90°

2) Use a calculator to find angle theta (θ). Give your answer to the nearest whole number.

a) tan θ = 1

b) sin θ = 0.866025403

c) cos θ = 0.819152044

d) cos θ = 0.927183854

e) sin θ = 1

f) cos θ = 1

3) From the triangles below, state which side is the opposite, adjacent and the hypotenuse.

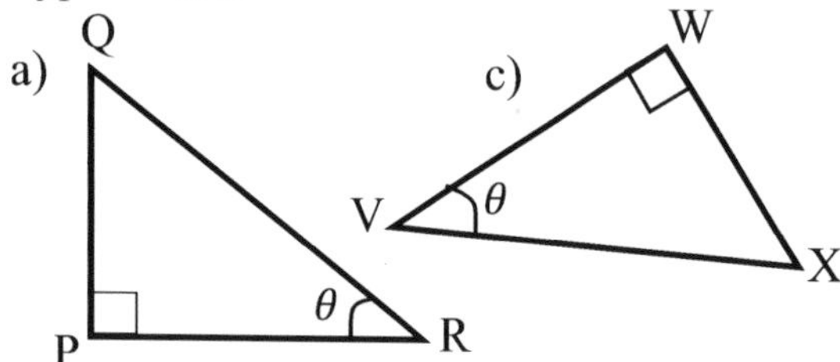

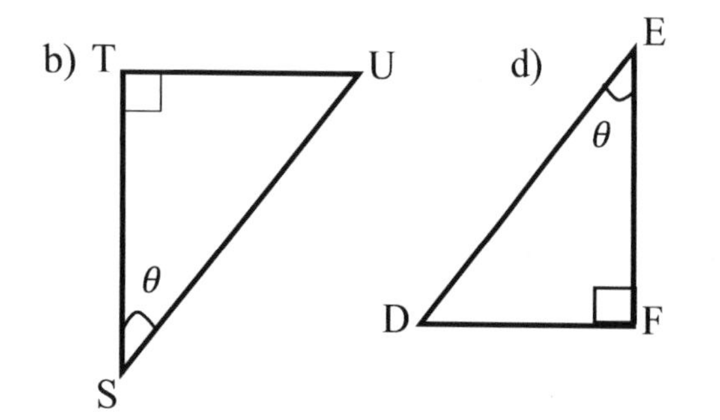

4) Work out these values to 3 decimal places.

a) 3 tan 30° d) 3.4 sin 46.9°

b) 4 sin 39° e) 10 sin 0°

c) 8 cos 54.8° f) $\frac{3}{4}$ tan 55°

5) For each of the triangles below, write sin θ, cos θ and tan θ as fractions.

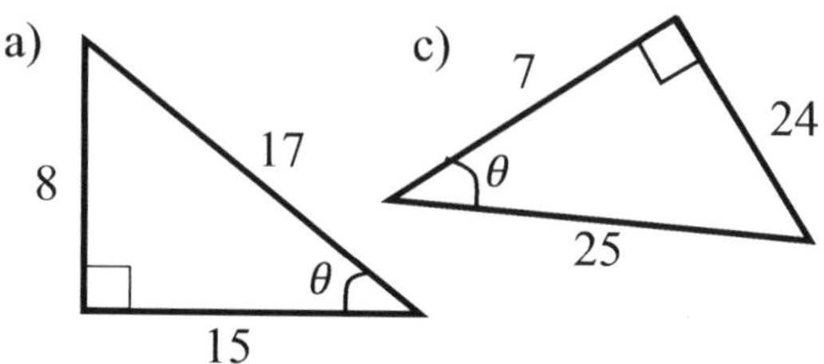

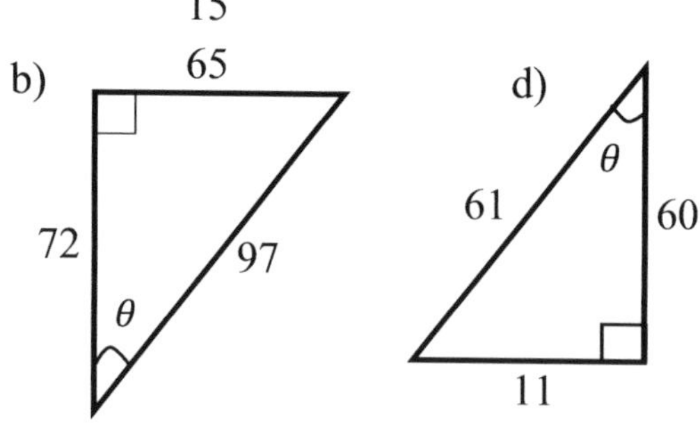

6) Calculate the values of the following and round your answers to 3 significant figures where possible.

a) $\frac{5}{\tan 5^{\circ}}$ b) $\frac{6}{\sin 30^{\circ}}$ c) $\frac{20}{\cos 77^{\circ}}$

7) Sin 30° = $\frac{1}{2}$. Using this as fact,

work out the exact values of

a) cos 60° d) tan 30°

b) tan 60° e) cos 30°

c) sin 60° f) $\sqrt{27}$ × tan 30°.

25.2 FINDING LENGTHS IN RIGHT-ANGLED TRIANGLES

Example 1: Work out the length of the side marked ***a*** in the triangle to 1 d.p.

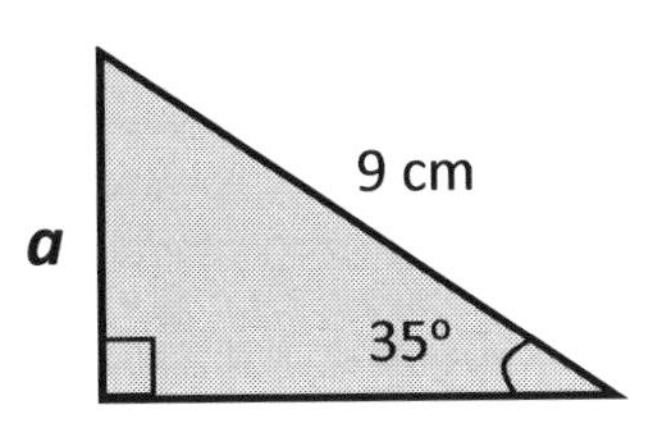

Solution: First, identify all the sides with opposite, adjacent and hypotenuse.

Opposite = a,   Hypotenuse = 9 cm

Adjacent: Not given

You must decide the trigonometric ratio (sin, cos or tan) to use in solving for the length, a.

However, adjacent is not relevant in the triangle above. Therefore any trigonometric ratio with adjacent **will not** be used.

Conversely, opposite and hypotenuse are relevant in finding the length of the above triangle. Therefore, any trigonometric ratio with opposite and hypotenuse **will be** used.

From SOH CAH TOA, the ratio with opposite and hypotenuse is **sine.** We, therefore, use SOH.

From $\sin\theta = \frac{opposite}{hypotenuse}$

$$\sin 35^{\circ} = \frac{a}{9}$$

$a = \sin 35^{\circ} \times 9 = 0.573576436 \times 9$

$= 5.2$ to 1 d.p

Example 2: Find the length of side x.

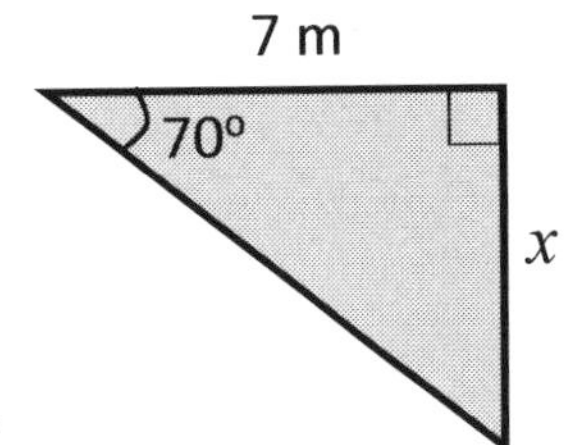

Solution:

Opposite and adjacent are relevant. Hypotenuse is not relevant.

From SOH CAH TOA, tangent (TOA) is the appropriate ratio to use.

$$\text{Tan } \theta = \frac{opposite}{adjacent}$$

$$\text{Tan } 70^{\circ} = \frac{x}{7}$$

$x = \tan 70^{\circ} \times 7 = 19.2$ m to 1 d.p.

Example 3: Find the length marked w.

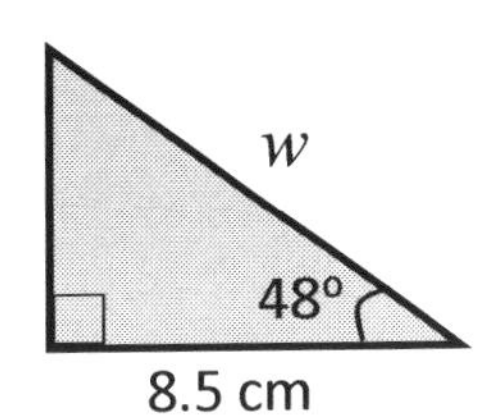

Solution: Adjacent and hypotenuse is relevant. We use cosine (CAH).

$$\text{Cos } \theta = \frac{adjacent}{hypotenuse}$$

$$\text{Cos } 48^{\circ} = \frac{8.5}{w}$$

Make w the subject of the formula.

$$w = \frac{8.5}{\cos 48^{\circ}}$$

$= 12.7$ to 1 d.p.

EXERCISE 25 B

1) Work out the length of the side marked x in each triangle to 1 decimal place. All the lengths are in metres.

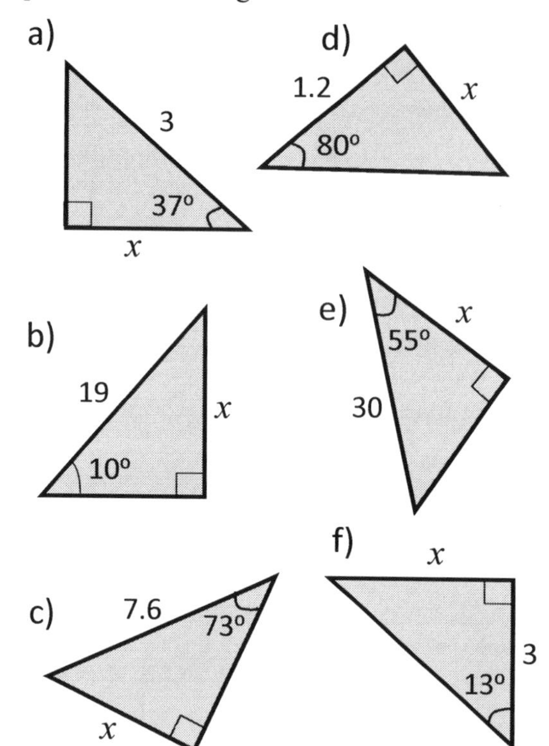

2) A 7.8 m plank leans against a tree as shown.

a) How far is the top of the plank from the ground?

b) How far is the bottom of the plank from the tree?

3) A container ship sails 1200 km on a bearing of 063°.

a) How far east has the ship sailed?

b) How far north has the ship sailed?

4) Find the length marked x in the triangles. All the lengths are in cm.

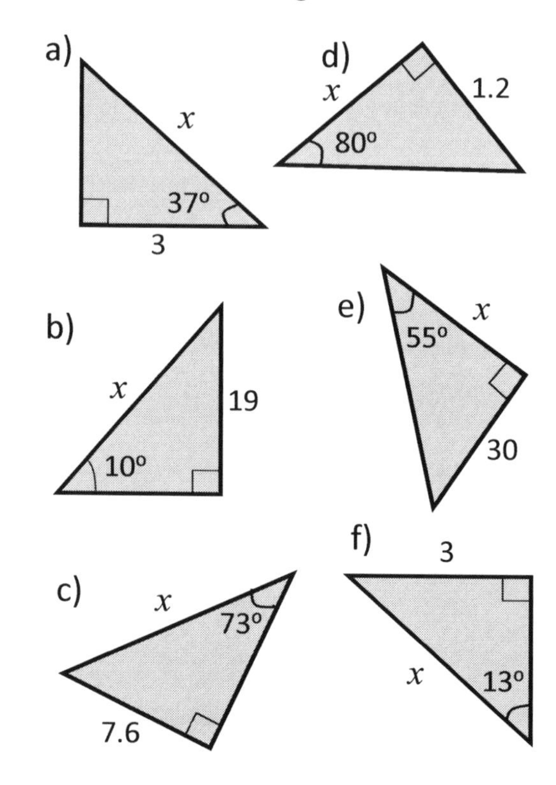

5) Find the missing lengths marked with letters. Round your answers to 3 significant figures where possible.

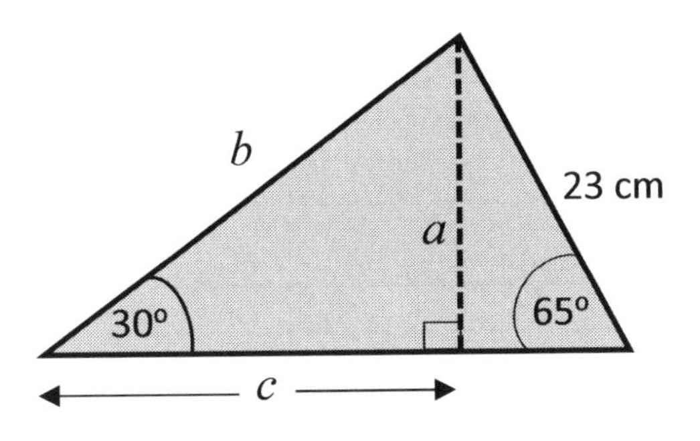

6)

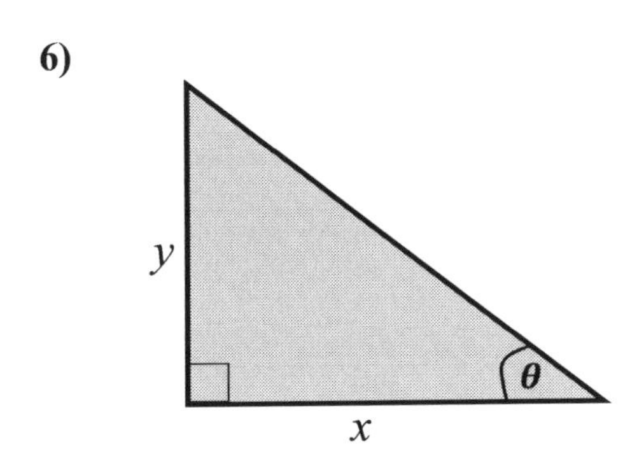

a) Work out the length, y in terms of x and θ.

b) Find the formula for the area of the triangle in terms of x and θ.

7) PQR is a triangle with the following:

$\angle RPQ = 90^{\circ}$

$\angle PRQ = 26.5^{\circ}$

QR = 19.3 m

a) Sketch the triangle.

b) Calculate the length of PQ.

c) Calculate the length of PR.

8)

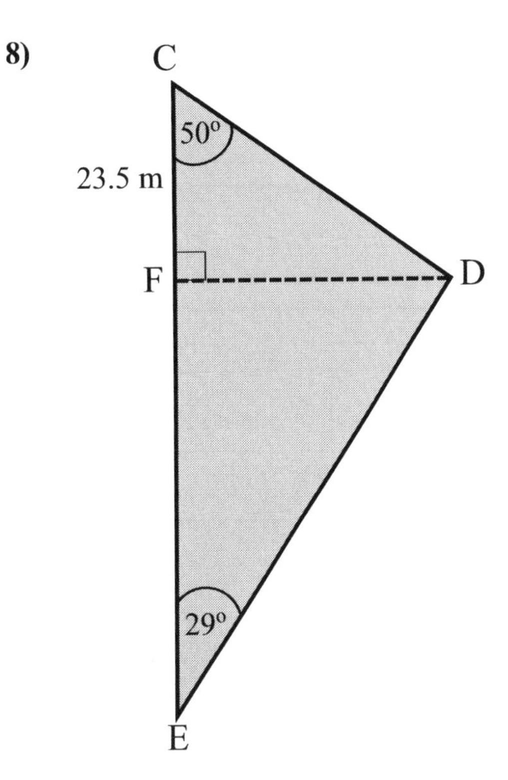

CDE is a triangle. Calculate

a) length DE

b) length DF

c) length EF.

9) In question 8 above, calculate the area of triangles

a) CDF

b) CDE

10) Work out the perimeter of triangle CDE in question 8 above.

25.3 FINDING ANGLES IN RIGHT-ANGLED TRIANGLES

In a right-angled triangle, if the lengths of any two sides are known, then the remaining angles can be found by using sine, cosine or tangent.

Example 1: Find the value of the angle marked θ in the triangle below to 1 decimal place.

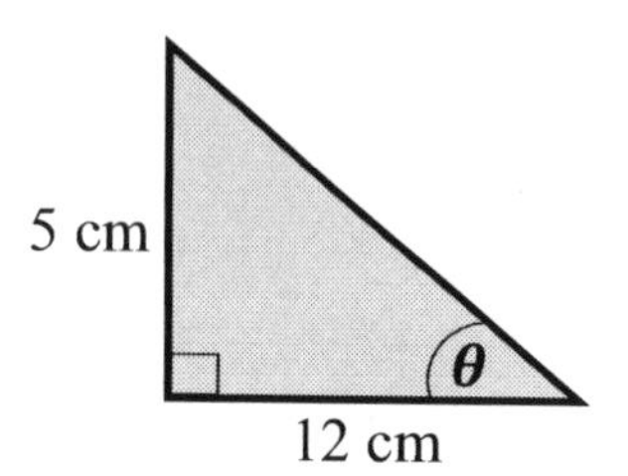

Solution:

We know the opposite (5 cm) and the adjacent (12 cm). Therefore, tangent (TOA) will be used.

$$\text{Tan } \theta = \frac{opposite}{adjacent} = \frac{5}{12} = 0.41666666$$

Using the SHIFT and TAN button on a calculator, $\theta = \tan^{-1} 0.41666666 = \mathbf{22.6^o}$

Example 2: Find the angle marked θ.

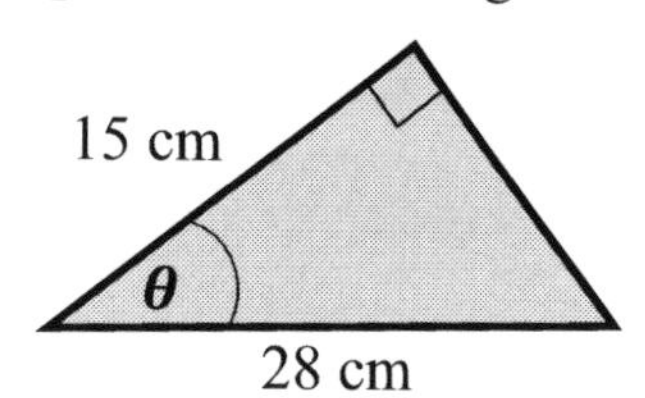

We have adjacent and hypotenuse; we use cosine (CAH). $\text{Cos } \theta = \dfrac{adjacent}{hypotenuse}$

$$\text{Cos } \theta = \frac{15}{28} = 0.535714285$$

$$\theta = \text{Cos}^{-1}\ 0.535714285 = \mathbf{57.6^o}$$

EXERCISE 25 C

1) Find the angles marked x in the triangles. All the lengths are in cm.

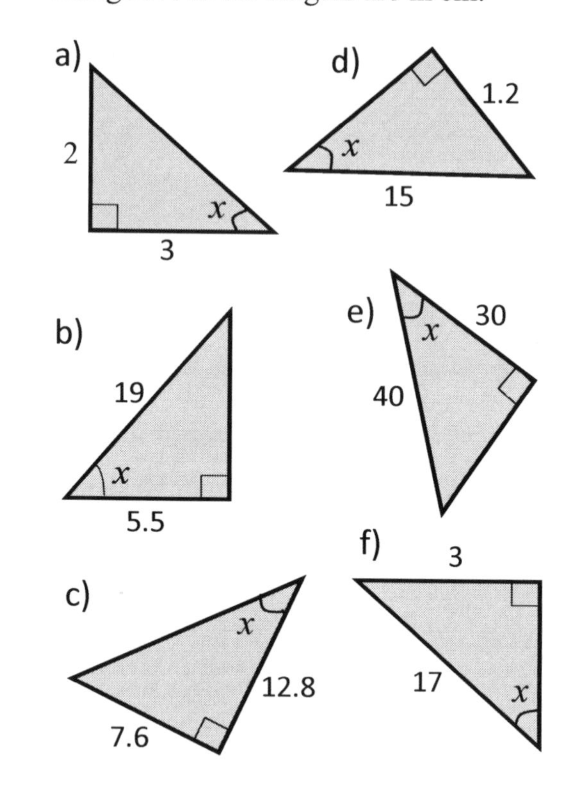

2)

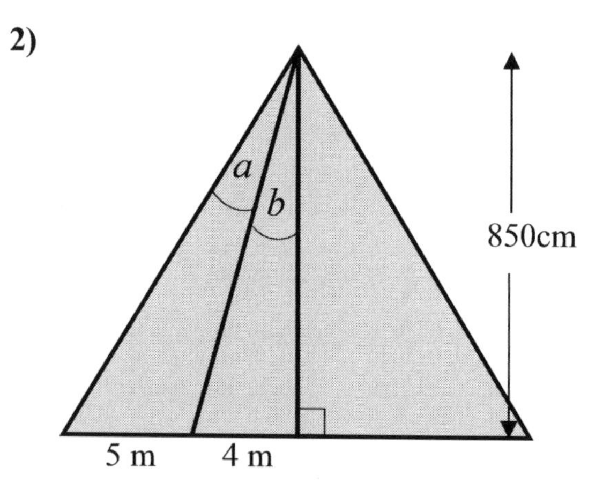

Work out angles ***a*** and ***b***.

25.4 EXACT TRIGONOMETRIC VALUES

In this section, we shall consider working out the exact values of sin, cos and tan ratios for 0°, 30°, 45°, 60° as well as the values of sin and cos for 90°.

Note: tan 90° is not defined and therefore, there is no value for it. The simple reason being that

$\tan\theta = \frac{\sin\theta}{\cos\theta}$ and if $\theta = 90^o$,

$\tan 90^o = \frac{\sin 90^o}{\cos 90^o} = \frac{1}{0}$. This is undefined and **not** allowed in maths when the denominator is zero. Hence, there is no tan ratio for 90°.

Pupils are expected to **memorise** the values of these ratios. However, some will forget and hence the reason for this section.

EXACT VALUES

Angle θ	**Sin θ**	**Cos θ**	**Tan θ**
0°	0	1	0
30°	$\frac{1}{2}$	$\frac{\sqrt{3}}{2}$	$\frac{1}{\sqrt{3}}$
45°	$\frac{1}{\sqrt{2}}$	$\frac{1}{\sqrt{2}}$	1
60°	$\frac{\sqrt{3}}{2}$	$\frac{1}{2}$	$\sqrt{3}$
90°	1	0	**Undefined**

WORKING OUT SINE, COSINE AND TANGENT RATIOS FOR 30° and 60°

An equilateral triangle of side 2 cm is drawn below. ∠ABC is bisected by BD to give 30° each.

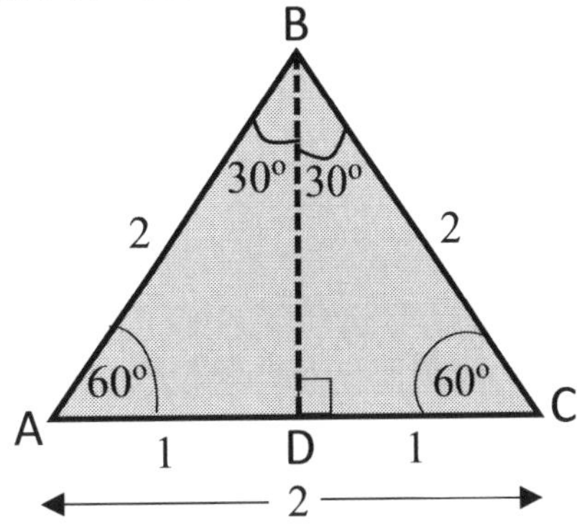

Considering triangle ABD, the length BD can be calculated using Pythagoras' theorem.

$BD^2 = 2^2 - 1^2 = 4 - 1 = 3$

$BD = \sqrt{3}$

Using SOH CAH TOA.

$\sin 30^o = \frac{opposite}{hypotenuse} = \frac{1}{2}$

$\sin 60^o = \frac{opposite}{hypotenuse} = \frac{\sqrt{3}}{2}$

$\cos 30^o = \frac{adjacent}{hypotenuse} = \frac{\sqrt{3}}{2}$

$\cos 60^o = \frac{opposite}{hypotenuse} = \frac{1}{2}$

$\tan 30^o = \frac{opposite}{adjacent} = \frac{1}{\sqrt{3}}$

$\tan 60^o = \frac{opposite}{adjacent} = \frac{\sqrt{3}}{1} = \sqrt{3}$

WORKING OUT SINE, COSINE AND TANGENT RATIOS FOR 45°

To show these trigonometric ratios, draw a right-angled isosceles triangle with the equal sides of 1 cm.

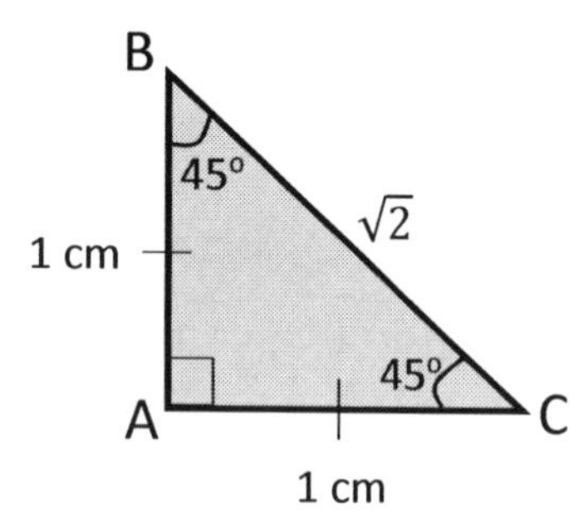

Using Pythagoras' theorem,

$BC^2 = 1^2 + 1^2 = 1 + 1 = 2$

$BC = \sqrt{2}$

Using SOH CAH TOA,

$\sin 45° = \frac{opposite}{hyopotenuse} = \frac{1}{\sqrt{2}}$

$\cos 45° = \frac{adjacent}{hypotenuse} = \frac{1}{\sqrt{2}}$

$\tan 45° = \frac{opposite}{adjacent} = \frac{1}{1} = 1$

Example 1: Work out the exact value of sin 60° + cos 60° without using a calculator.

Solution: $\sin 60° = \frac{\sqrt{3}}{2}$ and $\cos 60° = \frac{1}{2}$

$\sin 60° + \cos 60° = \frac{\sqrt{3}}{2} + \frac{1}{2} = \frac{\sqrt{3}+1}{2}$ ✓

EXERCISE 25 D

1) Without using a calculator, calculate the **exact values** of the following.

a) sin 30° + sin 60°

b) 2 tan 30°

c) cos 30° + tan 60°

d) sin 30° + cos 60° + sin 45°

e) 4 tan 60° + cos 30°

f) tan 60° ÷ tan 30°

g) Prove that $\sin 60° + \tan 30° = \frac{5\sqrt{3}}{6}$.

2) Work out the **exact values** of the missing lengths in metres.

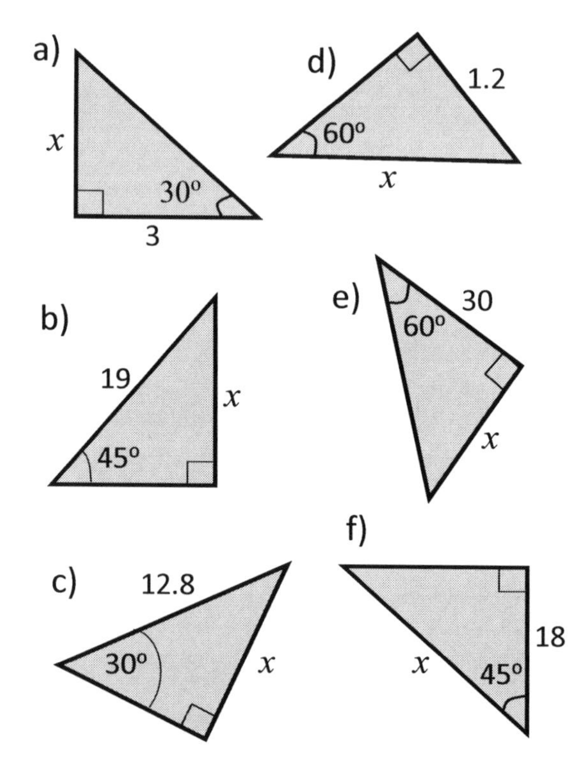

SINE 0° and 90°

A quick look at the sine graph will give the values of sin 0° and 90°.

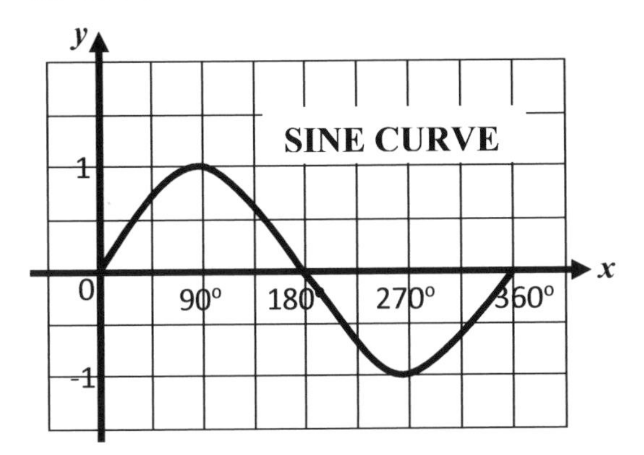

From the sine graph above, **sin 0° = 0** and sin **90° = 1**.

COSINE 0° and 90°

The cosine curve gives the value of cos 0° and cos 90°.

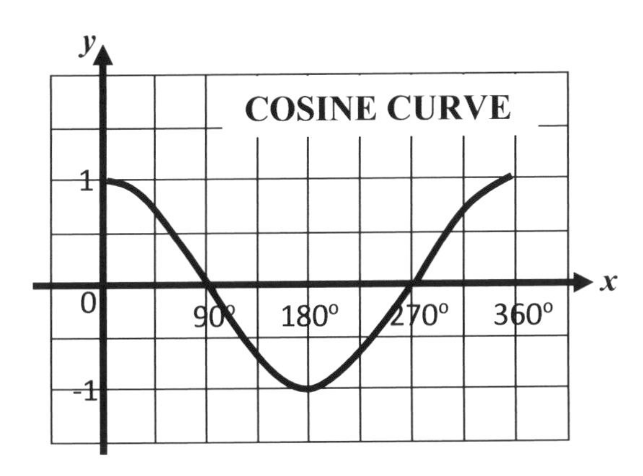

From the graph above, **cos 0° = 1** and **cos 90° = 0**.

SQUARE OF TRIGONOMETRIC RATIOS

$\text{Sin}^2 X = (\sin X)^2$

$\cos^2 X = (\cos X)^2$ and $\tan^2 X = (\tan X)^2$

Example 2: Find the value of $\sin^2 30°$.

Solution: $\sin 30° = \frac{1}{2}$

$\sin^2 30° = \left(\frac{1}{2}\right)^2 = \frac{1}{2} \times \frac{1}{2} = \mathbf{\frac{1}{4}}$

Example 2:

Find the value of $\tan^2 60° + \sin^2 45°$

Solution: $\tan 60° = \sqrt{3}$ and $\sin 45° = \frac{1}{\sqrt{2}}$

$\tan^2 60° = (\sqrt{3})^2 = \sqrt{3} \times \sqrt{3} = 3$

$\sin^2 45° = (\frac{1}{\sqrt{2}})^2 = \frac{1}{\sqrt{2}} \times \frac{1}{\sqrt{2}} = \frac{1}{2}$

$\tan^2 60° + \sin^2 45° = 3 + \frac{1}{2} = \mathbf{3\frac{1}{2}}$

EXERCISE 25E

1) Evaluate

a) $\sin^2 60°$ c) $\tan^2 30°$ e) $\tan^2 60°$

b) $\sin^2 45°$ d) $\cos^2 45°$ f) $\cos^2 60°$

2) Find the exact values of

a) $\tan^2 30° + \sin^2 60°$

b) $\tan^2 60° + \sin^2 60°$

c) $\sin^2 45° + \cos^2 60° + \sin^2 60°$

d) $\sin^2 30° + \cos^2 30°$

EXERCISE 25F (MIXED)

1) Calculate the value of:

a) the missing length, w
b) ∠CED.
c) Work out the area of triangle CDE.

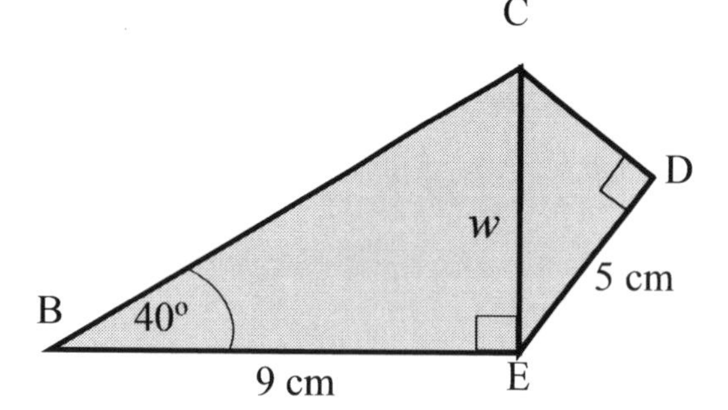

2) John looks out to sea from a cliff top at a height of 15 m. He sees a small boat 200 m from the cliffs. Work out the angle of depression.

3) Calculate the area of the triangle below.

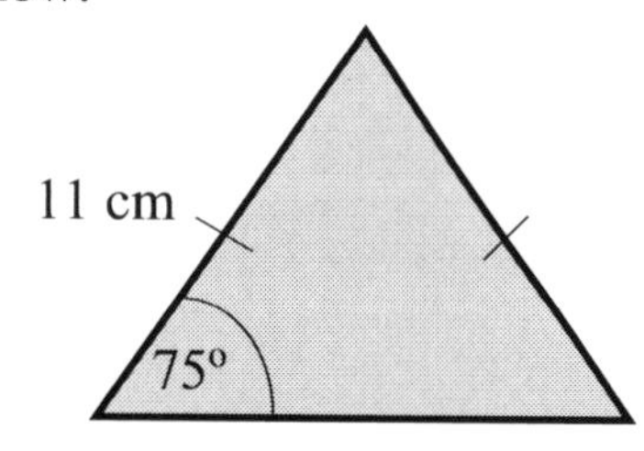

4) Arinze boarded a ship that sails on a bearing of 130° for 7 km.

a) How far east has the ship travelled?

b) How far south has the ship travelled?

5)

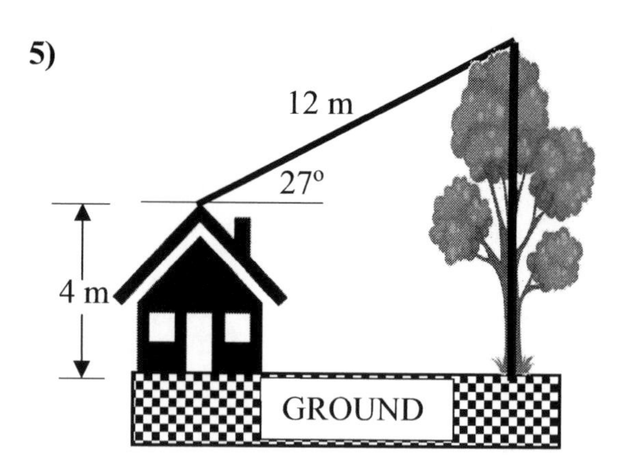

A rope 12 m long is tied from the highest point on a building to the top of a tree.

a) Work out the height of the tree.

b) Work out the angle the rope makes with the tree.

6) Work out the length marked *w*.

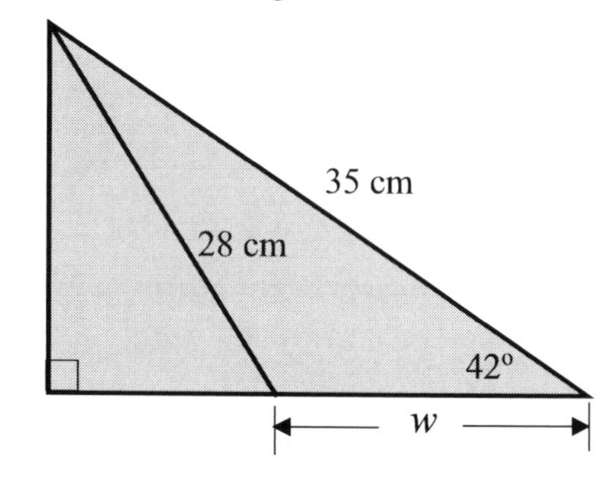

7)

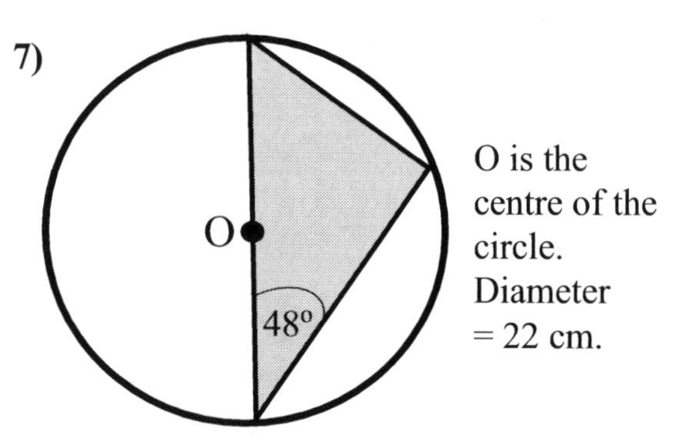

O is the centre of the circle. Diameter = 22 cm.

Calculate the area of the triangle.

8) The diagram below shows a Nigerian flag on top of a church. Work out the height of the flag.

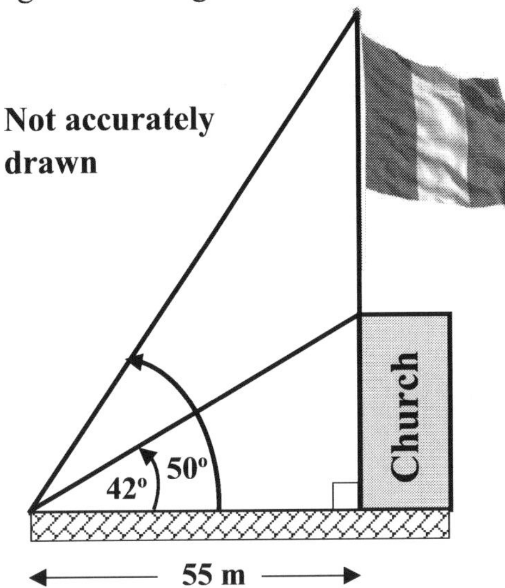

9) ABCD is a kite with the dimensions shown.

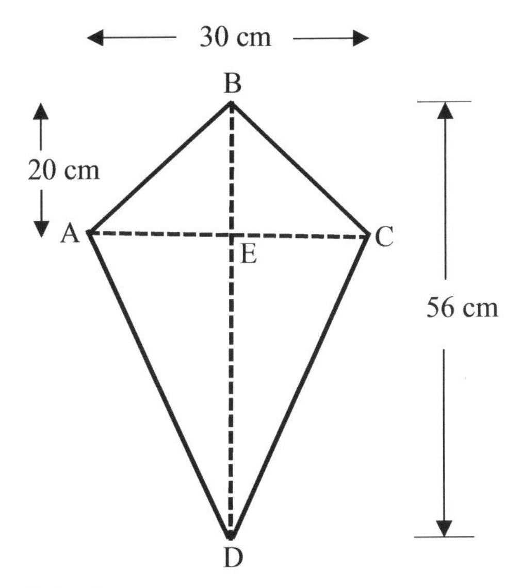

Calculate

a) length AD c) angle EDC

b) length BC d) angle ABE.

10) The height of a school building is 9.5 metres high. Clive looks at the school from a distance of 90 metres.

a) Sketch the diagram.

b) Work out the angle of elevation of the top of the school from Clive to the nearest degree.

11)

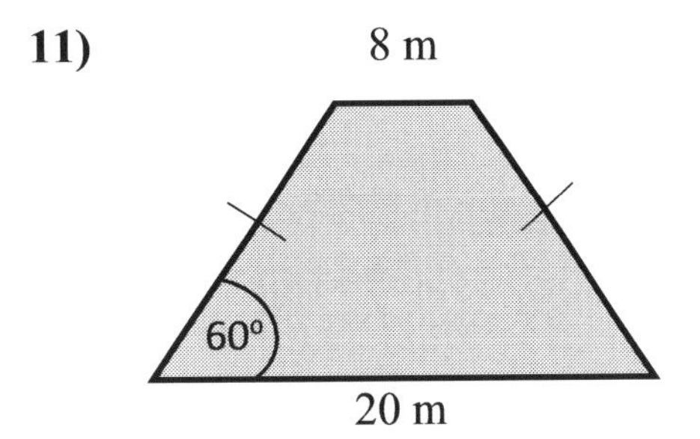

a) What is the mathematical name for the shape above?

b) Work out the exact area of this shape.

12)

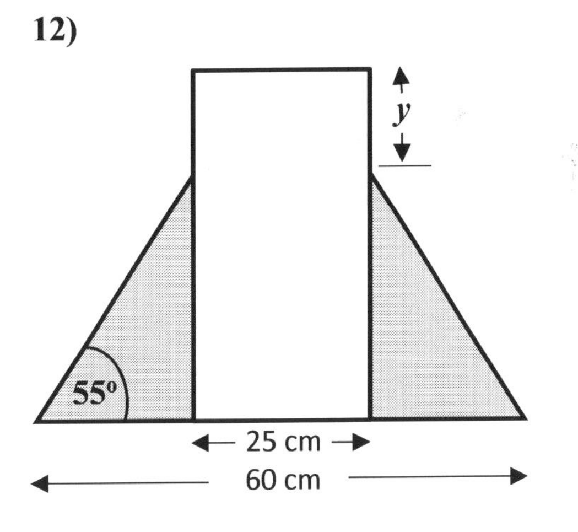

This shape is made of 2 identical triangles and a rectangle. Area of the rectangle is 750 cm^2.

Calculate the height, ***y*** to the nearest whole number.

25.5 THE SINE RULE

The **sine rule** works in any triangle and helpful in triangles that are not right-angled.

Sine rule is used for finding **a side** when a side and two angles are known in a triangle. It is also used to calculate **an angle** when two sides and an angle opposite one of the sides are known.

In the triangle below, the side opposite angle A is assigned length a, the side opposite angle B is assigned length b, and the side opposite angle C is assigned length c.

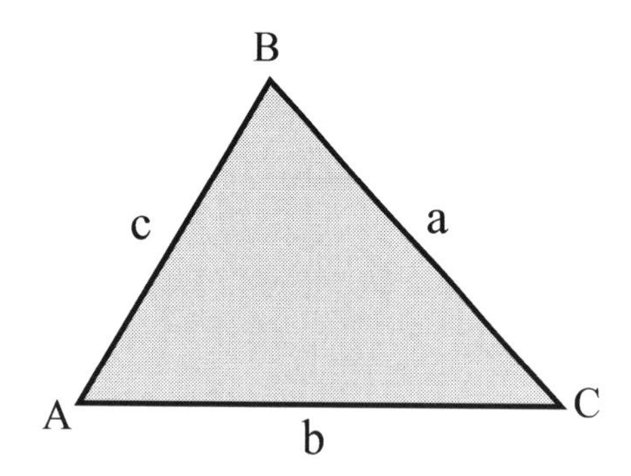

The sine rule states that:

$$\frac{a}{\sin A} = \frac{b}{\sin B} = \frac{c}{\sin C}$$

The format for the sine rule above is used when working out a length.

The preferred format for working out an angle is

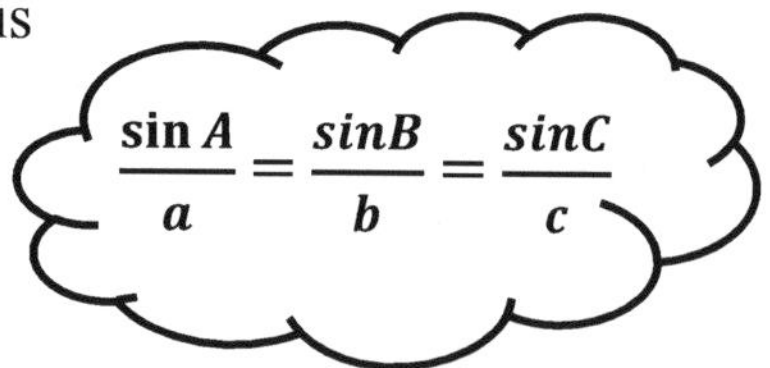

Example 1: Calculate the missing length. Round to 2 decimal places.

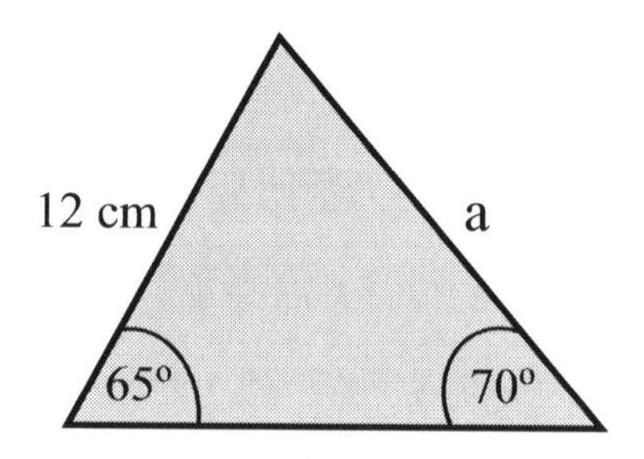

Solution:

Using the sine rule,

$$\frac{a}{\sin A} = \frac{b}{\sin B} = \frac{c}{\sin C}$$

The last part is not relevant. Therefore,

$$\frac{a}{\sin 65^{\circ}} = \frac{12}{\sin 70^{\circ}}$$

Rearrange to make ***a*** the subject.

$$a = \frac{12}{\sin 70^{\circ}} \times \sin 65^{\circ} = \mathbf{11.57} \text{ cm.}$$

Example 2: Work out the value of the missing angle.

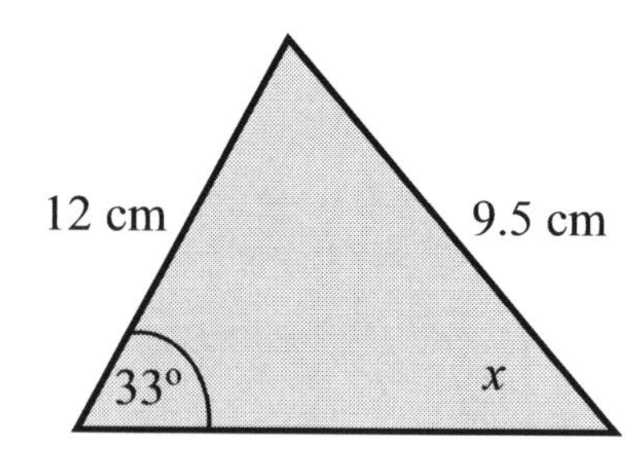

Solution: $\frac{\sin x}{12} = \frac{\sin 33^{\circ}}{9.5}$

$$\sin x = \frac{\sin 33^{\circ}}{9.5} \times 12 = 0.687965096$$

$$x = \sin^{-1} 0.687965096 = \mathbf{43.5^{\circ}} \text{ to 1 d.p}$$

Example 3: Calculate the obtuse angle **Q**, in the diagram below.

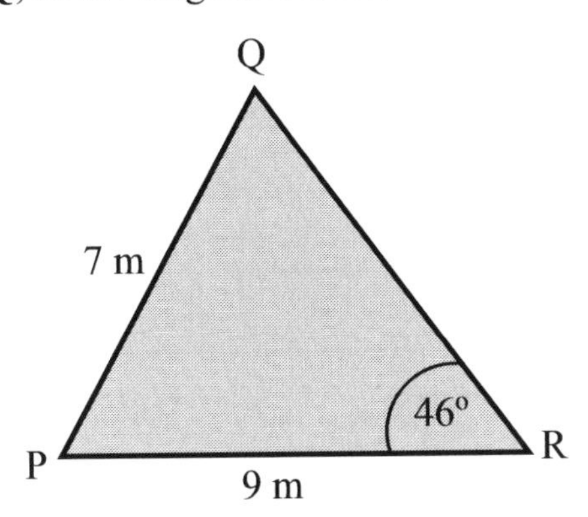

Solution: Work out angle Q using the sine rule. After that, find the obtuse angle by subtracting from 180°.

Using the sine rule $\frac{\sin Q}{9} = \frac{\sin 46^o}{7}$

$\sin Q = \frac{\sin 46^o}{7} \times 9 = 0.924865457$

$\angle Q = \sin^{-1} 0.924865457 =$ **67.6°**

However, the question mentioned "Obtuse" angle. 180 – 67.6 = **112.4°**

EXPLANATION:

From the sine graph, there are two solutions between 0° and 180°.

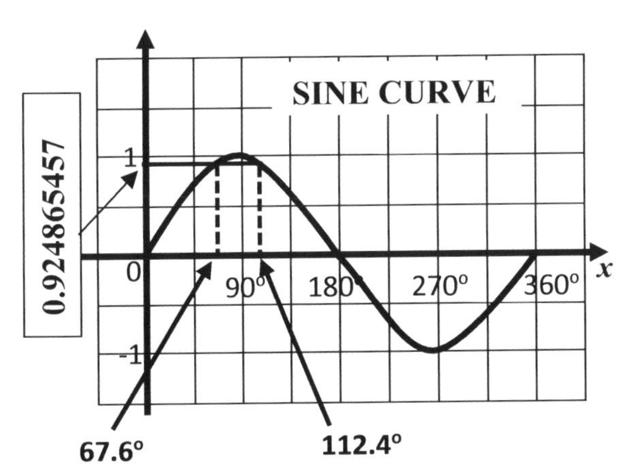

EXERCISE 25G

1) Work out the value of ***w*** in each of the following equations to 1 dp.

a) $\frac{w}{sin20^o} = \frac{5}{\sin 50^o}$ c) $\frac{11}{sin20^o} = \frac{w}{\sin 85^o}$

b) $\frac{w}{sin56^o} = \frac{4.8}{\sin 47^o}$ d) $\frac{4.5}{sin20^o} = \frac{w}{\sin 10^o}$

2) Calculate the length of each side marked with a letter. All lengths in **cm**. Give your answers to 2 decimal places where possible.

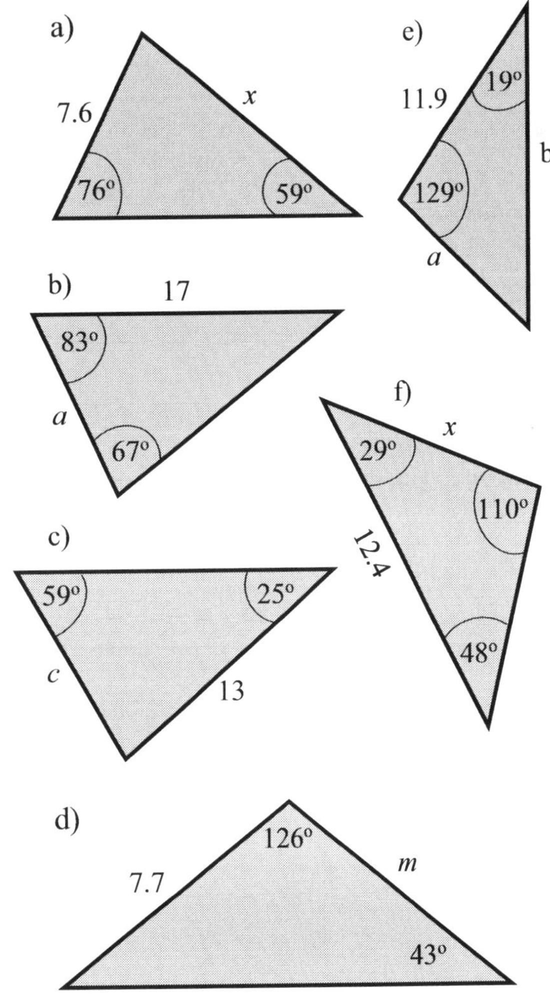

3) Find the unknown angles for each of the triangles. All lengths are in **metres**. Round to 1 decimal place.

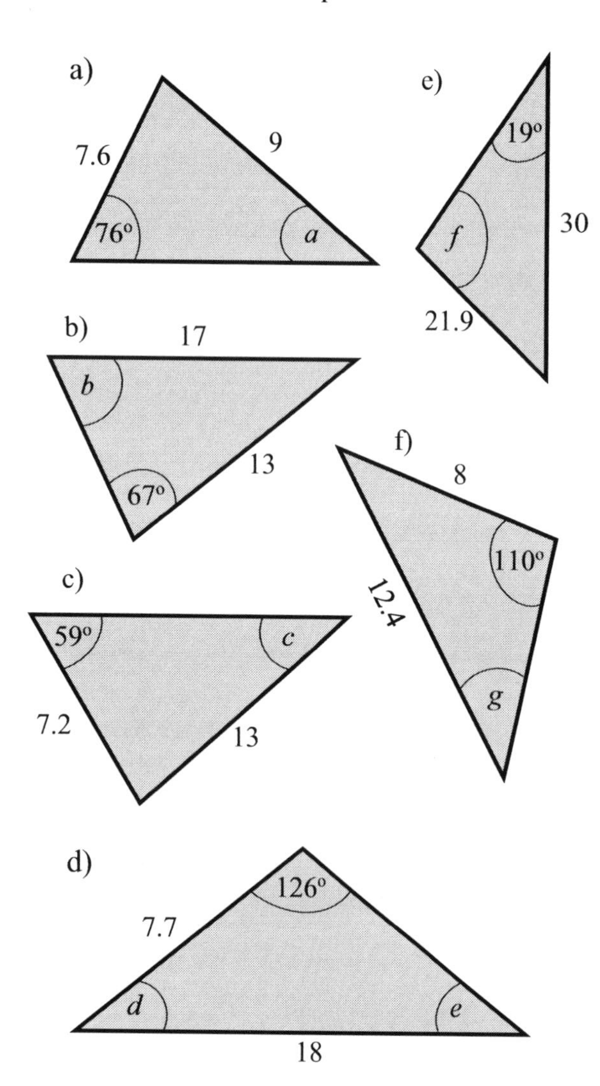

4) In Δ PQR, QR = 20 m, PR = 12 m and ∠QPR = 50°.

a) Sketch the triangle.

b) Explain why ∠ PQR must be less than 50°.

c) Calculate ∠PRQ.

25.6 THE COSINE RULE

Use the cosine rule in any triangle to:

a) find a side when two sides and the angle between them are known.

b) find an angle when you know all the three sides.

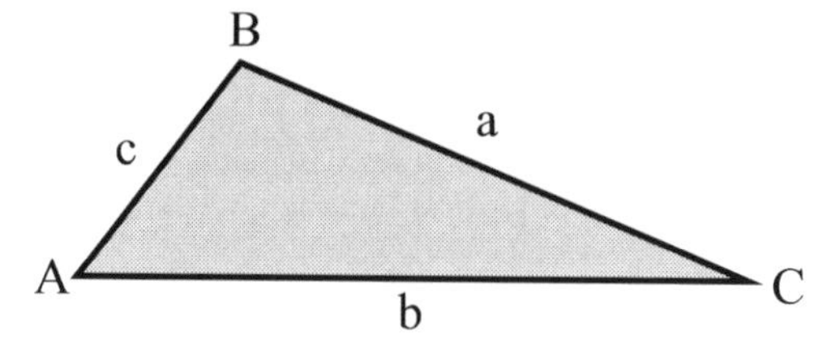

The cosine rule states that

$$a^2 = b^2 + c^2 - 2bc \cos A$$ ✓

Pupils are expected to memorise the cosine rule.

From the same reason,

$b^2 = a^2 + c^2 - 2ac \cos B$

$c^2 = a^2 + b^2 - 2ab \cos C$

You may use the formats above to find a length.

However, for ease of manipulations, the formats below may be used to find an angle.

$$\text{Cos } A = \frac{b^2 + c^2 - a^2}{2bc}$$

$$\text{Cos } B = \frac{a^2 + c^2 - b^2}{2ac}$$

$$\text{Cos } C = \frac{a^2 + b^2 - c^2}{2ab}$$

Example 1: Calculate the length of PQ and round your answer to three significant figures.

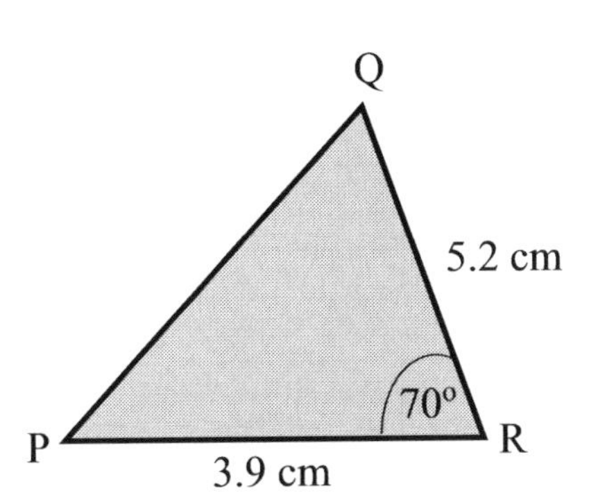

Solution: Cosine rule will be used. Notice that angle 70° is the angle between the two sides PR and QR.

Let length PQ = x

$x^2 = 3.9^2 + 5.2^2 - 2 \times 3.9 \times 5.2 \text{ Cos}70^\circ$

Work out the calculation in one step on your calculator. If not sure, put a bracket around the two parts and work it out as seen below.

$x^2 = (3.9^2 + 5.2^2) - (2 \times 3.9 \times 5.2 \cos 70^\circ)$

$x^2 = 42.25 - 13.87233701$

$x^2 = 28.37766299$

$x = \sqrt{28.37766299} = 5.327068893$ cm.

Therefore, PQ = **5.33 cm** to 3 s.f.

Note: DO NOT round any part of your calculations until you get to the final answer. You ONLY round your final answer based on the question. In this case: to three significant figures.

Example 2: Calculate the size of angle ***a*** and round the answer to 3 significant figures.

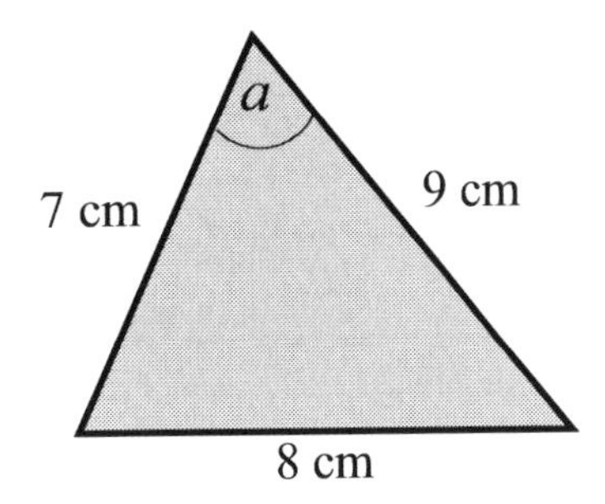

Solution: Use the cosine rule because three lengths are given.

$8^2 = 7^2 + 9^2 - 2 \times 7 \times 9 \cos a$

$64 = 49 + 81 - 126 \cos a$

$64 = 130 - 126 \cos a$

Take 130 away from both sides

$-66 = -126 \cos a$

Divide both sides by -126

$\frac{-66}{-126} = \cos a$

The two minus signs will cancel out.

$\cos a = \frac{66}{126} = 0.523809523$

$a = \cos^{-1} 0.523809523$

SHIFT and then COS button

$a =$ **58.4°** to three significant figures.

Alternatively, use the format

$\cos a = \frac{7^2 + 9^2 - 8^2}{2 \times 7 \times 9} = \frac{66}{126} = 0.52380...$

$a = \cos^{-1} 0.523809523$

a = **58.4°** to 3 s.f.

EXERCISE 25H

1) Find the length of the side marked with a letter and give your answers to three significant figures where possible. All lengths are in cm and not accurately drawn.

2) Work out the size of each angle marked with letters. All lengths are in metres. Round your answers to the **nearest degree**. Diagrams are not accurately drawn.

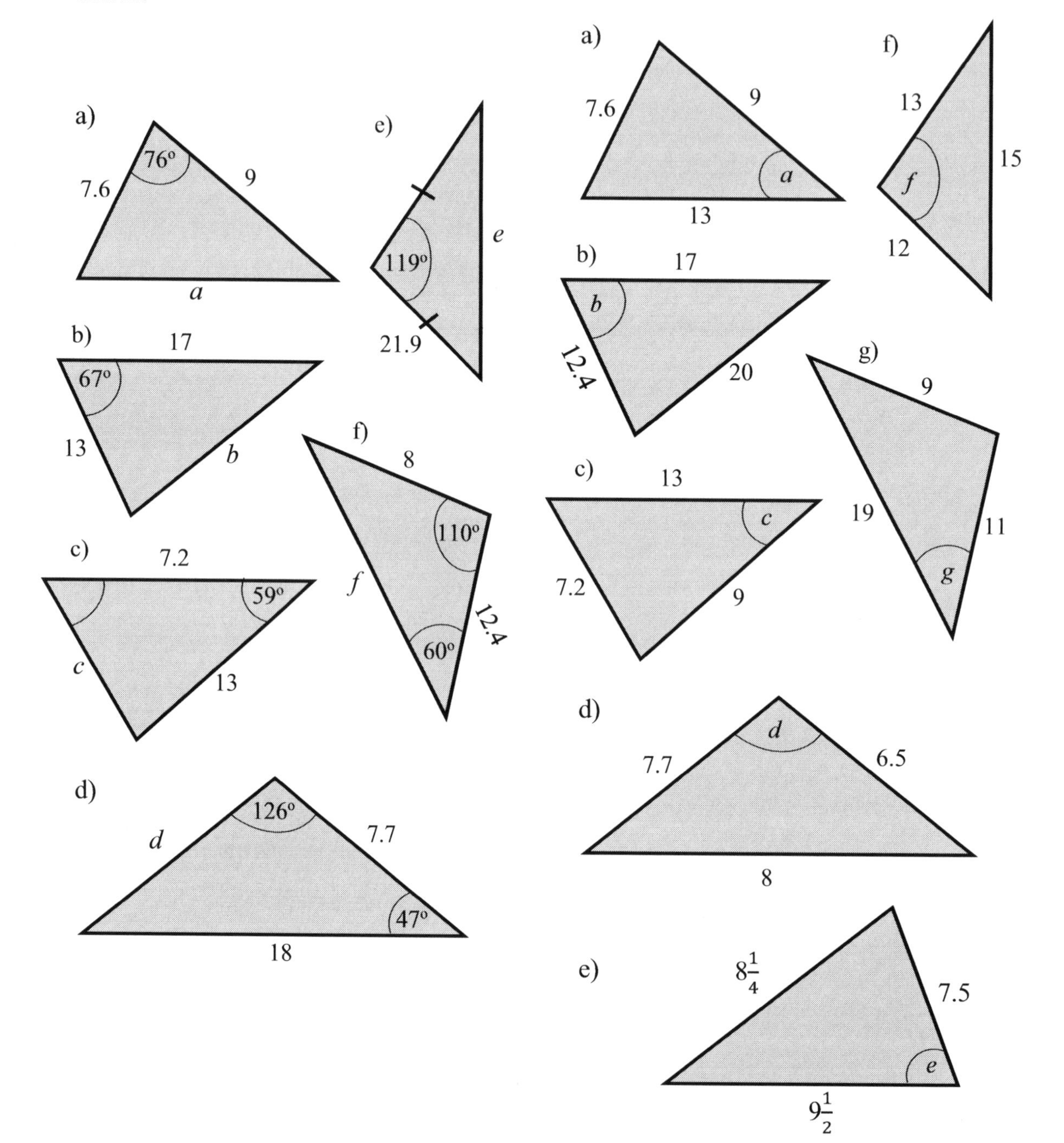

3) Work out the length of the diagonal BD to three decimal places.

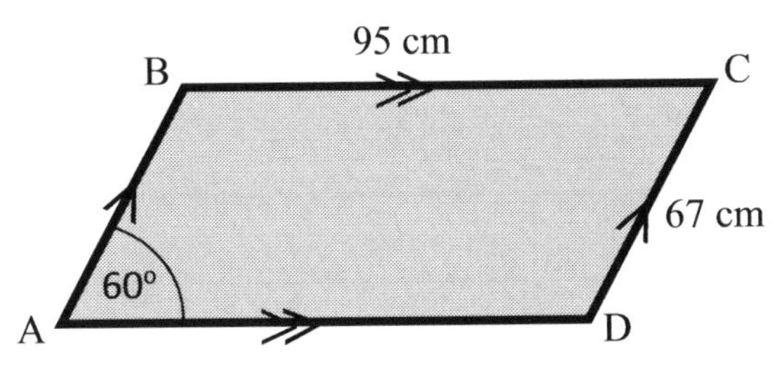

4) Calculate the missing lengths.

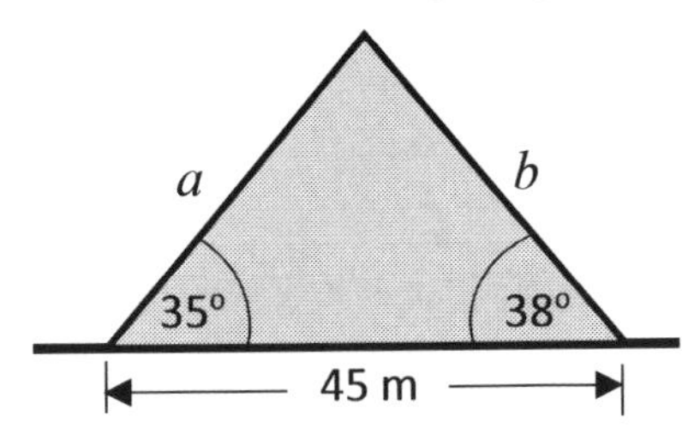

5) From triangle PQR, show that

$r^2 = p^2 + q^2 - 2pq \cos R$

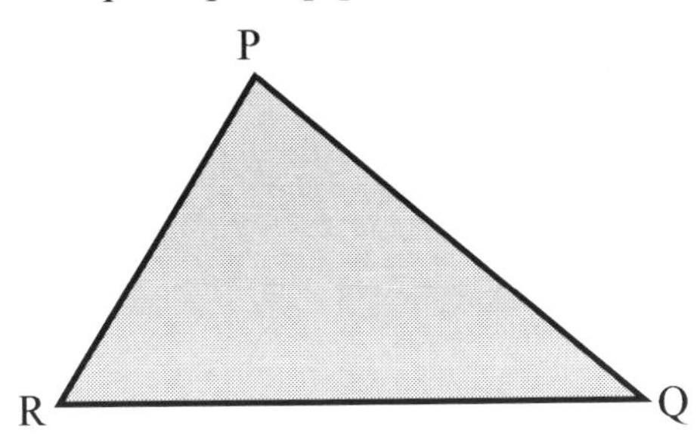

6) Town A is 35km due north of town B. Town C is 53km from town B, and 60 km from town A. Work out the bearing of town C from town B.

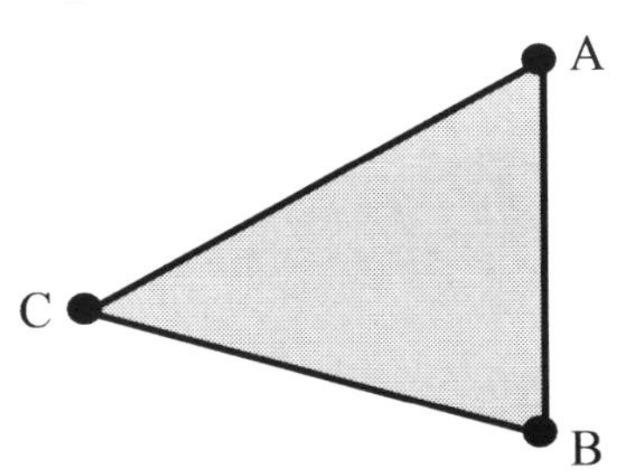

25.7 THE AREA RULE

The area of a triangle without a right-angle can be calculated using trigonometry. In the past, you have worked out the area of a right-angled triangle by using the formula:

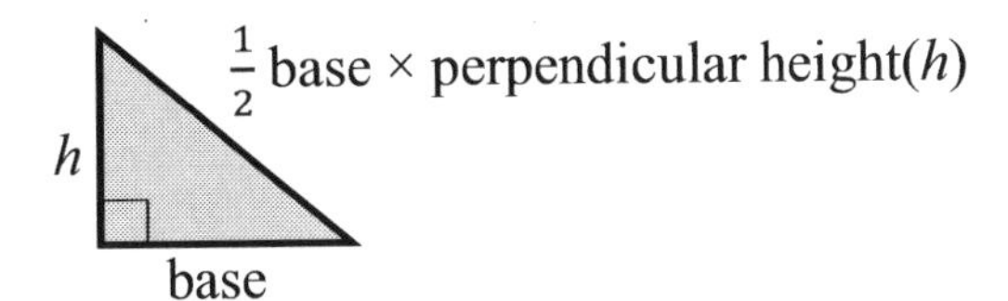

DERIVING THE FORMULA FOR AREA OF ANY TRIANGLE

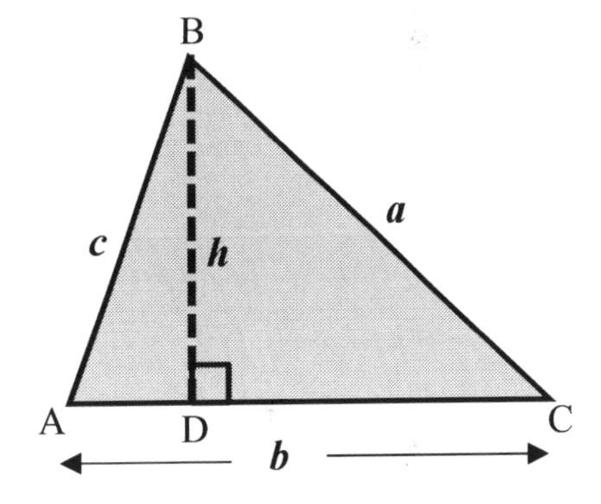

Draw a perpendicular from B to AC to make a right-angled triangle. Let height BD = $\boldsymbol{h}$.

From Δ BDC, $\sin C = \frac{Opposite}{Hypotenuse} = \frac{h}{a}$

and $h = a \sin C$.

But, the area of Δ ABC $= \frac{1}{2} \times$ base × height $= \frac{1}{2} \times b \times a \sin C$. In a simplified format, Area $= \boldsymbol{\frac{1}{2} ab \sin C}$ ✓

This is the formula for finding the area of any triangle if an angle is given.

Example 1: Calculate the area of triangle ABC. Give your answer to three significant figures.

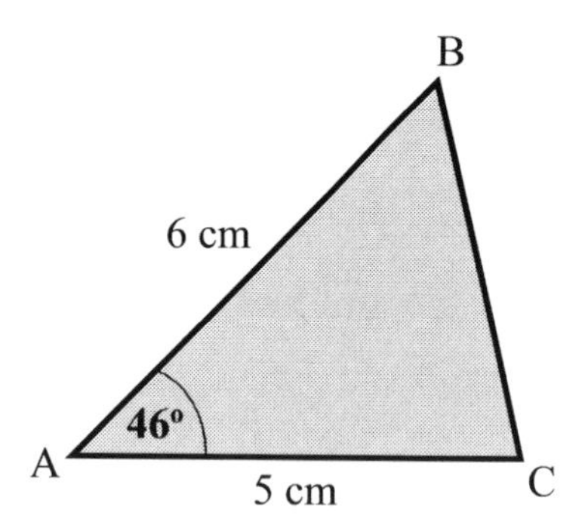

Solution: Area $= \frac{1}{2} \times 5 \times 6 \times \sin 46^\circ$

$A = 15 \times 0.7193398 = 10.8 \text{ cm}^2$ to 3 s.f.

NOTE: The angle used must be an included angle (angle between two known lengths).

Example 2: The area of the triangle below is 66.47 m². Work out the obtuse angle, w.

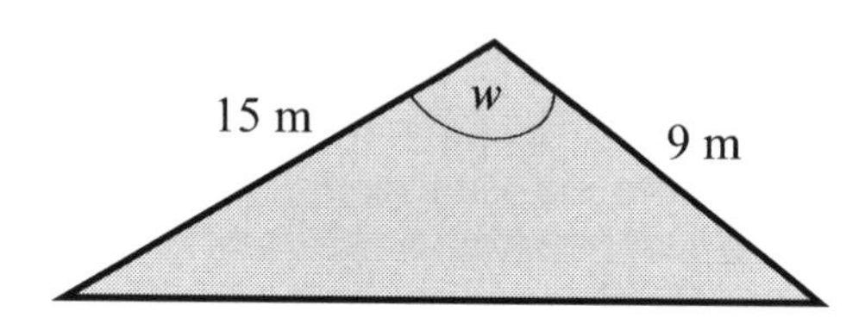

Solution: Area $= \frac{1}{2} \times 15 \times 9 \times \sin w$

$66.47 = 67.5 \sin w$

$\text{Sin } w = \frac{66.47}{67.5} = 0.98474074$

$w = \sin^{-1} 0.98474074 = 80^\circ$ …to the nearest degree.

Obtuse angle $= 180^\circ - 80^\circ =$ **100°**

EXERCISE 25I

1) Work out the area of these triangles. All lengths are in cm. Give your answer to 3 significant figures where necessary.

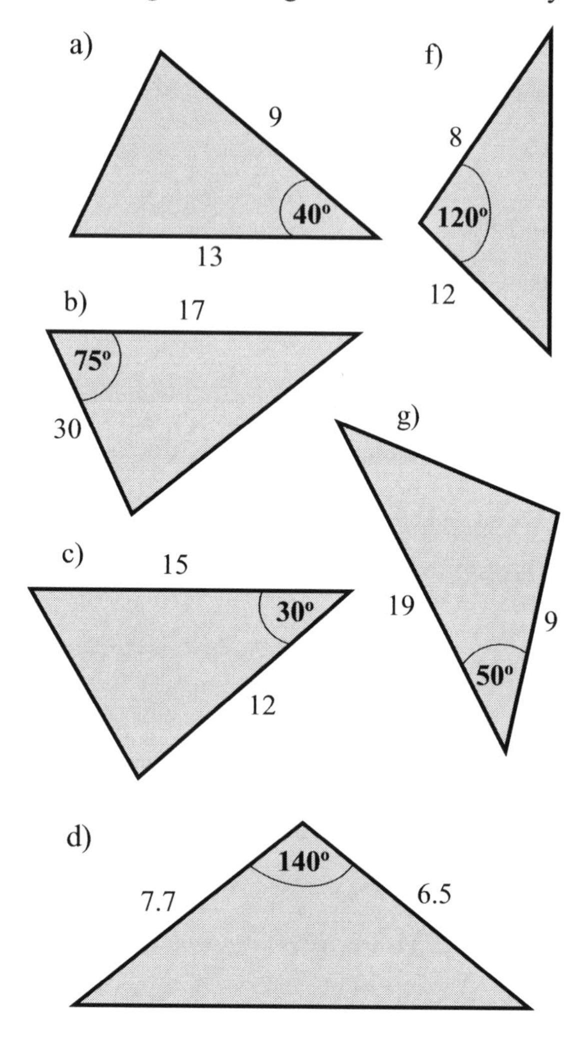

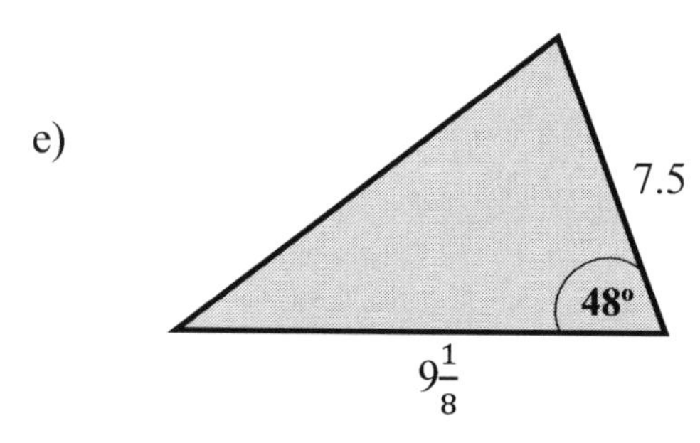

2) Work out the area of each triangle PQR with the following measurements.

a) $p = 2$ cm, $q = 5$ cm and $\angle R = 70^\circ$

b) $p = 8.7$ m, $q = 6.5$ m and $\angle R = 100^\circ$

c) $q = 7$ cm, $r = 13$ cm and $\angle P = 110^\circ$

3) Find the area of the quadrilateral ABCD.

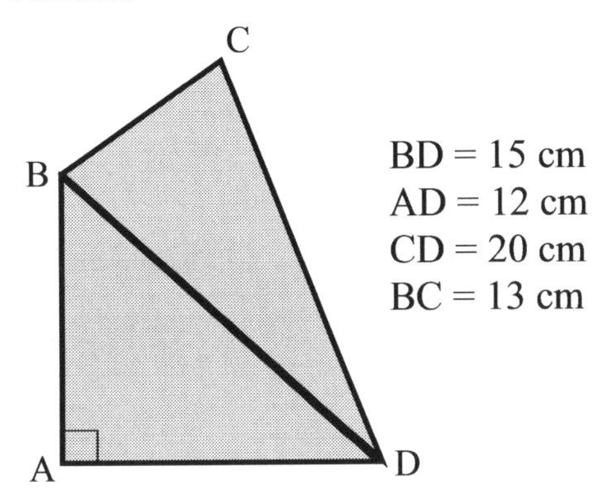

BD = 15 cm
AD = 12 cm
CD = 20 cm
BC = 13 cm

4) The diagram shows an isosceles trapezium PQRS. Work out the area of PQRS. All the lengths are in cm.

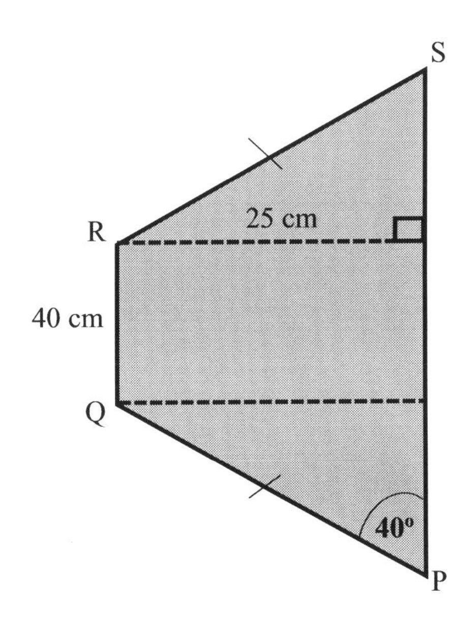

5) Show that the area of the quadrilateral below is $50\sqrt{3}$ cm^2.

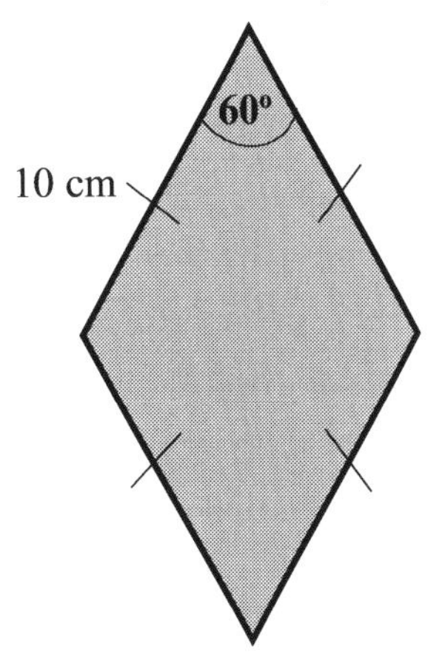

6) In ΔABC, AB = 8 cm and BC = 9 cm. The area of ΔABC is 27.58 cm^2. Show that $\angle$ ABC = 50°.

7) XYZ is a triangle with XY = 15 cm, YZ = 22 cm and $\angle XYZ = 42^\circ$.

a) Sketch the triangle.

b) Work out the perpendicular height from X to YZ to two decimal places.

c) Work out the area of ΔXYZ.

8)

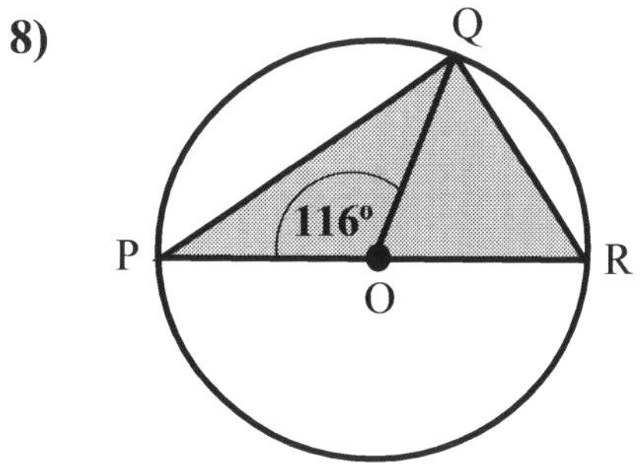

OR = 7.5 cm and O is the centre of the circle.

a) Work out the area of $\Delta\ PQR$.
b) Work out the area of the circle.
c)Work out the area of the unshaded area.

25.8 TRIGONOMETRY AND PYTHAGORAS' THEOREM IN THREE DIMENSIONS

In this section, we shall use the knowledge of Pythagoras' theorem and trigonometry to calculate missing lengths and angles in 3-D shapes.

Example 1:
Consider the cuboid below with the longest side (diagonal) AF.

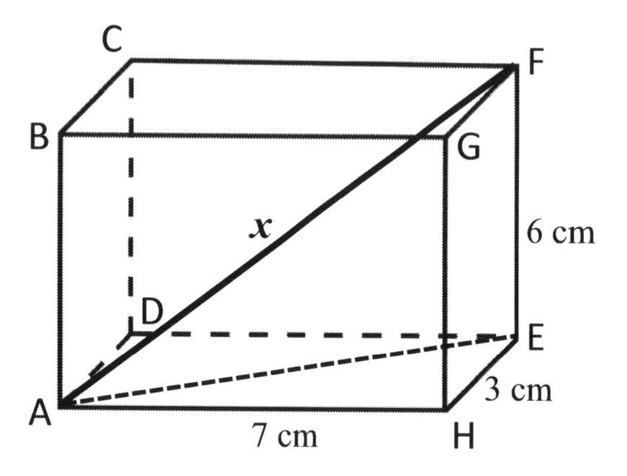

There are two options to work out the length of AF (x).

Option A: Apply Pythagoras' theorem twice by looking at rectangle ADEH (base).

Work out the diagonal AE, using triangle AHE since angle AHE = 90°. Applying Pythagoras' theorem:

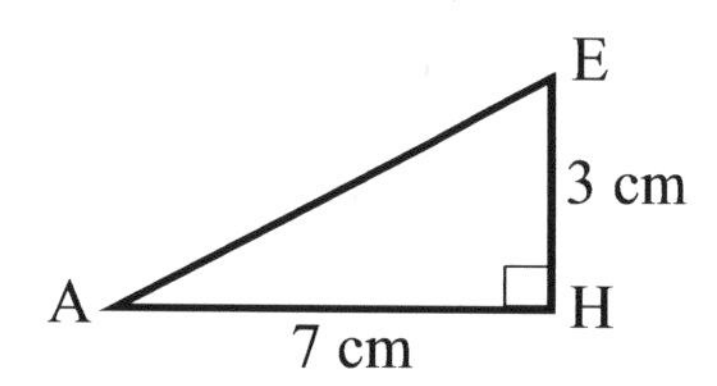

$AE^2 = 7^2 + 3^2 = 49 + 9 = 58$
$AE = \sqrt{58} = 7.62$ to 2 decimal places.

Now, consider triangle AEF and use Pythagoras' theorem to work out the value of x which is the longest diagonal. Remember: $\angle AEF = 90°$.

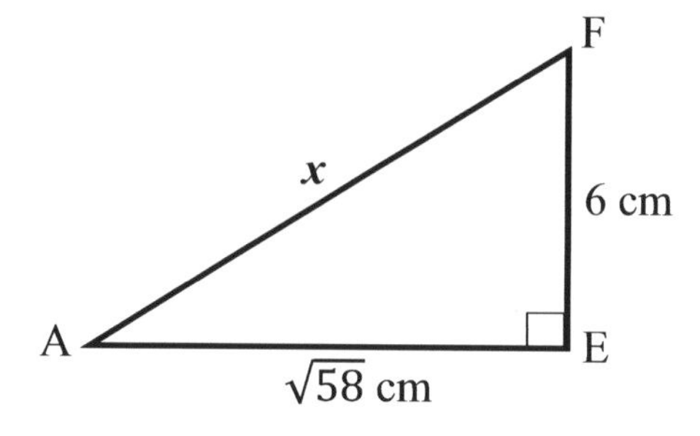

$x^2 = 6^2 + (\sqrt{58})^2$
$x^2 = 36 + 58 = 94$
$x = \sqrt{94} = 9.7$ cm to one decimal place.

The longest diagonal (AF) = 9.7 cm to one decimal place.

Option B: Use the 3-d version of Pythagoras' theorem which states that

$$\mathbf{x^2 = a^2 + b^2 + c^2}$$ ✓

$AF^2 = 7^2 + 3^2 + 6^2$
$x^2 = 7^2 + 3^2 + 6^2$
$x^2 = 49 + 9 + 36$
$x^2 = 94$
$x = \sqrt{94}$
$x = 9.7$ cm to one decimal place.

Any of the options will give the same answer.

Example 2: Calculate a) angle CHD and b) length CH.

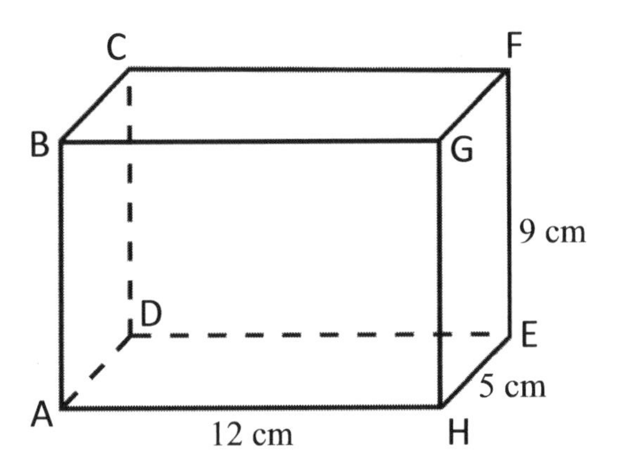

Solution: Let $\angle CHD = \theta$.

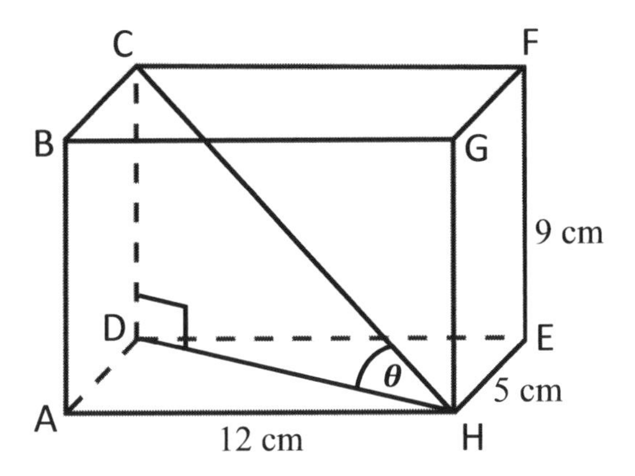

From rectangle ADEH, find the diagonal DH, using Pythagoras' theorem.

$DH^2 = 12^2 + 5^2 = 144 + 25 = 169$ cm

$DH = \sqrt{169} = 13$ cm.

Use DH = 13 cm in triangle CDH. 9 cm

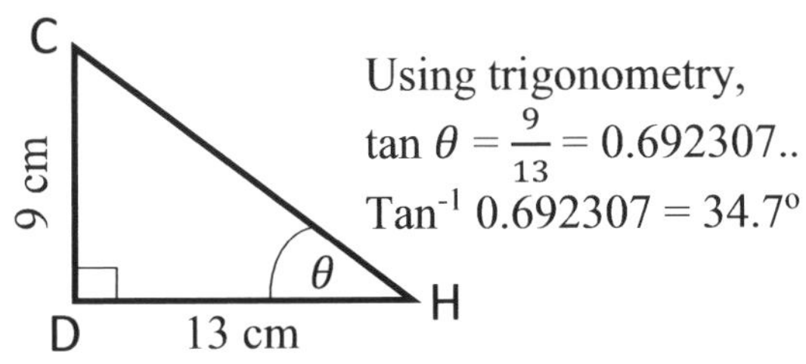

Using trigonometry,

$\tan\theta = \frac{9}{13} = 0.692307..$

$\text{Tan}^{-1}\ 0.692307 = 34.7^{\circ}$

Angle CHD = θ = **34.7°**

b)

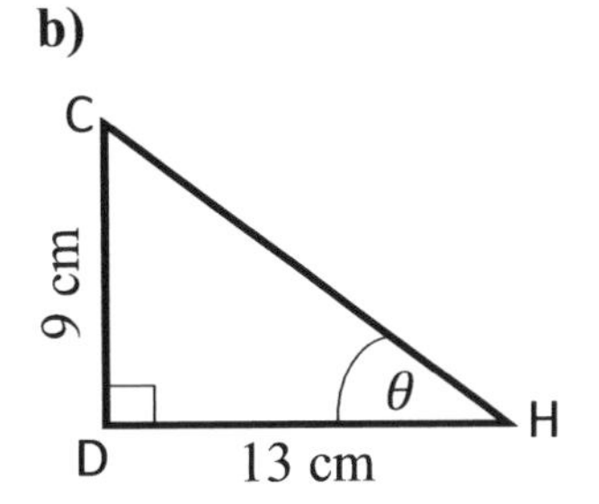

Using Pythagoras' theorem,

$CH^2 = 9^2 + 13^2$

$CH^2 = 81 + 169 = 250$

$CH = \sqrt{250}$ = **15.8 cm** to 1 d.p.

Alternatively, length *CH* could have been calculated from the 3-D version of Pythagoras' theorem since it is the longest diagonal.

From the original diagram,

$CH^2 = 12^2 + 5^2 + 9^2$

$CH^2 = 144 + 25 + 81 = 250$

$CH = \sqrt{250} = 15.8$ cm to 1 d.p.

PLANES AND ANGLES

A plane is simply a flat surface, mainly two dimensional. The surface of a table lies in a horizontal plane.

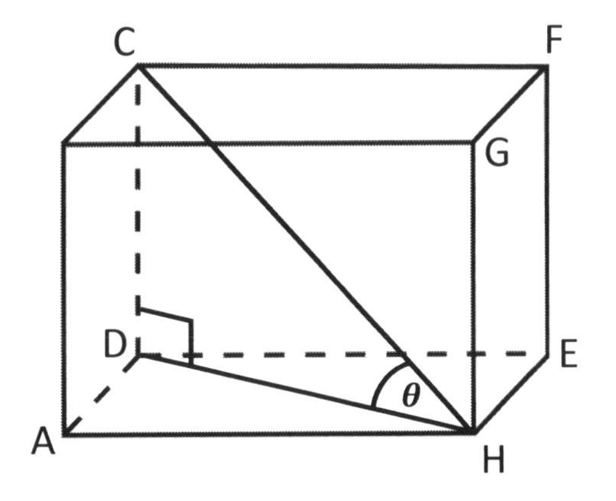

The base ADEH is in a horizontal plane, and triangle CDH is in a vertical plane.

Example 3: Work out the angle between the diagonal AF and the plane ABCD.

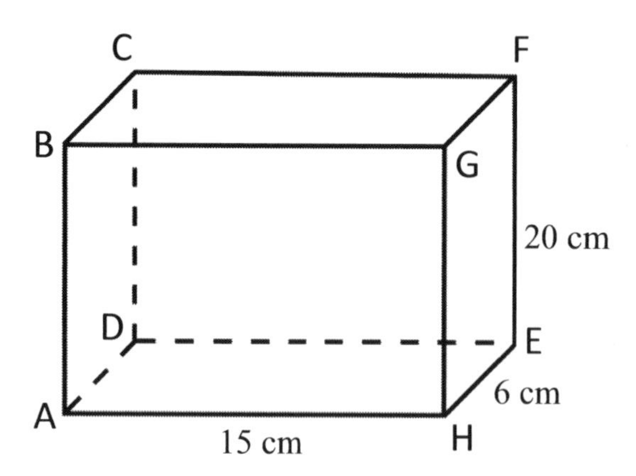

Solution: Draw line AF. The required angle line AF makes with the plane ABCD is shown as θ.

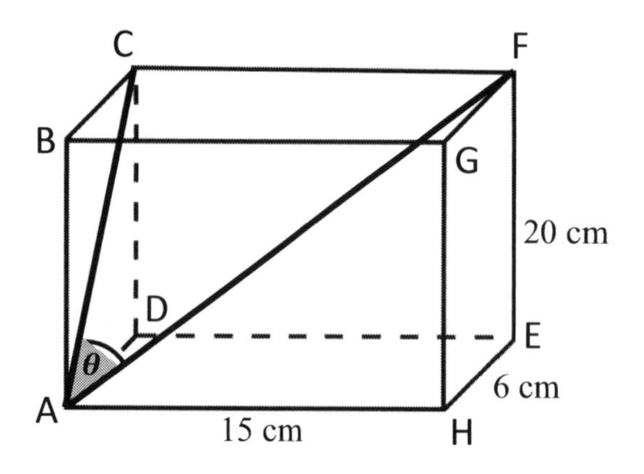

Using the 3-D version of Pythagoras' theorem, $AF^2 = 15^2 + 6^2 + 20^2$

$AF^2 = 225 + 36 + 400 = 661$

$AF = \sqrt{661} = 25.7$ cm to 1 d.p.

Using triangle ADC in the plane ABCD,

$AC^2 = 6^2 + 20^2$

$AC^2 = 36 + 400 = 436$

$AC = \sqrt{436} = 20.9$ cm to 1 d.p.

From triangle AFC,

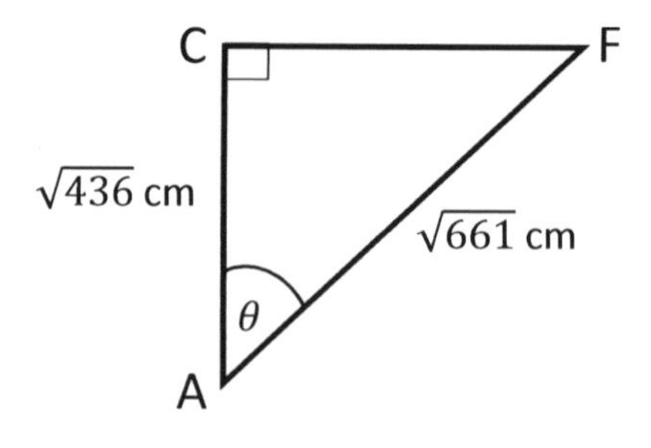

Using SOH CAH TOA,

$$\text{Cos } \theta = \frac{adjacent}{hypotenuse} = \frac{\sqrt{436}}{\sqrt{661}}$$

$$= \frac{20.88061302}{25.70992026} = 0.812161718$$

$\theta = \cos^{-1} 0.812161718 =$ **35.7°** to 1 d.p.

Example 4: ABCDE is a rectangular based pyramid. AB = AE = AD = AC = 25 cm. Calculate

a) the vertical height from A to the plane BCDE and

b) the angle between AB and the plane BCDE.

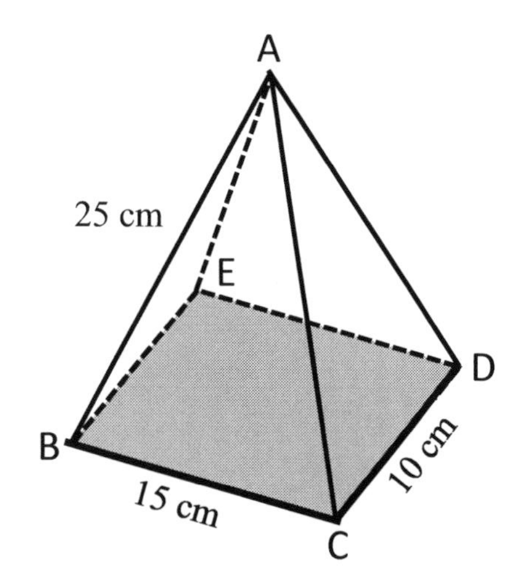

Solution: Insert a perpendicular line from the apex A to the base, BCDE.

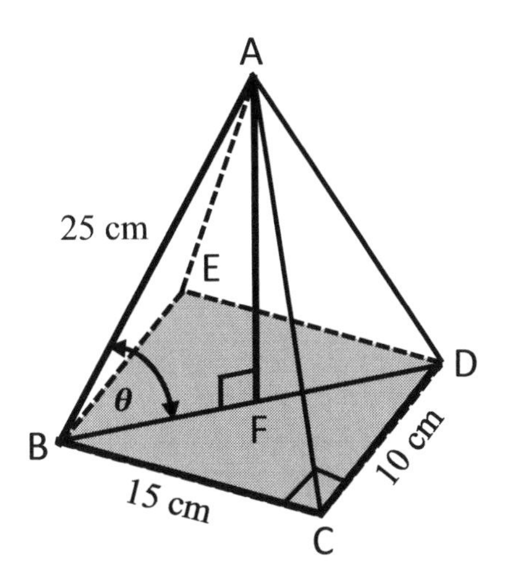

Using Pythagoras' theorem

$BD^2 = 15^2 + 10^2 = 225 + 100 = 325$

$BD = \sqrt{325} = 18.03$ cm to 2 dp.

$BF = \frac{1}{2} BD = 18.03 \div 2$

$= 9.01$ cm to 2 decimal places.

a) Using triangle AFB, the vertical height from A to the plane BCDE is AF.

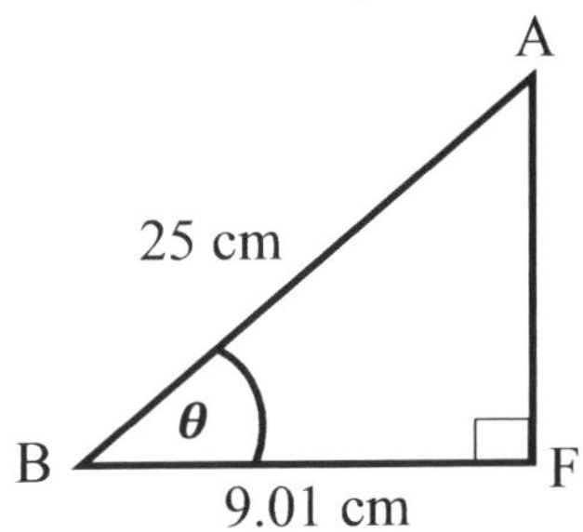

a) $AF^2 = 25^2 - 9.01^2 = 625 - 81.18$

$AF = \sqrt{543.82} =$ **23.32 cm** to 2 dp.

b) $\text{Cos } \theta = \frac{9.01}{25} = 0.3604$

$\theta = \cos^{-1} 0.3604 = 68.89°$ to 2.dp.

The angle between AB and the plane BCBE ≈ **69°** to the nearest degree.

NOTE: Always remember that most of the corners pupils encounter in 3-D trigonometric problems are right angles. The pupils must identify the right angles and apply Pythagoras' theorem or trigonometric ratios (sine, cosine and tangent – SOH CAH TOA) where appropriate.

Where possible, redraw the triangle needed for a particular question from the original diagram to avoid confusion. Draw a separate right-angled triangle and annotate accordingly.

Finally, most of the questions/problems require Pythagoras' theorem and trigonometric ratios. Look for the correct right-angled triangle and apply the correct method of Pythagoras' theorem and trigonometric ratio for that problem.

EXERCISE 25 J

1) The diagram is a cuboid.
a) Calculate the length of:
i) AE ii) AF iii) CE iv) CG
to two decimal places where possible.

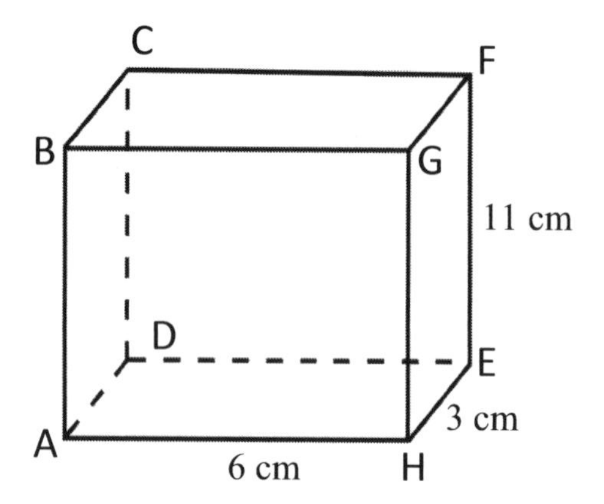

b) Work out the angle between the diagonal AF and the plane ADEH.
c) Work out the angle between the diagonal CH and the plane EFGH.

2) PU = 980 cm. Calculate i) the length RP. ii) the angle PS makes with the plane QRST. Leave your answers to 3 significant figures.

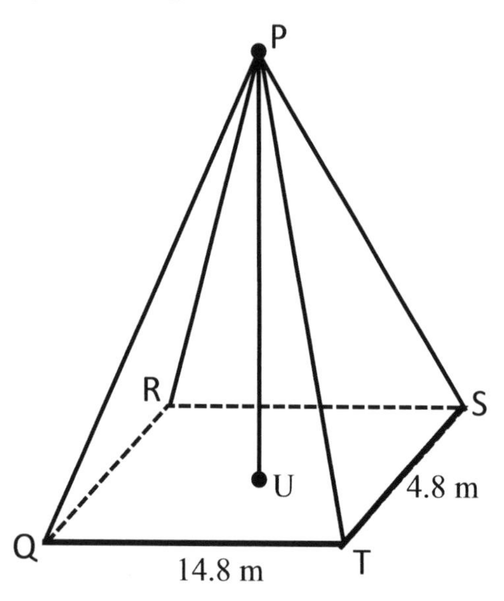

3)

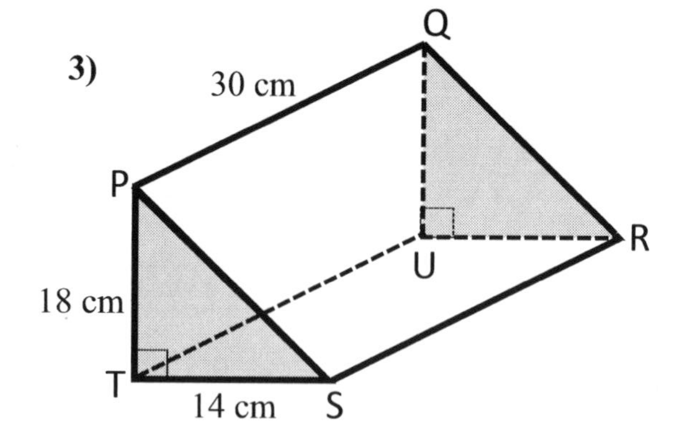

The cross-section of this prism is a right-angled triangle as shown. The other faces are rectangles.

a) Calculate the slant height TQ.
b) Calculate the angle the diagonal PR makes with the plane RSTU.
c) Calculate the volume of the prism.

4) A square-based pyramid PQRST is shown below, and each triangular face is an isosceles triangle.

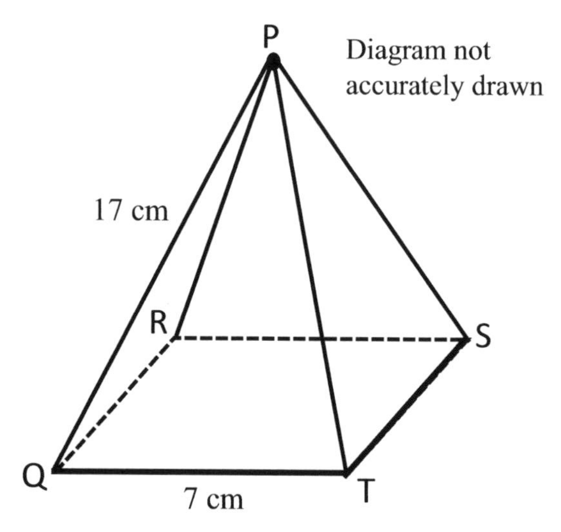

a) Calculate angle QPT.
b) Work out the area of triangle PQS, correct to 3 significant figures.
c) Calculate the total surface area of the pyramid.

26 Sectors, Arcs & Segments

This section covers the following topics:

- Introduction to sectors and arcs
- Segments

LEARNING OBJECTIVES

By the end of this unit, you should be able to:

a) Understand sectors and arcs
b) Work out the length of arcs
c) Calculate the area of sectors
d) Calculate the area of a segment

KEYWORDS

- Arc
- Sector
- Area
- Circumference
- Segment

26.1 LENGTH OF AN ARC

An **arc** is part of a circle and is the distance between the points where the radii meet the circumference.

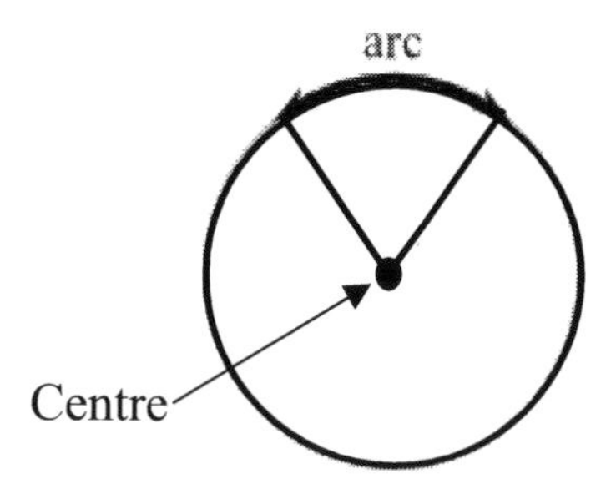

The smaller length is known as the **minor** arc while the bigger length is the **major** arc. The arc length is related to the angle at the centre.

The formula for the length of an arc is

Length of arc = $\frac{\theta}{360^0}$ × circumference

Length of arc = $\frac{\theta}{360^0} \times \pi\, d$

where $\boldsymbol{\theta}$ is the angle at centre and $\boldsymbol{d}$ is the diameter of the circle.

26.2 AREA OF A SECTOR

The sector is $\frac{\theta}{360}$ of s full circle.

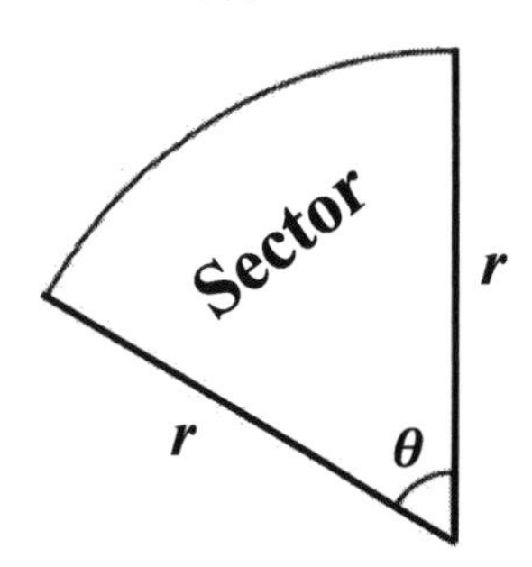

Therefore,

Area of a sector = $\frac{\theta}{360^0} \times \pi r^2$

Where $\boldsymbol{\theta}$ is the angle at centre and $\boldsymbol{r}$ is the radius of the circle.

Example 1: Work out
i) the length of arc BC
ii) the perimeter of the sector ABC and
iii) Area of the sector ABC. Give your answers to one decimal place.

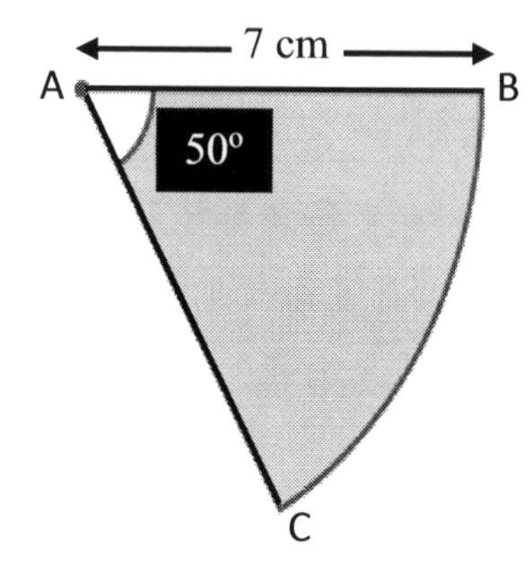

Solution:

i) Length of arc BC = $\frac{\theta}{360^0} \times \pi\, d$

Diameter = 7 × 2 = 14 cm

Length of arc = $\frac{50}{360^0} \times \pi \times 14$

= **6.1 cm** to 1 d.p.

ii) The perimeter is the distance around.
AB + BC + AC = 7 + 6.1 + 7 = **20.1 cm**

iii) Area of sector ABC = $\frac{\theta}{360^0} \times \pi r^2$

$= \frac{50}{360} \times \pi \times 7^2$

= **21.4 cm²** to 1 d.p ✓

Example 2

Work out the missing angle θ.

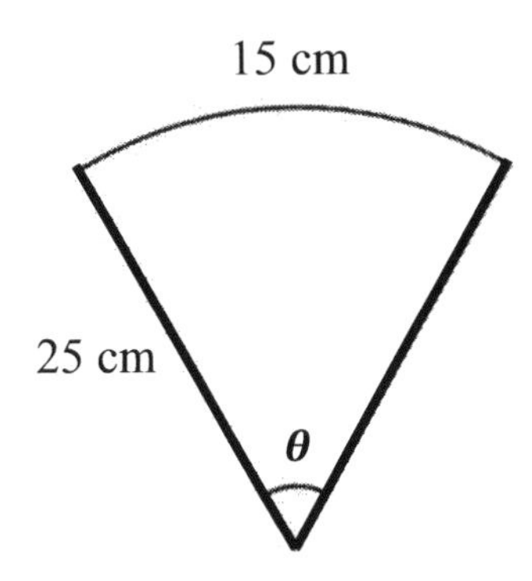

<u>Solution</u>

Length of arc $= \frac{\theta}{360^0} \times \pi\, d$

But radius = 25 cm, therefore diameter (d) = 50 cm.

$15 = \frac{\theta}{360^0} \times \pi \times 50$

$15 = \theta \times 0.436332313$

Make θ the subject of the formula.

$\theta = \frac{15}{0.436332313}$

$\theta =$ **34.4°** ✓

EXERCISE 26A

1) Calculate
i) the length of the arcs
ii) the perimeters
iii) the areas of the sectors.

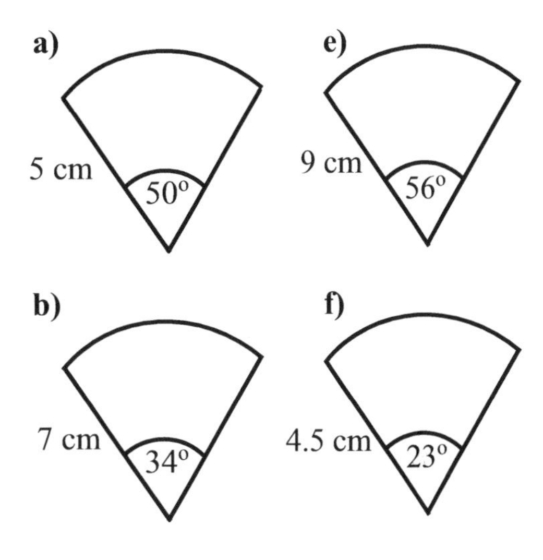

c) **g)**

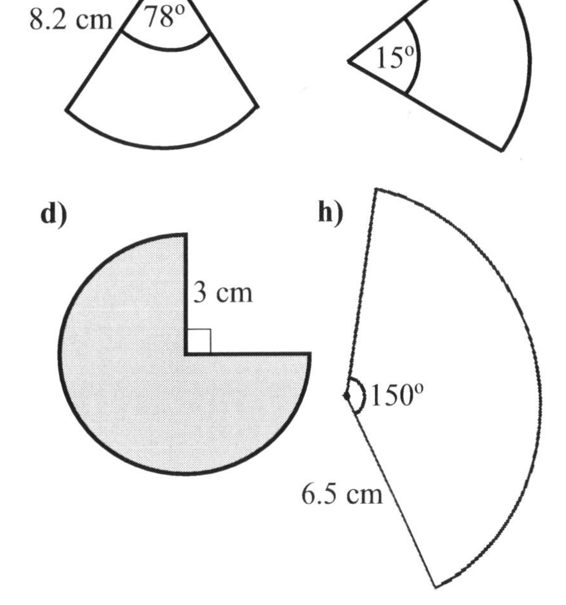

2) Work out the value of the angles marked θ.

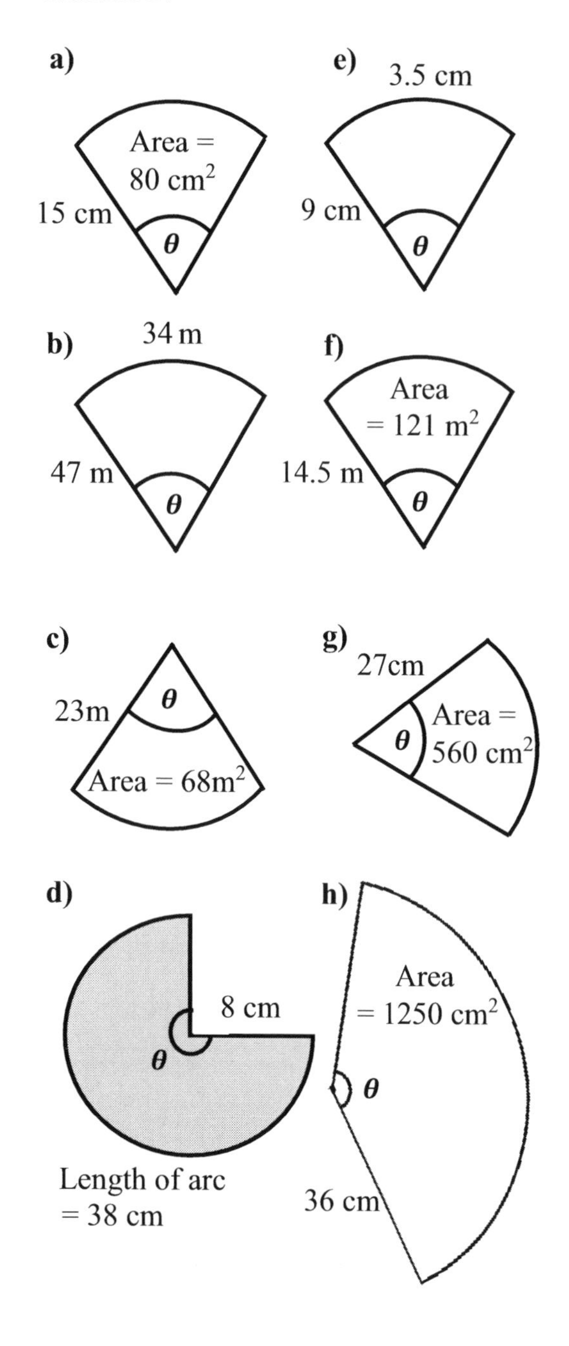

26.3 AREA OF A SEGMENT

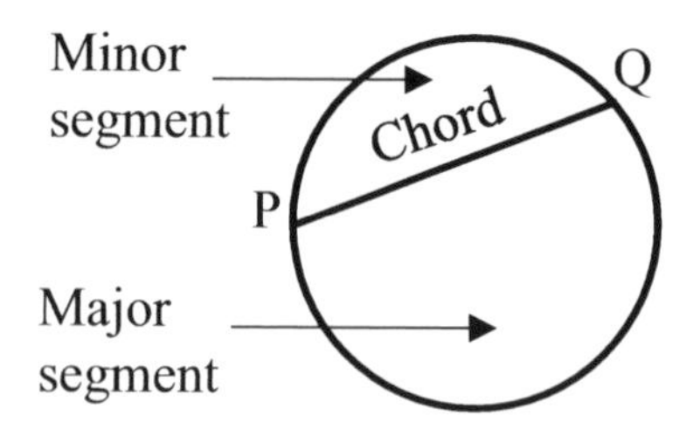

Chord PQ divides the circle into two segments: **minor segment** and **major segment**.

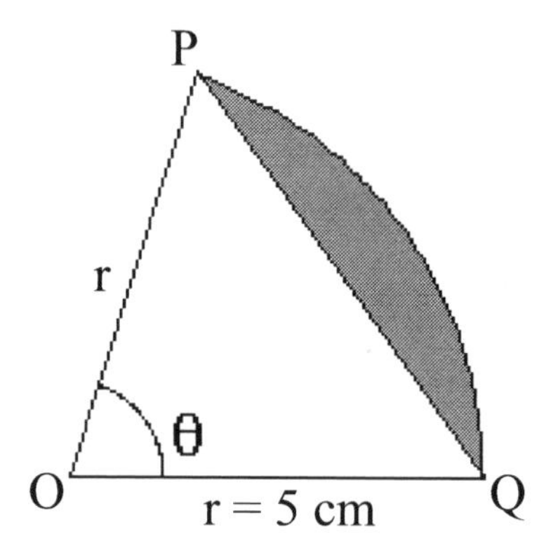

Area of segment
= area of sector OPQ – area of triangle.

Area of segment $= \left(\frac{\theta}{360^0} \times \pi r^2\right) - \left(\frac{1}{2} r^2 \sin\theta\right)$

Example 1: Calculate the area of the segment (shaded part).

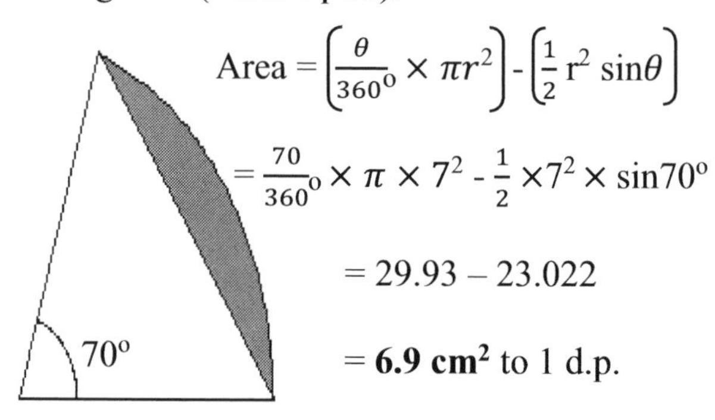

$$\text{Area} = \left(\frac{\theta}{360^0} \times \pi r^2\right) - \left(\frac{1}{2} r^2 \sin\theta\right)$$

$$= \frac{70}{360^0} \times \pi \times 7^2 - \frac{1}{2} \times 7^2 \times \sin 70^o$$

$$= 29.93 - 23.022$$

$= \mathbf{6.9\ cm^2}$ to 1 d.p.

Example 2:
Work out the area of the segment and write your answer in terms of π.

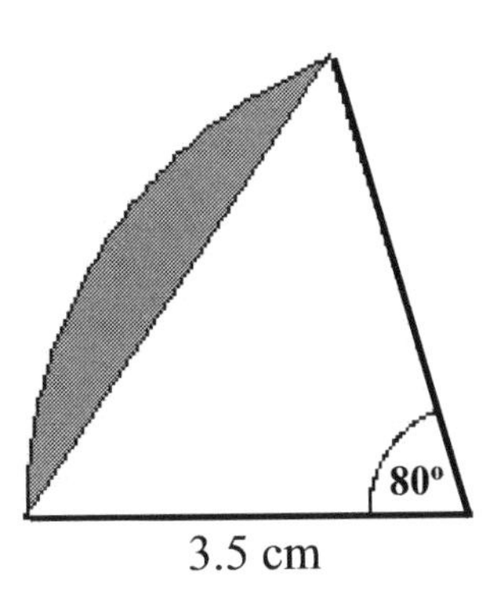

Solution:

4 cm

Area of segment $= \frac{\theta}{360^{o}} \times \pi r^2 - \frac{1}{2} r^2 \sin\theta$

To work this out in terms of π, **do not** multiply the value of π while performing the calculations for the area of the sector.

$$= \left(\frac{80}{360} \times \pi \times 3.5^2\right) - \left(\frac{1}{2} \times 3.5^2 \times \sin 80^{o}\right)$$

$$= \mathbf{2.722\pi - 6.03\ cm^2}$$

EXERCISE 26 B

1) Work out the area of each shaded segment and give your answers to 3 significant figures.

a)

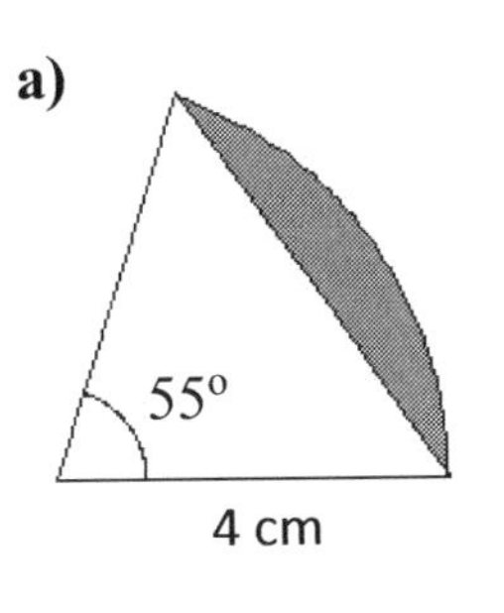

b)

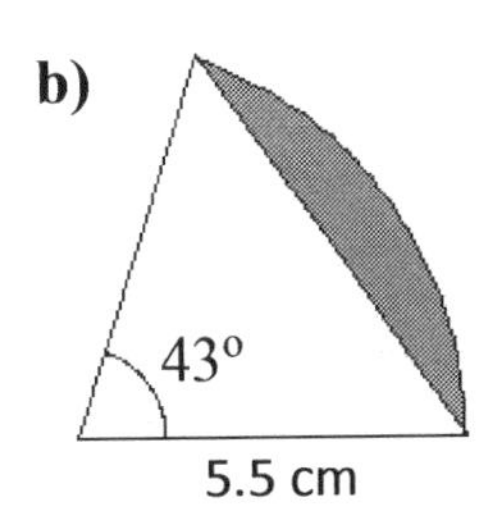

c)

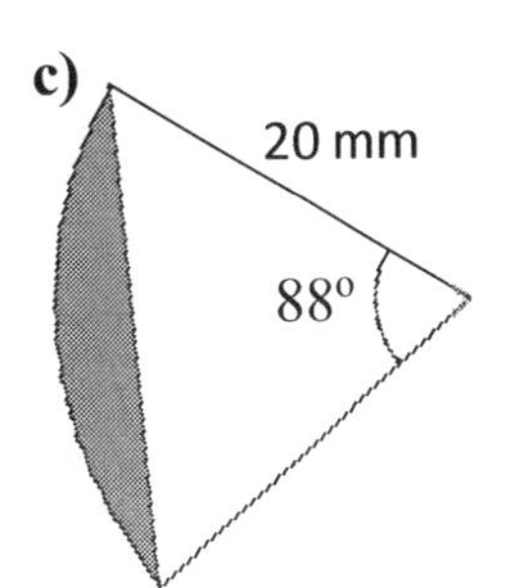

d)

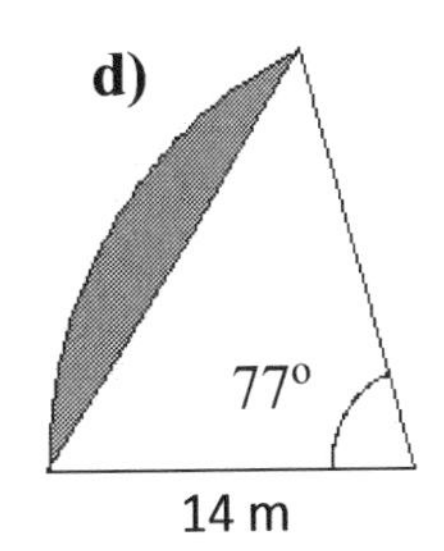

e)

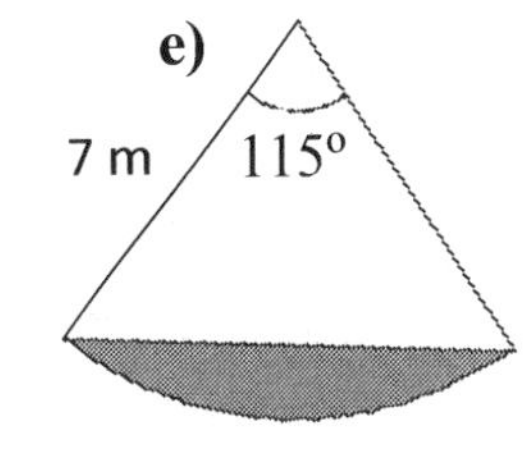

f)

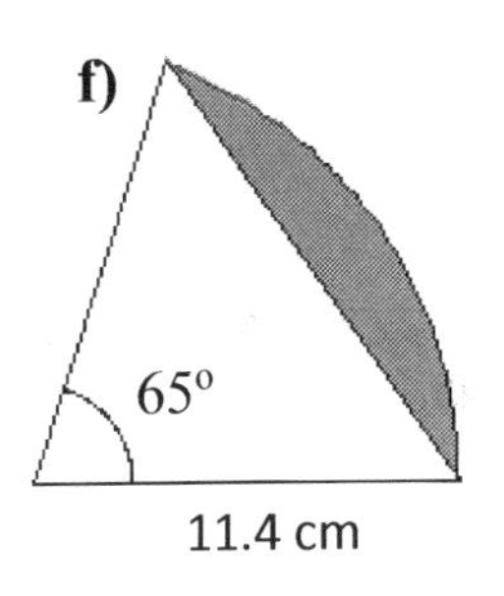

2) Find the area of the shaded part and round your answer to 1 decimal place. O is the centre of the circle.

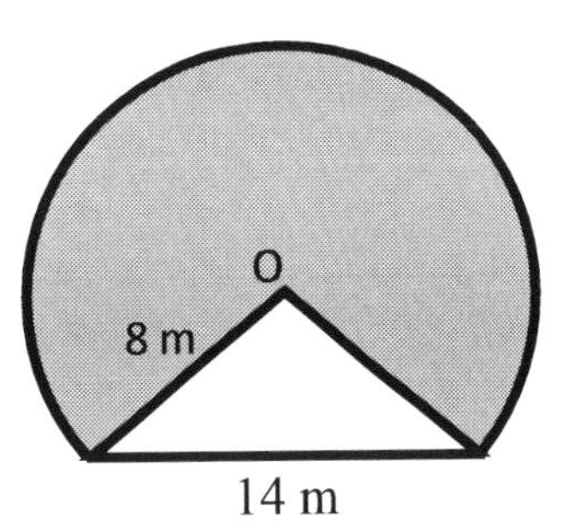

27 Circle theorems

This section covers the following topics:

- Properties of circles
- Circle theorems

LEARNING OBJECTIVES

By the end of this unit, you should be able to:

a) Use tangents and chord properties to solve circle problems
b) Solve geometrical problems using circle theorems
c) Prove circle theorems

KEYWORDS

- Circles
- Tangents
- Chords
- Circle theorems
- Proof

27.1 PROPERTIES OF CIRCLES

Parts of a circle are shown below. For letter O, is always the centre of the circle.

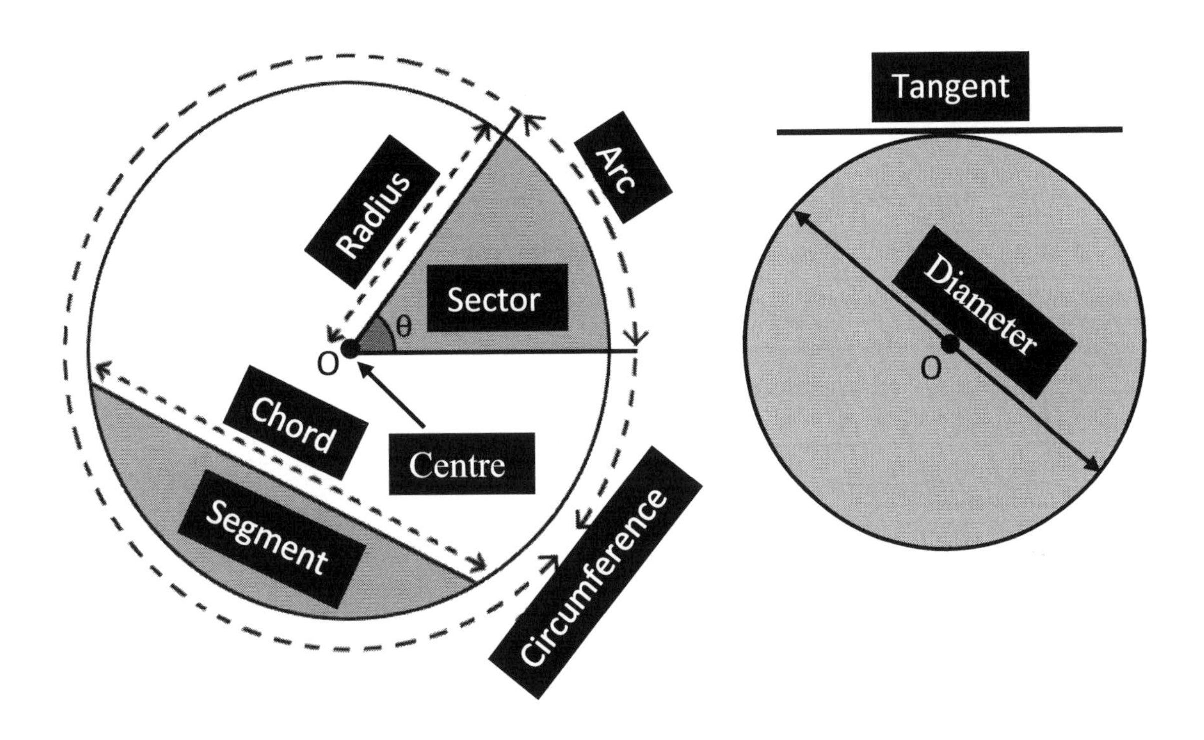

A LINE PERPENDICULAR TO A CHORD FROM THE CENTRE OF A CIRCLE BISECTS THE CHORD.

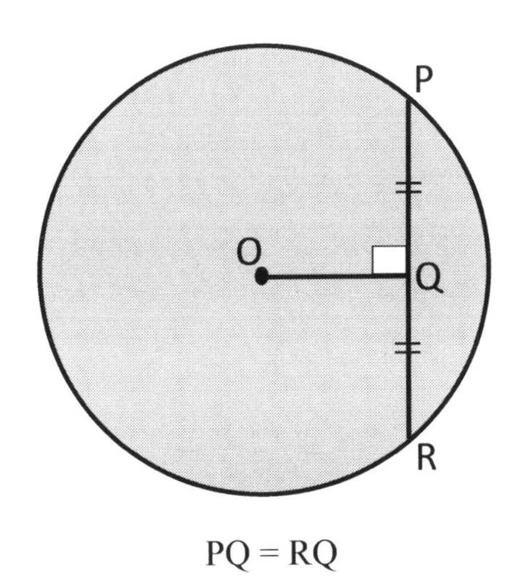

PQ = RQ

Example 1: A circle has a chord of length 8 cm and radius 5cm. Work out the distance from the centre of the circle to the midpoint of the chord.

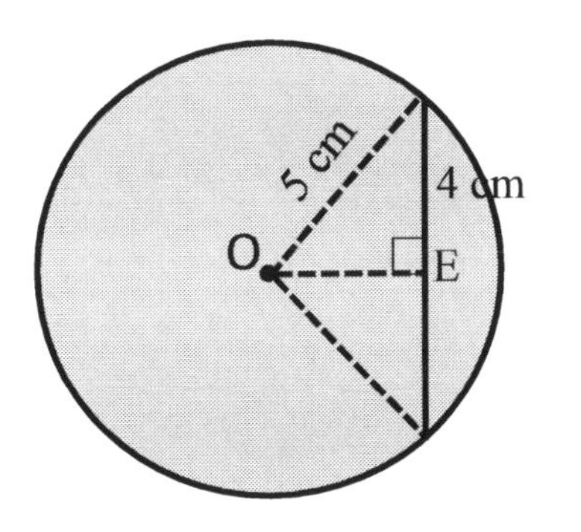

Using Pythagoras; theorem,

$OE^2 = 5^2 - 4^2$
$= 25 - 16$
$= 9$
$OE = \sqrt{9}$
$= \mathbf{3\ cm}$

Example 2: Prove that $\angle PQO = \angle RQO = 90^o$

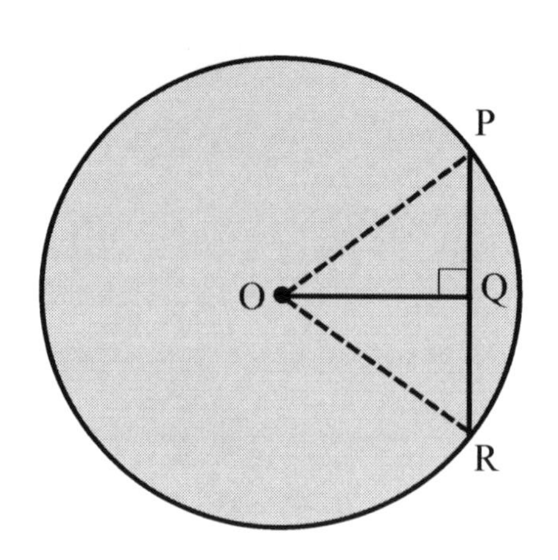

Solution:
Given: Chord PR on a circle with centre, O
Construction: Join OP and OR
Prove that $\angle PQO = \angle RQO = 90^o$

OP = OR ……radii
PQ = RQ ……line perpendicular to a chord from the centre, bisects the chord
OQ = OQ ……common

Triangle PQO ≡ triangle RQO ……SSS (congruent)
Therefore, $\angle$ PQO = $\angle$ RQO
However, $\angle$ PQO + $\angle$ RQO = 180° ……angle on a straight line
Therefore, $\angle$ PQO = $\angle$ RQO = (180 ÷ 2) = 90°

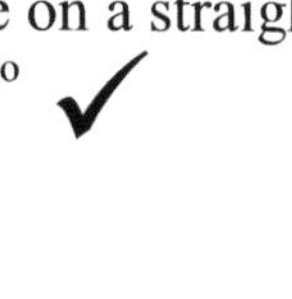

EXERCISE 27A

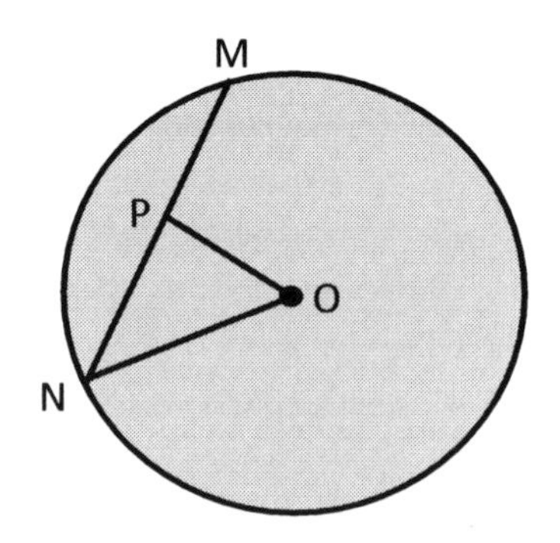

1) O is the centre of the circle and OP = 7cm.
MP = NP
Diameter = 20cm.
a) Calculate the length of the chord MN to 1 d.p.
b) Work out the area of triangle NOP.

2) If the radius of a circle is 9.5cm and the length of a chord of the circle is 14cm, calculate the vertical distance from the chord to the centre of the circle.

3) Prove that line OC bisects the line AB.
O is the centre of the circle.

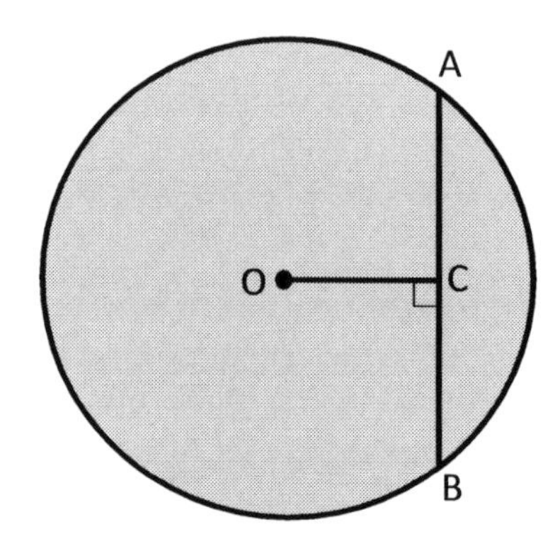

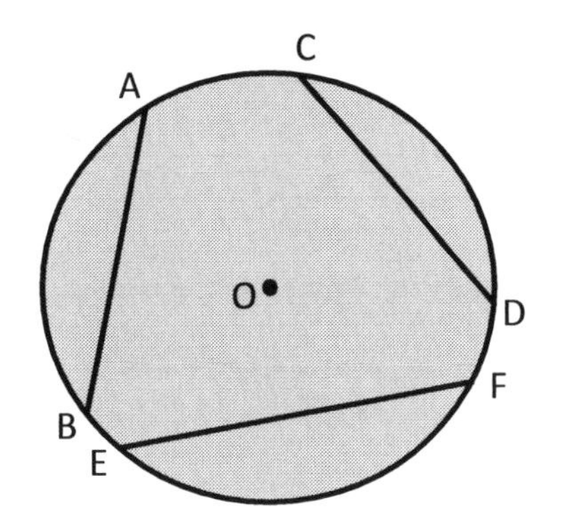

4) Three chords, AB, CD and EF, are inscribed in the circle with radius 9 cm and centre O. AB = 15 cm, CD = 12 cm and EF = 14 cm.

a) What is the distance between the chord EF and the centre of the circle?
b) Which of the chords is the closest to the centre, O of the circle? Show all working out.

TANGENTS FROM EXTERNAL POINTS ARE EQUAL IN LENGTH

QP and QR are tangents from external point, Q.
PQ = RQ

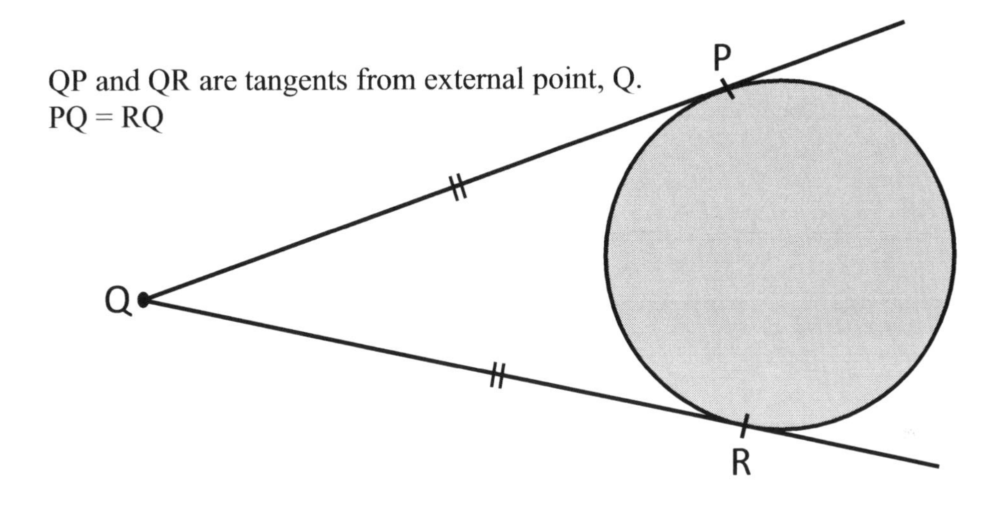

ANGLE BETWEEN A TANGENT AND A RADIUS IS 90°

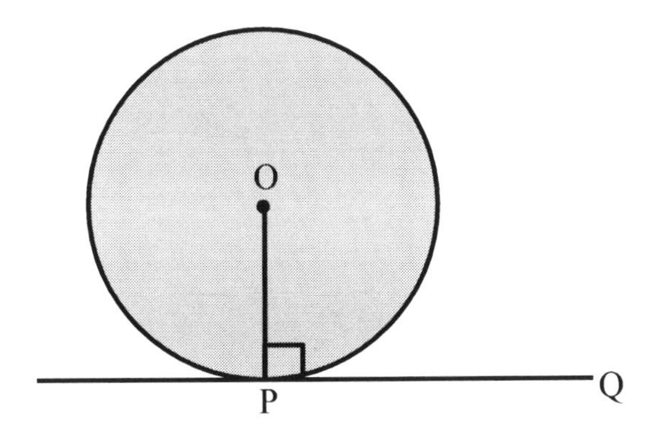

O is the centre of the circle.
Angle OPQ = 90°

O is the centre of the circle.

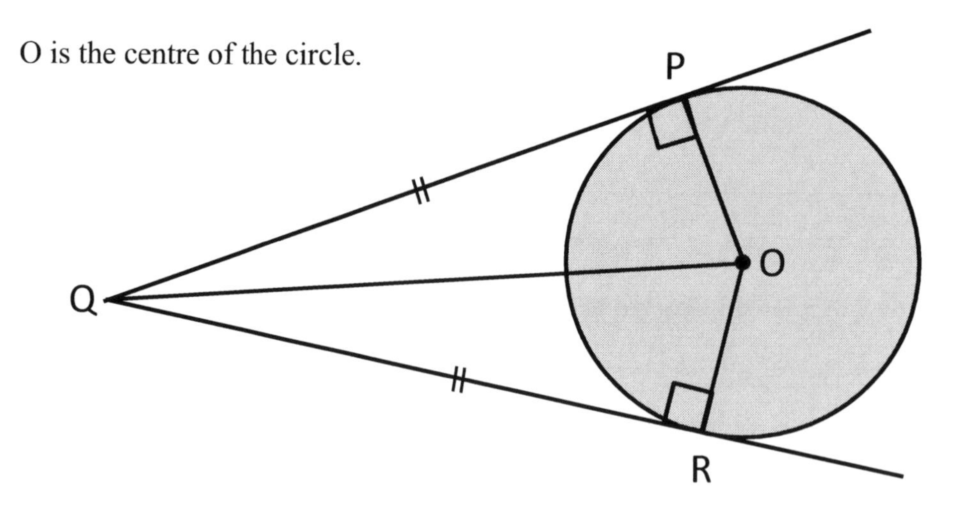

FACTS

QP = QR ……Tangents from external points
∠QPO = ∠QRO = 90° ……angle between a tangent and a radius is 90°.

Therefore, line QO bisects angle PQR and ∠PQO = ∠RQO.

Example 3: Calculate the missing angles ***a*** and ***b***. PR is a chord, and QP and QR are tangents. O is the centre of the circle.

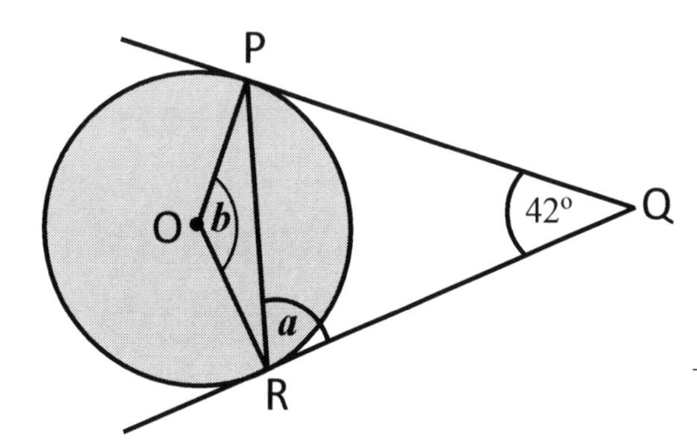

Solution:
QP = QR …tangents from external point

Therefore, triangle QPR is isosceles and the base angles QPR and QRP are equal.

180° – 42° = 138°
138 ÷ 2 = 69° and ***a*** = 69°

∠QPO = 90° …angle between a tangent and a radius. Likewise, ∠QRO = 90°. Since QPOR is a quadrilateral, interior angles add to 360°.
42° + 90° + 90° = 222°
Angle ***b*** = 360° – 222° = 138°

EXERCISE 27 B

1) Calculate the size of angles marked by letters. O is the centre of the circle.

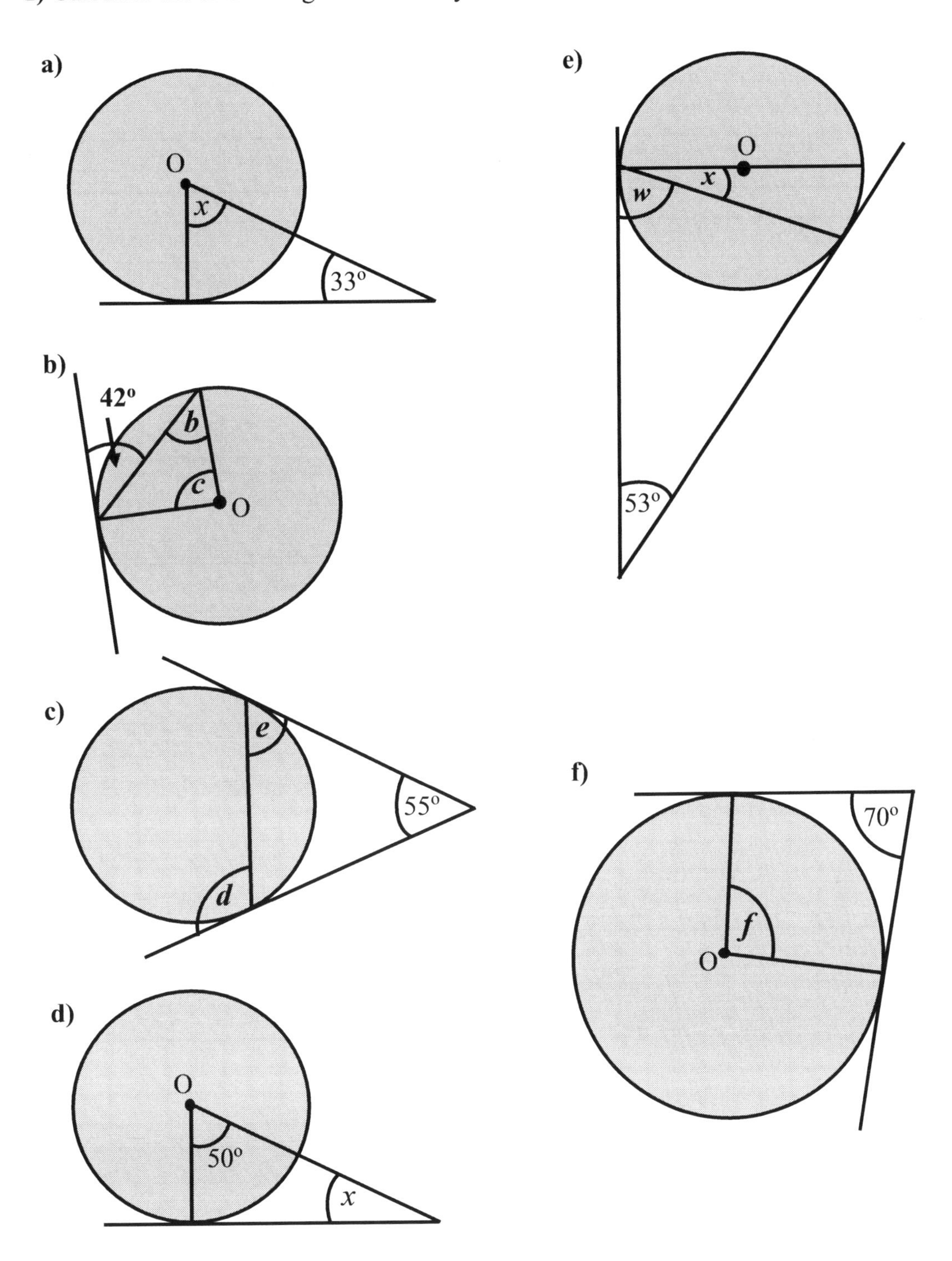

g)

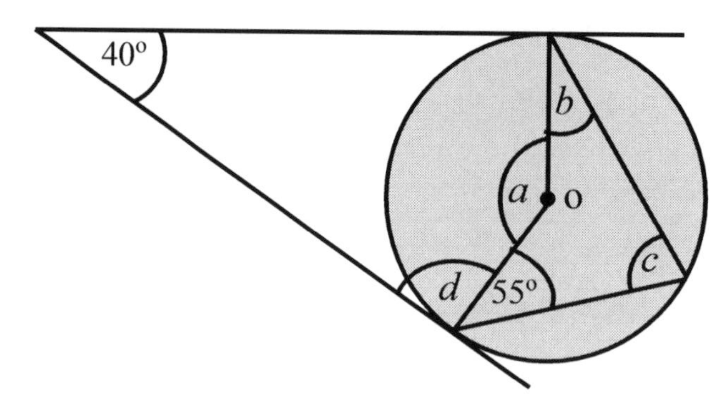

27.2 CIRCLE THEOREMS

Theorem 1: The angle subtended by an arc of a circle is twice the angle it subtends at any point on the remaining part of the circumference of the circle.
If the angle at B is 20°, the angle at centre, O will be $20 \times 2 = 40°$

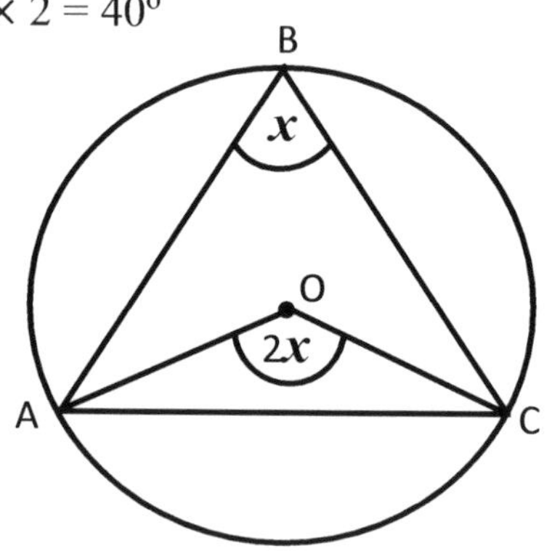

Example 1:

If the angle at B is 20°, the angle at centre, O will be

$20 \times 2 = 40°$

If $\angle AOC = 130°$, $\angle ABC = 130 \div 2 = 65°$

Example 2: Prove that $\angle AOC = 2 \times \angle ABC$

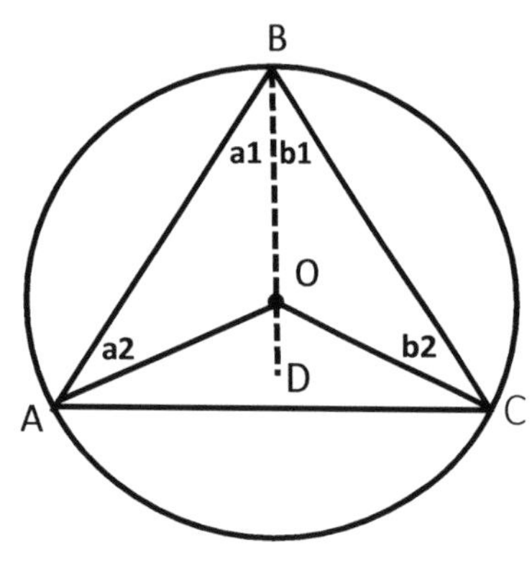

∠ AOC = 2 × ∠ ABC

Given: Circle ABC with centre O
Prove: $\angle AOC = 2 \times \angle ABC$
Construction: Join B and O with a straight line and extend to any point D

OC = OB ....radii
b1 = b2 ....base angles of isosceles triangle COB

∠COD = b1 + b2 ......exterior angle of triangle COB
Therefore, ∠COD = 2b1 ....b1 = b2
Similarly, ∠AOD = 2a1
∠AOC = ∠AOD + ∠COD
= 2a1 + 2b1 = 2 (a1 + b1)
= 2 × ∠ABC ✓

Theorem 2: Angles in the same segments of a circle are equal. Alternatively, angles subtended by the same arc are equal.

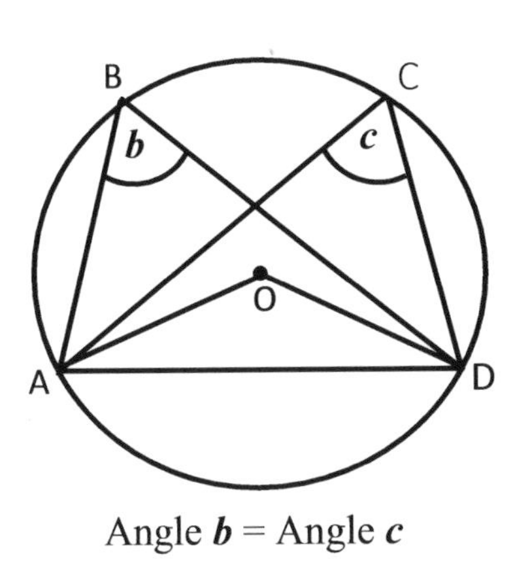

Angle ***b*** = Angle ***c***

If ∠ABD = 40°, ∠ACD is also 40°

Example 3: Prove that ∠ ABD = ∠ ACD

Solution:

Given: circle ABCD
Prove that ∠ ABD = ∠ ACD

∠ AOD = 2b …angle at centre is twice angle at circumference
∠ AOD = 2c …angle at centre is twice angle at circumference
Therefore, b = c ……. half of ∠ AOD

It means that ∠ABD = ∠ACD. ✓

EXERCISE 27 C

1) Calculate the sizes of the missing angles marked with letters. O is the centre of each circle.

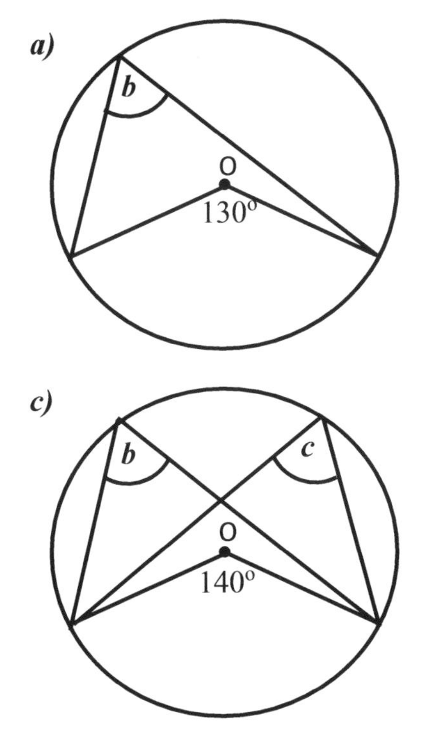

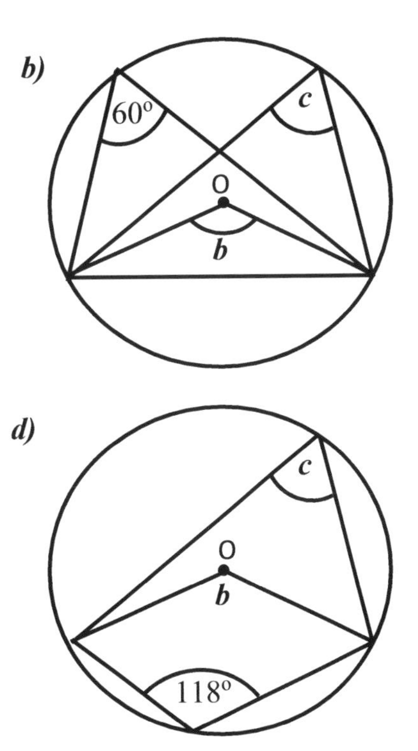

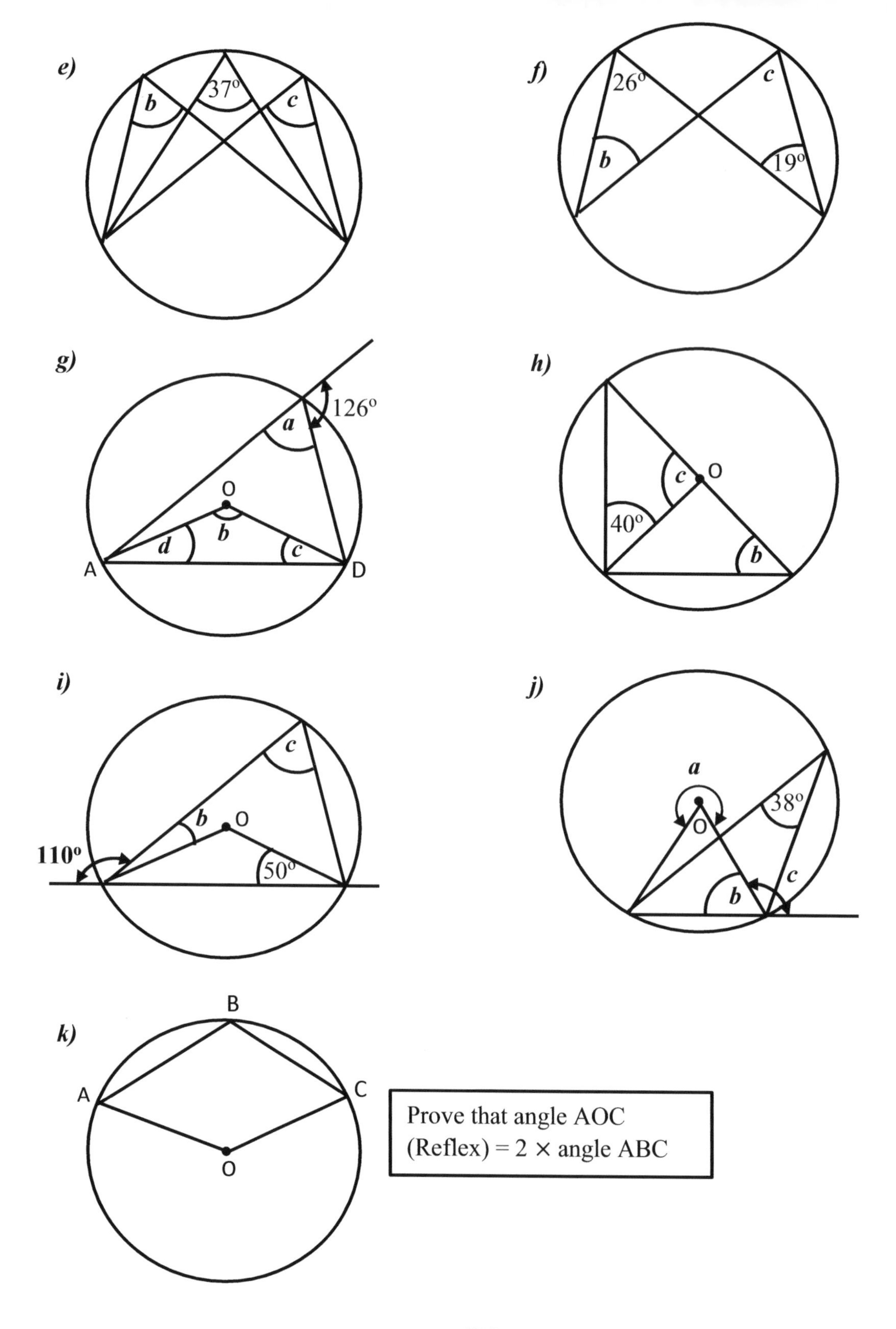

Prove that angle AOC (Reflex) = 2 × angle ABC

Theorem 3: Opposite angles of a cyclic quadrilateral are supplementary (add to 180°).

A **cyclic quadrilateral** is a quadrilateral inscribed in a circle. The vertices of the quadrilateral must touch the circumference of the circle at four different places.

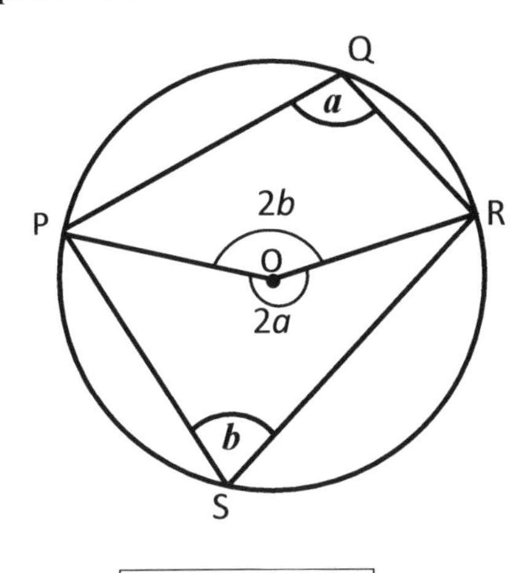

a + b = 180 °

Example 4: Prove that $\angle a + \angle b = 180°$

Solution:
Given: A cyclic quadrilateral
Prove: Let $\angle PQR = a$ and $\angle PSR = b$
Construction: Join P to O and R to O
O is the centre of the circle

$\angle POR = 2b$ …angle at centre is twice angle at circumference

$\angle POR$ (reflex) $= 2a$ …angel at centre is twice angle at circumference

Therefore, $2a + 2b = 360°$ ……angle at a point add to 360°

By factorising, $2(a + b) = 360°$
$a + b = 180°$ ✓

Example 5: Calculate the size of angles ***m*** and ***n***.

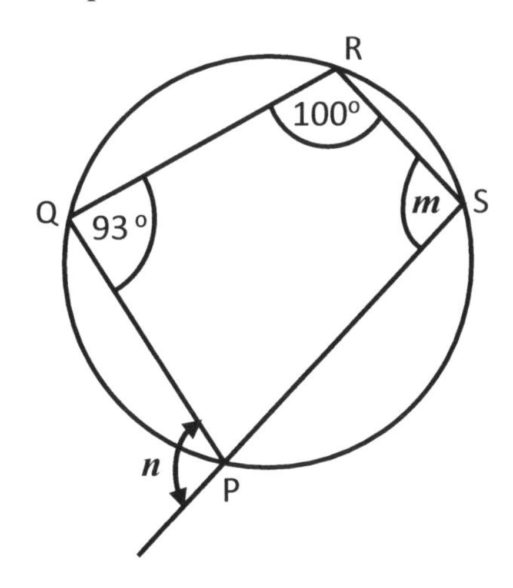

Solution:
PQRS is a cyclic quadrilaterals and angles Q and S are opposite each other.

Therefore, $m + 93° = 180°$ …Opposite angles of a cyclic quadrilateral.

$m = 180 - 93° =$ **87°**

Also, angle $P = 180° - 100° = 80°$
But angle $n = 180° - 80°$ …angles on a straight line.

Finally, **n = 100°**.

Theorem 4: The angle subtended at the circumference by a semicircle is 90°.

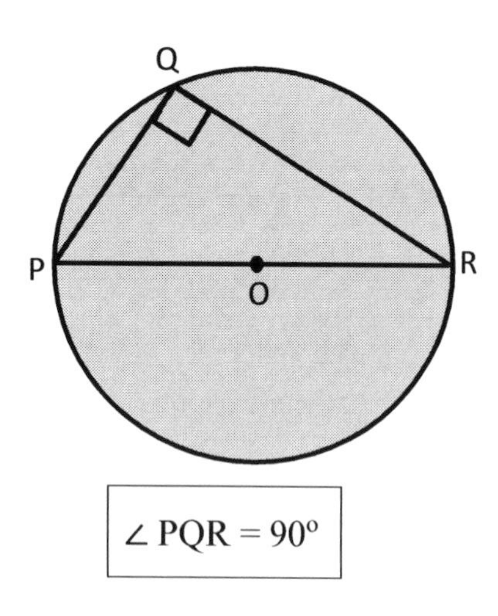

Prove that ∠ PQR = 90°

Given: Diameter PR with centre O and Q is a point on the circumference.

∠POR = 2 × ∠ PQR ……angle at centre is twice the angle at circumference.
Also, ∠ POR = 180° …angle on a straight line.

Therefore, 2 × ∠ PQR = 180°
This means that ∠ PQR = 90° ✓

Example 6: Work out the missing angles, m and n. O is the centre of the circle.

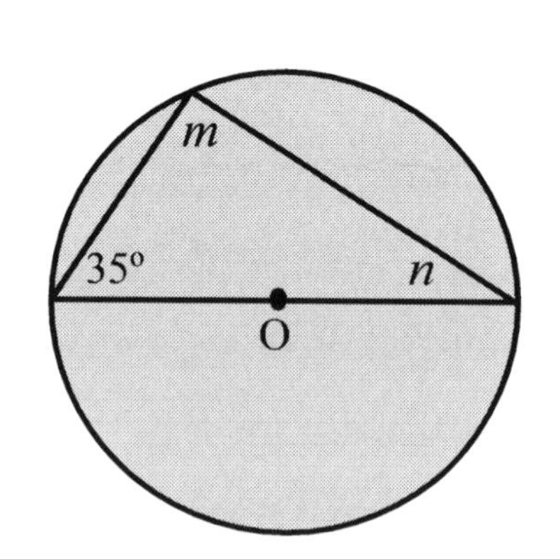

Solution:

$m = 90°$ ……angle in a semi-circle

$35° + 90° + n = 180°$ ……angles in triangles

$125° + n = 180°$

$n = 180 - 125 =$ **55°**

Example 7: Work out the missing angles marked with letters. O is the centre of the circle.

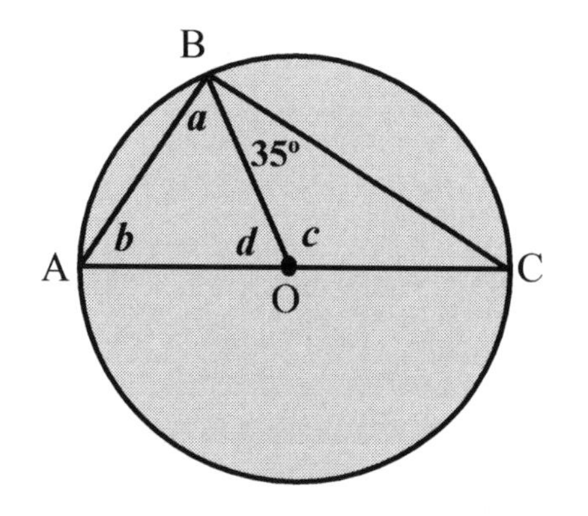

Solution:

a + 24° = 90° ………angle in a semicircle
a = 90° – 24° = 66°

b = a = 66° ……base angles of isosceles triangle AOB, since OA = OB (radii)

d = 180° – (66° + 66°) = 48°

c = 180° – 48° …angles on a straight-line AOC
c = 132°

EXERCISE 27D

1) Work out the missing angles. O is the centre of the circle.

a)

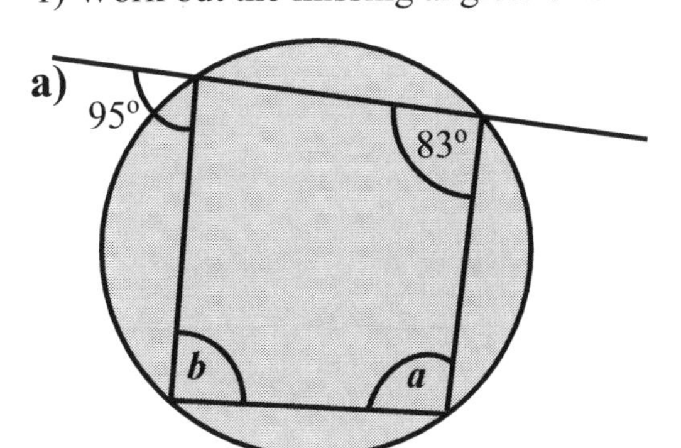

b)

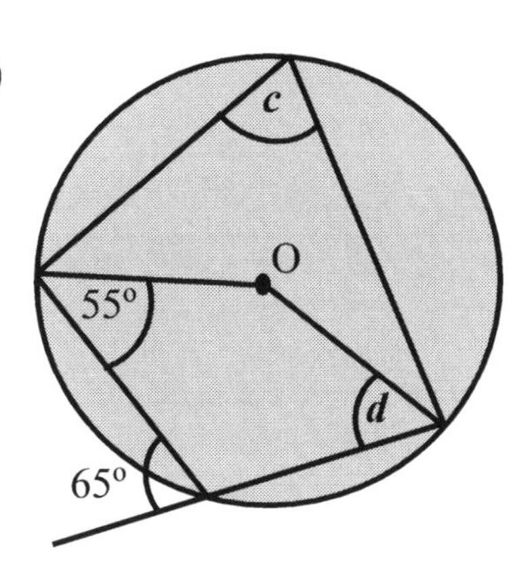

c)

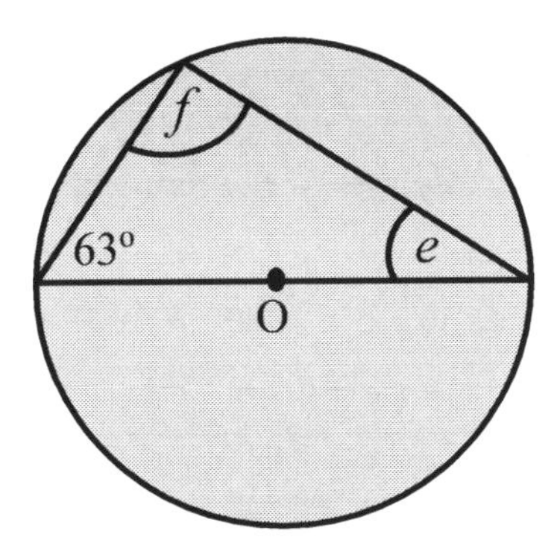

d)

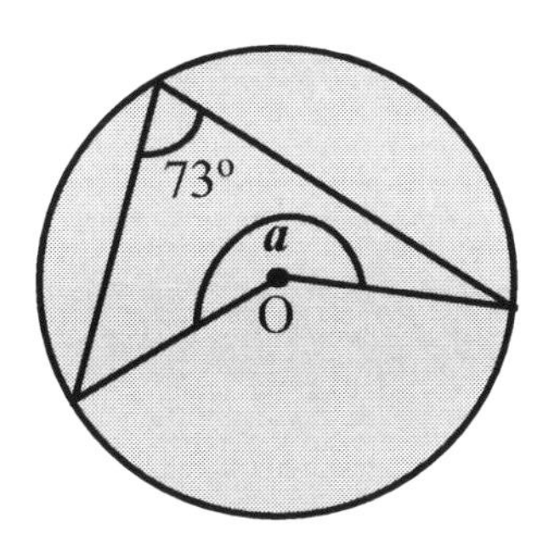

e)

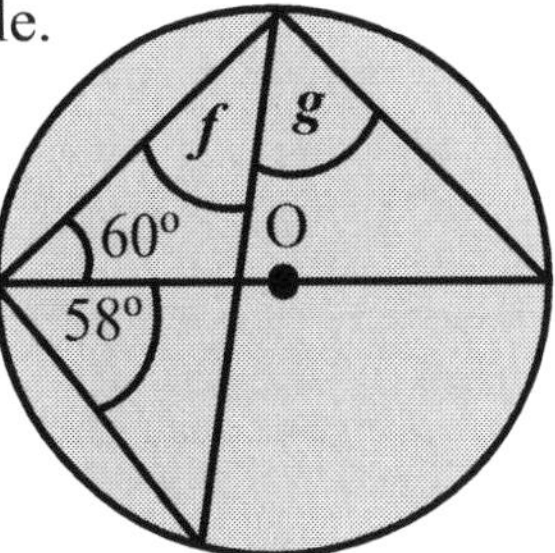

f)

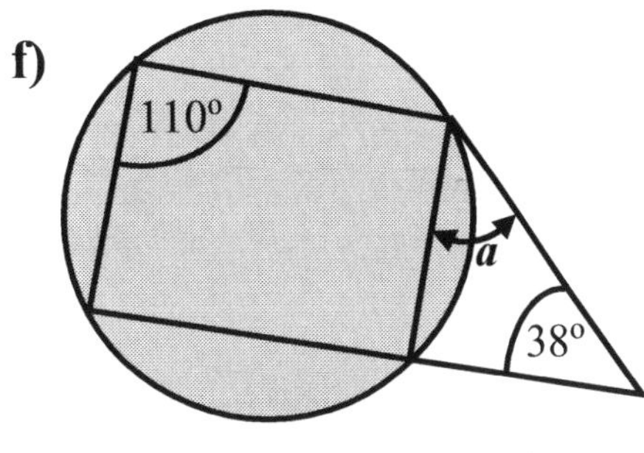

g)

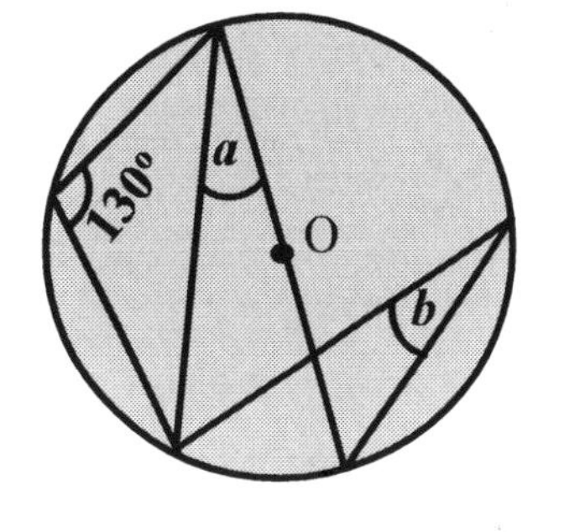

h) Prove that ∠ PQO = 30°

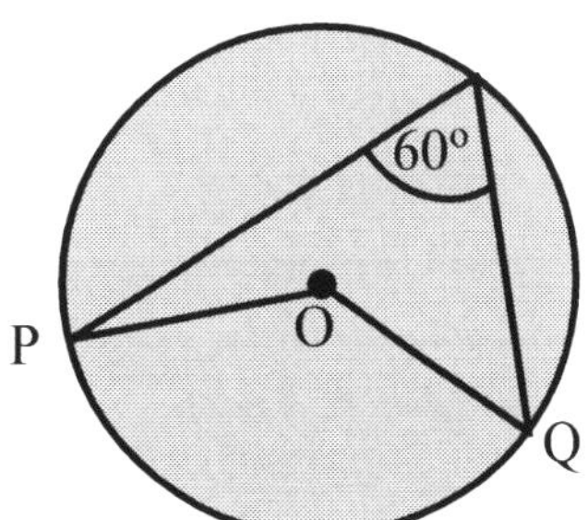

i) Prove that $x + y = 180°$

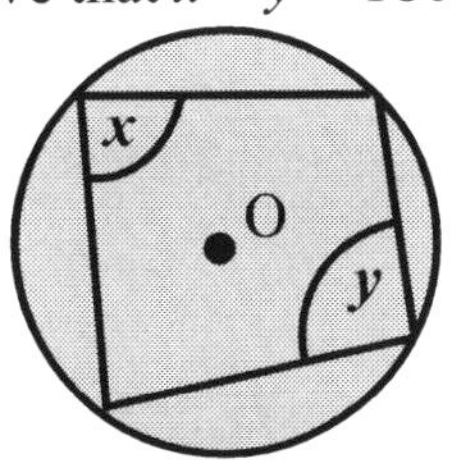

ALTERNATE SEGMENT THEOREM

The angle between a tangent and a chord through the point of contact is equal to the angle in the alternate segment.

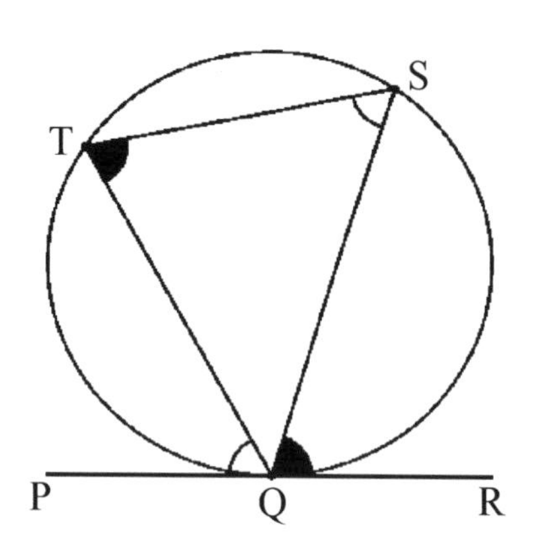

∠ QTS is known as the alternate segment of ∠RQS

Similarly, ∠ PQT = ∠ QST

∠QST is the alternate segment of ∠PQT.

∠ RQS = ∠ QTS ......alternate segment theorem

Example 1: Prove that ∠ RQS = ∠ QTS

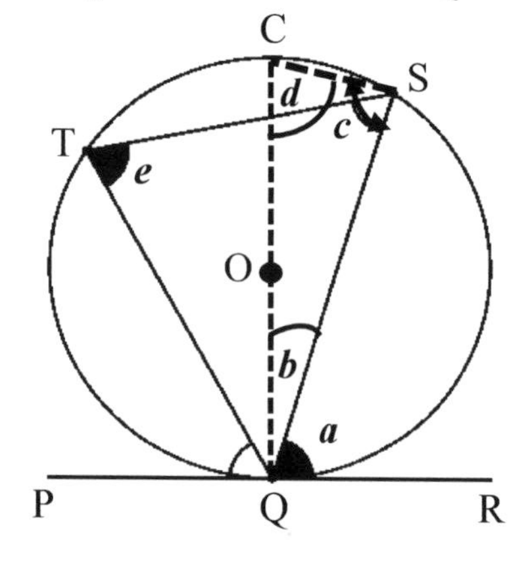

Solution:
Construction: Draw diameter QC and join line CS

Proof: a + b = 90°......tangent perpendicular to radius
c = 90° ...angle in a semi-circle
b + d = 90° .... sum of angles in a triangle
It then follows that a = d
Also, d = e .... angles in the same segment
a = e = d
Therefore, ∠ RQS = ∠ QTS.

Example 2: Calculate the size of angles a, b and c. Give reasons.

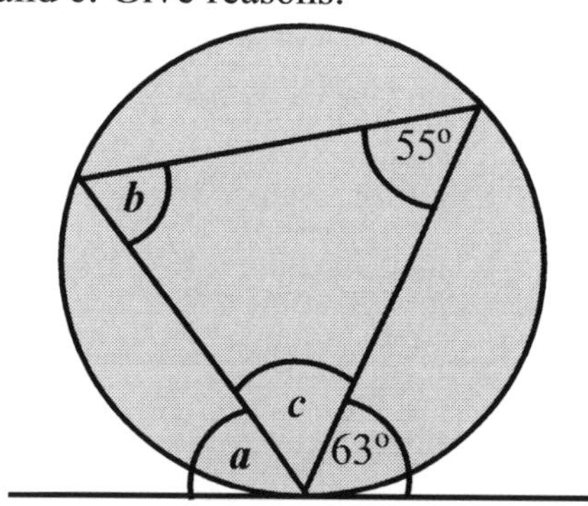

Solution:
a = 55° ......alternate segment theorem

b = 63° .... alternate segment theorem.

c + b + 55° = 180° ...angles in a triangle
c + 63° + 55° = 180°
c + 118° = 180°
c = 180° – 118°
c = 62°

EXERCISE 27E

Calculate the size of the missing angles in questions 1 to 5. All diagrams not drawn accurately. O is the centre of the circle.

1)

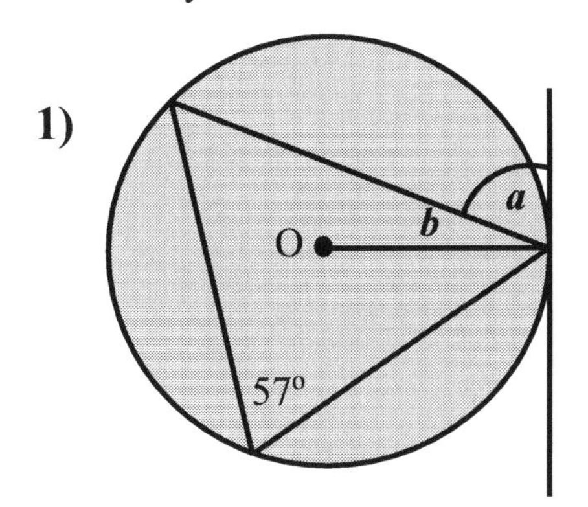

2)

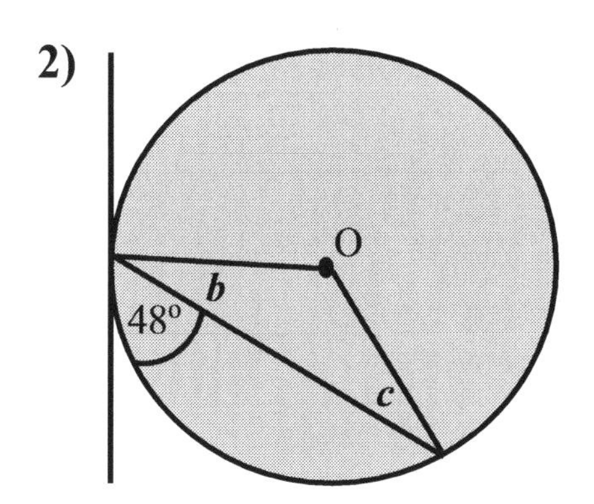

3)

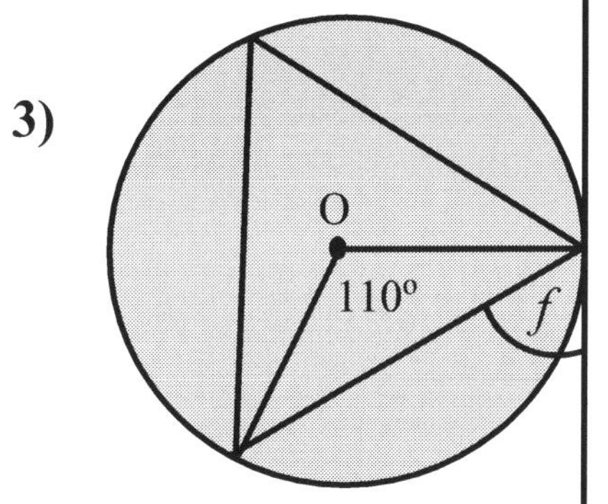

4)

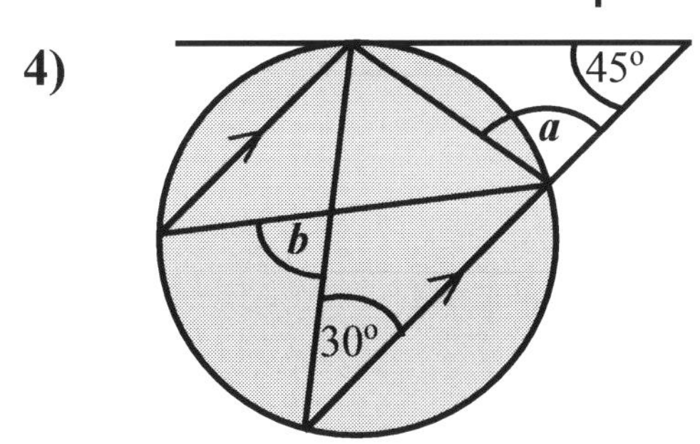

5)

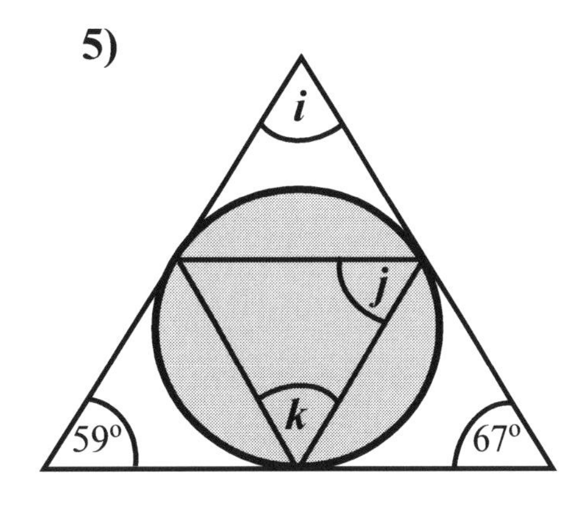

6) Show that m + 2n = 90°

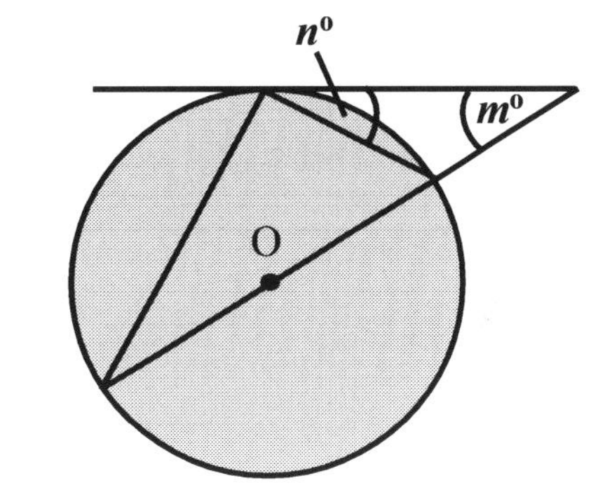

7) Prove that ∠ ABC = ∠ BEC

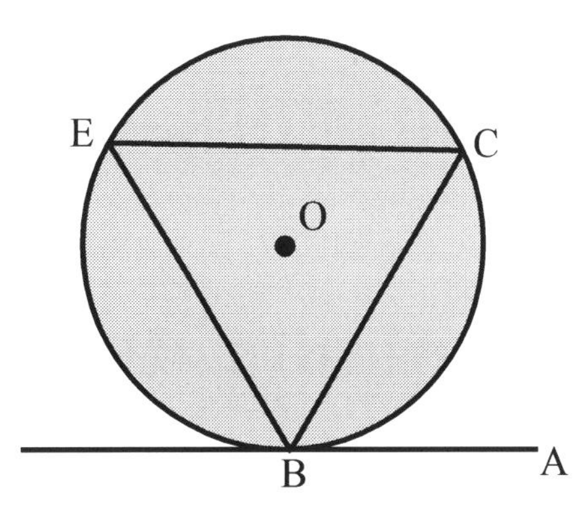

28 Cones, Pyramids and Spheres

This section covers the following topics:

- Volume of pyramids, cones and spheres
- Surface area of pyramids, cones and spheres

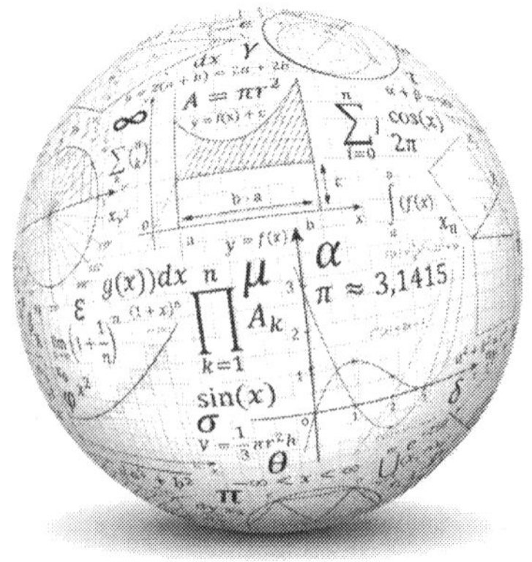

LEARNING OBJECTIVES

By the end of this unit, you should be able to:

a) Calculate the volume of cones, pyramids and spheres
b) Calculate the volume of a frustum of a cone
c) Calculate the surface area of cones, pyramids and spheres
d) Calculate the volume of compound solids

KEYWORDS

- Cone
- Frustum
- Pyramid
- Sphere
- Volume
- Surface area

CONES

A cone has a circular base and is known as a special pyramid.

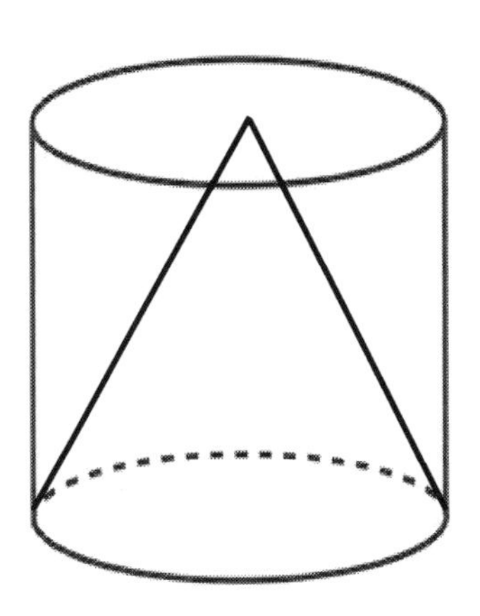

Three cones can fit into a cylinder

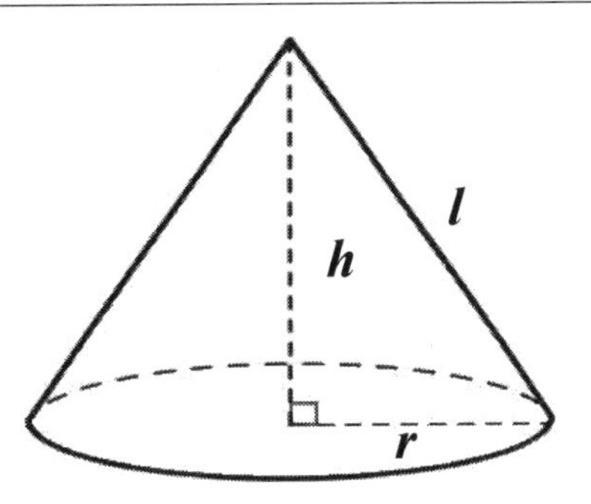

But, base area is a circle, which is πr^2. Therefore

Volume of a cone $= \frac{1}{3} \times$ base area $\times$ height

$= \boldsymbol{\frac{1}{3}\pi r^2 h}$

Curved surface area $= \pi \times$ radius $\times$ slant height

$= \boldsymbol{\pi r l}$

Surface area of a cone $= \boldsymbol{\pi r^2 + \pi r l}$

Where r = radius, h = vertical height and l = slant height

Example 1: Calculate the volume and total surface area of the cone.

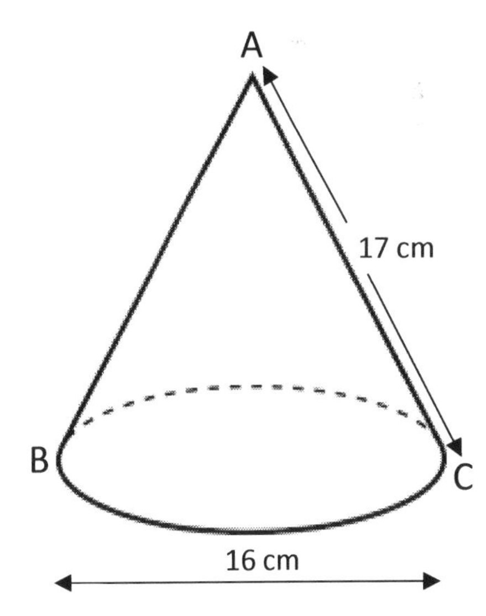

Solution: Volume of a cone $= \frac{1}{3}\pi r^2 h$

Radius = diameter ÷ 2 = 16 ÷ 2 = 8 cm.

Vertical height (h) is not given but using Pythagoras' theorem will give the vertical height.

$AC^2 = h^2 + DC^2$

$h^2 = AC^2 - DC^2$

$= 17^2 - 8^2$

$= 289 - 64 = 225$

$h = \sqrt{225} = 15\ cm$

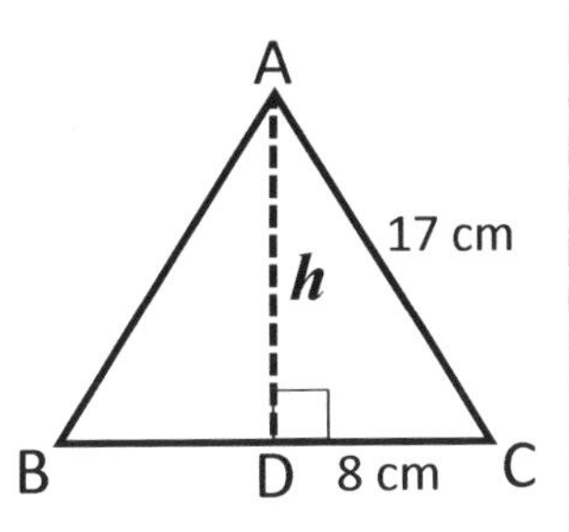

Volume of cone $= \frac{1}{3} \times \pi \times 8^2 \times 15$

$= 320\ \pi$ cm^3 or **1005.3 cm^3** to 1 d.p.

Surface area $= \pi r^2 + \pi r l = (\pi \times 8^2) + (\pi \times 8 \times 17)$

$= 200\ \pi$ cm^2 or **628.3 cm^2** to 1 d.p.

FRUSTUM OF A CONE

A **frustum** of a cone is a cone with the top cut off by making a slice parallel to the base of the cone. A frustum is what is left when the top of a cone is chopped off.

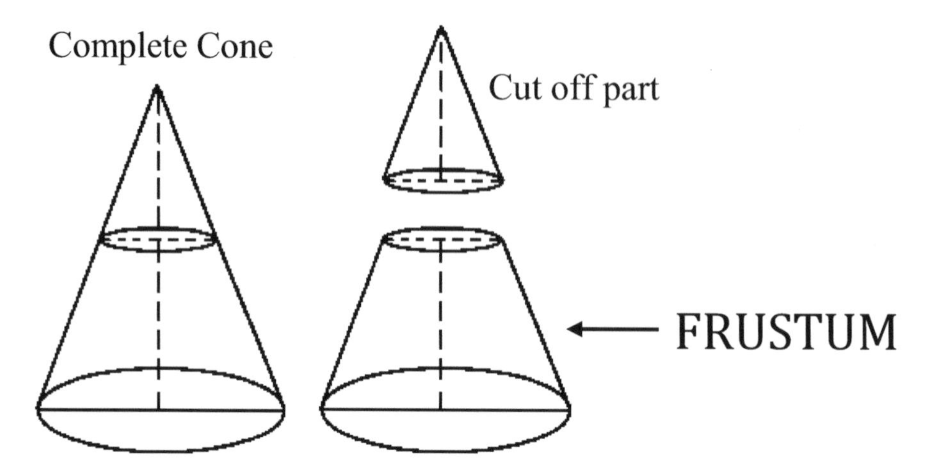

Volume of the frustum = Volume of the full cone – Volume of the cut-off part (cone).
Surface area of the frustum = Surface area of the whole cone – Surface area of the cut-off part

- Remember to add the area of the bottom and top circles when there is a need

Example 2: A cone is cut off, and the frustum shown below. Calculate the volume of the frustum.

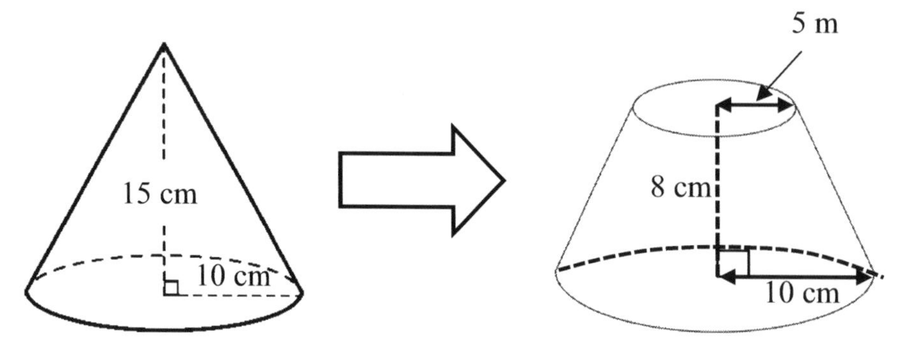

Solution: Cut off cone:

$15 - 8 = 7$ cm

7 cm 5 cm

Volume of big cone $= \frac{1}{3} \times \pi \times 10^2 \times 15 = 1570.796..$ cm^3

Volume of cut off cone $= \frac{1}{3} \times \pi \times 5^2 \times 7 = 183.259..$ cm^3

Volume of frustum = 1570.80 – 183.26 = **1387.5 cm³** to 1 d.p

VOLUME OF PYRAMIDS

Most pyramids have their names from its base. Examples of pyramids are:
Triangle-based pyramid, square-based pyramid or pentagonal pyramid.

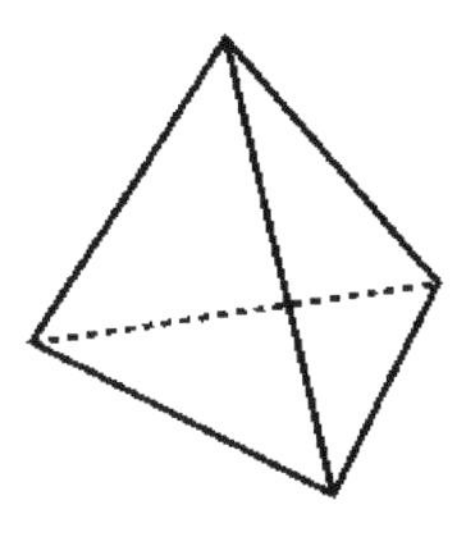

Triangle-based pyramid or tetrahedron

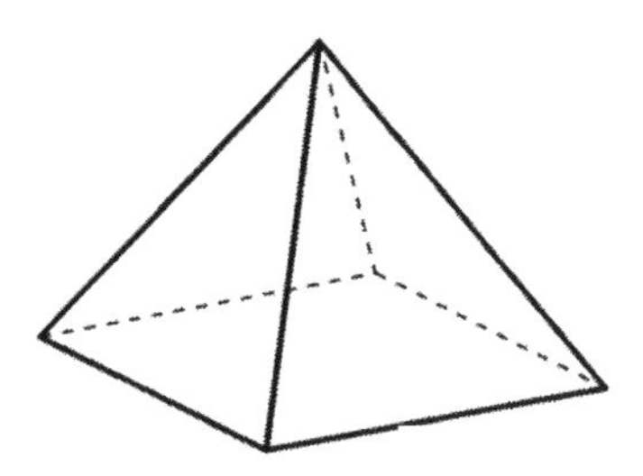

Square-based pyramid

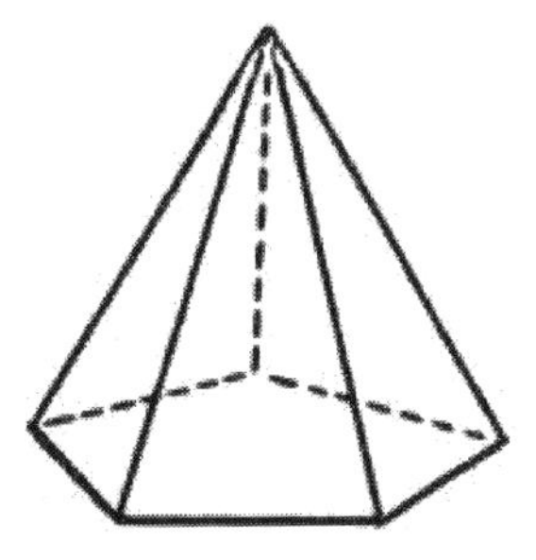

Pentagonal pyramid

The edges of any pyramid are all straight lines, and the base is a polygon. The edges meet at the vertex. As the volume of a cone, the volume of a pyramid is a third of the volume of its smallest surrounding prism.

Volume of a pyramid $= \frac{1}{3} \times$ base area $\times$ vertical height $= \frac{1}{3}\mathbf{A}\boldsymbol{h}$

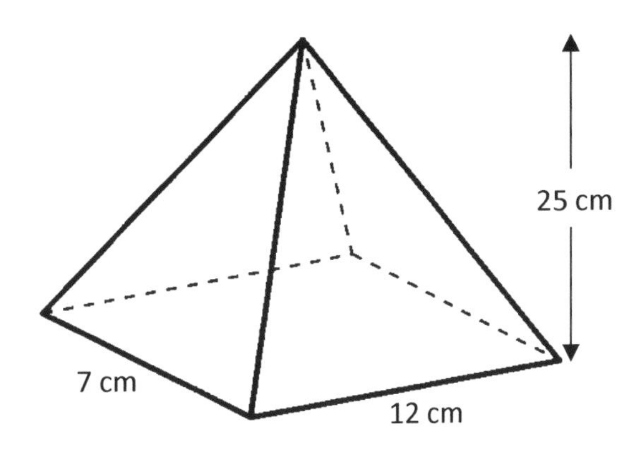

Example 3
Calculate the volume of the pyramid.

Solution:

Volume of pyramid
$= \frac{1}{3} \times$base area $\times$ vertical height

$= \frac{1}{3} \times 7 \times 12 \times 25$

$= 700$ cm^3

Example 4: A pyramid has a volume of 1000 cm^3, a square base of side *w* cm and a vertical height of 30 cm. Calculate the value of *w*.

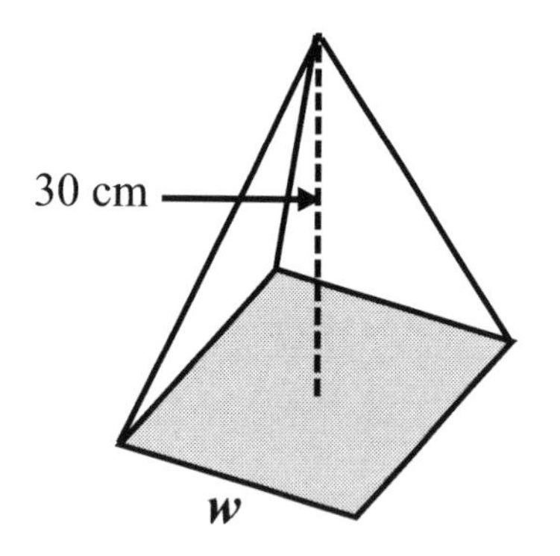

Solution:

Volume of pyramid

$V = \frac{1}{3} \times \text{base area} \times \text{vertical height}$

$1000 = \frac{1}{3} \times w \times w \times 30$

$1000 = 10\ w^2$

$w^2 = \frac{1000}{10} = 100$

$w = \sqrt{100} = \mathbf{10\ cm}$

SURFACE AREA OF PYRAMIDS

Adding together the areas of **all** the faces of a pyramid gives the total surface area.

Example 5: Calculate the total surface area of the pyramid.

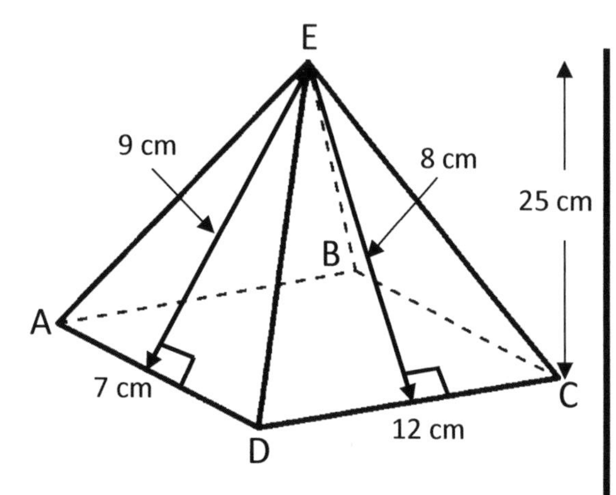

Solution:

Area of base ABCD = $7 \times 12 = 84$ cm^2

Area of triangle DEC = $\frac{1}{2} \times 12 \times 8 = 48$ cm^2

Area of triangle ABE = Area of triangle DEC = 48 cm^2

Area of triangle AED = = $\frac{1}{2} \times 7 \times 9 = 31.5$ cm^2

Area of triangle CEB = area of triangle AED = 31.5 cm^2

Total surface area

$= 84 + 48 + 48 + 31.5 + 31.5 = 243$ cm^2

VOLUME OF A SPHERE

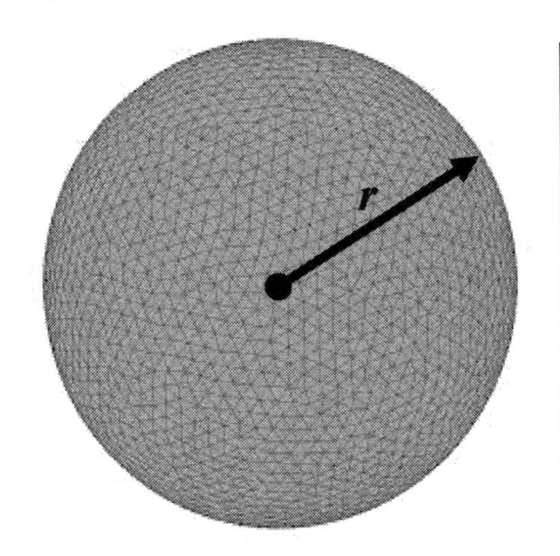

Example 6: Calculate the volume and surface area of a sphere with radius 7 cm.

Volume = $\frac{4 \times \pi \times 7^3}{3}$

$= \frac{4 \times \pi \times 343}{3} = 1436.8$ cm^3 to 1 d.p

Surface area = $4 \times \pi \times 7^2$
= 615.8 cm^2 to 1 d.p

The volume of a sphere = $\frac{4\pi r^3}{3}$

The surface area of a sphere = $4\pi r^2$

Where r = radius of the sphere

Example 7: The total surface area of a sphere is 4250 cm^2. Calculate the radius of the sphere.

Solution:

Surface area of sphere = $4\pi r^2$

$4\pi r^2 = 4250$

$4 \times \pi \times r^2 = 4250$

Make r^2 the subject

$r^2 = \frac{4250}{4 \times \pi} = 338.2042541$

$r = \sqrt{338.2042541}$

r = 18.4 cm to 1 decimal place

EXERCISE 28

1) Calculate the volume of the cones below. Round your answers to one decimal place. All lengths in cm.

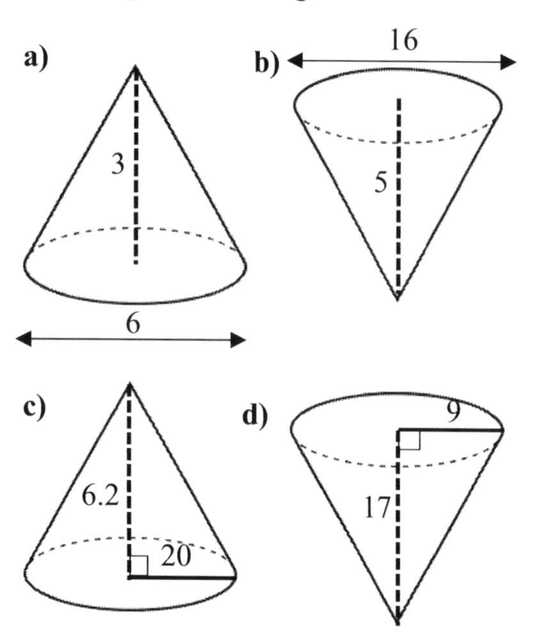

2) Calculate i) the curved surface area ii) the total surface area of the cones below. Round your answers to three significant figures. All lengths in cm.

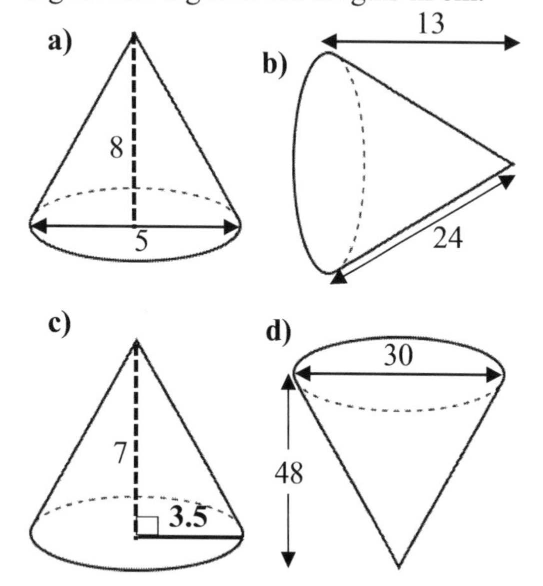

3) A pyramid has a volume of 320 cm^3, a square base of side $\boldsymbol{a}$ cm and a vertical height of 15 cm. Find the value of $\boldsymbol{a}$.

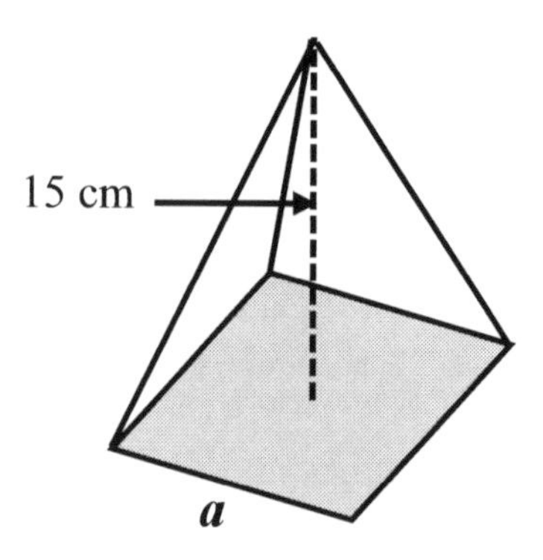

4) Calculate a) the volume b) the total surface area of the cone below.

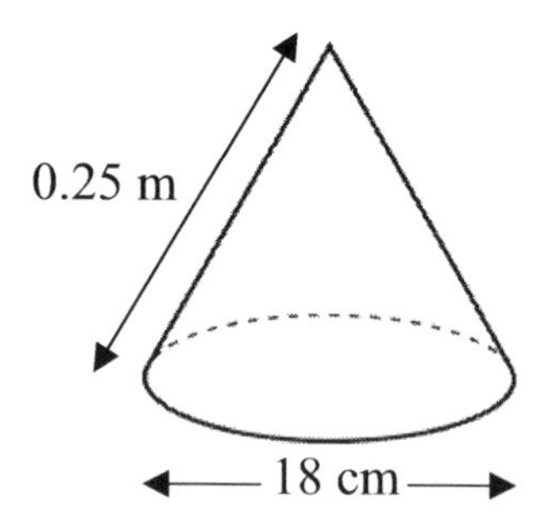

5) Calculate the a) volume b) surface area of the sphere to 2 decimal places.

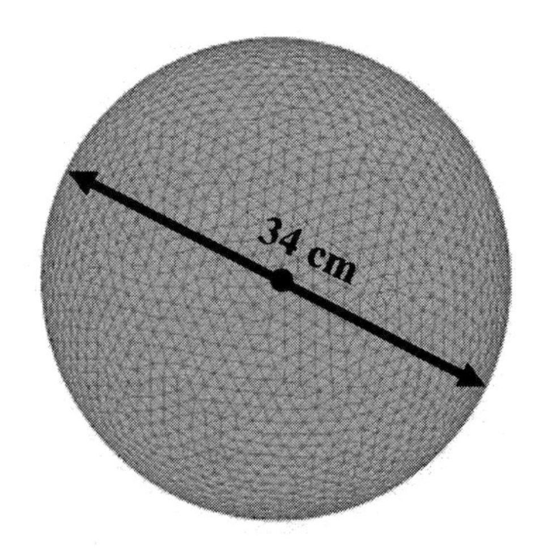

6) Work out the volume of the frustum below. Leave your answer to three significant figures.

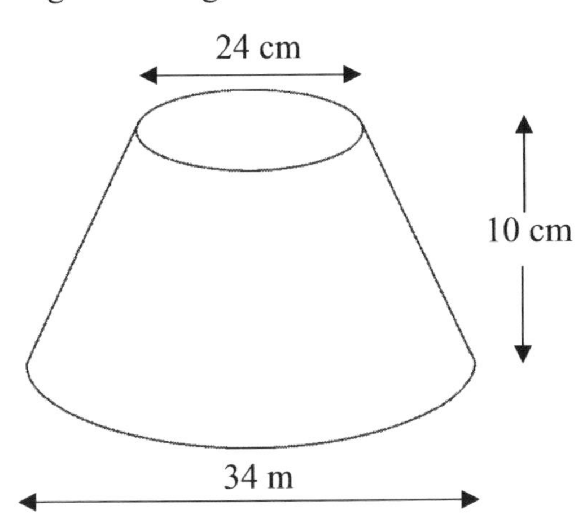

7) Calculate the volume of the solid below made from a hemisphere and a cone. Leave your answer to two decimal places.

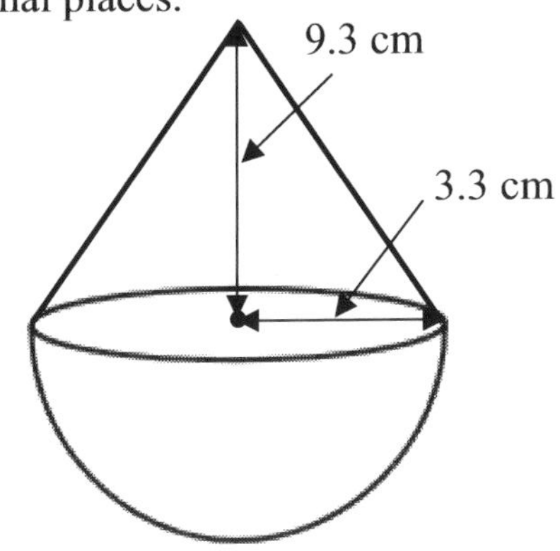

8) Calculate the surface area of the hemisphere below. Leave your answer to one decimal place.

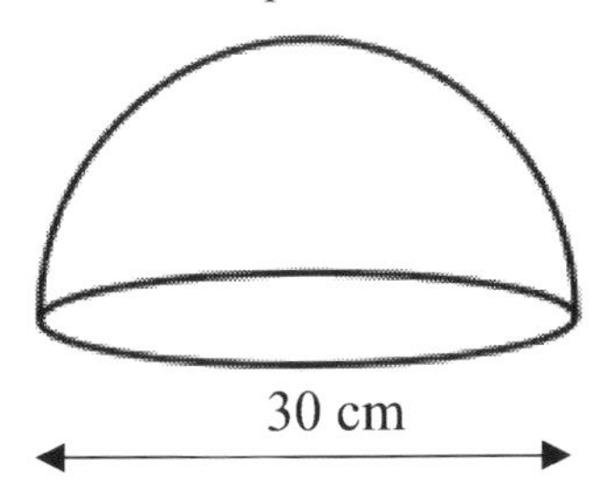

9) Calculate a) the volume b) the total surface area of the pyramid below.

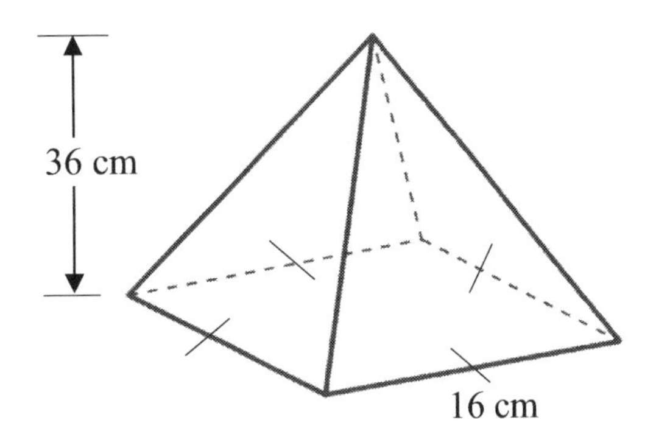

10) A cube is melted down and used to form a sphere as shown below.

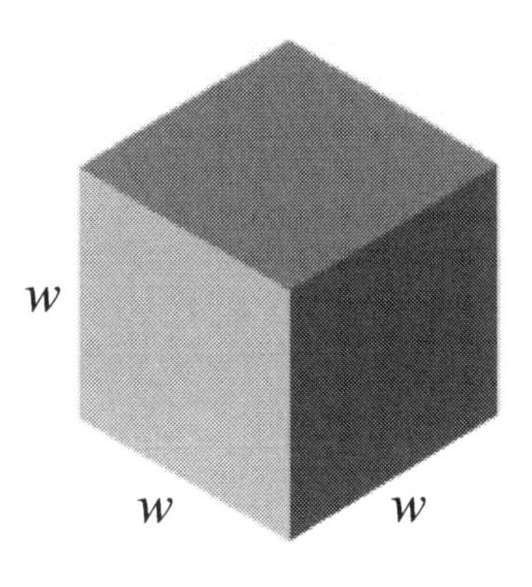

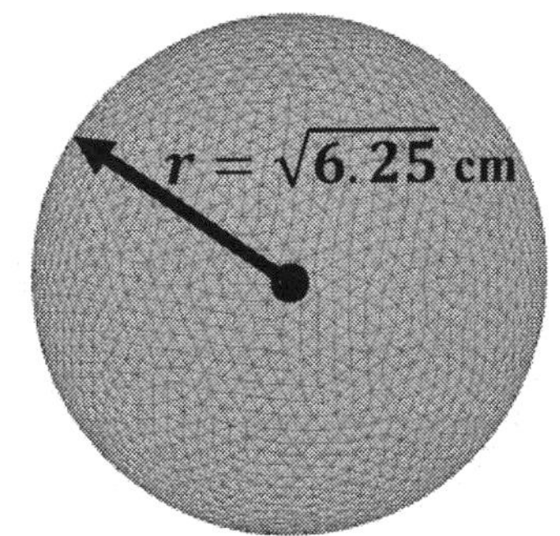

Assuming no material was lost in the process, calculate

a) the volume of the sphere

b) the surface area of the sphere

c) the length of the cube.

29 Similarity & Congruency

This section covers the following topics:

- Similar Shapes
- Congruency
- Area and volume scale factor
- Proof using congruency

LEARNING OBJECTIVES

By the end of this unit, you should be able to:

a) Find scale factor for similar shapes
b) Understand conditions for congruency
c) Calculate areas and volumes of similar objects
d) Prove that two shapes are congruent

KEYWORDS

- Scale factor
- Similarity
- Area scale factor
- Volume scale factor
- Congruent

29.1 SIMILAR SHAPES

Two or more shapes are similar if one shape is an enlargement of the other. Enlargement creates two similar shapes, and the corresponding lengths are in the same ratio. When two geometrical shapes are similar, the ratio of their corresponding sides is called the **scale factor**.

$$\text{Scale factor} = \frac{\textit{length of the side of one shape}}{\textit{length of corresponding side on other shape}}$$

Example 1: Find the scale factor and the missing length if the two shapes are similar.

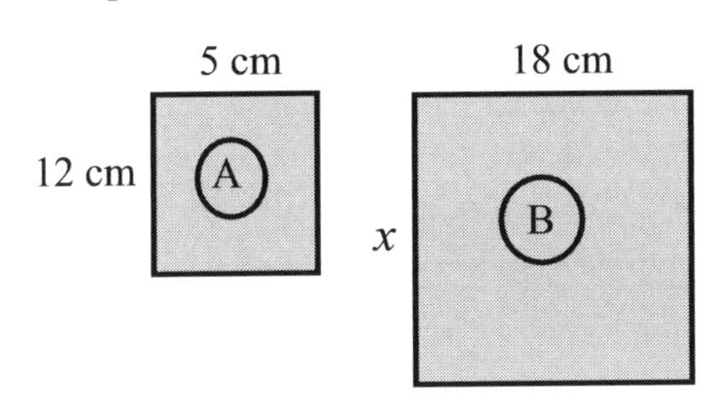

Solution:

To find the scale factor, look for the corresponding sides that are given. The sides of 5 cm and 18 cm are corresponding so,

Scale factor = $\frac{18}{5}$ = **3.6**

$x = 12 \times 3.6 =$ **43.2 cm**

CONDITIONS FOR SIMILARITY

If two shapes are similar, then

- The ratio of their corresponding sides is the same
- The corresponding angles must be the same

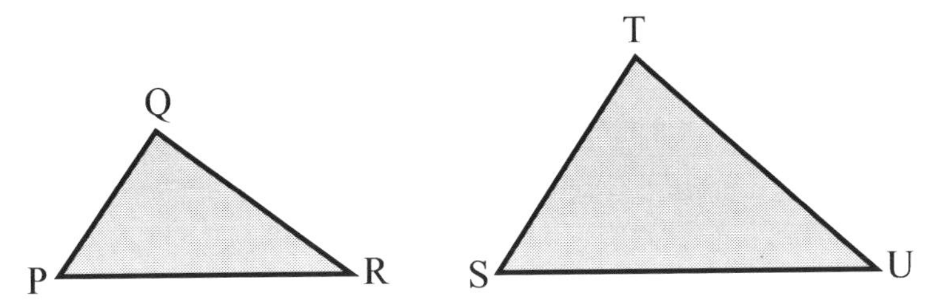

The corresponding sides are in the same ratio.

$\frac{TS}{QP} = \frac{SU}{PR} = \frac{TU}{QR} =$ Scale factor

Note: Identifying the corresponding sides is key in finding the correct ratio. The corresponding angles must be equal.

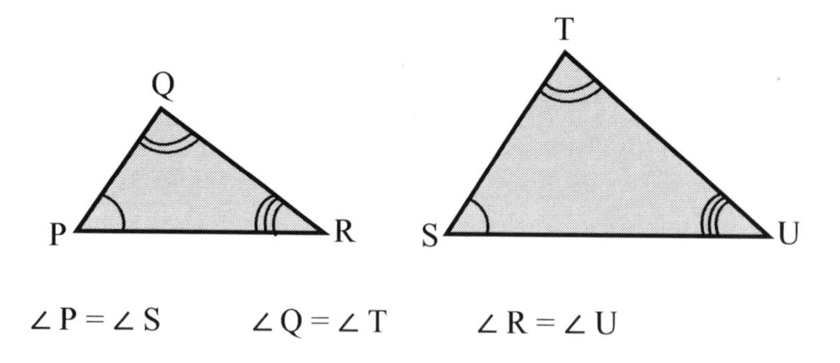

Example 2: The two triangles are similar. Work out the length of the side marked m.

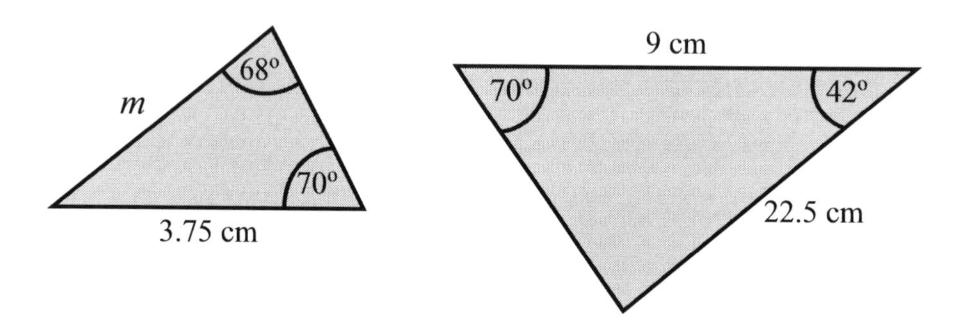

Solution: Work out the remaining angles and identify the corresponding lengths.

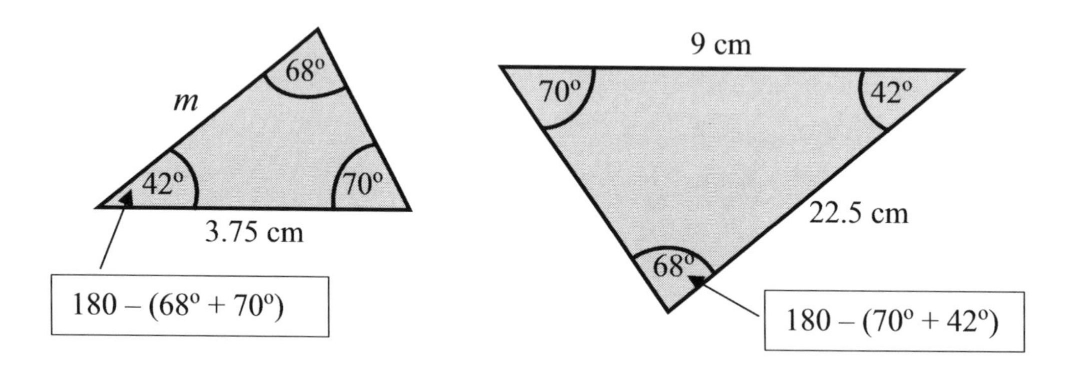

The corresponding sides (lengths) that are known are 3.75 cm and 9 cm.
Therefore, scale factor = 9 ÷ 3.75 = 2.4

Side 22.5 cm corresponds to the side of length, m.

Using the scale factor, m = 22.5 ÷ 2.4 = **9.375 cm**

Example 3:
a) Prove/show that triangle PST and triangle PQR are similar.
b) Work out the value of the missing lengths marked a and b.
All lengths are in cm.

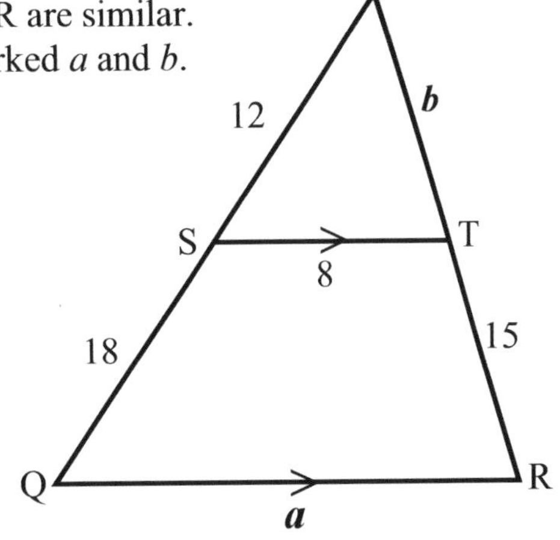

Solution:

a) $\angle$ PST = $\angle$ PQR ....corresponding angles
$\angle$ PTS = $\angle$ PRQ ....corresponding angles
$\angle$ SPT is common to both triangles
Therefore, Δ PST is similar to Δ PQR

b) Separate (split) the triangles for easy identification of corresponding sides.

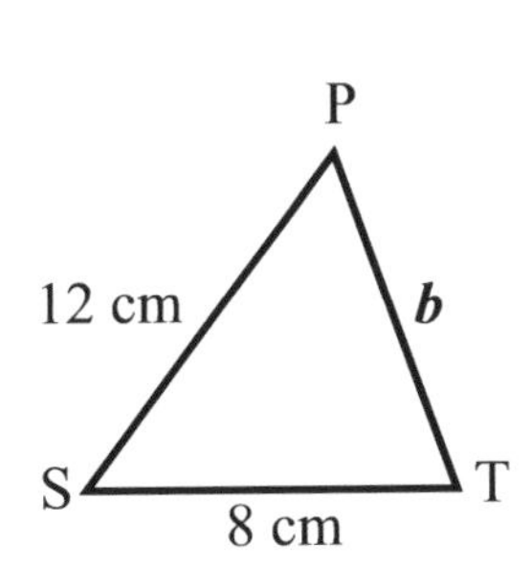

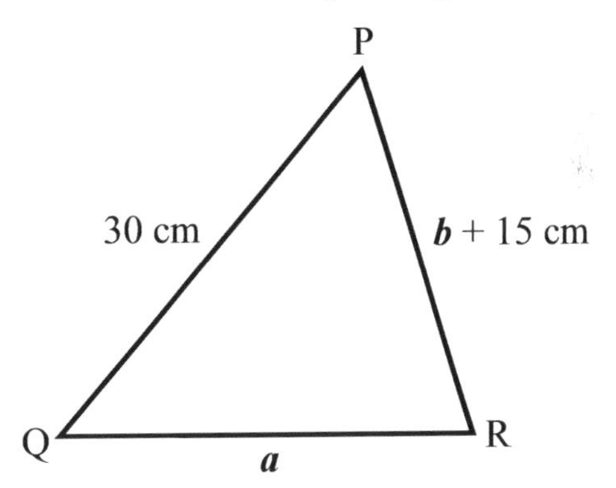

Scale factor = 30 ÷ 12 = 2.5
$a = 8 \times 2.5$
$a =$ **20 cm**

But, $b \times$ scale factor $= b + 15$

$2.5b = b + 15$
(Subtract b from both sides)

$1.5b = 15$

$b = \frac{15}{1.5}$

$b =$ **10 cm**

EXERCISE 29 A

1) Each pair of shapes is similar. Work out each of the lengths marked with letters. All lengths are in cm.

a)

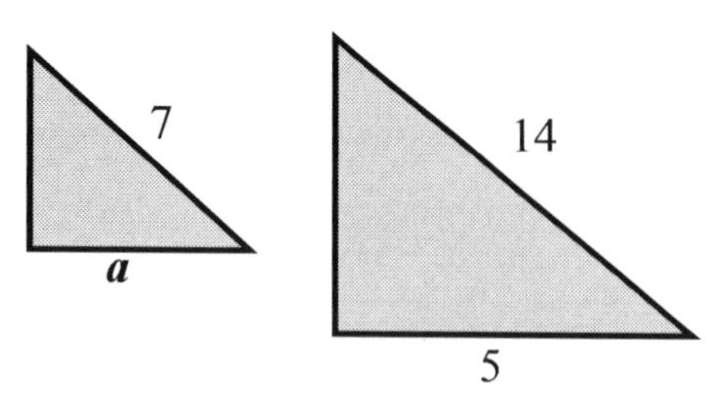

b)

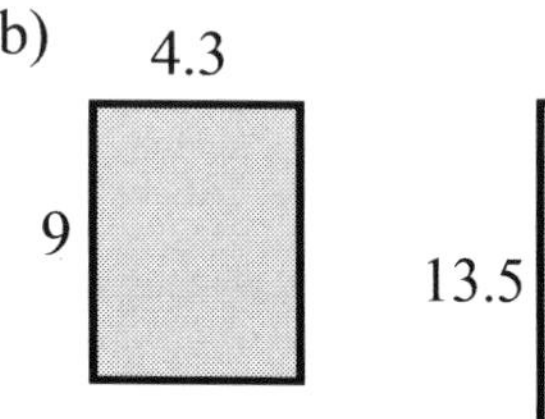

c)

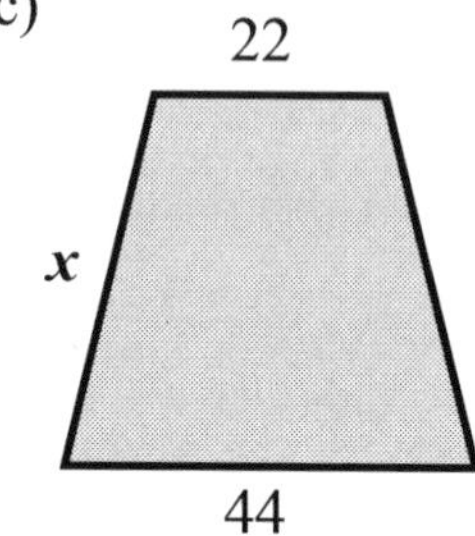

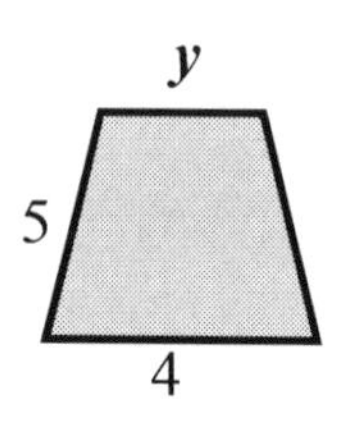

d)

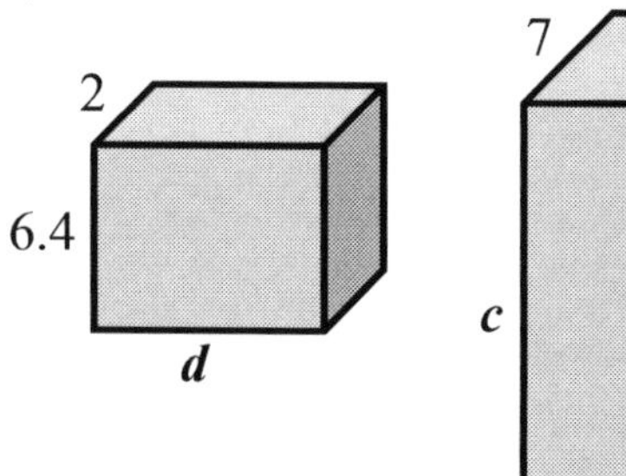

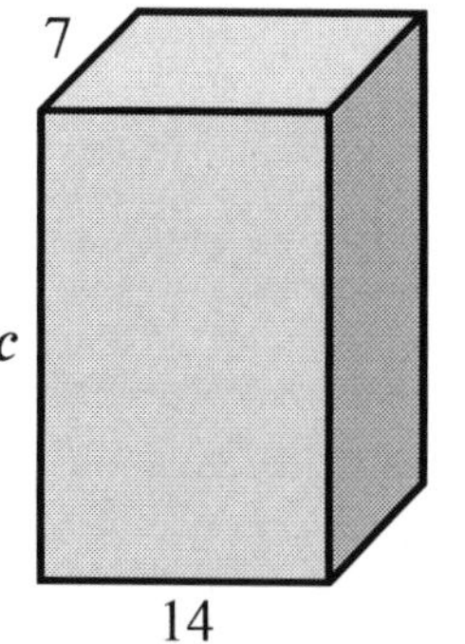

2)

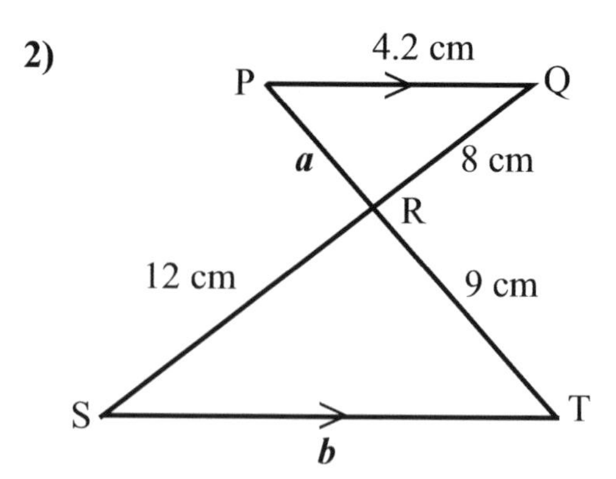

a) Show that triangles PQR and TSR are similar.
b) Calculate the lengths of a and b.

3)

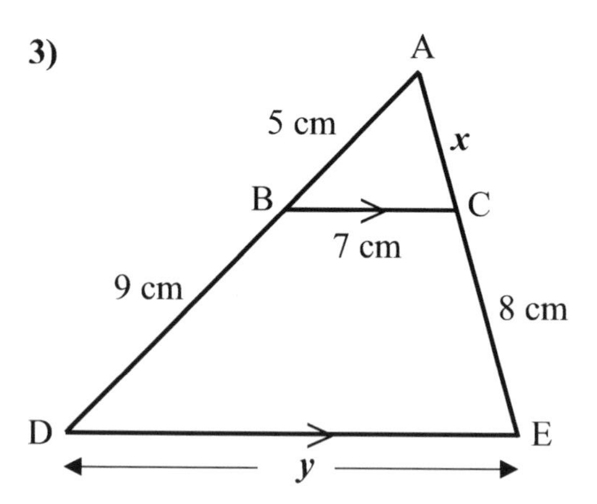

a) Explain why triangles ABC and ADE are similar.
b) Work out the lengths of x and y.

4) Match the pair of shapes that are similar. All lengths are in cm.

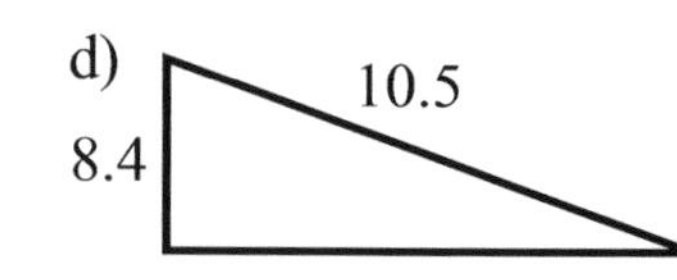

29.2 CONGRUENCY

When two or more shapes are the same shape and size, they are said to be congruent.

The shapes above are congruent. They are identical. It doesn't matter if one shape is rotated or turned over. The key is that the corresponding sides and corresponding angles are the same (identical).

When shapes are reflected, rotated or translated, they produce congruent images.

DIFFERENCE BETWEEN SIMILAR AND CONGRUENT SHAPES

An enlargement produces **similar** object(s). The created image is precisely the same shape but not the same size.

Reflected, translated and rotated objects produces **congruent** (same shape and size) images.

CONDITIONS FOR CONGRUENCY

The conditions below will always be true for congruent triangles.

1) SIDE, SIDE, SIDE (SSS)
Three sides are equal.

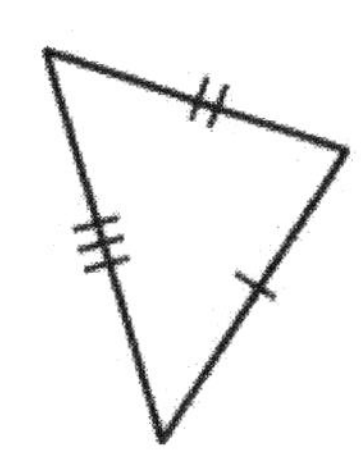

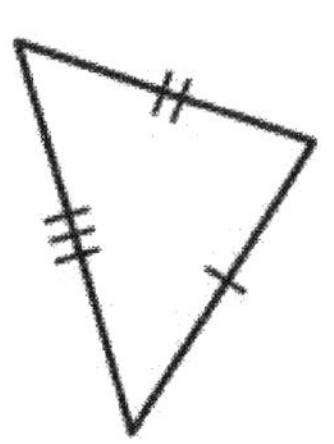

2) ANGLE, SIDE ANGLE (ASA)
or as **AAS**.

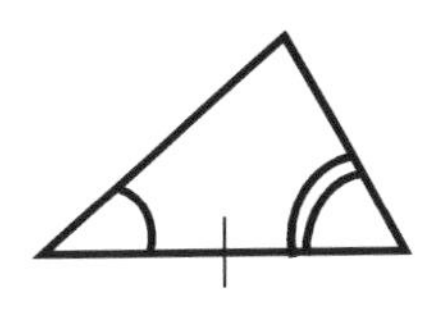

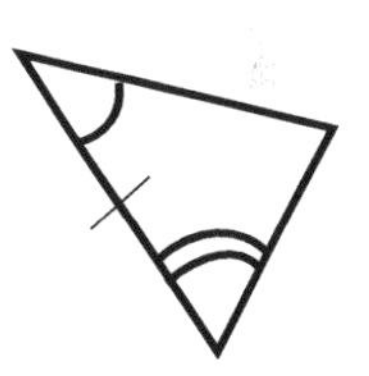

3) SIDE, ANGLE, SIDE (SAS)

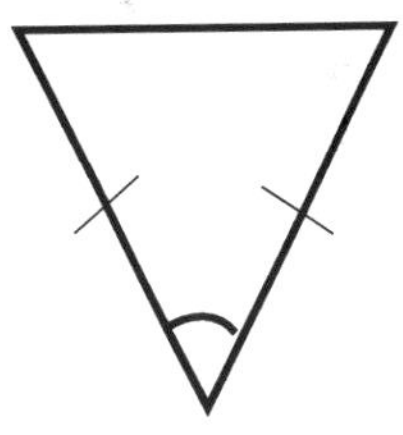

The angle must be included (made by the two lengths).

4) RIGHT-ANGLE, HYPOTENUSE SIDE (RHS)

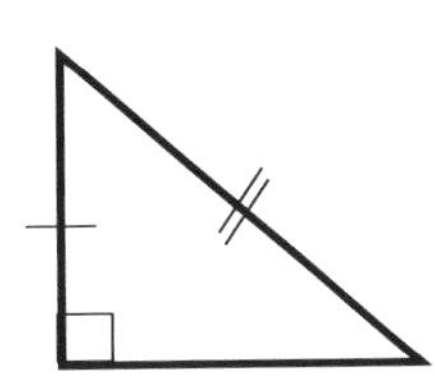

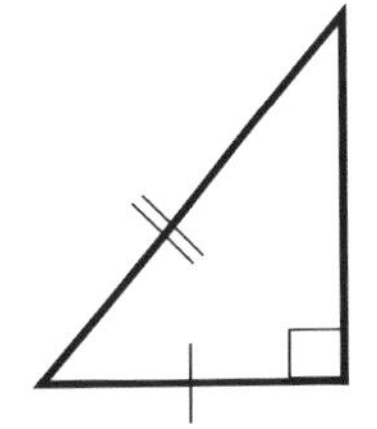

NOTE

1) Angle, Angle, Angle (AAA) is **not** a condition for congruency. It can produce similar but not congruent triangles.

2) Side, Side, Angle (SSA) is **not** a condition for congruency.

Trying to construct the triangle below will produce two triangles.

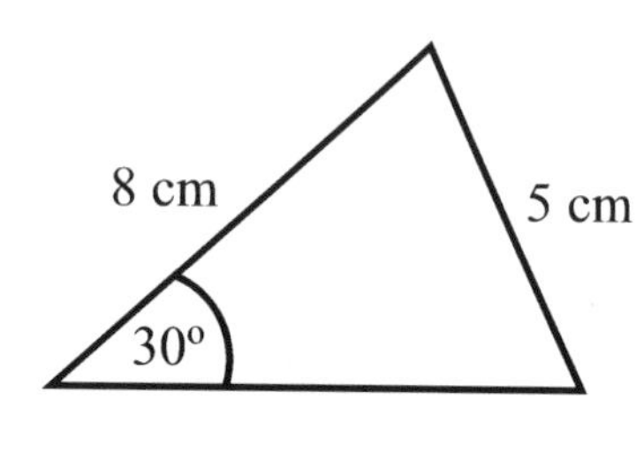

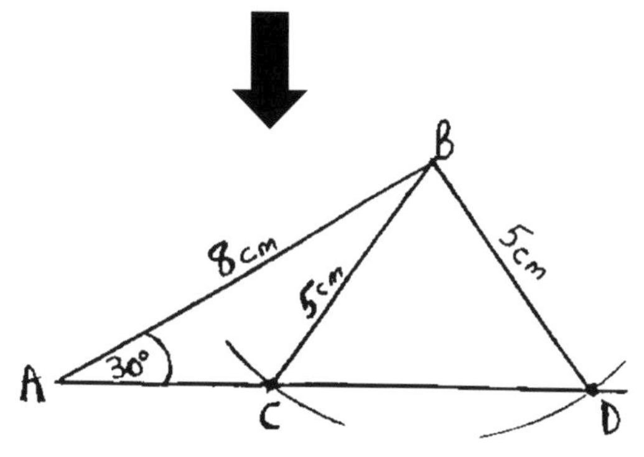

Triangles ABC and ABD will be produced.

Triangle ABC will have an acute angle at B while triangle ABD will have an obtuse angle at B.

This is **not** a condition for congruency since it has not produced identical triangles.

Example 1: Show that triangle ABC is congruent to triangle PQR.

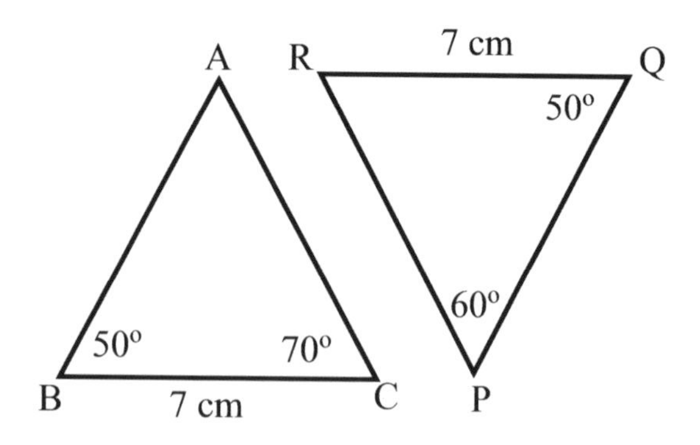

Solution:
Angle A = 180 – (50 + 70) = 60°

Likewise, angle R = 70°

All the angles are the same but not enough to prove congruency.

BC = QR = 7 cm
$\angle$ ABC = $\angle$ PQR = 50°
Also, $\angle$ BCA = $\angle$QRP = 70°

Therefore, triangle ABC is congruent (≡) to triangle PQR ....ASA

Example 2: State the condition of congruency for the pair of triangles.

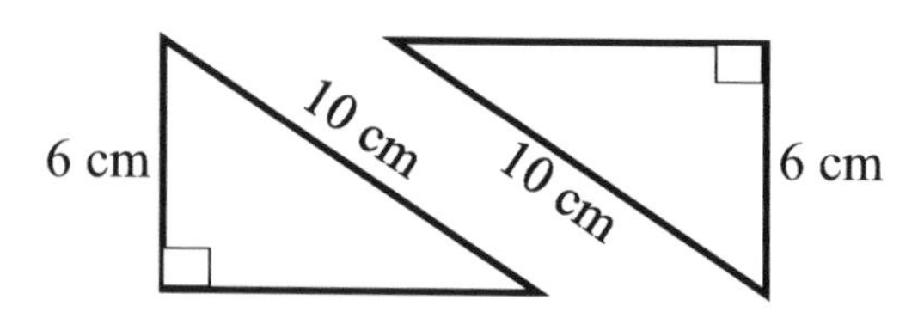

Hypotenuse and a side are equal to the corresponding sides. Therefore, the triangles are congruent…RHS

Example 3: Prove that triangles PQR and SRT are congruent.

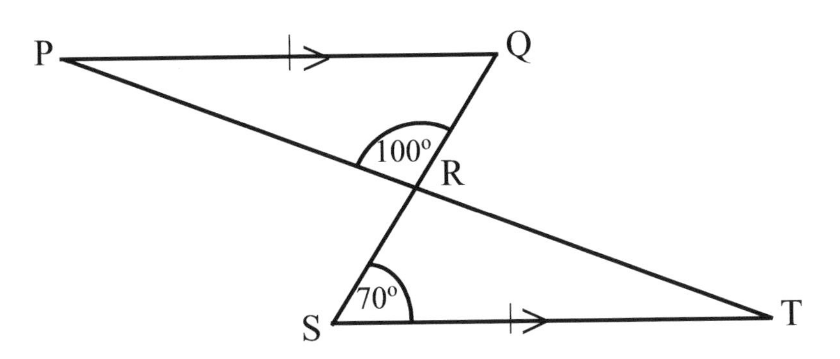

Solution:

∠ RQP = 70° ....alternate angles
∠ RPQ = 10° .... Angles in triangles add to 180° (from Δ PQR)

∠ SRT = 100° ....vertically opposite angles
∠ RTS = 10° ....alternate angles

Two angles and a corresponding side are equal. It shows that triangles PQR and SRT are congruent … (AAS)

EXERCISE 29B

1) From the pairs of shapes, state whether or not they are congruent. Give a reason.

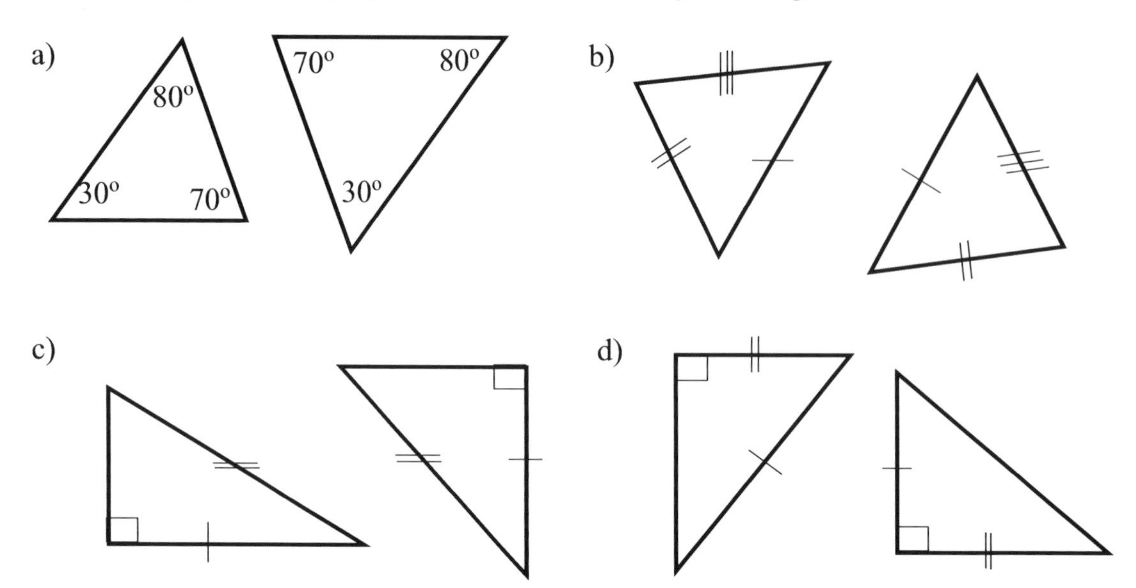

2) ABCD is a kite. Prove that triangles ABD and CBD are congruent.

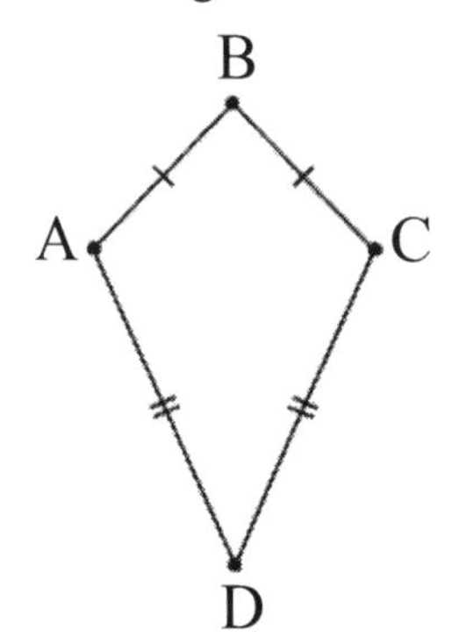

3) C is the midpoint of BD and C is also the midpoint of AE. BD and AE are straight lines. Prove that Δ ABC and Δ EDC are congruent.

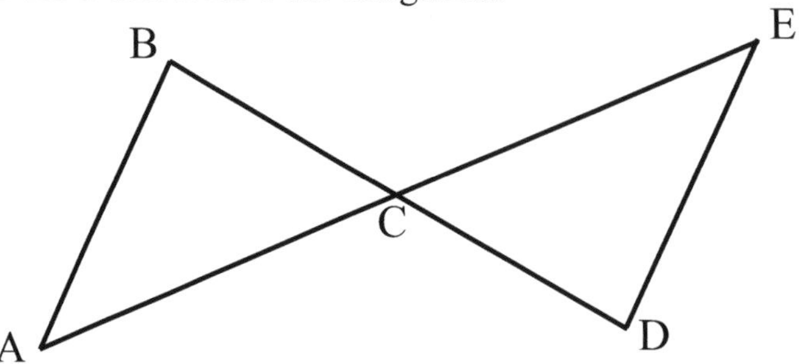

29. 3 AREAS OF SIMILAR OBJECTS

When linear scale factor (***k***) is known, the area of similar shapes can be calculated. Imagine a triangle enlarged by a scale factor of 2, as shown below.

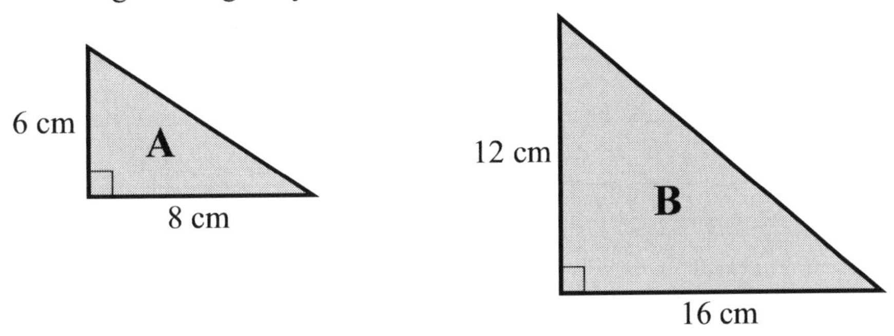

Area of triangle A = $\frac{8 \times 6}{2}$ = 24 cm^2 and area of triangle B = $\frac{16 \times 12}{2}$ = 96 cm^2.

Introducing the area scale factor, k^2

Example 1: Calculate the area of triangle B if the area of triangle A = 24 cm^2.

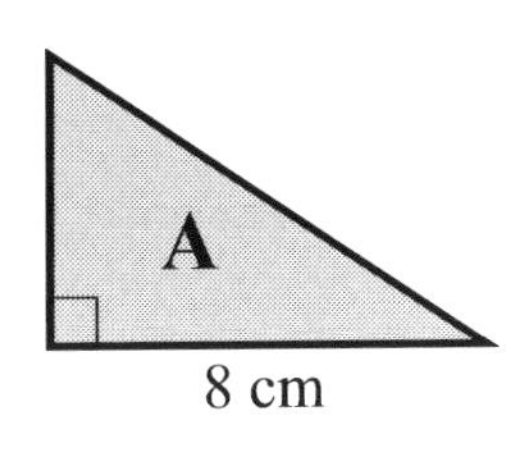

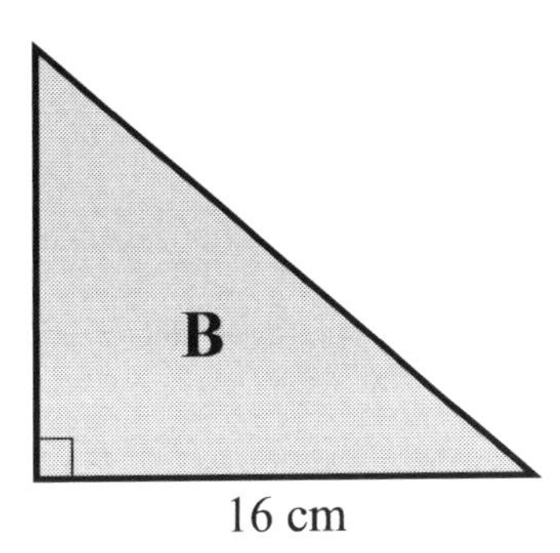

Solution
Use known corresponding lengths to work out the linear scale factor. The linear scale factor (k) for the shapes above is 16 ÷ 8 = **2**.

Area scale factor = $\mathbf{k^2}$. This is always the square of the linear scale factor. Therefore, area scale factor = 2^2 = 4.

To find the area of triangle B, multiply the area of triangle A by area scale factor, k^2. This is: 24 × 4 = **96 $\mathbf{cm^2}$**.

Example 2: Two similar rectangles are shown below. The area of the small rectangle is 20 cm^2. Calculate the area of the big rectangle.

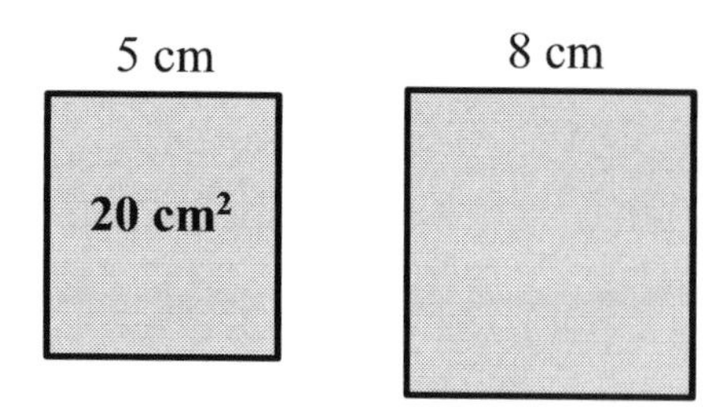

Solution:
Work out the linear scale factor (k).
$k = \frac{8}{5} = 1.6.$

Area scale factor = $k^2 = 1.6^2 = 2.56$
Area of big rectangle = $2.56 \times 20 =$ **51.2 cm²**

Example 3: A scale factor of 2.5 enlarges a trapezium.

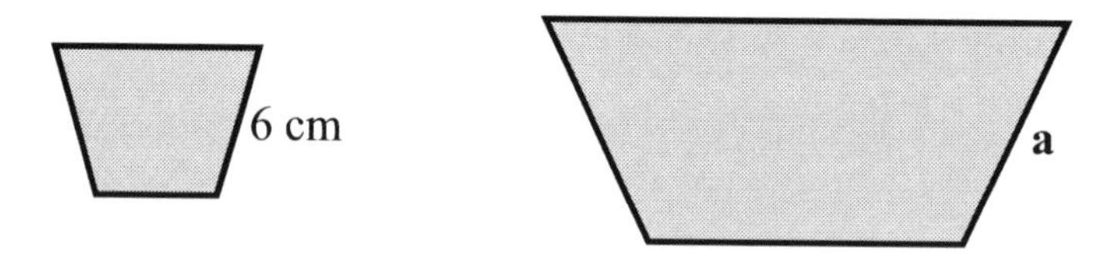

a) Calculate the missing length, **a**.
b) If the area of the big trapezium is 30 cm^2, work out the area of the small trapezium.

Solution:

a) Length of **a** = scale factor × 6
= 2.5 × 6 = 15 cm.

b) Area scale factor = k^2
$= 2.5^2 = 6.25$

Therefore, the area of small trapezium = area of big trapezium ÷ area scale factor (k^2).
= 30 ÷ 6.25
= **4.8 cm²**

29.4 VOLUMES OF SIMILAR OBJECTS

Imagine a cuboid enlarged by a scale factor of 3, as shown below.

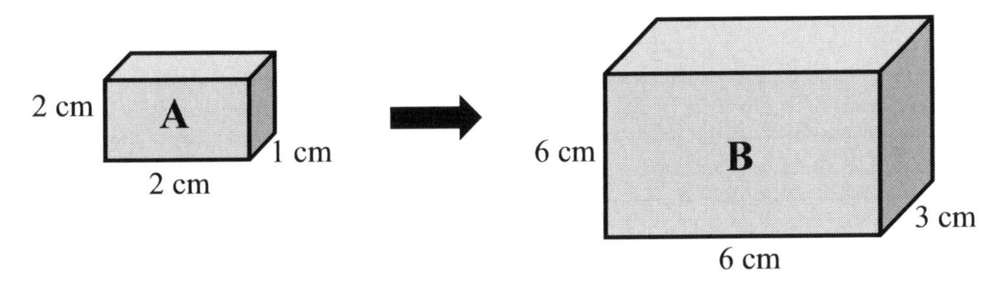

Volume of cuboid A = $2 \times 2 \times 1 = 4$ cm^3
Volume of enlarged cuboid B = $6 \times 6 \times 3 = 108$ cm^3

Introducing the volume scale factor, k^3

Example 1: The volume of cuboid A is 4 cm^3. Work out the volume of cuboid B.

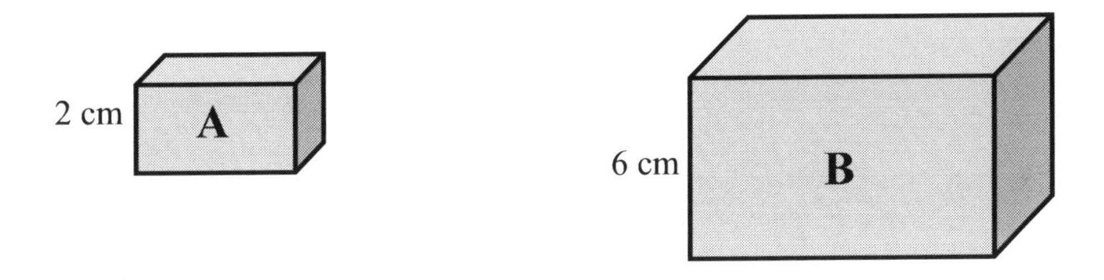

Solution: Work out the linear scale factor, k using corresponding lengths.

$k = \frac{6}{2} = 3.$

Volume scale factor = k^3. This is always the cube of the linear scale factor.
So, $k^3 = 3^3 = 3 \times 3 \times 3 = 27$.

Volume of cuboid B = volume of cuboid A × volume scale factor (k^3)
$= 4 \times 27 = 108$ cm^3.

Example 2: The radii of two circles are in the ratio 3:5.
Calculate the ratio of

a) the volumes of the circles
b) the area of the circles.

Solution:
a) Ratio of volumes = $3^3 : 5^3 = 27 : 125$
b) Ratio of areas = $3^2 : 5^2 = 9 : 25$

Example 3:

Iconic Concepts Limited produces two similar cylinders, A and B.

Cylinder A has a volume of 2500 cm^3.
Cylinder B has a volume of 8500 cm^3.

Calculate the height of cylinder B. Round your answer to one decimal place.

Solution:

First, find the scale factor of the enlargement. To do this, find the volume scale factor, k^3 and cube root the value.

$k^3 = \frac{8500}{2500} = 3.4$

$k = \sqrt[3]{3.4} = 1.5036945......$

Height of cylinder B = $42 \times 1.5036945......$
= 63.15517304
= 63.2 cm to 1 d.p

EXERCISE 29 C

1) The two shapes are similar. The area of the larger parallelogram is 34 m^2. Calculate the area of the smaller parallelogram to one decimal place. All the lengths are in metres.

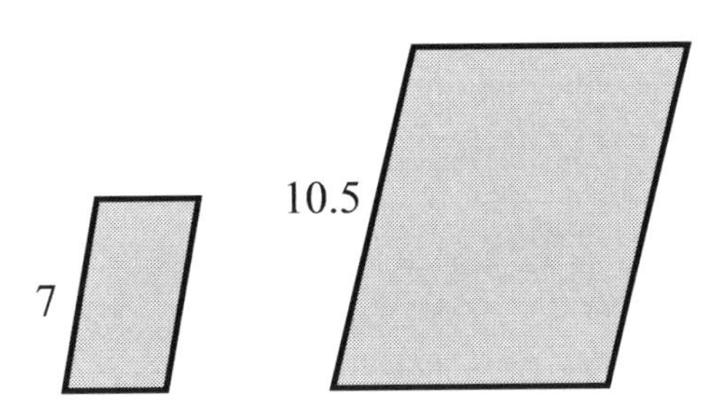

2) The rectangles below are similar. Work out the area of the larger rectangle.

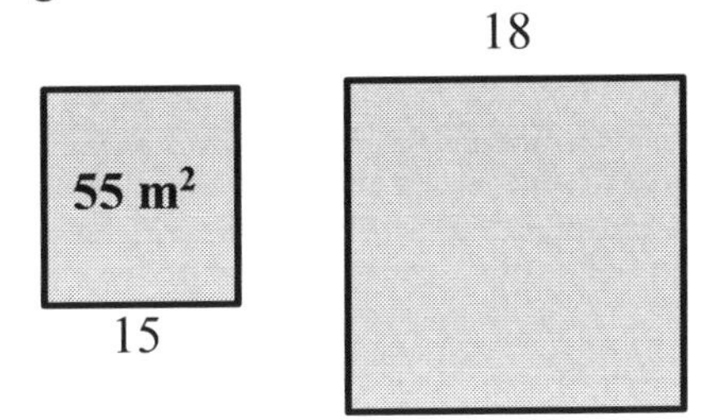

3)

The volume of the cold container is 785 cm^3 and the volume of the hot container is 175 cm^3. If the vertical height of the hot container is 17 cm, calculate the height of the cold container.

4) The ratio of the sides of two cylinders are in the ratio 1:7. Work out the ratio of:

a) the volumes of the cylinders

b) the areas of the cylinders.

5) The two cylinders below are similar. All lengths are in cm.

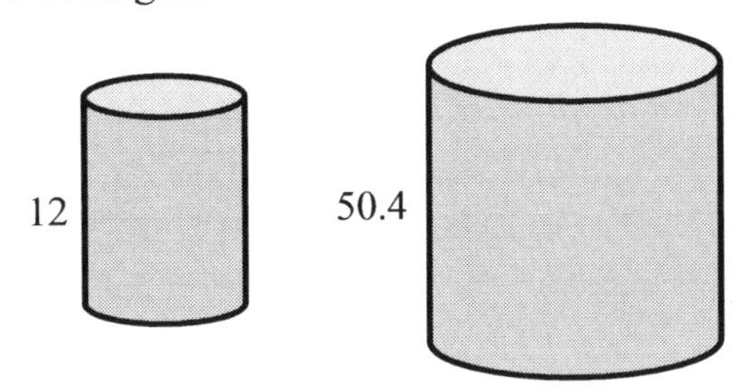

The area of cross-section of the small cylinder is 80 cm^2.

a) Calculate the cross-sectional area of the large cylinder.

b) The two cylinders are made of the same material. Calculate the mass of the small cylinder if the mass of the large cylinder is 250 g.

6) The two shapes are mathematically similar.

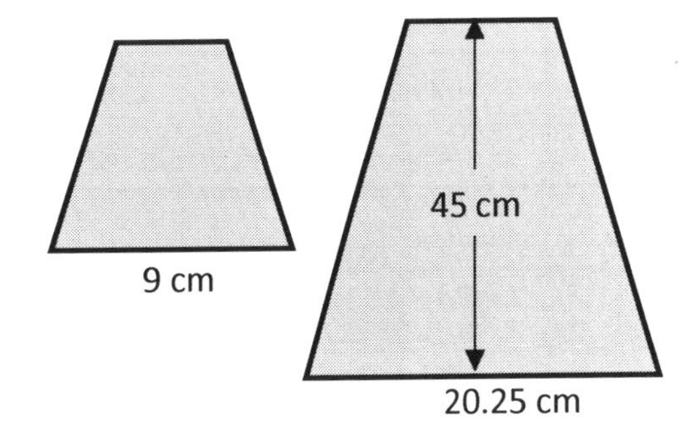

a) Work out the vertical height of the small trapezium.

b) The area of the small trapezium is 140 cm^2, calculate the area of the big trapezium.

30 Proportion/Variation

This section covers the following topics:

- Direct proportion
- Inverse proportion

LEARNING OBJECTIVES

By the end of this unit, you should be able to:

a) Write equations to solve direct proportion problems
b) Use equations to solve inverse proportion problems

KEYWORDS

- Direct proportion
- Inverse proportion
- Constant of proportionality

30.1 DIRECT PROPORTION

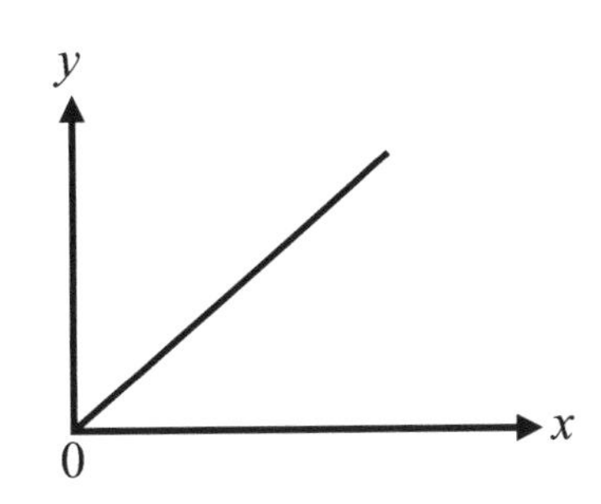

From the above graph, y is proportional to x. The above statement can be written as $\boldsymbol{y \propto x}$.

If the slope of the line is $\boldsymbol{k}$, then $y = kx$. The quantity k is called the **constant of proportionality.**

Rearranging the equation gives, $k = \frac{y}{x}$

Therefore, the ratio of y to x is always a constant. From the above, y is said to vary **directly** as x.

It follows that x increases in equal amounts as y.

x also decreases in equal amounts as y. For example, if x is doubled, y is also doubled. A glaring example is a relationship between the circumference (C) of a circle and its diameter (d).

$C = \pi \times$ diameter

It means that the circumference of a circle is directly proportional to its diameter.

Example 1:
y is directly proportional to x.
When $y = 12$, $x = 3$. Calculate the value of a) y when $x = 7$ b) x when $y = 20$.

Solution:
From the question, $y \propto x$. To introduce the equality sign, multiply x by the constant of proportionality, k.

This gives $y = kx$ .... (1)
Using the values given for y and x,
$12 = k \times 3$ and $k = 4$.

Rewrite equation 1 with the value of k.

$\boxed{y = 4x}$ .... (2)

Using equation 2,
a) $y = 4x$
$y = 4 \times 7 = \mathbf{28}$

b) From $y = 4x$, $20 = 4 \times x$
$x = \frac{20}{4} = \mathbf{5}$

Example 2: The volume (V) of a solvent in a jar is proportional to the height (h) of the jar. When the height of the solvent in the jar is 12 cm, the volume is 45 cm^3. Work out the volume of solvent when the height of solvent is 8 cm.

Solution: $V \propto h$ and $V = kh$

$45 = k \times 12$ and $k = \frac{45}{12} = 3.75$.
Rewrite the equation to give
$V = 3.75h$,
$V = 3.75 \times 8 = 30$

Volume of solvent = **30 cm^3**.

Example 3:
B varies directly as x^2.
When B = 75, x = 5. Work out
a) B when $x = 4$ b) x when B = 300.

<u>Solution</u>: $B \propto x^2$ and $B = kx^2$

Using the given numbers,
$75 = k \times 5^2$ and $k = \frac{75}{25} = 3$
Rewrite the equation to give
$B = 3x^2$

a) $B = 3 \times 4^2 = 3 \times 16 = \mathbf{48}$

b) From $B = 3x^2$,
$300 = 3 \times x^2$
$\frac{300}{3} = x^2$

$100 = x^2$
$x = \sqrt{100}$
$\mathbf{x = 10}$

EXERCISE 30A

1) W varies directly as T. If W = 30 and T = 6, calculate the value of a) W when T = 9   b) T when W = 17.5.

2) y is directly proportional to x.
When y = 20 and x = 8.

a) Write an equation in terms of y and x.

b) Calculate the value of

i) y when x = 0.6   ii) x when y = 40.

3) If c is directly proportional to d, complete the table below.

c	15		27	
d	5	7		10.5

4) **e** varies directly as the square of f.
Given that e = 13.5 when f = 1.5,

a) form an equation in terms of e and f.
b) Calculate the value of i) e when f = 4 and ii) f when e = 2.

5) y is in direct proportion to the cube of x. Given that y = 320 when x = 4, calculate
a) y when x = 2.5   b) x when y = 53.24.

6) a is directly proportional to the square root of b. Complete the table.

a		10	16	
b	1	25		100

7) p varies in direct proportion to the cube root of q. When $p = 6$, $q = 8$.
a) Form an equation in terms of p and q.
b) Calculate (i) p when $q = 125$
ii) q when $p = 9$.

30.2 INVERSE PROPORTION

If one quantity increases at the same rate as the other decreases, then the two quantities are in **inverse proportion**. If one quantity doubles, the other one halves.

We could say that y is inversely proportional to x or that y varies as the inverse of x.

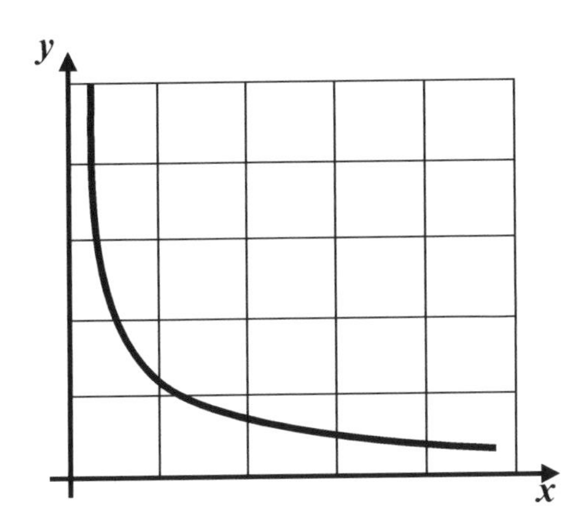

The above graph is a typical graph of y is inversely proportional to x.

We write it as:

$$y \propto \frac{1}{x}$$

Removing the proportional sign and adding the constant of proportionality gives $y = k \times \frac{1}{x}$

$$y = \frac{k}{x}$$

Example 1:
y varies as the inverse of x.
When x = 5, y = 2.
a) Write the equation linking x and y.
b) Work out the value of
i) y when x = 4   ii) x when y = 8.

Solution: $y \propto \frac{1}{x}$

To remove the proportionality sign, we multiply by the constant of proportionality, k.

$y = \frac{k}{x}$

Using the values given and substituting,
$2 = \frac{k}{5}$ and k = 2 × 5 = 10.

a) The equation is $\mathbf{y = \frac{10}{x}}$

b) i) $y = \frac{10}{4} = \mathbf{2.5}$

ii) $8 = \frac{10}{x}$ and $x = \frac{10}{8} = \mathbf{1.25}$

Example 2: w is inversely proportional to t^2. When w = 5, t = 2.
a) Form an equation linking w and t.
b) Work out
i) w when t = 2.5   ii) t when w = 4

Solution: $w \propto \frac{1}{t^2}$ so $w = \frac{k}{t^2}$

$5 = \frac{k}{2^2}$ so $k = 5 \times 2^2 = 5 \times 4 = 20$

a) the equation is $\mathbf{w = \frac{20}{t^2}}$

b) i) $w = \frac{20}{t^2} = \frac{20}{2.5^2} = \frac{20}{6.25} = \mathbf{3.2}$

ii) $4 = \frac{20}{t^2}$ and $4t^2 = 20$

$t^2 = 20 \div 4 = 5$

$t = \sqrt{5}$ or **2.24** to two decimal places.

EXERCISE 30B

1) y is inversely proportional to x.
When $y = 8$, $x = 3$.
a) Write an equation for y in terms of x.
b) Calculate the value of y when $x = 4$.
c) Calculate the value of x when $y = 10$.

2) w is inversely proportional to c.
When $w = 5.5$, $c = 2.5$.
a) Write an equation for w in terms of c.
b) Calculate the value of w when $c = 10$.
c) Calculate the value of c when $w = 8$.

3) Match each statement to the correct graph shown below.

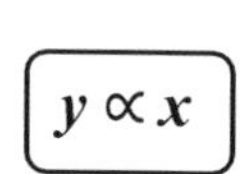

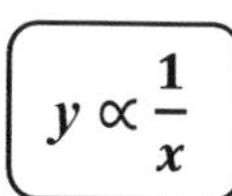

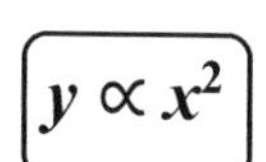

a)

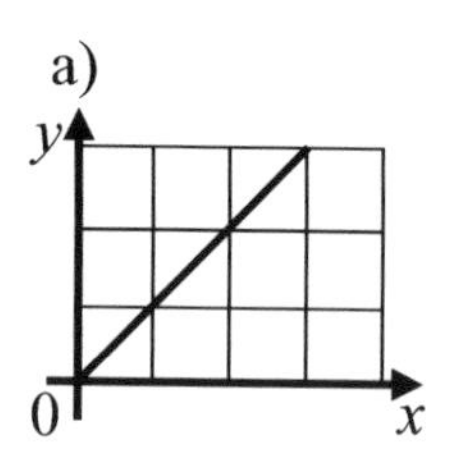

b)

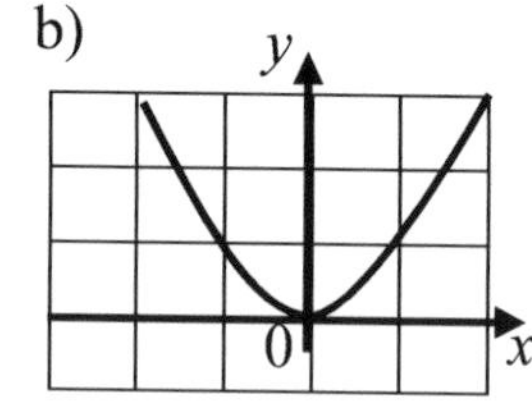

c)

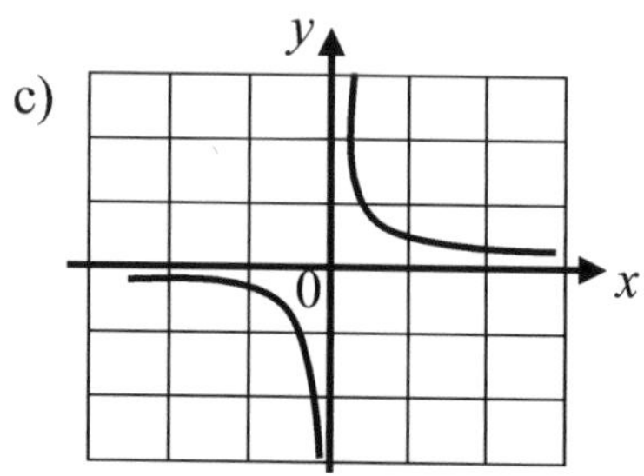

4) b is inversely proportional to c.
Complete the table.

b	1	6		15
c		2	4	

5) Below is a graph that shows two variables p and q that are inversely proportional to each other.
Find c, d and e.

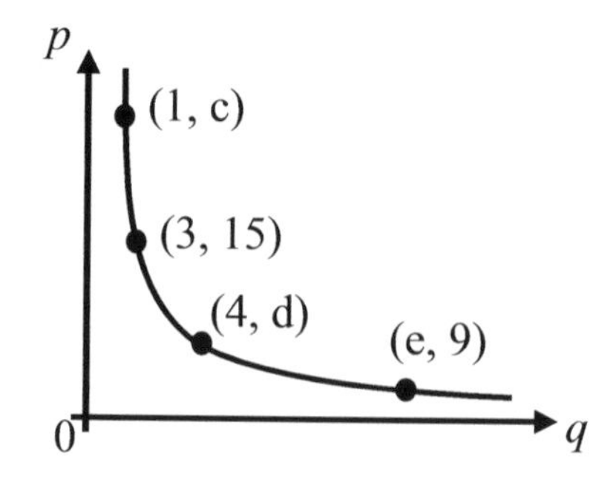

6) y is inversely proportional to $\sqrt{x}$.
When y = 50, x = 16.

a) Form an equation in term of y and x.

b) Calculate the value of y when x = 36.

c) Calculate the value of x when y = 18.75.

7) When oil is poured into a cylinder, the depth, d of the oil is inversely proportional to the square of the radius, r of the cylinder.

When $r = 7$ cm, $d = 20$ cm.

a) Write a formula for $\boldsymbol{d}$ in terms of r.

b) Calculate the depth of the oil in the cylinder when the radius is 4 cm.

c) Calculate the radius of the cylinder when the depth is 115 cm to 1 d.p.

31 Statistics

This section covers the following topics:

- Frequency Polygon
- Cumulative frequency diagrams
- Box plots
- Histograms

LEARNING OBJECTIVES

By the end of this unit, you should be able to:

a) Draw and interpret a cumulative frequency diagram
b) Draw and interpret box plots
c) Draw and solve problems involving histograms with unequal class intervals
d) Draw and interpret frequency polygons

KEYWORDS

- Frequency
- Cumulative frequency
- Box plot
- Histogram
- Frequency density

31.1 FREQUENCY POLYGONS

Discrete and continuous data can be represented in a frequency polygon. Frequency polygons are drawn by plotting points over the middle of each group. The points are then joined with straight lines.

Example 1:
Draw a frequency polygon for the information below.

Distance, in miles	Frequency
$0 < m \leq 5$	2
$5 < m \leq 10$	5
$10 < m \leq 15$	7
$15 < m \leq 20$	3
$20 < m \leq 25$	1

Solution: Use the middle value for each group.

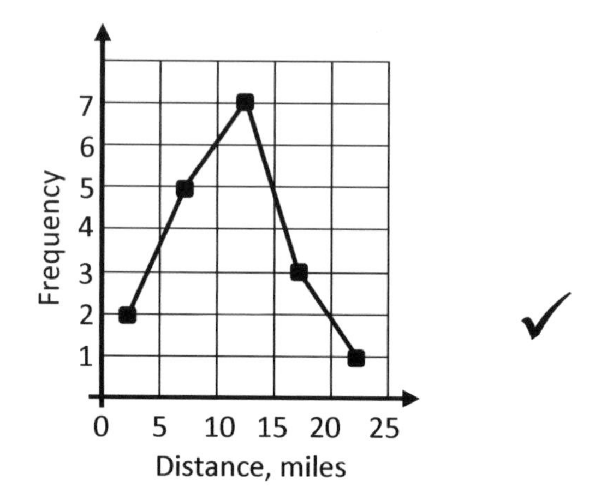

✓

EXERCISE 31A

1) Draw a frequency polygon for each of the following frequency tables.

a)

Time, t (mins)	Frequency
$0 < m \leq 5$	3
$5 < m \leq 10$	5
$10 < m \leq 15$	8
$15 < m \leq 20$	5
$20 < m \leq 25$	2

b)

weight, n (kg)	Frequency
$10 \leq n < 20$	1
$20 \leq n < 30$	9
$30 \leq n < 40$	17
$40 \leq n < 50$	12
$50 \leq n < 60$	5

2) Draw a frequency table from the frequency polygon shown below.

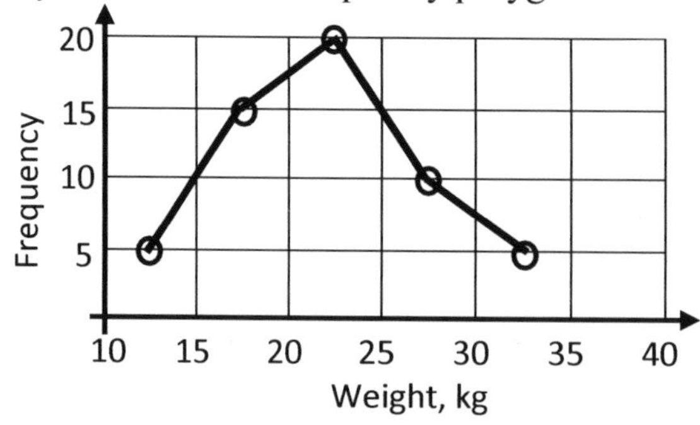

31.2 CUMULATIVE FREQUENCY DIAGRAMS

If the more detailed information is required about a set of data, cumulative frequencies come in handy. **Cumulative frequency** is the running total of the frequencies.

Example 1: a) Draw a cumulative frequency table from the frequency table below and b) draw a **cumulative frequency graph**.

Height, h, cm	Frequency
$10 < h \leq 20$	10
$20 < h \leq 30$	20
$30 < h \leq 40$	30
$40 < h \leq 50$	40
$50 < h \leq 60$	15
$60 < h \leq 70$	5

Total: → **120**

Solution: Extend the table to include cumulative frequency.

a)

Height (h)	Frequency	Cumulative frequency
$10 < h \leq 20$	10	**10** (the same)
$20 < h \leq 30$	20	$10 + 20 = 30$
$30 < h \leq 40$	30	$30 + 30 = 60$
$40 < h \leq 50$	40	$60 + 40 = 100$
$50 < h \leq 60$	15	$100 + 15 = 115$
$60 < h \leq 70$	5	$115 + 5 =$ **120**

Before drawing the cumulative frequency graph, check that the total frequency (120) is equal to the final cumulative frequency (120).

b) Now, rewrite the table to include only the numbers needed for drawing the graph.

Height, h, cm	Cumulative Frequency
$10 < h \leq 20$	10
$20 < h \leq 30$	30
$30 < h \leq 40$	60
$40 < h \leq 50$	100
$50 < h \leq 60$	115
$60 < h \leq 70$	120

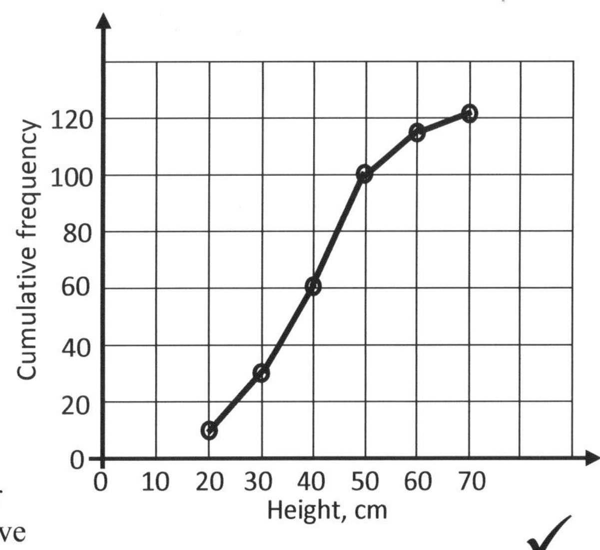

NOTE: You must join the points either by straight lines or freehand. Cumulative frequency is always shown on the vertical axis and is plotted at the upper end of the class interval.

MEDIAN, LOWER QUARTILE, UPPER QUARTILE

Cumulative frequency graphs are also useful in working out the median, lower and upper quartiles.

MEDIAN: This is the middle data value when listed in order of size.

LOWER QUARTILE: This is the value corresponding to a quarter of the way up the cumulative frequency axis. In short form, it is the $\frac{1}{4}n$th value.

UPPER QUARTILE: This is the value corresponding to three-quarters of the way up the cumulative frequency axis. It is the $\frac{3}{4}n$th value.

INTERQUARTILE RANGE: This is the difference between the lower and upper quartiles.

Example 2: From the cumulative frequency graph, estimate a) the median (M) b) the lower quartile (LQ) c) the upper quartile (UQ) d) the interquartile range (IQR).

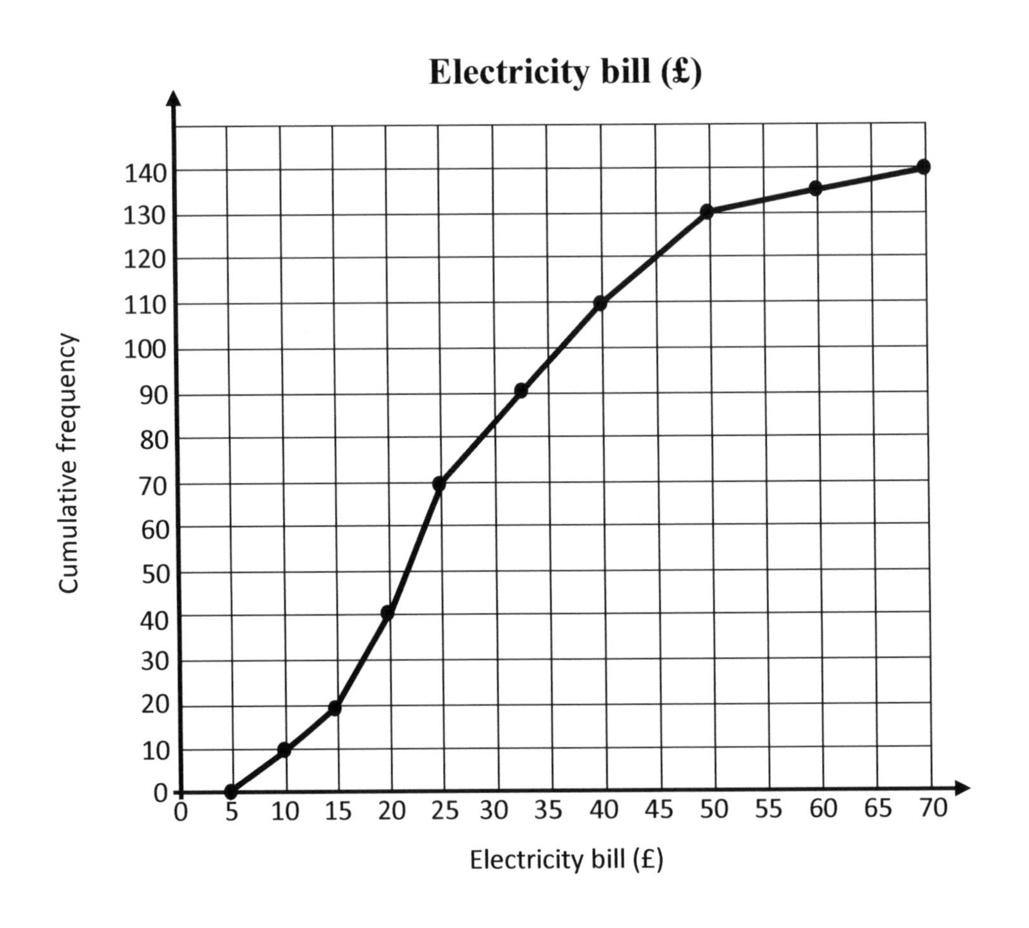

Solution: a) To find the median, divide 140 by 2. Median $= \frac{140}{2} = 70^{th}$ value. From the cumulative frequency axis, draw a horizontal line at 70 to meet the graph/curve. Then draw a vertical line to meet the electricity bill axis. Read off the value. The median is **£25.00**.

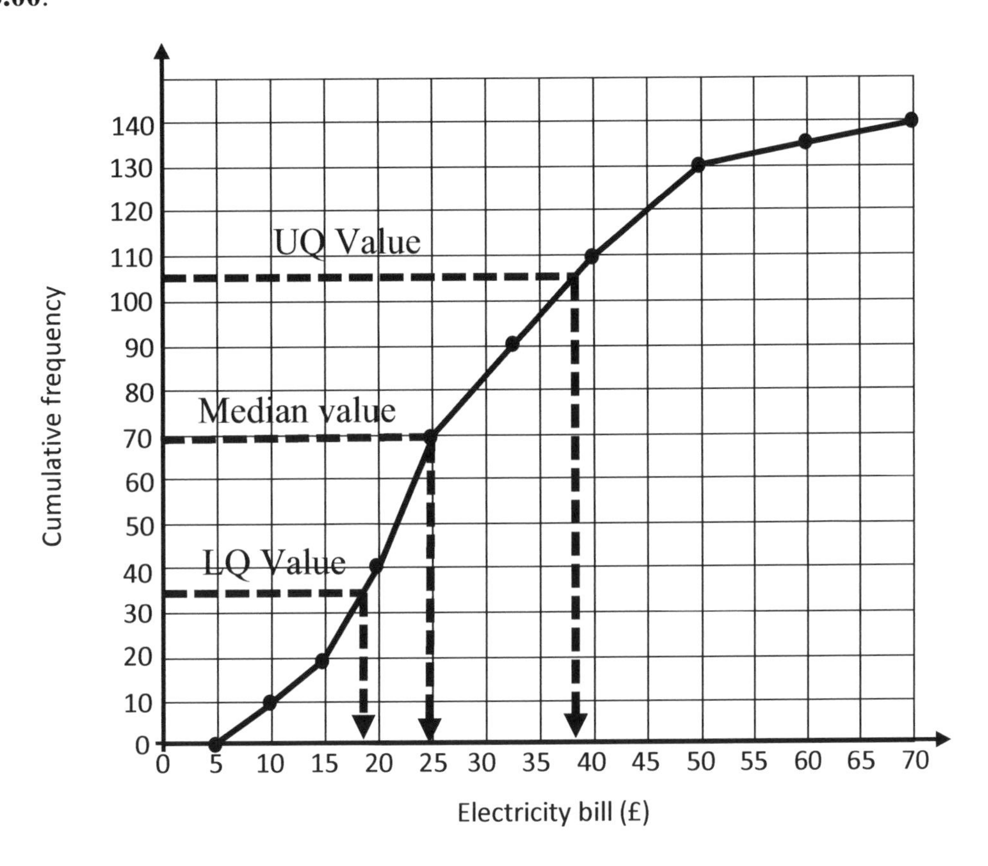

b) Lower quartile value $= \frac{1}{4}$ of $140 = 35^{th}$ value. Draw a horizontal line at 35 to meet the graph and then read off the value. Lower quartile = **£18.00**.

c) Upper quartile value $= \frac{3}{4}$ of $140 = 105^{th}$ value. Draw a horizontal line at 105 to meet the graph and then read off the value. UQ = **£38.00**.

d) Interquartile range = Upper quartile – Lower quartile
= £38 - £18
= **£20**

NOTE: Interquartile range eliminates extreme values and is the middle 50% of the data. It is a way of measuring how spread out the middle 50% of the data values are.

Example 3: Look at the cumulative frequency graph in example 2.
a) Customers who pay less than £22.50 must pay an extra 1% of electricity bill. Estimate the number of customers who will pay the additional 1%.
b) Work out the number of customers who spent £35 or more for the electricity bill.

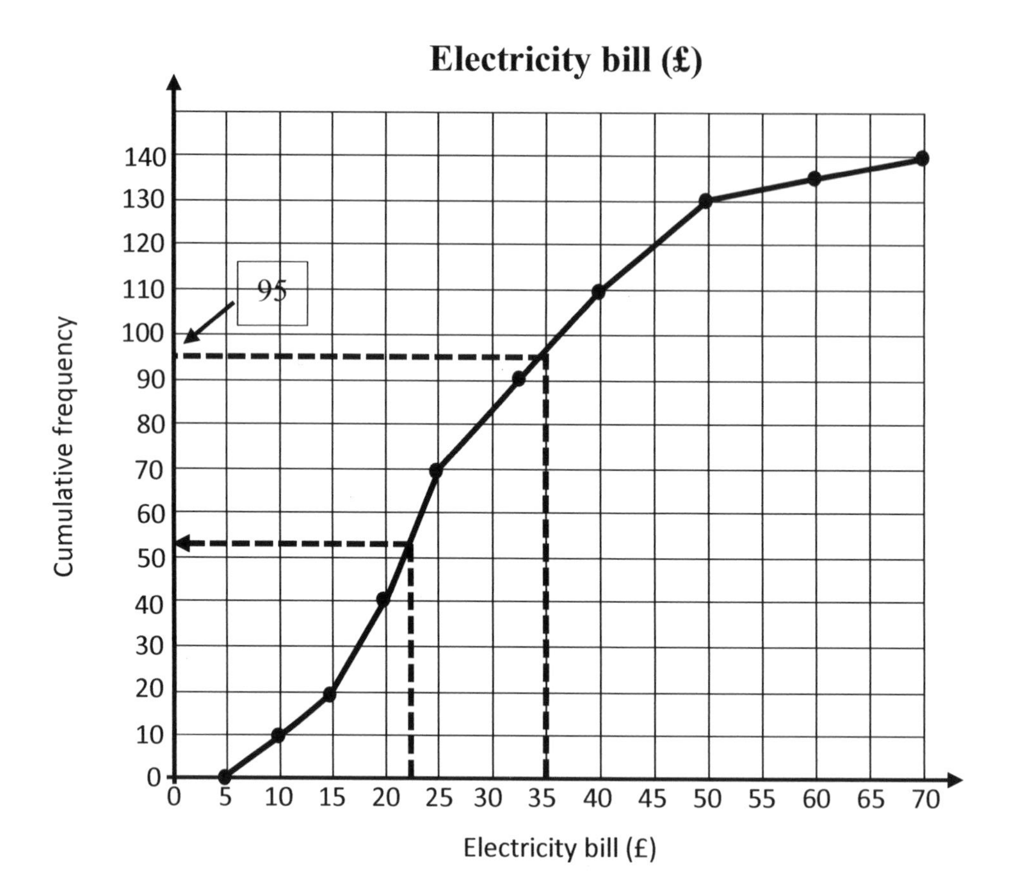

a) Draw a vertical line up from £22.50 mark on the electricity bill axis, to intersect the graph. At that point, draw a horizontal line to meet the cumulative frequency axis. From the chart, the number of customers to pay the 1% extra charge is **53**.

b) From 35 on the horizontal axis (electricity bill), draw a vertical line to meet the graph. Read across to the vertical axis. This value is 95.

The number of customers who spent £35 or more will be the difference between the highest number and 95.

140 – 95 = **45 customers**

EXERCISE 31B

1) Draw a cumulative frequency table and then draw the cumulative frequency graphs for each table.

a)

Height, cm	Frequency
$70 < h \leq 80$	7
$80 < h \leq 90$	20
$90 < h \leq 100$	30
$100 < h \leq 110$	35
$110 < h \leq 120$	20
$120 < h \leq 130$	8

b)

Total score, m	Frequency
$2 \leq h \leq 11$	7
$12 \leq h \leq 21$	11
$22 \leq h \leq 31$	22
$32 \leq h \leq 41$	14
$42 \leq h \leq 51$	8
$52 \leq h \leq 61$	3

2)

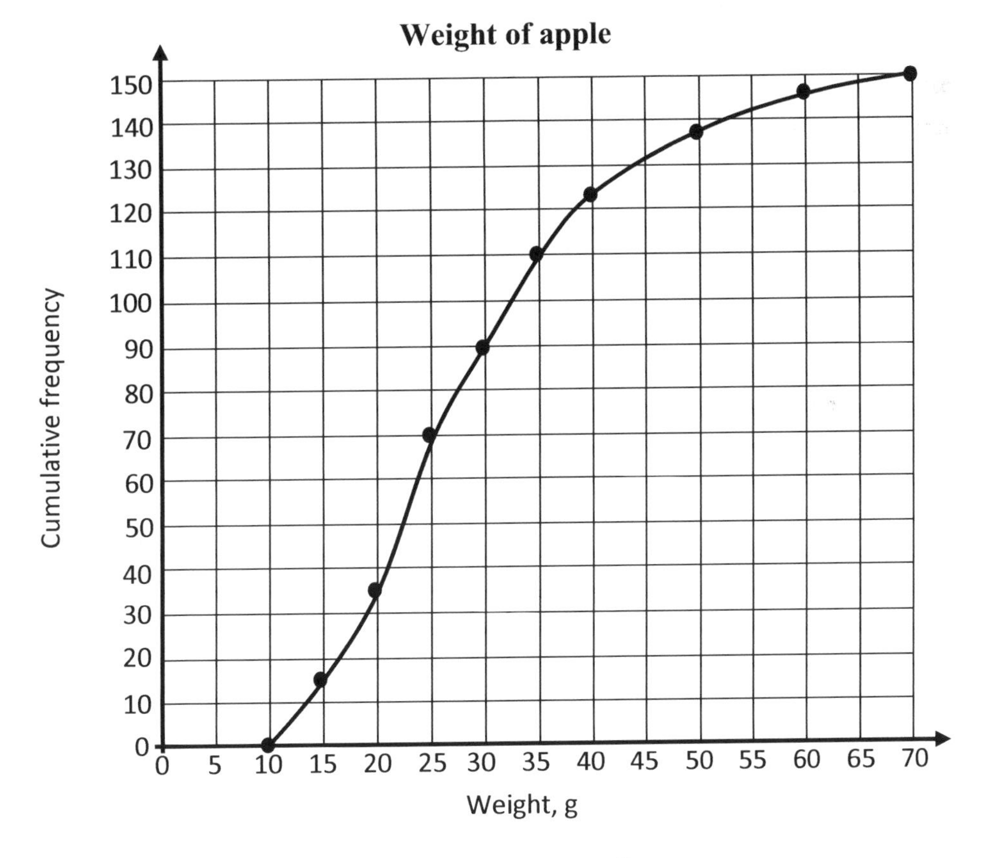

a) Calculate an estimate of the median and interquartile range.

b) Apples that weigh at least 30g are classified as good fruit.
i) What number of apples are classified as good fruit?
ii) What percentage of apples are classified as good fruits?

3) From question **1a** above and the related graph, estimate
a) the median b) the lower quartile c) the upper quartile d) the interquartile range.

4) From the cumulative frequency table below, work out an estimate of the mean height.

Height, h, cm	Cumulative Frequency
$10 < h \leq 20$	10
$20 < h \leq 30$	30
$30 < h \leq 40$	60
$40 < h \leq 50$	100
$50 < h \leq 60$	115
$60 < h \leq 70$	120

Hint: Work backwards to find the frequencies. After that, use the midpoints.

5)

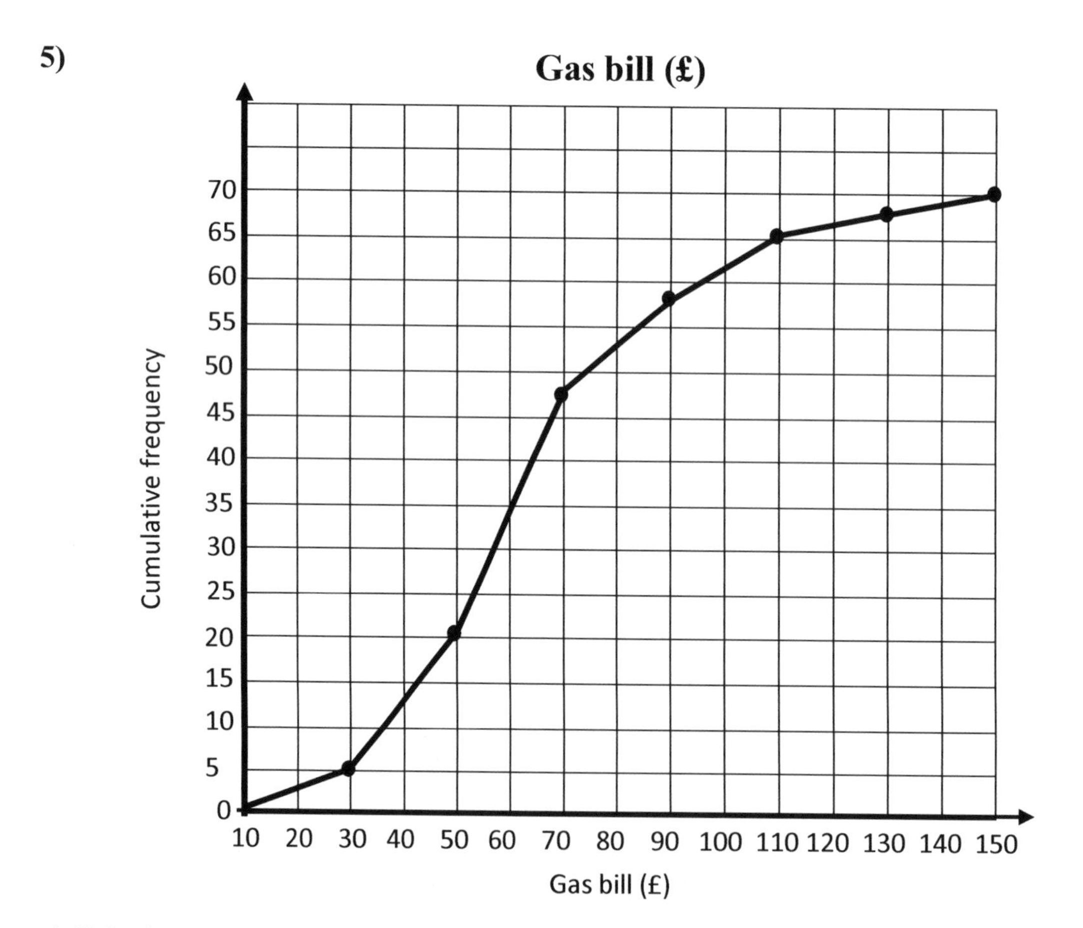

a) Calculate an estimate of the mean if the first class boundary is $10 \leq £ < 30$.
b) Calculate an estimate of the i) median ii) lower quartile iii) interquartile range.

31.3 BOX PLOTS

A box plot is another technique of displaying data. It is also used in comparing two or more data. Some call it box-and-whisker diagram. A typical box plot looks like this:

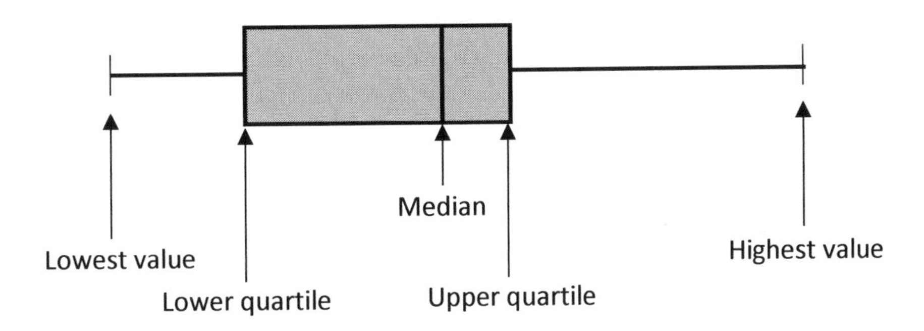

The median line can be anywhere in the box.

Example 4: Draw a box plot for the set of numbers below.

14 5 20 3 7 16 11 9 18 21 23

Solution: Arrange the numbers in order of size and work out the median, lower quartile, upper quartile. The lowest number is 3 while the highest number is 23.

(3) 5 (7) 9 11 (14) 16 18 (20) 21 (23)

Median is the middle number and is **14**.

Lower quartile can be easily read as the number halfway between the lowest number and the median. From the ordered numbers above, that number is **7**.

Upper quartile is the number halfway between the median and the highest number. That number is **20**. Now, the box plot would look like this:

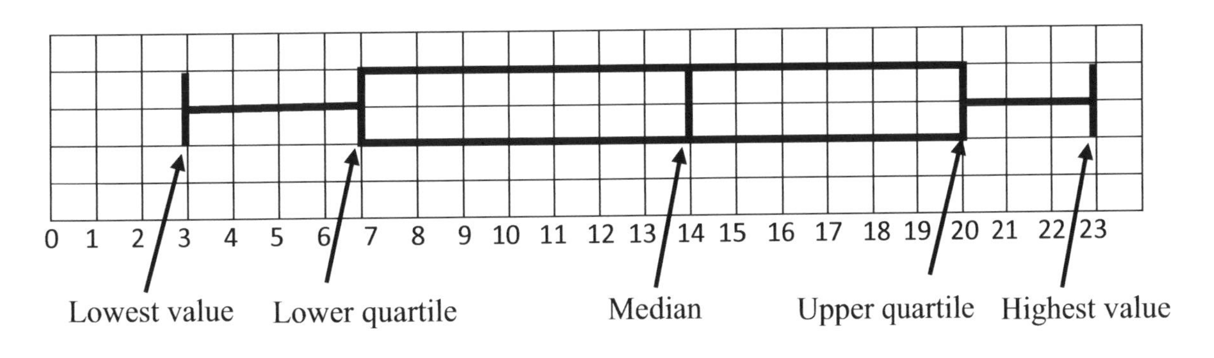

Example 5: Work out the lowest number, the median, the lower quartile, the upper quartile, the interquartile range and the maximum number of the numbers below.

5 10 15 2 13 28 30 25

Solution: Arrange in order of size.

2 5 10 13 15 25 28 30

Lowest number = **2**

Maximum number = **30**

Median is the middle number: $\frac{13+15}{2} = \frac{28}{2} = \mathbf{14}$

Lower quartile is the number halfway between the lowest number and median.

2 5 10 13 | 15 25 28 30

14 ← Median

Therefore, the number between 2 and 14 from the list above will lie between 5 and 10.
So, $\frac{5+10}{2} = \frac{15}{2} = 7.5$. Lower quartile = **7.5**

Upper quartile is the number between the median and the highest number.
Upper quartile = $\frac{25+28}{2} = \frac{53}{2} = \mathbf{26.5}$

Interquartile range = Upper quartile – lower quartile = 26.5 – 7.5 = **19**

POINTS TO REMEMBER

In a grouped data, the **actual** lowest and highest numbers are **not** known because the data is grouped. However, we may use the lowest and maximum possible values for the data.
In the grouped frequency table, the lowest possible time will be 10 minutes while the highest possible time will be 60 minutes.

weight, n (mins)	Frequency
$10 \leq n < 20$	1
$20 \leq n < 30$	9
$30 \leq n < 40$	17
$40 \leq n < 50$	12
$50 \leq n < 60$	5

BOX PLOTS FROM CUMULATIVE FREQUENCY GRAPHS

Box plots can be easily drawn from a cumulative frequency graph by finding the five points that make up the box plot.

Example 6: a) Draw box plot from the cumulative frequency graph below.
b) Calculate the interquartile range.

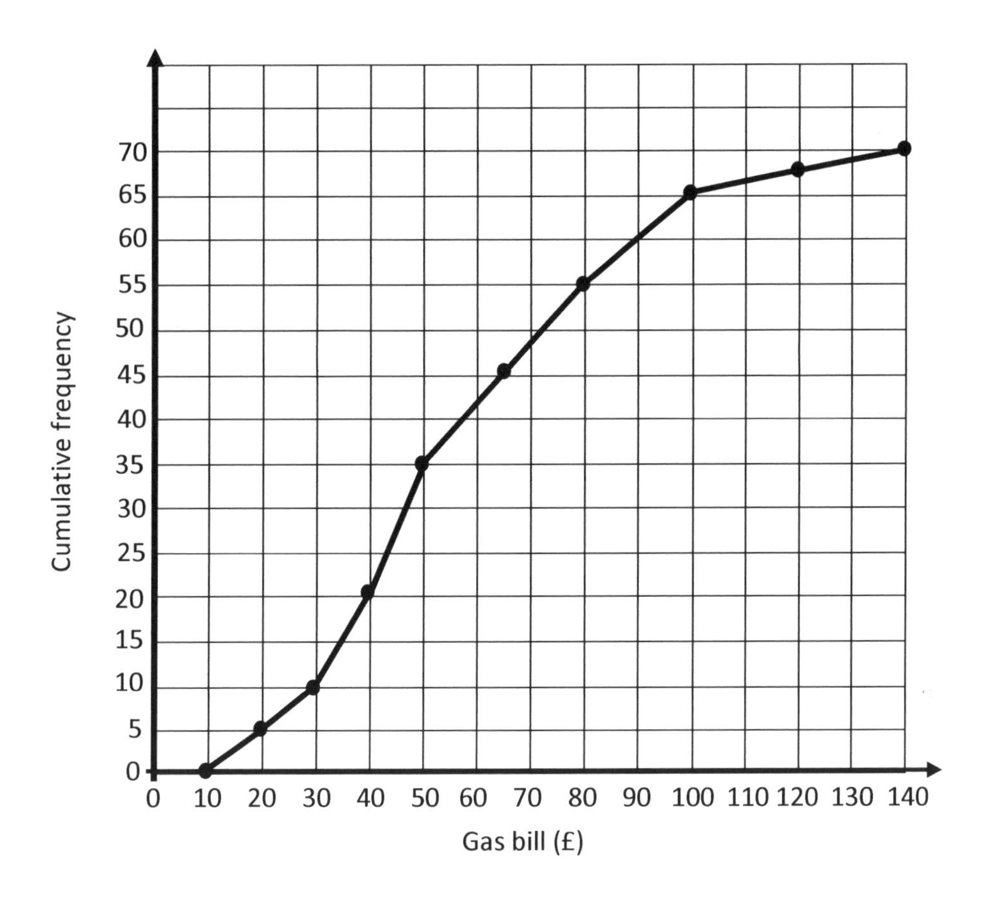

Solution: Find the lower quartile, the median, the upper quartile and then use the lowest number of 10 and the maximum number of 140. Then, draw the box plot.
Refer to example 2 above for the technique.

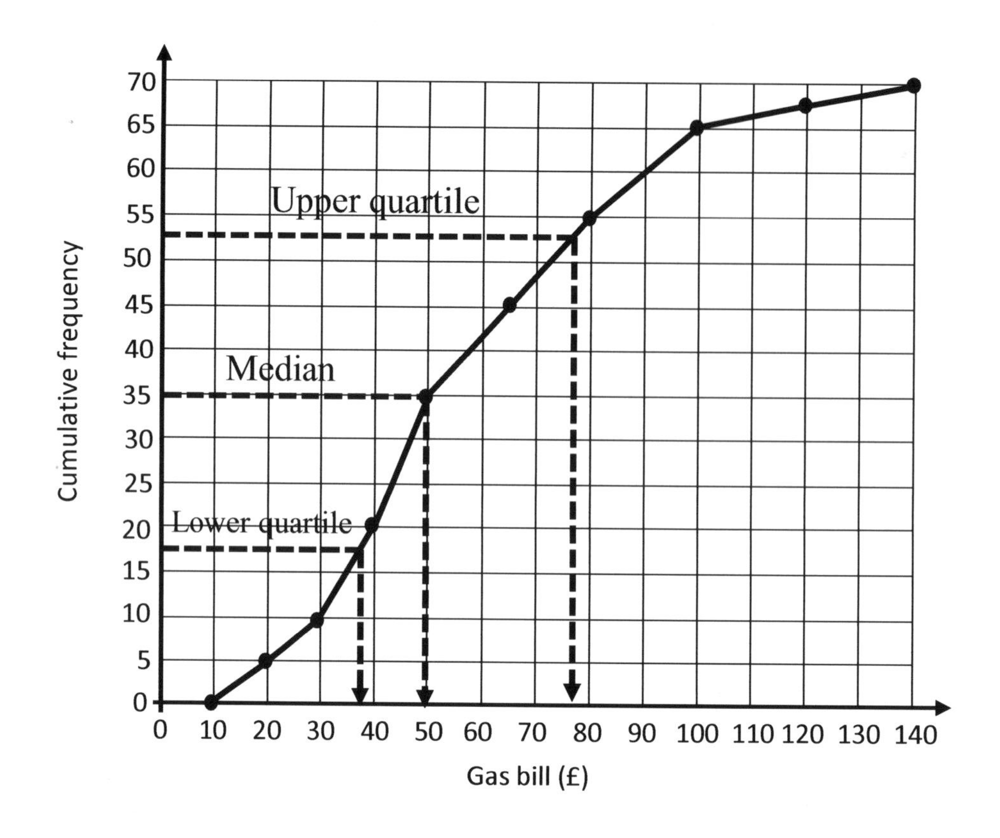

From the graph, median = £50, lower quartile = £37 and upper quartile = £77. Also, the lowest value is 10, and the maximum is 140.

The box plot looks like this:

a)

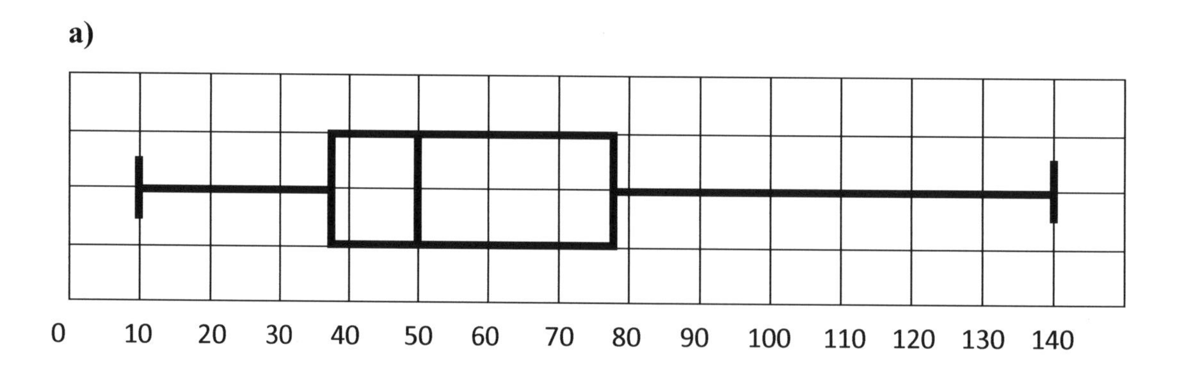

b) The interquartile range = Upper quartile – Lower quartile = 77 – 37 = **£40**

31.4 COMPARING DISTRIBUTIONS

Distributions can be compared if some statistical parameters are known. Two box plots or cumulative frequency graphs may be compared with each other as follows:

- Compare one or two measures of average like the median
- Compare the range or interquartile range.

Remember: Low range or low interquartile range means that the distribution is **more consistent** in relation to the compared distribution.

Example 7: The box plots show the times it took, to the nearest minute, for girls and boys in a maths class to complete an assignment. There are 20 girls and 12 boys.

a) Compare the box plots for each gender and give reasons.
b) How many pupils took more than 12 minutes to complete an assignment?

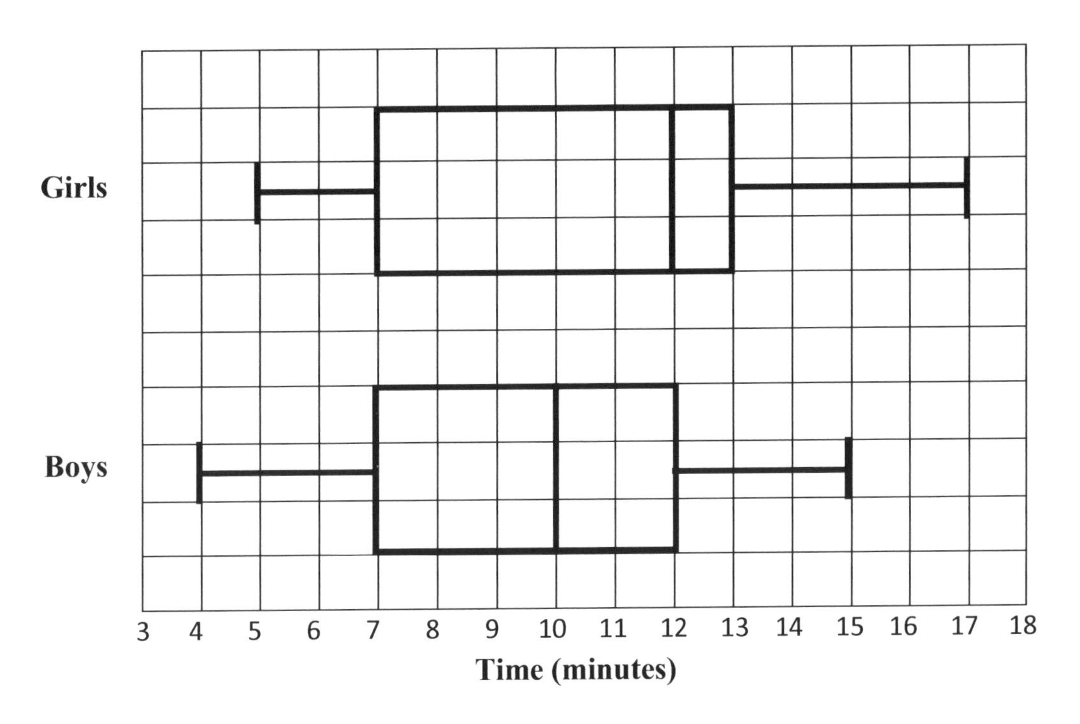

Solution:

a) The median for girls is 12 minutes while the median for boys is 10 minutes. It shows that **on average**, girls spent more time doing their assignment than boys. The range for boys is (15 – 4) = 11 and the range for girls is (17 – 5) = 12. Also, the interquartile range for boys is (12 – 7) = 5 and the interquartile range for girls is (13 – 7) = 6. It follows that the range and interquartile range for boys are slightly smaller than for girls, which shows that boys are **more consistent** than the girls.

b) Boys: 12 minutes signifies the upper quartile which is $\frac{3}{4}$ of 12 = 9. Therefore, the number of boys that took more than 12 minutes is 12 – 9 = **3**

Girls: 12 minutes signifies the median which is $\frac{1}{2}$ of 20 = 10. Therefore, the number of girls that took more than 12 minutes is 20 – 10 = **10.**

Finally, the total number of students (boys and girls) that took more than 12 minutes to do an assignment is 3 + 10 = **13 pupils**.

EXERCISE 31C

1) a) Work out the lowest number, the median, the lower quartile, the upper quartile, the interquartile range and the maximum number of the numbers below.

i) 15 6 30 4 8 17 12 10 19 22 24
ii) 4 9 14 1 12 27 35 24

b) Draw the box plots for i and ii.

2)

a)

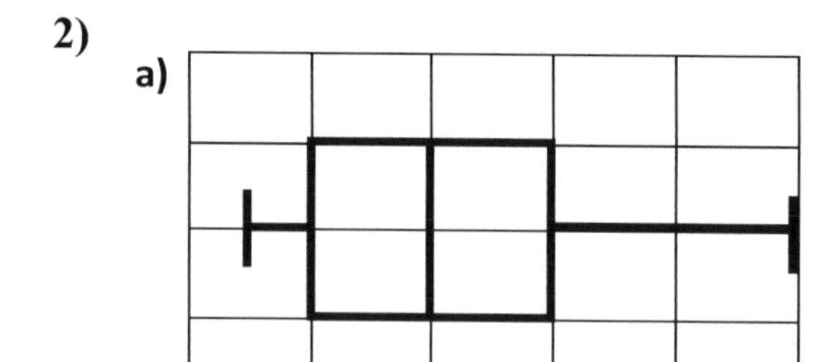

b)

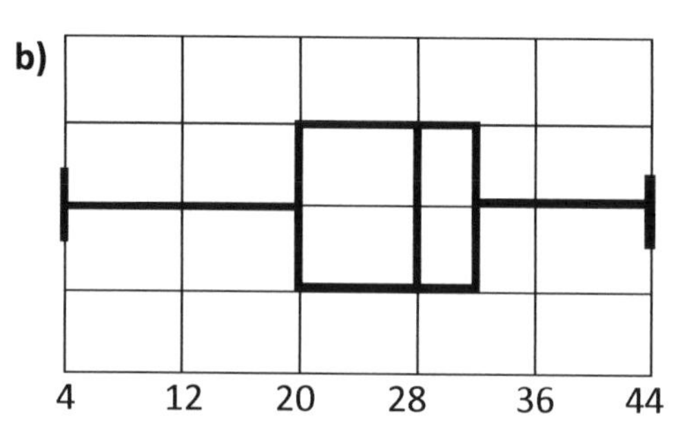

From the box plots above, find i) the lowest number ii) the median iii) the lower quartile iv) the highest number, v) the upper quartile vi) the range and vii) the interquartile range.

3) The times taken to complete a scrabble game are shown below.

a) Draw a cumulative frequency diagram
b) Draw a box plot
c) Use your cumulative frequency diagram to estimate the number of people who took less than 34 minutes to complete the game.

Time, mins	Frequency
$0 \leq t < 11$	5
$11 \leq t < 21$	15
$21 \leq t < 31$	32
$31 \leq t < 41$	40
$41 \leq t < 51$	20
$51 \leq t < 61$	9

4) The box plots show the percentage marks of pupils that sat for GCSE examinations in History and Mathematics. 60 pupils sat for History, and 120 sat for Maths.
a) Compare the box plots for each subject. Give reasons.
b) How many pupils scored more than 45% in the two subjects?
c) How many pupils scored 30% in History?

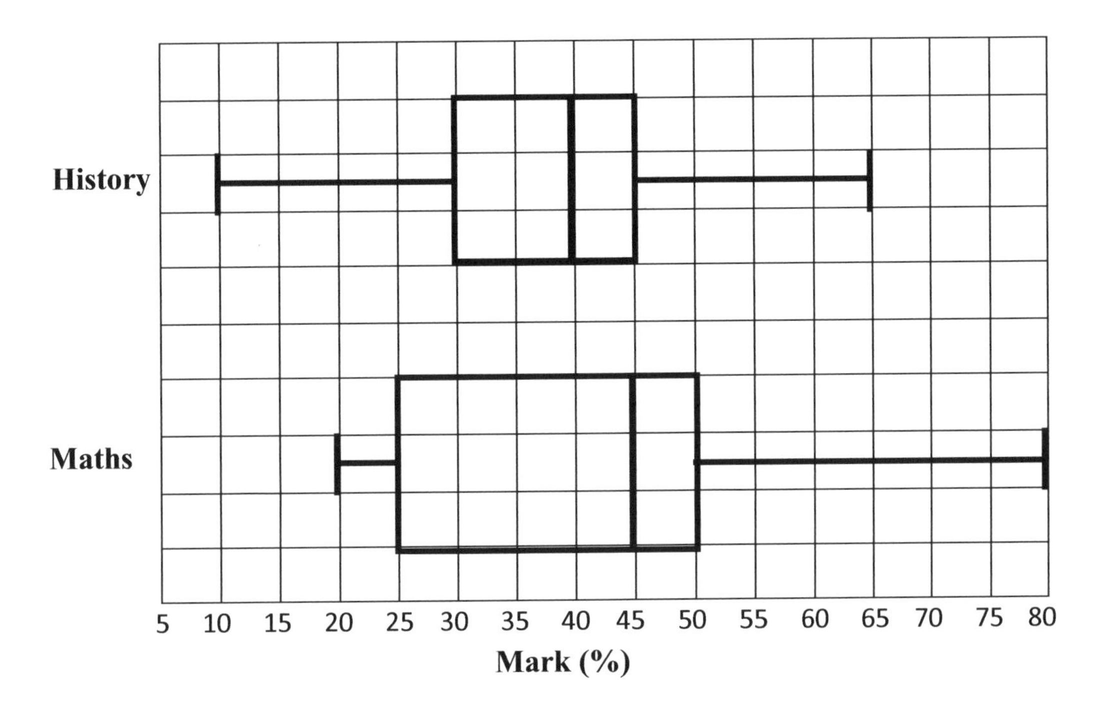

5) The cumulative frequency graphs show the weights of a sample of some males and females in a small community.

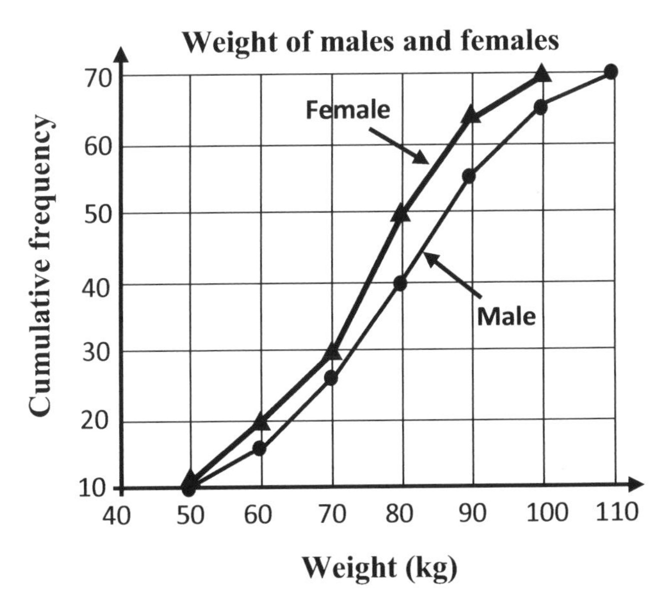

a) How many females were weighed?

b) How many males were weighed?

c) What is the estimate of the minimum and maximum weights of males?

d) Compare the male and female weights and give reasons for your answers.

31.5 HISTOGRAMS

Histograms are used to display continuous data. The bars from or in a histogram can have different widths unlike a simple bar chart with the same class width all through.

In this section, we shall consider plotting and interpreting a histogram.

Example 1: The frequency table shows the times taken to solve a mathematics puzzle. Draw a histogram to illustrate the data.

Time, mins	Frequency
$0 \leq t < 10$	5
$10 \leq t < 20$	15
$20 \leq t < 25$	30
$25 \leq t < 45$	40
$45 \leq t < 50$	20
$50 \leq t < 65$	9

Solution:
Find the **frequency density** from the table and plot against the class width. Frequency density must be in the vertical axis while the class width will be in the horizontal axis.
The **height** of the bar represents the frequency density while the **area** of the bar represents the **frequency**.

Frequency = frequency density × class width

$$\textbf{Frequency density} = \frac{\textbf{frequency}}{\textbf{class width}}$$

Time, mins	Frequency	Class width	Frequency density
$0 \leq t < 10$	5	$10 - 0 = 10$	$5 \div 10 =$ **0.5**
$10 \leq t < 20$	15	$20 - 10 = 10$	$15 \div 10 =$ **1.5**
$20 \leq t < 25$	30	$25 - 20 = 5$	$30 \div 5 =$ **6**
$25 \leq t < 45$	40	$45 - 25 = 20$	$40 \div 20 =$ **2**
$45 \leq t < 50$	20	$50 - 45 = 5$	$20 \div 5 =$ **4**
$50 \leq t < 65$	9	$65 - 50 = 15$	$9 \div 15 =$ **0.6**

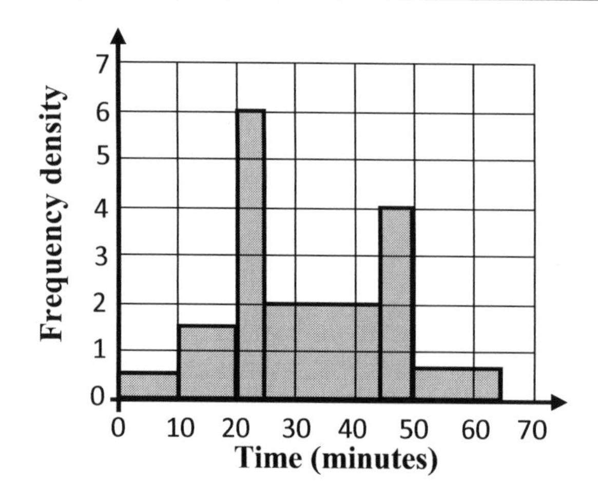

Example 2: The histogram shows the recorded speed of cars on M11 Essex motorway.

a) Calculate the frequency for each class interval.

b) What is the total sample size?

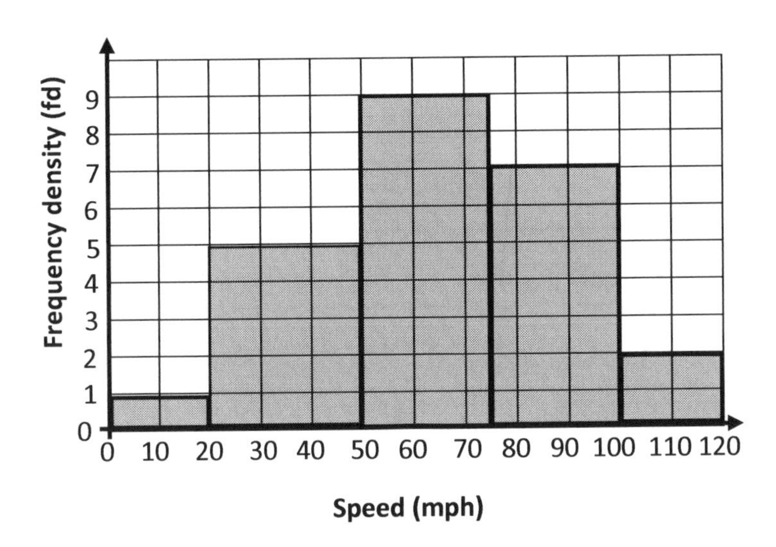

Solution: The frequency (f) for each class interval is the area of the bars (class width $\times fd$)

a) 1st class interval: Frequency = class width $\times$ frequency density = $20 \times 1 =$ **20.**

2nd class interval: Frequency = $30 \times 5 =$ **150**

3rd class interval: Frequency = $25 \times 9 =$ **225**

4th class interval: Frequency = $25 \times 7 =$ **175**

5th class interval: Frequency = $20 \times 2 =$ **40**

b) The total sample size is the total of all the frequencies.

$20 + 150 + 225 + 175 + 40 =$ **610**.

Example 3: An incomplete histogram is shown below. Complete the table and the histogram.

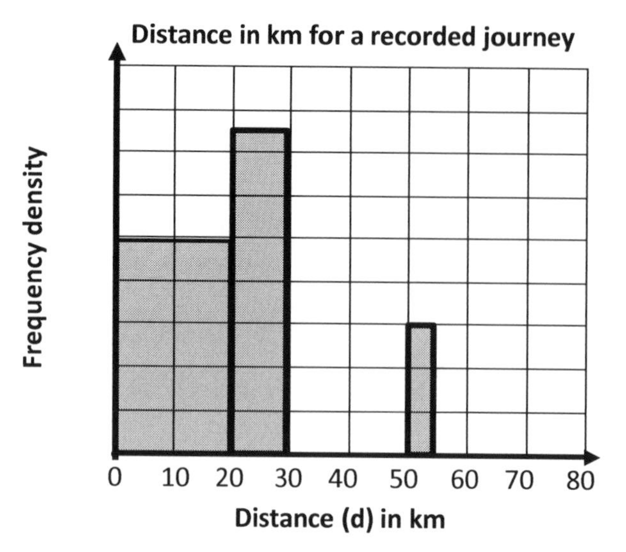

Distance (d) in km	Frequency
$0 < d \leq 20$	200
$20 < d \leq 30$	
$30 < d \leq 50$	100
$50 < d \leq 55$	
$55 < d \leq 75$	10

Solution: Use complete information to work out the frequency density. Using the class width of $0 < d \leq 20$ and the frequency of 200, frequency density $= \frac{frequency}{class\ width} = \frac{200}{20} = 10$
To get the scale on the frequency density axis, divide 10 by the number of divisions from 0. There are 5 divisions, therefore $10 \div 5 = 2$. It means that the vertical axis goes up in 2's. Label the frequency density axis accordingly.

Frequency for the class width of 20 – 30 is $10 \times 15 = 150$
Frequency for the class width of 50 – 55 is $5 \times 6 = 30$.

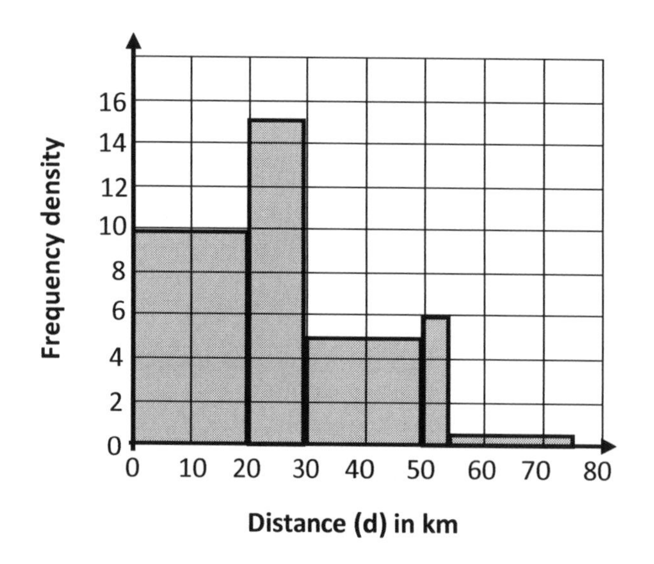

Distance (d) in km	Frequency
$0 < d \leq 20$	200
$20 < d \leq 30$	**150**
$30 < d \leq 50$	100
$50 < d \leq 55$	**30**
$55 < d \leq 75$	10

Example 4: The histogram illustrates the weights of some hospital equipment. Work out an estimate of the **median** weight.

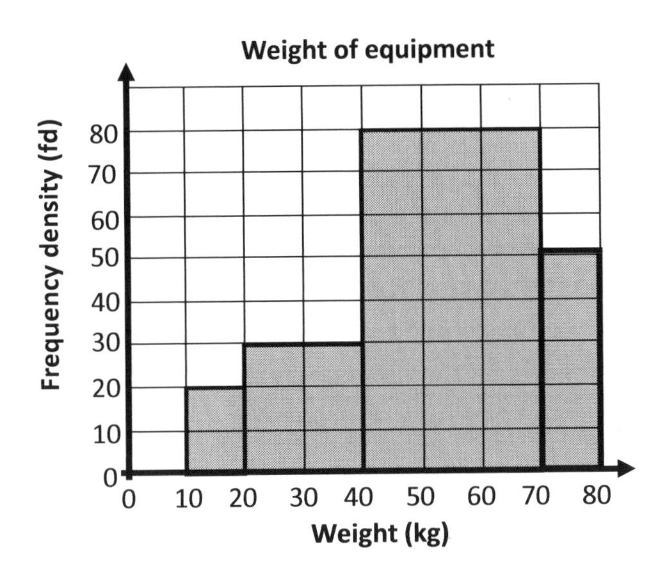

Solution: First, work out the total frequency by working out individual frequencies for each class width.

Frequency = frequency density × class width

$= (10 \times 20) + (20 \times 30) + (30 \times 80) + (10 \times 50)$

$= \quad 200 \quad + \quad 600 \quad + \quad 2400 \quad + \quad 500 = 3700$

The median is the $\frac{3700+1}{2} = 1850.5^{th}$ value and lies in the class of $40 < m \leq 70$.

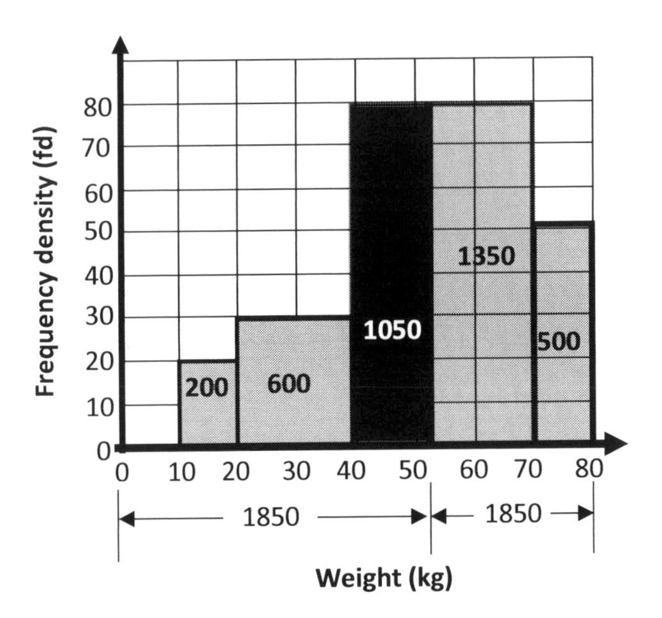

$200 + 600 = 800$

$1850 - 800 = 1050$. The portion of the bar in black is what we are interested in, since $200 + 600 + 1050$ will add to 1850 knowing that the median will lie in the 1850.5^{th} value.

Class width $= \frac{frequency}{frequency\ density}$

$= \frac{1050}{80} = 13.125$

Adding the lower-class boundary (40) to the new class width (13.125) will give the estimate of the median weight.

$40 + 13.125 =$ **53.13 kg** to 2 dp.

EXERCISE 31D

1) The histogram shows travelling times of some pupils. Draw a histogram to represent the data.

Time, mins	Frequency
$0 \leq t < 10$	15
$10 \leq t < 20$	10
$20 \leq t < 25$	25
$25 \leq t < 45$	40
$45 \leq t < 50$	20
$50 \leq t < 65$	9

2) The histogram shows the recorded speed of cars on a motorway.

a) Calculate the frequency for each class interval.

b) What is the total sample size?

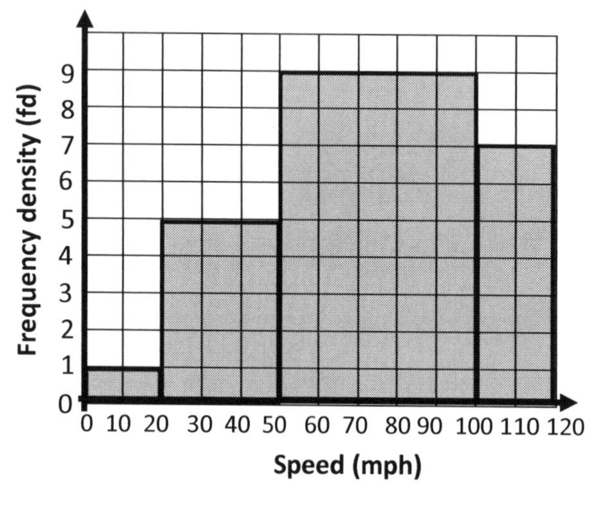

3) a) Copy and complete the table

Length (cm)	Frequency density	Frequency
$0 \leq l < 5$	8	
$5 \leq l < 8$		30
$8 \leq l < 12$		48
$12 \leq l < 17$	5	
$17 \leq l < 20$	10	

b) Draw a histogram to represent the information above.

c) Work out an estimate of the median length.

4) The histogram below shows the times taken to cook different meals.

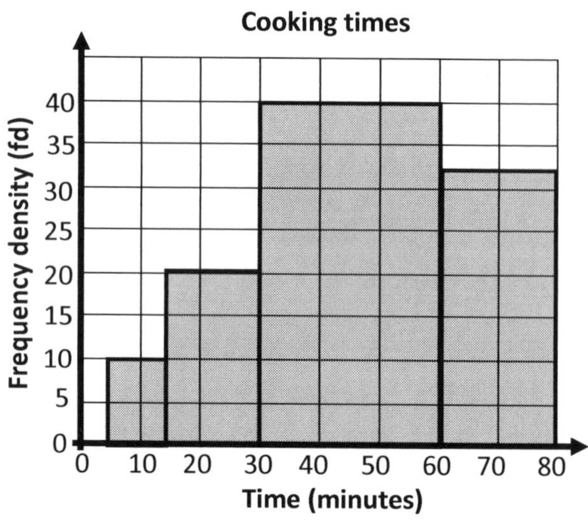

a) Copy and complete the table below.

Time, minutes	Frequency
$5 \leq t < 15$	
$15 \leq t < 30$	
	1200
$60 \leq t < 80$	

b) How many meals took less than 60 minutes to cook?

5) An incomplete histogram is shown below. Complete the table and the histogram.

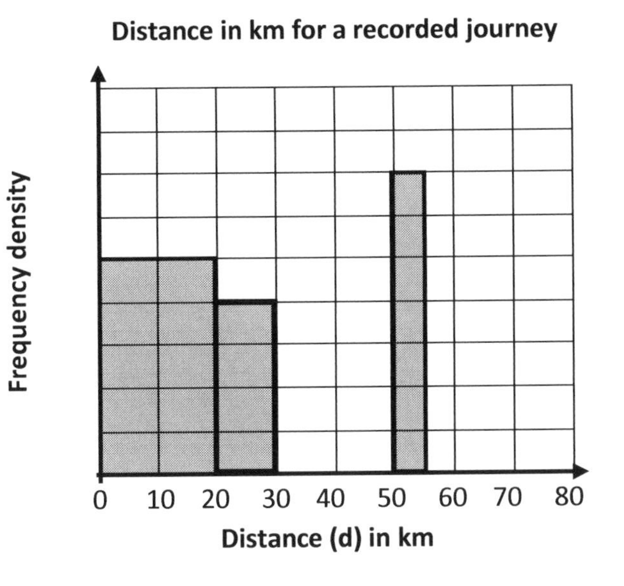

Distance (d) in km	Frequency
$0 < d \leq 20$	200
$20 < d \leq 30$	
$30 < d \leq 50$	100
$50 < d \leq 55$	
$55 < d \leq 75$	10

6) The histogram for the masses of a few bags of apples is shown below.

a) How many bags of apples are there in total?
b) How many bags of apples had a mass of between 40 and 70 grams?
c) As an estimate, how many bags of apples had a mass bigger than 65 grams?

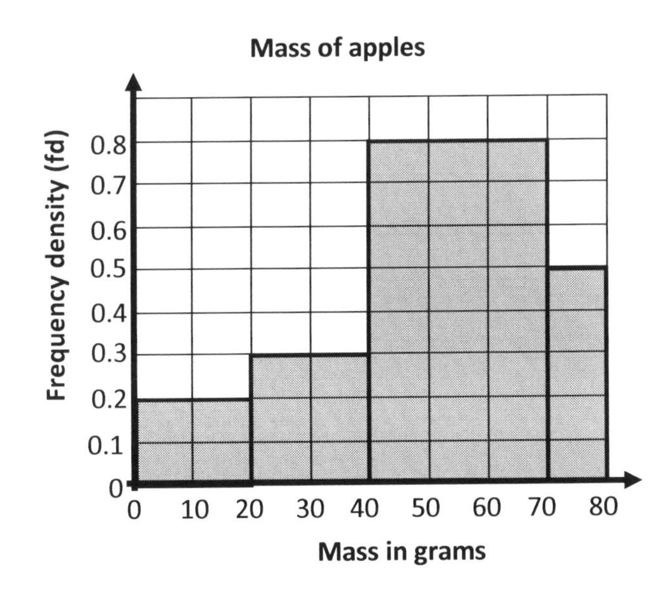

32 Other higher topics

This section covers the following topics:

- Recurring decimals
- Density
- Upper and lower bounds
- Venn Diagrams and Probability
- Transforming Graphs
- Trigonometric Graphs
- Cubic and reciprocal functions
- Exponential functions

LEARNING OBJECTIVES

By the end of this unit, you should be able to:

a) Understand recurring and terminating decimals
b) Solve problems involving density
c) Identify and solve problems comprising upper and lower bounds
d) Use Venn diagrams for calculating probabilities
e) Transform graphs
f) Understand and draw trigonometric graphs
g) Understand and draw exponential graphs
h) Understand and draw cubic graphs

KEYWORDS

- Recurring and terminating decimals
- Density
- Upper and lower bounds
- Venn diagram
- Reciprocal function
- Transformation
- Cubic function
- Trigonometric function
- Exponential functions

32.1 RECURRING DECIMALS

Recurring decimals are decimals that the digits go on forever.
For example: $\frac{4}{9} = 0.4444444444.......$ Notice that the number 4 goes on forever (recur).
As a standard, we write it as $0.\dot{4}$. The dot above the number shows which numbers recur.

$\frac{52}{99} = 0.525252\ldots = 0.\dot{5}\dot{2}$ and $\frac{235}{999} = 0.235235235\ldots = 0.\dot{2}3\dot{5}$

Recurring decimals are different from **terminating** decimals. Terminating decimals stop after a few decimal places. For example, $\frac{5}{8} = 0.625$.
0.625 is a terminating decimal.

RECURRING DECIMALS TO FRACTIONS

Example 1

Write the following recurring decimals to fractions.

a) $0.\dot{2}$ b) $0.\dot{2}\dot{3}$ c) $0.\dot{5}3\dot{4}$ d) $1.2\dot{5}$

Solutions:

a) Let $n = 0.222...$ (1)

Multiply equation (1) by 10 since the pattern recurs every one decimal place.

$10n = 2.222\ldots$ (2)

Now, subtract equation (1) from equation (2)

$$10n - n = 2.222 - 0.222$$
$$9n = 2$$
$$n = \frac{2}{9}$$ ✓

$0.\dot{2}$ is equivalent to the fraction, $\frac{2}{9}$

- Use a calculator to check that the answer is correct.

b) $0.\dot{2}\dot{3}$ to a fraction.

Let n = 0.2323... (1)

Multiply equation (1) by 100 since the pattern recurs every 2 decimal places.

100n = 23.2323… (2)

Now, subtract equation (1) from equation (2)

$$100n - n = 23.2323 - 0.2323$$
$$99n = 23$$
$$n = \frac{23}{99} \checkmark$$

$0.\dot{2}\dot{3}$ is equivalent to the fraction $\frac{23}{99}$

c) $0.\dot{5}3\dot{4}$ to a fraction.

Let n = 0.534534... (1)

Multiply equation (1) by 1000 since the pattern recurs every 3 decimal places.

1000n = 534.534534… (2)

Now, subtract equation (1) from equation (2)

$$1000n - n = 534.534534 - 0.534534$$

$$999\ n = 534$$

$$n = \frac{534}{999} \checkmark$$

$0.\dot{5}3\dot{4}$ is equivalent to the fraction $\frac{534}{999}$

d) $1.2\dot{4}$ to a fraction.

Let n = 1.2444... (1)

Multiply equation (1) by 10 since the pattern recurs every 1 decimal place.

10n = 12.4444… (2)

Now, subtract equation (1) from equation (2)

$$10n - n = 12.4444 - 1.2444$$
$$9n = 11.2$$
$$n = \frac{11.2}{9}$$

To get rid of the decimal point, multiply numerator and denominator by 10.

$$n = \frac{112}{90} = \frac{56}{45} \checkmark$$

$1.2\dot{4}$ is equivalent to the fraction $\frac{56}{45}$

EXERCISE 32A

1) Work out the fractions which are equivalent to the recurring decimals below.

a) 0.33333…
b) 0.77777…
c) 0.2727272…
d) 0.56565656…
e) 0.1234123412341234……
f) $0.1\dot{4}\dot{3}$
g) $1.\dot{2}3\dot{6}$
h) $0.3\dot{6}$
i) $0.00\dot{8}$

32.2 DENSITY

The **density** of a substance is the mass of the substance per unit volume.

$$\text{Density} = \frac{\text{mass}}{\text{volume}}$$

The unit of density depends on the unit of mass and volume for that measurement. However, it could be in grams per cubic centimetre (g/cm^3), kg/m^3 or other units based on mass and volume.

Example 1: The mass of an object is 7500g with a volume of 425 cm^3. Calculate the density of the object to two decimal places.

Solution: Using the formula,

$$\text{Density} = \frac{\text{mass}}{\text{volume}}$$

$$= \frac{7500}{425}$$

$= \mathbf{17.65}$ g/cm^3 to 2 d.p.

Example 2:
The density of a substance is 300.52 kg/m^3. Calculate the mass of the object if its volume is 4.5 m^3.

Solution: $\text{Density} = \frac{\text{mass}}{\text{volume}}$

Rearranging the equation gives

Mass = density × volume
= 300.52 × 4.5
= **1352.34 kg**

NOTE: Use the triangle to remember the formula

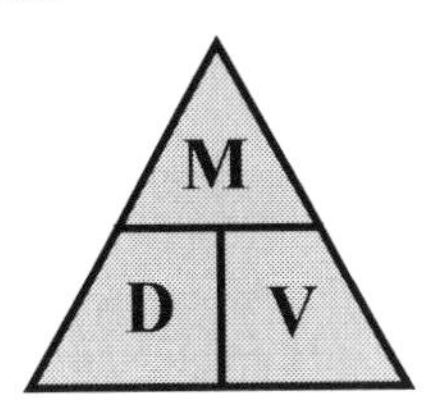

$$\text{Density (d)} = \frac{mass\ (M)}{volume\ (V)}$$

$$\text{Volume (v)} = \frac{mass\ (M)}{density\ (D)}$$

$$\text{Mass (m)} = density\ (D) \times volume\ (V)$$

EXERCISE 32B

1) Tony weighs a 35 cm^3 container with a mass of 322 g. Calculate the density of the container.

2) A sphere has a radius of 5 cm and mass of 320 g.

a) calculate the volume of the sphere.
b) Calculate the density of the sphere.

3) The concrete has a mass of 3 kg.

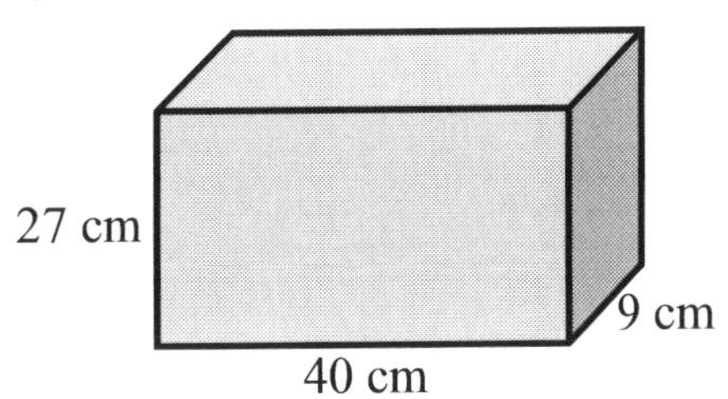

a) Work out the volume of the concrete.
b) Calculate the density of the concrete.
c) Calculate the mass of an 850 cm^3 concrete of the same make-up in grams.

32.3 UPPER AND LOWER BOUNDS

FOR CONTINUOUS DATA

Continuous data can take any value. Measurements for lengths and weights are examples of continuous data.

Most measuring instruments are not accurate enough not to have errors.

A 38 cm stick measured to the nearest centimetre could be anything from 37.5 cm up to but not including 38.5 cm (38.499 999 999 … cm) long. The number line below explains the situation.

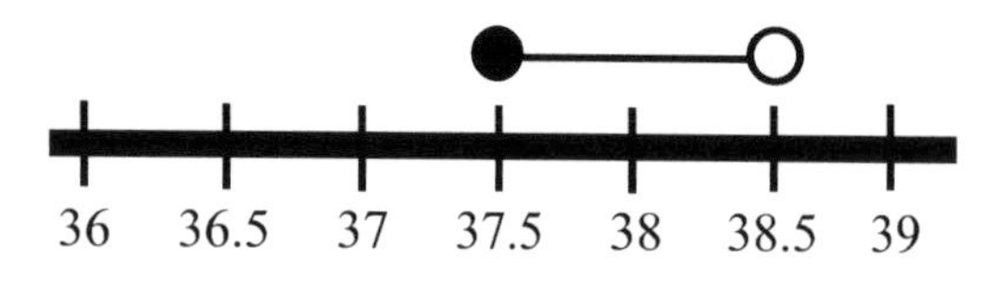

Anything below 37.5 will be rounded down to 37. If it were 38.5, it would be rounded up to 39.

The above information is best represented using inequality symbols. If n represents the length of the stick, then the possible measurements can be written as

$37.5 \text{ cm} \leq n < 38.5 \text{ cm}$.

The inequality above can be explained as follows: the length is greater than or equal to 37.5 cm but less than 38.5 cm.

The **lower bound** is 37.5 cm, and we use 38.5 cm as the **upper bound** because 38.499 999 999… is only 0.000 000 001 away from 38.5 cm.

Lower bound: this is the smallest value a number can be when given to a specified accuracy.

Upper bound: this is the largest value a number can be when given to a specified accuracy.

The difference between the upper and lower bounds is known as **error interval**.

Example 1: The width of a laptop is 38 cm measured to the nearest cm. Write down the range of possible widths between which the actual width of the laptop lies.

Solution: 1 cm ÷ 2 = 0.5 cm
Lower bound = 38 – 0.5 = 37.5 cm
Upper bound = 38 + 0.5 = 38.5 cm

Range: $\mathbf{37.5 \text{ cm} \leq w < 38.5 \text{ cm}}$
where w is the width of the laptop.
This is also known as the error interval.

The range of possible lengths means that the width will be greater than or equal to 37.5 cm but **less than** 38.5 cm.

Example 2: The dimensions of the rectangle below are length 5.5 cm and width 2.4 cm, both measured to the nearest 0.1 cm.

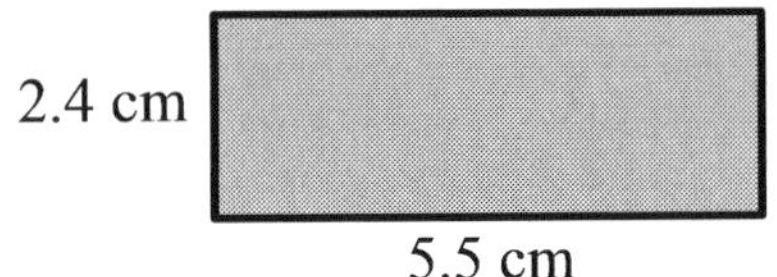

Calculate
a) the lower bound for the length
b) the upper bound for the length
c) the error interval for length
d) the lower bound for the width
e) the upper bound for the width
f) the maximum possible area of the rectangle.

Solution:
0.1 cm ÷ 2 = 0.05 cm

a) Lower bound for 5.5 cm
= 5.5 – 0.05 = **5.45 cm**

b) upper bound for 5.5 cm
= 5.5 + 0.05 = **5.55 cm**

c) the error interval for 5.5 cm is
$\mathbf{5.45\ cm \leq \ell < 5.55\ cm}$

d) lower bound for 2.4 cm
= 2.4 – 0.05 = **2.35 cm**

e) upper bound for 2.4 cm
= 2.4 + 0.05 = **2.45 cm**

f) The maximum possible area is the upper bound of length × upper bound of width = 5.55 cm × 2.45 = **13.5975 cm²**

Example 3: Find the lower and upper bounds for a length of 0.6 cm measured to the nearest one decimal place.

Solution: One decimal place is 0.1.
0.1 ÷ 2 = 0.05.

Lower bound = 0.6 – 0.05 = 0.55 cm
Upper bound = 0.6 + 0.05 = 0.65 cm

Example 4: Find the lower and upper bounds for a length of 0.34 cm measured to three decimal places.

Solution: Three decimal places is 0.001.
0.001 ÷ 2 = 0.0005.

Lower bound = 0.34 – 0.0005
= 0.3395 cm

Upper bound = 0.34 + 0.0005
= 0.3405 cm

Example 5: Find the lower and upper bounds for a length of 19.34 cm measured to three significant figures.

Solution: 3 significant figures in this case corresponds to 0.1. Therefore, 0.1 ÷ 2 = 0.05.
Lower bound = 19.34 – 0.05 = 19.29 cm
Upper bound = 19.34 + 0.05 = 19.39 cm

Example 6: Find the lower and upper bounds for 48 cm rounded to two significant figures.

Solution: 2 significant figures, in this case, corresponds to 1. Therefore, 1 ÷ 2 = 0.5.
Lower bound = 48 - 0.5 = 47.5 cm
Upper bound = 48 + 0.5 = 48.5 cm

FOR DISCRETE DATA

Data that can have only certain values are called **discrete data**. They are usually whole numbers. There are some exceptions like shoe sizes which can be in fractions or decimals.

Number of pupils in a school are discrete data. We can have 200 pupils but not $200\frac{1}{2}$ pupils.

Example 7: There are 1200 year 8 pupils at Stewards Academy, Harlow in 2017 academic year, counted to the nearest 100.

Find the lower and upper bounds for the number of year 8 pupils.

Solution: This is discrete data and so must take a certain value.

$100 \div 2 = 50$.

Lower bound = 1200 – 50 = 1150
Upper bound = 1200 + 49 = 1249.

If n is the number of pupils, this can be represented as:

$$1150 \leq n \leq 1249$$

or

$$1150 \leq n < 1250$$

The number of pupils in the school could be as low as 1150 because it rounds to 1200 or as many as 1249 as it rounds down to 1200.

Lower bound = **1150**
Upper bound = **1249**.

EXERCISE 32 C

1) Find the lower and upper bounds for each of these quantities.

a) 45 cm rounded to the nearest cm
b) 1200 cm to the nearest 100 cm
c) 67 rounded to 2 sf
d) 0.08 km rounded to 1 dp
e) 8.8 kg rounded to 2 sf
f) 56.34 cm rounded to 3 sf
g) 7.7 m measured to the nearest 0.1 m
h) 540 pens measured to nearest 10
i) 12350 people, to the nearest 1000

2) The weight of a table is 8.3 kg to the nearest kg. What is the least and greatest weight it could be?

3) The dimensions of the rectangle below are length 4.5 cm and width 2.2 cm; both measured to the nearest 0.1 cm.

Calculate
a) the lower bound for the length
b) the upper bound for the length
c) the error interval for length
d) the lower bound for the width
e) the upper bound for the width
f) the maximum possible area of the rectangle.

4) A cube has a side of length 3.8 cm to 2 significant figures.
What is the smallest possible volume of the cube?

32.4 ABSOLUTE AND PERCENTAGE ERROR

Absolute error is the error encountered due to the lack of precision of most of the measuring instruments. It is the difference between the measured value and theoretical value/**nominal value**. The nominal value is the value calculated if there were no errors.

$$\text{Percentage error} = \frac{\textit{absolute error}}{\textit{nominal value}} \times 100\%$$

Example 1: The length of a field should be 99.5 m but is 98.4 m.
Calculate a) the absolute error
b) the percentage error.

Solution:
a) Absolute error = 99.5 – 98.4 = **1.1 m**.
b) Percentage error $= \frac{1.1}{99.5} \times 100\%$
$= \mathbf{1.11\ \%}$ to 2 dp.

Example 2: The measurements for the triangle below is correct to 2 significant figures.

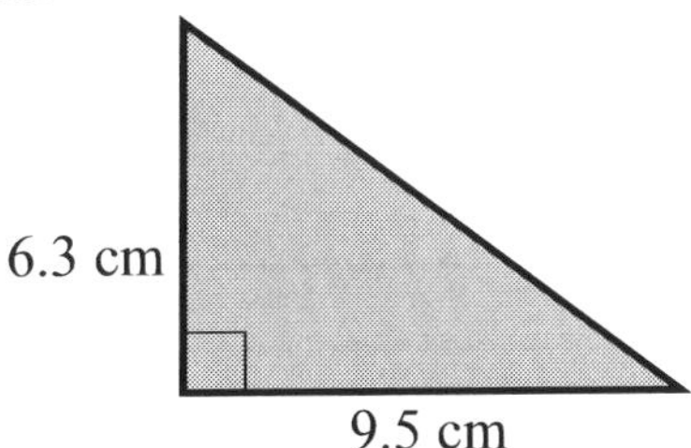

Work out
a) the nominal area of the triangle
b) the least and greatest possible areas of the triangle
c) the maximum absolute error
d) the maximum percentage error.

Solution:

a) Nominal area $= \frac{9.5 \times 6.3}{2} = \mathbf{29.925\ cm^2}$

b)
For the base of the triangle,
lower bound = 9.5 – 0.05 = 9.45 cm
Upper bound = 9.5 + 0.05 = 9.55 cm

For the vertical height of the triangle,
Lower bound = 6.3 – 0.05 = 6.25 cm
Upper bound = 6.3 + 0.05 = 6.35 cm

The least possible area
$= \frac{9.45 \times 6.25}{2} = \mathbf{29.53125\ cm^2}$

The greatest possible area
$= \frac{9.55 \times 6.35}{2} = \mathbf{30.32125\ cm^2}$

c) The maximum absolute error
$= 30.32125 - 29.925 = \mathbf{0.39625\ cm^2}$

d) The maximum percentage error

$= \frac{\textit{maximum absolute error}}{\textit{Nominal value}} \times 100\%$

$= \frac{0.39625}{29.925} \times 100\%$

= **1.3%** to 2 significant figures.

EXERCISE 32D

1) A box of chocolates weighs 1.45 kg. It should weigh 1.50 kg. Calculate the percentage error.

2) The length of a luxury curtain should be 1.92 m but measured 2.1 m.
Calculate
a) the absolute error
b) the percentage error.

3) The measurements for the triangle below is correct to one decimal place.

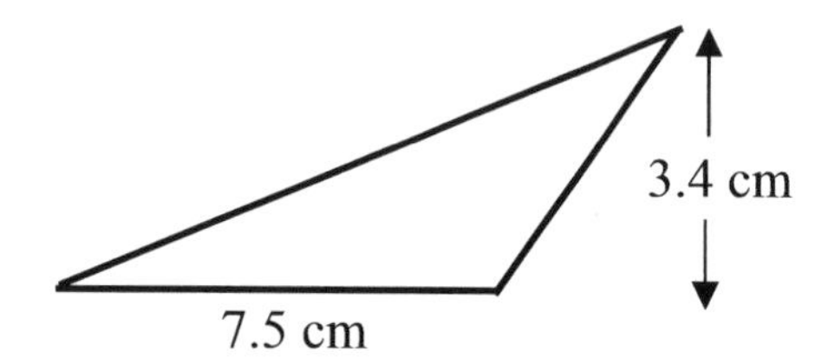

Work out
a) the nominal area of the triangle
b) the least and greatest possible areas of the triangle
c) the maximum absolute error
d) the maximum percentage error.

4)

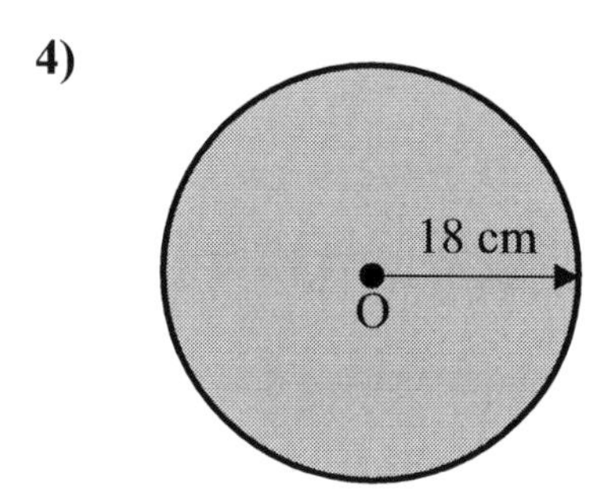

O is the centre of the circle, and the dimensions are measured to 2 significant figures.
Calculate the **maximum** percentage error of the area of the circle, to 2 decimal places.

32.5 VENN DIAGRAM

In this section, we shall learn how to draw a Venn diagram, write notations using special symbols and calculate probabilities from Venn diagrams.

SETS

A **set** is a collection of numbers or objects.

If we want to represent a set of prime numbers less than 15 in a set represented by P, it should look like this:

P = $\{\mathbf{2, 3, 5, 7, 11, 13,}\}$

P represents the set of prime numbers less than 15. Note that capital letters are mostly used to represent a set. In the set above, 2, 3, 5, 7, 11 and 13 are all **elements** within the set.

The **universal set** is the set containing **all** the elements. It is represented as $\boldsymbol{\varepsilon}$.
We could also have a set of no element(s). It is called the **empty set**, ∅.

NOTATIONS

There are basically three special notations used in Venn diagrams.

1) A ∩ B
This means the intersection of A and B. It is the overlapping part of the Venn diagram.

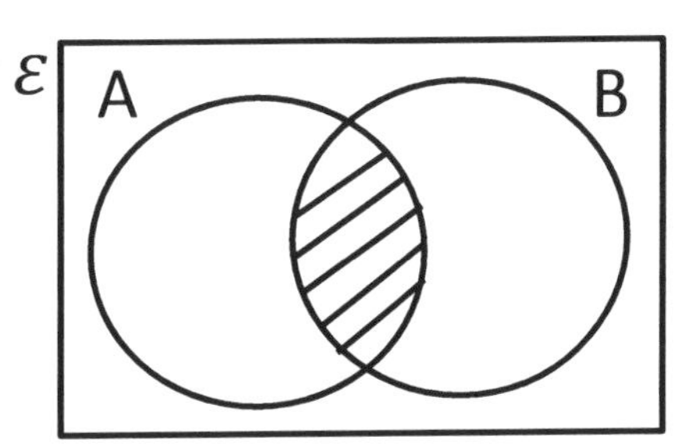

2) A ∪ B
This means the union of A and B.
It is everything in A and B and is represented by the shaded part in the diagram.

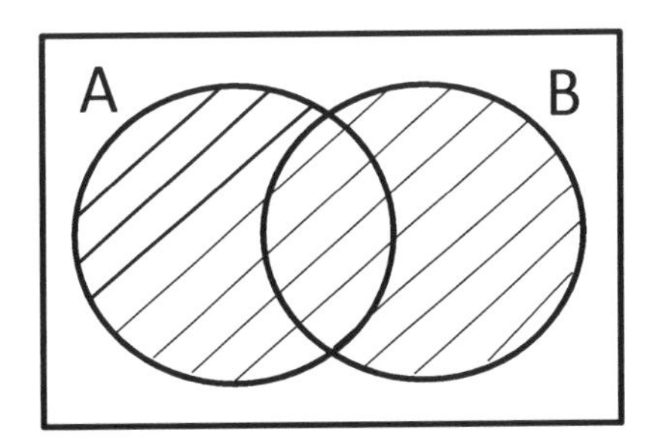

3) A′
This is the complement of A.
It is everything **not** in A. It is the
Shaded part.

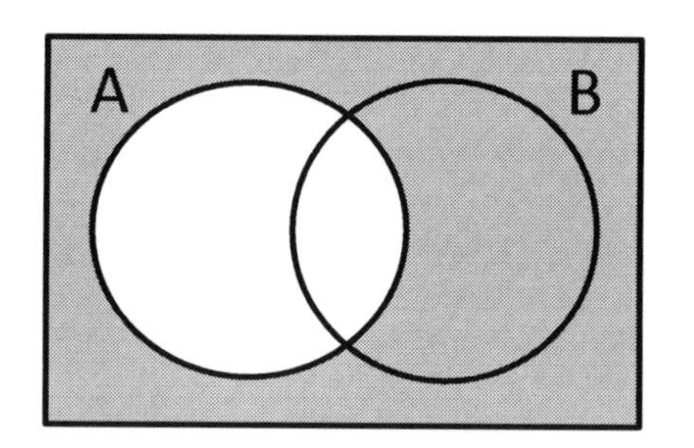

Example 1: $\varepsilon = \{2, 4, 6, 8, 10, 12, 14, 16\}$, A = $\{multiples\ of\ 4\}$ and B = $\{2, 5, 8, \}$

a) Draw a Venn diagram to illustrate the information.

b) Work out i) A ∪ B   ii) A ∩ B   iii)) A′ ∪ B   iv) (A ∪ B)′   v) A′

Solutions:

a) Venn diagram

$\varepsilon = \{2, 4, 5, 6, 8, 10, 12, 14, 16\}$
A = $\{Multiples\ of\ 4\} = \{4, 8, 12, 16\}$
B = $\{2, 5, 8, \}$

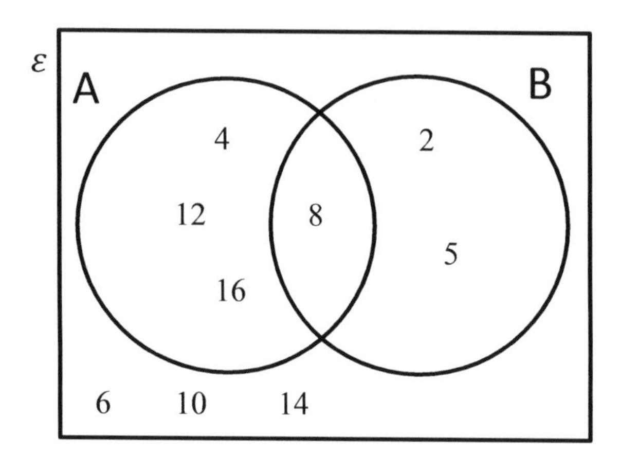

b) i) A ∪ B = $\{\mathbf{2, 4, 5, 8, 12, 16}\}$
These are all numbers in both A and B.

ii) A ∩ B = $\{\mathbf{8}\}$
This is the number in the intersection part.

iii) A′ ∪ B = $\{\mathbf{2, 5, 6, 8, 10, 14}\}$
These are all numbers excluding 4, 12 and 16.

iv) (A ∪ B)′ = $\{\mathbf{6, 10, 14}\}$
This is the complement of A Union B.
These are all the numbers **not** in A and B.

v) A′ = $\{\mathbf{2, 5, 6, 10, 14}\}$
This is the complement of A. They are all the numbers **not** in A.

PARTS AND DESCRIPTION OF THE VENN DIAGRAM

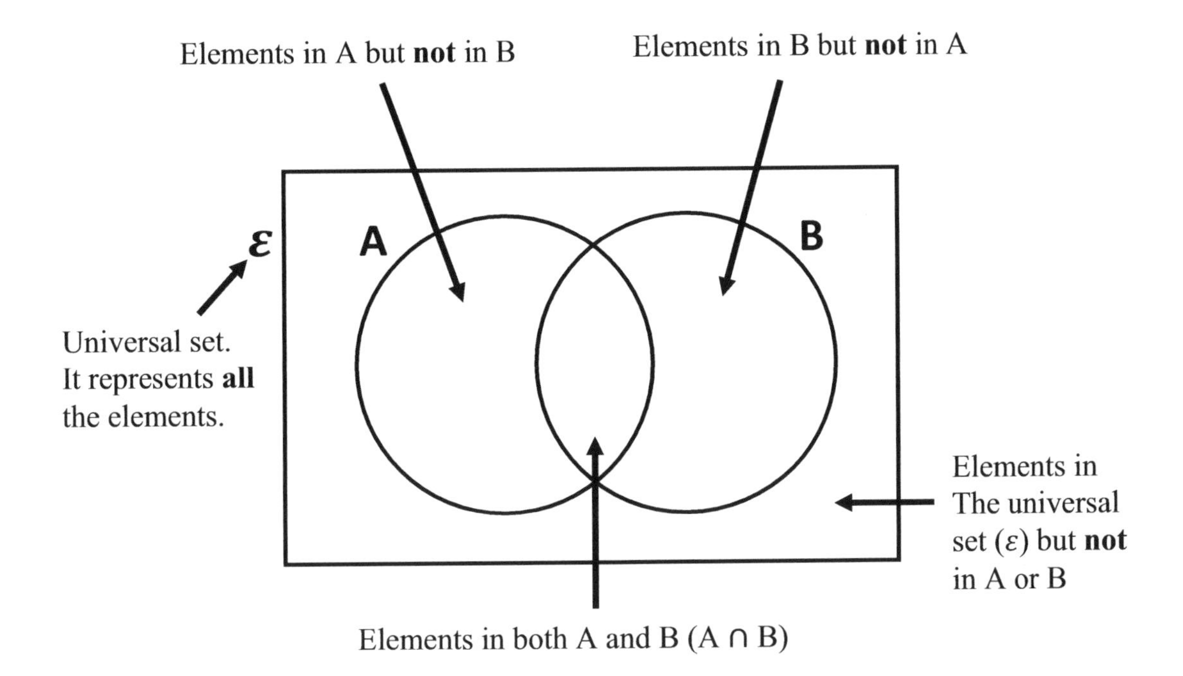

32.6 PROBABILITY AND VENN DIAGRAM

We may use the Venn diagram to calculate probabilities as follows.

$$P(A) = \frac{\textit{Number of elements in set A}}{\textit{Total number of elements in } \varepsilon}$$

Probability of A

Also:

P (A ∩ B) = P (A *and* B)

P (A ∪ B) = P (A *or* B)

P (A′) = P (*not* A)

Example 2: The Venn diagram shows the number of pupils that are right-handed (R) and the number of pupils that are left-handed (L) in a class.

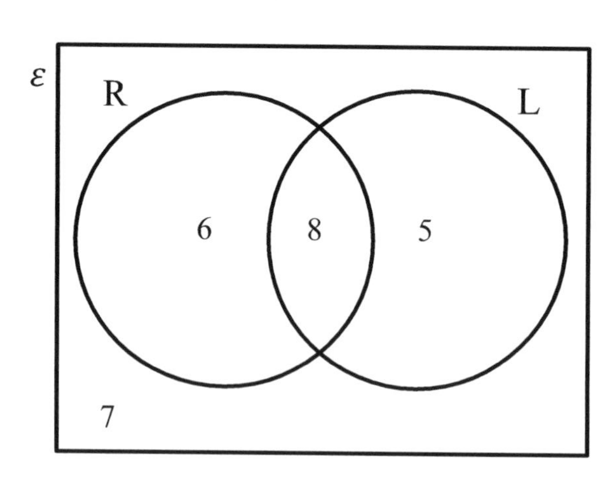

a) What is the total number of pupils in the class?

b) What is the probability that a pupil chosen at random is right handed?

Work out

c) $P(R \cap L)$
d) $P(R \cup L)$
e) $P(R \cup L)'$
f) $P(R')$
g) $P(R' \cup L)$
h) $P(R \cap L)'$

<u>Solution:</u>

a) Total number of pupils = 6 + 8 + 5 + 7 = **26 pupils.**

b) 6 + 8 = 14...........14 pupils are right handed therefore, $P(R) = \frac{\mathbf{14}}{\mathbf{26}}$

c) $P(R \cap L)$ is worked out by looking at the intersection part of the Venn diagram.

$P(R \cap L) = \frac{\mathbf{8}}{\mathbf{26}}$

d) 6 + 8 + 5 = 19. Therefore, $P(R \cup L) = \frac{\mathbf{19}}{\mathbf{26}}$

e) $P(R \cup L)'$ means the probability that the pupils are neither right nor left-handed. This is also $1 - P(R \cup L) = 1 - \frac{19}{26} = \frac{\mathbf{7}}{\mathbf{26}}$

f) R' means not in R or not right handed. This is 5 + 7 = 12. Therefore, $P(R') = \frac{\mathbf{12}}{\mathbf{26}}$

g) 8 + 5 + 7 = 20. Therefore, $P(R' \cup L) = \frac{\mathbf{20}}{\mathbf{26}}$

h) $(R \cap L)'$ means everything that **is not** in the intersection part of the Venn diagram. Everything that is not 8 is 6 + 5 + 7 = 18. Therefore, $P(R \cap L)' = \frac{\mathbf{18}}{\mathbf{26}}$

EXERCISE 32E

1) List all the elements of the following sets.

a) P – the first six multiples of 7
b) Q – factors of 48
c) R – prime numbers between 6 and 24
d) S – double-digit cube numbers

2) The Venn diagram shows information about pupils playing flute and guitar at a music festival.

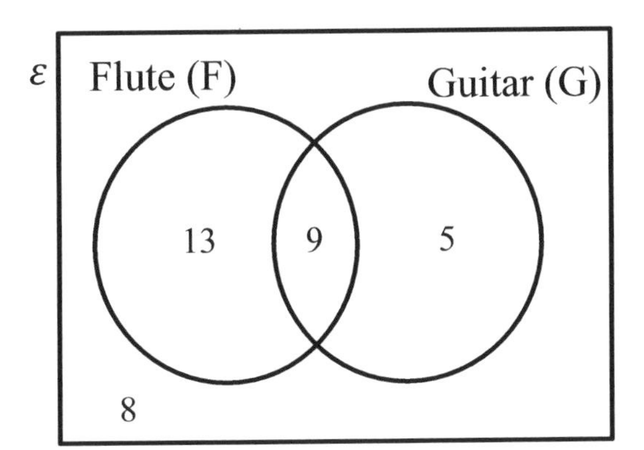

a) How many pupils were present at the festival?
b) How many pupils played the flute?
c) How many pupils played the guitar?
d) How many pupils did not play flute?
e) How many pupils played the guitar but not the flute?

If a pupil is chosen at random, find

f) P $(F \cap G)$
g) P (F)
h) P $(F \cup G)$
i) P $(F \cup G)'$
j) P $(F' \cup G)$
k) P $(F \cap G)'$

3) $\mathcal{E} = \{1, 2, 3, 4, 5, 6, 7, 8, 9, 10, 11, 12, 13\}$
$A = \{2, 3, 7, 9\}$
$B = \{1, 5, 7, 9, 13\}$

a) Represent the information in a Venn diagram.
b) Use the Venn diagram to find:
i) P (B)
ii) P (A')
iii) P $(A \cap B)$
iv) P $(A \cup B)'$

4) If

$\mathcal{E} = \{x : x \text{ is a positive integer}, x \leq 19\}$,
list all the elements of:

a) $B = \{x : x \text{ is even}\}$
b) $C = \{x : x \text{ is a multiple of } 4\}$
c) $D = \{x : x \text{ is a prime number}\}$

5) Using set notations, describe the shaded area in each Venn diagram.

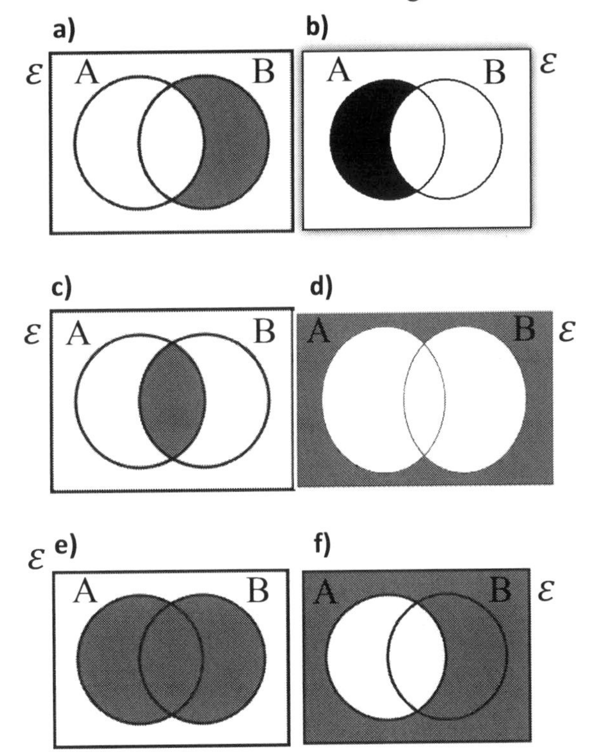

Example 3: There are 20 pupils in a class. They are each asked whether they have a dog (D) or a hamster (H).
15 say they have a dog.
8 say they have a hamster. Two pupils have neither of the pets.
Draw a Venn diagram to represent the information.

Solution: $15 + 8 = 23$
$20 - 2 = 18$
$23 - 18 = 5$ (overlap)

$15 - 5 = 10$ pupils $\quad 8 - 5 = 3$ pupils

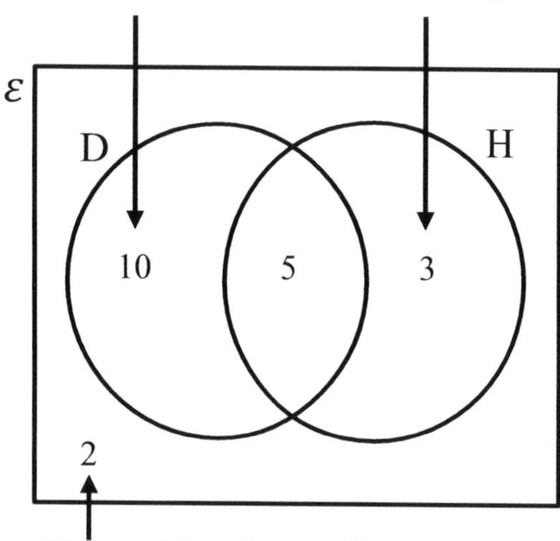

Have neither Cat nor Hamster

Example 4:
Complete the Venn diagram for pupils with equipment during an examination.

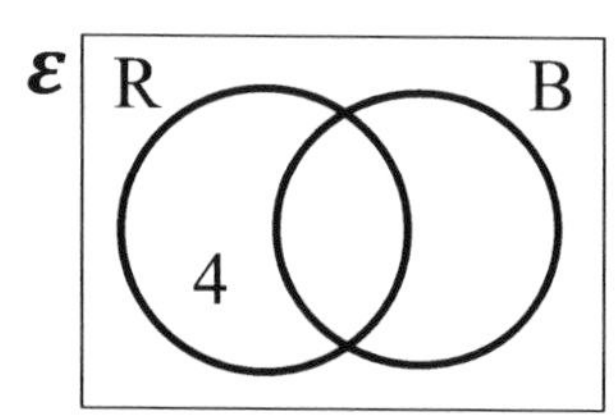

ε = 24 equipment,
R = Red pen
B = Blue pen

16 have red pens, 14 have blue pens, 4 have red pens but not blue pens.

Solution:

From 16 with red pens, $16 - 4 = 12$
From 14 with blue pens, $14 - 12 = 2$

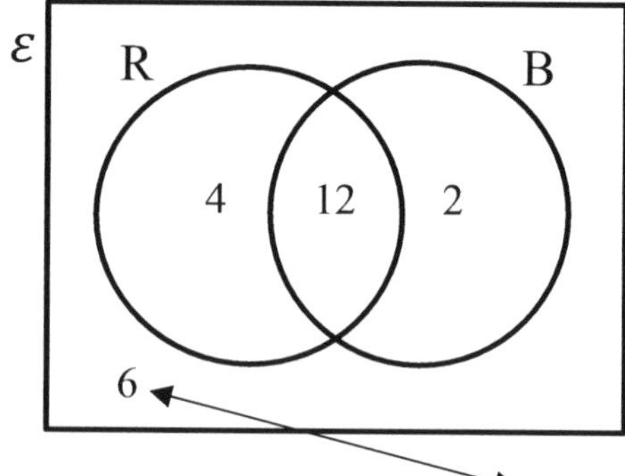

$4 + 12 + 2 = 18$ so, $24 - 18 = 6$

Example 5: An employer has **80** workers. 45 workers are in ICT (I) department. 50 workers are right-handed (R). 15 workers are not right handed and not in ICT department.
a) Draw a Venn diagram
b) Find the number of workers who are right-handed and in ICT department.
c) Work out the probability that a worker is right handed and in ICT department.

Solution: Let the number of workers who are right-handed and in ICT department be $\boldsymbol{n}$. $\varepsilon = 80$

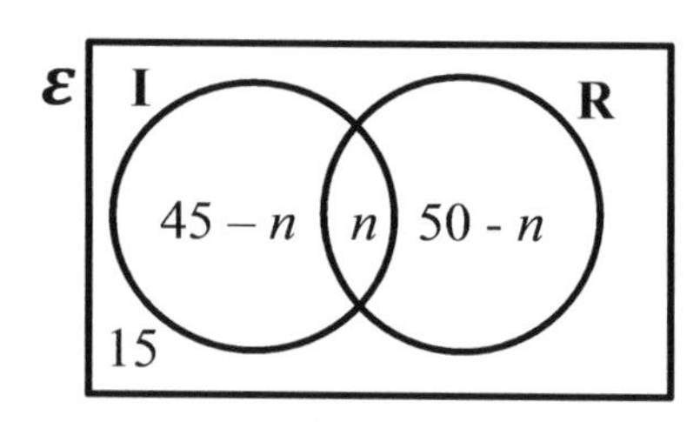

$45 - n + n + 50 - n + 15 = 80$
$110 - n = 80$
$110 - 80 = n$
$\boldsymbol{n} = 30$

a) Substituting the value of n in the Venn diagram gives

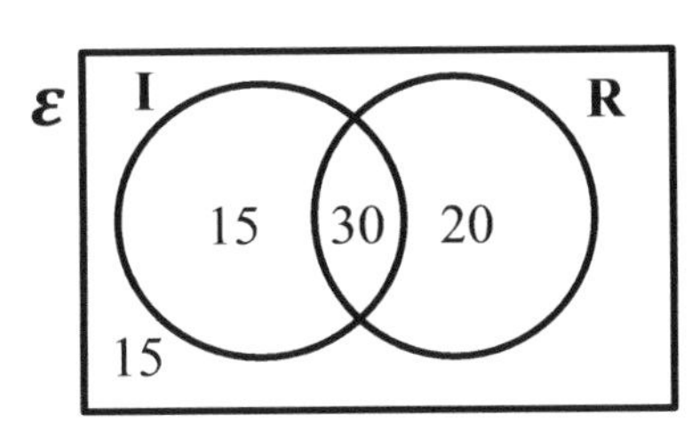

b) n = 30 workers

c) P (I and R) = $\frac{30}{80}$ or 0.375

Example 6: In a sixth-form class, 7 study both maths (M) and physics (P). 14 study physics but not maths and 5 study neither subject. The overall class has 32 students.

a) Draw a Venn diagram to represent the information above.

b) How many students study maths.

c) Find the probability that a student studies physics.

Solution: Let x represent the number of students who study maths only. $\varepsilon = 32$.

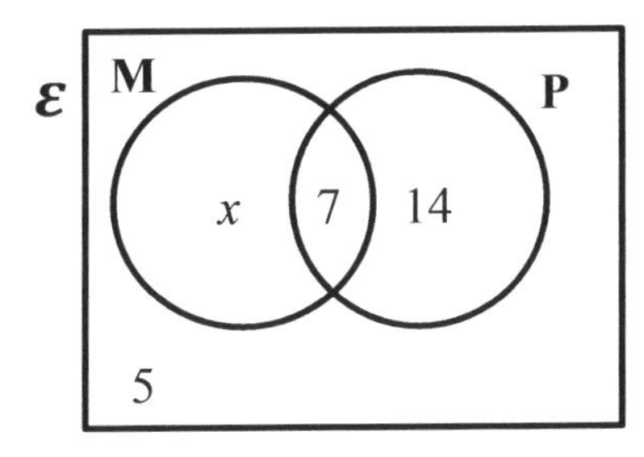

$x + 7 + 14 + 5 = 32$, so $x = 6$

a)

ε M P
6 7 14
5

b) 6 + 7 = 13 students

c) 7 + 14 = 21

P(P) = $\frac{21}{32}$

Example 7: There are 25 pupils in a class. They are each asked whether they have a dog (D) or a hamster (H).
18 say they have a dog.
10 say they have a hamster. 3 pupils have neither of the pets.

a) Draw the Venn diagram.

b) Find the probability that a pupil chosen at random from the class owns a dog **given that** they own a hamster.

Solution: 18 + 10 = 28
25 – 3 = 22
28 – 22 = 6 (overlap)
Dog only: 18 – 6 = 12
Hamster only: 10 – 6 = 4

a)

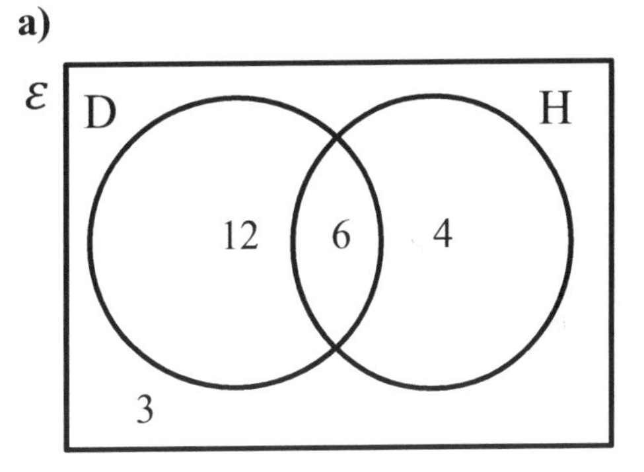

b)

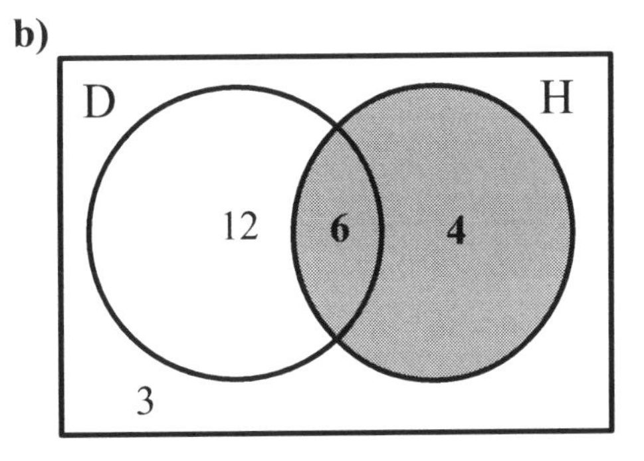

6 + 4 = 10

Therefore, the probability = $\frac{6}{10}$

Example 8: 92 students in a sixth form college.

40 study Biology (B)
50 study Physics (P)
25 study English (E)

10 study both Biology and Physics
12 study both Physics and English
8 study both Biology and English

All the pupils study one or more of these subjects.
a) Draw a Venn diagram to illustrate the information.
b) How many pupils study all three subjects?
c) What is the probability that a pupil picked at random studies all subjects mentioned?

Solution: Sketch the Venn diagram. Let the number of students that studies all three subjects be x.

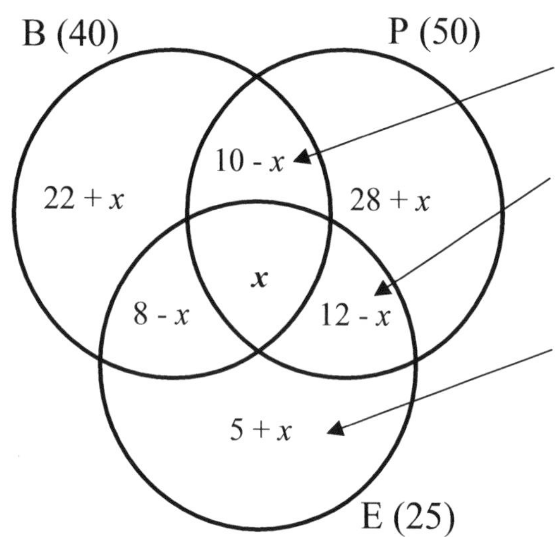

10 students study Biology and Physics. Therefore, this portion is $10 - x$

12 students study Physics and English. Therefore, this portion is $12 - x$
Do the same to the remaining part to get $8 - x$

This area $= 25 - (8 - x + x + 12 - x)$
$= 25 - (20 - x)$
$= 5 + x$
Do the same for other areas to get $22 + x$ and $28 + x$ respectively

a)

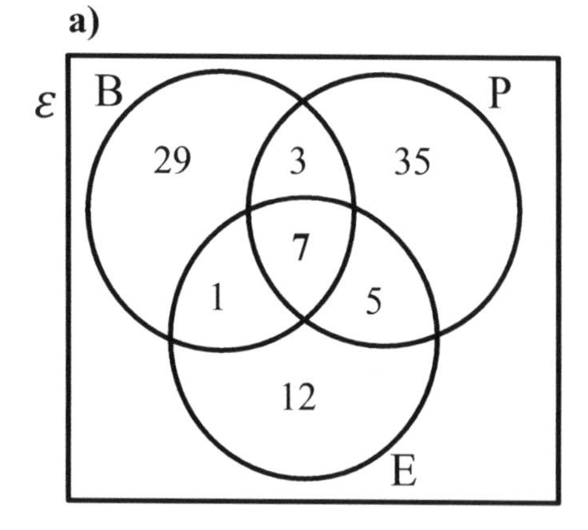

b) Add all the numbers and letters in the Venn diagrams. This will equal 92.
$22 + x + 10 - x + x + 8 - x + 28 + x + 12 - x + 5 + x = 92$. This reduces to $85 + x = 92$.
$x = 92 - 85 = 7$

Therefore, **7 students** study all three subjects.

c) $\frac{7}{92}$

EXERCISE 32F

1) In a class of 27 pupils, 20 have a cat, 9 have a dog. 6 pupils have neither.

a) Draw a Venn diagram to represent the above information.
b) Calculate the probability that a pupil chosen at random from the class owns
i) a cat and a dog   ii) a dog
c) Find the probability that a pupil chosen at random from the class owns a dog, given that they own a cat.

2) P(X) = 0.35 and P (Y) = 0.12.
Write down a) P (X′)   b) P (Y′)

3) An employer has **110** workers. 65 workers are in media (M) department. 60 workers are left-handed (L). 20 workers are not left handed and not in the media department.
a) Draw a Venn diagram
b) Find the number of workers who are left-handed and in media department.
c) Work out the probability that a worker is left-handed and not in the media department.

4) The number of elements in the complement of B = n (B′) = 27.

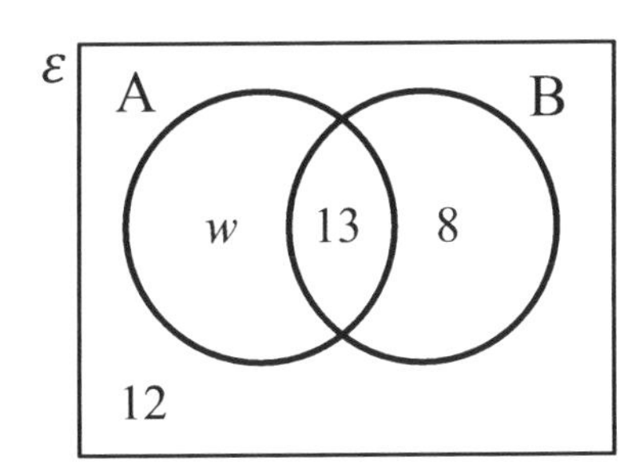

Work out
a) w b) n (A ∪ B)   c) n(A ∪ B′).

5) The Venn diagram below shows some probabilities.

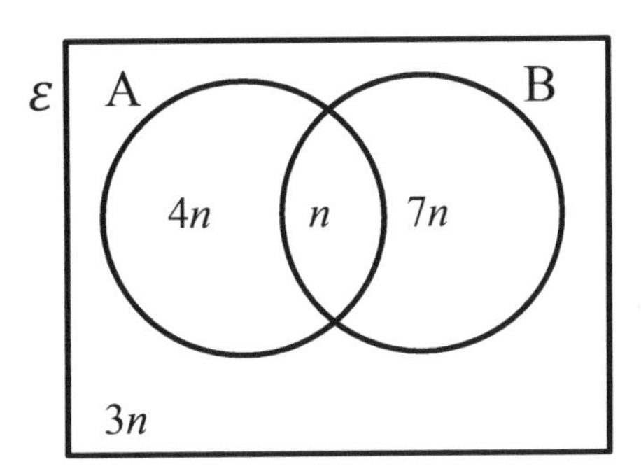

Calculate:
a) P (A)   b) P (B)   c) P (A′)   d) P (B′)

e) P (A ∪ B)   f) P (A ∩ B′)   g) P (A′ ∩ B)

6) 120 pupils in a school are in year 11.

60 study Geography
55 study History
45 study French

15 study both Geography and History
18 study both History and French
12 study both Geography and French

All the pupils study one or more of these subjects.

a) Draw a Venn diagram to illustrate the information.

b) How many pupils study all three subjects?

c) What is the probability that a pupil chosen at random studies all subjects mentioned?

32.7 CUBIC AND RECIPROCAL GRAPHS

A **cubic** function has the highest index number as 3. It will always include a power of 3. Examples of cubic numbers include: $x^3 + 4x + 8$, $3x^3 - 2x + 5$.

Example 1: a) Plot the graph of $y = x^3 + 2x + 1$, x values from -4 to 4.
b) Use your graph to find the value of y when x = 2.5.

Solution: Draw a table of values for the coordinates.
When x = -4, $y = (-4)^3 + 2(-4) + 1 = -71$

x	-4	-3	-2	-1	0	1	2	3	4
y	-71	-32	-11	-2	1	4	13	34	73

a)

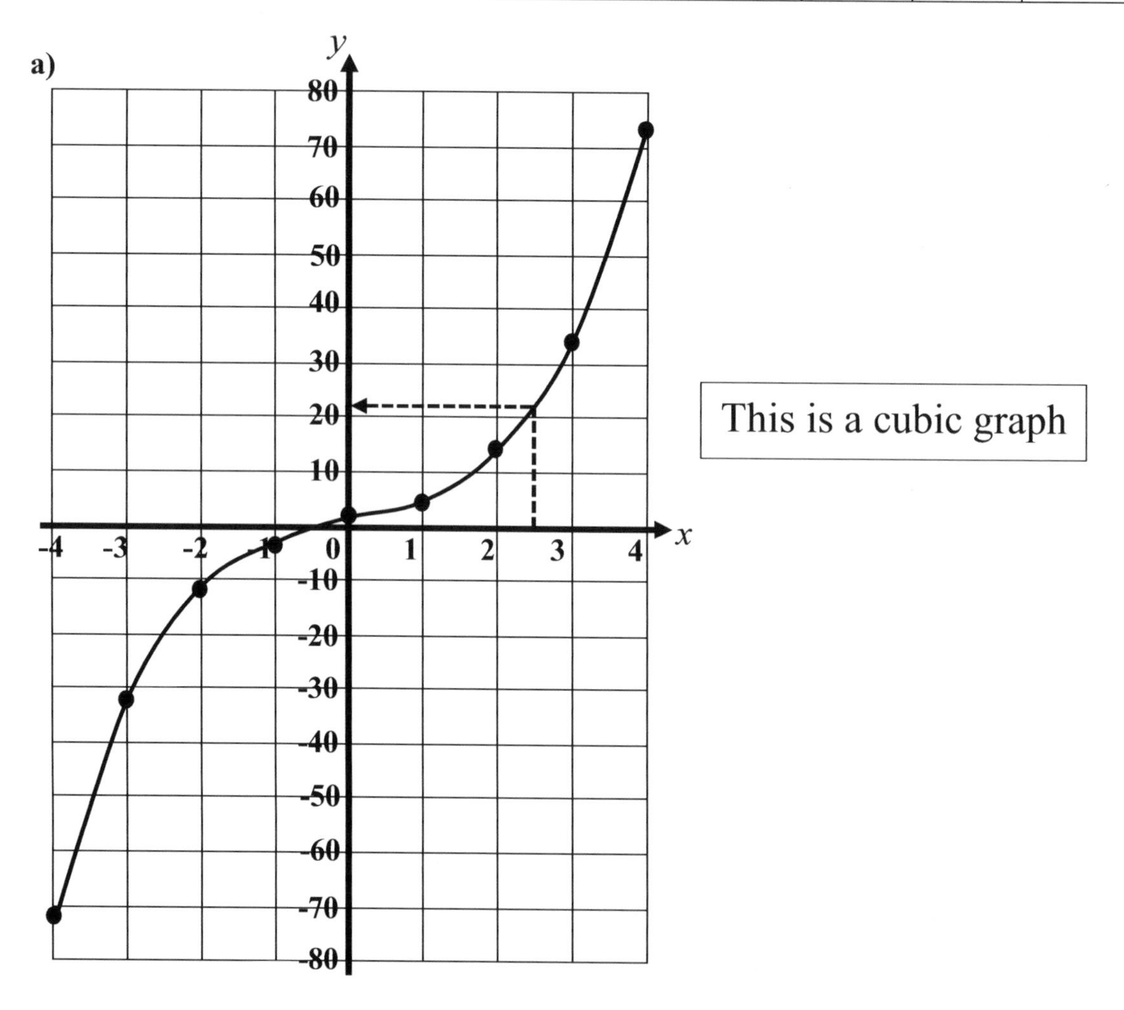

b) From 2.5 on the x-axis, draw a vertical line to meet the graph. Then, read off the value on the y-axis. Therefore, when x = 2.5, y is ≈ **21.5**

RECIPROCAL GRAPHS

A **reciprocal** function is in the form $\frac{1}{a}$ or a^{-1}. $y = \frac{1}{x}$ and $y = \frac{5}{x}$ are examples of reciprocal functions. In a reciprocal function, the graph **never** touches the axes. It gets closer but not touching. A sketch of the reciprocal function of $y = \frac{1}{x}$ is shown below.

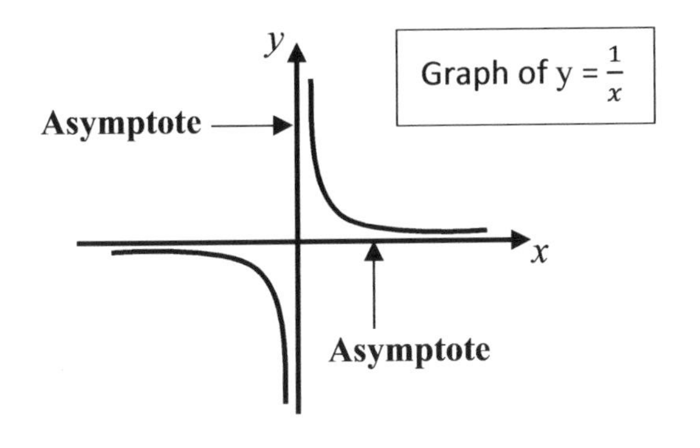

When a graph gets closer to a line but never touches it, that line is called an **asymptote**.

In a reciprocal graph, **two** asymptotes are visible.

For the graph of $y = \frac{1}{x}$, the equations of the asymptotes are $x = 0$ and $y = 0$.

Example 2: a) Draw the graph of $y = \frac{2}{x}$. Choose suitable values for the x-axis.

b) Write down the equations of the asymptotes.

Solution:

x	-5	-4	-3	-2	-1	-0.5	0.5	1	2	3	4
y	$-\frac{2}{5}$	$-\frac{1}{2}$	$-\frac{2}{3}$	-1	-2	-4	4	2	1	$\frac{2}{3}$	$\frac{1}{2}$

From $y = \frac{2}{x}$, when $x = 4$, $y = \frac{2}{4} = \frac{\mathbf{1}}{\mathbf{2}}$ or **0.5**

a)

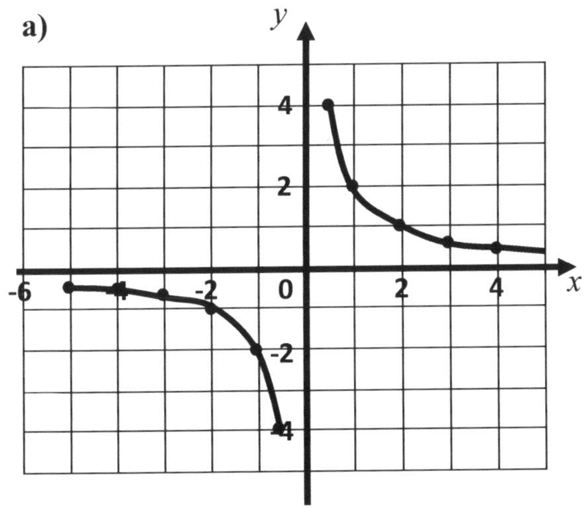

b) The asymptotes are:
$x = 0$ and $y = 0$

EXERCISE 32G

1) Copy and complete the table of values for each equation.

a) $y = x^3 + 1$

x	-4	-3	-2	-1	0	1	2	3
y								

b) $y = x^3 + x - 3$

x	-4	-3	-2	-1	0	1	2	3
y								

c) $y = \frac{4}{x}$

x	-5	-4	-3	-2	-1	-0.5	0.5	1	2	3	4
y											

d) $y = -\frac{4}{x}$

x	-5	-4	-3	-2	-1	-0.5	0.5	1	2	3	4
y											

2) a) Draw the graphs of the functions in question **1a** and **1c** only.
b) Show the asymptotes and write down their equations for **1c** only.

3) Copy and complete the table for $y = \frac{x}{x+2}$.

x	-3	-2	-1	$-\frac{1}{2}$	$-\frac{1}{5}$	$\frac{1}{5}$	1	2	3
$x + 2$	-1			$1\frac{1}{2}$					5
$y = \frac{x}{x+2}$	3			-0.33					0.6

4) Copy and complete the table for the function $y = (x + 3)^3$.

X	-4	-3	-2	-1	0	1	2	3	4
Y		0				64		216	

32.8 EXPONENTIAL FUNCTIONS

An **exponential function** has the form $\mathbf{y = a^x}$ where a is a positive constant and x can vary. Some examples of exponential functions are: $y = 5^x$ and $y = (0.25)^x$.
Two possible graphs (shapes) for exponential functions.

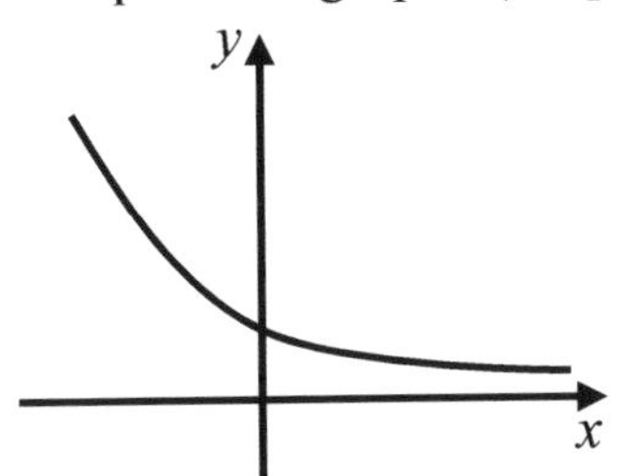

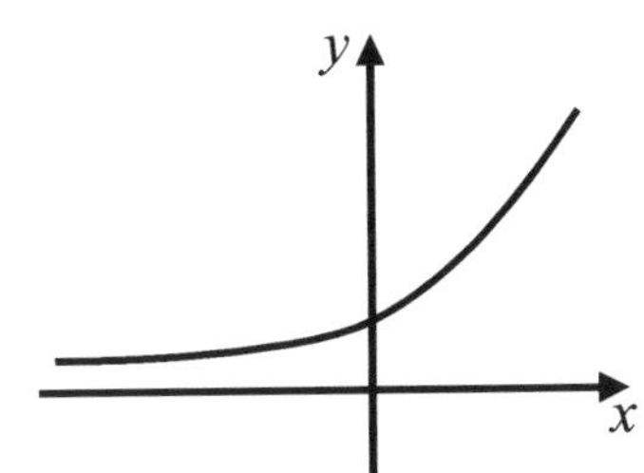

x is increasing while y is decreasing (like a negative slope). The curve approaches the x-axis but never touches it. The x-axis is called the asymptote and the equation of the asymptote is $\mathbf{y = 0}$.	x is increasing while y is increasing (like a positive slope). The curve approaches the x-axis but never touches it. The x-axis is called the asymptote and the equation of the asymptote is $\mathbf{y = 0}$.

Looking at the form of all exponential functions of the form $y = a^x$, it would suggest that the graph must pass through the coordinate point (0,1). This is explained as $y = a^0 = 1$.

Example 1: If $y = 4^x$ is an exponential function, a) draw its graph for x values from -3 to +4. b) Use your graph to estimate the value of x when y = 9.5.

Solution: Draw a table of values.

x	-3	-2	-1	0	1	2
$y = 4^x$	0.02	0.06	0.25	1	4	16

a)

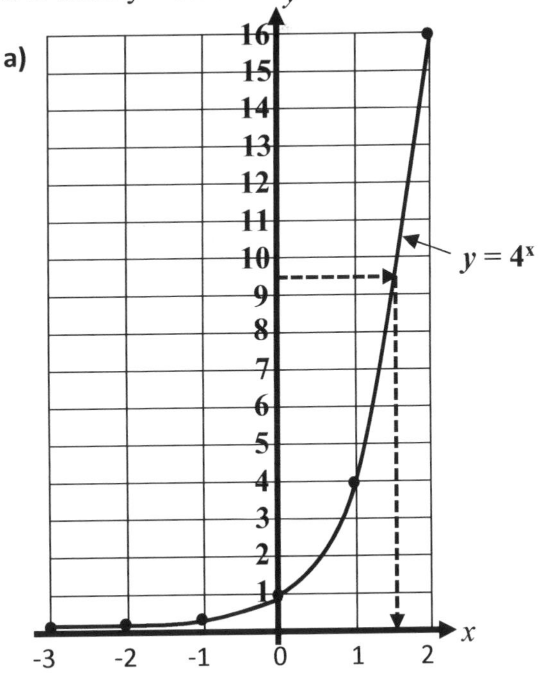

b) From the graph, when y = 9.5, $x \approx 1.6$.

EXERCISE 32H

1) a) Copy and complete the table for $y = 3^x$.

x	-4	-3	-2	-1	0	1	2	3
$y = 3^x$	0.01						9	

b) Draw the graph of $y = 3^x$.
c) Use your graph to estimate the value of x when $y = 2.5$.

2) Draw a table of values for the graphs of a) $y = 5^x$ b) $y = 5^{-x}$. x values from -3 to 3.
c) At your leisure time, draw the graphs of $y = 5^x$ and $y = 5^{-x}$.

3) A radioactive substance decays by the exponential function

$$B = 60 \times 2^{-t}$$

where B is the amount of the radioactive material and t is the time in years.

a) Copy and complete the table below.

t	0	1	2	3	4	5	6	7	8
B		30						0.47	

b) Draw the exponential graph of the decay over the eight years.

c) Use the graph to estimate the amount of radioactive material remaining after $4\frac{1}{2}$ years.

4) a) Copy and complete the table for the function $y = (\frac{2}{5})^x$.

x	-2	-1	0	1	2	3	4
$y = (0.4)^x$		2.5		0.4			

b) At your leisure time, draw the graph of $y = (\frac{2}{5})^x$, for $-2 \leq x \leq 4$.

32.9 SINE, COSINE AND TANGENT GRAPHS

This section deals with recognising and drawing the graphs of $y = \sin\theta$, $y = \cos\theta$, and $y = \tan\theta$.

THE SINE GRAPH ($y = \sin\theta$)

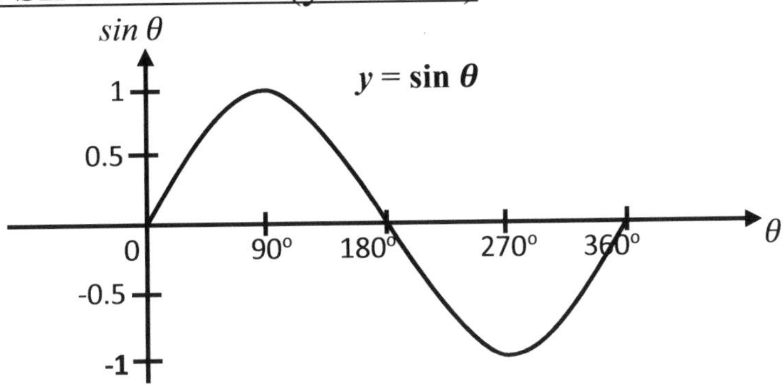

CHARACTERISTICS OF THE SINE GRAPH

1) The highest value of y is **+1**. It occurs when $\theta = 90°, 450°, -270°, \ldots$
2) The minimum value of y is **-1**. It occurs when $\theta = -90°, 270°, -450°, \ldots$
3) The graph repeats itself every 360°. It shows a **period** of 360°.

Below is the extended graph of $y = \sin\theta$ to show other values after each period.

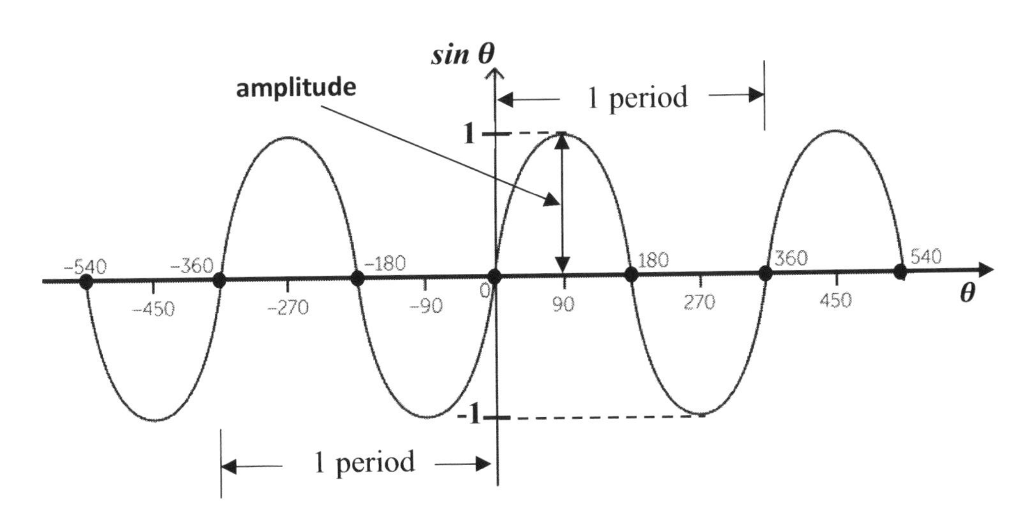

POINTS TO NOTE: $\sin(-\theta) = -\sin\theta$, $\sin(180° + \theta) = -\sin\theta$ and $\sin(180° - \theta) = \sin\theta$.

THE COSINE GRAPH ($y = \cos\theta$)

Below is a graph of $y = \cos\theta$. It has a maximum value of **+1** and a minimum value of **-1** just like the sine graph.

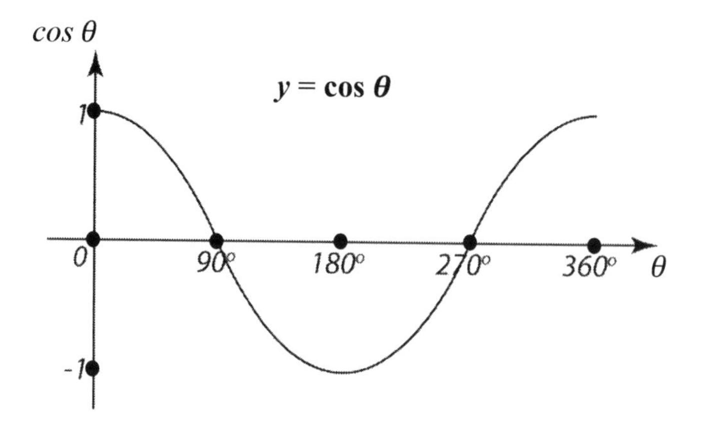

The cosine graph repeats itself every 360° and its **period** is 360°. The maximum value of +1 occurs when $\theta = 0°, 360°, -360°, \ldots$ and the minimum value of -1 occurs when $\theta = 180°, -180°, 540°, \ldots$

THE TANGENT GRAPH ($y = \tan\theta$)

The graph will never touch the 90° line. Type in tan 90° in your calculator. What do you notice? This topic will be explained in detail in subsequent Whiz-kid, Mathematics Series (A levels).

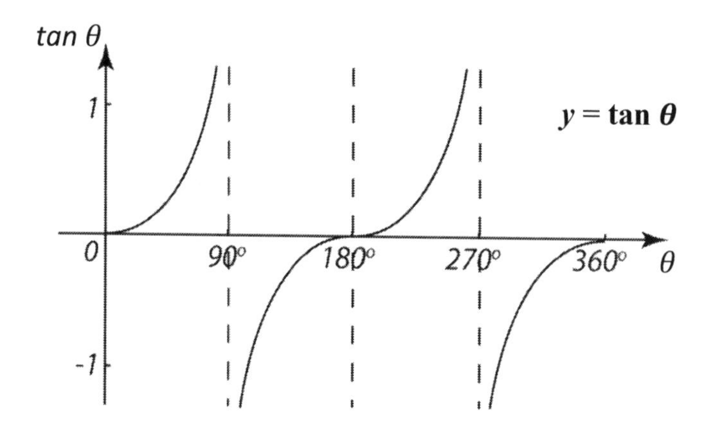

The tan graph repeats itself every 180o and the period is 180o. It is also infinite at $x\ (\theta) = \pm 90°, \pm 270°, \ldots$

EXERCISE 32I

1) a) Copy and complete the table and give your values to two decimal places.

θ	0°	30°	45°	60°	90°	135°	180°
sin θ							
cos θ							
tan θ							

b) Draw the graph of $y = \sin\theta$ for x values in the interval $0° \leq \theta \leq 360°$.
c) Draw the graph of $y = \cos\theta$ for x values in the interval $0° \leq \theta \leq 360°$

2) From the graph of $y = \sin\theta$ in question **1b**, estimate the angle when
i) $\sin\theta = 0.5$ ii) $\sin\theta = 0.87$

3) From the graph of $y = \cos\theta$ in question **1c**, estimate the angle when
i) $\cos\theta = 0.5$ ii) $\cos\theta = 0.86602$.

4) a) Copy and complete the table.

x	$2x$	$2\sin 2x$
-180°		
-135°	-270°	2
-90°		
-45°		
0°		
45°		
90°	180°	
135°		
180°		0

b) Draw the graph of $y = 2\sin 2x$.
c) Work out the minimum and maximum values of $2\sin 2x$ and the values of x at which they occur.

32.10 TRIGONOMETRIC EQUATIONS

To any trigonometric equation, there is uncountable (infinite) number of solutions due to the symmetry and properties of the curves.

Using a calculator will give only one value and students should respond accordingly when prompted.

Example 1:
a) Solve the equation $10\sin x = 5$.
b) Also, state other values of x in the interval $0° \leq \theta \leq 360°$.

Solution:

a) $10\sin x = 5$
$\text{Sin } x = \frac{5}{10} = \frac{1}{2}$ or 0.5.
$x = \sin^{-1} 0.5 = 30°$

b) Using the symmetry of the sine curve $y = \sin x$,

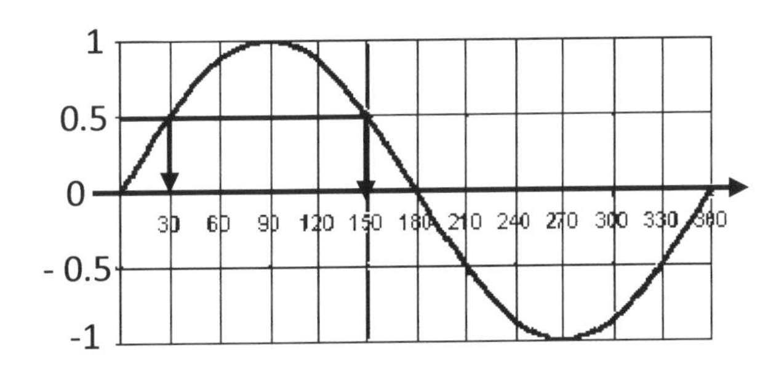

From 0.5, draw a horizontal line to cut the graph at two points. Read off the values to give **30°** and **150°**.

Check with a calculator. Sin 30° and 150° is = 0.5.

EXERCISE 32J

1) Solve the equation 4 sin $x = 2$ and give all the values of x in the range $-360^\circ \leq x \leq 360^\circ$.

2) a) Sketch the graph of y = 4 cos x, x values from 0° to 360°.

b) Solve the equation 4 cos $x = 1$ and give all the values of x between 0° and 360°.

3) For the interval, $0^\circ \leq \theta \leq 360^\circ$, find all the solutions of the equations

a) sin $\theta = 0.6$ c) sin $\theta = -0.2$

b) cos $\theta = 0.6$ d) cos 0.965

4) a) Solve the trigonometric equation 8 sin 2x = 3, giving your answer to the nearest degree.

b) Find two solutions to the equation 8 sin $2x = 3$ in the interval $0^\circ \leq \theta \leq 180^\circ$.

5) a) Use the graph to find i) cos 135° ii) cos 27°. b) Solve the equation cos $x = 0.4$.

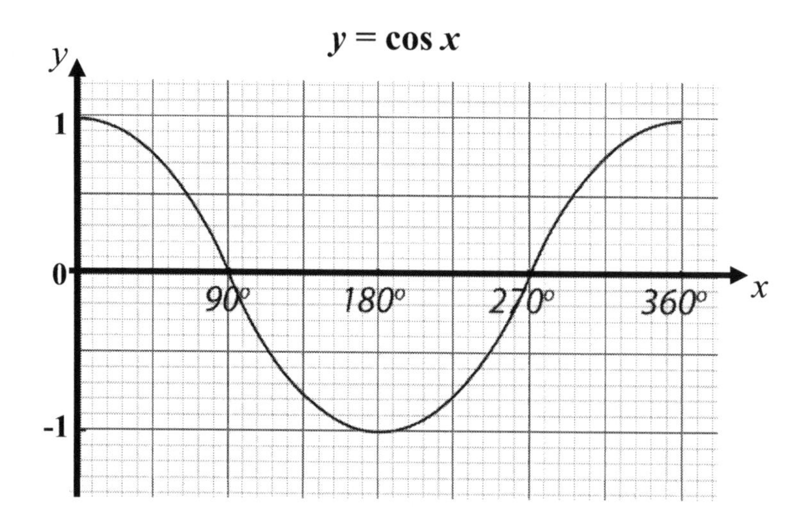

32.11 TRANSFORMING GRAPHS

You should have encountered basic transformations (reflections, rotations, translations and enlargements) in earlier years as a mathematics student.

This section deals with translations and its applications when manoeuvring from one function to another. It will involve some stretches along the ***y*** and *x*-axes depending on the functions to be translated/transformed. For ease of understanding, the graph of $y = x^2$ will be used to demonstrate different transformations. However, the rules apply to **any shape** or function to be transformed.

RULES FOR TRANSFORMATIONS

1) **f(x) + c** moves the function **c** units upwards. See graph ($y = x^2 \longrightarrow y = x^2 + 3$)
The graph of $y = x^2$ has moved vertically upwards 3 units by the vector $\begin{pmatrix} 0 \\ 3 \end{pmatrix}$.

2) **f(x) – c** moves the function **c** units downwards. See graph ($y = x^2 \longrightarrow y = x^2 - 8$)
The graph of $y = x^2$ has moved down 8 units by the vector $\begin{pmatrix} 0 \\ -8 \end{pmatrix}$.

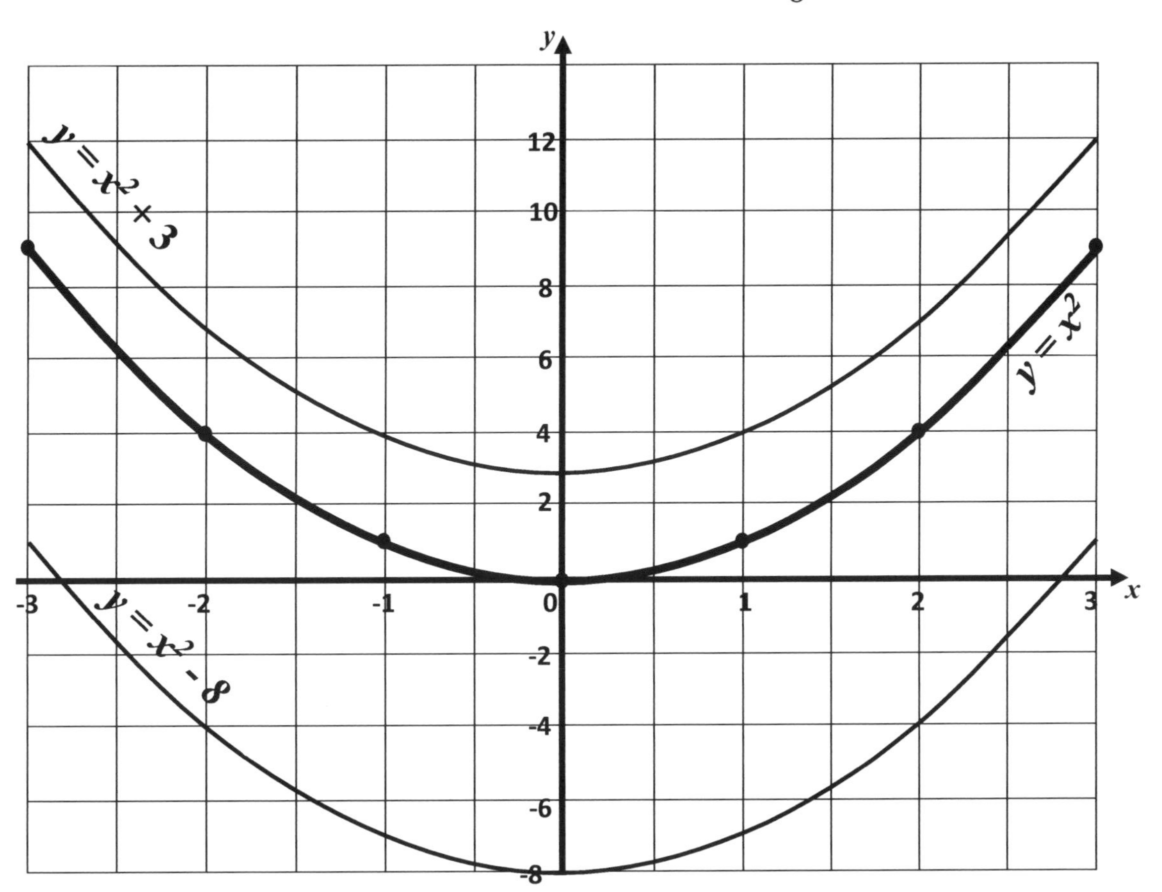

3) **f (x + c)** moves the function **c** units to the left. From the diagram below, the graph of $y = x^2$ is translated 1 unit to the **left** by the vector $\begin{pmatrix} -1 \\ 0 \end{pmatrix}$ to become $y = (x + 1)^2$.

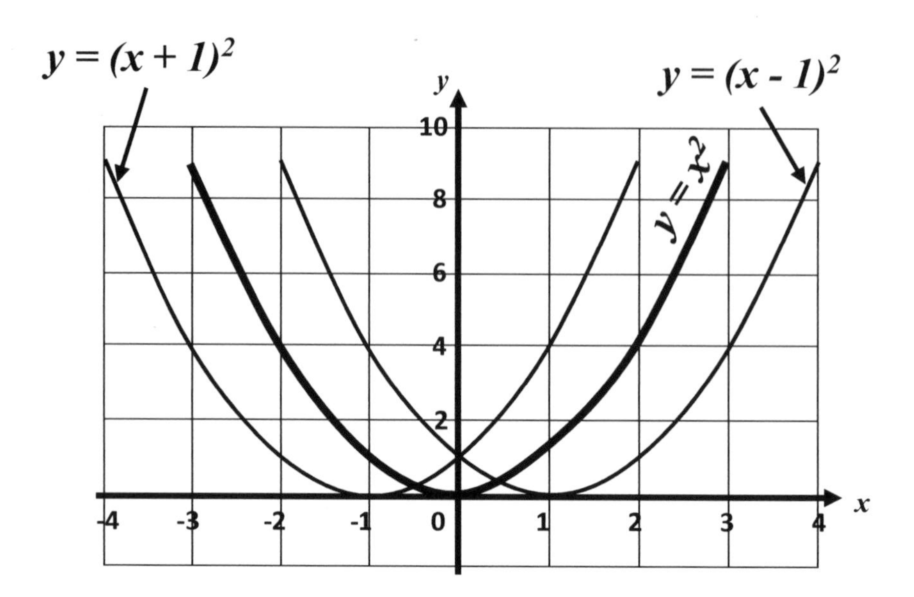

4) **f (x – c)** moves the function **c** units to the right. From the diagram above, the graph of $y = x^2$ is translated 1 unit to the **right** by the vector $\begin{pmatrix} 1 \\ 0 \end{pmatrix}$ to become $y = (x - 1)^2$.

5) If a function is multiplied by a constant, we talk of **stretches** instead of translations.

Using the function $y = x^2$ as a reference, sketch the graph of $y = 2x^2$ on the same set of axes.

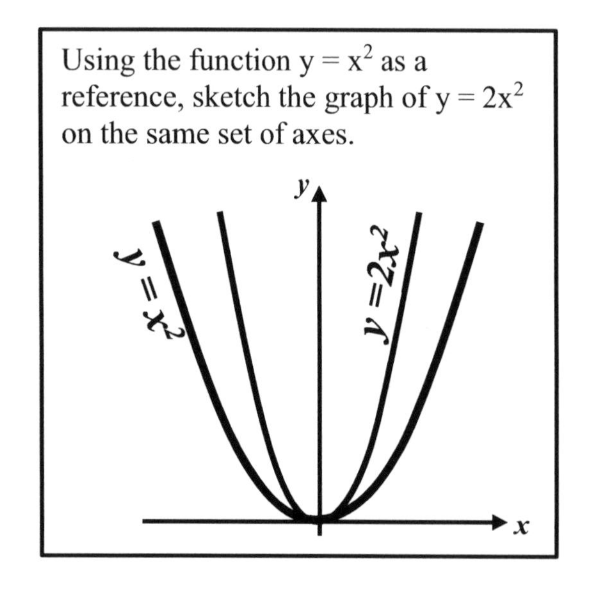

The curve of $y = x^2$ has been stretched out in the direction of the y-axis by a scale factor of 2 to give the graph of $y = 2x^2$.

The y-coordinates of $y = 2x^2$ are 2 times bigger when compared with the y- coordinates of $y = x^2$.

Note to students: If in doubt about the shape of the transformed function, draw the graph by choosing sensible x and y values.

Also, the function $y = x^2$ can be reduced horizontally using a fractional scale factor. Again, using $y = x^2$ as a reference, sketch the graph of $y = (4x)^2$.

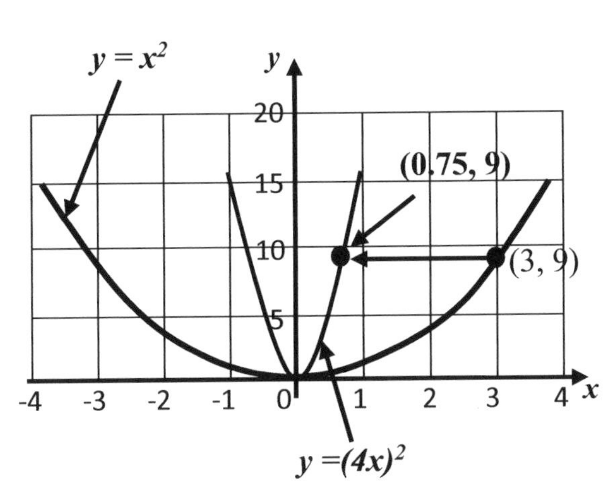

The diagram is the sketch of $y = (4x)^2$. $(4x)^2 = 4x \times 4x = 16x^2$. The y-coordinates of $y = x^2$ has been multiplied by 16 to give the corresponding y-coordinates of $y = (4x)^2$.

The implication is that the x-coordinates of $y = x^2$ is a quarter $(\frac{1}{4})$ compared to that of $y = (4x)^2$.

It therefore moves the graph of $y = x^2$ inwards (towards the y-axis) to produce the graph of $y = (4x)^2$.

EXERCISE 32K

1a) Draw the graph of $y = x^2$, x values From -3 to 3.

b) On the same graph, draw the graph of
i) $y = x^2 + 1$ ii) $y = x^2 - 3$ iii) $y = (x - 2)^2$

c) Describe the transformations that moves the graph of $y = x^2$ to the graphs in question **1b**.

2a) Sketch the graph of $y = x^2$.
b) Sketch the graph of $y = (x + 0.5)^2$.
c) Describe the transformation that takes the graph of **a** to the graph of **b**.

3) Sketch the graphs of the functions
a) $y = x^2$ and b) $y = 3x^2$ on the same graph.
c) Explain in full how the graph of $y = 3x^2$ is obtained from that of $y = x^2$.

4) Explain the transformation that takes the graph of $y = x^2$ to a) $y = 6x^2$ b) $y = (2x)^2$.

5) When translated by the vectors
a) $\binom{0}{2}$ b) $\binom{0}{-5}$ c) $\binom{1}{0}$,

the graph of $y = x^2$ moves to a different position. Sketch the graphs to illustrate the movements.

6) Explain fully the transformations that will take the function $y = f(x)$ to
a) $y = f(x) - 3$ b) $y = 3f(x) + 2$

7) The graph of $y = \sin x$ is shown below.

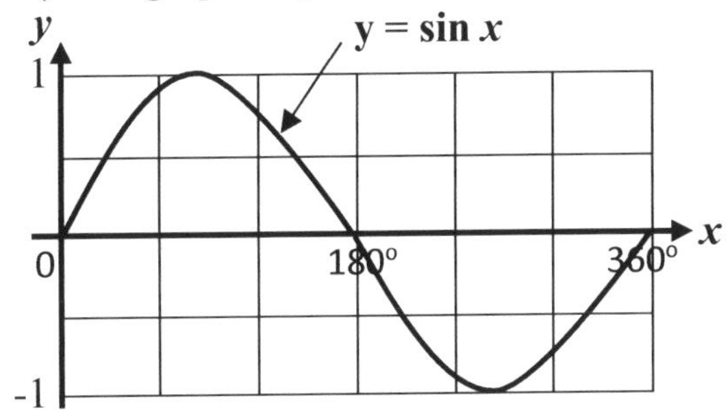

a) Sketch the graphs of
i) $y = \sin x + 1$ ii) $y = \sin x - 2$
iii) $2 \sin x + 2$
b) What is the period of $y = \sin x + 1$?

ANSWERS

ANSWERS

EXERCISE 1A

1a) 1, 2, 3, 4
b) 2, 4, 6, 8
c) 5, 10, 15, 20
d) 7, 14, 21, 28
e) 9, 18, 27, 36
f) 10, 20, 30, 40
g) 13, 26, 39, 52
h) 17, 34, 51, 68
i) 20, 40, 60, 80

2a) 2, 12, 16, 22, 30, 36, 54, 70
b) 12, 30, 36, 54
c) 12, 16, 36
d) 5, 30, 70
e) 7, 49, 70

3) Any number that fits the requirements, e.g. 9, 15, ...
4) None
5) Examples: 10, 20...
6) 49 and so on
7) 8 and 20
8a) 33 and 55
b) 27 and 63
c) 16
9) YES, because 6 x 3 = 18
10) NO, Sanusi is incorrect. The first multiple of a number is the number itself. Therefore, the first five multiples of 4 are: 4, 8, 12, 16, and 20

11a) 7, 14, 21, 28, 35
b) 3, 6, 9, 12, 15, 18, 21, 24, 27, 30

1	2	3	4	5	6	7
8	9	10	11	12	13	14
15	16	17	18	19	20	21
22	23	24	25	26	27	28
29	30	31	32	33	34	35
36	37	38	39	40	41	42
43	44	45	46	47	48	49

c) e.g. 21
12) 12 and 30
13) 18, 36, 63

EXERCISE 1B

1a) 4 b) 6 c) 35
d) 70 e) 24 f) 30
g) 60 h) 20
2a) 12 b) 210 c) 240
d) 56 e) 315 f) 84
g) 90 h) 300

3) 360 minutes

4a) 2 × 2 × 3 × 3 × 3
b) 2 × 2 × 3 × 3 × 5 × 5
c) 2 × 2 × 2 × 3 × 3 × 3 × 5 × 5
d) 5 × 5 × 7 × 7 × 7 × 11 × 11
f) 2 × 2 × 3 × 3 × 3 × 3 × 4 × 4 × 5 × 5 × 5

5) Jude is wrong. LCM of 3 and 6 is 6 and not (3 × 6= 18)
6) 560
7) 40
8) 4 cups

EXERCISE 1C
1)
a) 1, 3

b) 1,2,4

c) 1,2,7,14

d) 1,2,3,4,6,8,12,24

e) 1,2,3,5,6,10,15,30

f) 1,5,7,35

g) 1,2,3,4,6,9,12,18,36

h) 1,2,3,4,6,8,12,16,24,48

i) 1,2,4,7,8,14,28,56

j) 1,2,3,4,6,7,12,14,21,28,42,84

k) 1,2,4,5,10,20,25,50,100

l) 1,2,3,4,5,6,8,10,12,15,20,24,30,40, 60,120

2)
a) 3, 6
b) 3, 7
c) 3, 6, 9
d) 3, 9
e) 3, 5, 6
f) 3, 6, 9

3) 1 or 5
4) Could be 2, 6 or 18
5) 5, 10 and 25
6) 1 and 5
7) 4 and 18
8) Yes, Anthony is correct. 7 × 8 = 56
9) 72

EXERCISE 1D

1)
a) 3, 7, 19

b) 2, 5, 13

c) 31, 41

d) 37, 59, 97

e) 11, 17

2)
a) 2

b) 2 and 3

c) 17 and 2

d) 2 and 5

3) 2 or 3
4) 2 or 3

5a and b

1	2	3	4	5	6	7
8	9	10	11	12	13	14
15	16	17	18	19	20	21
22	23	24	25	26	27	28
29	30	31	32	33	34	35
36	37	38	39	40	41	42
43	44	45	46	47	48	49

5c) 71
5d) Yes, 71 is a prime number because it has only two factors, 1 and 71.

EXERCISE 1E

1a) $2 \times 3 \times 5$
b) 5×11
c) $2 \times 5 \times 7$
d) $2 \times 2 \times 2 \times 2 \times 3 \times 3$
$= 2^4 \times 3^2$
e) $2 \times 2 \times 2 \times 2 \times 3 \times 3 \times 5 = 2^4 \times 3^2 \times 5$
f) $2 \times 2 \times 5 \times 47$
$= 2^2 \times 5 \times 47$
2a) 2, 5 and 7
b) $2 \times 2 \times 5 \times 7$
$= 2^2 \times 5 \times 7$
3a) $2 \times 3 \times 5^2$
b) $2^2 \times 5^3$
c) $2 \times 3 \times 5 \times 17$
4a) 2×3
b) 3×7
c) 11×13
5a) 3, 3, 5
b) 3, 7, 2, 5
6a) $2^3 \times 3^3$
b) $2^2 \times 5 \times 7^4$
c) $3^4 \times 5^2$
7) NO, he is incorrect. For example;

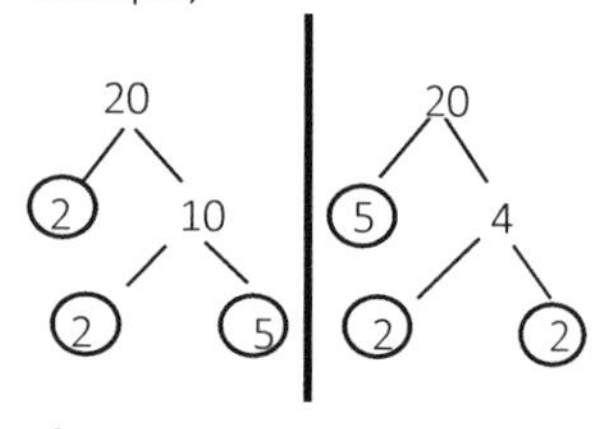

$2^2 \times 5$ $2^2 \times 5$

8a) $2^2 \times 3 \times 5^2 \times 11$
b) $2^2 \times 5 \times 7 \times 11$

EXERCISE 1F

1a)
6: 1 2 3 6
10: 1 2 5 10

HCF = 2

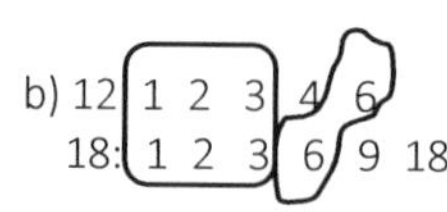

HCF = 6

c) 5: 1 5
25: 1 5 25
HCF = 5

d) 27: 1 3 9 27
49: 1 7 49
HCF = 1

e) 30: 1,2,3,5,6,10,15,30
76: 1,2,4,19,38,76
HCF = 2

f) 36: 1,2,3,4,6,9,12,18,36
48: 1,2,3,4,6,8,12,16,24,48
Common factors are 1,2,3,4,6 and12
HCF = 12

2a) 20 b) 30 c) 140

3a $3 \times 5 \times 5 \times 5 \times 7 \times 7$
$= 3 \times 5^3 \times 7^2$
3b $2 \times 2 \times 3 \times 3 \times 3$
$= 2^2 \times 3^3$
3c $5 \times 7 \times 13$
4a) 6 b) 5
5a) 8 b) 9 c) 30 d) 5
6a) 1,2,4 b) 1,2,4
c) 1,3,9 d) 1,2,4
7) 35 cm by 35 cm
8) No, Kenechukwu is incorrect. The HCF of 25 and 50 is 25.

EXERCISE 1G

1a) YES, last digit is even
b) YES, last digit is even
c) NO, last digit (7) is not 0 or even
d) NO, last digit (9) is not 0 or even
e) YES, last digit is 0
f) NO, last digit (5) is not 0 or even

2a) YES, sum of digits is divisible by 3. $1 + 2 = 3$ and $3 \div 3 = 1$
2b) YES, same reason as above
2c) NO, $3 + 4 = 7$ and 7 is not divisible by 3
2d) NO, $4 + 6 = 10$ and 10 is not divisible by 3
2e) YES, $1 + 8 + 0 = 9$ and 9 is divisible by 3
2f) YES, $6 + 5 + 4 = 15$ and 15 is divisible by 3

3) YES, last digit is 5.
$5 \times 2 = 10$. $28794 - 10 = 28784$ and $28784 \div 7 = 4112$

4a) Yes, b) Yes, c) No
d) No e) Yes, f) Yes
5) NO, the last digit is not 0 or 5
6) No, the last three digits (451) is not divisible by 8.
7) Yes, last digit is 0 and even
8a) Yes b) No

EXERCISE 1H

1a) 1 or 64
b) 1 or 125
c) 13
d) 3
2) 55
3a) 81 b) 2.25 c) 169
d) 121
4a) 15 b) 17 c) 21
d) 2.5
5) NO..because two identical numbers cannot multiply to give a negative number.
6) 109 7) 85
8a) 125 b) -64 c) 1/8
d) 8/125
9a) 4 b) 8 c) 343
d) 1.728
10a) 121 b) 7 c) 25 d) 1022
11a) 64 b) 22 c) 98 d) 65
e) 2 f) 1.6 g) 0.44 h) 7
i) 14.96 j) 5
k) a ... 2, b)... -14 c).. - 30

EXERCISE 1I

1a) 5^3 b) 6^5 c) 13^7
d) $4^3 \times 7^2$
2a) 1 b) 8 c) 289 d) 1000
e) 1 f) 125000 g) 484
h) 27000
3a) 40 b) 40 c) 255 d) 16 e) 77 f) 16 g) -63 h) 81
4a) 2^8 b) 4^7 c) 10^9 d) 17^{11}
e) 3^1 f) 9^9 g) e^8 h) y^16
i) w^5 j) $6d^{10}$ k) $105h^8$ l) y^4
5a) 4^3 b) 3^7 c) n^2 d) c^6
e) y^6 f) $3w^5$ g) 5^4 h) $3w^8$
i) 3d j) 5^{-6}
6a) ¼ b) 1/3 c) 1/25
d) 1/2197
7a) 5^{-2} b) 5^{-1} c) 10^2 d) 5^5
e) n f) 8^5 g) 11^6 h) 15^{-12}
8a) false b) true c) true
d) false e) true f) false
9a) $12x^6$ b) $5a^4$ c) $16x^2y^2$
d) $25x^2$ e) $8p^2y$

EXERCISE 1 J

1) 2×10^2 2) 5×10^2
3) 4×10^3 4) 1.2×10^5
5) 6.5×10^2 6) 2.345×10^3
7) 3.037×10^3 8) 1×10^4
9) 3.4×10^5 10) 3.9×10^4
11) 4×10^1 12) 5.4×10^1
13) 9.89×10^2
14) 1.345×10^3 15) 3×10^3
16) 6×10^4 17) 1×10^9
18) 7.893×10^6
19) 9.05×10^3
20) 9.999×10^3 21) 2.1×10^8
22) 1.68×10^{-24}
23) 2×10^5 m/s 24) b
25) 4×10^{12}

EXERCISE 1K

1) 2×10^{-2} 2) 3×10^{-3}
3) 5×10^{-5} 4) 1.23×10^{-4}
5) 8.09×10^{-2} 6) 7×10^{-6}
7) 6.4×10^{-4} 8) 4.44×10^{-5}
9) 1.2×10^{-2} 10) 4.5×10^{-2}
11) 3.23×10^{-3} 12) 9×10^{-3}
13) 8.4×10^{-5}
14) 6.128×10^{-2}
15a) 0.000018 b) 0.0067
c) 0.000205 d) 0.073
e) 0.89 f) 0.00502

EXERCISE 1L

1a) 2×10^3 b) 2.8×10^3
c) 3.5×10^2 d) 6×10^1
e) 2.5×10^4 f) 1.6×10^3
g) 2.86×10^8 h) 2.25×10^{10}
i) 8×10^3 j) 2.76×10^{-11}
2) 6×10^{17} b) 1.5×10^{-5}
c) 2.00003×10^{11}
d) 1.99997×10^{11}
3a) 8.6×10^4, 2.3×10^5, 1.5×10^8
3b) 2.5×10^{-6}, 3.7×10^{-3}, 6.5×10^{-3}, 6.9×10^{-3}
3c) 6.1×10^{-9}, 7.1×10^{-9}, 3.4×10^{-6}, 4.5×10^{-6}
4a) 2×10^3 b) 5×10^2
c) 4×10^{-2} d) 2×10^{-9}
e) 3×10^1 f) 2×10^{-12}
g) 4.035×10^{-4}
h) 8.778×10^{-3}
5a) 2.78×10^6 km
5b) 2.43×10^{13} km^2
6) 6.05×10^{-8} m^2
7a) 5.9736×10^{24} kg
7b) 1.1036×10^{24} kg
7c) 460.2 million km^2
7d) 1.989×10^{30} kg
8a) 60000000000 cm^3
8b) 6×10^{10} cm^3
8c) 1.2×10^7 cm^2

CHAPTER 1 REVIEW SECTION

1a) 3,6,9,12,15
b) 4,8,12,16,20
c) 14,28,42,56,70
2a) 1, 13
b) 1,2,3,4,6,9,12,18,36
c) 1,2,4,8,13,26,52,104
3a) 7 e) None
b) 7,9
c) 6, 72
d) 2, 7, 23
4) $2 \times 5 \times 5 \times 5$
$= 2 \times 5^3$
5a) 9 b) 100
6a) 14 b) 350 c) 5460
7a) LCM = 132
HCF = 66
b) LCM = 1680
HCF = 140
8a) $2 \times 3^3 \times 5$ b) 7
9) Azubuike is wrong. LCM of 2 and 4 is **4**
10) 9
11) 165 seconds
12a) $2 \times 3 \times 3 \times 3$
b) 2×3^3
c) 18
13) 60
14) Henry is not correct. 1 is not a prime number since it has only one factor. Also, 2 is the only even prime number with factors 1 and 2
15) 37, 41, 43, 47, 53
16a)

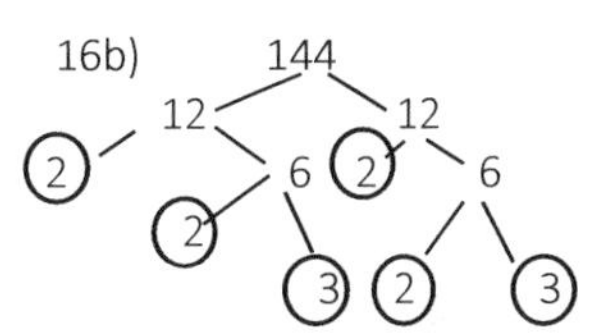

16b) 144
12 12
2 6 2 6
2 3 2 3

17) Yes.
1+8+ 9 + 3 + 5 + 7 = 33
33 ÷ 3 = 11 (Whole number)

18) $2 \times 3 \times 5 \times 5 \times 5$ which implies 2, 3 and 5
19a) 47 b) 54
c) 16 or 32
d) 12 or 16
e) 16
20)

x	1	4	9	16	25	144
x^2	1	16	81	256	625	20736
$\sqrt{x}$	1	2	3	4	5	12
$-\sqrt{x}$	-1	-2	-3	-4	-5	-12

21a) $(-3)^2$ b) None, they are the same. c) 4^2 d) 2^3 e) 0.3^2
22a) 7^{10} b) 3^8 c) 12^{-4} d) 4^{-7} e) a^{-16}
23a) 3.45×10^2
b) 2.8×10^4
c) 8.98×10^9 d) 1.2×10^{-3}
e) 9.8×10^{-6} f) 4.05×10^{-1}
g) 8.732×10^3 h) 6×10^9
i) 9.15×10^{-5}
24a) 6000 b) 0.42
c) 0.00007 d) 0.0000066
e) 0.00094 f) 0.014
25a) 1.52×10^9
b) 1.505×10^{-5} c) 2×10^{-2}
d) 5×10^{-4} e) 9.6×10^8
f) 5.67×10^4

EXERCISE 2A

1) ¼ 2) 2/4 3) 7/12
4) 4/6 5) 1/5 6) 5/6
7)

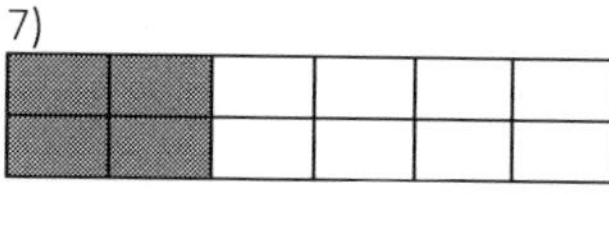

8)

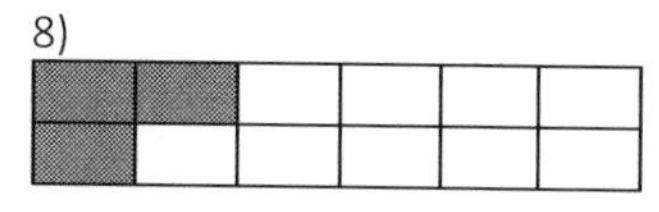

9)

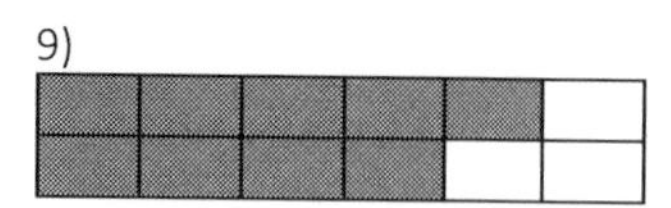

10) $1\frac{1}{2}$ 11) $2\frac{2}{6}$ or $2\frac{1}{3}$
12) $3\frac{2}{11}$ 13) $3\frac{2}{40}$ or $3\frac{1}{20}$
14) $\frac{8}{5}$ 15) $\frac{7}{3}$ 16) $\frac{32}{5}$
17) $\frac{68}{7}$ 18) 12 19) 14
20) 6 21) 14 22) 21
23) 5 24) 35 25) 5
26) 1 27) 2 and 4
28) 3, 4, 5

EXERCISE 2B

1a) $\frac{1}{2}$, b) $\frac{2}{3}$ c) $\frac{1}{3}$
d) $\frac{2}{11}$ e) $\frac{1}{4}$ f) $\frac{2}{9}$
g) $\frac{5}{24}$ h) $\frac{5}{8}$ i) $\frac{1}{3}$
j) $\frac{3}{10}$ k) $\frac{4}{15}$ l) $\frac{11}{20}$
m) $\frac{1}{5}$ n) $\frac{2}{5}$ o) 1

2a) $\frac{12}{15}$ and $\frac{10}{15}$
b) $\frac{5}{30}$ and $\frac{24}{30}$
c) $\frac{6}{8}$ and $\frac{4}{8}$
3a) $1\frac{3}{5}$ b) $4\frac{1}{7}$
c) $18\frac{1}{3}$ d) $1\frac{5}{6}$
e) $1\frac{4}{5}$ f) $4\frac{1}{4}$
g) $10\frac{12}{13}$ h) $13\frac{5}{15}$

4a) $\frac{4}{9}$ b) $\frac{3}{7}$ c) $\frac{5}{6}$
d) 0 e) $\frac{7}{12}$ f) $\frac{14}{45}$
g) $\frac{13}{15}$ h) 0 i) $\frac{26}{45}$
j) $\frac{34}{40} = \frac{17}{20}$ k) $\frac{9}{14}$
l) $\frac{47}{70}$ m) $1\frac{1}{12}$
n) $1\frac{1}{10}$ o) $1\frac{3}{20}$
5a) $\frac{2}{3}$ b) $\frac{8}{9}$ c) $\frac{1}{3}$
d) $\frac{2}{17}$ e) $\frac{7}{13}$ f) $\frac{4}{19}$

6a) $\frac{5}{6}, \frac{2}{3}, \frac{1}{2}$
b) $\frac{7}{8}, \frac{1}{2}, \frac{1}{4}$
c) $3\frac{2}{3}, 2\frac{2}{3}, 2\frac{1}{3}$

7a) ¼ b) ¼ c) ¼
d) ¼
8) $\frac{7}{15}$
9a) $1\frac{1}{19}$ m b) $\frac{12}{19}$ m
10) $\frac{5}{45}$. In its lowest form, $\frac{5}{45} = \frac{1}{9}$ but the rest are $\frac{4}{15}$ to their lowest terms.
11a) 3 b) $2\frac{1}{2}$ c) 1
d) 1 e) $2\frac{1}{2}$

EXERCISE 2C

1) $3\frac{1}{3}$ 2) $4\frac{1}{15}$ 3) 8
4) 3 5) $6\frac{1}{2}$ 6) 5
7) $1\frac{37}{105}$ 8) $3\frac{13}{21}$ 9) 0
10) $4\frac{13}{15}$ 11) $5\frac{5}{12}$

EXERCISE 2D

1a) $\frac{2}{15}$ b) $\frac{4}{35}$ c) $\frac{18}{28} = \frac{9}{14}$
d) $\frac{10}{54} = \frac{5}{27}$ e) $\frac{9}{25}$ f) $\frac{7}{32}$
g) $\frac{40}{63}$ h) $\frac{3}{96} = \frac{1}{32}$ i) $\frac{12}{63}$
j) $\frac{12}{105}$ k) $\frac{24}{40} = \frac{3}{5}$
l) $\frac{4}{400} = \frac{1}{100}$

2) 12 m 3) $\frac{12}{55}$ m² 4) $\frac{1}{6}$
5) P: $\frac{3}{21}$ m² Q: $\frac{15}{63}$ m²
R: $\frac{5}{63}$ m² S: $\frac{25}{189}$ m²

6a) $\frac{1}{27}$ b) $\frac{6}{60} = \frac{1}{10}$
c) $\frac{36}{143}$
7a) Nnaemeka is wrong. Correct answer should be $\frac{5}{18}$
b) Nnaemeka added the numerators and denominators instead of multiplying them.

EXERCISE 2E

1)
a) 1 b) 4 c) 9 d) 21
e) 8 f) 10 g) 50 h) 15
i) 40 j) 55 k) 350 l) 1

2a) ₦40 b) ₦50 c) $50
d) 6 kg e) 40 f) ₦1200
g) £15.50 h) 1020 kg

3a) £40
b) Arinze is correct. $\frac{4}{5}$ of £40 = £32 and £32 is more than £30

EXERCISE 2F

1a) $\frac{7}{4}$ b) $\frac{28}{5}$ c) $\frac{103}{9}$
d) $\frac{52}{3}$
2a) $1\frac{14}{25}$ b) $3\frac{31}{35}$ c) $4\frac{14}{25}$
d) $4\frac{24}{50}$ e) $5\frac{1}{4}$ f) $17\frac{1}{7}$
g) $110\frac{1}{4}$ h) $6\frac{19}{25}$
3a) $1\frac{10}{21}$ b) $6\frac{3}{25}$ c) $17\frac{2}{3}$
d) $60\frac{3}{4}$
4) $3\frac{1}{10}$
5a) $27\frac{2}{5}$ kg b) 29 kg

EXERCISE 2G

1a) 2 b) $1\frac{1}{5}$ c) $1\frac{1}{6}$
d) $4\frac{2}{7}$ e) 14 f) $1\frac{2}{3}$
g) 21 h) $7\frac{21}{45}$ i) $11\frac{1}{9}$
j) $1\frac{97}{153}$ k) $2\frac{8}{81}$ l) $3\frac{9}{77}$
2a) 15 b) 21 c) 27
d) 45
3a) 60 b) 80 c) 110
d) 200
4) $1\frac{1}{2}$ m 5) $12\frac{3}{5}$ m
6) 9 7) $2\frac{26}{235}$

CHAPTER 2 REVIEW SECTION

1a) $\frac{2}{3}$ b) $\frac{1}{3}$ c) $\frac{1}{3}$
2) $\frac{3}{12} = \frac{1}{4}$

3) $\frac{1}{3}$

4a) $\frac{10}{40}$ b) $\frac{15}{40}$ c) $\frac{32}{40}$

d) $\frac{36}{40}$

5a) $\frac{15}{7}$ b) $\frac{13}{8}$ c) $\frac{37}{3}$

d) $\frac{229}{11}$

6a) $\frac{5}{7}$ b) $1\frac{5}{21}$ c) $1\frac{17}{42}$

d) $\frac{65}{77}$

7a) $\frac{1}{2}$ b) $\frac{1}{3}$ c) $\frac{3}{5}$

d) $\frac{6}{7}$

8a) 12 b) 60 c) 72 d) 12

9a) $2\frac{1}{3}$ b) $5\frac{3}{4}$ c) $8\frac{3}{7}$

d) $10\frac{4}{20} = 10\frac{1}{5}$

10a) $\frac{8}{17}$ b) $\frac{5}{9}$ c) $6\frac{3}{5}$

d) $3\frac{1}{3}$ e) $1\frac{5}{9}$ f) $\frac{5}{6}$

g) $5\frac{29}{54}$ h) $1\frac{44}{81}$

11a) 12 b) 27 kg c) ₦300

d) 140

12a) Q = $\frac{6}{77}$ m^2 R = $\frac{8}{77}$ m^2

S= $\frac{36}{77}$ m^2

12b) 1 m^2 12c) Square

Same lengths for the four sides.

I metre each

13a) 2 b) 3 c) $9\frac{9}{35}$

14a) $19\frac{5}{6}$ kg b) $38\frac{1}{2}$ kg

c) $5\frac{1}{2}$ kg d) ₦115 500

EXERCISE 3A

1a) 0.6 b) 0.8 c) 0.3 d) 0.1

e) 0.28 f) 0.125 g) 0.25

h) 0.625 i) 0.06 j) 0.76

k) 0.75 l) 0.8 m) 0.35

n) 0.48 o) 0.6 p) 0.5 q) 0.1

r) 0.875 s) 0.02 t) 0.8

2a) $\frac{1}{50}$ b) $\frac{3}{100}$ c) $\frac{1}{10}$ d) $\frac{3}{20}$

e) $\frac{9}{50}$ f) $\frac{1}{5}$ g) $\frac{1}{4}$ h) $\frac{3}{10}$ i) $\frac{7}{20}$

j) $\frac{23}{50}$ k) $\frac{1}{2}$ l) $\frac{11}{20}$ m) $\frac{3}{5}$ n) $\frac{7}{10}$

o) $\frac{3}{4}$ p) $\frac{79}{100}$ q) $\frac{4}{5}$ r) $\frac{81}{100}$ s) $\frac{9}{10}$

t) $\frac{49}{50}$

EXERCISE 3B

1a) 25% b) 20% c) 40%

d) 30% e) 80% f) 15%

g) 60% h) $37\frac{1}{2}$% i) 35%

j) 36% k) 32% l) 60%

m) 24% n) 64% o) 65%

p) 10% q) 50% r) 50%

s) 25% t) 5%

2a)

i) $\frac{7}{10}$ ii) 70% iii) $\frac{3}{10}$ iv) 30%

2b) i) $\frac{3}{25}$ ii) 12% iii) $\frac{22}{25}$

iv) 88%

2c)

i) $\frac{1}{2}$ ii) 50% iii) $\frac{1}{2}$ iv) 50%

EXERCISE 3C

1) $\frac{5}{7}$ 2) $\frac{4}{20} = \frac{1}{5}$ 3) $\frac{5}{200} = \frac{1}{40}$

4) $\frac{30}{300} = \frac{1}{10}$

5a) $\frac{3}{13}$ b) $\frac{1}{4}$ c) $\frac{1}{30}$ d) $\frac{17}{50}$

e) $\frac{5}{48}$ f) $\frac{2}{125}$ g) $\frac{1}{150}$

EXERCISE 3D

1a) 40% b) 25% c) 20%

d) $33\frac{1}{3}$ %

2a) 10% b) 50% c) 25%

d) 100%

3) $1\frac{2}{3}$% 4) 15% 5) 30%

6a) 62.5% b) 37.5%

7a) 20% b) 50% c) No, Okonkwo is wrong. Uduak scored 70%

CHAPTER 3 REVIEW SECTION

1) 27%

2a)

2b)

2c)

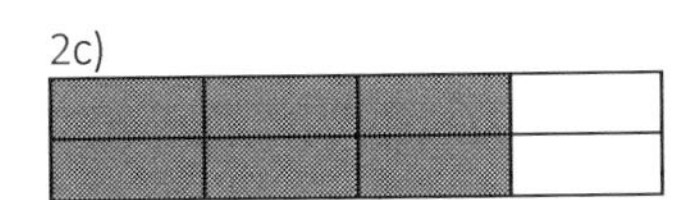

3) $\frac{1}{50}$

4a) 28.8% b) $\frac{1}{4} \times 100 = 25\%$

c) 50%

5a) 50% b) 60% c) 70%

d) 15%

6a) Maths – 68%

Economics – 5%

English - 60% Physics – 35%

6b) 5%, 35%, 60%, 68%

7)

Fractions	%	Decimals
$\frac{2}{5}$	40%	0.4
$\frac{7}{20}$	35%	0.35
$\frac{3}{5}$	60%	0.6
$\frac{6}{25}$	24%	0.24

8) 0.65, $\frac{3}{5}$, 51%, $\frac{1}{2}$

9) $\frac{8}{25} \times 100 = 32\%$

$\frac{8}{25}$ is bigger

10a) 75% b) 25%

EXERCISE 4A

1a) 3 b) 0.4 c) 6 d) 0.8 e) 1.6 f) 4 g) ₦8 h) ₦30
i) 6 j) 0.8 k) 12 l) 1.6 m) 16 n) 18 o) 16 p) 50
2a) 1 b) 7 c) 60 d) 2 e) 1.9 f) 0.24 g) 1220
h) 28 kg

EXERCISE 4B

1a) 1.3 b) 1.05 c) 16 d) 24 e) 6.93 f) 7.7 g) 5.2
h) ₦21 i) £33 j) 675 k) 1.2 l) 28 m) 24.4 kg
n) 62.3 o) ₦460 p) 0.36
2a) £1.89 b) 1.792 c) 5.6 d) 80.5 e) 0.75 f) 9
g) 2.38 kg h) ₦325

EXERCISE 4C

1a) 22 b) 88 c) ₦220 d) ₦374 e) 924 f) 3300
2a) 23 b) 92 c) ₦230 d) ₦391 e) 966 f) 3450
3a) 32 kg b) 73.6 litres c) £98.40 d) $146.42
e) 3200 g f) 196 cm 4) ₦2500 5) £3150
6) b 50 to 70
7) For £15, sale price = £9.75
For £32, sale price = £20.80
For £45, sale price = £29.25
8) 1.94 kg 9) 33.3% 10) ₦5610 11) ₦5220
12) 82.8 kg 13) £772.20

EXERCISE 4D

1a) £21 b) £320 c) ₦2400 d) ₦2500 e) ₦900
F) $412.50 g) $490 h) ₦9000
2a) ₦125 b) ₦5125
3a) £1750 b) £5250
4) Bank A: £1500
5a) ₦840 000 b) ₦4 340 000 c) ₦120 555.56

EXERCISE 4E

1a) £16 b) £96 2) £90 3a) £420 b) £14 400
c) £1080 4a) £70 b) £420

PROBLEM SOLVING

1a) £193 b) i) £393 ii) £2 2a) £3400 b) £6000
c) £9400 3) £40 4a) £2100 b) £375 c) £600
d) £2475

EXERCISE 4F

1a) £2970.52 b) £3059.64 c) £3443.65
2) £779.14 3a) £7334.57 b) £20834.57
4) £5324.46 5a) 4 years
5b) 215 456 800.80 6a) Niamh
6b) Niamh will pay £141.55 more.

EXERCISE 4G

1) £710.78 2) 246.57 cm^3 3) £500
4) A £64 B £80
5a) £483.50 b) £96.70

CHAPTER 4 REVIEW SECTION

1a) 13 b) 0.34 c) 180 d) ₦1260
2) ₦3600 3) 67.5kg 4a) 25% 4b) 4%
4c) 4.2% 5) 10% 6) ₦1068.48
7a) ₦24 500 7b) ₦350 000 7c) ₦31 500
8a) £26 b) £156 9) £114
10a) £360 b) £18 000 c) £1320
11a) £7334.57 b) £20834.57
c) £29386.67

EXERCISE 5A

1a) 22 b) 11 c) 47 d) 8 e) 4 f) 5
g) 18 h) 22 i) 44 j) 9 k) 39 l) 3
m) 18 n) 22 o) 14 p) 10 q) 15
r) 11 s) -2 t) 300
2a) 5 + (6÷ 2) = 8 b) 21 ÷ (3+4) = 3
c) 2 × (7-4) = 6 d) 8 – (4÷4) = 7
e) 9 + (3+3) × 2 = 21
f) (40-8) × 7 = 224
3a) 12w + 10 b) 3 + 8y c) 47c
d) 5 + 3x e) 5n – 1 f) 7w – 2
g) 18 h) 12m + 10 i) 47p – 3
j) 4w + 5 k) 51x + 2 l) 10x – 2
m) 27p – 9 n) 24 – 2n o) 35k – 21
p) 10 q) 44b r) 24g^2 – 3
s) -2x t) 28v

EXERCISE 5B

1a) 20 b) 30 c) 13 d) 7 e) 10 f) 5
g) 100 h) 60.25 i) ¼ j) 18 k) 130
l) 5
2a) 21 b) 33 c) 37 d) 49 e) 0 f) 70
g) 42 h) -6 i) 21 j) 9 k) 49 l) 20
3a) i) 22 cm ii) 10 m 3b) i) 18 cm
ii) 9m c) i) 36 cm ii) 19.5 m
d) i) 11 cm ii) 6.6 m
4a) 18 b) 116 c) 10 d) 140 e) 3
f) -24 g) 2 h) 10
5a) 25 b) 3 c) 31 d) 36 e) 2 f) -10
6a) 10 b) 270 c) 63 d) 169
7a) 10 b) 4.25 c) 1/8 d) -5.75

EXERCISE 5C

1a) -9 b) -6 c) -2 d) 13 e) -10 f) 24 g) -103 h) -9 i) -27
j) 0 k) 1 l) -58
2a) -2 b) -11 c) 28 d) 1 e) -1 f) 8 g) 95 h) -21 i) -48 j) 4
k) 34 l) 19
3a) -18 b) -18 c) -30 d) 6 e) -6 f) -24 g) -13 h) -24 i) 9
j) 2 k) 1 l) 34

EXERCISE 5D

1a) -5 b) 0 2) 2 3a) 50 b) -30 4) 19 5) 4 6a) 201 b) 170
7a) 81.225 b) 6
8) $4y^3 + 20 = 128$, $10y - 5 = 25$, $3y^3 = 81$, $5 - 2y = -1$, $2(y - 1) = 4$
$6y - 3y = 9$
9a) 5/6 b) 5 10a) 1 b) -7.6

EXERCISE 5E

1a) c = 2 b) c = 13 c) c = 15 d) x = 1 e) x = 11 f) x = 91
g) x = 18 h) w = 10 i) w = 22 j) w = -9 k) u = -13 l) x = - 16
m) u = -7 n) r = 53 o) g = -5 p) w = 23 q) y = -16 r) r = 64
s) y = 30 t) n = 22 u) e = -7 v) x = 24 w) c = 10 x) x = 6.8
y) x = 7.8 z) u = 344

EXERCISE 5F

1a) x = 2 b) x = 3 c) x = 6 d) x = 7 e) x = 11 f) w = 3 g) w = 9
h) c = -2 i) c = -2 j) w = -4 k) x = 8 l) x = -10 m) x = 7 n) n = 3
o) n = 42 p) x = 30 q) w = -18 r) n = 3 s) x = 7 t) n = -9
u) n = -2 v) m = 3 w) w = 4.5 x) z = 1/3 y) n = 10 z) y = -94

EXERCISE 5G

1a) c = 4 b) c = 2 c) x = 4 d) w = 2 e) x = 1 f) x = -1 g) x = 3
h) w = 1 i) y = 6 j) x = 2 k) u = -1 l) x = -1 m) y = 2 n) y = 10
o) n = 6 p) v = 3 q) x = 2 r) x = 3 s) c = 5 t) w = 2 u) x = 8
v) n = 7 w) w = 8 x) x = 3 y) c = 4 z) x = -1

EXERCISE 5H
1a) n = 15 b) n = 40 c) n = 10 d) n = 5 e) n = -9 f) n = 300
g) n = -63 h) n = 5 i) n = 14 j) n = 120 k) n = 3 l) n = 78
m) n = 32 n) n = 4 o) n = 2 p) n = 4/5 q) n = -4 r) n = -18
s) n = 5 t) n = 4 u) n = 5 v) n = 1 w) n = 1 x) n = 4 y) n = 28
z) n = -8.
EXTENSION QUESTIONS
1) 9 cm 2) 36 cm^2 3a) d = -15 b) w = -10 c) n = 1
EXERCISE 5I
1) 3 2) 1 3) 5 4) 12 5) 3 6) 1 7) -1/8 8) $8\frac{1}{3}$ 9) 1 10) 5
11) $1\frac{21}{44}$ 12) 2 13) 2 14) 1 15) 10 16) -4 17) 3 18) -1
19) 3/7 20) $14\frac{8}{21}$
EXERCISE 5J
1) 4 2) 3 3) $-1\frac{1}{6}$ 4) 2 5) 3 6) -2 7) $-1\frac{7}{9}$ 8) $3\frac{2}{7}$ 9) $-3\frac{1}{2}$ 10) 3
11) $2\frac{3}{4}$ 12) 7 13) 3/8 14) ½ 15) -5 16) 6 17) 2 18) 1/3
19) $10\frac{2}{3}$ 20) $2\frac{3}{4}$ 21)0 22) -1/2 23) -4/7 24) $-1\frac{2}{5}$

EXERCISE 5K

1) 2/3 2) 4 3) 3 4) 2 5) ½ 6) -2 7) 9
8) -1 9) -2/5 10) 18 11) 2 12) 16
13) 10 14) -2 15) -7

EXERCISE 5L

1) 2x + 10 = 44, x = 17 2) 7(x + 3) = 70, x = 7
3) $\frac{3x + 6}{3} = 2x$, x = 2
4) $\frac{x - 6}{4} = x - 4$, $x = \frac{10}{3}$

5) 2(3x + 2) = 2x + 12, x = 2
6) 21, 22, 23
7a) t + 10 years b) t + 35 years
c) t – 5 years d) 25 year
8a) 4x + 6 = 22 b) 4
c) Length = 9 cm, width = 2 cm d) 18 cm²
9) £15 000 10) $\frac{x + 38}{2} = 10x$, x = 2
11a) 7(x + 6) – 40 = 16, 7x + 2 = 16
11b) x = 2 12) 2(x – 9) = $\frac{x}{5}$, x = 10
13a) 3w + 30 = 180°, w = 50°
13b) 4w + 40 = 360°, w = 80°
13c) 4w + 80 = 180°, w = 25°
14a) 4x + 6 = 450 b) 111, 112, 113, 114
15a) 3x + 28 b) 3x + 28 = 49 c) x = 7
15d) 10 cm, 12 cm and 27 cm

EXERCISE 5M

1) x > -1 2) x ≤ 4 3) x ≥ 0 4) x ≥ −1
5) -1 ≤ x < 5 6) x < -1 and x ≥ 2
7) x ≤ 0 and x > 3

EXERCISE 5N

1a) x < 5 b) x < 3 c) x > 1 d) x ≤ −2
e) y ≥ 1
2a)

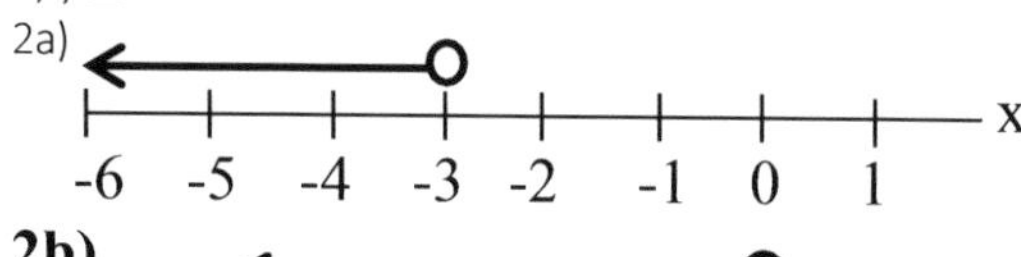

2b)

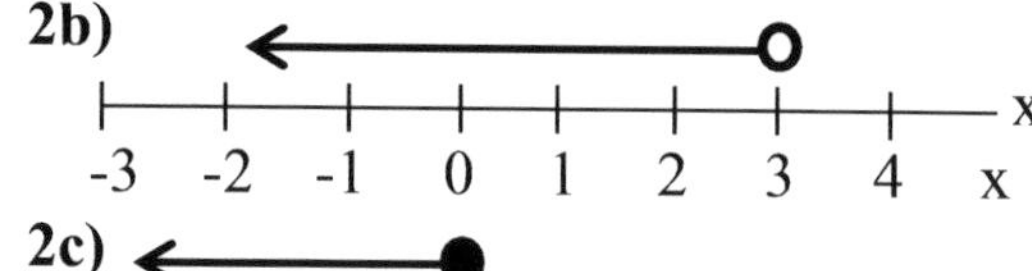

2c)

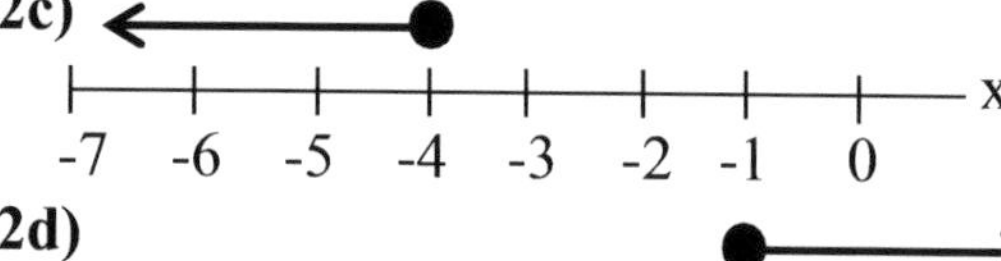

2d)

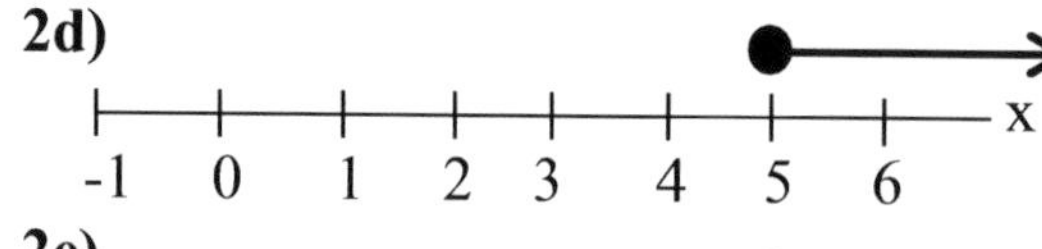

2e)

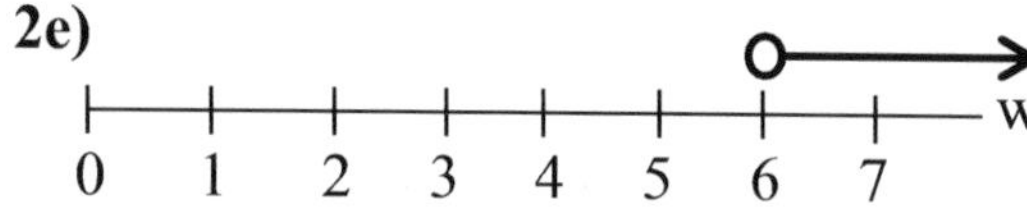

2f)

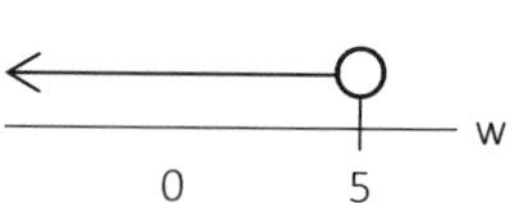

3a) True b) False c) True d) True e) True f) True
g) False h) False
4a) 2 b) -1 c) 5 d) -9 5a) 1 b) 1 c) -1 d) -4

EXERCISE 5O

1) x < 5 2) x < 5 3) x ≤ −6 4) x ≥ 12 5) x > 1
6) x > 1/3 7) x < -2 8) x ≥ −2 9) x < -13 10) x > $2\frac{1}{2}$
11) n ≤ 4 12) w > ½
13) n ≥ −2

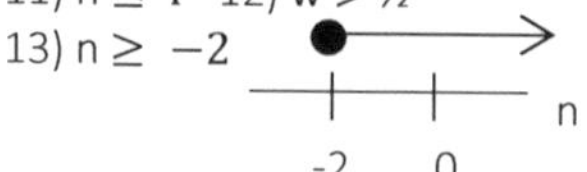

14) n ≥ −8

15) n < 3

16) n < 6

17) 1 < x < 2

18) x > 1/3

19) x > - 13

20) -13 < x ≤ 5

21) -2 ≤ x < 1

22) -4 < x < 6

23) -1, 0   24) 6, 7, 8, 9, 10   25) Not possible
26) x < 5 and x ≥ −1............ -1, 0, 1, 2, 3, 4
27) x > 2 and x < 5................ 3, 4
28) x > -3 and x < 6 ....-2, -1, 0, 1, 2, 3, 4, 5

EXERCISE 5P

1) x > 3

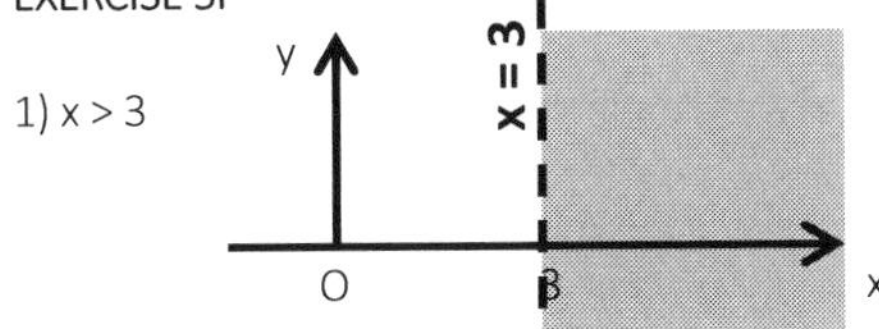

2) x ≤ 5

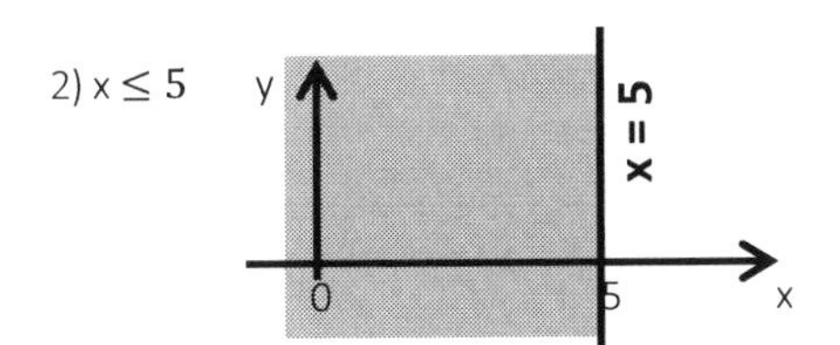

3) x > -1

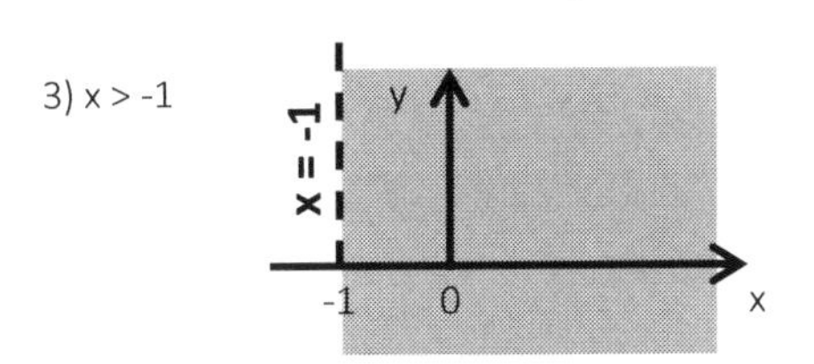

4) y > 5

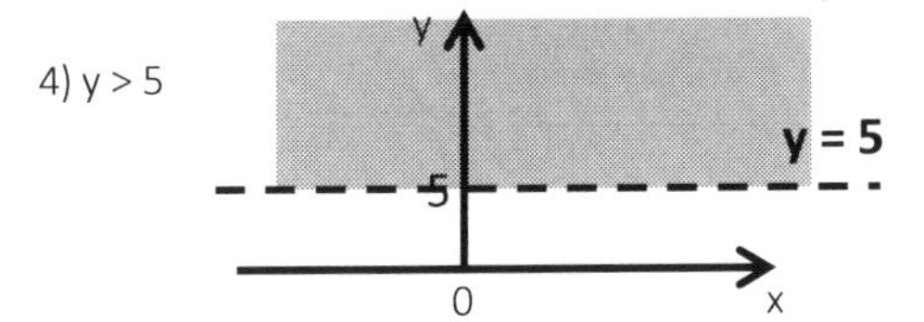

5) y > x + 2

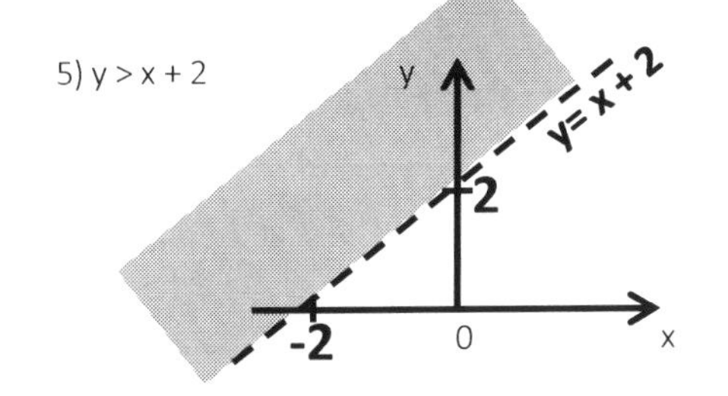

6) x + y ≤ 3

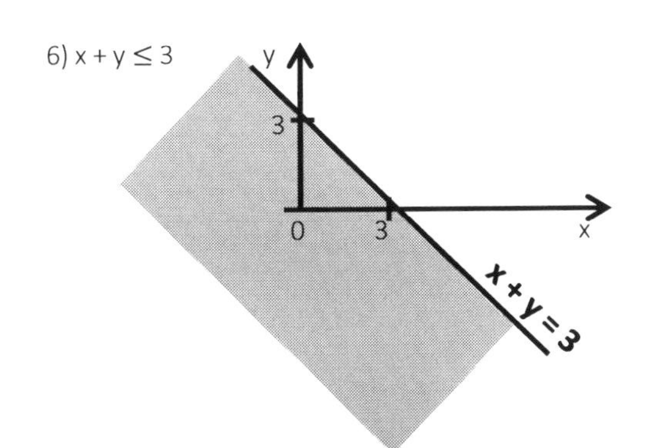

7) x + y ≥ −2

8) y > 3 − 2x

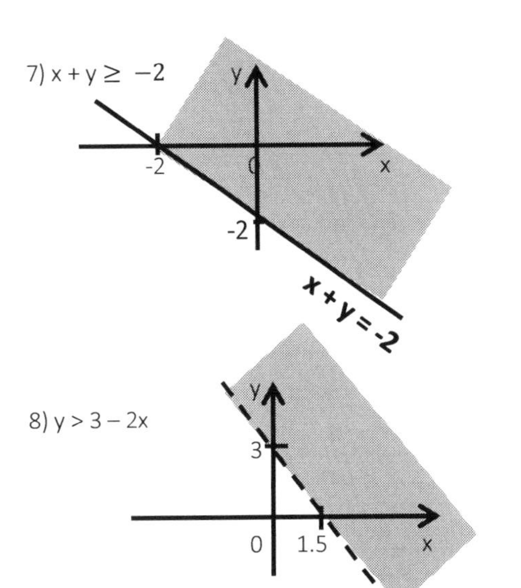

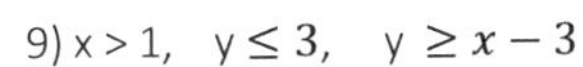
9) x > 1,   y ≤ 3,   y ≥ x − 3

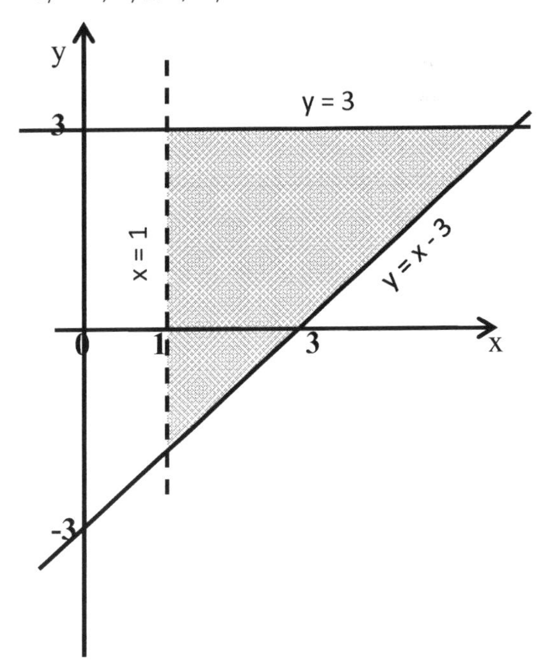

10) x ≥ 2   11) x < 5   12) y ≤ 2 and x < 3
13) x + y ≥ 3   14a) s + c ≤ 8, s ≥ 3, c ≥ 1
14b)   c

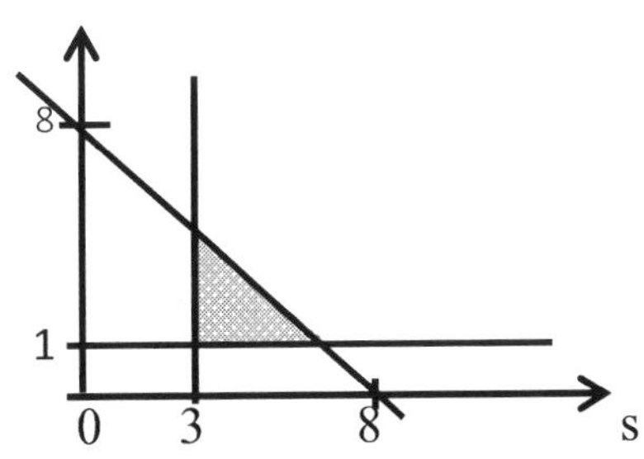

CHATPER 5 REVIEW SECTION

1a) 16 b) -3 c) 7
2a) c – d + e – f b) 13w + 6n
3a) f + 4b b) 7a – 3d + 2e c) $8w^2$ d) $3a^2b$
4a) 12 b) 44 c) 23
5a) 3 b) 7 c) 4 d) 9
6a) n = 2 b) n = 11 c) n = 27 d) x = 4 e) x = -9
f) x = -1 g) y = 14 h) x = 10
7a) n = 15 b) x = 70 c) x = 7 d) x = -45
e) x = -60 f) x = 18 g) y = 48 h) x = 3
8a) 3 b) -4 c) 3 d) -3 e) 2/3 f) -1/7 g) -2
h) 4/9
9) $10/3 = 3\frac{1}{3}$ 10a) 8x + 14 = 30 10b) x = 2
10c) Length = 10 cm, width = 5 cm 10d) 50 cm^2
11a) < b) > c) > d) > e) <

12a)

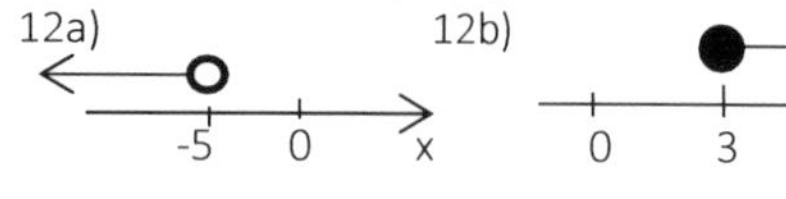

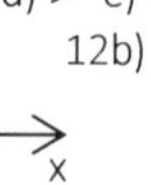

12b)

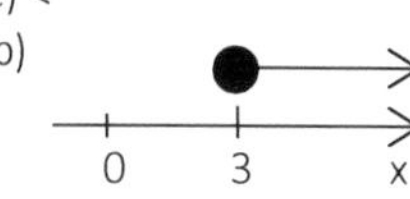

c)

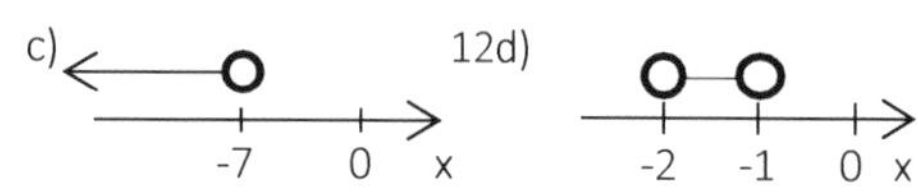

12d)

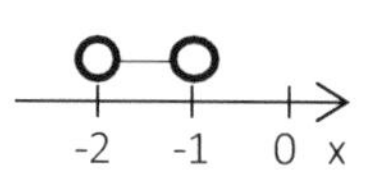

12e)

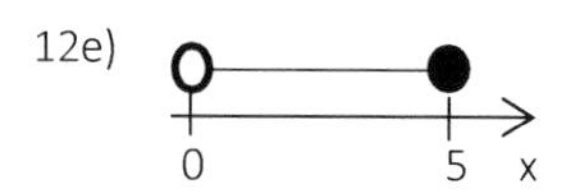

13a) $x > 1\frac{1}{6}$ b) $x \leq -10$ c) $w \geq -7$
d) a < -25 e) -8 < x < 2
14a) 4, 5, 6, 7, 8 b) -4, -3, -2, -1, 0, 1, 2
c) 3, 4, 5, 6 d) -2, -1, 0

15) $y \geq 2x - 3$

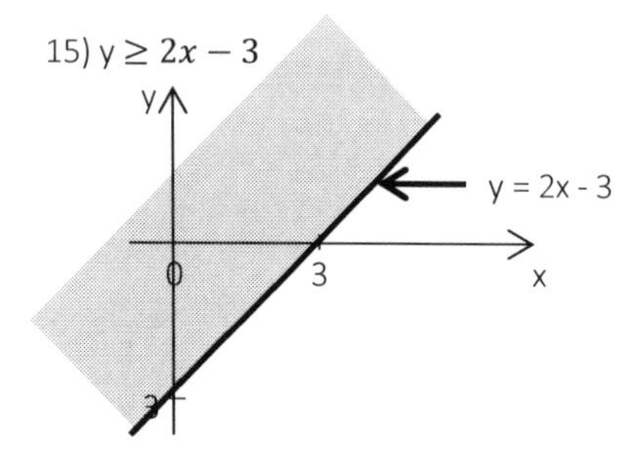

16a) $x \geq 3$ b) x + y < 3 c) y < x or x > y
d) $y \geq 2$

EXERCISE 6A

1a) 30° b) 10° c) 145° d) 44° e) 60°
f) 100° g) 50° h) 141° i) a= 27°,
b = 153°, c = 153° j) x = 30° k) 125°
l) w = 90°, y = 47° m) 215°
n) n = 90°, x = 40°

EXERCISE 6B

1a) a = 100° 2) b = 40°
3) c = 42°, d = 42°
4) e = 70°, f = 70°, g = 110°
5) h = 75°, i = 30°
6) j = 60°, k = 60°, l = 60°
7) m = 32°, n = 50°, o = 82°, p = 98°
8) q = 60°, r = 60°, s = 60°, t = 120°, u = 120°
9) v = 33°
10a) scalene.............three different angles
b) Right-angled..........have a 90° angle
c) Isosceles......two equal lengths
d) Equilateral......all sides the same
e) Isosceles......two base angles are equal
f) Equilateral.......three equal sides
11a) 25° b) 25° c) 59° d) 56° e) 84°
12) NO, Obiora is incorrect. 54 + 85 + 42 = 181° but angles in a triangle add up to 180°.
13a) 94° b) 30° c) 81° d) 10° e) 40° f) 60° g) 90°
h) 81° i) 39° j) 55°
14a) 9w = 180 b) w = 20° c) 100° d) 80
15a) i) 55° ii) 70° iii) 18°
15b) i) Isosceles triangle because two base angles are the same (55° and 55°).
ii) Scalene triangle because all three angles are different.

EXERCISE 6C

1a) Rhombus b) Trapezium c) Rectangle d) Square e) kite
f) Parallelogram g) Trapezium
2) 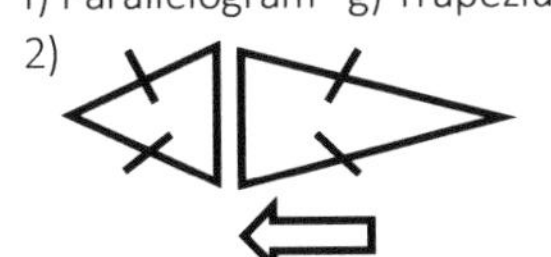 3) 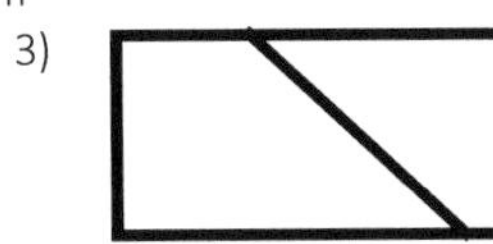

4) A square has all the corners at 90° while the corners of a rhombus are not 90°.
5) No. A quadrilateral has four sides, but shape B has three sides.
6a) Trapezium b) Rhombus c) Kite d) Square e) Rectangle
f) Parallelogram
7) No, Kunle is incorrect. A trapezium has only a pair of parallel sides. The shape drawn is a parallelogram.

EXERCISE 6D

1a) p = 141° b) w = 65°, x = 95° c) a = 93°, b = 75°, c = 92°
d) f = 90° e) d = 43° 2) 140° 3a) 3n = 156 , n = 52°
3b) 8x = 280, x = 35° 3c) 6x = 210, x = 35° 3d) 10c = 360, c = 36°

EXERCISE 6E

1) NO. A polygon has straight sides, but a circle is curved.
2) Any four quadrilaterals, for example; Pentagon – 540°,
Hexagon – 720°, Heptagon – 900°, Octagon – 1080°
3) Dodecagon – 1800°
For questions 4 - 7): Students to construct circles accurately, using the radii given. They must use a pencil, ruler and a pair of compasses for the construction. Teachers/parents to check the accuracy of the circles drawn.
8) No, Okoro is incorrect. A circle with a diameter of 20 cm must have a radius of 10 cm. *2 × radius = diameter*

EXERCISE 6F

1a) a = 47°........ Alternate angles are equal
b) b = 107°...... alternate angles are equal
c) c = 38° .....Angle on a straight line where
142° lies is 38° (180 – 142). Therefore,
Angle C = 38° ....alternate angles. Other
Mathematical reasons are also encouraged.
d) d = 102°... corresponding angles are equal
e) e = 113° ....alternate angles
f = 70° ....corresponding angles
f) g = 96° ....corresponding angles
g) h = 30° ....alternate angles
i = 47° ....angles on a straight line
j = 133° ....corresponding angles
h) k = 110°
Angle next to K is 70° ....Corresponding angles.
Angle on a straight line = 180 – 70 = 110°
i) l = 40° ....angles on a straight line
n = 14° .... corresponding angles
m = 166° ... from (180 – 14) = 166°, then
corresponding angles are equal.
m = 40° ......alternate angle to l
j) p = 58° ...co-interior angles add up to 180°

CHAPTER 6 REVIEW SECTION

1a) x = 85° b) a = 109°, b = 71°, c = 71°
d = 109° c) e = 63°, f = 117°, g = 63°, h = 54°
i = 63°
2a) j = 92°, k = 110°, l = 70° b) m = 95°
c) n = 48°, o = 89°, p = 48°
3a) 3 b) CD c) AB d) Regular hexagon
e) 6 f) obtuse

EXERCISE 7A

1) Teacher's supervision
2a) 9.4 cm b) 7.1 cm c) 8.1 cm d) 5.1 cm
3a) 22 cm b) 44 cm c) 110 km d) 132 cm
e) 176 m f) 6.3 km
4a) 25 cm b) 90 cm c) 3.6 m d) 94 cm
e) 35.7 cm
5) 131.9 cm 6) 502.4 cm 7) 2.2 cm
8) 159 rev.

EXERCISE 7B

1a) 21 cm^2 b) 24 cm^2 c) 22 cm^2 d) 4 cm^2
2) A = 3cm^2, B = 6 cm^2, C = 2 cm^2, D = 3.5 cm^2
3) P = 2 cm^2, Q = 11.5 cm^2
4) Could be 4 cm and 2 cm or 8 cm and 1 cm.
5) 6 m 6) 127 cm^2 7a) 22 m^2 b) ₦37 300
8)

Area	Length of side
	1 cm
	3 m
64 cm^2	
18.49 cm^2	
	9 cm

EXERCISE 7C
1a) 15 cm^2 b) 6 cm^2 c) 24 cm^2 d) 14 cm^2
e) 17.5 cm^2 f) 18 cm^2
2) 63.5m^2 3) 6 cm
4) For example; 14 cm and 4 cm or any two lengths that will multiply to give 56.
5a) 18 cm^2 b) 12 cm^2 c) 14 cm^2
d) 240 cm^2 e) 40 cm^2
6a) 18 cm^2 b) 20 cm^2 c) 28 cm^2
d) 75 cm^2 e) 36 cm^2 f) 175 cm^2
7a) 5 m b) 5 cm c) 5 m d) 2 m
8a) 3.52 cm^2 b) 171 cm^2 c) 30 cm^2
d) 153 cm^2 e) 67.5 cm^2 f) 6.3 cm^2
g) 19.22 cm^2

EXERCISE 7D

1a) 80 cm^2 b) 45 cm^2 c) 19.5 cm^2 d) 45.5 cm^2 e) 21.2 cm^2
f) 17.5 cm^2 g) 48 cm^2 h) 117 cm^2
2a) 6 m b) 5 m c) 7 m d) 20 m

EXERCISE 7E

1a) 28 cm^2 b) 36 cm^2 c) 126.5 cm^2 d) 148 cm^2 e) 24 cm^2
f) 58 cm^2 g) 90 cm^2 2) 66 cm^2 3) $4 \times 12 = 48$ m^2, $5 \times 6 = 30$ m^2...........48 + 30 = 78m^2 is the area.
4a) 85 m^2 b) 88 m^2 c) 125 m^2

EXERCISE 7F

1a) 129 m^2 b) 41.5 m^2 c) 51.5 m^2 d) 81 m^2 e) 88 m^2
f) 40.5 m^2

EXERCISE 7G

1a) 12.6 cm^2 b) 78.6 cm^2 c) 95.1 m^2 d) 314.3 cm^2
e) 201.1 km^2 f) 113.1 mm^2
2a) 140 m b) 15386 m^2
3)

Diameter (m)	Radius (m)	Area m^2
	7	154
	21	1386
42		1386
28		616

4a) 14.1 m^2 b) 50.2 m^2 c) 100.5 m^2 d) 379.9 m^2
5a) 218.5 m^2 b) 74.1 m^2

EXERCISE 7H

1a) 13.8 cm^2 b) 109.9 cm^2 2a) 19.63 m^2 2b) 58.88 m^2
2c) 16.13 m^2 3) 21.5% 4) 314 cm^2 5a) 471 cm^2
5b) 66.67% 6) 36.25 m^2 7) 176.7 m^2

CHAPTERS 7 REVIEW SECTIONS

1a) 7 b) 25 2) times 10
3)

Input	9	30	6	15	45	39
Output	15	50	10	25	75	65

4a)

8	3	10
9	7	5
4	11	6

4b)

9	4	11
10	8	6
5	12	7

5) 11.4 cm 6) 37.84 am
7a) 66 cm b) 264 cm 8) 237 cm^2
9a) 15 cm^2 b) 24 cm^2 c) 27 cm^2
d) 100 cm^2 10) 6 cm
11a) 3 cm^2 b) 52.5 cm^2
c) 92 cm^2 d) 420 cm^2
12a) 7 m b) 5 m c) 4 m d) 20 m
13a) 142 cm^2 b) 72 cm^2
14a) 7.1 cm^2 b) 314.3 cm^2
c) 173.3 cm^2 d) 44.2 cm^2 e) 285.2 cm^2
15a) 294.6 cm^2 b) 60%
16) ₦3 142 8 57.14 17) 7626 m^2

g) 2, 1, 0 h) 6, 8, 10 i) 5, 0, -5
j) 11, 14, 17 k) 8, 11, 16 l) 10, 13, 18
m) 17, 37, 57 n) 2, 8, 18 o) 2, 11, 26
p) 14, 26, 46 q) 11, 9, 7 r) 1, $\frac{1}{4}$, $\frac{1}{9}$

2a) 40, 44, 48, 52, 56 b) 7, 12, 17, 22, 27
c) 13, 11, 9, 7, 5 d) 2, -1, -4, -7, -10
e) -7, 2, 11, 20, 29
3) 4 and 10
4a) add 5 each time b) 6 c) 481
5a) 7, b) 11, c) 391
6a) 108 b) 972
7) 17, 22, 42
8a) 13, 15, 17 b) -2, -1, 0
c) 1, 4, 9 d) 6, 18, 36

EXERCISE 8A

1a) 8, 9 b) 15, 17 c) 21, 26 d) 49, 97
e) 19, 23 f) -2, -8 g) 15, 20 h) 6.6, 6.5
i) 32, 64 j) 19, 26 k) 81, 243 l) $\frac{5}{162}, \frac{6}{486}$
2a) add 1 each time
b) add 2 each time
c) add 5 each time
3) Could be
i) 1, 8, 15, 22, 29 ii) 5, 12, 19, 26, 33
4a) add 5 each time b) 4 c) 49
5a) 4, 10, 16, 22 b) 47, 62, 77, 92
c) 2, -5, -12, -19 d) 3, 9, 27, 81
e) 200, 100, 50, 25
6) 88, 5, 88, 616, 1672
7a) i) multiply difference by 2 and add to get the next term. ii) 127
7b) i) Subtract 45 ii) 90, 45
7c) i) Subtract 3 ii) 83
7d) i) Multiply by 10 ii) 3, 3000
7e) i) Difference increases by 1 each then, then add to the next number ii) 18
8) 3 086 358 025

EXERCISE 8B

1a) 2, 4, 6 b) 14, 28, 42 c) 4, 6, 8
d) 2, 5, 8 e) 5, 9, 13 f) 1, 4, 9

EXERCISE 8C

1a) 2n + 3 b) 3n + 8 c) n + 3 d) -5n + 35 e) 4n – 3 f) 3n + 6
g) 7n h) 4n – 4 i) 2n – 3 j) -3n + 20 k) 50n + 450
2)

Term number	4n	Term
1	4	7
2	8	11
3	12	15
4	16	19
70	280	283

3a) 2n + 2 b) -3n + 36 c) 2n d) 3n – 1 e) n + 2 f) n^2
4a) 15 cm
4b)

Pattern 4

Pattern 5

4c)

Pattern number	1	2	3	4	5	6
Perimeter (cm)	5	10	15	20	25	30

4d) 5n
4e) 500 cm
4f) 630
5a) 14 rods
5b) 3n - 1
5c) 20
5d)

Pattern number	1	2	3	50
Number of rods	2	5	8	149

EXERCISE 8D

1i) a, b, e ii) c, d, f
iii) c:71 and 97
d: 79 and 105
f: 101 and 123
iv
a) $2n - 1$ b) $-2n + 13$
c) $2n^2 + 4n + 1$ d) $2n^2 + 4n + 9$
e) $10n + 10$ f) $n^2 + 11n + 21$

2a) $n^2 + 1$ b) $n^2 + 5$ c) $2n^2 + 1$
d) $2n^2 - 2$ e) $3n^2 + 3$ f) $2n^2 + n + 1$
g) $2n^2 + 3n - 1$ h) $3n^2 + n - 1$
i) $2n^2 + n$ j) $6n^2$
3) 10005 4) 9, 14, 21
5a) 20 squares b) $n^2 + n$ c) 90300 squares

6a)
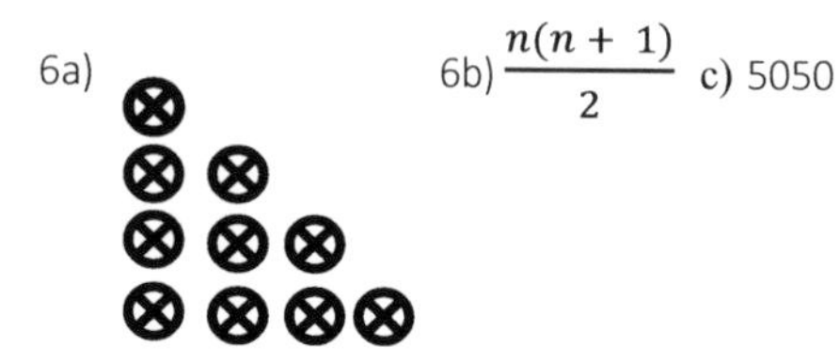
6b) $\frac{n(n+1)}{2}$ c) 5050

EXERCISE 8E
1a) 4, 7, 12 b) 0, 3, 8 c) 11, 14, 19
d) 5, 8, 13 e) 4, 13, 28 f) 5, 28, 87
g) 4, 10, 20 h) -4, 2, 12 i) 5, 12, 23
j) 3, 15, 33
2a) 9809 b) 14 492

CHAPTER 8 REVIEW SECTION

1a) 19 b) $k + 3$ c) $k - 3$
2a) $3n^2 + 3$ b) 120003 3) 40, 121, 364, 1093
4a) i) $3n + 2$ ii) 62 b) i) $-10n + 80$ ii) -120
c) i) $2n^2 - 2$ ii) 798 d) i) $n^2 + 30$ ii) 430
e) i) $\frac{n+2}{2n+3}$ ii) $\frac{22}{43}$ f) i) $n(n + 2)$ ii) 440

EXERCISE 9A

1a) 1 : 3 b) 1 : 1 c) 1 : 2
2a) 1 : 2 b) 1 : 2 c) 2 : 1 d) 2 : 1 e) 2 : 1 f) 1 : 8
g) 19 : 30 h) 1 : 3 i) 3 : 2 j) 2 : 3 k) 7 : 11 l) 5:6
m) 1:4:10 n) 4:3:1 o) 1:2:7 p) 1:3:1 q) 2:3:6
r) 6:2:3
3a) 2:3 b: 3:5 c) 1:2 d) 1:3 e) 1:4 f) 1:10
g) 1:4 h) 1:2 4) 6:1 5) 18:7
6) 56 blue and 42 red pens 7) $\frac{16}{33}$
8a) 2:1 8b) 1:5 8c) 1: 2000 d) 20:1

9) 12: 30, 30: 75, 2:5, 42:105, 18:45, 2:5

EXERCISE 9B

1a) £12 and £24 b) £8 and £28 c) £15 and £21
2a) £60 and £360 b) £168 and £252
c) £240 and £180
3a) Mark received £3600 b) £4800 c) 70%
4) £240 5a) 100°, 140°, 40°, 80° b) 100°
6) 1:3 7) 13:22
8) A £20, £70, £40 B £40, £30, £60
C £55, £25, £50 D £39, £91 E £65, £65
F) £52, £78

EXERCISE 9C

1a) £450 b) £5400 2) £45 3) ₦132
4a) 1/5 b) ½ c) 1/5 d) ½
5a) 3/5 b) £2000 6) £400 7a) 1/3 b) 2/3
c) 23 men and 46 women

EXERCISE 9D

1) 5 hours 2a) 40 days b) $\frac{120}{w}$ days 3) 6 hours
4a) 3:2 b) 27 red books

EXERCISE 9E

1) $\frac{150}{10}$ = £15/hour 2a) 10 m/s b) 600 m/min
3) £800/month

EXERCISE 9F

1) 2100 km 2) 2 400 000 m 3) 1200 km
4) 3600 km
5a) 10000 cm × 6000 cm = 100 m × 60 m
5b) Area = 6000 m^2 6) 7 cm

EXERCISE 10A
1a) $2x + 6$ b) $3x + 12$ c) $4x + 8$ d) $5x - 15$
e) $7x - 28$ f) $40 - 10x$ g) $24 + 8x$ h) $12x + 60$
i) $8x - 80$ j) $99x + 990$ k) $60 - 12x$ l) $a^2 + 3a$
m) $w^2 + 7w$ n) $3x^2 + 15x$ o) $m^2 + 9m$
p) $42 - 18c$ q) $4d^2 + 8d$ r) $7a - a^2$ s) $2x + 4c - 18$
t) $0.5x^2 + 2x$ u) $3x - 9$ v) $0.18c + 2.7$

EXERCISE 10B

1a) $-2x - 4$ b) $-8x - 24$ c) $-3x - 30$ d) $-5x - 10$
e) $-9x - 36$ f) $-x - 2$ g) $-4x - 8$ h) $-7x - 49$
i) $-10x - 60$ j) $-20x - 100$ k) $-42 - 6x$ l) $-3x + 12$
m) $-9x + 9$ n) $-4x + 8$ o) $-8x + 40$ p) $-35 + 7x$
q) $-a^2 - 3a$ r) $-w^2 - 10w$ s) $-36w^3 - 30w^2$
t) $-3v^2 + 2v$ u) $-28w^3 - 24w^2$ v) $-a^2 - 4an + 2a$

EXERCISE 10C

1a) $5x + 9$ b) $9x + 27$ c) $12x + 52$ d) $2x + 8$
e) $3x + 11$ f) $-x - 7$ g) $x + 4$ h) $5x + 6$ i) $12y - 16$
j) $4y + y^2$ k) $30t + 2$ l) $6x + 18$ m) $40d - 26$
n) $11d + 1$ o) $2x^2 - 3x$ p) $-4d^2 - 9d$ q) $5n^2 - 15n$
r) $-98w + 26$ s) $-6s^2 + 5s + 59$

EXERCISE 10D

1) $12x + 8$ 2a) $2c + 18$ 2b) $c + 15$ 2c) $3c + 14$
3a) $f^2 - 2f$ b) f^2 c) $6f^2 - 8f$ d) $f^3 - 2f^2$
4a) $3x$ b) $2x + 6$ c) $10x + 36$ d) $90x$
e) $80x - 36$ f) $69x$

EXERCISE 10E

1a) $x^2 + 4x + 3$ b) $x^2 + 7x + 10$ c) $x^2 + 10x + 21$
d) $n^2 + 8n + 16$ e) $n^2 + 15n + 54$ f) $x^2 + x - 2$
g) $x^2 - 4x - 21$ h) $w^2 - 6w + 5$ i) $w^2 - 9w - 10$
j) $c^2 - 3c - 108$
2a) $12m^2 + 36m + 15$ b) $6x^2 + 6y^2 + 20xy$
c) $x^2 - 16x + 64$ d) $a^2 + 2ab + b^2$
e) $28c^2 - 41c + 15$ f) $3w^3 - 2w^2 - 16w$
3a) $8y + 16$ b) $4y^2 + 16y + 16$ c) 16

EXERCISE 10F

1a) $4(x + 4)$ b) $3(2x + 3)$ c) $4(3x + 4)$
d) $5(a + 2)$ e) $9(x + 10)$ f) $3(2x - 1)$
g) $9(x - 2)$ h) $5(x - 6)$ i) $15(2x - 3)$
j) $7(m - 7)$ k) $7(4n + 3)$ l) $7(x - 2b + 3c)$
m) $2n(2 - 3y)$ n) $2(6x + 7)$ o) $16(y - 1)$
p) $bcd(a - 1)$ q) $5(m + 4n - 1)$ r) $75(k - 6)$

EXERCISE 10G

1a) $x(x + 5)$ b) $4t(t - 4)$ c) $c(bc - d)$ d) $7p(p + 4)$
e) $cd(c + d)$ f) $f(f^2 - 2)$ g) $s(p + q - pr)$
h) $x(11x + 1)$ i) $df(df - mndf + 1)$
j) $3x(x - 3 + 9x^2)$ k) $7y(2 - y + 4y^2)$ l) $(c+f)(d+e)$
m) $(x + 3)(b + c)$ n) $-3(10y^2 + 3)$

EXERCISE 10H

1a) 1, 2, 4, c, 2c, 4c
b) 1, 2, 3, 4, 6, 12, t, 2t, 3t, 4t, 6t, 12t
c) 1, a, c, ac d) 1, 7, x, 7x, y, 7y, 7xy, xy
e) 1, y, y^2 f) 1, 2, 3, 6, x, 2x, 3x, 6x, y, 2y, 3y, 6y, xy, 2xy, 3xy, 6xy g) 1, 2, 5, 10, x, 2x, 5x, 10x, x^2, $2x^2$, $5x^2$, $10x^2$ h) 1, 5, 7, 35, e, 5e, 7e, 35e, f, 5f, 7f, 35f, ef, 5ef, 7ef, 35ef

EXERCISE 10I

1a) qr b) n c) $5c^2$ d) 30mn e) b f) 5 g) 6j
h) n i) 5ab j) 6ny

EXERCISE 10J

1a) 12p b) cd c) 15tu d) $7n^2$ e) $8cd^2$ f) $30p^2q^2$
g) $10ab^2c$ h) amn i) $105mn^2$

EXERCISE 10K

1) 2t 2) $\frac{2x}{y}$ 3) 2/7 4) c/5 5) 2bc 6) $\frac{7a}{2}$ 7) 24y
8) $\frac{df}{eg}$ 9) p/2 10) $\frac{75}{10ps} = \frac{15}{2ps}$ 11) 7x 12) q/5
13) $\frac{b}{9a}$ 14) $\frac{10}{3s}$ 15) x/3 16) 6 17) $w^2/2y$
18) $d^2/8c$ 19) 2/ap 20) $4cd/e^3$ 21) 9w/4y
22) $\frac{g+1}{g-1}$ 23) $\frac{y+2}{3}$ 24) 3x

EXERCISE 10L

1) 4x/7 2) 7n/6 3) 3x/7 4) $\frac{1+n}{y}$ 5) $\frac{5xy}{4w}$
6) $\frac{4c-d}{6}$ 7) $\frac{12x+tu}{4u}$ 8) $\frac{23}{2y}$ 9) $\frac{cx+30}{10c}$ 10) $\frac{15y-7x}{21}$

11) $\frac{7c+3}{10}$ 12) $\frac{8b-15}{15}$ 13) $\frac{28dg-5f}{7g}$ 14) $\frac{79u+v}{9}$

15) k/9 16) $\frac{4x-8}{8} = \frac{1}{2}x - 1$ 17) $\frac{3x+11}{7}$

CHAPTERS 9 & 10 REVIEW SECTIONS

1a) 11 : 10 b) 1 : 7 c) 1 : 14 d) 1 : 4
2) 4 : 1
3) 18:15, 36:30, 48:40, 36:30, 24:20, 42:35
4) Isabel - 15 oranges, Edward – 25 oranges
Angela – 35 oranges
5) £45 6a) $-4x - 20$ 6b) $-3x + 12$
6c) $-9s^2 + 5s + 16$ 7a) $wd(w + d)$ b) $(a+3)(b+c)$
c) $4ry(4y + 2r - 1)$
8a) HCF = 15wx, LCM = $30w^2x$
8b) HCF = 5ny, LCM = $25n^2y^2$
9a) 6bc b) 1/2ps
10a) 2x/3 b) 5n/3 c) $\frac{21c-3d}{21}$

EXERCISE 11

1a) ½ b) ¼ c) 7 d) 9/5 e) – ½ f) – 7/5
g) – 1/11 h) 14/9 i) 7/4 j) 1/12 k) – 1/5
l) -12 m) 11/2 n) -1/4 0) 17/3

2a) – 10/9 b) – 2/3 c) – 4/5 d) – 1/100
e) 7/36 f) 7/26 g) – 2/5 h) – 2/9 i) w j) 7/n
3) Andrew is WRONG. Reciprocal of 6 is 1/6 = 0.16666.. Reciprocal of ¼ is 4. 4 is greater than 0.166666....
4a) 17/3 b) 3/17 c) – 3/17 d) 1

EXERCISE 12A

1a)

x	1	2	3	4	5
y	4	7	10	13	16
C	1,4	2,7	3,10	4,13	5,16

1b) Graph of y = 3x + 1 drawn as a straight line
c) (0,1)
2a) (-1,2), (0,3), (2,5),(6,9)
2b and c) Students drawing
3a – f) Student's drawings of straight line graphs
4a)

d	0	5	13	20
c	0	15	39	60

4b)

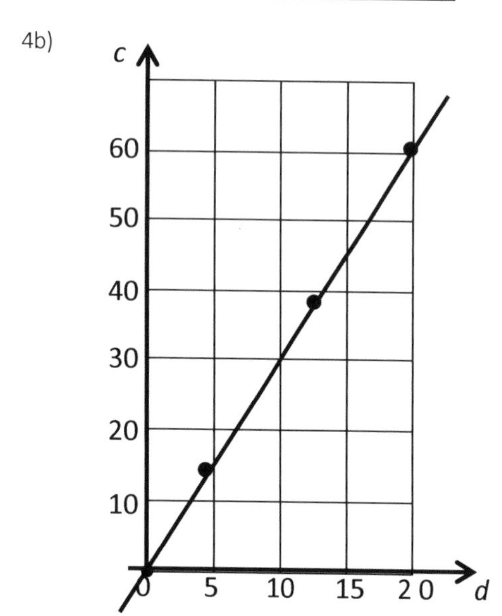

4c)

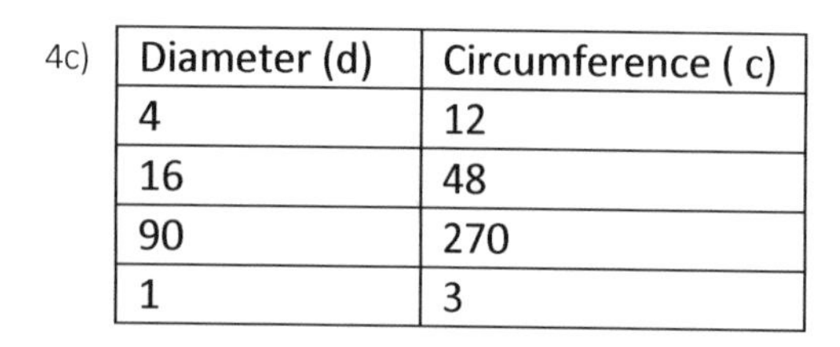

Diameter (d)	Circumference (c)
4	12
16	48
90	270
1	3

5a) P (2, -2) Q(-1, 1)

5b)

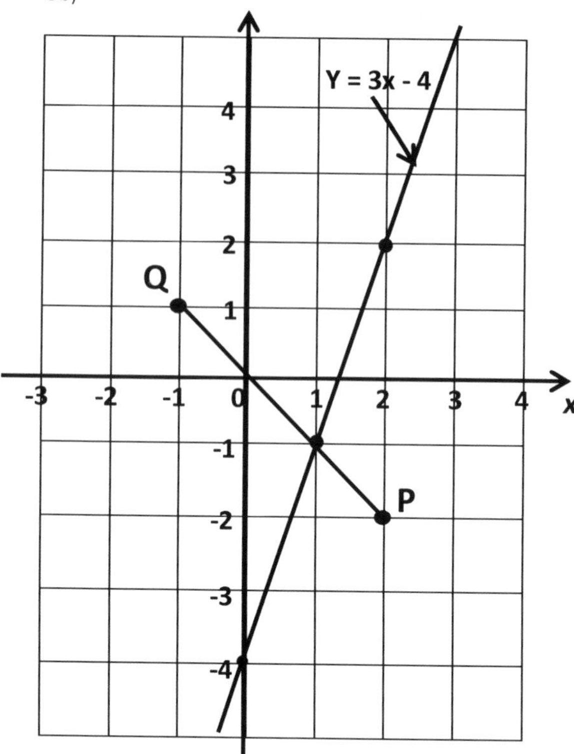

5c) See diagram 5d) (1, -1)

EXERCISE 12B

1) A x = -3 B x = 2 C x = 4 D y = 3 E y = -3
F x = 0 G y = 0

2a)

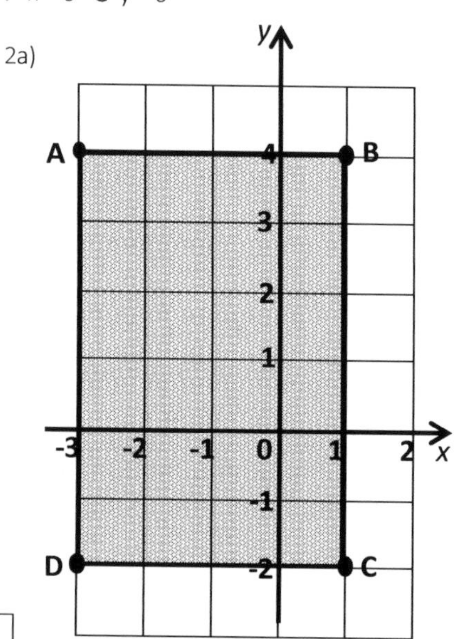

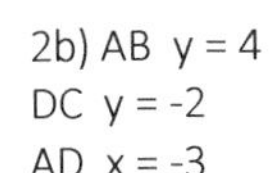

2b) AB y = 4
DC y = -2
AD x = -3
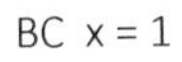
BC x = 1

3a)

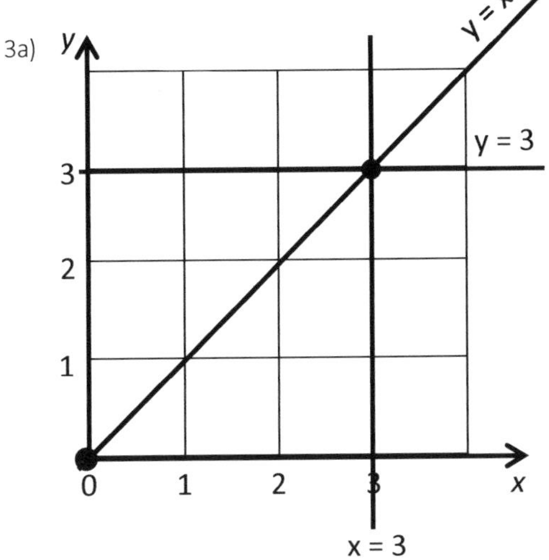

3b) (3, 0) (3, 3) and (0, 0)   3c) 4.5 cm^2

4a)

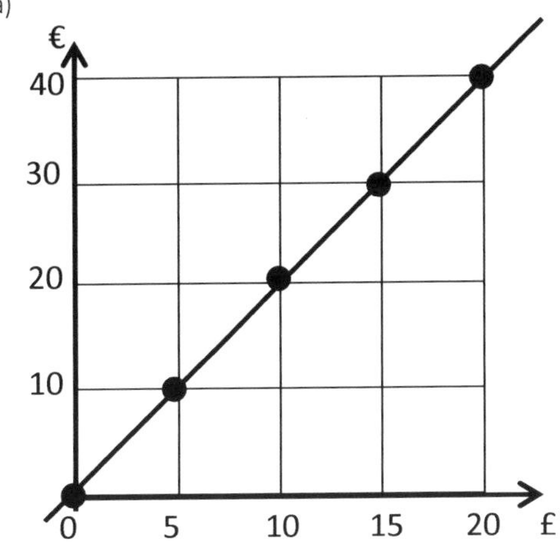

4b) i) €12 ii) €28 iii) €34
4c) i) £2.50 ii) £11 iii) £18.50

EXERCISE 12C

1a) 1 b) 1 c) 6 d) 2 e) 3/2 f) 3 g) -5
h) – 10/7 i) 4 j) -1 k) 1 l) – ¼ m) 4
n) -2 o) 2 p) 2
2a) 1 b) ½ c) – 3/2 d) 2 e) ½ f) – ½ g) – 5/3
3) AB 4 BC – 4/3 DC 5/6 AD – 3/8
4a) 1 b) -2 c)1 d) 7/5

EXERCISE 12D

1a) (4, 6) b) (1, 4) c) (3.5, 4) d) (2, 3),
e) (- 0.5, 0.5), f) (1, -2)
2) A (2, -3) B (3.5, 1) C (4, 5) D (-3, 4)
E (-6, 3.5) F (- 4.5, - 3.5)
3) (-4, 5)
4a) AB (1, 4) BC (3, 4) AC (3, 2) 4b) 8 cm^2

EXERCISE 12 E

1a) $y = 4x + 4$ b) $y = -5x + 5$ c) $y = 4x – 6$
d) $y = \frac{2}{3}x + 4$ e) $y = -4x – 5$ f) $-\frac{5}{6}x – 4$
2a) $y = -\frac{2}{3}x + 4$ b) $y = \frac{4}{5}x – 2$ c) $y = 3x - 1$

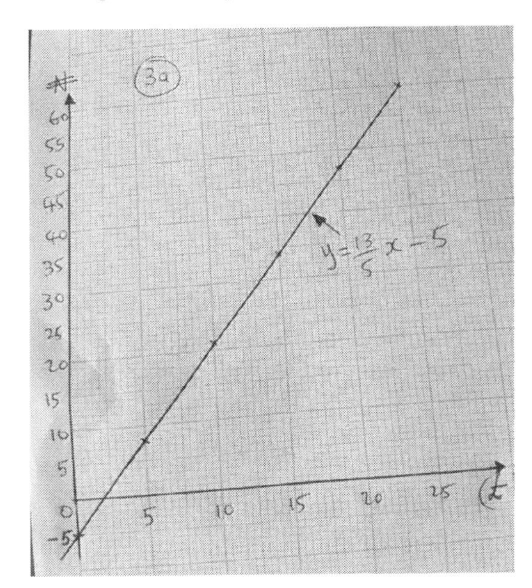

3b) $y = \frac{13}{5}x – 5$

4) A ➡ $y = -\frac{1}{2}x – 2$ B ➡ $y = -\frac{2}{7}x + 2$
C ➡ $y = \frac{1}{3}x + 3$ D ➡ $y = x + 5$
E ➡ $x = - 6$ F ➡ $y = -x - 5$

EXERCISE 12F

1a) Q and S are parallel b) P and R are perpendicular
c) T is neither parallel nor perpendicular

2) Student's drawing and the gradient is 2. Student's drawing of a line perpendicular to
$y = 2x + 1$ with a gradient of -1/2.
3) 5 4) $y = x + 4$

EXERCISE 12G

1a)

x	-3	-2	-1	0	1	2	3
x^2	9	4	1	0	1	4	9
y	13	8	5	4	5	8	13

1b) Student's own drawing
2a,b,c) Student's own drawing
3a) Student's own drawing
3b) i) x = 1.5 ii) x = - 2.5

EXTENSION WORK

4) Student's drawings
5a) Graph of $y = x^2 + 1$

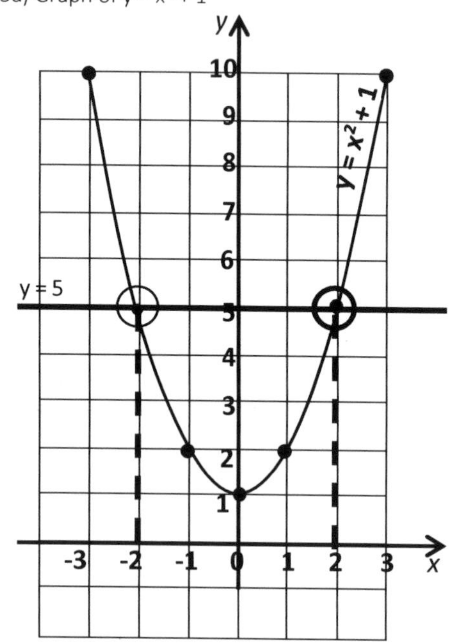

5b) See diagram above.
5c) x = -2 and a = 2

CHAPTERS 11 AND 12 REVIEW SECTIONS

1a) 11/3 b) 3/11 c) – 3/11 d) 1

2a)

x	1	2	3	4	5
y	4	9	14	19	24
C	1,4	2,9	3,14	4,19	5,24

2b) (0, -1) 3a) 1 3b) y = x - 4
4a) (1, 4) 4b) (- 0.5, 0.5)
5a) $y = -\frac{5}{3}x + 7$ b) $y = 4x - 1$
6) $4 \times -\frac{1}{4} = -1$. Two lines are perpendicular if the product of their gradients is -1.

EXERCISE 13A

1a) 4 b) 9 c) 6 d) 4 e) 13.25
2a) ₦3 600 b) ₦240
3a) 1270 g b) 423.33 g
4) 287.5 g 5a) 6 b) $13\frac{1}{3}$ c) $5\frac{3}{4}$
d) 5 e) 6 m f) $1\frac{1}{2}$ 6) 37%
7a) 27 b) 8 8) 7 9a) 5.2 b) 5.8 c) 5.3 d) 2.7

EXERCISE 13B

1a) 6 b) 2.5 c) 4 d) 5 e) 40 f) 4.5
2a) 19 b) 20 c) Chuba's marks were more consistent. 3) 20°C 4) 10 5) 0.9 6a) 5.75 b) 60

EXERCISE 13C

1a) 1 b) 3 c) 7 and 12 d) 9 e) No mode 2) 8
3) Blue 4) No mode 5a) 13 b) -8 c) 3.5 d) $\frac{2}{5}$
6a) Blue b) π c) Dog and Cat

EXERCISE 13D

1a) 9 b) 12 c) 30 d) 80 2) 2 3) 36 cm
4a) 40 b) 16 c) 5.5 d) 0.4

EXERCISE 13E

1a) 115 b) 110 c) 15 d) 110
2a) 4 b) 5 c) 4 d) 3.6 3a) 1.2 b) 40 matches
4a) 61 b) 56 c) 99 d) 2 e) 6
5a) 0 b) 1.9 c) 5 d) 1

EXERCISE 13F

1a) 17 b) 2 c) 4 d) 2 2) Dog 3a) $\frac{80}{360} = \frac{2}{9}$ 3b) C4
3c) i) 72 students ii) C6 : 20 Students C4: 24 students
Altogether 44 students
4a) 28 b) 28 c) Friday d) 138 apples
5a) Akin b) 10 c) 10.5 d) 20
6a) ₦45 000 b) No mode c) ₦48 000 d) The average salary is ₦48 000 which is less than ₦49 000 stated in the advert. Also, the Carpenter's salary was advertised as ₦40 000.
7) i) No title ii) No label on both axes iii) Bars are of different widths.
8a) LAGOS

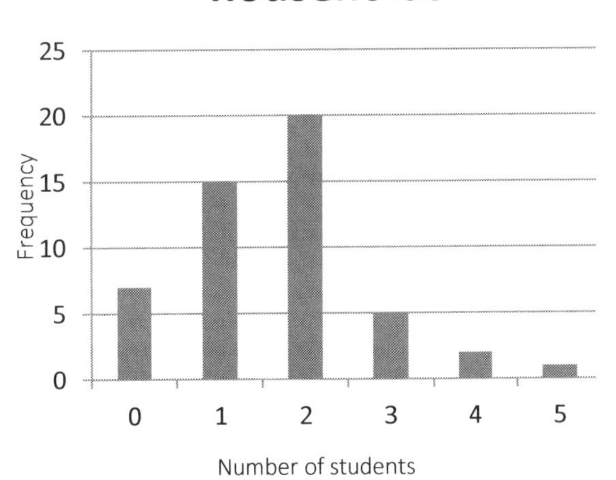

ENUGU

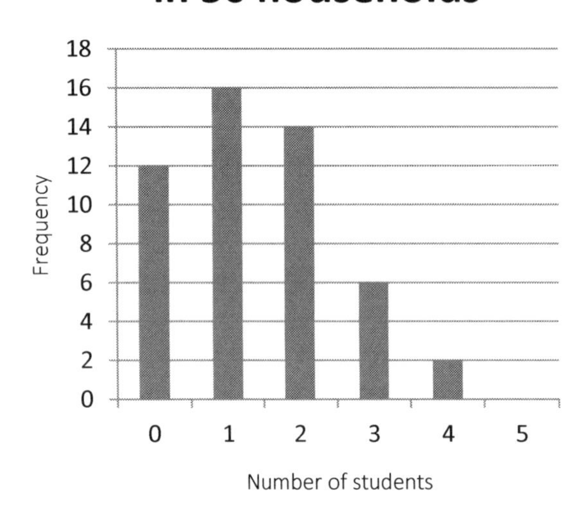

b) For Lagos:
Mean = 1.7 ≈ 2 students
Median = 2 students
Mode = 2 students
Range = 5 students

For Enugu:
Mean = 1.4 ≈ 1 student
Median = 1 student
Mode = 1 student
Range = 4 students

c) From the averages calculated, it shows that fewer students were living in households in Enugu than in Lagos. Also, the spread of students per household is bigger in Lagos than in Enugu (From their ranges).

EXERCISE 13G

1a) 21 – 25 b) 16 – 20 c) 16.7g
2a) $55 \leq w < 60$ b) $60 \leq w < 65$ c) 56.2 kg
d) Midpoints are used instead of actual numbers.
3a) 63.25 mm b) 63.67 mm c) Mean is almost the same 4a) 2.7 – 2.8 b) 2.7 – 2.8

EXERCISE 14A

1a) 10 cm b) 11.40 cm c) 9.22 cm d) 9.43 cm
e) 5.39 cm f) 19.10 cm g) 16.28 cm h) 36.06 cm

EXERCISE 14B

1a) 6 m b) 10.72 m c) 4.58m d) 5.20 m e) 6.71 m
f) 12.69 m g) 4.80 m h) 34.64 m

EXERCISE 14C

1) 7.2 cm 2) 11.3 m 3) 19.1 cm to 1 d.p 4) 8.31 cm
5) 13.9 m 6a) 21.2 cm b) y = 5 cm

EXERCISE 14D

1) P ⟶ 4.47 units, Q ⟶ 2 units, R ⟶ 6.40 units
S ⟶ 7.21 units, T ⟶ 3 units
2a) 5 units b) 5 units c) $\sqrt{29}$ units d) $\sqrt{18}$ units
e) $\sqrt{61}$ units

EXERCISE 15A

1i) 050° ii) 230° 2i) 280° ii) 100° 3a) 280°
b) 310° c) 250° 4a) i)298° ii) 118° b) i)50°
ii) 230° 5a) 120° b) 300° c) 30° d) 210°

6) Accurate construction.

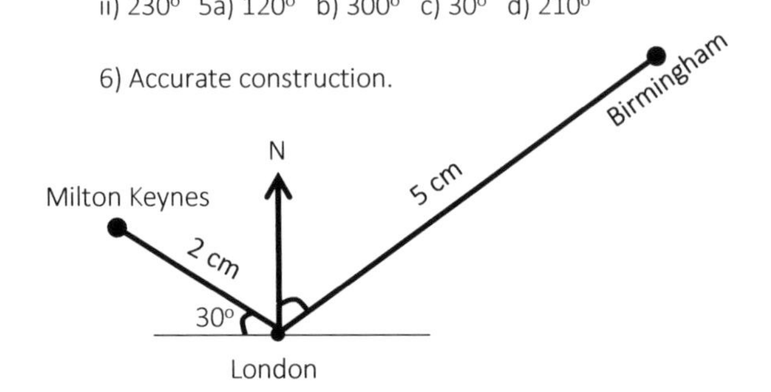

7a) 763.22 km
7b) Accurate diagram drawn by student and should look like this:

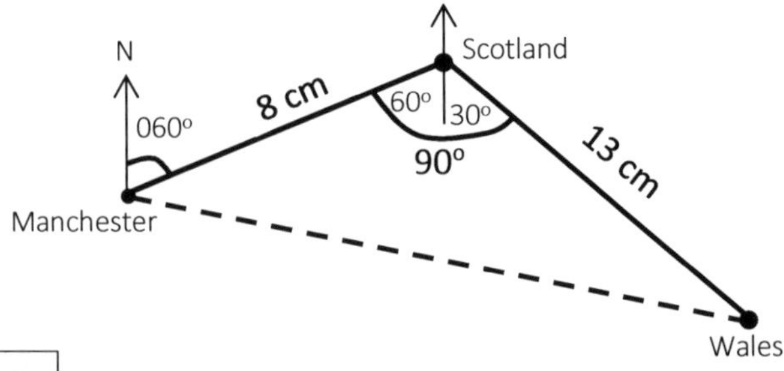

7c) Measure the distance from Wales to Manchester. (Should be around 15.26 cm or 763.22 km)
7d) They are almost the same.

EXERCISE 16A

1) $x = 5$ and $y = 2$
2) $x = 8$ and $y = 5$
3) $x = 20$ and $y = 5$
4) $x = 7$ and $y = -1$
5) $x = 3$ and $y = 1$
6) $x = 5$ and $y = 2$
7) The left side will be zero. The two lines if plotted would be parallel because they have the same gradients (2) and there won't be a point of intersection. Therefore, it is impossible to solve the simultaneous equation.
8) a) $t - h = 37$ and $t + h = 67$
   b) Tom is 52 years and Harry, 15 years.
   c) 780 years

EXERCISE 17A

1a) $x^3 + 7x^2 + 14x + 8$
b) $x^3 + 6x^2 + 11x + 6$
c) $x^3 - 9x^2 + 24x - 20$
d) $bcd + 3bd - c^2d - 3cd - 2bc - 6b + 2c^2 + 6c$
e) $w^3 - 10w^2 + 26w - 35$

2a) $30x^3 + 116x^2 + 82x + 12$
b) $125x^3 - 300x^2 + 240x - 64$

3a) $24w^3 + 38w^2 - 112w - 30$
b) $70w^2 - 8w - 38$
c) 42 cm by 8 cm by 10 cm

4) $g = 6$ and $c = 1$

EXERCISE 17B

1a) $(a - 2)(a + 2)$
b) $(a - 3)(a + 3)$
c) $(x - 10)(x + 10)$
d) $(x - 12)(x + 12)$
e) $(x - 15)(x + 15)$
f) $(r - 1)(r + 1)$
g) $(k - 9)(k + 9)$
h) $(w - 13)(w + 13)$
i) $(w - 18)(w + 18)$
j) $(m - 27)(m + 27)$

2) a) $(4m - 1)(4m + 1)$
b) $(5x - 7)(5x + 7)$
c) $(10t - 3)(10t + 3)$
d) $(20w - 8)(20w + 8)$
e) $(10x - 5)(10x + 5)$
f) $5(2x - 3)(2x + 3)$

3a) $x^2 - 81$
b) $(x - 9)(x + 9)$

4) Mark is **incorrect**. $(x - 3)(x + 3)$ will give $x^2 - 9$ and not $x^2 + 9$.

5a) $25x^2 - 400$
b) $(5x - 20)(5x + 20)$

EXERCISE 17C

1a) $(x + 2)(x + 2)$
b) $(x + 1)(x + 5)$

c) $(x + 3)(x + 4)$
d) $(x + 2)(x + 7)$
e) $(x + 6)(x + 8)$

2a) $(x - 5)(x + 1)$
b) $(x - 3)(x + 4)$
c) $(x - 2)(x + 7)$
d) $(x - 3)(x + 7)$
e) $(x - 8)(x + 6)$

3a) $(w - 11)(w + 5)$
b) $(w - 12)(w + 4)$
c) $(w - 6)(w + 10)$
d) $(w - 8)(w - 8)$
e) $(w - 6)(w - 7)$

4) There are no two numbers that will multiply to give 10 but will also add to give 3.

5) $(w^6 - 600)(w^6 + 3)$

EXERCISE 17D

1a) $(x + 3)(5x + 1)$
b) $(x + 4)(2x + 7)$
c) $(2x + 3)(5x + 4)$
d) $(x + 10)(4x + 1)$
e) $(x + 5)(6x + 1)$

2a) (4x + 3) (2x – 4)
b) (2x – 3) (3x – 5)
c) (x – 2) (4x – 5)
d) (7x – 3) (6x + 5)
e) (5x + 5) (6x – 2)
3) (3x + 2)
4a) (5w + 4) (6w – 3)
b) 22w + 2
c) 376 cm

EXERCISE 17E

1a) x = -3 or x = $-\frac{1}{5}$

b) x = -4 or x = -3.5
c) x = -1.5 or x = $-\frac{4}{5}$

d) x = -10 or x = $-\frac{1}{4}$

e) x = -5 or x = $-\frac{1}{6}$

2a) x = $-\frac{3}{4}$ or x = 2

b) x = 1.5 or x = $\frac{5}{3}$

c) x = 2 or x = $\frac{5}{4}$

d) x = $\frac{3}{7}$ or x = $-\frac{5}{6}$

e) x = -1 or x = $\frac{1}{3}$
3a) a = 2 and - 2
b) a = 3 and -3
c) x = 10 and - 10
d) x = 12 and – 12
e) x = 15 and – 15
f) r = 1 or -1
g) k = 9 or - 9
h) w = 13 or - 13
i) w = 18 or - 18
j) m = 27 or - 27
4a) w = 1 and w = - 5
b) w = 12 and w = - 4
c) w = 6 and w = - 10
d) w = 8
e) w = 6 and w = 7

EXERCISE 17F

1) x = $\frac{-9 \pm \sqrt{73}}{2}$

2) x = $\frac{-4 \pm \sqrt{12}}{2}$

3) x = $\frac{-7 \pm \sqrt{73}}{2}$

4) x = $\frac{-5 \pm \sqrt{37}}{2}$

5) x = $\frac{3 \pm \sqrt{17}}{2}$

6) x = $\frac{-7 \pm 5}{4}$

7) x = $\frac{-8 \pm \sqrt{88}}{6}$

8) x = $\frac{5 \pm \sqrt{97}}{12}$

9) x = $\frac{-5 \pm \sqrt{73}}{8}$

10) x = $\frac{1 \pm \sqrt{29}}{2}$

EXERCISE 17G

1) x = 0.772 and x = - 7.77

2a) x = 3.19 or x = -2.19
b) x = 2.11 or x = - 2.61
c) x = - 1.70 or x = - 5.30
d) y = 5.54 or y = - 0.541
e) x = 5.34 or x = - 0.0936
f) x = 12.3 or x = 5.68

3) The two numbers are 0.65 and 7.65

4) 17 and 18

5) 5 m

EXERCISE 17 H

1) $b^2 - 4ac < 0$, therefore, no solution.

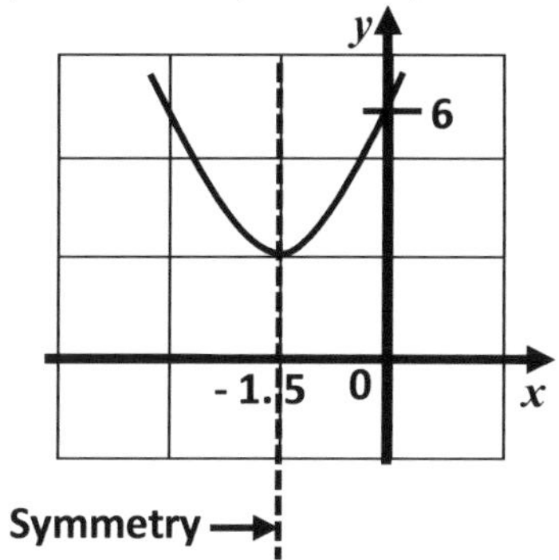

2) $b^2 - 4ac > 0$, therefore, two solutions.

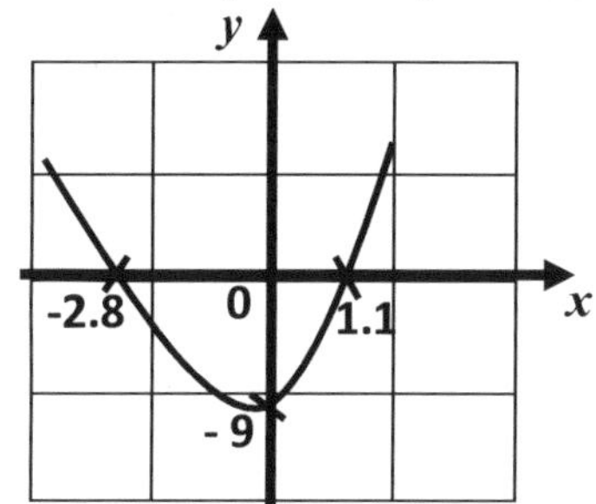

3) $b^2 - 4ac > 0$, therefore, two solutions.

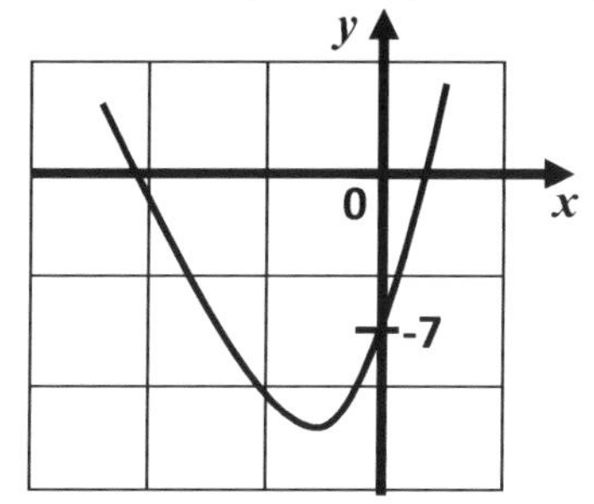

4) $b^2 - 4ac = 0$, therefore, one solution.

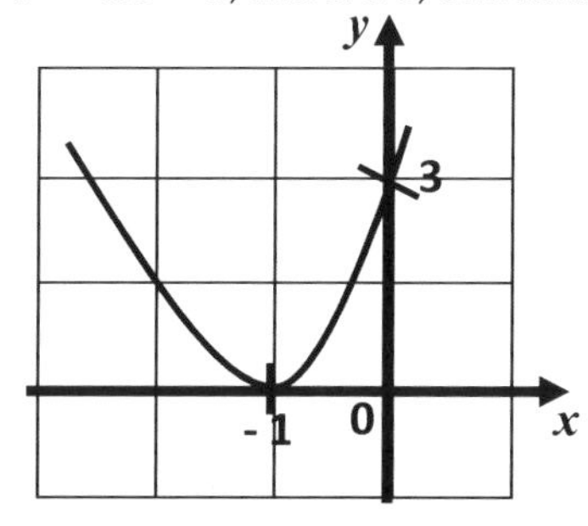

5) $b^2 - 4ac = 0$, therefore, one solution.

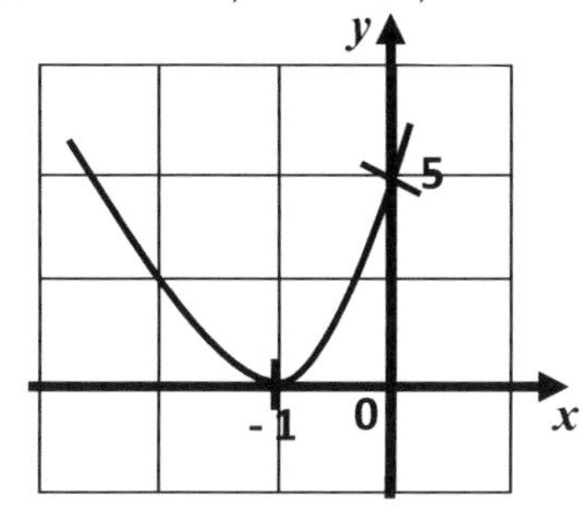

6) $b^2 - 4ac < 0$, therefore, no solution.

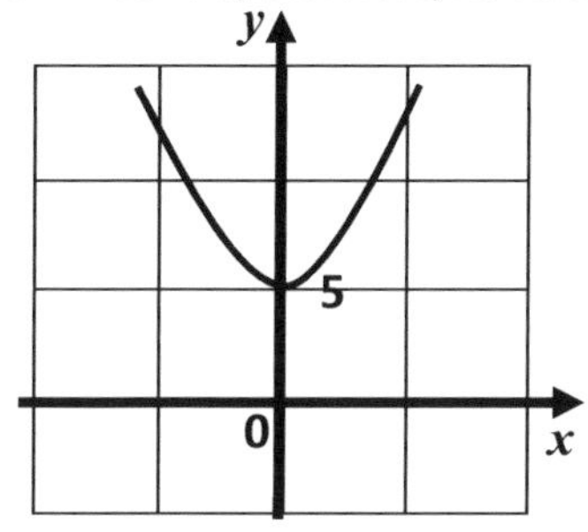

7) $b^2 - 4ac > 0$, therefore, two solutions.

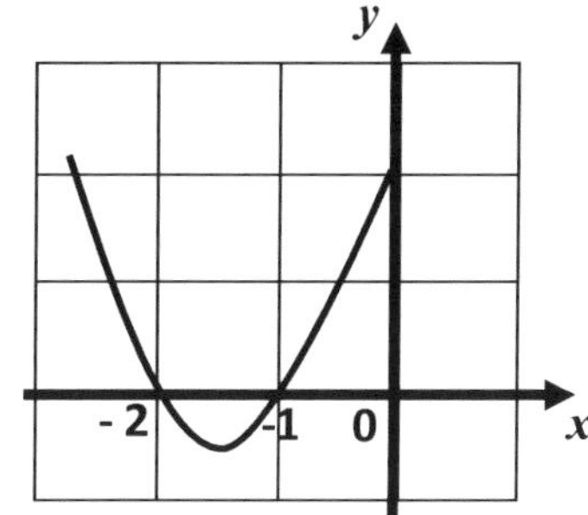

8) $b^2 - 4ac > 0$, therefore, two solutions.

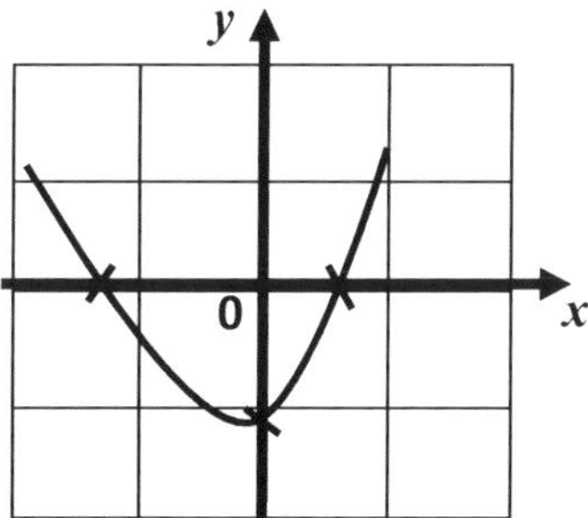

9a) P = - 3.56 and Q = 0.56

9b) x = - 1.5

EXERCISE 17I

1) $(x + 4)^2 – 14$
2) $(x + 5)^2 – 18$
3) $(x + 3)^2 – 18$
4) $(x – 3)^2 – 10$
5) $(x – 8)^2 – 59$
6) $(x + 1)^2 – 6$
7) $(x + 2.5)^2 – 1.25$
8) $(x – 3.5)^2 – 21.25$
9) $(x + 10)^2 - 93$
b = 10 and c = -93
10) $(x – 3.5)^2 – 7.25$
b = -3.5 and c = -7.25
11) $(x – 1.5)^2 – 8.25$
b = -1.5 and c = - 8.25
12) $(x + 7)^2 - 41$
b = 7 and c = -41
13) $(x + 3)^2 – 7$
b = 3 and c = -7
14) $(x – 2)^2 - 8$
b = -2 and c = -8
15) $(x + 1)^2 + 4$
b = 1 and c = 4
16) $(x + 3)^2 + 4$
b = 3 and c = 4
17) p = 4 and q = 24

EXERCISE 17J

1a) $m = \pm\sqrt{19} - 4$
b) $x = \pm\sqrt{21} - 4$
c) $x = \pm\sqrt{5} + 2$
d) $x = \pm\sqrt{19} - 5$
e) $x = \pm\sqrt{11} + 3$
f) $x = \pm\sqrt{27} - 6$

2a) c = 1 or c = -3
b) x = 0.58 or x = -8.58
c) x = 4.24 or x = -0.24
d) x = -0.64 or x = -9.36
e) x = 6.32 or x = -0.32
f) y = -0.80 or y = -11.20

EXERCISE 17K

1) $3(x + 1)^2 + \frac{2}{3}$

2) $5(x + 1.5)^2 – 1.25$
3) $4(x – 0.75)^2 – 1.25$
4) $7(x – 1)^2 - 1$
5) $0.5(n + 5)^2 – 6.5$
6) $0.75(k – 2)^2 + 3$

EXERCISE 17L

1) x = -1 or -2
2) x = 1.31 or 0.19
3) x = 1.38 or 0.62
4) x = -1.39 or -8.61

EXERCISE 17M

1) $2(x + 1.5)^2 + 2.5$
2) $3(x – 1.5)^2 – 5.75$
3) $4(b + 0.5)^2 + 2$
4) $7(c – 1.5)^2 – 7.75$
5) $5(x – 1)^2 - 10$
6) $9(x + 0.5)^2 – 7.25$

7)
1) a = 2, p = 1.5, q = 2.5
2) a = 3, p = -1.5, q = -5.75
3) a = 4, p = 0.5, q = 2
4) a = 7, p = -1.5, q = -7.75
5) a = 5, p = -1, q = -10
6) a = 9, p = 0.5, q = -7.25

8i)

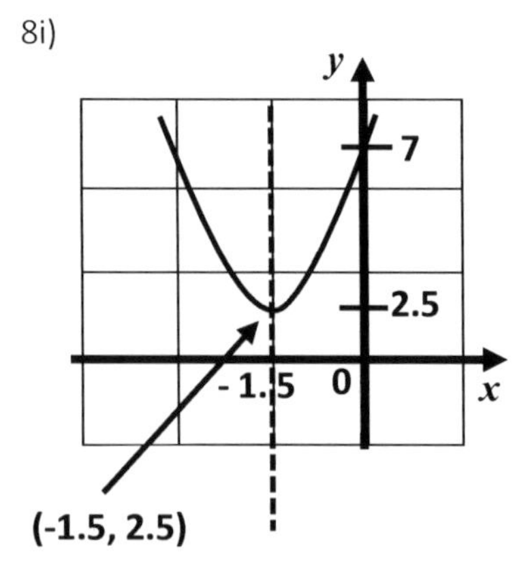

8ii)

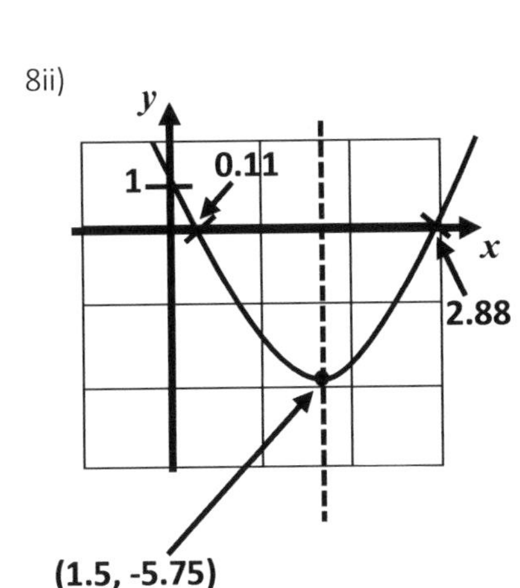

9i)
1) (-1.5, 2.5)
2) (1.5, -$\frac{23}{4}$)
3) (-0.5, 2)
4) 1.5, -7.75)
5) (1, -10)
6) (-0.5, 7.25)

9ii)
1) x = -1.5
2) x = 1.5
3) x = -0.5
4) x = 1.5
5) x = 1
6) x = -0.5

10a) $y^2 - 12y + 20 = 0$
b) y = 2, y = 10
c) 10 cm, 8 cm and 6 cm

11a) No solution
b) x = 0.11 or 2.89
c) No solution
d) x = 0.45 or 2.55
e) x = -0.41 or 2.41
f) x = 0.4 or -1.4

12a) c = -1 and d = 6
b) x = 2.414 and x = -0.414
c) (1, 6)
d) x = 1

EXERCISE 17N
1) (1, 1) (-1, 1)
2) (2, 4) (-2, 4)
3) (3, 9) (-1, 1)
4) (3, 9) (-4, 16)
5) (4, 16) (-3, 9)
6) (5, 60) (-3, 4)
7) (1, 0) (-4, 15)
8) (3, 4) (-3, -4)
9) (0, 5) (-4, 3)
10) (4.16, 4.32) (-0.96, -5.92)

EXERCISE 17O
1a) w = c - 7
b) $\frac{y+2}{6} = w$
c) w = $\frac{y-t}{a}$

d) w = $\frac{2}{b}$

e) w = $\frac{3t+5f}{6}$

f) w = $\frac{k+7f}{7}$

g) w = $\sqrt{k - 18}$
h) w = $\sqrt{5m - 5y}$
i) w = $\frac{a^2-k}{7}$
j) w = $\sqrt{\frac{b+10}{c}}$

k) w = $\sqrt{9t + 9y}$
l) w = $\frac{3aw-ac-7d}{7}$

2a) c = $\frac{y-wg+wt}{5g-5t}$

b) c = $\frac{mr-m}{1+r}$

c) c = $\frac{g^2w}{25n^2}$

d) c = $\sqrt{\frac{e+5r}{h-p}}$

e) c = $\frac{aby}{n+by}$
f) 7

EXERCISE 18A

1) 9, 21, 45, 93, 189

2a) $x^3 + 7x - 3 = 0$
$x(x^2 + 7) - 3 = 0$
$x(x^2 + 7) = 3$
$x = \frac{3}{x^2+7}$

2b) 0.42

2c) $x^3 + 7x - 3 = 0$
$x^3 + 7x = 3$
$x^3 = 3 - 7x$
$x = \sqrt[3]{3 - 7x}$

3a) $y(y - 3)$
b) $y^2 - 3y - 64 = 0$
c) $y_{n+1} = \sqrt{3y_n + 64}$
d) $y = 9.64$

4a) $x^3 - 8x + 7 = 0$
$x^3 + 7 = 8x$
$x^3 = 8x - 7$
$x = \sqrt[3]{8x - 7}$

4b) $x_{n+1} = \sqrt[3]{8x_n - 7}$
4c) 2.19
5) $x = 3.19$
6) Does not exist

EXERCISE 19A

1) $x < 6$ and $x > -6$

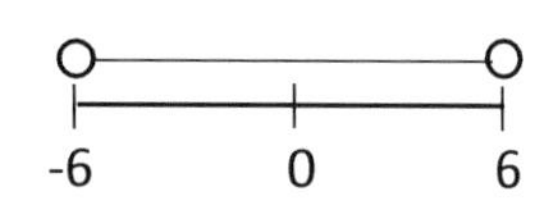

2) $x < 2$ and $x > -2$

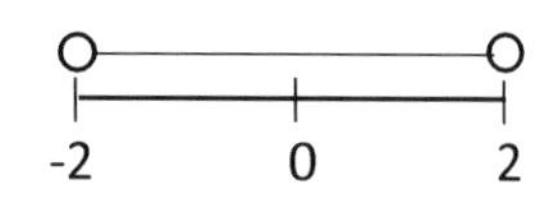

3) $x < 10$ and $x > -10$

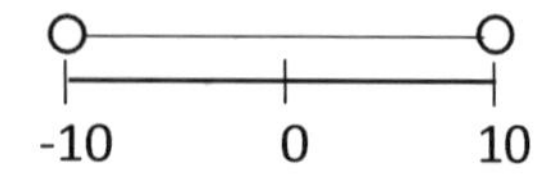

4) $x > 6$ and $x < -6$

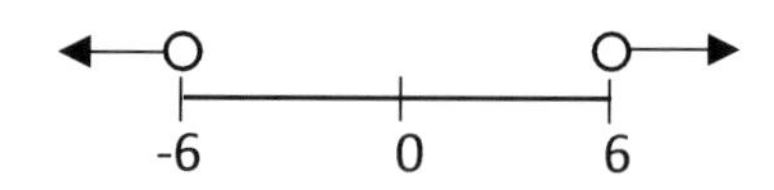

5) $x > 2$ and $x < -2$

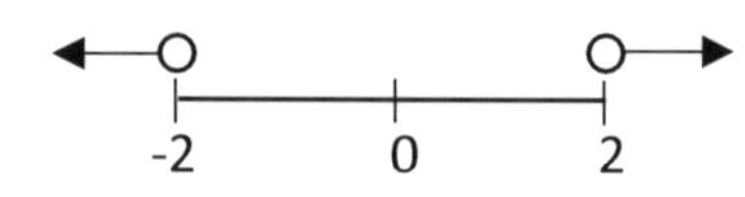

6) $x > 10$ and $x < -10$

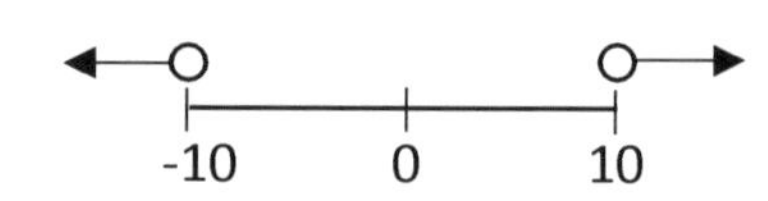

7) $x \leq 8$ and $x \geq -8$

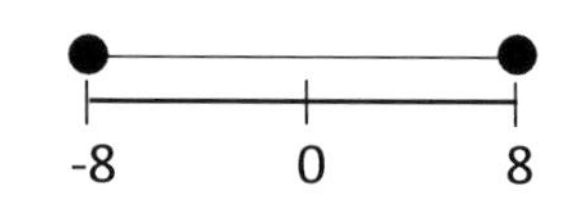

8) $x \leq 7$ and $x \geq -7$

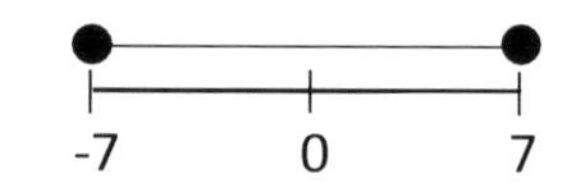

9) $x \leq 10$ and $x \geq -10$

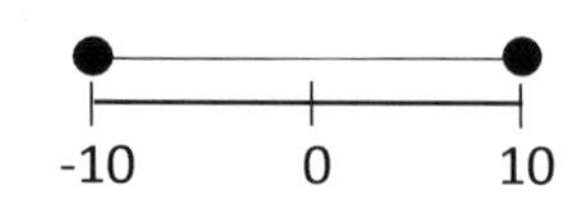

10) $x \geq 8$ and $x \leq -8$

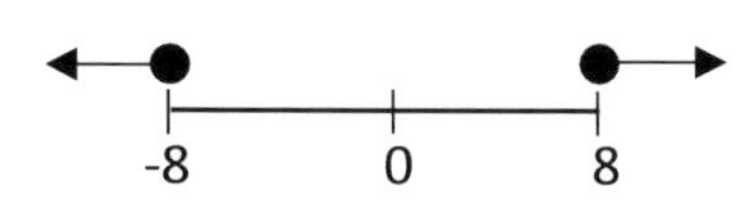

11) $x \geq 1$ and $x \leq -1$

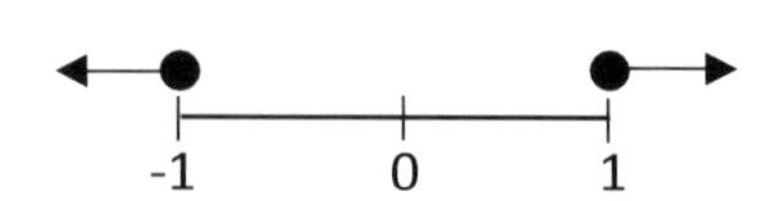

12) $x \geq 10$ and $x \leq -10$

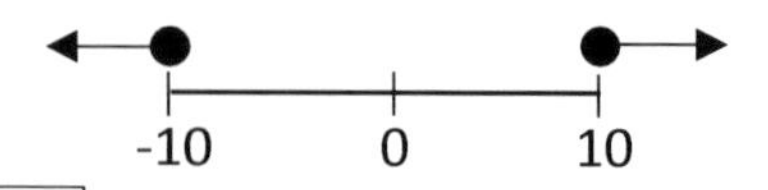

EXERCISE 19B

1) $x < 4$ and $x > -4$

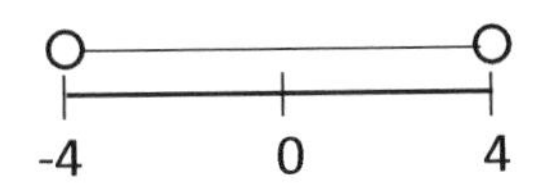

2) $x < 6$ and $x > -6$

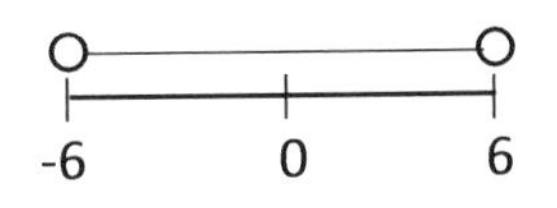

3) $x > 8$ and $x < -8$

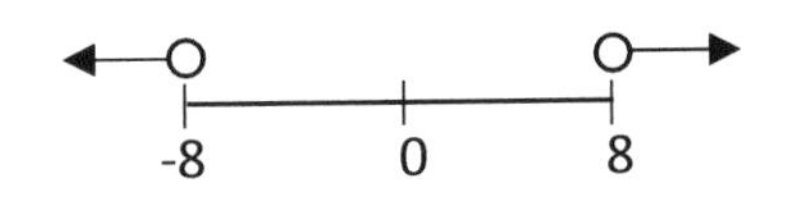

4) $x \leq 7$ and $x \geq -7$

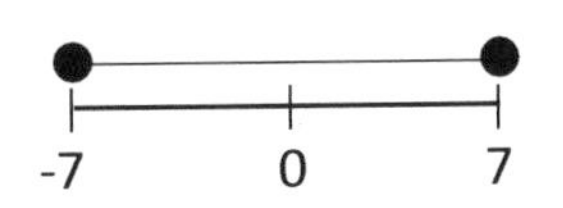

5) $x < 9$ and $x > -9$

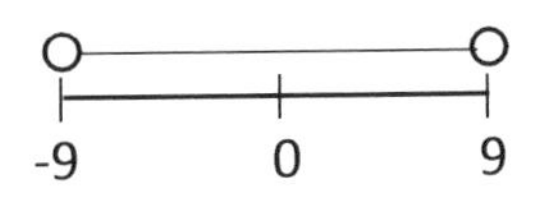

6) $x \leq 4$ and $x \geq -4$

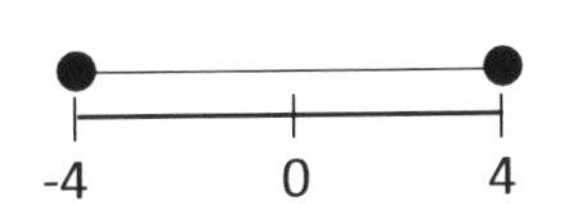

7) $x > 4$ and $x < -4$

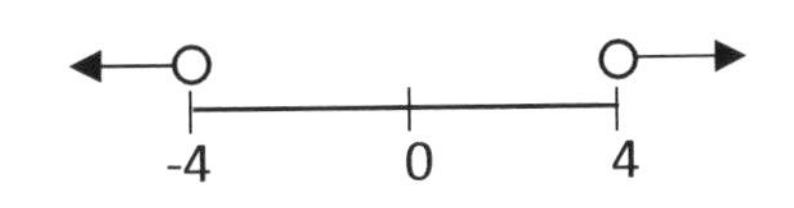

8) $x \leq 10$ and $x \geq -10$

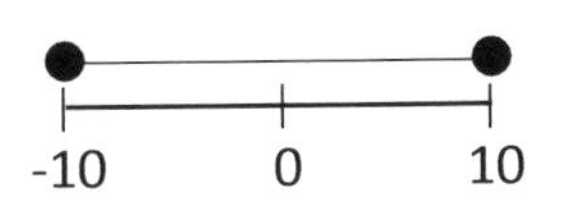

9) $x < 5$ and $x > -7$

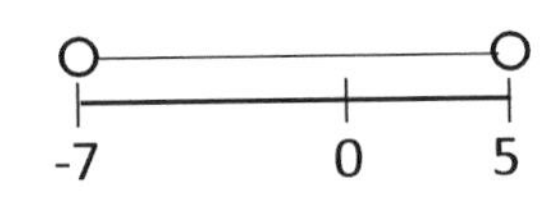

10) $x \geq 1$ and $x \leq -6$

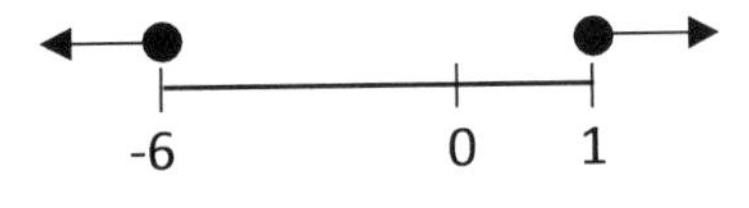

11) $x \leq 10$ and $x \geq -3$

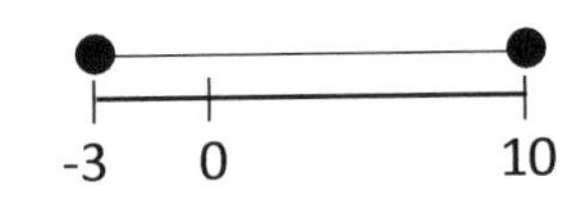

12) $x > 4$ and $x < -5$

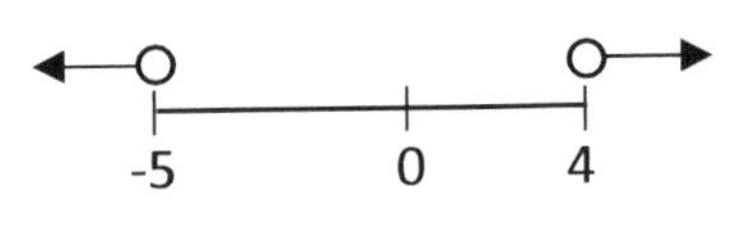

13) $x < 4$ and $x > -8$

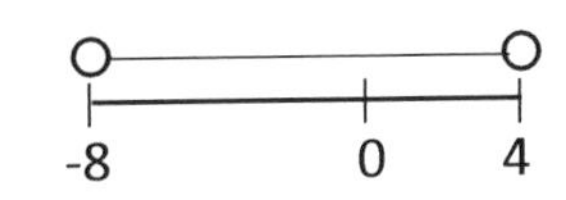

EXERCISE 20A

1a) $\sqrt{21}$

b) $\sqrt{30}$

c) 5

d) $7\sqrt{7}$

e) 6

f) $2\sqrt{33}$

g) $8\sqrt{5}$

h) $36\sqrt{28}$

i) $5\sqrt{2}$

j) 5

k) $\frac{\sqrt{5}}{4}$

l) $18\sqrt{21}$

m) $24\sqrt{2}$

n) $\frac{\sqrt{30}}{2}$

o) $1\frac{4}{5}$

2a) $2\sqrt{2}$
b) $3\sqrt{2}$
c) $2\sqrt{10}$
d) $5\sqrt{3}$
e) $4\sqrt{7}$
f) $2\sqrt{7}$
g) $8\sqrt{5}$

3a) $ax\sqrt{by}$
b) $pc\sqrt{dq}$

c) $\frac{a}{p}\sqrt{\frac{b}{q}}$

4a) c = 3 or $\sqrt{9}$
b) c = $3\sqrt{7}$
c) c = 4

EXERCISE 20B

1) $5\sqrt{6}$
2) $6\sqrt{5}$
3) $2\sqrt{10}$
4) $11\sqrt{11}$
5) $9\sqrt{2}$
6) $8\sqrt{7}$
7) $8\sqrt{3}$
8) $5\sqrt{5}$
9) $5\sqrt{5}$
10) $20\sqrt{6}$

EXERCISE 20C

1) $\frac{\sqrt{6}}{6}$

2) $\frac{2\sqrt{5}}{5}$

3) $\frac{7\sqrt{11}}{11}$

4) $\frac{\sqrt{5}}{10}$

5) $\frac{\sqrt{10}}{10}$

6) $\frac{10-2\sqrt{2}}{23}$

7) $\frac{3\sqrt{5}+5}{5}$

8) $\frac{7\sqrt{3}+18}{3}$

9) $\frac{11\sqrt{5}-15}{5}$

10) $\frac{7\sqrt{5}-9}{4}$

11) $\frac{15-5\sqrt{2}+3\sqrt{3}-\sqrt{6}}{7}$

12) $\frac{4+2\sqrt{2}+2\sqrt{7}+14}{2}$

EXERCISE 20D
1) $4\sqrt{2}+\sqrt{6}$
2) $3-4\sqrt{3}$
3) $6\sqrt{15}+10\sqrt{5}$
4) $12\sqrt{3}+9$
5) $\sqrt{10}-\sqrt{6}-\sqrt{15}+5$
6) $9+2\sqrt{14}$
7) $5\sqrt{5}-11$
8) $35+10\sqrt{21}-\sqrt{14}-2\sqrt{6}$

EXERCISE 20E
1) $3\sqrt{2}$
2a) $3\sqrt{5}$
b) $8\sqrt{2}$
c) $8\sqrt{5}$
d) -12.5
3a) $(5\sqrt{5}+\sqrt{15})$ cm²
b) $(2\sqrt{3}+2\sqrt{5}+10)$ cm
4a) $(216-72\sqrt{5})\pi$ cm²
b) $6(\sqrt{20}-2)\pi$ cm
5) $40\sqrt{3}$ m

EXERCISE 21A

1a) $\binom{2}{1}$
b) $\binom{-4}{-1}$
c) $\binom{0}{-3}$
d) $\binom{4}{2}$
e) $\binom{3}{-2}$
f) $\binom{-4}{0}$
g) $\binom{3}{4}$
h) $\binom{2}{2}$
i) $\binom{-2}{5}$
j) $\binom{2}{-2}$

2) a and d
3a)

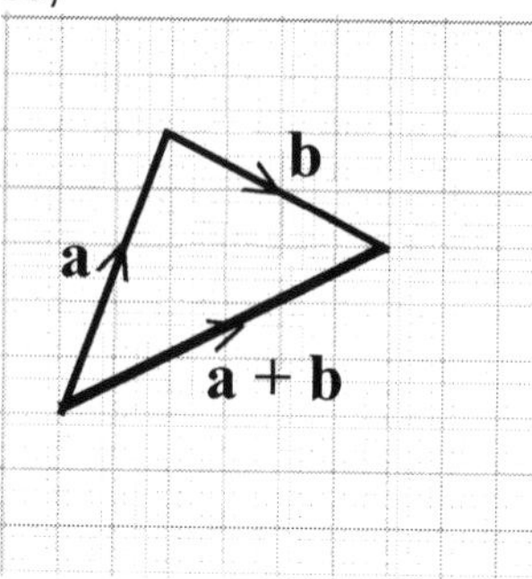

3b)

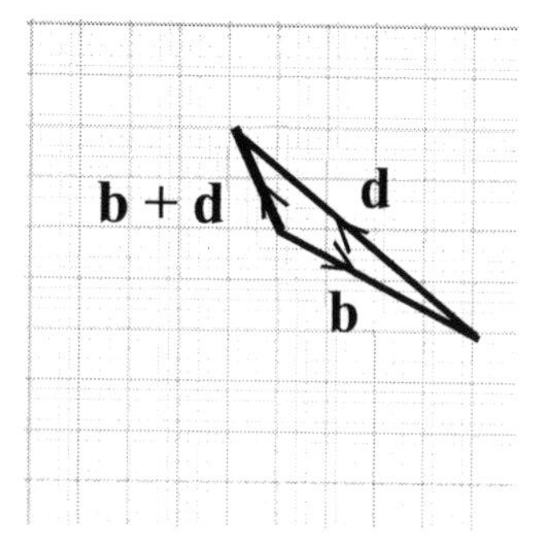

3c)

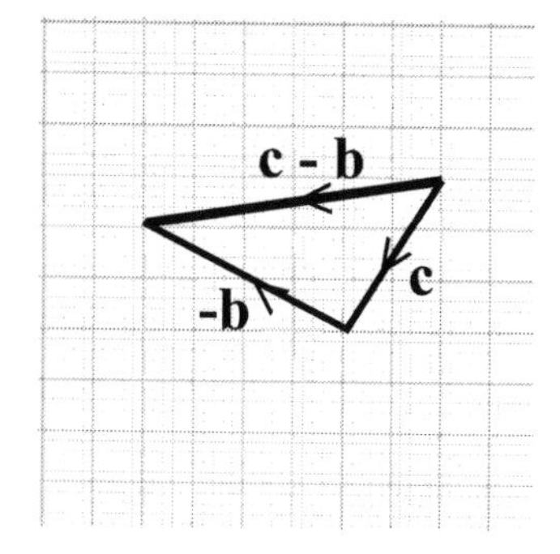

3d)

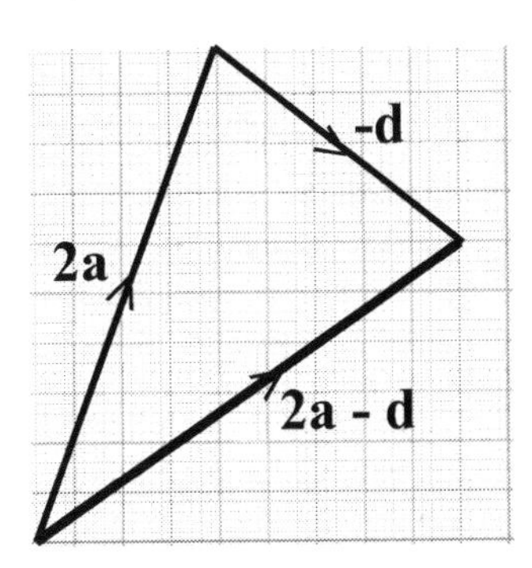

3e)

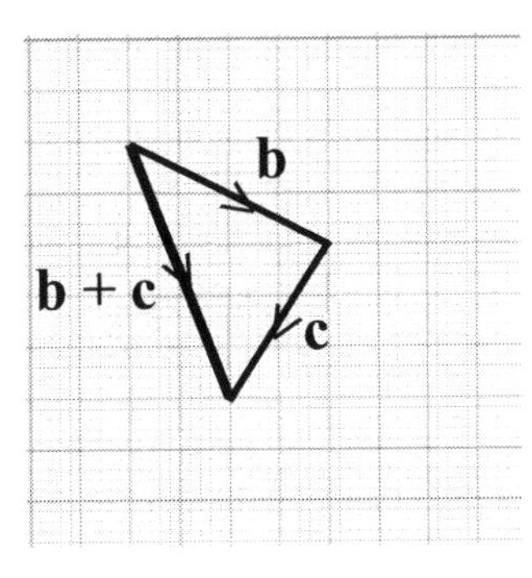

4a) magnitude = $\sqrt{53}$
direction = 74°

4b) magnitude = 3
direction = 90°

4c) magnitude = $\sqrt{34}$
direction = -30.96°

5a) $\begin{pmatrix}-7\\14\end{pmatrix}$

b)) $\begin{pmatrix}7\\-14\end{pmatrix}$

c) (0.5, 2)

6a) i) $\begin{pmatrix}2\\1\end{pmatrix}$

ii) $\begin{pmatrix}2\\3\end{pmatrix}$

iii) $\begin{pmatrix}-5\\5\end{pmatrix}$

iv) $\begin{pmatrix}-9\\1\end{pmatrix}$

v) $\begin{pmatrix}9\\4\end{pmatrix}$

vi) $\begin{pmatrix}4\\-1\end{pmatrix}$

vii) $\begin{pmatrix}-1\\4\end{pmatrix}$

viii) $\begin{pmatrix}7\\6\end{pmatrix}$

7a) p = 4, q = 8
b) p = 3, r = 2
c) p = 4, r = -30
d) p = 3, q = 7
e) p = -2, q = 9

8a) Parallelogram
b)

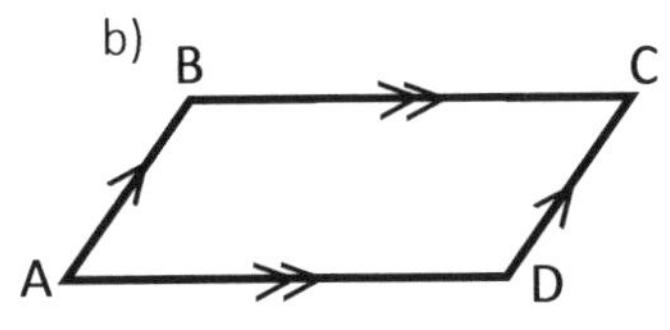

c) AB is parallel to DC and BC is parallel to AD.

EXERCISE 21B
1a)

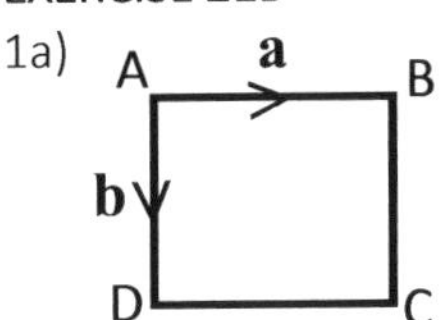

bi) a
ii) b
iii) a + b
iv) a - b
v) -a

2a) b - a
b) $\frac{5}{8}$ b - a
c) $\frac{3}{8}$ b

d) - $\frac{3}{8}$ b

e) – (b – a)
or – b + a
or a – b

3a) a + b
b) a + b
c) 2b
d) 2a

4a) a - b
b) $\frac{2}{5}\boldsymbol{a} + \frac{3}{5}\boldsymbol{b}$

5) $\overrightarrow{EA}$ = 1.5 a – 0.5b

$\overrightarrow{FE}$ = 0.5a - $\frac{1}{8}\boldsymbol{b}$

No, $\overrightarrow{EA}$ and $\overrightarrow{FE}$ have no common multiple. Therefore, the points are not all on a straight line.

6a) i) a + b
ii) $\frac{3}{4}(\boldsymbol{a}+\boldsymbol{b}) = \frac{3}{4}\boldsymbol{a} + \frac{3}{4}\boldsymbol{b}$

iii) $\frac{1}{4}(\boldsymbol{a}+\boldsymbol{b}) = \frac{1}{4}\boldsymbol{a} + \frac{1}{4}\boldsymbol{b}$

b) $\overrightarrow{NK} + \frac{1}{4}\boldsymbol{b} = \frac{1}{4}\boldsymbol{a} + \frac{1}{4}\boldsymbol{b}$

$\overrightarrow{NK} = \frac{1}{4}\boldsymbol{a} + \frac{1}{4}\boldsymbol{b} - \frac{1}{4}\boldsymbol{b}$

$\overrightarrow{NK} = \frac{1}{4}\boldsymbol{a}$

Since $\overrightarrow{BC}$ = a is a multiple of $\overrightarrow{NK}$, $\overrightarrow{NK}$ is parallel to $\overrightarrow{BC}$.
7a) i) z - x
ii) y - z
iii) 0.5y + 0.5z
iv) 0.5x – 0.5y
v) y – 0.5x
7b) $\overrightarrow{AD} + \frac{1}{2}x = \frac{1}{2}y$

$\overrightarrow{AD} = \frac{1}{2}y - \frac{1}{2}x$ = 0.5(y – x)

$\overrightarrow{BC} = \frac{1}{2}z - \frac{1}{2}x + \frac{1}{2}y - \frac{1}{2}z$

$\overrightarrow{BC} = \frac{1}{2}y - \frac{1}{2}x$ = 0.5(y – x)
The vectors are equal and in the same direction. They are parallel.

7c) Rhombus
8a) $8x + 10y$
b) 6.25
9a) b - a
b) a
c) a + b
d) b – 2a
e) b
f) b - a

EXERCISE 22A

1a) 24 b) -11 c) 4 d) 17
2a) 4 b) -1 c) -1.8 d) -2.95
3a) 275 b) -200 c) 90
d) 18.75
4a) 27 b) 195 c) 3 d) 1
e) ± 4 5a) 123 b) -1697
c) 123 d) -3
6) 11, 12, 13, 14

EXERCISE 22B
1a) $x + 7$
b) $\frac{x-9}{2}$
c) $\sqrt{x+1}$
d) $3x - 2$
e) $\frac{3-5x}{x}$
f) $\sqrt[3]{x+10}$

2a) 15 b) -0.5 c) 3 d) 22
e) -4.625
f) $\sqrt[3]{18} = 2.62$ to 2 d.p.

3) $\frac{9x+7}{5-2x}$
4a) $\frac{4}{x-3}$

4b) $\frac{17x-3}{3} = \frac{17x}{3} - 1$
4c) $\frac{13}{x}$

4d) $\frac{q+px}{rx-p}$

5a) $\frac{5x+1}{4-2x}$

5b) $-\frac{13}{3}$

EXERCISE 22C
1a) $5x^2 - 16$
b) $25x^2 - 10x - 2$
c) $44 - 5x^2$
d) $25x - 6$

2a) $\frac{4}{3}$ b) 6 c) $\frac{2}{3}$ d) $-\frac{2}{9}$
3a) $49x - 168$ b) $27 - 7x$
c) $21 - 7x$ d) 18.6 e) 42
4) $2x^2 - 3x - 2$

EXERCISE 23
1a) Student's own drawing
b) Student's own drawing of a circle with radius of 3.
c) diameter = 6
d) x – axes: (-3, 0) and (3, 0)
y – axes: (0, -3) and (0, 3)

2a)
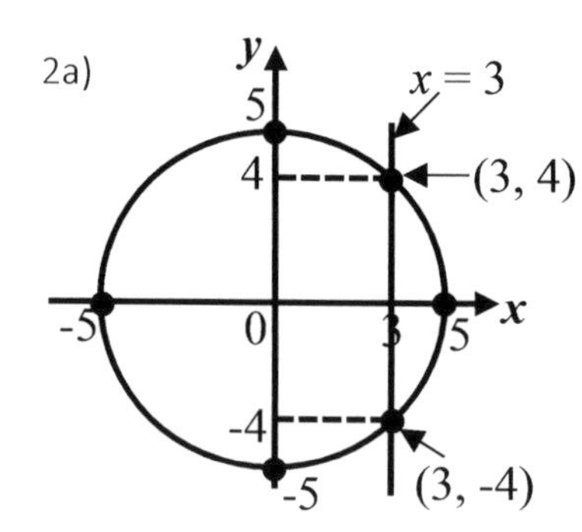

2b) See diagram above.
2c) $x^2 + y^2 = 25$
2d) x – axes: (5, 0) and (-5,0)
y – axes: (0, -5) and (0, 5)
2e) (3, 4) and (3, -4)

3a) Student's own drawing of a circle with radius of 8 units.
3b) Student's own drawing
3c) x = - 3.12 and x = 1.92
3d) 3
3e) Student's own drawing
3f) x = - 4.26 and x = 0.66

4a) $\frac{1}{2}$ 4b) $y = \frac{1}{2}x + 10$

5) y = 11 and x = 11

6a) $x^2 + y^2 = 9$
6b) $x^2 + y^2 = 169$
6c) $x^2 + y^2 = 5$
6d) $x^2 + y^2 = 27$
6e) $x^2 + y^2 = 15.21$

7a) $y = -\frac{7}{24}x + \frac{625}{24}$

7b) $y = \frac{7}{24}x + \frac{625}{24}$

8) $x^2 + y^2 = 40.44$
9a) $2\sqrt{17}$
9b) $y = -\frac{1}{4}x - 4.25$
9c) $y = \frac{3}{\sqrt{8}}x + \frac{17}{\sqrt{8}}$

10a) $(0, 61)$ and $(0, -61)$
10b) $(11, 60)$ and $(11, -60)$
10c) $(-11, 60)$ and $(11, 60)$

EXERCISE 24A

1a) 12.5 m/s b) 25 m/s
c) 875 m d) 125 m
e) 17.5 m/s
2) 145 km/h
3a) 120 km
b) i) 25 km/h^2 ii) 6.7 km/h^2
4a) 37.5 m/s
4b) P: Positive gradient with increasing velocity. Speed (velocity) increased from 0 m/s to 30 m/s for 10 seconds.
Q: This section is horizontal signifying that speed is constant at 30 m/s for 10 seconds (20 – 10 = 10).
R: Positive gradient with increasing velocity (30 m/s to 45 m/s) which is 15 m/s for 10 seconds (30 – 20= 10).
S: Decelerates for 20 seconds (50 – 30 = 20).
4c) Bike was decelerating for 20 seconds.
4d) $\frac{45}{20} = 2.25$ m/s^2

EXERCISE 24B

1) 175 km and under-estimation. (Teacher's guidance recommended)
2a) 40 m/s b) 80 m/s
c) 330 m/s
3a) 190 km
b) under-estimation.
4a) 105 km
b) over-estimation
5) 192.5 m

EXERCISE 24C

1a) 10 hrs. This is the highest point on the curve before deceleration.
1b) i) 12.5 km/h ii) 5 km/h
1c) 12.5 km/h
2a) 2.7 m/s^2 2b) 1.75 m/s^2
2c) t = 20. This is the highest point on the curve before deceleration.
3a) 6.7 km/h^2 3b) 365 km
4a) 15 km/h 4b) 0 km/h
4c) 13.333...km/h
5a)

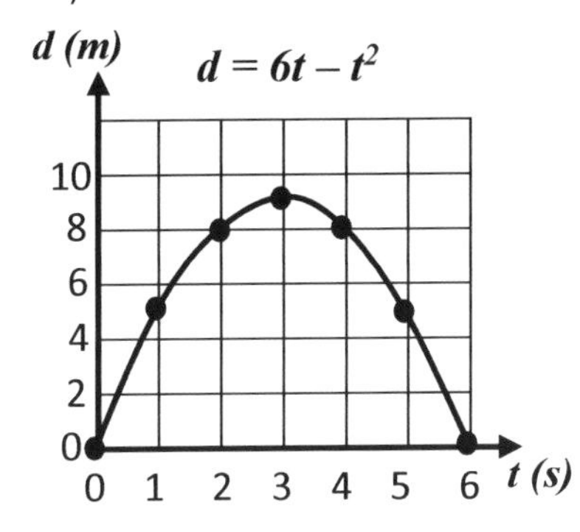

5b) The gradient is zero at the maximum point (3,9). Therefore, the time when gradient is zero is 3 seconds.

5c) Maximum speed is when t = 0 or at t = 6. This occurs when the curve is steepest. By drawing a tangent at t = 0, the gradient is 6. Therefore, the maximum speed is 6 m/s.

5d) At point (4, 8), the gradient is negative, as the line is going down. The gradient will be -2 and it can be said that the speed is moving down with a speed of 2 m/s.

6a) 0.5 m/s^2
6b) Zero
6c) ≈ 1575 m (teacher's guidance recommended).

7a)

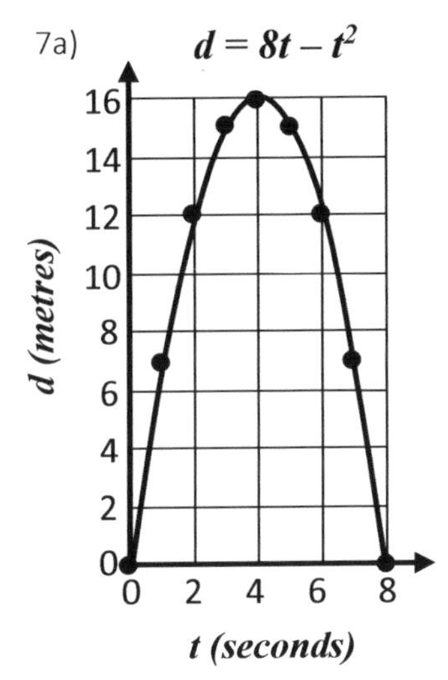

7b) Maximum speed is when t = 0 or at t = 8.
$\frac{dy}{dx} = 8 - 2t$
When t = 0, speed (gradient) = 8 m/s.

7c) i) $\frac{dy}{dx} = 8 - 2t$
When t = 2, the gradient/speed = 4 m/s.

ii) $\frac{dy}{dx} = 8 - 2t$
when t = 4, speed = 0 m/s.

EXERCISE 25A

1a) 0.26 b) 0.56 c) 4.01
d) 0.95 e) 0.69 f) 0
2a) 45° b) 60° c) 35°
d) 22° e) 90° f) 0°
3a) PQ = Opposite
PR = Adjacent
QR = Hypotenuse
3b) TU = Opposite
TS = Adjacent
SU = Hypotenuse
3c) WX = Opposite
WV = Adjacent
VX = Hypotenuse
3d) DF = Opposite
EF = Adjacent
DE = Hypotenuse
4a) 1.732 b) 2.517
c) 4.611 d) 2.483
e) 0.000 f) 1.071

5a) $\sin\theta = \frac{8}{17}$
$\cos\theta = \frac{15}{17}$
$\tan\theta = \frac{8}{15}$

5b) $\sin\theta = \frac{65}{97}$
$\cos\theta = \frac{72}{97}$
$\tan\theta = \frac{65}{72}$

5c) $\sin\theta = \frac{24}{25}$
$\cos\theta = \frac{7}{25}$
$\tan\theta = \frac{24}{7}$

5d) $\sin\theta = \frac{11}{61}$
$\cos\theta = \frac{60}{61}$
$\tan\theta = \frac{11}{60}$
6a) 57.2 b) 12 c) 88.9
7a) $\frac{1}{2}$ b) $\sqrt{3}$ c) $\frac{\sqrt{3}}{2}$
d) $\frac{1}{\sqrt{3}}$ e) $\frac{\sqrt{3}}{2}$ f) 3

EXERCISE 25B

1a) 2.4 m b) 3.3 m c) 7.3 m
d) 6.8 m e) 17.2 m f) 0.7 m
2a) 7.3 m b) 2.7 m
3a) 1069.2 km b) 544.8 km
4a) 3.8 cm b) 109.4 cm
c) 7.9 cm d) 0.2 cm
e) 21 cm f) 13.3 cm

5a) a = 20.8 cm
b = 41.7 cm
c = 36.1 cm
6a) $y = x \tan\theta$
6b) $A = \frac{1}{2}x^2 \tan\theta$

7a)
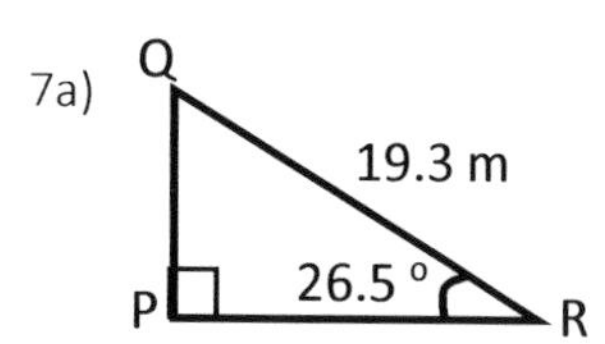

7b) 8.6 m 7c) 17.3 m
8a) 57.8 m b) 28 m
c) 50.5 m
9a) 329 m^2 b) 1036 m^2
10) 168.3 m

EXERCISE 25C

1a) 33.7° b) 73.2° c) 30.7°
d) 45.9° e) 41.4° f) 10.2°
2a) 21.4° b) 25.2°

EXERCISE 25D

1a) $\frac{1+\sqrt{3}}{2}$ b) $\frac{2}{\sqrt{3}}$ c) $\frac{3\sqrt{3}}{2}$

d) $1\frac{1}{\sqrt{2}}$ e) $9\frac{\sqrt{3}}{2}$ f) 3

g) $\frac{\sqrt{3}}{2} + \frac{1}{\sqrt{3}}$

$= \frac{\sqrt{3}(\sqrt{3})}{2 \times \sqrt{3}} + \frac{1 \times 2}{2\sqrt{3}} = \frac{3}{2\sqrt{3}} + \frac{2}{2\sqrt{3}}$
$= \frac{5}{2\sqrt{3}}$
Rationalise the denominator
$= \frac{5}{2\sqrt{3}} \times \frac{\sqrt{3}}{\sqrt{3}} = \frac{5 \times \sqrt{3}}{2\sqrt{3} \times \sqrt{3}}$
$= \frac{5\sqrt{3}}{6}$

2a) $\sqrt{3}$ b) $\frac{19\sqrt{2}}{2}$ c) 6.4
d) $\frac{4\sqrt{3}}{5}$ e) $30\sqrt{3}$ f) $18\sqrt{2}$

EXERCISE 25E

1a) $\frac{3}{4}$ b) $\frac{1}{2}$ c) $\frac{1}{3}$ d) $\frac{1}{2}$ e) 3 f) $\frac{1}{4}$
2a) $1\frac{1}{12}$ b) $3\frac{3}{4}$ c) 1.5 d) 1

EXERCISE 25F

1a) 7.6 cm b) 48.5°
c) 14.15 cm^2
2) 4.3° to 1 d.p.
3) 30.25 cm^2
4a) 5.4 km b) 4.5 km
5a) 9.45 m b) 63°
6) 10.7 cm 7) 120.3 cm^2
8) 16.02 m
9a) 39 cm b) 25 cm c) 22.6°
d) 36.9°
10a)

9.5 m
90 m

10b) 6°
11a) Trapezium
11b) $84\sqrt{3}$ m^2
12) 5 cm

EXERCISE 25G

1a) 2.2 b) 5.4 c) 32.0 d) 2.3
2a) 8.60 cm b) 9.23 cm
c) 6.41 cm d) 2.15 cm
e) a = 7.31 cm b) = 17.45 cm
f) 9.81 cm
3a) 55.0° 3b) 44.7° 3c) 28.3°
3d) e = 20.2° d = 33.8°
3e) 26.5° f) 37.3°

4a)
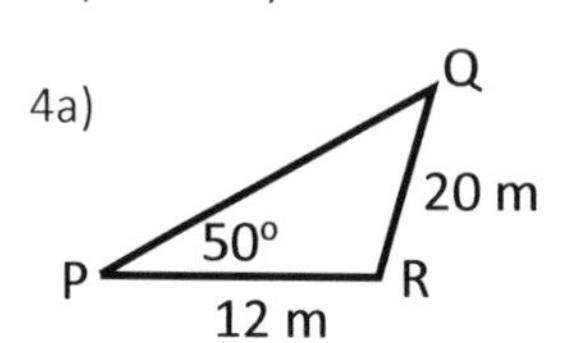

4b) ∠PQR must be less than 50° because the 12 m length facing angle PQR is less than the 20 m length facing 50°. Therefore, angle facing the bigger length (in this case, 50°) must be greater than the angle facing the smaller length.

4c) 102.6°

EXERCISE 25H

1a) 10.3 cm b) 16.9 cm
c) 11.2 cm d) 13.9 cm
e) 37.7 cm f) 16.9 cm
2a) 35° b) 84° c) 32° d) 68°
e) 57° f) 74° g) 16°
3) 84.552 cm
4) a = 28.97 m to 2 d.p
b = 27 m to the nearest whole number.

5)
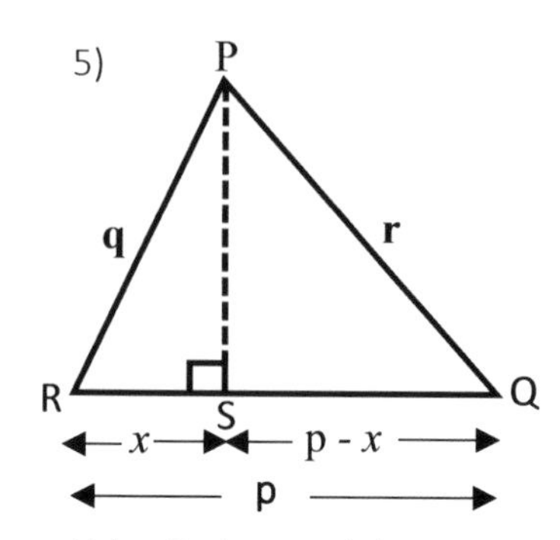

Using Pythagoras' theorem from ΔPQS,
$r^2 = (p - x)^2 + (PS)^2$
...but PS = q sin R
and x = q cos R
Therefore,
$r^2 = (p - q \cos R)^2 + (q \sin R)^2$
$r^2 = p^2 - 2pq \cos R + q^2 \cos^2 R + q^2 \sin^2 R$
$r^2 = p^2 - 2pq \cos R + q^2(\cos^2 R + \sin^2 R)$
...but $\cos^2 R + \sin^2 R = 1$
$r^2 = p^2 - 2pq \cos R + q^2$
$\mathbf{r^2 = p^2 + q^2 - 2pq \cos R}$

6) 276.7° to 1 d.p

EXERCISE 25I

1a) 37.6 cm^2 b) 246 cm^2
c) 45 cm^2 d) 16.1 cm^2
e) 25.4 cm^2 f) 41.6 cm^2
g) 65.5 cm^2
2a) 4.7 cm^2 b) 27.8 m^2
c) 42.8 cm^2 3) 151.5 cm^2
4) 1744.8 cm^2

5) Area of top triangle

$= \frac{1}{2} \times 10 \times 10 \times \sin 60^\circ$

$= 50 \times \frac{\sqrt{3}}{2} = 25\sqrt{3}$

Since the two triangles are identical, area of top triangle multiply by 2 will give the area of the quadrilateral. Therefore,

$2 \times 25\sqrt{3} = 50\sqrt{3}$ cm^2.

6) Area $= \frac{1}{2} \times 8 \times 9 \times Sin B$

$27.58 = 36 \times sin B$

Sin B $= \frac{27.58}{36} = 0.7661111$

B $= \sin^{-1} 0.7661111$

B = 50°

7a)

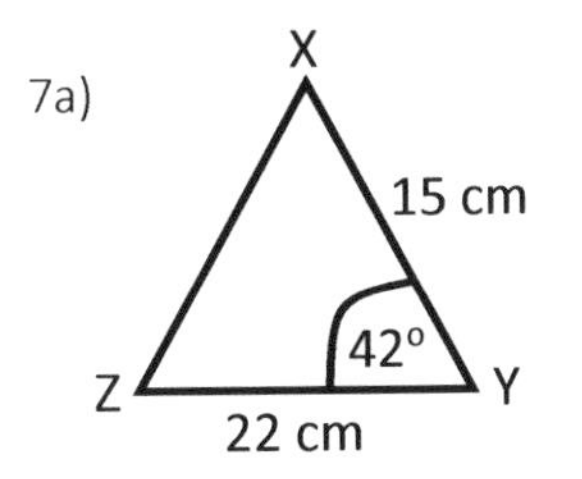

7b) 10.04 cm 7c) 110.4 cm^2

8a) 50.6 cm^2 b) 176.7 cm^2

c) 126.2 cm^2

EXERCISE 25J

1a) i) 6.71 cm ii) 12.88 cm

iii) 12.53 cm iv) 6.71 cm

2i) 9.39 m ii) 51.6°

3a) 34.99 cm b) 28.53°

c) 3780 cm^3

4a) 23.8° b) 80.5 cm^2

c) 281.9 cm^2

EXERCISE 26A

1a) i) 4.36 cm ii) 14.36 cm

iii) 10.9 cm^2

1b) i) 4.15 cm ii) 18.15 cm

iii) 14.54 cm^2

1c) i) 11.16 cm ii) 27.56 cm

iii) 45.77 cm^2

1d) i) 14.14 cm ii) 20.14 cm

iii) 21.21 cm^2

1e) i) 8.80 cm ii) 26.80 cm

iii) 39.58 cm^2

1f) i) 1.81 cm ii) 10.81 cm

iii) 4.06 cm^2

1g) i) 3.93 cm ii) 33.93 cm

Iii) 29.45 cm^2

1h) i) 17.02 cm ii) 30.02 cm

iii) 55.31 cm^2

2a) 40.7° b) 41.4° c) 14.7°

d) 272.2° e) 22.3° f) 65.9°

g) 88° h) 110.5°

EXERCISE 26B

1a) 1.13 cm^2 b) 1.04 cm^2

c) 107 mm^2 d) 36.2 m^2

e) 27.0 m^2 f) 14.8 cm^2

2) 132.9 m^2

EXERCISE 27A

1a) 14.3 cm b) 25 cm^2

2) 6.42 cm to 1 d.p

3) Join OB and OA.

OB = OA ....radii

BC = CA since C is the midpoint of AB.

OC is common.

Therefore, triangles OCB and OCA are congruent.....SSS.

∠OCB = ∠ OCA = 180°

......angles on a straight line

Then, ∠OCB = ∠OCA = 90°

4a) 5.7 cm

b) AB is 4.7 cm to the centre. EF is 5.7 cm to the centre and CD is 6.7 cm to the centre. Therefore, chord AB is the closest to the centre as it has the shortest distance from the centre.

EXERCISE 27B

1a) x = 57°

b) b = 48°, c = 84°

c) d = 117.5° e = 62.5°

d) x = 40°

e) w =63.5°, x = 26.5°

f) f = 110°

g) a = 140°, b = 15°

c = 70° d) 90°

EXERCISE 27C

1a) b = 65°

b) b = 120°, c = 60°

c) b = c = 70°

d) b = 124°, c = 62°

e) b = c = 37°

f) b = 19°, c = 26°

g) a = 54°, b = 108°,

c = d = 36°

h) b = 50°, c = 100°

i) b = 20°, c = 40°

j) a = 284°, b = 52°

c = 128°

k)

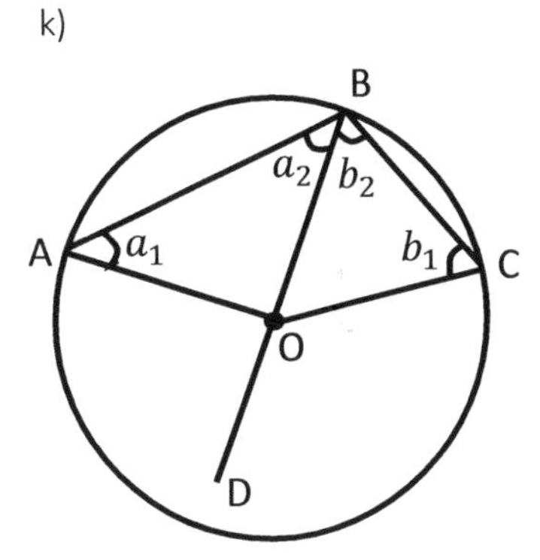

Join BO and extend to any point D.

OA = OB ...radii

$a_1 = a_2$ .....base angles of isosceles triangle

∠AOD = $a_1 + a_2$ ...exterior angles of triangle AOB

Therefore, ∠AOD = $2a_2$

(since $a_1 = a_2$)

Similarly, ∠COD = $2b_2$

$= 2a_2 + 2b_2$

$= 2(a_2 + b_2)$

= 2 × ∠ABC

EXERCISE 27D

1a) a = 95° and b = 97°

b) c = 65° and d = 60°

c) e = 27° and f = 90°

d) a = 214°

e) f = 32° and g = 58°

f) a = 32°

g) a = 40° and b = 40°

h)

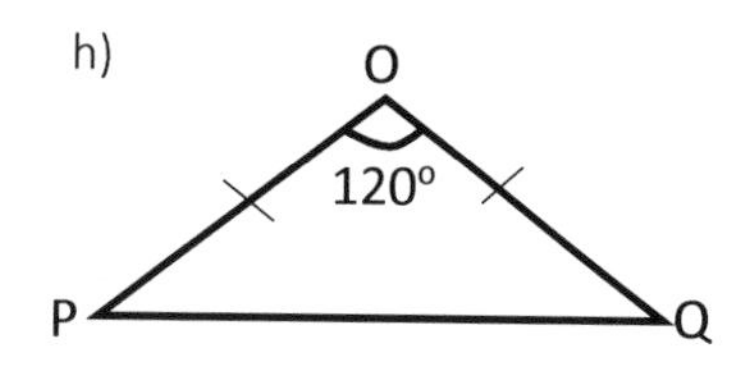

Base angles of isosceles triangles are equal.
180° – 120° = 60°
60° ÷ 2 = 30°
∠PQO = 30°

i)

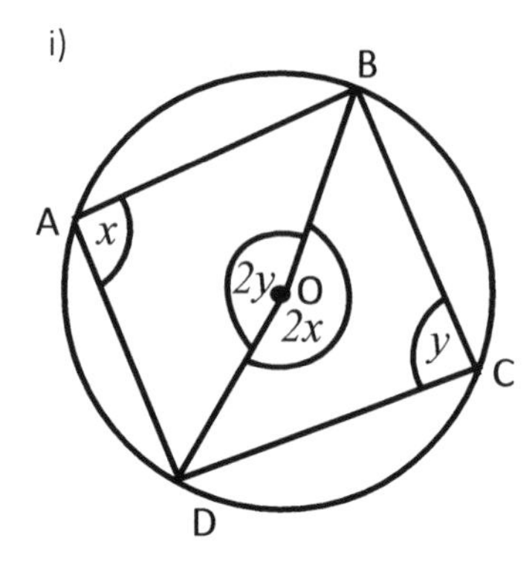

Join D to O and B to O.
∠DOB = 2y ...angle at centre is twice angle at circumference
∠DOB (reflex) = 2x ...angle at centre is twice angle at circumference
Therefore, 2x + 2y = 360° ......angles at a point.
Factorising 2(x + y) = 360°
x + y = 180°
Therefore,
∠DAB + ∠DCB = 180°

EXERCISE 27E

1) a = 57°, b = 33°
2) b = 42°, c = 42°
3) f = 55°
4) a = 105°, b = 60°
5) i = 54°, j = 60.5°, k = 63°

6)

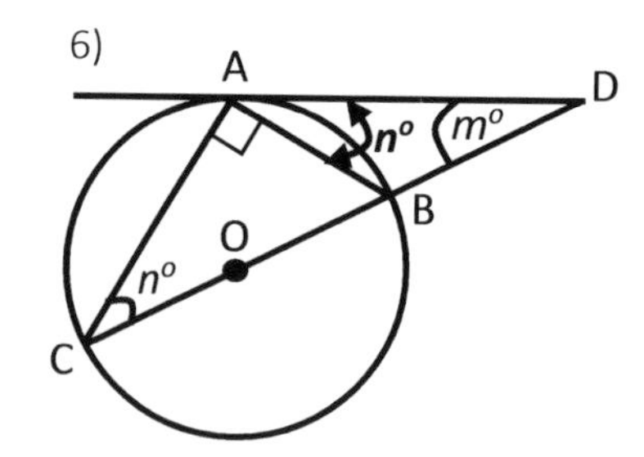

∠CAB = 90° ...angle in a semi-circle
∠ACB = n° ......alternate segment theorem
From ΔCAD,
m + n + 90° + n = 180°
2n + m + 90° = 180°
2n + m = 90°

7)

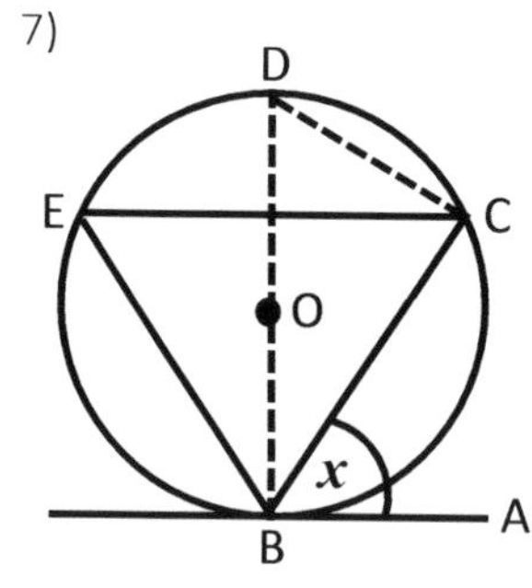

Draw diameter BD. Join DC.
Let ∠ABC = x
∠DBA = 90°.........radius
Perpendicular to tangent
∠DBC = 90° - x
∠DCB = 90° ......angle in a semi-circle
∠CDB + ∠DCB + ∠DBC = 180° ....angles in a triangle
∠CDB + 90° + 90° – x = 180°
∠CDB + 180° – x = 180°
∠CDB = x
But, ∠CDB = ∠BEC ...angles in the same segment
Therefore, ∠BEC = x and
∠ABC = ∠BEC = x

EXERCISE 28

1a) 28.3 cm³ b) 335.1 cm³
c) 2597.0 cm³ d) 1442.0 cm³

2a) i) 65.8 cm² ii) 85.5 cm²
b) i) 1520 cm² ii) 2800 cm²
c) i) 86.1 cm² ii) 125 cm²
d) i) 2370 cm² ii) 3080 cm²
3) 8 cm
4a) 1978.4 cm³ b) 961.3 cm²
5a) 20579.53 cm³
b) 3631.68 cm²
6) 6670 cm³
7) 181.32 cm³
8) 1413.7 cm²
9a) 3072 cm³
b) 1436.1 cm²
10a) 65.4 cm³
b) 78.5 cm²
c) 4.03 cm to 2 d.p.

EXERCISE 29A

1a) 2.5 cm b) 6.45 cm
c) x = 55 cm, y = 2
d) c = 22.4 cm, d = 4 cm

2a)
∠PQR = ∠TSR ......alternate angles
∠QPR = ∠STR ...alternate angles
∠PRQ = ∠TRS...vertically opposite angles
Since their angles are equal, Δs PQR and TSR are similar

2b) a = 6 cm, b = 6.3 cm
3a) ∠ABC = ∠ADE ...corresponding angles
∠ACB = ∠AED ...corresponding angles
∠A = ∠A ...common
Therefore, ΔABC is similar to ΔADE
3b) x = 4.44.. cm, y = 19.6 cm
4a) a and b are similar
b and d are similar

EXERCISE 29B

1a) Not congruent. AAA is not a condition for congruency.
b) Congruent (SSS).
c) Congruent (RHS).
d) Not congruent. The hypotenuse is different for both triangles and cannot be RHS.
2) AB = CB ...sides of a kite
AD = CD ...sides of a kite
BD is common to both Δs.
Therefore,
$\Delta ABD \equiv \Delta CBD$...SSS

3) BC = DC since C is the midpoint of BD
AC = EC since C is the midpoint of AE.
$\angle ACB = \angle ECD$...vertically opposite angles
Therefore,
$\Delta ABD \equiv \Delta EDC$ ...SAS

EXERCISE 29C

1) 15.1 m^2 2) 79.2 m^2
3) 28 cm to the nearest whole number
4a) 1 : 343 b) 1 : 49
5a) 1411.2 cm^2 b) 3.4 g
6a) 20 cm b) 708.75 cm^2.

EXERCISE 30A

1a) 45 b) 3.5
2a) y = 2.5 x
2b) i) 1.5 ii) 16
3)

c	15	21	27	31.5
d	5	7	9	10.5

4a) e = 6f^2
b) i) 96 ii) 0.58 to 2 d.p.
5a) 78.125 b) 2.2

6)

a	2	10	16	20
b	1	25	64	100

7a) $p = 3\sqrt[3]{q}$
b) i) 15 ii) 27

EXERCISE 30B

1a) $y = \frac{24}{x}$ b) 6 c) 2.4
2a) $w = \frac{13.75}{c}$
2b) 1.375 c) 1.72
3) graph (a) $y \propto x$
graph (b) $y \propto x^2$
graph (c) $y \propto \frac{1}{x}$
4)

b	1	6	3	15
c	12	2	4	0.8

5) c = 45 d = 11.25 e = 5
6a) $y = \frac{200}{\sqrt{x}}$
6b) 33.333... 6c) 113.8
7a) $d = \frac{980}{r^2}$
7b) 61.25 cm c) 2.9 cm

EXERCISE 31A

1a)

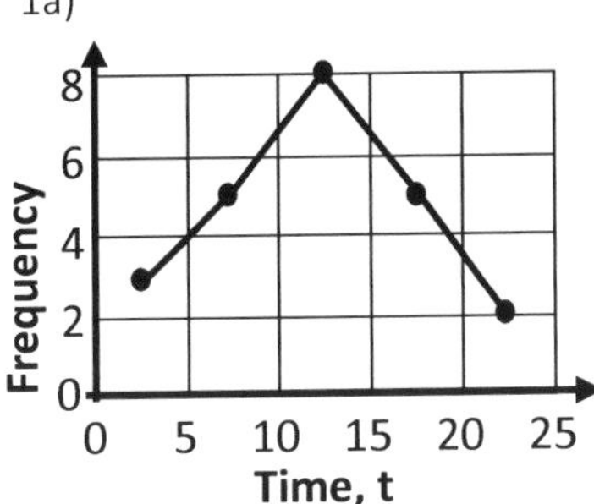

1b)

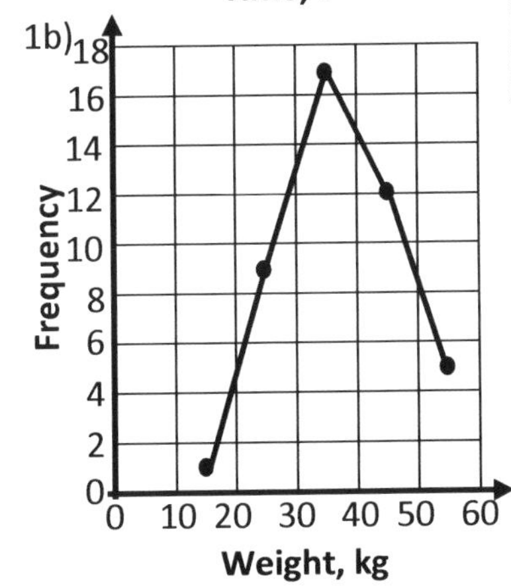

2)

Weight, n (kg)	Frequency
$10 \le n < 15$	5
$15 \le n < 20$	15
$20 \le n < 25$	20
$25 \le n < 30$	10
$30 \le n < 35$	5

EXERCISE 31B

1a)

Height, cm	CF
$70 < h \le 80$	7
$80 < h \le 90$	27
$90 < h \le 100$	57
$100 < h \le 110$	92
$110 < h \le 120$	112
$120 < h \le 130$	120

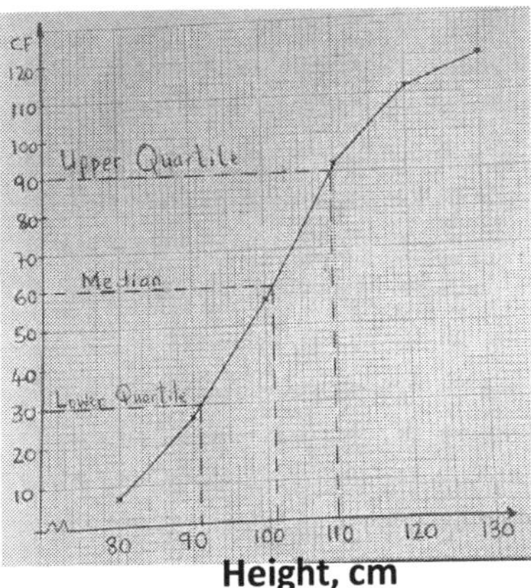

1b)

Total score, m	CF
$2 \le h \le 11$	7
$12 \le h \le 21$	18
$22 \le h \le 31$	40
$32 \le h \le 41$	54
$42 \le h \le 51$	62
$52 \le h \le 61$	65

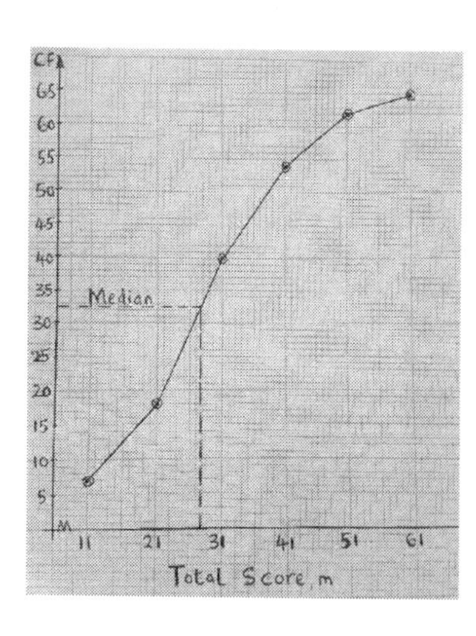

2a) median = 27 g (±2)
Interquartile range = 15 g (±2)
2b) i) 60 apples ii) 40%
3a) median = 101 cm
b) lower quartile = 91 cm
c) upper quartile = 109 cm
d) interquartile range = 18 cm
4) 38.75 cm
5a) £65
b) i) £61 ii) £45 iii) £35

EXERCISE 31C

1a)
i) lowest number = 4
median = 15
lower quartile = 8
upper quartile = 22
interquartile range = 14
maximum number = 30

ii) lowest number = 1
median = 13
lower quartile = 6.5
upper quartile = 25.5
interquartile range = 19
maximum number = 35

1b)

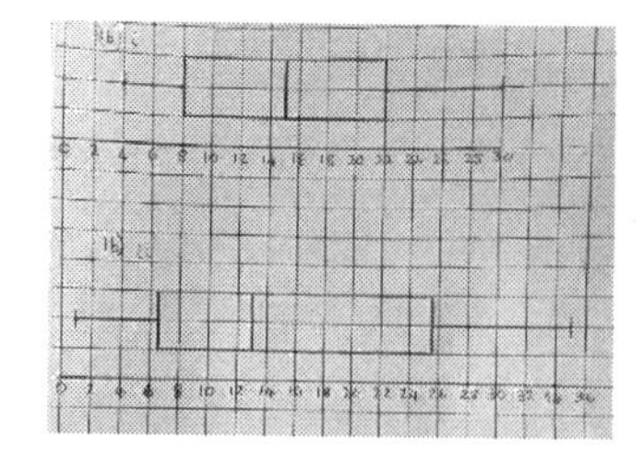

2a)
i) lowest number = 2.5
ii) median = 10
iii) lower quartile = 5
iv) highest number = 25
v) upper quartile = 15
vi) range = 22.5
vii) interquartile range = 10

2b)
i) lowest number = 4
ii) median = 28
iii) lower quartile = 20
iv) highest number = 44
v) upper quartile = 32
vi) range = 40
vii) interquartile range = 12

3a)

Time, mins	CF
$0 \leq t < 11$	5
$11 \leq t < 21$	20
$21 \leq t < 31$	52
$31 \leq t < 41$	92
$41 \leq t < 51$	112
$51 \leq t < 61$	121

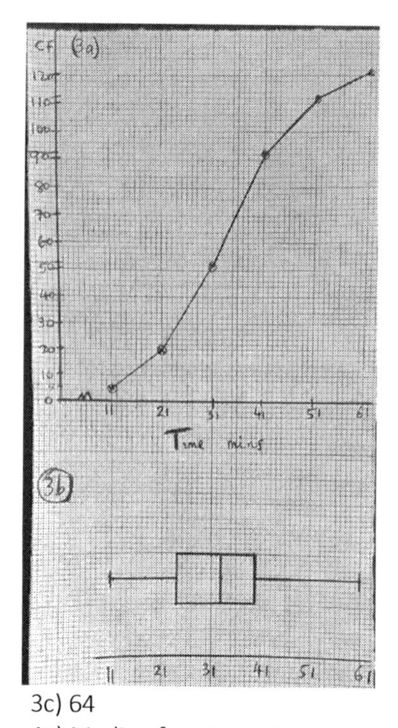

3c) 64
4a) Median for History is 40% and the median for Maths is 45% showing that on average, pupils scored better in maths.
Also, the interquartile range for History is 15% (45 – 30) and that for Maths is 25% (50 – 25). This shows that the scores for History were more consistent.

5a) 70 females
5b) 70 males
5c) Minimum weight of males = 50 kg and maximum weight for males = 110 kg.
5d) i) Median for males (estimate) = 77 kg.
Median for females (estimate) = 73 kg. This shows that on average, males weighed more than females. ii) The range for males is 60 kg and the range for females = 50 kg. It means that the weights for females were more consistent.

EXERCISE 31D

1)

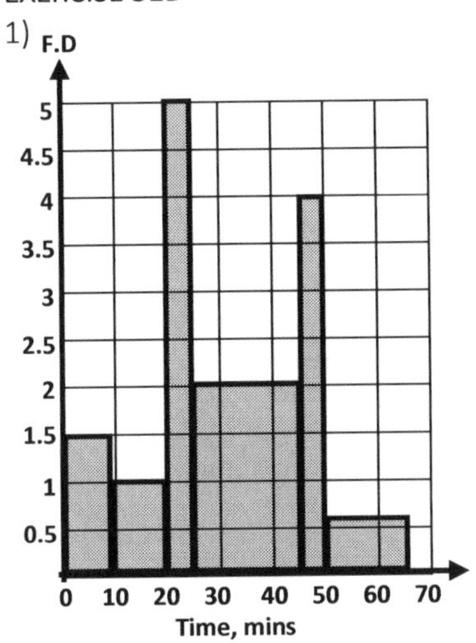

2a) 20, 150, 450, 140

2b) 760

3a)

Length (cm)	FD	Frequency
$0 \le t < 5$	8	40
$5 \le t < 8$	10	30
$8 \le t < 12$	12	48
$12 \le t < 17$	5	25
$17 \le t < 20$	10	30

3b)

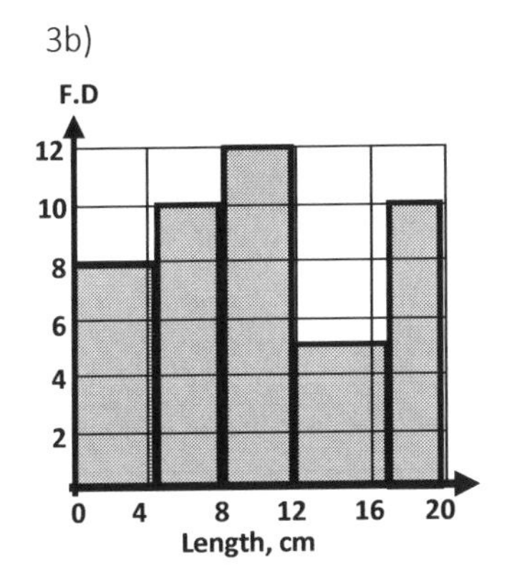

3c) 9.42 cm to 2 d.p.

4a)

Time, minutes	Frequency
$5 \le t < 15$	100
$15 \le t < 30$	300
$30 \le t < 60$	1200
$60 \le t < 80$	650

4b) 1600 meals

5)

Distance (d) in km	Frequency
$0 < d \le 20$	200
$20 < d \le 30$	80
$30 < d \le 50$	100
$50 < d \le 55$	70
$55 < d \le 75$	10

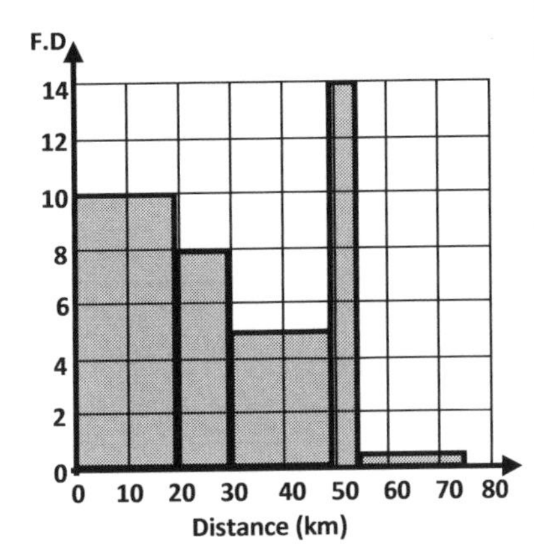

6a) 39 6b) 24 6c) 9

EXERCISE 32A

1a) $\frac{3}{9}$ b) $\frac{7}{9}$ c) $\frac{27}{99}$ d) $\frac{56}{99}$
e) $\frac{1234}{9999}$ f) $\frac{71}{495}$ g) $\frac{1235}{999}$

h) $\frac{11}{30}$ i) $\frac{2}{225}$

EXERCISE 32B

1) 9.2 g/cm^3

2a) 523.6 cm^3 to 1 d.p.

b) 0.6 g/cm^3 to 1 d.p.

3a) 9720 cm^3

b) 0.3 g/cm^3 to 1 d.p.

c) 262.4 g

EXERCISE 32C

1a) Lower bound = 44.5 cm
Upper bound = 45.5 cm

b) LB = 1150, UB = 1250

c) LB = 66.5, UB = 67.5

d) LB = 0.03, UB = 0.13

e) LB = 8.75, UB = 8.85

f) LB = 56.29, UB = 56.39

g) LB = 7.65, UB = 7.75

h) LB = 535, UB = 545

i) LB = 11850, UB = 12849

2) Least weight = 7.8 kg
Greatest weight = 8.8 kg

3a) LB = 4.45 cm

b) UB = 4.55 cm

c) $4.45 \le l < 4.55$

d) LB = 2.15 cm

e) UB = 2.25 cm

f) 10.2375 cm^2

4) 52.7 cm^3

EXERCISE 32D

1) 3.333..%

2a) 0.18 m b) 9.4% to 1 d.p

3a) 12.75 cm^2

b) Least possible area = 12.47875 cm^2
Greatest possible area = 13.02375 cm^2

c) Maximum absolute error = 0.27375 cm^2

d) Maximum percentage error = 2.1 % to 1 d.p.

4) 5.63%

EXERCISE 32E

1a) P = {7,14,21,28,35,42}

b) Q= {1,2,3,4,6,8,12,16,24,48}

c) R = {7,11,13,17,19,23}

d) S = {27,64}

2a) 35 pupils b) 22 pupils

c) 14 pupils d) 13 pupils

e) 5 pupils f) $\frac{9}{35}$ g) $\frac{22}{35}$

h) $\frac{27}{35}$ i) $\frac{8}{35}$ j) $\frac{22}{35}$ k) $\frac{26}{35}$

3a)

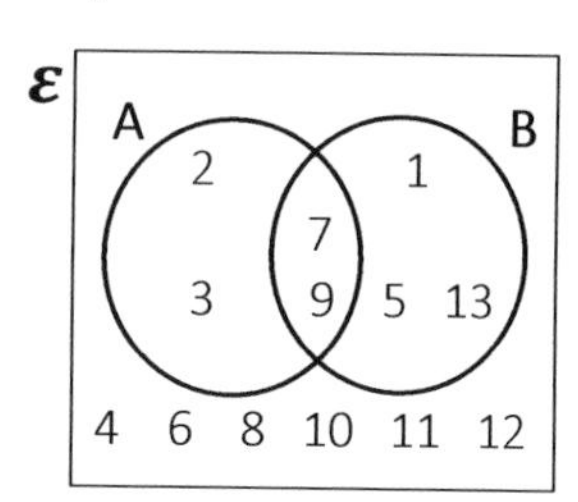

3b) i) $\frac{5}{13}$ ii) $\frac{9}{13}$ iii) $\frac{2}{13}$ iv) $\frac{6}{13}$

4a)
B = {2,4,6,8,10,12,14,16,18}
C = {4,8,12,16}
D = {2,3,5,7,11,13,17,19}

5a) $A' \cap B$
b) $A \cap B'$
c) $A \cap B$
d) $(A \cup B)'$
e) $A \cup B$
f) A'

EXERCISE 32F

1a) c = Cat and D = dog

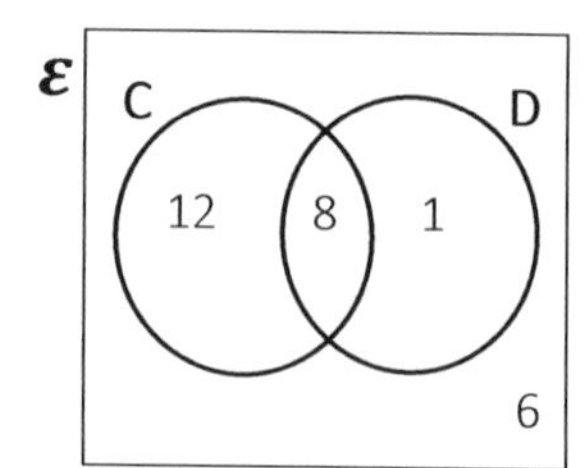

b) i) $\frac{21}{27} = \frac{7}{9}$ ii) $\frac{9}{27} = \frac{1}{3}$
c)) $\frac{8}{20} = \frac{2}{5}$

2a) 0.65 b) 0.88
3a)

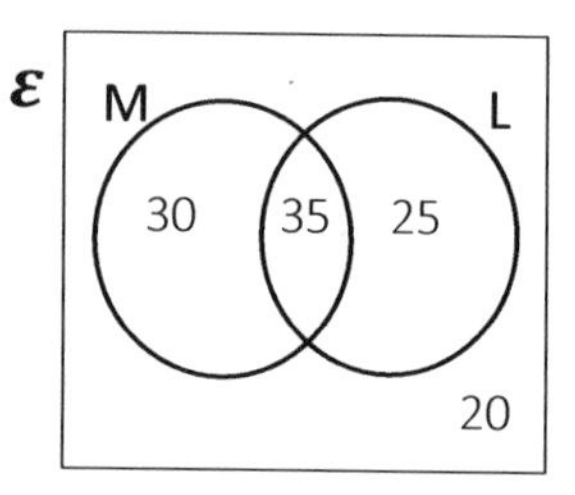

Where M = media dept and
L = left handed workers

b) 35 workers
c) $\frac{25}{110} = \frac{5}{22}$

4a) w = 15 b) 36 c) 40
5a) $\frac{5}{15}$ b) = $\frac{8}{15}$ c) $\frac{10}{15}$
d) $\frac{7}{15}$ e) $\frac{12}{15}$ f) $\frac{4}{15}$ g) $\frac{7}{15}$

6a) G = Geography,
H = History and F = French.

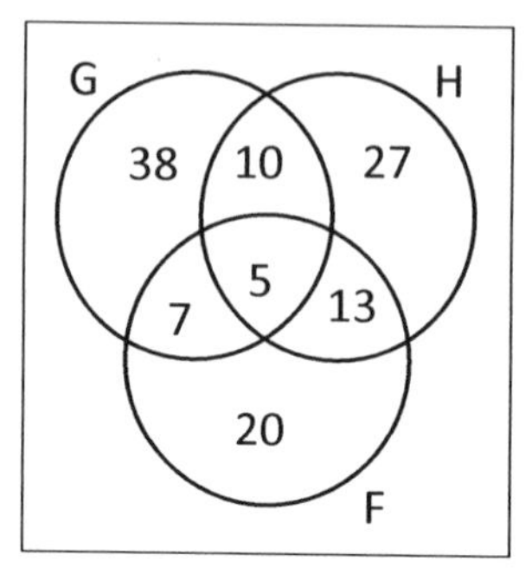

6b) 5 students c) $\frac{5}{120}$

EXERCISE 32G

1a) $y = x^3 + 1$

x	-4	-3	-2	-1	0	1	2	3
y	-63	-26	-7	0	1	2	9	28

1b) $y = x^3 + x - 3$

x	-4	-3	-2	-1	0	1	2	3
y	-71	-33	-13	-5	-3	-1	7	27

1c) $y = \frac{4}{x}$

x	-5	-4	-3	-2	-1	$-\frac{1}{2}$	0.5	1	2	3	4
y	-0.8	-1	-1.3	-2	-4	-8	8	4	2	1.3	1

1d) $y = -\frac{4}{x}$

x	-5	-4	-3	-2	-1	$-\frac{1}{2}$	0.5	1	2	3	4
y	0.8	1	1.3	2	4	8	-8	-4	-2	-1.3	-1

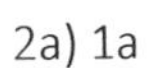

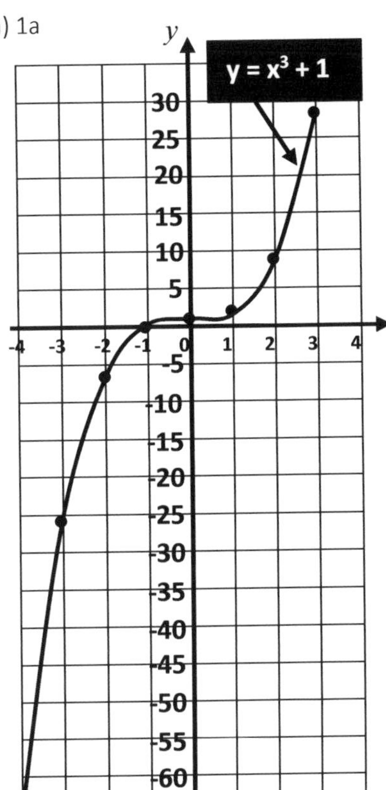

3) $y = \dfrac{x}{x+2}$

x	-3	-2	-1	-0.5	-0.2	0.2	1	2	3
x + 2	-1	0	1	1.5	1.8	2.2	3	4	5
y	3	-	-1	-0.33	-0.11	0.09	0.33	0.5	0.6

4)

x	-4	-3	-2	-1	0	1	2	3	4
y	-1	0	1	8	27	64	125	216	343

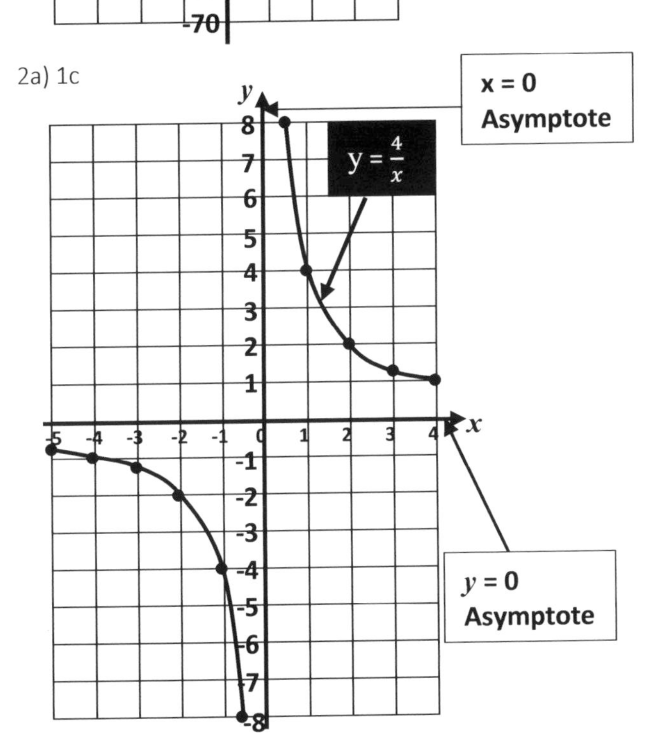

EXERCISE 32H

1a) $y = 3^x$

x	-4	-3	-2	-1	0	1	2	3
y	0.01	0.04	0.11	0.33	1	3	9	27

1b)

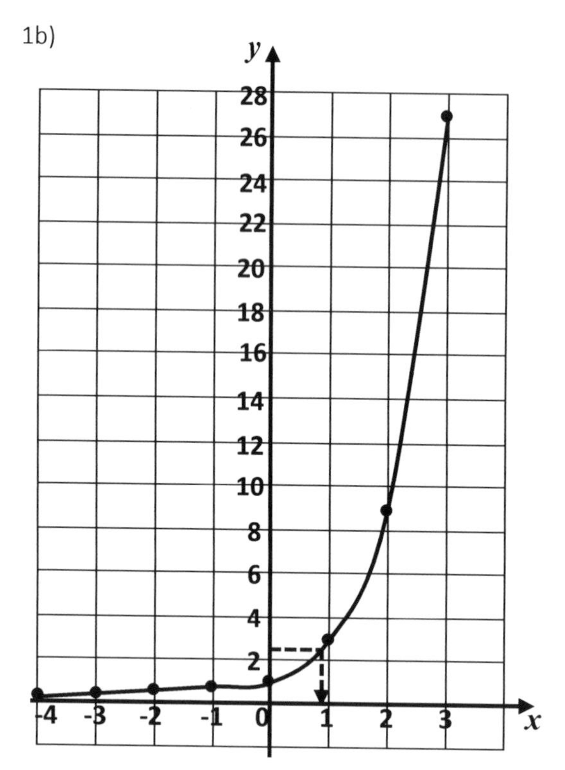

$x = 0.84$

2a) $y = 5^x$

x	-3	-2	-1	0	1	2	3
y	0.008	0.04	0.2	1	5	25	125

2b) $y = 5^{-x}$

x	-3	-2	-1	0	1	2	3
y	125	25	5	1	0.2	0.04	0.008

2c) Student's correct graph.

3a)

t	0	1	2	3	4	5	6	7	8
B	60	30	15	7.5	3.75	1.88	0.94	0.47	0.23

3b)

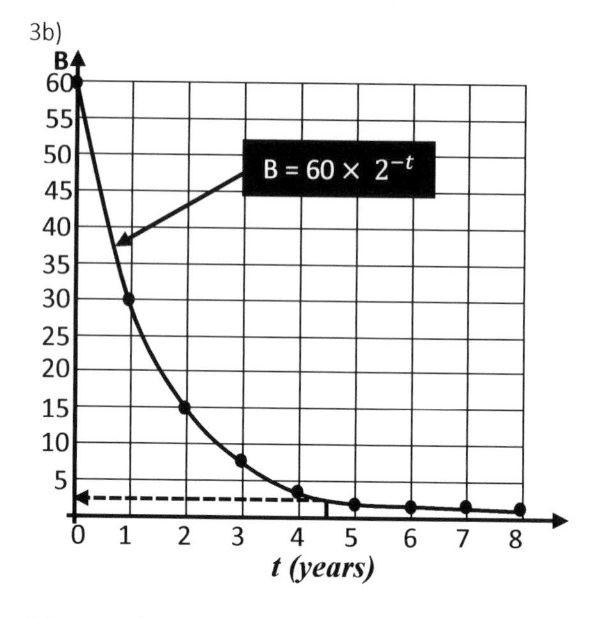

3c) Accept between 2 and 2.8.

4a) $y = (\frac{2}{5})^x$

x	-2	-1	0	1	2	3	4
y	6.25	2.5	1	0.4	0.16	0.06	0.03

4b) Student's own correct graph.

EXERCISE 32I

1a)

θ	0°	30°	45°	60°	90°	135°	180°
sinθ	0	0.5	0.71	0.87	1	0.71	0
cosθ	1	0.87	0.71	0.5	0	-0.71	-1
tanθ	0	0.58	1	1.73	-	-1	0

1b)

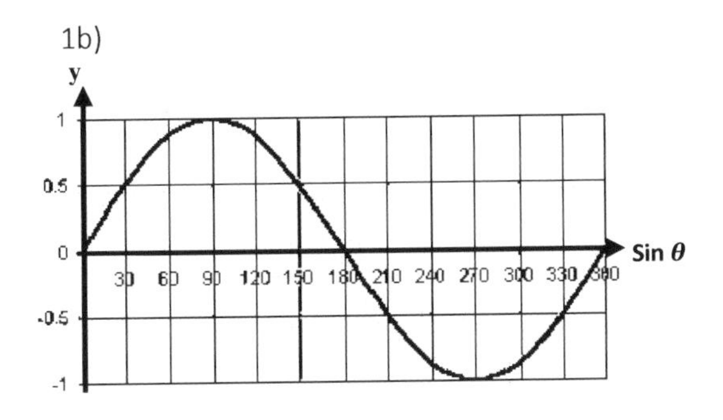

1c)

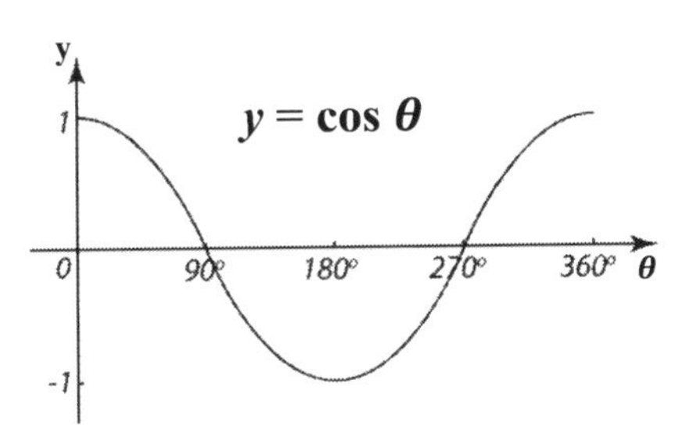

2) i) 30° ii) 60°

3) i) 60° ii) 30°

4a)

x^o	$2x^o$	$2\sin 2x$
-180	-360	0
-135	-270	2
-90	-180	0
-45	-90	-2
0	0	0
45	90	2
90	180	0
135	270	-2
180	360	0

4b)

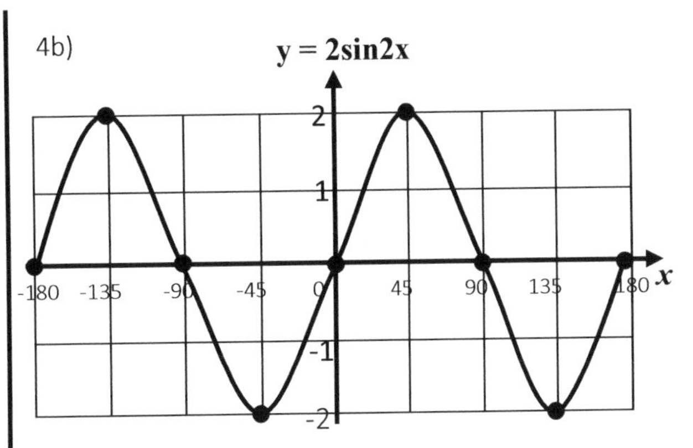

4c) Minimum values i) (-45, -2) and (135, -2)

Maximum values i) (-135, 2) and (45, 2)

EXERCISE 32J

1) x = 30°, 150°, -210° and -330°

2a)

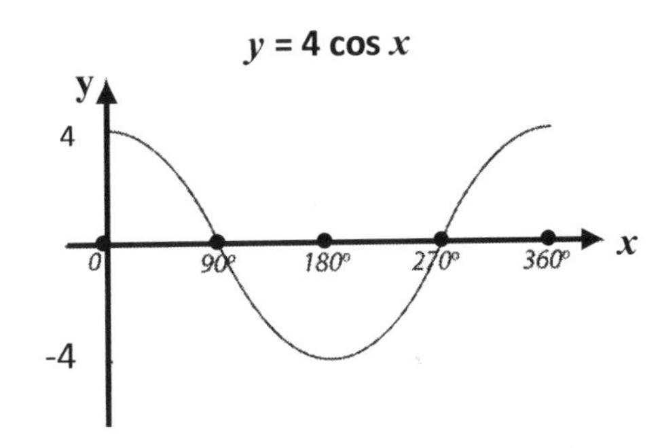

2b) x = 75.5° and 284.5°

3a) 37° and 143°

3b) 53° and 307°

3c) 348° and 192°

3d) 15° and 345°

4a) 11° b) 11° and 169°

5a) i) -0.707 ii) 0.891

5b) 66.4°

EXERCISE 32K

1a)

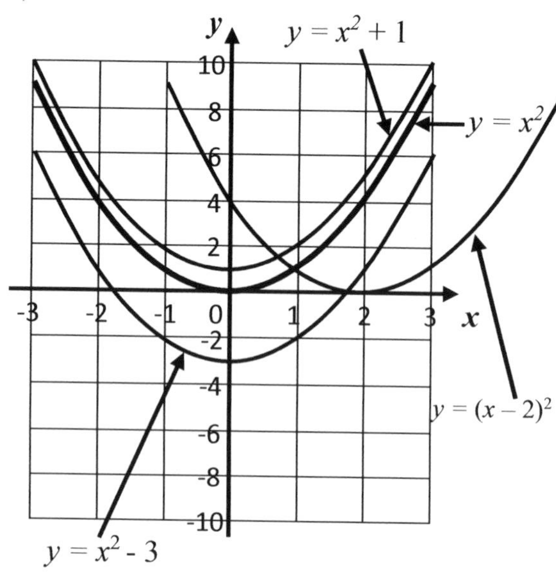

1b) See diagram above

1c) i)

$y = x^2$ has been translated by the vector $\begin{pmatrix}0\\1\end{pmatrix}$ to give $y = x^2 + 1$.

ii) $y = x^2$ has been translated by the vector $\begin{pmatrix}0\\-3\end{pmatrix}$ to give $y = x^2 – 3$.

iii) $y = x^2$ has been translated by the vector $\begin{pmatrix}2\\0\end{pmatrix}$ to give $y = (x – 2)^2$.

2a)

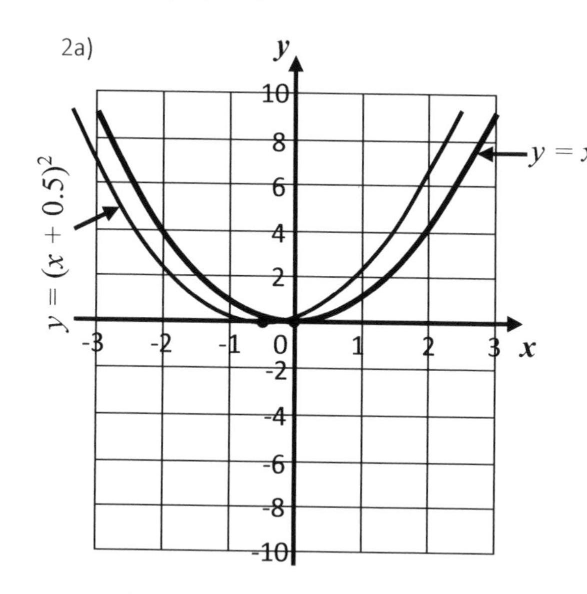

2b) see diagram

2c) $y = x^2$ has been translated by the vector $\begin{pmatrix}-0.5\\0\end{pmatrix}$ to give $y = (x + 0.5)^2$.

3a and 3b)

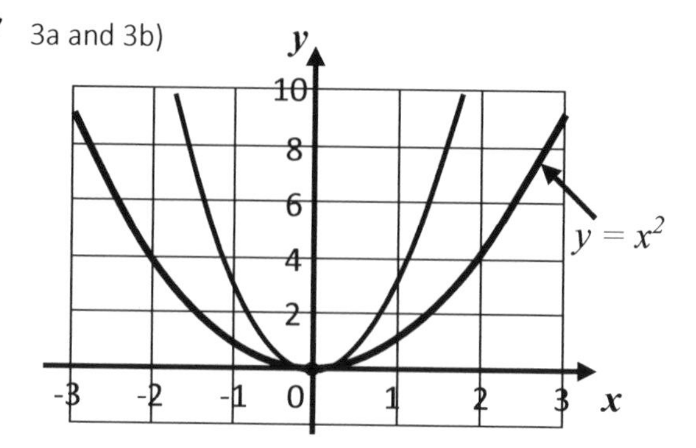

3c) The curve of $y = x^2$ has been stretched out in the direction of the y-axis by a scale factor of 3 to give the graph of $y = 3x^2$.
Also, the y-coordinates of $y = 3x^2$ are 3 times bigger when compared with the y- coordinates of $y = x^2$.

4a) The curve of $y = x^2$ has been stretched out in the direction of the y-axis by a scale factor of 6 to give the graph of $y = 6x^2$.

4b)
The x-coordinates of $y = x^2$ is a quarter $(\frac{1}{2})$ compared to that of $y = (2x)^2$. It therefore moves the graph of $y = x^2$ inwards (towards the y-axis) to produce the graph of $y = (2x)^2$.

5)

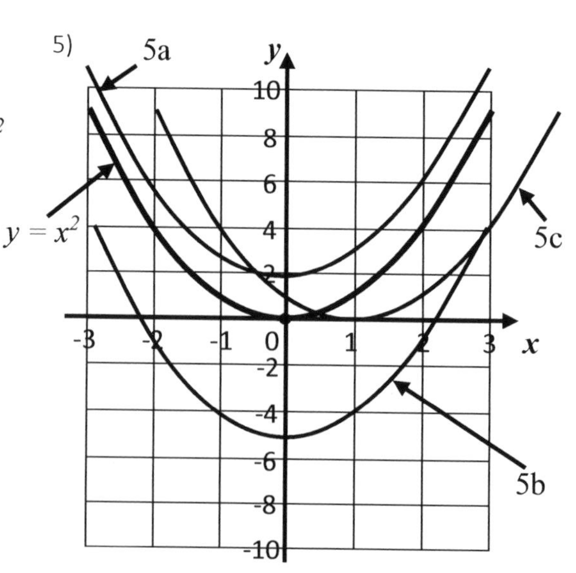

6a) Translation 3 units down or by the vector $\begin{pmatrix} 0 \\ -3 \end{pmatrix}$.

6b) A stretch parallel to the y-axis by a scale factor of 3 and translation 2 units up by the vector $\begin{pmatrix} 0 \\ 2 \end{pmatrix}$.

Question 7a

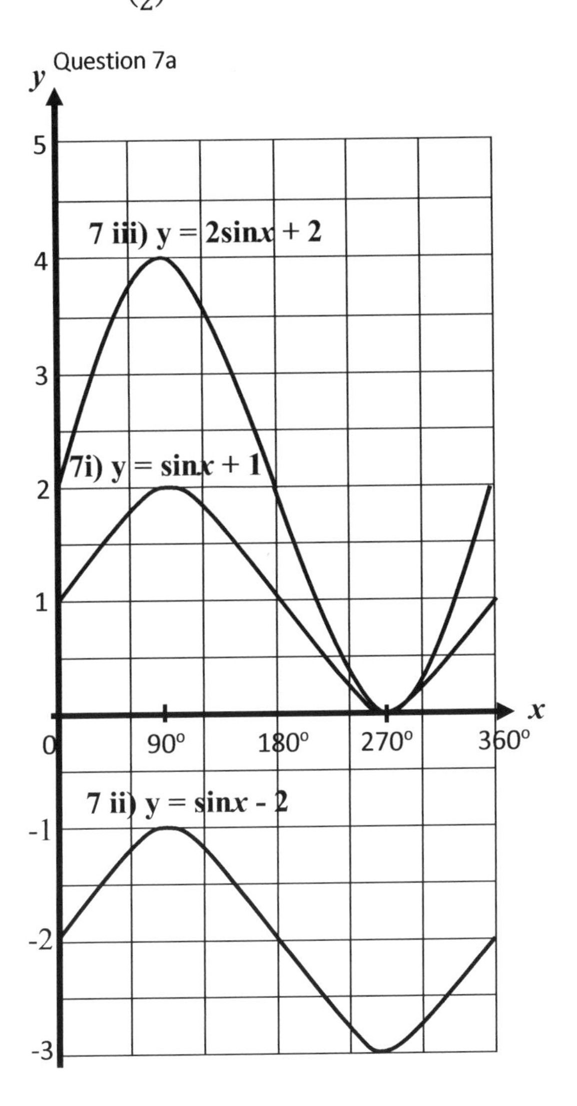

Question 7b) The period is 360°.

INDEX

Y